# Introduction
## to
# Plant Biotechnology
### Third Edition

Third Edition

# Introduction
# to
# Plant Biotechnology

## H.S. Chawla

Genetic & Plant Breeding Department
G.B. Pant University of Agriculture & Technology
Pantnagar, India

**Oxford & IBH Publishing Co. Pvt. Ltd**
**New Delhi**

Oxford & IBH Publishing Company Pvt. Ltd.
113-B Shahpur Jat
*Asian Village Side*
New Delhi 110 049, India

*Fax:* (011) 4151 7559
*Email:* oxford@oxford-ibh.in

© 2009, Copyright Reserved

Last Reprint 2009

ISBN 978-81-204-1732-8

Printed at Chaman Enterprises, New Delhi.

# Preface to the Third Edition

In view of the developments in the field of biotechnology, the scope of this textbook has been broadened in this revised edition, but the basic philosophy remains the same. Protocols are given in the chapters where laboratory exercises are essential for the students. References have been quoted at the end of each chapter. Emphasis is placed on application of a technique, its contribution and impact on crop improvement.

Chapters on plant tissue culture have been suitably amended wherever needed but the basic aspects remain the same except for few additions. Chapters on genetic material, organization of DNA in the genome and basic techniques for understanding the recombinant DNA technology remains the same.

The recombinant DNA technology part in this revised edition has been broadened to a large extent. Chapter on gene transfer methods has been expanded with emphasis on gene silencing and development of marker free transgenics. A new chapter on chloroplast and mitochondrion DNA transformation has been introduced. Chapter on Transgenics in Crop Improvement has been revised with some recent examples of transgenics developed for various characters and its implications have been added as new chapters. Impact of Recombinant DNA Technology chapter details the status of biotech crops and arguments made in favor and against the GM crops in detail and in other role of rDNA technology in other fields of biotechnology. Biosafety issues on biotech crops are taking a centre stage and so many concerns have been raised. Great emphasis has been laid on a chapter on biosafety and regulatory framework. Issues and controversies arisen at different times on biosafety have been given. Regulatory frameworks for the release of transgenic crops and biotech products in different countries of the world have been explained. Chapters on Bioinformatics and Intellectual Property Rights have been expanded to great length. IPR issues as such and with respect to biotechnology and protection mechanism for plants in the form of breeders rights have been explained. Molecular markers and marker assisted selection chapter has been expanded to give more information on some of the new molecular markers, genetic fingerprinting and further updated by giving some recent examples. Other chapters on gene cloning, *in vitro* mutagenesis, transposons and gene tagging, gene isolation and genomics have been amended, updated wherever some lacunas were there. Metabolomics portion has been added in the chapter on genomics.

I am grateful to the reviewers of the previous editions. It enabled me to improve this book. I am thankful to my well wishers, colleagues and innumerable students who have made great contributions directly or indirectly by giving suggestion for improvement in one way or the other for this revised edition.

Nothing can move without the will of God. I am lucky and fortunate enough to have the blessings of my gracious Almighty God to complete the task of writing this book. I am grateful to my elders for giving me strength and support, and to my wife and children, Komaljit and Jasmit for their perseverance.

January, 2009

H.S. Chawla

# Preface to the First Edition

Plant biotechnology has emerged as an exciting area of plant sciences by creating unprecedented opportunities for the manipulation of biological systems. We are seeing the genes and genomes of a wide range of different organisms being manipulated by the use of new techniques for the benefit of man. One of the key techniques in genetic engineering is gene transfer, which encompasses a variety of methods for returning cloned genes to cells and to generate transgenic plants. Cell and tissue culture are the innovative breeding techniques applied to meet the increasing need for improved crop varieties. Tissue culture techniques can shorten the time and can lessen the labor and space requirements needed to produce a new variety. Cell and plant tissue culture and recombinant DNA technology constitutes an important aspect of plant biotechnology. Further, to understand gene technology it has become essential to understand the basic structure of gene and its organization in plant cell.

Courses on biotechnology are offered at various levels of undergraduate and graduate studies in various departments of Botany, Genetics, Plant Breeding, Horticulture, Plant Pathology, Entomology, Plant Science, Biotechnology and Bioscience. A good understanding of genetic engineering and plant tissue culture at the undergraduate level has become very critical for realizing the full potential of biotechnology. There are a number of books which deal with specialized aspects of plant tissue culture, cloning of genes and genome organization, but, I expect it will be useful to have a book describing the basic aspects of gene and genome organization in plant cells, basic tissue culture techniques and the fundamentals of cloning, gene transfer approaches and molecular markers. Basic tissue culture courses usually include history of the technique, laboratory organization, nutrition media, micropropagation, organ culture, cell suspension culture, anther culture, somatic fusion, secondary metabolite production and cryopreservation. I have given information on variability generated by tissue culture as somaclonal variation in one of the chapters. Gene cloning, gene transfer techniques, genome mapping and molecular markers have been described in relation to plants. A chapter on intellectual property rights has been included to give basic information on various aspects of patenting, copyright and plant breeders right.

Plant breeders are striving to meet the challenge of increased production by developing plants with higher yield, resistance to pests, diseases and weeds and tolerance to various abiotic stresses. I have tried to give suitable examples of transgenics developed for various characters in one of the chapters so that a student is aware of the impact of biotechnology on crop improvement. In most of the chapters protocols for conducting laboratory exercises have been given. A very important point is that in most chapters emphasis is placed on application of a technique and its contribution and impact on crop improvement.

I would very much appreciate receiving your suggestions, criticism and research contributions (as reprints) which relate to different aspects relevant to this book. It will be most helpful during the penetration of a revised edition. Please mention the errors you find with page numbers and describe mistakes. I'll highly appreciate your assistance in this regard.

I am thankful to Mr. Vijay Upadhaya and Mr. Fahim for their patience and agreeing to my suggestions for making diagrams. I am thankful to innumerable students who have made great contributions in one way or the other during the preparation of manuscript.

I am thankful to my elders whose continuous support and inspiration led me to complete this work. I am fortunate in having a family which understands the preoccupation that goes with such projects. I am grateful to my wife and to my children, Komaljit and Jasmit for their perseverance and help.

March, 2000                                                                                        H.S. Chawla

# Abbreviations

| | |
|---|---|
| 2,4-D | 2,4-Dichlorophenoxyacetic acid |
| IAA | Indole-3-acetic acid |
| IBA | Indole-3-butyric acid |
| NAA | Naphthaleneacetic acid |
| pCPA | *p*-Chlorophenoxyacetic acid |
| BAP or BA | Benzylamino purine or benzyladenine |
| 2iP | 2-Isopentenyladenine |
| Kin | Kinetin |
| Zea | Zeatin |
| ABA | Abscisic acid |
| A.t. | *Agrobacterium tumefaciens* |
| $B_5$ | Gamborg's medium |
| DMSO | Dimethylsulfoxide |
| GA | Gibberellic acid |
| kb(p) | Kilobase (pairs) |
| LS | Linsmaier and Skoog medium |
| Mb | Megabase |
| Mdal | Magadalton |
| MS | Murashige and Skoog medium |
| ng | Nanogram |
| PAGE | Polyacrylamide gel electrophoresis |
| pg | Picogram |
| Ri | Root inducing plasmid of *Agrobacterium rhizogenes* |
| SH | Schenk and Hildebrandt medium |
| Ti | Tumor inducing plasmid of *Agrobacterium tumefaciens* |

# Contents

# PART I

# *Plant Tissue Culture*

# 1

# Introduction

Plants are the key to life on earth as they directly supply 90% of human calorie intake, and 80% of the protein intake, the remainder being derived from animal products, although these animals have also derived their nutrition from plants. Of the three thousand plant species which have been used as food by man, the world now depends mainly on around twenty crop species for the majority of its calories, with 50% being contributed by eight species of cereals. Minerals and vitamins are supplied by a further thirty species of fruits and vegetables. Most important of the staple foods are the cereals, particularly wheat and rice, with more than one-third of all cultivated land used to produce these two crops.

As the population continues to expand, concern has been expressed over the finite number of people world agriculture can support. An estimate of how many people could be supported by world food production can be obtained from consideration of the average daily calorie requirements and the net amount of grain yield from well managed areas. It has been calculated that the earth can support about 15 billion people on a strictly vegetarian diet, or five billion on a mixed diet.

The farming practices and crops cultivated today have developed over a relatively short time span. Crop plants of today have changed in a number of ways so that they now bear very little resemblance to their wild type ancestors. These changes have come about through selection, either conscious or unconscious, for traits which are advantageous to the people growing the crops. Thus modern wheat does not disperse their seeds or legumes do not have pods which burst open. Today, varieties are the result of generation of plants cultivated under ideal conditions from the man´s point of view.

From the beginning of crop cultivation to the late nineteenth century, all improvements in the species used were brought about by those who were directly involved, i.e. farmers themselves. In the following 100 years, the laws of genetic inheritance and rules governing species variation were laid down by Mendel, Darwin and others which redefined the breeding techniques by making them predictable and therefore quicker, more precise and more productive. Furthermore, despite the implementation of breeding techniques, the time taken to produce and test varieties is an important limiting consideration.

## NEW TECHNOLOGIES

Over the past few years a number of methodologies have come that would seem to have much more to offer in terms of advancing current research in the plant sciences, and exploiting the knowledge gained to develop new crops. The first of these areas is concerned with manipulation and subsequent growth of cells, tissues, organs and naked plant cells (protoplasts) in tissue culture. The second field is genetic engineering or recombinant DNA (rDNA) technology which has grown out very fast from the work initially carried out on microorganisms. With genetic engineering, scientists have more exact methods for breeding better livestock and crop varieties. This technology allows for the detailed manipulation of genes. These two areas of research have in recent years become associated with the general field of biotechnology and are potentially applicable to a wide variety of plant species, as well as offer a precision in manipulating genetic material. Recombinant DNA techniques in particular have already contributed much to the elucidation of basic mechanisms in plants at the molecular level.

Our knowledge of the structure and expression of the plant genome has come largely through the use of recombinant DNA or gene cloning techniques. This technology allows the isolation and characterization of specific pieces of DNA and by cloning the DNA sequences into bacterial cells, they can be replicated to yield large quantities for analysis. In addition to supplying much basic information concerning gene structure and expression, recombinant-DNA technology provides the opportunity for specifically manipulating genetic material, and moving such material among different organisms.

## ORIGIN OF BIOTECHNOLOGY

The term Biotechnology was coined by Karl Ereky, a Hungarian engineer in 1919. The origin of biotechnology can be traced back to prehistoric times when microorganisms were already used for processes like fermentation, formation of yoghurt and cheese from milk, vinegar from molasses, production of butanol and acetone from starch by *Clostridium acetobutylicum* or the production of antibiotics like penicillin from *Penicillium notatum*. However, biotechnology got a boost in the 1970's with the discovery of restriction enzymes which led to the development of a variety of gene technologies and is thus considered to be the greatest scientific revolution of this century. Biotechnology thus consists of a variety of techniques, designed to genetically improve and/or exploit living systems or their components for the benefit of man. Infact, biotechnology is the product of interaction between sciences of biology and technology. It is the technological exploitation and control of biological systems. Attempts have been made to define biotechnology and it has been interpreted in different ways by different groups of workers. However, the following definitions seem to be most appropriate.

i.  The application of science and engineering in the direct or indirect use of living organisms, or parts or products of living organisms, in their natural or modified form.

ii. The integrated use of biochemistry, microbiology and engineering sciences in order to achieve technological (industrial) application of the capabilities of the microorganisms, cultured tissue cells and parts thereof (European Federation of Biotechnology).

iii. The controlled use of biological agents such as microorganisms or cellular components for beneficial use (US National Science Foundation).

The term genetic engineering refers to a number of new techniques involving the transfer of specific genetic information from one organism to another. These techniques do not rely on sexual methods, but instead involve genetic manipulation at the cellular and molecular levels. These are the nonsexual methods for gene transfer.

An important aspect of all plant biotechnology processes is the culture of either the microorganisms or plant cells or tissues and organs in artificial media. One of the problems in conventional plant breeding is that the range of organisms among which genes can be transferred is severely limited by species barriers. The new technologies provide a better approach for defining and manipulation of targets, and species-specific barriers are broken. They do not replace plant breeding but provide methods capable of achieving objectives not possible by other means.

# HISTORY

The last three decades have seen a very rapid rise in the number of plant scientists using the techniques of organ, tissue and cell cultures for plant improvement. The term "plant tissue culture" broadly refers to the *in vitro* cultivation of plants, seeds, plant parts (tissues, organs, embryos, single cells, protoplasts etc.) on nutrient media under aseptic conditions.

During the 1800s, the cell theory, which states that the cell is the basic structural unit of all living creatures, was very quick to gain acceptance. However, the second portion of the cell theory states that these structural units are distinct and potentially totipotent physiological and developmental units, failed to gain universal acceptance. The skepticism associated with the latter part was because of the inability of scientists such as Schleiden and Schwann to demonstrate totipotency in their laboratories. It was in 1902 that the well-known German plant physiologist, Gottlieb Haberlandt (1854–1945), attempted to cultivate plant tissue culture cells *in vitro*. He is regarded as the father of plant tissue culture. He clearly stated the desirability of culturing the isolated vegetative cells of higher plants. He stated: "To my knowledge, no systematically organized attempts to culture isolated vegetative cells from higher plants in simple nutrient solutions have been made. Yet the results of such culture experiments should give some interesting insights into the properties and potentialities which the cell as an elementary organism possesses. Moreover, it would provide information about the interrelationships and complementary influences to which cells within the multicellular whole organism are exposed". Haberlandt started his experiments in 1898 using single cells isolated from the palisade tissue of *Lamium purpureum*, pith cells from petioles of *Eicchornia crassipes,* grandular hair of *Pulminaria* and *Utrica*, stamen hair cells of *Tradescantia,* Stomatal guard cells of *Ornithogalum*, and the other plant materials. He grew them on Knops salt solution with glucose, sucrose and peptone. Gottlieb Haberlandt (1902) developed the concept of *in vitro* cell culture. He was the first to culture isolated, fully differentiated cells in a nutrient medium containing glucose, peptone, and Knop's salt solution. Haberlandt realized that asepsis was necessary when culture media are enriched with organic substances metabolised by microorganisms. In his cultures, free from microcontamination, cells were able to synthesize starch as well as increase in size and survived for several weeks. However, Haberlandt failed in his goal to induce these cells to divide. Despite drawbacks, he made several predictions about the requirements for cell division under experimental conditions in 1902, which have been confirmed with the passage of time. Haberlandt is thus regarded as the father of tissue culture.

Hanning (1904) initiated a new line of investigation involving the culture of embryogenic tissue. He excised nearly mature embryos of some crucifers (*Raphanus sativus, R. landra, R. caudatus,* and *Cochlearia donica*) and successfully grew them to maturity on mineral salts and sugar solution. Winkler (1908) cultivated segments of string bean and observed some cell divisions but no proliferation.

In the early 1920s workers again attempted to grow plant tissues and organs *in vitro*. Molliard in 1921 demonstrated limited success with the cultivation of plant embryos and subsequently Kotte (1922a, 1922b), a student of Haberlandt in Germany and, independently Robbins (1922a,b) were successful in the establishment of excised plant root tips *in vitro*. However, in 1934, the pioneering work of growing excised roots of tomato *in vitro* for periods of time without theoretical limits was demonstrated by White (1934). However, his investigation led to the fundamental discovery by W.J. Robbins and S.M. Bartley in 1937 and 1939 that vitamins $B_1$ (thiamine) and $B_6$ (pyridoxine) were the root growth factors.

Initially, White used a medium containing inorganic salts, yeast extract and sucrose, but later yeast extract was replaced by three B-vitamins viz. pyridoxine, thiamine and nicotinic acid. Snow (1935) demonstrated that indole acetic acid (IAA—a growth substance discovered by Went in 1926) stimulated cambial activity. Following these observations in the same year of 1934, Gautheret who was engaged in experimentation with excised root tips and the cultivation of cambial tissue removed under aseptic conditions from *Salix capraea*, *Populus nigra* and other trees on Knop's solution containing glucose and cysteine hydrochloride and recorded that they proliferated for a few months. White (1939) reported similar results in the cultures from tumor tissues of the hybrid *Nicotiana glauca* x *N. langsdorffii*. These studies continued, and in 1939, Gautheret, Nobecourt and White, published independently studies on the successful cultivation for prolonged periods of cambial tissues of carrot root and *Acer pseudoplatanus* (Gautheret, 1939), tobacco (White, 1939) and carrot (Nobecourt, 1939). These were the first true plant tissue cultures in the strict sense of prolonged cultures of unorganized materials. The methods and media now used are in principle modifications of those established by these three pioneers. All the three workers used meristematic cells to generate continuously growing cultures. Philip R. White (1943, 1954, 1963) and Roger J. Gautheret (1959) can be credited with providing a significant impetus to the field with the publication of their authoritative handbooks.

From 1939 to 1950, experimental work with root cultures drew attention to the role of vitamins and growth regulators in plant growth and advanced the knowledge of the shoot-root relationship. Skoog (1944), Skoog and Tsui (1951) demonstrated that adenine stimulates cell division and induces bud formation in tobacco

tissue even in the presence of IAA (which normally acts as a bud inhibitor). This convinced Skoog and collaborators those nucleic acids which contain substances such as adenine influence tissue proliferation.

In 1955, Skoog and collaborators (Miller *et al.,* 1955) finally isolated from autoclaved yeast extract, a derivative of adenine (6-furfyl aminopurine), named *kinetin*. A substance with kinetin-like properties was also detected in young maize endosperm (Miller, 1961), which was isolated by Letham (1963) and named *zeatin*. It was also verified that a similar substance called *ribosylzeatin* occurred in coconut milk (Letham, 1974). Now many synthetic as well as natural compounds with kinetin-like activity are known which show bud-promoting properties. These substances are collectively called *cytokinins* and are used to induce divisions in cells of highly mature and differentiated tissues. The work of Miller and Skoog (1953) on bud formation from cultured pith explants of tobacco led to the discovery of kinetin. In 1952, Steward initiated work on cultured carrot explants and used coconut milk as a nutrient (Steward *et al.*, 1952) that ultimately led to the discovery of embryogenesis (Steward *et al.*, 1958).

An important breakthrough was achieved in 1965 when Vasil and Hildebrandt observed that colonies arising from cloning of isolated cells of the hybrid *Nicotiana glutinosa* x *N. tabacum* regenerated plantlets.

Sandford *et al.* (1948) initiated studies on single cell cultures by demonstrating division in animal cells using conditioned media (media in which tissue has been growing for sometime). In 1953, Muir reported that if fragments of callus of *Tagetes erecta* and *Nicotiana tabacum* are transferred to liquid culture medium and the medium is agitated on a reciprocal shaker, then the callus fragments break up to give a suspension of single cells and cell aggregates (Muir, 1953). The following year Muir and associates (1954) applied the 'conditioning principle' to induce division in isolated single cells of *Tagetes* and tobacco growing in shake (suspension) cultures. A single isolated cell was placed on top of a filter paper positioned on top of a callus mass. About 8% of these isolated single cells on the filter paper multiplied and formed colonies. An important technique of cloning large number of single cells of higher plants was developed by Bergmann (1960). Jones *et al.* (1960) designed a microculture method using hanging drops of free cells in conditioned medium. This facilitated continuous observation of cells growing in cultures.

J.B. Routien and L.G. Nickel of Pfizer & Co. Inc., New York, got the US Patent in 1952 on their claim that plant cells could be grown in liquid nutrient medium and had the potential to produce useful compounds. The gestation period was 30 year when the first commercial production of a natural product (Shikonin) by plant cell suspension culture was achieved. In a recent breakthrough, commercial production of paclitaxel from *Taxus* cell cultures in bioreactors has been made possible by Phyton Gesellschaft fur Biotechnik mbH, Germany (Phyton, 2002), which will go a long way in treating cancer.

The differentiation of whole plants in tissue cultures may occur via shoot and root differentiation, or alternatively the cells may undergo embryogenic development to give rise to somatic embryos. Differentiation of plants from callus cultures has often been suggested as a potential method for rapid propagation. Ball in 1946 successfully raised transplantable whole plants of *Lupinus* and *Tropaeolum* by culturing their shoot tips with a couple of leaf primordia. The practical usefulness of shoot meristem technique is credited to Morel and Martin (1952), who for the first time recovered virus-free Dahlia plants. However, the glaring early example of saving valuable plant varieties from extinction due to virus infection through shoot meristem culture is of *Solanum tuberosum* varieties King Edward and Arran Victory by B. Kassanis in 1957. However, the foundation of commercial plant tissue culture was laid by G.M. Morel (1960) for a million-fold increase in clonal multiplication of an orchid, *Cymbidium*.

The release of protoplasts from root tip cells using a fungal cellulase in 0.6M sucrose was reported by Cocking in 1960. Protoplasts released by cell wall degrading enzymes have now been prepared from many plant tissues. The most universally used high-salt medium was developed by Skoog and his students (Murashige and Skoog, 1962). In addition to mineral salts, media contain an energy source, vitamins and growth regulators.

Another landmark in the development of Plant tissue culture has been the discovery of haploids, when Guha and Maheshwari (1964) obtained haploid embryos from *Datura innoxia*. In an earlier effort to produce haploids, however, W. Tulecke (1953) could obtain colonies of cells originating from cultured pollen grains of *Ginkgo biloba* and subsequently androgenic haploid plant of *Nicotiana* were actually produced by J.P. Bourgin and J.P. Nitsch in 1967. Although haploids have been produced in a number of Plant species, the goal of creating new improved varieties of *Oryza sativa*, *Triticum aestivum* and *N. tabacum* was realized in China, as described by H. Hu in 1978.

In 1970s, restriction enzymes were discovered and technique was developed by Lobban and Kaiser for joining the two restriction fragments by ligase. Paul Berg working at Stanford University, was the first to make a recombinant DNA molecule combining DNA from SV40 virus with that of lambda virus in 1972. However, realizing the dangers of his experiments, he terminated it immediately. He proposed, in what is known as "Berg Letter", one year moratorium on such research in order that safety concerns could be worked out. In 1973 Stanley Cohen of Stanford University and Herbert Boyer of University of California produced world's first recombinant DNA organism. This set the foundation of modern biotechnology and genetic engineering. Genetic engineering became a reality when a man made insulin gene was used to manufacture a human protein in bacteria. The 1980s saw the development of various genetic transformation techniques which revolutionized the rDNA technology and led to the development of transgenic plants for various crops.

It was in mid 1970s that techniques to sequence DNA were developed. Sanger *et al.* (1977) and Maxam and Gilbert (1977) were the first to report gene sequencing techniques. Sanger *et al.* (1977) sequenced virus φx 174 comprising 5375 nucleotides that code for 10 proteins. This was followed by report on complete DNA sequence of simian virus 40 (SV40).

Molecular breeding work started in 1980s with the development of restriction fragment length polymorphism technique by Botstein and coworkers in 1980. The technique of DNA fingerprinting based upon the sequence of bases in segments of DNA was introduced in 1986 by Jeffreys, to identify individuals and it was first used as an evidence in the court room in USA in 1987. The discovery of polymerase chain reaction (Mullis *et al.*, 1986) was another landmark and proved to be a unique process that brought about a new class of DNA markers, the random amplified polymorphic DNA (RAPD) marker (Williams *et al.*, 1990; Welsh and McClelland, 1990), sequence characterized amplified region (Patran and Michelmore, 1993), Single nucleotide polymorphism (Jordan and Humphries, 1994), retrotransposon based insertional polymorphism (Flavell *et al.*, 1998). A large number of molecular markers have been reported that led to the development of methodologies for mapping of genes on the chromosomes and in molecular marker assisted selection.

Sequencing of DNA began to be performed in laboratories in 1978. The available technology was limited to a manual approach based on utilization of radioactive isotopes, only small regions were sequenced, about the size of a gene of several thousand bases. In 1986, Japanese scientists sequenced the whole series of chloroplast genomes (Shinozaki *et al.*, 1986).

But for successful sequencing projects the scientific community had to wait for the arrival of the first automatic sequencers. It was in 1995 that Fleischmann *et al.*, from TIGR (The Institute for Genomic Research, Washington) published the sequence of the first bacterial genome, *Haemophilus influenzae* (1.8 Mb). The genome of yeast (16 Mb) was sequenced by a public international consortium (Goffeau, 1997) and the genome of a multicellular eukaryote, *Caenorhabditis elegans* was published in 1998 (The *C. elegans* Sequencing Consortium, Sanger Centre, U.K. and Washington University, USA). In recent years entire genomes of a number of prokaryotes and Arabidopsis, rice, wheat, maize, *Medicago* and Lotus genomes have either been sequenced or draft sequences published.

Some of the contributions have been mentioned below.

1902 First attempt of plant tissue culture (Haberlandt).

1904 Embryo culture of selected crucifers attempted (Hannig).

1922 Asymbiotic germination of orchid seeds *in vitro* (Knudson).

1922 *In vitro* culture of root tips (Robbins).

1925 Use of embryo culture technique in interspecific crosses of *Linum* (Laibach).

1934 Successful culture of tomato roots (White).

1939 Successful establishment of continuously growing callus cultures (Gautheret, Nobecourt and White).

1941 Use of coconut milk containing a cell division factor for the first time in *Datura* (van Overbeek).

1944 *In vitro* adventitious shoot formation in tobacco (Skoog).

1946 Raising of whole plants of *Lupinus* and *Tropaeolum* by shoot tip culture (Ball).

1952 Use of meristem culture to obtain virus free dahlias (Morel and Martin).

1952 First application of micrografting (Morel and Martin).

1952 Got the U S Patent on their claim that plant cells could be grown in liquid nutrient medium and had the potential to produce useful compounds (Routien and Nickel of Pfizer & Co. Inc., New York).

1953 Production of haploid callus of the gymnosperm *Ginkgo biloba* from pollen (Tulecke).

1954 First plant from a single cell (Muir *et al.*).

1955 Discovery of kinetin a cell division hormone (Miller *et al.*).

1957 Discovery of the regulation of organ formation by changing the ratio of auxin: cytokinin (Skoog and Miller).

1958 Regeneration of somatic embryos *in vitro* from the nucellus of *Citrus* ovules (Maheshwari and Rangaswamy).

1959 Regeneration of embryos from callus clumps and cell suspensions of *Daucus carota* (Reinert, Steward).

1959 Publication of first handbook on plant tissue culture (Gautheret).

1960 First successful test tube fertilization in *Papaver rhoeas* (Kanta).

1960 Use of the microculture method for growing single cells in hanging drops in a conditioned medium (Jones *et al.*).

1960 Enzymatic degradation of cell walls to obtain large number of protoplasts (Cocking).

1960 Filtration of cell suspensions and isolation of single cells by plating (Bergmann).

1962 Development of Murashige and Skoog nutrition medium (Murashige and Skoog).

1964 Production of first haploid embryos from pollen grains of *Datura innoxia* (Guha and Maheshwari).

1967 Androgenic haploid plants of *Nicotiana* were produced (Bourgin and Nitsch).

1968 Restriction endonuclease term coined to a class of enzymes involved in cleaving DNA (Meselson and Yuan).

1970 Selection of biochemical mutants *in vitro* by the use of tissue culture derived variation (Carlson).

1970 First achievement of protoplast fusion (Power *et al.*).

1970 Discovery of first restriction endonuclease from *Haemophillus influenzae* Rd. It was later purified and named *Hind*II (Kelly and Smith).

1970 In certain cancer causing animal virus genetic information flow in reverse form. Discovery of reverse transcriptase (Temin)

1971 First restriction map using *Hind*II enzyme to cut circular DNA of SV 40 into 11 specific fragments was prepared (Nathans).

1971 Regeneration of first plants from protoplasts (Takebe *et al.*).

1972 First report of interspecific hybridization through protoplast fusion in two species of *Nicotiana* (Carlson *et al.*).

1972 First recombinant DNA molecule produced using restriction enzymes (Jackson *et al.*).

1972 Joining of two restriction fragments regardless of their origin produced by the same restriction enzyme by the action of DNA ligase (Mertz and Davis).

1973 Use of Lobban and Kaiser technique to develop hybrid plasmid—insertion of *Eco*RI fragment of DNA molecule into circular plasmid DNA of bacteria using DNA ligase. Gene from African clawed toad inserted into plasmid DNA of bacteria (Cohen *et al.*).

1974 Discovery that the Ti plasmid is the tumor inducing principle of *Agrobacterium* (Zaenen *et al.*, Larebeke *et al.*).

1975 Positive *in vitro* selection of maize callus cultures resistant to T toxin of *Helminthosporium maydis* (Gengenbach and Green).

1975 High resolution 2 Dimensional gel electrophoresis procedure developed that led to the development of proteomics (O'Farrel).

1976 Shoot initiation from cryopreserved shoot apices of carnation (Seibert).

1977 Successful integration of the Ti plasmid DNA from *A. tumefaciens* in plants (Chilton *et al.*).

1977 A method of gene sequencing based on degradation of DNA chain (Maxam and Gilbert).

1977 Discovery of split genes (Sharp and Roberts).

1978 Somatic hybridization of tomato and potato resulting in pomato (Melchers *et al.*).

1979 Co-cultivation procedure developed for transformation of plant protoplasts with *Agrobacterium* (Marton *et al.*).

1980 Commercial production of human insulin through genetic engineering in bacterial cells (Eli Lilly and Co.).

1980 Restriction fragment length polymorphism (RFLP) technique developed.

1980 Studies on the structure of T-DNA by cloning the complete *EcoR*I digest of T37 tobacco crown gall DNA into a phage vector, thus allowing the isolation and detailed study of T-DNA border sequences (Zambryski *et al.*).

1981 Introduction of the term somaclonal variation (Larkin and Scowcroft).

1982 Incorporation of naked DNA by protoplasts resulting in the transformation with isolated DNA (Krens *et al.*).

1984 Transformation of tobacco with *Agrobacterium*; transgenic plants developed (De Block *et al.*, Horsch *et al.*).

1986 Genetic fingerprinting technique developed for identifying individuals by analyzing polymorphism at DNA sequence level (Jeffreys).

1986 TMV virus-resistant tobacco and tomato transgenic plants developed using cDNA of coat protein gene of TMV (Powell-Abel *et al.*).

1986 Nucleotide sequencing of tobacco chloroplast genome (Shinozaki *et al.*).

1986 The discovery of polymerase chain reaction (Mullis *et al.*).

1987 Development of biolistic gene transfer method for plant transformation (Sanford *et al.*, Klein *et al.*).

1987 *Bt* gene from bacterium (*Bacillus thuringiensis*) isolated (Barton *et al.*).

1990 The Human Genome Project is formally launched.

1990 Random amplified polymorphic DNA (RAPD) technique developed (Williams *et al.*, Welsh and McClelland).

1991 DNA microarray system using light directed chemical synthesis system developed (Fodor).

1995 The Institute for genomic research reported the complete DNA sequence of *Haemophilus influenzae* (Fleischmann *et al.*).

1995 DNA finger printing by amplified fragment length polymorphism (AFLP) technique developed (Vos *et al.*).

1997 Sequencing of *E. coli* genome (Blattner *et al.*).

1997 The genome of yeast (16 Mb) was sequenced by a public international consortium (Goffeau).

1998 The genome of a multicellular organism (*Caenorhabditis elegans*) sequenced (*C. elegans* sequencing consortium).

2001 Sequencing of human genome successfully completed (Human Genome project Consortium and Venter *et al.*).

2002 Draft sequence of rice genome (*Oryza sativa* L. *ssp. japonica*) (Goff *et al.*, 2002).

2003 Draft sequence of rice genome (*Oryza sativa* L. *ssp. indica*) (Yu *et al.*, 2002).

2005 Map based sequence of the rice genome (International Rice Genome Sequencing Project).

# REFERENCES

Ball, E. 1946. Development in sterile culture of stem tips and adjacent regions of *Tropaeolum majus* L. and of *Lupinus albus* L. *Am. J. Bot.,* **33:** 301–318.

Barton *et al.* 1987. *Bacillus thurigiensis* δ-endotoxin expressed in transgenic *Nicotiana tabacum* providse resistance to lepidoptera insects. *Plant Physiology* **85:** 1103–1109.

Bergmann, L. 1960. Growth and division of single cells of higher plants *in vitro. J. Gen. Physiol.* **43:** 841–851.

Bergner, A.D. (1921). Cited by Han, H. 1987. Application of pollen derived plants in crop improvement: Natesh, S. *et al.* (eds). Biotechnology in Agriculture, Oxford and IBH Publishing Co. Pvt. Ltd, New Delhi, pp. 155–168.

Blattner, F.R., Plunkett, G., Bloch, C.A., Perna, N.T., Burland, V., Riley, M., Collado-Vides, J., Glasner, J.D., Rode, C.K. and Mayrew, G.F. 1997. The complete genome sequence of *Escherichia coli* K-12. *Science* **227:** 1453–1474.

Bourgin, J.P. and Nitsch, J.P. 1967. Obtentionde Nicotiana haploides a partir d´etamines cultivers *in vitro. Ann. Physiol. Veg. (Paris)* **9:** 377–382.

Carlson, P.S. 1970. Induction and isolation of auxotrophic mutants in somatic cell culture of *Nocatiana tabacum. Science, N.Y.* **168:** 487–489.

Carlson, P.S., Smith, H.H. and Dearing, R.D. 1972. Parasexual interspecific plant hybridization. Proc Natl. Acad. Sci. *U.S.A.* **69:** 2292–2294.

Chilton, M.D., Drummond, M.H., Merlo, D.J., Sciaky, D., Montoya, A.L., Gordon, M.P. and Nester, E.W. 1977. Stable incorporation of plasmid DNA into higher plant cells: the molecular basis of crown gall tumorigenesis. *Cell* **11:** 263–271.

Cocking, E.C. 1960. A method for the isolation of plant protoplasts and vacuoles. *Nature.* **187:** 962–963.

Cohen *et al.* 1973. Construction of biologically functional Bacterial Plasmids *in vitro. PNAS* **70.**

De Block, M., Herrera-Estrella, L., van Montagu, M., Schell, J. and Zambryski, P. 1984. Expression of foreign genes in regenerated plants and their progeny. *EMBO. J.* **3:** 1681–1689.

Flavell, A.J., Knox M.R., Pearce S.R. and Ellis T.H.N. 1998. Retrotransposon-based insertion polymorphisms (R.B.I.P.) for high throughput marker analysis. *Plant J.* **16:** 643–650.

Fleischmann, R.D., Adams, M.D., White, O., Clayton, R.A., Kirkness. E.F., Kerlavage A.R., Bult, C.J., Tomb, J.K, Dougherty, B.A., Merrick, J.M., McKenney, K., Sutton, G., Fitzhugh, W., Flelds, C., Gocayne, J.D., Scoot, J., Shirley, R., Liu. L., Glodek, A., Kelley, J.M., Weidman, J.K., Phillips, C.A., Spriggs, T., Hedblom, E., Cotton, M.D., Utterback, T.R., Hanna, J.L., Geoghagen, N.S.M., Gnehm, C.L., McDonald, L.A., Small, K.V., Fraser, C.M., Smith, H.O. and Venter, J.C. 1995. Whole-genome random sequencing and assembly of *Haemophilus influenzae* Rd. *Science* **269:** 131–134.

Fodor, S.P.A. 1997. Massively parallel genomics. *Science* **270:** 393–394.

Gengenbach, B.G. and Green, C.E. 1975. Selection of T-cytoplasm maize callus resistant to Helminthosporium maydis race T pathotoxin. *Crop Sci.* **15:** 645–649.

Gautheret, R.J. 1939. Culture du tissu cambial. C.R. Acad. Sci. (Paris) **198:** 2195–2196.

Gautheret, R.J. 1939. Sur la possibilitie de realiser la culture indefinie des tissue de tubercules de carotte. *C.R. Acad. Sci.* Paris **208:** 118–120.

Gautheret, R.J. 1959. *La culture des tissus vegetaux: Techniques et realisations.* Masson, Paris.

Goff, S. *et al.* 2002. A draft sequence of the Rice genome (*Oryza sativa* L ssp. *Japonica*). *Science* **296:** 92–100.

Goffeau, A. *et al.* 1997. The Yeast genome directory. *Nature* **387** (suppl.) 1–105.

Guha, S. and Maheshwari, S.C. 1964. *In vitro* production of embryos from anthers of *Datura. Nature* **204:**497.

Haberlandt, G. 1902. Culturversuchemit isolierten Pflanzenzellen. *Sitzungsber. Math. Naturwiss. Kl. Kais. Akad. Wiss. Wien* **111:** 69–92.

Hannig, E. 1904. Zur Physiologie pflanzlicher embryonen. I. Uber die cultur von Cruciferen-Embryonen ausserhalb des Embryosacks. *Bot. Ztg.* **62:** 45–80.

Horsch, R.B., Fraley, R.T., Rogers, S.G., Sanders, P.R., Lloyd, A. and Hoffman, N. 1984. Inheritance of functional foreign gene in plants. *Science* **223:** 496–498.

International Rice Genome Sequencing Project. 2005. The map-based sequence of the rice genome. *Nature* **436:** 793–800.

Jackson *et al.* 1972. A Biochemical Method for Inserting New Genetic Information into SV40 DNA: Circular SV40 DNA molecules containing Lambda phage genes and the Galactose operon of *E. Coli. Proc. Nat. Acad. Sci.* USA **69:** 2904.

Jones, L.E., Hildebrandt, A.C., Riker, A.J. and Wu, J.H. 1960. Growth of somatic tobacco cells in microculture. *Am. J. Bot.* **47:** 468–475.

Jordan, S.A. and Humphries, P. 1994. Single nucleotide polymorphism in Exon 2 of the B.C.P. Gene on 7931–935, Hum Mol Genet **3:** 1915.

Kanta, K. 1960. Intra-ovarian pollination in *Papaver rhoeas* L. *Nature* **188:** 683–684.

Klein, T.M., Wolf, E.D., Wu, R. and Sanford, J.C. 1987. High velocity microprojectiles for delivering nucleic acids into living cells. *Nature* **327:** 70–73.

Knudson, L. 1922. A symbiotic germination of orchid seeds. *Bot. Gaz.* **73:** 1–25.

Kotte, W. 1922a. Wurzelmeristem in Gewebekultur. *Ber. Deut. Ges.* **40:** 269–272.

Kotte, W. 1922b. Kulturversuche mit isolierten Wurzelspitzen. Beitr. Allg. Bot. **2:** 413–434.

Krens, F.A., Molendijk, L., Wullems, G.J., and Schilperoort, R.A. 1982. *In Vitro* transformation of plant protoplasts with Ti- plasmid D.N.A. *Nature (London)* **296:** 72–74.

Laibach, F. 1925. Das Taubverden von Bastardsmen und die kunsliche Aufzucht fruh absterbender Bastardembryonen. *Z. Bot.* **17:** 417–459.

Larkin, P.J. and Scowcroft, W.R. 1981. Somaclonal variation—a novel source of variability from cell cultures for plant improvement. *Theor. Appl. Genet.* **60:** 197–214.

Letham, D.S. 1963. Zeatin, a factor including cell division isolated from *Zea mays. Life Sci.* **2:** 569–579.

Letham, D.S. 1974. Regulators of cell division in plant tissues. The cytokinins of coconut milk. *Physiol. Plant.* **32:** 66–70.

Maheshwari, N. 1958. *In Vitro* culture of excised ovules of *Papaver somniferum. Science N.Y.* **127:** 342.

Marton *et al.* 1979. *In vitro* transformation of cultured cells from *Nicotiana tabacum* by *Agrobacterium tumefaciens. Nature* **277:** 129–131.

Maxam, A.M. and Gilbert W. 1977. A new method for sequencing D.N.A. *Proc. Natl. Acad. Sci. U.S.A.* **74:** 560–564.

Melchers, G., Sacristan, M.D. and Holder, A.A. 1978. Somatic hybrid plants of potato and tomato regenerated from fused protoplasts. *Carlsberg Res. Commun.* **43:** 203–218.

Meselson, M. and Yuan, R. 1968. D.N.A. restriction enzyme from *E. coli. Nature* **217:** 1110–14.

Miller, C. and Skoog, F. 1953. Chemical Control of Bud Formation in Tobacco Stem Segments. *American Journal of Botany* **40 (10):** 768–773.

Miller, C.O. 1961. Kinetin related compounds in plant growth. *Ann. Rev. Plant Physiol.* **12:** 395–408.

Miller, C.O., Skoog, F., Okumura, F.S., Von Saltza., M.H. and Strong, F.M.1955. Structure and synthesis of kinetin. *J. Am. Chem. Soc.* **77:** 2662–2663.

Morel, G. 1960. Producing virus free cymbidium. *Am. Orchid Soc. Bull.* **29:** 495–497.

Morel, G. and Martin, C. 1952. Guerison de pommes de terre atteintics de maladie a virus. *C. R. Acad. Agric. Fr.* **41:** 471–474.

Muir, W.H. 1953. Cultural conditions favouring the isolation and growth of single cells from higher plants *In Vitro.* Ph.D. Thesis, Univ. of Wisconsin, U.S.A.

Muir, W.H., Hildebrandt, A.C.and Riker, A.J. 1954, Plant tissue cultures produced from single isolated plant cells. *Science* **119:** 877–878.

Mullis, K., Faloona, F., Scharf, S., Saiki, R., Horn, G. and Erlich, H., 1986. Specific enzymatic amplification of D.N.A. *in vitro*: the polymerase chain reaction. *Cold Spring Harbor Symposia on Quantitative Biology* **51:** 263–273.

Murashige, T. and Skoog, F. 1962. A revised medium for rapid growth and bioassays with tobacco tissue cultures. *Physiol. Plant.* **15:** 473–497.

Nathans, D. and Danna, K. 1971. Specific clevage of simian virus 40 DNA by restriction endonuclease of *Haemophilus influenzae. Proc. Nat. Acad. Sci.* USA. **68(12):** 2913–2917.

Nobecourt, P. 1939. Cultures en serie de tissus vegetaux sur milieu artificiel. *C. R. Seanc. Soc. Biol.* **205:** 521–523.

Nobecourt, P. 1939. Sur la perennite *et* l'augmentation des cultures de tissus vegetaux.

O'Farrell, P.H. 1975. High resolution two-dimensional electrophoresis of proteins. *J. Biol. Chem.* **250:** 4007–4021.

Paran, I. and Michelmore, R.W. 1993. Development of reliable PCR based markers linked to downy mildew resistance gene in lettuce. *Theor. Appl. Genet.* **85:** 985–993.

Powell-Abel, P.A., Nelson, R.S., De, B., Hoffman, N., Rogers, S.G., Fraley, R.T. and Beachy, R.N. 1986. Delay of disease development in transgenic plants that express tobacco mosaic virus coat protein gene. *Science* **232:** 738–743.

Power, J.B., Cummins, S.E. and Cocking, E.C. 1970. Fusion of isolated plant protoplasts. *Nature* **225:** 1016–1018.

Rangaswami, N.S. 1958. Culture of nucellus tissue of citrus *in vitro. Experienta* **14:** 11–12.

Reinert, J. 1959. Uber die kontrolle der morphogenese und die induktion von advientive embryonen an gewebekulturen aus Karotten. *Planta* **58:** 318–333.

Robbins, W.J. 1922a. Cultivation of excised root tips under stzerile conditions. *Bot. Gaz.* **73:** 376–390.

Robbins, W.J. 1922b. Effect of autolyzed yeast and peptone on growth of excised corn root tips in the dark. *Bot. Gaz.* **74:** 57–79.

Sandford, K.K., Earle, W.R. and Likely, G.D. 1948. The growth *in vitro* of single isolated tissue cells. *JNCL, Natl. Cancer Inst.* **9:** 229–246.

Sanfod, J.C., Klein, T.M., Wolf, E.D. and Allen, N.J. 1987. Delivery of substances into cells and tissues using a particle bombardment process. *J. Part Sci. Techn.* **6:** 559–563.

Sanger, F., Nicklen, S., Coulson, A.R., 1977 D.N.A. sequencing with chain- terminating inhibitors. *Proc Natl Aad Sci USA* **74 (12):** 5463–5467.

Seibert, M., 1976. Shoot initiation from carnation shoot apices frozen to −196°C. *Science (N.Y.)* **191:** 1178–1179.

Shinozaki, K., Ohme, M., Tanaka, M., Wakasugi, T., Hayashida, N., Matsubayashi, T., Zaita, N., Chunwongse, J., Obokata, J., Yamaguchi-Shinozaki, K., Ohto, C., Torazawa, K., Meng, B.Y., suugita, M., Deno, H. and Sugiura, M. 1986. The complete nucleotide sequence of the tobacco chloroplast genome: its gene organization and expression, *EMBO J.* **5:** 2043–2049.

Skoog, F. 1944. Growth and organ formation in tobacco tissue cultures. *Am. J. Bot.* **31:** 19–24.

Skoog, F. and Miller, C.O. 1957. Chemical regulation of growth and organ formation in plant tissues cultivated *in vitro*. In: *Biological Action of Growth Substances. Symp. Soc. Exp. Biol.* **11:** 118–131.

Skoog, F. and Tsui, C. 1951. Growth substances in the formation of buds in plant tissues. In: Plant Growth Substances (Skoog, F., ed). University of Wisconsin Press, Madison, WI, pp. 263–285.

Smith, H.O. and Kelly, T.J. 1970. A restriction enzyme from *Haemophilus influenzae II*. *Journal of Molecular Biology.* **51 (2):** 379–391.

Snow, R. 1935. Activation of cambial growth by pure hormones. New Phytol. **34:** 347–360.

Steward, F.C., Caplin, S.M. and Miller, F.K. 1952. Investigations on growth and metabolism of plant cell. I. New techniques for the investigation, metabolism, nutrition and growth in undifferentiated cells. *Ann. Bot.* **16:** 57–77.

Steward, F.C., Mapes, M.O. and Mears, K. 1958. Growth and organized development of cultured cells. II. Organization in cultures grown from freely suspended cells. *Am. J. Bot.* **45:** 705–708.

Takebe, I., Labib, G. and Melchers, G. 1971. Regeneration of whole plants from isolated mesophyll protoplasts of tobacco. *Naturwissen.* **58:** 318–320.

Temin, H.M. 1970. RNA dependent DNA polymerase in virions of Rous sarcoma virus. *Nature* **226:** 1211–1213.

The C. *elegans* Sequencing consortium. 1998. Genome sequence of the nematode, *C. elegans*: a platform for investigation biology. *Science* **282:** 2012–2018.

Tulecke, W. 1953. A tissue derived from the pollen of *Ginkgo biloba*. *Science (N.Y.)* **117:** 599–600.

Van Larebeke, N., Genetello, C., Schell, J., Schilperoort, R.A. and Schell, J. 1974. Large plasmid in *Agrobacterium tumefaciens* essential for crown gall-inducing ability. *Nature* **255:** 742–743.

Van Overbeek, J., Conklin, M.E., and Blakslee, A.F., 1941. Factors in coconut milk essential for growth and development of very young *Datura* embryos. *Science* **94:** 350–351.

Vasil, V. and Hildebrandt, A.C. 1965. Differentiation of tobacco plants from single isolated cells in microcultures. *Science* **150:** 889–890.

Venter, J.C. *et al.* 2001. The sequence of the human genome. *Science* **291:** 1304–1351.

Vos, P., Hogers, R., Bleeker, M., Reijans, M., Van de Lee T., Hornes, M., Fritjers, A., Pot, J., Peleman, J., Kuiper, M. and Zabeau, M. 1995. AFLP: a new technique for DNA fingerprinting. *Nucl. Acids Res.* **23:** 4407–4414.

Welsh, J. and McClelland, M. 1990. Fingerprinting genomes using PCR with arbitrary primers. *Nucl. Acids Res.* **8:** 7213–7218.

Went, F.W. 1926. On growth accelerating substances in the coleoptile of *Avena sativa. Proc. K. Ned. Akad. Wet. Ser. C* **30:** 10.

White, P.R. 1934. Potentially unlimited growth of excised tomato root tips in a liquid medium. *Plant Physiol.* **9:** 585–600.

White, P.R. 1939. Potentially unlimited growth of excised plant callus in an artificial nutrient. *Am. J. Bot.* **26:** 59–64.

White, P.R. 1943. Morphogenesis of *Cymbidium atropurpureum in vitro*. In: *A Handbook of Plant Tissue Culture*. Jacquess Cattell Press, Tempe, Arizona.

White, P.R. 1954. *The Cultivation of Animal and Plant Cells*. Ronald Press Co., New York.

White, P.R. 1963. *The Cultivation of Animal and Plant Cells*. Ronald Press Co, New York.

Williams, J.G.K., Kubelik, A.R., Livak, K.J., Rafalsk, J.A. and Tingey, S.V., 1990. DNA polymorphisms amplified by arbitrary primers are useful as genetic markers. *Nucl. Acids Res.* **18:** 6531–6535.

Winker, H. 1908. Besprechung der Arbeit G. Haberlandt's Kultur Versuche mit isolierten Pflanzenzellen. Bot. Z. **60:** 262–264.

Yu, J. *et al.* 2002. A draft sequence of the Rice genome (*Oryza Sativa L ssp. indica*). *Science* **296:** 79–91.

Zaenen, I., Van Larebeke, N., Teuchy, H., Van Montagu, M. and Schell, J. 1974. Supercoiled circular DNA in crown gall inducing *Agrobacterium* strains. *J. Mol. Biol.* **86:** 107–127.

Zambryski, P., Holster, M., Kruger, K., Depicker, A., Schell, J., Van Mongagu, M. and Goodman, H. 1980. Tumor DNA structure in plant cells transformed by A. *tumefactions Science* **209:** 1358–1391.

# 2

# Laboratory Organization

It is generally accepted that the term "plant tissue culture", broadly refers to the *in vitro* cultivation of all plant parts, whether a single cell, a tissue or an organ, on nutrient medium under aseptic conditions. The underlying principles involved in plant tissue culture are very simple. Firstly, a plant part from the intact plant and its inter-organ/inter-tissue is isolated. Secondly, it is necessary to provide the plant part with an appropriate environment in which it can express its intrinsic or induced potential. This means that a suitable culture medium and proper culture conditions must be provided. Finally, the above procedures must be carried out aseptically. A facility for plant tissue culture operations involving any type of *in vitro* procedures must include certain essential elements. The overall design in any laboratory organization must focus on maintaining aseptic conditions. The specific design can vary, but an effective laboratory organization for plant tissue culture must include certain elements irrespective of whether used by a scientist or group of workers. The size and proportion of different parts of the facility will be dependent upon the function, purpose and size of the operations. The basic rule essential in the planning of an efficient and functional laboratory facility is the order and continuous maintenance of cleanliness because all operations require aseptic conditions.

A general guideline for setting up a facility is to focus on a design that different units and activities need to be arranged to make operational steps possible with the least amount of cross-traffic. Thus a tissue culture facility should have the following features.

*Washing facility:* An area with large sinks (some lead lined to resist acids and alkalies) and draining area is necessary. The conventional method for cleaning laboratory glassware involves chromic acid-sulphuric acid soak followed by thorough washing with tap water and subsequent rinsing with distilled water. However, due to the corrosive nature of acids, it is recommended that for routine procedures glassware should be soaked in a 2% detergent cleaner for 16 h followed by washing with 60–70°C hot tap water, and finally distilled water. The cleaned glassware should then be dried at 150–200°C in a convection-drying oven for 1–4 h. Only highly contaminated glassware should be first treated in a chromic acid bath for 16 h. Some precautions must be taken during acid cleaning: (i) handling of acid and processing of glassware in the acid should be done in a fume hood; (ii) never add water to acid but for dilution acid should be added slowly to the water while stirring; (iii) after acid cleaning, if the solution in the acid becomes dark colored, it should be discarded. However, with the availability of a wide range of reusable plasticware, it is recommended that these should be washed with mild detergents followed by a rinse with tap water and then distilled water.

The following precautions should be followed:

1. Contents of containers should be discarded immediately after completion of an experiment.
2. Contaminated containers should be autoclaved, contents discarded in a waste bottle or container and then washed. The waste bottle containing autoclaved contaminated media should not be emptied in the laboratory sink.
3. Flasks or beakers used for agar-based media should be rinsed immediately after dispensing the media into culture vessels to prevent drying of the residual agar in the beaker prior to washing.

*General laboratory and media preparation area*: This part is a central section of the laboratory where most of the activities are performed. The general laboratory section includes the area for media preparation and autoclaving of media. It also includes many of the activities that relate to the handling of tissue culture materials. The area to be set-aside for media preparation should have ample storage and bench space for chemicals, glassware, culture vessels, closures and other items needed to prepare media. The following equipment items are essential:

Laminar air flow hood
Autoclave (available as horizontal or upright) and/or pressure cookers
Refrigerator
Freezer
Balances, preferably electronic
pH meter
Water distillation unit
Magnetic stirrer (with hot plate)
Ovens
Water bath with temperature control
Hot plate/ gas plate

Microwave oven for heating agar media
Centrifuge tabletop
Vortex
Shaker, gyratory with platform and clips for different size flasks
Dissecting microscope
Lab carts and Trays

Some basic points should be followed:

• A laboratory should have an inventory and a complete up-to-date record of all the equipment and their operating manuals.
• Up-to-date record of all the chemicals including the name of manufacturer and the grade.
• All chemicals should be assigned to specific areas preferably by arranging alphabetically.
• Special handling or storage procedures should be posted in the records so that retrieving of chemicals is easy because chemicals need storage at different temperatures.
• Strong acids and bases should be stored separately.
• Solvents such as chloroform, alcohol, phenol which are volatile or toxic in nature must be stored in a fume hood.
• Hygroscopic chemicals must be stored in desiccators to avoid caking.
• Chemicals kept in refrigerators or freezers should be arranged either alphabetically or in smaller baskets.

*Transfer area*: Tissue culture techniques can be successfully carried out in a very clean laboratory, dry atmosphere with some protection against air-borne microorganisms. But it is advisable that a sterile dust-free room should be available for routine transfer and manipulation work. The laminar airflow cabinet is the most common accessory used for aseptic manipulations. It is cheaper and easier to install than a transfer room. The cabinet should be designed with horizontal airflow from the back to the front, and should be equipped with gas cocks if gas burners are to be used. Vertical flow sterile cabinet is recommended if pathogens or *Agrobacterium*-mediated transformation work is to be carried out. In the airflow cabinets, air is forced into a cabinet through small motor into the unit first through a coarse filter, where large dust particles are separated and then subsequently passes through a bacterial 0.3 μm HEPA (high efficiency particulate air) filter, it flows outwards (forward) over the working bench at a uniform rate. The air coming out of the fine filter is ultra clean (free from fungal or bacterial contamination) and its velocity (27 ± 3 m/min) prevents the microcontamination of the working area by a worker sitting in front of the cabinet. It has been observed that a 0.3 μm HEPA filter shows 99.97 to 99.99% efficiency in not allowing the bacteria to pass through. One or more laminar airflow cabinets can be housed in such a transfer area. The roof of some of the cabinet houses an ultraviolet (UV) germicidal light often used to sterilize the interior of chamber. These are turned on about 30 min prior to using the chamber but must be turned off during operations. Even so, before starting work, it is desirable to swab down the inside of cabinet and especially the table top with 70% alcohol.

***Culturing facilities:*** Plant tissue cultures should be incubated under conditions of well-controlled temperature, illumination, photoperiod, humidity and air circulation. Incubation culture rooms and commercial available incubator cabinets, large plant growth chambers, and walk-in environmental rooms satisfy this requirement. These facilities can also be constructed by developing a room with proper air-conditioning, perforated shelves to support culture vessels, fluorescent tubes and a timing device to set light dark regimes (photoperiods) which are the standard accessories. In addition, a dark area simply closed off from the rest of room with thick black curtains is necessary if some cultures require continuous darkness. This type of culture room can also be used to house a variety of liquid culture apparatus including gyratory and reciprocating shakers or a variety of batch and continuous bioreactors etc. For most culture conditions the temperature control should be

adequate to stay within ± 1°C in a range from 10 to 32°C. The photoperiod may be set according to the specific types of culture. Lighting is adjusted in terms of quantity and photoperiod duration by using automatic clocks. Cultures are generally grown in diffused light (less than 1 klx) but provisions to keep them at higher light intensity and also in dark must be made. Provision of humidity range of 20–98% with ± 3% should be made.

***Light units:*** Lux or foot candles terms have been used but more acceptable terms are:

Joule per square meter ($J/m^2$) is an expression of radiant energy per unit area.

Watts per square meter is ($W/m^2$) used to express irradiance.

The preferred term for photosynthetically active radiation units is to use photon flux density which is μmol photons/ $m^2/s$.

A common term for photon flux is based on 1mol of photons being equal to 1 Einstein (E);

Thus, a unit of photon flux density may then be expressed as μE/ $m^2/s$.

***Greenhouses:*** Greenhouses are required to grow regenerated plants for further propagation and for growing plants to maturity. This facility is required as a transitional step of taking plant materials from culture containers present in the controlled room to the field. Thus in the greenhouse, plants are acclimatized and hardened before being transferred to the field conditions. The plants are grown in the greenhouse to develop adequate root systems and leaf structures to withstand the field environment. The greenhouse should be equipped with cooling and heating systems to control temperature. For maximum use, the houses should have an artificial lighting system. The lights should include a mixture of fluorescent and incandescent lights or contain lamps designed to provide balanced wavelengths of light for plant growth and photosynthesis. The installation of a misting system is recommended on benches in the greenhouse that will be used for acclimating the plants.

# LABORATORY AND PERSONAL SAFETY

Every new worker should be given some orientation before starting work. It should include instructions on how to operate various equipment such as pH meter, balance, laminar airflow, microscope, gas burners etc. Special instructions should be given on potential for fire, broken glass, chemical spills or accidents with sharp-edged instruments. The following regulations/ instructions should be followed:

1. Toxic chemicals should be handled with appropriate precautions and should be discarded into separately labeled containers:

    Organic solvents (e.g. alcohol, chloroform, etc.)

    Mercury compounds (e.g. mercuric chloride)

    Halogens and mutagenic chemicals

2. Broken glass and scalpel blades must be disposed into separate marked containers.

3. Pipettes, tips, Pasteur pipettes and other items used for genetically modified organisms and various materials used in the pathogen work. It should be first collected in autoclavable bags, autoclaved and then disposed off.

4. A pipettor should always be used for pipetting any solution.

5. First-aid kits should be placed in every laboratory and the staff should know its location and how to use its contents.

6. Fire extinguishers should be provided in each laboratory. Some portable extinguishers should also be placed at convenient sites for easy handling by amateurs.

# 3

# Nutrition Medium

The greatest progress towards the development of nutrition medium for plant cells grown in culture took place in 1960s and 1970s. The basic nutritional requirements of cultured plant cells are very similar to those utilized by plants. However, the nutritional composition varies according to the cells, tissues, organs and protoplasts and also with respect to particular plant species. Thus in a discussionth on nutrition media it is important to recognize what type of culture, i.e. callus, cell, organ, or protoplast is to be studied and what are the objectives and purposes of the investigation.

Different types of culture have unique requirements of one or more nutritional components. A wide variety of salt mixtures have been reported in various media. A nutrient medium is defined by its mineral salt composition, carbon source, vitamins, growth regulators and other organic supplements which include organic nitrogen, acids and complex substances. When reference is made to a particular medium, the intention is to identify only the salt composition unless otherwise specified. Any number and concentration of amino acids, vitamins, growth regulators or organic supplements can be added in an infinite variety of compositions to a given salt composition to achieve the desired results.

## FACILITIES AND EQUIPMENTS

Autoclave and/or pressure cookers
Balances, preferably electronic
Centrifuge tabletop
Cotton plugs, aluminium foil, brown sheets etc.
Culture vessels (flasks, tubes, petri dishes, jars, etc.)
Desiccators
Dissecting microscope
Filter units
Flasks, beakers, pipettes, etc.
Freezer
Hot plate/gas plate
Lab carts/trolley, trays, etc.
Laminar airflow hoods
Magnetic stirrer (with hot plate)
Microwave oven
Ovens
pH meter
Refrigerator
Shaker
Vortex
Water bath with temperature control
Water distillation unit

# UNITS FOR SOLUTION PREPARATION

The concentration of a particular substance in the media can be expressed in various units that are as follows:

## Units in weight

It is represented as milligram per liter (mg/l)

$10^{-6}$ = 1.0 mg/l or 1 part per million (ppm)
$10^{-7}$ = 0.1 mg/l
$10^{-9}$ = 0.001 mg/l or 1 mg/l

## Molar concentration

A molar solution (M) contains the same number of grams of substance as is given by its molecular weight.

1 molar (M) = the molecular weight in g/l
1 mM = the molecular weight in mg/l or $10^{-3}$M
1 mM = the molecular weight in mg/l or $10^{-6}$M or $10^{-3}$mM.

## Conversion from milli molar (mM) to mg/l

For example, the molecular weight of auxin 2,4-D = 221.0
1 M 2,4-D solution consists of 221.0 g/l
1 mM 2,4-D solution consists of 0.221 g/l = 221.0 mg/l
1 $\mu$M 2,4-D solution consists of 0.000221 g/l = 0.221 mg/l

## Conversion from mg/l to mM

The molecular weight of $CaCl_2 \cdot 2H_2O$
= 40.08 + 2 × 35.453 + 4 × 1.008 + 2 × 16
= 147.018
(the atomic weights of Ca, Cl, H and O being 40.08, 35.453, 1.008 and 16.0 respectively).
If 440 mg/l of $CaCl_2 \cdot 2H_2O$ is to be converted into mM, then

$$\text{The number of mM } CaCl_2 \cdot 2H_2O = \frac{\text{No. of mg } CaCl_2 \cdot 2H_2O}{\text{Molecular weight of } CaCl_2 \cdot 2H_2O}$$

$$= \frac{440}{147.018} = 2.99 \text{ mM}$$

Thus, 440 mg/l $CaCl_2 \cdot 2H_2O$ = 2.99 mM

# MEDIA COMPOSITION

A number of basic media are listed in Table 3.1. The MS medium of Murashige and Skoog (1962) salt composition is very widely used in different culture systems. In the development of this medium and also other media, it was demonstrated that not only the presence of necessary nutrients but also the actual and relative concentrations of various inorganic nutrients are of crucial significance. The $B_5$ medium of Gamborg *et al.* (1968) or its derivatives have been useful for cell and protoplast culture. This medium was originally designed for callus and suspension cultures but it has also been effectively used for plant regeneration. It differs from MS medium in having much lower amounts of nitrate in the form of ammonium. The $N_6$ medium was developed for cereal anther culture and is used successfully with other types of cereal tissue culture (Chu, 1978). Any success with a medium is in all probability due to the fact that the ratios as well as concentrations most nearly match the optimum requirements for the cells or tissues for growth and/or differentiation. The nutritional milieu generally consists of inorganic nutrients, carbon and energy sources, vitamins, phytohormones (growth regulators), and organic supplements which include organic nitrogen, acids and complex substances.

## Inorganic nutrients

Mineral elements are very important in the life of a plant. For example, calcium is a component of the cell wall, nitrogen is an important part of amino acids, proteins, nucleic acids, and vitamins, and magnesium is a part of chlorophyll molecules. Similarly iron, zinc and molybdenum are parts of certain enzymes. Besides, C, H, N, and O, 12 other elements are known to be

**Table 3.1:** Composition of plant tissue culture media (values expressed in mg/l ).

| Constituent | Nitsch & Nitsch (1956) | White (1963) | Murashige & Skoog (1962)– MS | Gautheret (1942) | Gamborg' s 1968)–B$_5$ | Chu (1978)–N$_6$ |
|---|---|---|---|---|---|---|
| KCl | 1500 | 65 | – | – | – | – |
| MgSO$_4$·7H$_2$O | 250 | 720 | 370 | 125 | 250 | 185 |
| NaH$_2$PO$_4$·H$_2$O | 250 | 16.5 | – | – | 150 | – |
| CaCl$_2$·2H$_2$O | – | – | 440 | – | 150 | 166 |
| KNO$_3$ | 2000 | 80 | 1900 | 125 | 2500 | 2830 |
| CaCl$_2$ | 25 | – | – | – | – | – |
| Na$_2$SO$_4$ | – | 200 | – | – | – | – |
| NH$_4$NO$_3$ | – | – | 1650 | – | – | – |
| KH$_2$PO$_4$ | – | – | 170 | 125 | – | 400 |
| Ca(NO$_3$)$_2$·4H$_2$O | – | 300 | – | 500 | – | – |
| (NH$_4$)2SO$_4$ | – | – | – | – | 134 | 463 |
| NiSO$_4$ | – | – | – | 0.05 | – | – |
| FeSO$_4$·7H$_2$O | – | – | 27.8 | 0.05 | – | 27.8 |
| MnSO$_4$·4H$_2$O | 3 | 7 | 22.3 | 3 | – | – |
| MnSO$_4$·H$_2$O | – | – | – | – | 10 | 3.3 |
| KI | – | 0.75 | 0.83 | 0.5 | 0.75 | 0.8 |
| CoCl$_2$·6H$_2$O | – | – | 0.025 | – | 0.025 | – |
| Ti(SO$_4$)$_3$ | – | – | – | 0.2 | – | – |
| ZnSO$_4$·7H$_2$O | 0.5 | 3 | 8.6 | 0.18 | 2 | 1.5 |
| CuSO$_4$·5H$_2$O | 0.025 | – | 0.025 | 0.5 | 0.025 | – |
| BeSO$_4$ | – | – | – | 0.1 | – | – |
| H$_3$BO$_3$ | 0.5 | 1.5 | 6.2 | 0.05 | 3 | 1.6 |
| H$_2$SO$_4$ | – | – | – | 1 | – | – |
| Na$_2$MoO$_4$·2H$_2$O | 0.025 | – | 0.25 | – | 0.25 | – |
| Fe$_2$(SO$_4$)$_3$ | – | 2.5 | – | – | – | – |
| EDTA disodium salt | – | – | 37.3 | – | – | 37.3 |
| EDTA- Na ferric salt | – | – | – | – | 43 | – |
| m-inositol | – | – | 100 | – | 100 | – |
| Thiamine | – | 0.1 | 0.1 | 0.1 | 1.0 | 1.0 |
| Pyridoxine | – | 0.1 | 0.5 | 0.1 | 1.0 | 0.5 |
| Nicotinic acid | – | 0.5 | 0.5 | 0.5 | 1.0 | 0.5 |
| Glycine | – | 3 | 2 | 3 | – | – |
| Cysteine | – | 1.0 | – | – | 10 | – |
| Sucrose | 34,000 | 20,000 | 30,000 | 30,000 | 20,000 | 30,000 |

essential for plant growth. According to the recommendations of International Association for Plant Physiology, the elements required by plants in concentration greater than 0.5 mmol/l are referred to as macro/major elements and those required in concentration less than that are micro/ minor elements. A variety of salts supply the needed macro/major- and micro/minor nutrients that are the same as those required by the normal plant. The following are required in macro- or millimole quantities: N, K, P, Ca, S, and Mg. The essential micronutrients which are required in micromolar concentrations include iron (Fe),

manganese (Mn), boron (B), copper (Cu), zinc (Zn), iodine (I), molybdenum (Mo) and cobalt (Co). For achieving maximum growth rate, the optimum concentration of each nutrient can vary considerably. The mineral composition of culture medium is defined precisely by the equilibrium of the concentrations of different ions in a solution. When mineral salts are dissolved in water, they undergo dissociation and ionization. The active factor in the medium is the ions of different types rather than the compounds. Therefore, a useful comparison between the two media can be made by looking into the total concentrations of different types of ions in them. To choose a mineral composition and then compare their different ionic balances, one uses ionic concentrations expressed in milliequivalents per litre (Table 3.2). For most purposes a nutrient medium should contain from 25 to 60 mM inorganic nitrogen. The cells may grow on nitrate alone, but often there is a distinct beneficial effect and requirement for ammonium or another source of reduced nitrogen. Besides, nitrate alone in the medium drifts the pH towards alkalinity. Adding a small amount of an ammonium compound together with nitrate checks this drift. Nitrate is used in the range of 25–40 mM and ammonium in the range of 2–20 mM. The response to ammonium varies from inhibitory to essential depending upon the tissue and the purpose of culture. In case of amounts in excess of 8mM of ammonium or if grown solely on this source of nitrogen, then citrate, malate, succinate or another TCA cycle acid should be present. Most plants prefer nitrate to ammonium, although the opposite is also true in some cases. Potassium is required at concentrations of 2 to 26 mM. Potassium is generally supplied as the nitrate or as the chloride form and sodium cannot be substituted for potassium. A concentration of

1–3mM calcium, sulphate, phosphate and magnesium is usually adequate. Iron is generally added as a chelate with ethylene diamine tetra acetic acid (EDTA). In this form iron remains available up to a pH of 8.0.

## Carbon and Energy Source

The standard carbon source without exception is sucrose but plant tissues can utilize a variety of carbohydrates such as glucose, fructose, lactose, maltose, galactose and starch. In the cultured tissues or cells, photosynthesis is inhibited and thus carbohydrates are needed for tissue growth in the medium. The sucrose in the medium is rapidly converted into glucose and fructose. Glucose is then utilized first, followed by fructose. Sucrose is generally used at a concentration of 2–5%. Most media contain myo-inositol at a concentration of *ca.* 100mg/l, which improves cell growth.

## Vitamins

Normal plants synthesize the vitamins required for growth and development. But plant cells in culture have an absolute requirement for vitamin $B_1$ (thiamine), vitamin B (nicotinic acid) and vitamin $B_6$ (pyridoxine). Some media contain pantothenic acid, biotin, folic acid, p-amino benzoic acid, choline chloride, riboflavine and ascorbic acid. The concentrations are in the order of one mg/l. Myo-inositol is another vitamin used in the nutrient medium with a concentration of the order of 10–100mg/l.

## Growth regulators

Hormones now referred to as growth regulators are organic compounds that have been naturally synthesized in higher plants, which influence growth and development. They are usually active at a site different from where they are produced and are only present and active in very small quantities. Apart from natural compounds, synthetic compounds have been developed which correspond to the natural ones. There are two main classes of growth regulators that are of special importance in plant tissue culture.

**Table 3.2:** Composition of mineral solutions used in *in vitro* culture

| Ions | Gautheret | Gamborg | Murashige & Skoog | White |
|------|-----------|---------|-------------------|-------|
| $NO_3^-$ | 5.49 | 25.0 | 39.4 | 3.34 |
| $H_2PO_4^-$ | 0.92 | 1.1 | 1.3 | 0.12 |
| $SO_4^-$ | 1.0 | 4.0 | 3.0 | 8.64 |
| $Cl^-$ | – | 2.0 | 6.0 | 0.88 |
| Total anions (mEq/l) | 7.41 | 32.1 | 49.7 | 12.98 |
| $K^+$ | 2.17 | 25.0 | 20.1 | 1.68 |
| $NH_4^+$ | – | 2.0 | 20.6 | – |
| $Na^+$ | – | 1.1 | – | 2.92 |
| $Ca^{++}$ | 4.24 | 2.0 | 6.0 | 2.54 |
| $Mg^{++}$ | 1.0 | 2.0 | 3.0 | 5.84 |
| Total cations (mEq/l) | 7.41 | 32.1 | 49.7 | 12.98 |
| Total anions + cations (103 M/l) | 11.70 | 60.2 | 93.3 | 17.45 |

**Table 3.3:** Characteristics of growth regulators and vitamins

| Name | Chemical formula | Molecular weight | Solubility |
|---|---|---|---|
| p-Chlorophenoxy acetic acid | $C_8H_7O_3Cl$ | 186.6 | 96% alcohol |
| 2,4-Dichlorophenoxy acetic acid | $C_8H_6O_3Cl_2$ | 221.0 | 96% alcohol, heated lightly |
| Indole-3 acetic acid | $C_{10}H_9NO_2$ | 175.2 | 1N NaOH/96% alcohol |
| Indole-3 butyric acid | $C_{12}H_{13}NO_2$ | 203.2 | 1N NaOH/96% alcohol |
| α-Naphthalene acetic acid | $C_{12}H_{10}O_2$ | 186.2 | 1N NaOH/96% alcohol |
| β-Naphthoxy acetic acid | $C_{12}H_{10}O_3$ | 202.3 | 1N NaOH |
| Adenine | $C_5H_5N_5 \cdot 3H_2O$ | 189.1 | $H_2O$ |
| Adenine sulphate | $(C_5H_5N_5)_2 \cdot H_2SO_4 \cdot 2H_2O$ | 404.4 | $H_2O$ |
| 6-Benzyl adenine | $C_{12}H_{11}N_5$ | 225.2 | 1N NaOH |
| N-isopentenylamino purine | $C_{10}H_{13}N_5$ | 203.3 | 1N NaOH |
| Kinetin | $C_{10}H_9N_5O$ | 215.2 | 1N NaOH |
| Zeatin | $C_{10}H_{13}N_5O$ | 219.2 | 1N NaOH/1N HCl, heated lightly |
| Gibberellic acid | $C_{19}H_{22}O_6$ | 346.4 | Alcohol |
| Abscisic acid | $C_{15}H_{20}O_4$ | 264.3 | 1N NaOH |
| Folic acid | $C_{19}H_{19}N_7O_6$ | 441.4 | 1N NaOH |
| Colchicine | $C_{22}H_{25}NO_6$ | 399.4 | $H_2O$ |

These are the auxins and cytokinins, while others viz. gibberellins, abscisic acid, ethylene, etc. are of minor importance. Some of the naturally occurring growth regulators are the auxin indole acetic acid (IAA) and the cytokinins zeatin and isopentenyladenine (2 iP), while others are synthetic growth regulators. The characteristics of growth regulators are listed in Table 3.3.

## Auxins

A common feature of auxins is their property to induce cell division, cell elongation and formation of callus. Auxin has a clear rhizogenic action i.e. induction of adventitious roots. It often inhibits adventitious and axillary shoot formation. At low auxin concentrations, adventitious root formation predominates, whereas at high auxin concentrations, root formation fails to occur and callus formation takes place. Auxin is present in sufficient concentration in the growing shoot tips or flowering tips of plants to ensure multiplication and elongation. Auxin circulates from the top towards the base of the organs with a polarity strongly marked in young organs. The compounds most frequently used are highly effective 2,4-dichlorophenoxy acetic acid (2,4-D), naphthalene acetic acid (NAA), indole acetic acid (IAA), indole butyric acid (IBA), 2,4,5-trichlorophenoxy acetic acid (2,4,5-T), p-chlorophenoxy acetic acid (pCPA) and picloram (4-amino-3,5,6-trichloropicolinic acid).

## Cytokinins

Cytokinins are derivatives of adenine and have an important role in shoot induction. Cytokinins also have a clear effect on cell division. These are often used to stimulate growth and development. They usually promote cell division if added together with an auxin. Cytokinins have a clear role in organogenesis where they stimulate bud formation. They are antagonistic to rhizogenesis. At higher concentrations (1 to 10 mg/l), adventitious shoot formation is induced but root formation is generally inhibited. Cytokinins promote axillary shoot formation by decreasing apical dominance. The most frequently used cytokinins are 6-benzyl amino purine (BAP) or 6-benzyladenine (BA), N-(2-furfurylamino) 1-H-purine-6-amine (kinetin), 6-(4-hydroxy-3methyl-trans-2-butylamino) purine (zeatin), and isopentenyladenine (2 iP). Zeatin and 2-iP are naturally occurring cytokinins while BA and kinetin are synthetically derived cytokinins. Subsequently it was discovered that N,N'-diphenyl urea (DPU), thidiaziron, N-2-chloro-4-puridyl-N-phenyl urea (CPPU) and other derivatives of diphenyl urea show the cytokinin type activity.

Stock solutions of IAA and kinetin are stored in amber bottles or bottles covered with a black paper and kept in dark since they are unstable in light.

## Other growth regulators

Gibberellins, although found in all plants and fungi, are unevenly distributed. The sites of synthesis are very young, unopened leaves, active buds, root tips and embryos. Gibberellins are normally used in plant regeneration. $GA_3$ is essential for meristem culture of some species. In general, gibberellins induce elongation of internodes and the growth of meristems or buds *in vitro*. Gibberellins usually inhibit adventitious root as well as shoot formation. During organogenesis, gibberellins are antagonistic. They seem to oppose the phenomenon of dedifferentiation.

Abscisic acid is an important growth regulator for induction of embryogenesis. This is a growth inhibitor, which seems to be synthesized when a plant is under difficult conditions. It has a favorable effect on abscission.

Ethylene is a gaseous compound identified a long time ago, but its functions as a growth regulator were not evident. Ethylene is produced by cultured cells, but its role in cell and organ culture is not known. The practical use of ethylene, which is difficult in a gaseous state, made great progress after the discovery of 2-chloroethane phosphoric acid. This product, when applied in a powder form, penetrates the tissue where it liberates ethylene.

## Organic Supplements

### Organic nitrogen

Cultured cells are normally capable of synthesizing all the required amino acids, but it is often beneficial to include organic nitrogen in the form of amino acids such as glutamine and asparagine and nucleotide such as adenine. For cell culture, it is good to add 0.2 to 1.0 g/l of casein hydrolysate or vitamin-free casamino acids. In some cases L-glutamine (up to 8 mM

or 150 mg/l) may replace the casein hydrolysate. The amino acids when added should be used with caution, since they can be inhibitory. The amino acids included in the media and amount in mg/l are: glycine (2), aspargine (100), tyrosine (100), arginine (10) and cysteine (10). Sometimes adenine sulphate (2–120 mg/l) is added to the agar media for morphogenesis.

### Organic acids

Plant cells are not able to utilize organic acids as a sole carbon source. Addition of TCA cycle acids such as citrate, malate, succinate or fumarate permits growth of plant cells on ammonium as the sole nitrogen source. The cells can tolerate a concentration of upto 10mM of the acid. Pyruvate may also enhance growth of cells cultured at low density.

### Complex substances

A variety of extracts *viz*. protein hydrolysate, yeast extract, malt extract, coconut milk, orange and tomato juice have been tested. With the exception of protein hydrolysate and coconut milk, most others are used as the last resort. If used at higher concentration, the complex substances may adversely affect cell growth. It is advisable to make a preliminary test in a range of 0.1 to 1.0 g/l to assess its effect on growth. Coconut milk is commonly used at 2–15% (v/v). The present trend is, however, towards fully defined media and the use of complex mixtures is losing favour.

### Anti browning compounds

Many plants are rich in polyphenolic compounds. After tissue injury during dissection, such compounds will be oxidized by polyphenol oxidases and the tissue will turn brown or black. The oxidation products are known to inhibit enzyme activity, kill the explants, and darken the tissues and culture media, a process which severely affects the establishment of explants Activated charcoal is generally acid washed and neutralized (if not done) at concentrations of 0.2 to 3.0% (w/v) in the medium where phenol like compounds are a problem for *in vitro* growth of cultures. It can adsorb toxic brown/black pigments and stabilizes pH. It also helps to reduce toxicity by removing toxic substances produced during the culture and permits unhindered cell growth. Besides activated charcoal, polyvinylpyrrolidone (250–1000 mg/l), citric acid and ascorbic acid (100 mg/l each), thiourea or L-cysteine are also used to prevent oxidation of phenols.

Phloroglucinol, a phenolic compound, is sometimes added to inhibit the enzyme IAA oxidase responsible for the breakdown of IAA.

## Gelling Agents

Agar, a seaweed derivative, is the most popular solidifying agent. It is a polysaccharide with a high molecular mass and has the capability of gelling media. Solublized agar forms a gel that can bind water and adsorb compounds. The higher the agar concentration, the stronger is the binding of water. Plant tissue culturists often use Difco Bacto agar at a concentration of 0.6 to 1.0% (w/v), although other forms of agar (agarose, phytagar, flow agar, etc.) are also becoming popular. *In vitro* growth may be adversely affected if the agar concentration is too high. With higher concentrations, the medium becomes hard and does not allow the diffusion of nutrients into the tissues. Impurities, if present in the agar can be removed for critical experiments on nutrient studies. The procedure involves washing agar in double distilled water for atleast 24h, rinsing in ethanol and then drying at 60°C for 24h. Agar has several advantages over other gelling agents as: (i) agar gels do not react with media constituents; (ii) agar is not digested by plant enzymes and remains stable at all feasible incubation temperatures. Besides agar, the following alternatives are also available.

i. Alginate can be used for plant protoplast culture.
ii. Gelrite at 0.2% can be used for solidification of media. Gelrite gels are remarkably clear in comparison to those formed by agar. Gelrite requires both a heating cycle and

the presence of divalent cations ($Mg^{++}$ or $Ca^{++}$) for gelation to take place.

iii. Synthetic polymer biogel P200 (polyacrylamide pellets) or a starch polymer can be used.

iv. Agargel: A mixture of agar and synthetic gel has been developed by Sigma Company. It has the properties of both synthetic gel and agar.

v. Gelatin at higher concentration (10%) has been tried but has limited use because it melts at low temperature (25°C).

vi. The FMC Corporation has developed a highly purified agarose called Sea plaque (k) which can be used in the recovery of single protoplasts from cultures.

The advantages of synthetic gelling compounds is that they form clear gels at relatively low concentrations (1.25–2.5 g/l) and are valuable aids for detection of contamination and also root formation during the culture. Whether explants grow better on agar or other supporting agents, depends on the tissue and species.

Liquid media with a support can also be used instead of solid media. These include

i. liquid medium without agar using clean foam plastic, glasswool or rockwool as support;
ii. filter paper-bridge which is hung in a liquid medium;
iii. growth on a liquid medium containing glass beads;
iv. viscose sponge underneath filter paper as a carrier for a liquid medium instead of agar.

## pH

pH determines many important aspects of the structure and activity of biological macromolecules. pH is the negative logarithm of the concentration of hydrogen ions. Nutrient medium pH ranges from 5.0 to 6.0 for suitable *in vitro* growth of explant. pH higher than 7.0 and lower than 4.5 generally stops growth and development. The pH before and after autoclaving is different. It generally falls by 0.3 to 0.5 units after autoclaving. If the pH falls appreciably during plant tissue culture (the medium becomes liquid), then a fresh medium should be prepared. It should be known that a starting pH of 6.0 could often fall to 5.5 or even lower during growth. pH higher than 6.0 give a fairly hard medium and a pH below 5.0 does not allow satisfactory gelling of the agar.

# PROTOCOLS

## 3.1: General Methodology for Medium Preparation

*Preparation of macro- and micronutrient stock solutions*

Stock solutions of macro- and micronutrients, vitamins and growth regulators are prepared in distilled or high purity demineralized water. Chemicals should be of the highest grade. Usually the stock solutions of macro and micronutrients are prepared as 10x and 100x concentrations as shown in Table 3.4. These nutrient solutions can

be dispensed in plastic bags with zipper seals and stored frozen (e.g. 10x macronutrient solution dispensed into a bag containing 100ml of solution for preparation of 1l medium). While dissolving the nutrients in water, one compound is added at a time to avoid precipitation. For macronutrient stock solution, calcium chloride is dissolved separately in water and then added to the rest of the solution to avoid precipitation.

*Preparation of vitamin and growth regulator stock solutions*

All the growth regulators are not soluble in water. Solubility of different growth regulators is given in Table 3.3. The compound should be dissolved in few-ml of solvent and then water slowly added to make up to the requisite volume. Concentrations of compounds can be taken as mg/l or in molarity.

**Table 3.4:** Preparation of stock solutions of Murashige and Skoog (MS) medium

| Constituent | Concentration in MS medium (mg/l) | Concentration in the stock solution (mg/l) | Volume to be taken/ liter of medium |
|---|---|---|---|
| **Macronutrients (10x)** | | | |
| $NH_4NO_3$ | 1650 | 16500 | 100 |
| $KNO_3$ | 1900 | 19000 | |
| $MgSO_4 \cdot 7H_2O$ | 370 | 3700 | |
| $CaCl_2 \cdot 2H_2O$ | 440 | 4400 | |
| $KH_2PO_4$ | 170 | 1700 | |
| **Micronutrients (100x)** | | | |
| $H_3BO_3$ | 6.2 | 620 | 10 |
| $MnSO_4 \cdot 4H_2O$ | 22.3 | 2230 | |
| $ZnSO_4 \cdot 7H_2O$ | 8.6 | 860 | |
| KI | 0.83 | 83 | |
| $Na_2MoO_4 \cdot 2H_2O$ | 0.25 | 25 | |
| $CuSO_4 \cdot 5H_2O$ | 0.025 | 2.5 | |
| $CoCl_2 \cdot 6H_2O$ | 0.025 | 2.5 | |
| **Iron source** | | | |
| Fe-EDTA- Na salt | 40 | Added fresh | |
| **Vitamins** | | | |
| Nicotinic acid | 0.5 | 50 mg/100 ml | 1 |
| Thiamine HCl | 0.1 | 50 mg/100 ml | 0.2 |
| Pyridoxine HCl | 0.5 | 50 mg/100 ml | |
| Myo-inositol | 100 | Added fresh | 1 |
| **Others** | | | |
| Glycine | 2 | 50 mg/100 ml | 4 |
| Sucrose | 30,000 | Added fresh | |
| Agar | 8000 | Added fresh | |
| pH 5.8 | | | |

**Table 3.5:** Preparation of stock solutions of growth regulators

| Compound | Abbreviation | mg/100 ml (1mM) |
|---|---|---|
| **Auxins** | | |
| 2,4-Dichlorophenoxy acetic acid | 2,4-D | 22.1 |
| Indole-3 acetic acid | IAA | 17.5 |
| Indole-3 butyric acid | IBA | 20.32 |
| α-Naphthalene acetic acid | NAA | 18.62 |
| β-Naphthoxy acetic acid | NOA | 20.23 |
| 2,4,5-Trichlorophenoxy acetic acid | 2,4,5-T | 25.56 |
| p-Chlorophenoxy acetic acid | pCPA | 18.66 |
| Picloram | PIC | 24.12 |
| **Cytokinins** | | |
| Adenine | Ade | 18.91 |
| Benzyl adenine or benzyl amino purine | BA or BAP | 22.52 |
| N-isopentenylamino purine | 2-iP | 20.33 |
| Kinetin | KIN | 21.52 |
| Zeatin | ZEA | 21.92 |
| **Others** | | |
| Gibberellic acid | $GA_3$ | 34.64 |
| Abscisic acid | ABA | 26.43 |
| Colchicine | Col | 39.94 |

**Concentration in mg/l**: It is preferable to dissolve 50 mg/100 ml to give a concentration of 0.5 mg/ml.

**Concentration in mM**: The growth regulator solutions can be prepared as 1mM for 100 ml as shown in Table 3.5. Concentration of growth regulators in mg/l converted to mM are also given in Table 3.6. If a culture medium is to contain 10 mM of the growth regulator (e.g. 2,4-D), then

1 mM = 221 mg/l or 0.221 mg/ml

The amount in 100 ml stock solution = 0.221 × 100 ml =22.1 mg

10 µM = 2210 mg or 2.210 mg

The required volume of stock solution to be added is 10 ml (22.1/10 = 2.210 mg).

Add each component according to the list, including growth regulators, by using correctly sized graduated cylinders or pipettes or balance.

Add water to just below the final volume.

**Table 3.6:** Molarity concentrations of some commonly used plant growth regulators

| Concentration used | BA | Kinetin | Zeatin | 2,4-D | IAA | IBA |
|---|---|---|---|---|---|---|
| Mol. Weight | 225.3 | 215.2 | 219.2 | 221.0 | 175.2 | 203.2 |
| mg/l | µM | µM | µM | µM | µM | µM |
| 0.1 | 0.44 | 0.46 | 0.46 | 0.45 | 0.57 | 0.49 |
| 0.2 | 0.89 | 0.93 | 0.91 | 0.90 | 1.14 | 0.98 |
| 0.3 | 1.33 | 1.39 | 1.36 | 1.36 | 1.71 | 1.48 |
| 0.4 | 1.78 | 1.86 | 1.82 | 1.81 | 2.28 | 1.97 |
| 0.5 | 2.22 | 2.32 | 2.28 | 2.26 | 2.85 | 2.46 |
| 0.6 | 2.66 | 2.79 | 2.73 | 2.71 | 3.42 | 2.95 |
| 0.7 | 3.11 | 3.25 | 3.19 | 3.17 | 3.99 | 3.44 |
| 0.8 | 3.55 | 3.72 | 3.65 | 3.62 | 4.57 | 3.94 |
| 0.9 | 3.99 | 4.18 | 4.11 | 4.07 | 5.14 | 4.43 |
| 1.0 | 4.40 | 4.60 | 4.56 | 4.52 | 5.71 | 4.90 |
| 2.0 | 8.90 | 9.30 | 9.12 | 9.05 | 11.42 | 9.80 |
| 5.0 | 22.20 | 23.20 | 22.81 | 22.62 | 28.54 | 24.60 |
| 10.0 | 44.0 | 46.0 | 45.62 | 45.25 | 57.08 | 49.00 |

Adjust the pH of the medium to the required value (e.g. pH 5.8 for MS) by drop-wise addition of 1N KOH or 1N HCl with constant stirring.

Preparation of 1N HCl: Commercially available HCl has minimum assay of 35.4%, specific gravity = 1.18, M.W. = 36.46

Conversion of liquid into solid = minimum assay × specific gravity 35.4 × 1.18 = 41.7 g

Thus 100 ml of HCl contains 41.77 g

$$\text{Normality} = \frac{\text{Grams in one liter}}{\text{Equivalent weight}}$$

E.wt. = M.W./acidity or basicity

Therefore, E.wt of HCl = 36.46/1 = 36.46

1N HCl = 417.7/ 36.46 = 11.5 ml of HCl to make volume of one liter

Make up the solution to the final volume.

Add the needed amount of agar (e.g. 6–8 g/l) or any other gelling agent. Heat the solution while stirring until the agar is dissolved. The solution becomes transparent when the agar is completely dissolved. Distribute the medium in glass or polypropylene vessels and plug it with cotton or cover it with aluminium foil.

The culture medium is sterilized in an autoclave for 20 min at 121°C at 15 p.s.i (105 kPa). If filter-sterilized hormones/compounds are to be added, then these are added to the autoclaved medium. Medium is cooled in laminar airflow to 50–60° C, the hormones are added by Millipore or any other filter assembly using 0.22 mm filter, and then dispensed into sterile vessels. After cooling, the media are preferably stored at 4–10°C.

It is advisable to mark each vessel to show the precise medium and date of preparation. The culture medium should be used 3–4 days after preparation, so that if it is not properly sterilized, contamination will start appearing and this medium can be discarded.

## REFERENCES

Chu, C.C. 1978. The N6 medium and its applications to anther culture of cereal crops. In: *Proc. Symp. Plant Tissue Culture,* Science Press, Peking, pp. 43–45.

Gamborg, O.L., Miller, R.A. and Ojima, K. 1968. Nutrition requirements of suspension cultures of soybean root cells. *Expt. Cell. Res.* **50:** 151–158.

Gautheret, R.J. 1942. *Manuel technique de culture des tissus vegetaux.* Masson, Paris, pp. 392.

Murashige, T. and Skoog, F. 1962. A revised medium for rapid growth and bioassays with tobacco tissue cultures. *Physiol. Plant.* **15:** 473–497.

Nitsch, J.P. and Nitsch, C. 1956. Auxin dependent growth of excised *Helianthus tuberosus* tissues. *Am. J. Bot.* **43:** 839–851.

White, P.R. 1963. *The Cultivation of Animal and Plant Cells.* Ronald Press Co, New York.

## FURTHER READING

Evans, D.A., Sharp, W.R., Ammirato, P.V. and Yamada, Y. 1983. *Handbook of Plant Cell Culture.* Macmillan Publishing Company, New York.

Pierik, R.L.M. 1987. *In vitro culture of Higher Plants.* Martinus Nijhoff Publishers, Dordrecht.

Reinert, J. and Bajaj, P.P.S. (eds.) 1977. *Applied and Fundamental Aspects of Plant Cell, Tissue and Organ Culture.* Springer-Verlag, Berlin.

Street, H.S. (ed.)1973. *Plant Tissue and Cell culture.* Blackwell Scientific Publications, Oxford.

Thorpe, T.A. 1981. *Plant Tissue Culture: Methods and Applications in Agriculture.* Academic Press Inc., New York.

# 4

# Sterilization Techniques

Sterilization is a procedure used for elimination of microorganisms. The maintenance of aseptic (free from all microorganisms) or sterile conditions is essential for successful tissue culture procedures. The need for asepsis requires that all culture vessels, media, and instruments used in handling tissues, as well as explant itself be sterilized. The importance is being laid on keeping the air, surface and floors free of dust. All operations are carried out in laminar airflow sterile cabinets. An obvious precaution is not to share these cabinets and areas of tissue culture work being carried out with other microbiologists and pathologists. If tissue culture work itself requires screening against pathogens or isolation of toxins etc., then the pathological work area must be separated. The cabinets for plant tissue culture work should have a horizontal airflow from back towards the front, while vertical airflow cabinets are preferable for working with pathogens or *Agrobacterium*-mediated transformation.

The objective in this chapter is to explain specific procedures/guidelines for preparing sterile media, containers/vessels, explant materials and for maintaining sterile conditions.

In general different sterilization procedures can be grouped under three categories.

1. Preparation of sterile media, containers and small instruments

2. Maintenance of aseptic conditions
3. Preparation of sterilized explant material

## PREPARATION OF STERILE MEDIA, CONTAINERS AND SMALL INSTRUMENTS

The most popular method of sterilizing culture media and apparatus is by autoclaving the material using steam/dry sterilization.

### Steam Sterilization

Most nutrient media are sterilized by using an autoclave. Autoclaving is a method of sterilizing with water vapour under high pressure. Nearly all microbes are killed on exposure to the superheated steam of an autoclave. It is normal practice now to sterilize glassware and other accessories *viz.* cotton plugs, plastic caps, pipettes, filters, gauge etc. in a commercial available autoclave. The standard conditions for autoclaving media are 121°C with a pressure of 15 psi for 20 min. These conditions are followed for test tubes or other containers containing between 20–50 ml of nutrient media and also for plastic labware which can be repeatedly autoclaved such as propylene, polymethyl pentene, Tfzel, ETPE and Teflon FEP. However, some polycarbonates show loss of mechanical strength with repeated autoclaving. An autoclave has a temperature range of 115–135°C. Good sterilization relies on time, pressure, temperature and volume of the object to be sterilized. The following guidelines must be followed:

Test tubes/flasks containing between 20–50 ml nutrient media—20 min. at 121°C, 15 psi.

Flasks containing between 50–500 ml nutrient media—25 min. at 121°C, 15 psi.

Flasks containing between 500–5000 ml nutrient media—35 min. at 121°C, 15 psi.

It must be realized that heat penetration is very important in an autoclave and large volumes must be sterilized for longer periods, as the heat will take longer to penetrate than with smaller volumes. The advantages of an autoclave are speed, simplicity and destruction of viruses, while disadvantages are change in pH by 0.3–0.5 units, components can separate out and chemical reactions can occur resulting in a loss of activity of media constituents.

*Units of energy and pressure*

Pounds per square inch (psi) is very commonly used but it can also be expressed as Joules per kilogram (J/kg) which is an expression of energy per unit of mass and should be used.

Pascal units are used to express pressure. 1J/kg = 1kPa (kilo Pascal)

Thus, 1psi = 6.9kPa or 15psi = ~ 105kPa

*Precaution for material in the bottles*: Bottles when autoclaving should not be tightly packed and their tops should be loose. After autoclaving, these are kept in the laminar airflow and tops of bottles are tightened when they are cool.

During and after autoclaving, the following points should be considered:

i. The pH of the media is lowered by 0.3–0.5 units.

ii. Autoclaving at too high temperature can caramelize sugars, which may be toxic.

iii. Volatile substances can be destroyed by the use of an autoclave.

iv. Autoclaving for too long periods can precipitate salts, and depolymerize the agar.

v. Care should be taken to use the correct duration and temperature.

vi. For sterilization of liquid material (media, water) wet or steam sterilization employing either autoclave or domestic pressure cooker can be used but with a precaution of slow exhaust.

### Dry Sterilization

The sterilization of glassware and metallic instruments can be carried out in dry heat for 3h at 160–180°C. An exposure of 160°C dry heat for 2h is regarded as equivalent to moist heat (steam) sterilization at 121°C for 15 min. Dry goods (instruments, empty glassware) can either be wrapped in aluminium foil, brown paper or sealed metal containers to maintain sterility.

Present day autoclaves come with both dry and steam sterilization programs. If material to be autoclaved is dry *viz.* instruments, glassware, pipettes, tips, etc. then dry cycle in an autoclave should be used. Domestic pressure cookers are very useful when small amount of media is to be sterilized.

Disadvantages of dry heat sterilization are poor circulation of air and slow penetration of heat. Therefore, sterilization by autoclaving is always recommended.

### Filter Sterilization

Some growth factors, amino acids, vitamins, and toxins are heat labile and get destroyed during autoclaving with the rest of nutrient medium. Such chemicals are therefore sterilized by filtration through a sieve or filtration assembly using filter membranes of 0.45 to 0.22 μm. A range of bacteria proof membrane filters (Millipore-Millipore Co., Nalgene-Nalgene Co., Sartorius-Sartorius Filters, Inc., etc.) and associated equipment are available for sterilization of different volumes of liquid in the range of 1–200 ml. Most filters membranes are of cellulose acetate and/or cellulose nitrate, and are available in pre-sterilized, plastic disposable units. During filter sterilization, all the particles, microorganisms and viruses that are bigger than the pore diameter of the filter used are removed. Filter sterilization is often used if a thermolabile substance (e.g. X) is needed in the medium (e.g. A). The basic medium without X substance is first autoclaved in a flask, while the medium is still liquid (i.e. ~50°C), the substance X is injected with the hypodermic syringe fitted to a membrane filter. These operations are carried out in the laminar airflow hood. The medium containing X is then stirred and dispensed into pre-sterilized containers. For larger volumes, filter sterilization can be carried out using a vacuum filtering setup marketed by various companies.

### Ultra Violet Sterilization

Sterilization of nutrient media by irradiation (via gamma rays) is hardly ever used for tissue culture media preparation because it is extremely expensive as compared to the usual method of autoclaving. The disposable plastic wares are generally UV-radiation sterilized. The sterilized autoclaved media are then dispensed into these sterilized containers in the laminar airflow cabinet. Radiation treatment as such is generally not used in research laboratories. Also UV irradiation may generate PCR inhibitors and ruin plasticware and pipettes.

## MAINTENANCE OF ASEPTIC CONDITIONS

The following procedures/techniques are employed during aseptic transfer:

### Alcohol Sterilization

It is essential that worker hands be relatively aseptic during manipulation work. A wash with an antibacterial detergent followed by spraying of 70% ethanol on hands is quite effective. The laminar airflow cabinets should also be sprayed with 70% ethanol before use.

### Flame sterilization

This method is used for sterilization of instruments that are continuously used during manipulation work. Instruments are soaked in 70–80% alcohol followed by flaming on a burner in the laminar airflow hood, which should be carried out repeatedly while aseptic manipulation work is in progress. This is essential even if instruments have been subjected to prior dry or steam sterilization.

Culture containers should be covered as quickly as possible after an operational step is completed.

Talking in the hood should be avoided. If necessary, turn your face either left or right, but the hands should remain inside the cabinet.

Set up all materials including containers, media, etc. in one side of the cabinet. It should not disturb the airflow pattern. Material should not come in the path of transfer work to be carried out in the cabinet.

## STERILIZATION OF EXPLANT

### Chemical Sterilization

Plant material can be surface sterilized by a variety of chemicals. It is the eradication of microorganisms with the aid of chemicals. The type and concentration of chemical sterilant to be used and exposure time must be decided upon empirically. Some of the commonly used chemical sterilants and their effectiveness is shown in a comparative form in Table 4.1. It is frequently seen that over-zealous sterilization leads not only to the complete removal of all microorganisms but is also lethal to the plant tissue. It is therefore important to determine the optimal conditions for each tissue.

1. 1% solution of sodium hypochlorite—NaClO, commercial bleach: It is generally available with 5% active chlorine content, so 20% can be used for normal sterilization.
2. Calcium hypochlorite $Ca(ClO)_2$: This comes in a powder form. It is mixed well with water,

**Table 4.1:** A comparison of the effectiveness and properties of common surface sterilants for explant

| Sterilizing Agent | Concentration used | Ease of Removal | Treatment time (min) | Remarks |
|---|---|---|---|---|
| Sodium hypochlorite | 1–1.4%[a] | +++ | 5–30 | Very effective |
| Calcium hypochlorite | 9–10% | +++ | 5–30 | Very effective |
| Hydrogen peroxide | 10–12 % | +++++ | 5–15 | Effective |
| Bromine water | 1–2% | +++ | 2–10 | Very effective |
| Silver nitrate | 1% | + | 5–30 | Effective |
| Mercuric chloride | 0.01–1% | + | 2–10 | Satisfactory |
| Antibiotics | 4–50 mg/l | ++ | 30–60 | Effective |

[a] Common usage rate is 20% v/v of commercial solution

allowed to settle and the clarified filtered supernatant solution is used for sterilization. Generally, 4–10g/100ml of $Ca(ClO)_2$ is used. Calcium hypochlorite enters the plant tissue slowly as compared to sodium hypochlorite. It can be stored only for a limited period of time because of deliquescent (takes up water) nature.

3. 1% solution of bromine water.

4. 0.01 to 1% solution of mercuric chloride: It is dissolved in water to make the solution. Mercuric chloride is an extremely toxic substance for plants, so rinsing must be very thorough. It should be disposed off in a separately marked container.

5. Alcohol: 70% alcohol is used for sterilization of plant material by dipping them for 30 sec to 2 min. Generally, alcohol alone is not sufficient to kill all the microorganisms and the plant material after alcohol treatment is treated with another chemical sterilant.

6. 10% hydrogen peroxide.

7. 1% silver nitrate solution.

8. 4–50 mg/l of antibiotic solution.

It is important that the plant surface be properly wetted by the sterilizing solution. A 30 sec immersion in 70% (v/v) ethanol or the addition of a few drops of liquid detergent (e.g. Teepol, Tween 20) to the other chemical sterilant solutions during use will ensure effectiveness of the sterilizing agent. Explants after treatment with sterilants must be thoroughly rinsed with several changes of sterile distilled water because

retention of such noxious chemicals will seriously affect the establishment of culture.

## PROTOCOLS

### 4.1: Sterilization of Seeds

1. Take seeds in a tea strainer/nylon net pouch/ muslin cloth and washed in running tap water. If seeds are highly infected then first wash them in a beaker with few drops of liquid detergent.

2. After washing, seeds are treated with 70% alcohol for 30 sec to 2 min in a beaker. The alcohol is then decanted.

3. Then treat seeds with 20–40% solution of sodium hypochlorite solution (preferably 40%) containing 2–3 drops of Tween 20 for 10–20 min with stirring. Alternatively, treat the seeds with 0.1 to 1% solution of mercuric chloride for 2–10 min or treat the seeds with 4–10% saturated solution of calcium hypochlorite for 10–20 min.

4. In a sterile laminar airflow hood, decant the chemical sterilant and wash the seeds thoroughly 5x with sterile distilled water.

### 4.2: Sterilization of Buds, Leaf, Stem, Roots, Tubers, Scales, etc.

1. Treat the tissue for 30 sec to 2 min in 70% alcohol by submerging in it. Tubers, roots and bulbs should be thoroughly washed in detergent and running tap water before alcohol treatment.

2. Tissue is then treated with 20% sodium hypochlorite solution (1% active chlorine content) for 5–10 minutes. Concentration and/or time can be varied depending upon the tissue and degree of contamination.

3. In a sterile laminar airflow hood, decant the chemical sterilant and wash the tissue thoroughly 3–5x with sterile distilled water.

## 4.3: Sterilization of Tissue for Immature Embryos, Ovules, and Flower Buds for Anther Culture

1. The whole enclosing tissue is surface sterilized, e.g. ovary, developing seeds of wheat/barley/maize/rice etc. for immature embryos and flower buds for anther culture.

2. Tissue is submerged in 20% sodium hypochlorite solution containing 2–3 drops of Tween 20 for 5–10 min with stirring.

3. In a sterile laminar airflow hood, decant the chemical sterilant and wash the tissue thoroughly 3–5x with sterile distilled water.

4. Dissect-out/excise the required explant.

## FURTHER READING

Evans, D.A., Sharp, W.R., Ammirato, P.V. and Yamada, Y., 1983. *Handbook of Plant Cell Culture*. Macmillan Publishing Company, New York.

Pierik, R.L.M., 1987. *In vitro culture of Higher Plants*. Martinus Nijhoff Publishers, Dordrecht.

Reinert, J. and Bajaj, P.P.S. (eds.) 1977. *Applied and Fundamental Aspects of Plant Cell, Tissue and Organ Culture*. Springer-Verlag, Berlin.

Street, H.S. (ed.)1973. *Plant Tissue and Cell culture*. Blackwell Scientific Publications, Oxford.

Thorpe, T.A., 1981. *Plant Tissue Culture: Methods and Applications in Agriculture*. Academic Press Inc., New York.

Vasil, I.K., 1984. *Cell Culture and Somatic Cell Genetics of Plants: Laboratory Procedures and Their Applications*. Academic Press, Inc., New York.

# 5

# Types of Culture

A plant consists of different organs and each organ is composed of different tissues, which in turn are made up of different individual cells. Plant tissue culture refers to the *in vitro* cultivation of plants, seeds, plant parts (tissues, organs, embryos, single cells, protoplasts) on nutrient media under aseptic conditions. The procedures of plant tissue culture have developed to such a level that any plant species can be regenerated *in vitro* through several methodologies. Unlike animal cells, plant cells, even highly mature and differentiated, retain the ability to change to a meristematic state and differentiate in to a whole plant if it has retained an intact membrane system and a viable nucleus.

In plant tissue culture, more often we use an explant (an excised piece of differentiated tissue or organ) to initiate their growth in culture. The non-dividing, differentiated, quiescent cells of the explant when grown on a nutrient medium first undergo changes to achieve the meristematic state. The phenomenon of mature cells reverting to a meristematic state and forming undifferentiated callus tissue is termed as dedifferentiation. Since the multicellular explant comprises cells of diverse types, the callus derived will be heterogeneous. The ability of the component cells of the callus to differentiate into a whole plant or a plant organ is termed as redifferentiation. These two phenomena of dedifferentiation and redifferentiation are inherent in the capacity of a plant cell, and thus giving rise to a whole plant is described as cellular totipotency. Generally, a callus phase is involved before the cells can undergo redifferentiation leading to the regeneration of a whole plant. The dedifferentiated cells can rarely give rise to whole plants directly without an intermediate callus phase.

## CYTODIFFERENTIATION

Cell differentiation is the basic event of development in higher organisms. The cells in a callus are parenchymatous in nature; the differentiation of these cells into a variety of cells is required during redifferentiation of cells into whole plants which is conveniently referred to as cytodifferentiation. *In vitro* and *in vivo,* the main emphasis in plant cytodifferentiation has been laid on vascular tissue differentiation (xylem, phloem), particularly the xylem elements. In an intact plant, tissue differentiation goes on in a fixed manner that is characteristic of the species and the organ. Callus cultures, which lack vascular elements offer a valuable system for the study of the effect of various factors that control cellular totipotency through cytological, histological and organogenic differentiation.

## Vascular Differentiation

One of the efficient systems for the study of cytodifferentiation *in vitro* is *xylogenesis*. Xylogenesis is the differentiation of parenchyma into cells that have localized secondary wall thickenings as seen in the xylem of vascular plants. These xylem-like cells have been variously named as wound vessel members, vessel elements, tracheids and tracheary elements. In general the tracheary element term has been used for such differentiated cells. The formation of tracheary elements helps in understanding the mechanism of differentiation in higher plant cells because (1) the morphological characteristics of tracheary elements such as annular, spiral, reticulate and pitted secondary wall thickenings enable us to distinguish differentiated cells from undifferentiated cells; (2) the formation of tracheary elements can be induced in tissue and cell cultures of many species; (3) specific biochemical events leading to deposition of cell wall polysaccharides and lignin make it possible to trace marker proteins associated with the process of cytodifferentiation and (4) due to loss of nucleus the tracheary element loses its capacity for de- or redifferentiation.

For the study of cytodifferentiation, it is desirable that explants ideally should have cells in a uniform physiological state so that differentiation occurs synchronously. Callus, which shows stable characteristics under specific conditions after subculture through many successive passages, is a suitable material for cytodifferentiation. The advantage of callus is that it is composed of a fairly homogeneous mass of cells and can be proliferated in large amounts under known culture conditions. In a callus system, tracheary elements often pre-exist and this necessitates selection of a maintenance medium that does not support the formation of these cells while subculturing the callus. The disadvantages are: (i) many cultures lose their potential for differentiation during continual subculture; (ii) tracheary elements are formed as discreet nests or nodules giving the appearance of a tissue, rather than as scattered groups or single elements and thus difficult to analyse the process of cytodifferentiation as distinct from tissue formation. Cells in suspension culture are ideally a good system for study of cytodifferentiation as cells receive more homogeneous stimuli in a defined medium. Theoretically, the percentage and synchrony of differentiation may be expected to be higher in cells of suspension culture. Vascular element formation has been reported in protoplasts also but the system needs to be improved because of the low rate of tracheary element formation and their asynchroneity.

*Cytological and cytochemical aspects*: The studies during various stages of cell differentiation reveal some early cytological changes. Cells elongate due to synthesis of metabolites and thickenings appear in the cell wall, which are mostly reticulate or helical in pattern. Secondary wall thickenings include cellulose microfibrils, xylan, lignin, hemicellulose, pectin and protein. Cell autolysis is the general feature after secondary wall thickenings due to loss of nucleus and cytoplasmic contents in cultured cells. This process, leading towards cell death, seems to be programmed at the beginning of secondary wall thickening. Vacuoles in the tracheary elements may function as lysosomes at some stage and release hydrolytic enzymes, resulting in disruption of the intracellular structures.

*Physiological aspects*: Phytohormones, particularly auxins, are reported to effect vascular differentiation quantitatively as well as qualitatively. Some evidence also points towards the involvement of cytokinins and gibberellins in the process of xylogenesis. Several reports suggest that an exogenous auxin is essential for cytodifferentiation. The optimum concentration varies with respect to the species. However, cytokinin is an absolute requirement for vascular differentiation in *Zinnia* mesophyll cells, soybean callus and root cortical explants. In most other explants (e.g., Jerusalem artichoke tubers, carrot roots), exogenous cytokinin appears to act only in the stimulation of auxin-induced tracheary element differentiation. Sometimes a high cytokinin and auxin ratio in the medium favors both xylem and phloem differentiation (sycamore callus) or different combinations of phytohormones may bring about different patterns of xylem formation (lettuce pith explants) in cultures. These results appear to show that auxin is a principal limiting factor of vascular differentiation *in vitro* and that cytokinin acts in the succeeding process to the onset of cytodifferentiation caused by auxin. The requirement of GA in the medium for cellular differentiation is not absolutely necessary.

Sucrose, in addition to its role as an energy source, may act as a regulator of xylem differentiation in cultured tissues. The relative amount of xylem and phloem tissues in the callus varies considerably under the varying concentrations of sucrose in combination with the auxin. Generally, auxin at low concentration (0.05mg/l) stimulates xylogenesis and there is an inverse relationship between the auxin concentration and degree of xylem differentiation. In studies on *Syringa*, keeping the concentration of IAA auxin at 0.05mg/l constant and sucrose level at 1% induced little xylem formation; 2% sucrose favoured better xylem formation with little or no phloem; 2.5–3.5% sucrose resulted in both xylem and phloem differentiation and with 4% sucrose, phloem was formed with little or no xylem. Besides sucrose, other sugars such as glucose, fructose, maltose and trehalose have also been used to stimulate vascular differentiation

Observations on various experiments deduce that a reduction in the total nitrogen ($NH_4^+$ and $NO_3^-$) content of the medium in cultures contributes to an increase in the percentage of vascularization without affecting cell division in tissues and cells. Besides, light and temperature and other physical factors have a pronounced effect on vascular differentiation.

*Cell Division and Cytodifferentiation*: The *in vitro* system is a good tool for studying the relationship between cell division and differentiation. Since the chemical factors (auxin, cytokinin, sugar) involved in vascular differentiation are generally the same as those regulating cell division, it appears that cell division and cytodifferentiation are two sides of the same coin.

The first report suggesting a close relationship between tracheary element differentiation and the cell cycle showed that inhibitors of DNA synthesis (mitomycin C, fluorodeoxyuridine and colchicines) prevented cytodifferentiation in cultured stem segments of *Coleus* in which cell division preceded the differentiation. Such observation gave the impression that cell division is essential for cell differentiation. On the contrary, there are several reports suggesting that tracheary differentiation occurs without cell division.

From the point of view of cytodifferentiation, it may be interesting to consider the type of cell cycle preceeding the origin of a new cell. The cell cycle may be either of the *quantal* or *proliferative* type. In a quantal cell cycle the daughter cells that form differ from the mother cell while the proliferative cell cycle generates copies of the mother cell. Cells originating from the quantal cycle are unlikely to differentiate, whereas a cell formed by the proliferative cycle need not undergo further division for differentiation. The cytodifferentiation of tracheary elements is an irreversible specialization of cells accompanied by the loss of nuclei and cell contents at maturity. This rules out dedifferentiation or redifferentiation of a cell that has once differentiated to a tracheary element or is in the progression of cytodifferentiation.

Recent advances on vascular differentiation have led to the identification of a number of vascular specific genes. Demura and Fukuda (1994) isolated 3 cDNA clones of tracheary element genes (TED2, TED3, and TED4) expressing in mesophyll cells of Zinnia during their redifferentiation into tracheary elements.

# ORGANOGENIC DIFFERENTIATION

In nature, totipotentiality of somatic cells has been observed in several taxa where stem, leaf and root pieces are able to differentiate into shoots and roots. *In vitro* studies have indicated that totipotentiality is not restricted to few species; most plant species if provided with appropriate conditions would show differentiation. For the regeneration of a whole plant from a cell or callus tissue, cytodifferentiation is not enough and there should be differentiation leading to shoot bud or embryo formation. This may occur either through organogenesis or somatic embryogenesis.

Organogenic differentiation is an outcome of the process of dedifferentiation followed by redifferentiation of cells. Dedifferentiation favors unorganized cell growth and the resultant development callus has meristems randomly divided. Most of these meristems, if provided appropriate *in vitro* conditions, would redifferentiate into shoot buds and roots. Mostly, the whole plant regeneration from cultured cells may occur either through shoot-bud differentiation or somatic embryogenesis and thus establishing the totipotency of somatic cells. The cell(s) of stem, leaf and root cuttings of several taxa are able to directly differentiate into shoots and roots, leading to the establishment of new individuals under *in vivo* conditions. Shoot bud formation is organogenesis and various factors affect the process.

Shoot-bud differentiation in cultured tissues is dependent on the auxin/cytokinin ratio in the medium. Skoog and Miller (1957) rejected the concept of organ forming substances (rhizocaulines and caulocalines) proposed by Went (1938) and instead suggested that organ formation is controlled by quantitative inter-action of different substances in their proportion rather than absolute concentration of substances.

Cytokinin (adenine or kinetin) in the medium leads to the promotion of bud differentiation and development. Kinetin is 30,000 times more potent than adenine. The shoot-forming effect is modified by auxin (IAA and NAA), which at lower concentrations (5μM IAA) suppresses the differentiation of buds in the tobacco callus. About 15000 molecules of adenine or 2 molecules of kinetin are required to neutralize the inhibitory effect of 1 molecule of IAA on shoot-bud differentiation. A relatively high concentration of auxin favours cell proliferation and root differentiation while higher levels of cytokinin promote bud differentiation. Thus, root-shoot differentiation is a function of quantitative interaction of auxin and cytokinin rather than absolute concentration. Not only growth regulators but increased levels of phosphate ($PO_4^-$) in the medium is reported to counteract the inhibitory effect of auxin and promotes bud formation in the absence of cytokinin. Casein hydrolyste or tyrosine also induces kinetin type bud formation even in the presence of higher levels of IAA in the medium.

The requirement for exogenous auxin and cytokinin in the process of bud differentiation varies with the tissue system and apparently depends on the endogenous levels of the auxin and cytokinin growth regulators in the tissue. Endosperm cultures require cytokinin alone or in combination with a very low level of auxin for bud differentiation. Polyamines have been shown to be associated with induction of cell division, growth and differentiation of bacterial, animal and plant cells. However, their mechanism of action is not known. Other cytokinins which influence the induction of shoot-buds include BAP, 2-iP, 6-tetrahydropyrane-adenine (SD 8339) and zeatin.

*Anatomy of Shoot-bud differentiation*: Under conditions favoring unorganized growth, the meristems in a callus are random and scattered. Transfer of callus pieces to conditions supporting organized growth leads to the formation of *meristemoids* (also termed 'nodules' or 'growth centres'). The meristemoids are localized clusters of cambium-like cells which may become vascularised due to the appearance of tracheidal cells in the centre. These are the site for organ formation in the callus and can form roots or shoots.

In shoot bud organogenesis, a monopolar structure that has a connection with the pre-existing vascular tissue within the callus, while in embryogenesis a bipolar structure with no vascular connection with the maternal callus tissue or explant is formed. Regeneration of plants via organogenesis and embryogenesis has been discussed in micropropagation (see Chapter 6).

## TYPES OF CULTURE

Plant tissue culture, which covers all types of aseptic plant culture should be used in a restricted sense and it is possible to distinguish it into various types of cultures (Fig. 5.1).

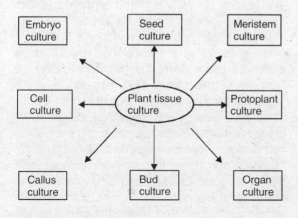

**Fig. 5.1:** Types of culture.

Seed culture — Culture of seeds *in vitro* to generate seedlings/plants

Embryo culture — Culture of isolated mature or immature embryos

Organ culture — Culture of isolated plant organs. Different types can be distinguished, e.g. meristem, shoot tip, root culture, anther culture

Callus culture — Culture of a differentiated tissue from explant allowed to dedifferentiate *in vitro* and a so-called callus tissue is produced

Cell culture — Culture of isolated cells or very small cell aggregates remaining dispersed in liquid medium

Protoplast culture — Culture of plant protoplasts, i.e. cells devoid of their cell walls

### Seed culture

Seed culture is an important technique when explants are taken from *in vitro*-derived plants and in propagation of orchids. Sterilizing procedures are needed for plant materials that are to be used directly, as explant source can cause damage to the tissues and affect regeneration. In that case, culture of seeds to raise sterile seedlings is the best method. Besides, in several species during *in vivo* conditions, seeds do not germinate well, e.g. orchids. It has been reported that orchids live in symbiotic relationship with the fungus from the moment of germination. Symbiosis is the association of two organisms to their mutual advantage. The research carried out by Knudson around 1922 showed that seeds of *Cattleya, Laelia, Epidendrum,* and many other orchids were able to germinate asymbiotically *in vitro*. In nature germination of orchid seedlings is dependent on a symbiotic relationship with a fungus. However, *in vitro* it is possible to be independent of the fungus by substituting its action with a nutrient medium, and this is known as asymbiotic germination. Orchid seeds are propagated vegetatively as well as generatively. With vegetative propagation, the progeny is identical to the parent plants. However, with generative propagation (by seed), identical progeny is rarely obtained. If seeds from a cultivated orchid are used (mainly obtained from a cross and strongly heterozygous), the progeny will be extremely heterogeneous and seldom identical to the starting material. In principle, cultivated orchids can only be propagated vegetatively. Orchid cloning *in vivo* is a very slow process. Thus seeds can be germinated *in vitro* and meri-cloning (vegetative propagation by meristem culture) is then carried out on a large scale.

Most of the orchids are sown *in vitro* because

i. Orchid seeds are very small and contain very little or no food reserves. Their small size (1.0 to 2.0 mm long and 0.5–1.0 mm wide) makes it very likely that they can be lost if sown *in vivo*, and the limited food reserves also make survival *in vivo* unlikely. The seed consists of a thickened testa, enclosing an embryo of about 100 cells. The embryo has a round or spherical form. Most orchid seeds are not differentiated: there are no cotyledons, roots and no endosperm. The cells of the embryo have a simple structure and are poorly differentiated;

ii. sowing *in vitro* makes it possible to germinate immature orchid embryos, thus shortening the breeding cycle; and

iii. germination and development take place much quicker *in vitro* since there is a conditioned environment and no competition with fungi or bacteria.

Orchid seeds imbibe water via the testa and becomes swollen. After cell division has begun, the embryo cracks out of the seed coat. A protocorm-like structure is formed from the clump of cells and on this a shoot meristem can be distinguished. Protocorm has a morphological state that lies between an undifferentiated embryo and a shoot. Protocorms obtained by seed germination have many close similarities with those produced from isolated shoot tips; the term protocorm like-bodies has been introduced while cloning orchids by meristem culture. The vegetative propagation of orchids follows culture of seeds, transformation of meristem into protocorm-like bodies, the propagation of protocorms by cutting them into pieces and the development of these protocorms to rooted shoots.

A large number of factors influence the germination and growth of orchids. The mineral requirement of orchids is generally not high and a salt poor medium of Knudson (1946) and Vacin and Went (1949) are good. Some of the orchids require darkness (*Paphiopedilum ciliolare*) for germination while others require low irradiance. Sugar is extremely important as an energy source, especially for those that germinate in darkness. Regulators are usually not necessary for seed germination, and their addition often leads to unwanted effects like callus formation, adventitious shoot formation etc.

## Embryo culture

Embryo culture is the sterile isolation and growth of an immature or mature embryo *in vitro*, with the goal of obtaining a viable plant. The first attempt to grow the embryos of angiosperms was made by Hannig in 1904 who obtained viable plants from *in vitro* isolated embryos of two crucifers *Cochleria* and *Raphanus* (Hannig, 1904). In 1924, Dietrich grew embryos of different plant species and established that mature embryos grew normally but those excised from immature seeds failed to achieve the organization of a mature embryo (Dietrich, 1924). They grew directly into seedlings, skipping the stages of normal embryogenesis and without the completion of dormancy period. Laibach (1925, 1929) demonstrated the practical application of this technique by isolating and growing the embryos of interspecific cross *Linum perenne* and *L. austriacum* that aborted *in vivo*. This led Laibach to suggest that in all crosses where viable seeds are not formed, it may be appropriate to excise their embryos and grow them in an artificial nutrient medium. Embryo culture is now a well-established branch of plant tissue culture.

There are two types of embryo culture

**Mature embryo culture:** It is the culture of mature embryos derived from ripe seeds. This type of culture is done when embryos do not survive *in vivo* or become dormant for long periods of time or is done to eliminate the inhibition of seed germination. Seed dormancy of many species is due to chemical inhibitors or mechanical resistance present in the structures covering the embryo, rather than dormancy of the embryonic tissue. Excision of embryos from the testa and culturing them in the nutrient media may bypass such seed dormancy. Some species produce sterile seeds, which may be due to incomplete embryo development. Embryo culture procedures may yield viable seedlings. Embryos excised from the developing seed at or near the mature stage are autotrophic and grow on a simple inorganic medium with a supplemental energy source.

**Immature embryo culture/embryo rescue:** It is the culture of immature embryos to rescue the embryos of wide crosses. This is mainly used to avoid embryo abortion with the purpose of producing a viable plant. Wide hybridization, where individuals from two different species of the same genus or different genera are crossed, often leads to failure. There are several barriers, which operate at pre- and post-fertilization levels to prevent the successful gene transfer from wild into cultivated species. The pre-fertilization barriers include all factors that hinder effective fertilization, which is usually due to inhibition of pollen tube growth by the stigma or upper style. Post-fertilization barriers hinder or retard the development of the zygote after fertilization and normal development of the seed. This frequently results from the failure of the hybrid endosperm to develop properly, leading to starvation and abortion of the hybrid embryo or results from embryo-endosperm incompatibility where the endosperm produces toxins that kills the embryo.

Raghavan (1976) has discussed evidence, which suggests that embryos of inviable hybrids possess the potential for initiating development, but are inhibited from reaching adult size with normal differentiation. Endosperm development precedes and supports embryo development nutritionally, and endosperm failure has been implicated in numerous cases of embryo abortion. Endosperm failure generally results in abnormal embryo development and eventual starvation. Thus, isolation and culture of hybrid embryos prior to abortion may circumvent these strong post-zygotic barriers to interspecific and intergeneric hybridization. The production of interspecific and intergeneric hybrids is the most conspicuous and impressive application of embryo rescue and culture technique, particularly for subsequent valuable gene transfer from wild species.

The underlying principle of embryo rescue technique is the aseptic isolation of embryo and its transfer to a suitable medium for development under optimum culture conditions. With embryo culture there are normally no problems with disinfecting. Florets are removed at the proper time and either florets or ovaries are sterilized. Ovules can then be removed from the ovaries. The tissue within the ovule, in which the embryo is embedded, is already sterile. For mature embryo culture either single mature seeds are disinfected or if the seeds are still unripe then the still closed fruit is disinfected. The embryos can then be aseptically removed from the ovules. Utilization of embryo culture to overcome seed dormancy requires a different procedure. Seeds that have hard coats are sterilized and soaked in water for few hours to few days. Sterile seeds are then split and the embryos excised.

The most important aspect of embryo culture work is the selection of medium necessary to sustain continued growth of the embryo. In most cases a standard basal plant growth medium with major salts and trace elements may be utilized. Mature embryos can be grown in a basal salt medium with a carbon energy source such as sucrose. But young embryos in addition require different vitamins, amino acids, and growth regulators and in some cases natural endosperm extracts. Young embryos should be transferred to a medium with high sucrose concentration (8–12%) which approximate the high osmotic potential of the intracellular environment of the young embryo-sac, and a combination of hormones which supports the growth of heart-stage embryos (a moderate level of auxin and a low level of cytokinin). Reduced organic nitrogen as aspargine, glutamine or casein hydrolysate is always beneficial for embryo culture. Malic acid is often added to the embryo culture medium. After one or two weeks when embryo ceases to grow it must be transferred to a second medium with a normal sucrose concentration, low level of auxin and a moderate level of cytokinin which allows for renewed embryo growth with direct shoot germination in many cases. In some cases where embryo does not show shoot formation directly, it can be transferred to a medium for callus induction followed by shoot induction. After the embryos have grown into plantlets *in vitro*, they are generally transferred to sterile soil and grown to maturity.

*Applications of embryo culture*

1. *Prevention of embryo abortion in wide crosses*: Successful interspecific hybrids have been seen in cotton, barley, tomato rice, legume, flax and well known intergeneric hybrids include wheat × barley, wheat × rye, barley × rye, maize × *Tripsacum, Raphanus sativus* × *Brassica napus*. Distant hybrids have also been obtained via embryo rescue in *Carica* and *Citrus* species. Embryo rescue technique has been successfully used for raising hybrid embryos between *Actidinia deliciosa* × *A. eriantha* and *A.deliciosa* × *A. arguata*. Embryo rescue from a cross between an early Japanese cv. Sunago Wase and a Chinese cv. Yuhualu obtained 'Zaoxialu' an

extra early maturing peach cultivar. It matures in 55 days in Hanzhou, China. Some of the hybrid plants raised by embryo culture have recombined desirable genes such as earliness, disease and pest resistance and are listed in Table 5.1. Embryo culture is also used in crosses between diploids and tetraploids.

2. *Production of haploids*: Embryo culture can be utilized in the production of haploids or monoploids. Kasha and Kao (1970) have developed a technique to produce barley monoploids. Interspecific crosses are made with *Hordeum bulbosum* as the pollen parent, and the resulting hybrid embryos are cultured, but they exhibit *H. bulbosum* chromosome elimination resulting in the monoploids of the female parent *H. vulgare*.

3. *Overcoming seed dormancy*: Embryo culture technique is applied to break seed dormancy. Seed dormancy can be caused by numerous factors including endogenous inhibitors, specific light requirements, low temperature, dry storage requirements and embryo immaturity. These factors can be circumvented by embryo excision and culture. Seed dormancy in *Iris* is due to the presence of stable chemical inhibitor in the endosperm. American basswood (*Tilia americana*) seed is borne within a tough indehiscent pericarp where the resistance is mechanical. Thus, there is difficulty in

**Table 5.1:** Resistance traits transferred to cultivated species through embryo rescue technique

| Crossing species | Resistance traits(s) |
| --- | --- |
| *Lycopersicon esculentum* x *L. peruvianum* | Virus, fungi and nematodes |
| *Solanum melongena* x *S. khasianum* | Brinjal shoot and fruit borer (*Leucinodes arbonalis*) |
| *Solanum tuberosum* x *S. etuberosum* | Potato leaf roll virus |
| *Brassica napus* x *Raphanobrassica* | Shattering resistance |
| *Brassica oleracea* x *B. napus* | Triazine resistance |
| *Brassica napus* x *B. oleracea* (Kale) | Cabbage aphid |
| *Triticum aestivum* x *Thinopyrum scirpeum* | Salt tolerance |
| *Hordeum sativum* x *H. vulgare* | Powdery mildew and spot blotch |
| *Hordeum vulgare* x *H. bulbosum* | Powdery mildew |
| *Oryza sativa* x *O. minuta* | Blast (*Pyricularia grisea*) and Bacterial blight (*Xanthomonas oryzae*) |

germinating seeds of this species. By excising embryos, germination occurred without delay at all stages of seed maturity. A potential use of the technique is in the production of seedlings from seed of naturally vegetatively propagated plants such as bananas (*Musa balbisiana*) and *Colocasia* whose seeds do not germinate in nature, probably due to recalcitrant dormancy. Embryo culture techniques may thus be capable of producing viable seedlings in these species.

4. *Shortening of breeding cycle*: There are many species that exhibit seed dormancy that is often localized in the seed coat and/or in the endosperm. By removing these inhibitions, seeds germinate immediately. Seeds sometimes take up water and $O_2$ very slowly or not at all through the seed coat, and so germinate very slowly if at all, e.g. Brussels sprouts, rose, apple, oil palm and iris. Hollies (*Ilex*) are important plants for Christmas decorations. *Ilex* embryos remain in the immature heart-shaped stage though the fruits have reached maturity. Under proper germination conditions, three years are required for seeds of *I. opaca* from mature berries to complete their embryonic development and to begin germination. The excised embryo go through all the developmental sequences *in vitro* and plants can be obtained in 2–3 weeks time. *Rosa* normally takes a whole year to come to flowering and through embryo culture it has been possible to produce two generations in a year.

5. *Prevention of embryo abortion with early ripening stone fruits*: Some species produce sterile seeds that will not germinate under appropriate conditions and will eventually decay in soil, e.g. early ripening varieties of peach, cherry, apricot, plum. Seed sterility may be due to incomplete embryo development, which results in the death of the germinating embryo. In crosses of early ripening stone fruits, the transport of water and nutrients to the yet immature embryo is sometimes cut off too soon resulting in abortion of the embryo. Macapuno coconuts are priced for their characteristic soft endosperm which fills the whole nut. These nuts always fail to germinate because the endosperm invariably rots before germinating embryo comes out of the shell. Embryo culture has been practised as a general method in horticultural crops including avocado, peach, nectarine and plum. Two cultivars 'Goldcrest peach' and 'Mayfire nectarine' have resulted from embryo culture and are now commercially grown.

6. Embryos are excellent materials for *in vitro* clonal propagation. This is especially true for conifers and members of Gramineae family.

7. Germination of seeds of obligatory parasites without the host is impossible *in vivo*, but is achievable with embryo culture.

**Root Culture**

Roots have been cultured since the early days of tissue culture research. Kotte (Germany) and Robbins (USA) started work on root culture in 1922. They postulated that a true *in vitro* culture could be made easier by using meristematic cells, such as those that operate in the root tip or bud. Small excised root tips of pea and maize cultivated by Kotte in a variety of nutrients containing salts of Knop's solution, glucose and several nitrogen compounds (such as asparagines, alanine, and yeast extract) grew successfully for two weeks. Robbins, working independently, maintained his maize root tip cultures for a longer period by subculturing them, but the growth gradually diminished and the cultures were ultimately lost. An important breakthrough for continuously growing root tip cultures came from White in 1934, who initially used yeast extract in a medium containing inorganic salts and sucrose but later replaced yeast extract by three B vitamins, namely, pyridoxine, thiamine and nicotinic acid. White's synthetic medium later proved to be one of the basic media for a variety of cell and tissue cultures.

Root-tip cultures are generally maintained in an agitated liquid medium with an appropriate auxin and they extended during culture, providing numerous roots. This simple and stable organ culture system was used by plant physiologists for many years and provided useful insights into plant root growth and mineral nutrition.

Solanaceous species produce tropane alkaloids in the roots. In cell cultures of most such species, only trace levels of alkaloids are produced. However, with the regeneration of roots from the undifferentiated calli, the production of alkaloids increased dramatically. Thus interest in root cultures for photochemical production developed in the 1980s, and root cultures of *Hyoscyamus, Duboisia, Atropa* were found to produce the alkaloids characteristic of the species.

However, due to the slow growth of roots (as compared to cell suspension cultures), and due to the difficulty in culturing roots in many species, interest in them as a source of secondary products has been limited.

## Callus culture

Callus is basically a more or less non-organized tumor tissue which usually arises on wounds of differentiated tissues and organs. Thus, in principle, it is a non-organized and little-differentiated tissue. The cells in callus are of a parenchymatous nature. When critically examined, callus culture is not homogeneous mass of cells, because it is usually made up of two types of tissue: differentiated and non-differentiated. Explant tissue is a differentiated tissue (roots, stem, leaves, flowers, etc.) which is used as a starting material for callus induction. These explant tissues generally show distinct planes of cell division, cell proliferation and organization into specialized structures such as vascular systems. If there are only differentiated cells present in an isolated explant, then dedifferentiation must take place before cell

division can occur. Parenchyma cells present in the explant usually undergo this differentiation. If the explant already contains meristematic tissue when isolated, then this can divide immediately without dedifferentiation taking place. Dedifferentiation plays a very important role, enabling mature cells in an explant isolated from an adult plant to be redetermined. In this process, adult cells are temporarily able to revert from the adult to the juvenile state. The rejuvenated cells have a greater growth and division potential and under special circumstances are able to regenerate into organs and/or embryos.

Callus formation takes place under the influence of exogenously supplied growth regulators present in the nutrient medium. The type of growth regulator requirement and its concentration in the medium depends strongly on the genotype and endogenous hormone content of an explant. These requirements can be put into three categories:

1. Auxin alone (especially in monocotyledons)
2. Cytokinin alone
3. Both auxin and cytokinin (carrot)

If the callus is difficult to induce, or if juvenile callus is needed, then immature embryos or seedlings or parts of these are used. It should be taken into account that the type of starting material (juvenile or adult) and the original position of the explant in the plant reflects the endogenous growth regulator level which have an important influence on processes such as cell division and organ and embryo formation.

Many other factors are important for callus formation: genotype, composition of the nutrient medium, physical growth factors (light, temperature, etc.). The Murashige and Skoog (1962) mineral medium, or modifications of this are often used. Sucrose or glucose (2–4%) is usually employed as the sugar source. The effect of light on callus formation is dependent on the plant species; light may be required in some cases and darkness in other cases. A temperature of 22–28°C is normally advantageous for callus formation.

Callus tissue induced from different plant species may be different in structure and growth habit: white or coloured, soft (watery) or hard, friable (easy to separate into cells) or compact. The callus growth within a plant species may also depend on factors such as the original position of the explant within the plant, and the growth conditions.

After callus induction, the callus is grown further on a new medium which is referred to as subculturing. When subcultured regularly on agar medium, callus cultures will exhibit an S-shaped or sigmoid pattern of growth during each passage (See Fig. 7.1). There are five phases of callus growth:

1. Lag phase, where cells prepare to divide.
2. Exponential phase, where the rate of cell division is highest.
3. Linear phase, where cell division slows but the rate of cells expansion increases.
4. Deceleration phase, where the rates of cell division and elongation decreases.
5. Stationary phase, where the number and size of cells remain constant.

Callus growth can be monitored by fresh weight measurements, which are convenient for observing the growth of cultures over time in a non-destructive manner. Dry weight measurements are more accurate than fresh weight, but this method requires sacrifice of the samples. Mitotic index measurement of cell division rates require extensive sampling to reduce sample error and are not easy to perform.

**Organ culture**

It is an isolated organ grown *in vitro*. It can be given different names depending upon the organ used as an explant. For example, meristem or shoot tip culture (see Chapter 6), root culture, nucellus, endosperm, ovule culture, anther culture for production of androgenic haploids while ovule and ovary culture for *in vitro* production of gynogenic haploids (see Chapter 8). The culture of plant organs results in three types of *in vitro* culture (De Fossard *et al.*, 1977).

*Organized*: The culture of whole plants (embryos, seeds) and organ has been termed as organized culture. In this, characteristic organized structure of a plant or the individual organ is maintained. If the organized structure is not broken down, then progeny arises which are identical to the original plant material (e.g. meristem culture).

*Non-organized*: If cells and/or tissues are isolated from an organized part of a plant, de-differentiate and are then cultured, a non-organized growth in the form of a callus tissue results. If the callus disperses into clumps of cells (aggregates) and/or single cells results, it is referred to as suspension culture. Non-organized cultures have very low genetic stability.

*Non-organized/organized*: This type of culture is intermediate between the above two types. Cells in an isolated organ or tissue, first dedifferentiate and then form callus tissue which then redifferentiate to form organs (roots or shoots) or embryos. Thus organized structures can develop from non-organized cultures either through special techniques or spontaneously. In this the progeny are often not completely identical to the original plant material.

## NUCELLUS CULTURE

Nucellus culture has been utilized to study factors responsible for the formation of adventive embryos. Normally the embryo originates from the zygote, but sometimes in plants such as citrus and mango, the embryos arise adventitiously from cells of the nucellus or integument. Rangaswamy (1958) studied for the first time nucellar embryony *in vitro*. He excised the nucellus from unpollinated ovules of *Citrus microcarpa*, a natural polyembryonic species and cultured on a modified White medium supplemented with casein hydrolysate. The nucellus proliferated into a callus mass, and differentiated into embryo like structures termed pseudobulbils, which eventually developed into plants. Rangan *et al.* (1968) demonstrated that it is possible to induce nucellar embryogenesis in monoembryonic citrus species as well. Adventive embryos were successfully initiated in nucellus cultures of *C. grandis*, *C. limon* and *C. reticulata* on MS medium supplemented with either malt extract (500 mg/l) or a combination of adenine sulphate (25 mg/l) + NAA (0.5 mg/l) + orange juice (5%). In these species nucellar explants did not produce callus or pseudobulbils, but gave rise directly to highly organized multiple embryos.

Pollination and fertilization are generally considered essential prerequisites for the induction of nucellar embryogenesis. However, in later studies, embryogenesis could be induced *in vitro* from nucelli of unfertilized ovules of *C. sinensis* and *C. aurantifolia*. MS medium is generally adequate for both poly and mono embryonic varieties. A high sucrose concentration of 5% is generally recommended. Medium should be supplemented with complex substances rich in amino acids such as casein hydrolysate, malt extract or yeast extract. In some systems malt extract can be replaced by a combination of adenine sulphate and kinetin. In general it appears that auxins are not essential for citrus proliferation particularly from immature fruit tissues. But auxin together with cytokinin may stimulate embryogenesis.

In a procedure of nucellus culture for direct embryogenesis in citrus immature green fruits 2.5–3.0 cm in diameter are harvested from trees 100–120 days after pollination. After sterilization, a shallow incision around the circumference of the fruit is given and ovules are excised. With the aid of microscope, the zygotic embryos along with nucellus are rolled away from the integument. At the micropylar end, it is pierced and embryo is pushed and the entire nucellus can be placed on the media.

Reports for induction of nucellar embryogenesis in other species have been very few. Mullins and Srinivasan (1976) succeeded in inducing nucellar embryony in ovule cultures of *Vitis vinifera*, a normally monoembryonic species. Unfertilized ovules grown on a media supplemented with BA and B-naphthoxy acetic acid formed a nucellar callus which subsequently produced embryoids. These embryoids when transferred to a semi-solid media containing GA and 2iP produced plantlets. Eichholtz *et al.* (1979) reported the formation of adventive embryos in nucellus cultures of apple (*Malus domestica*). Micropylar halves of nucellus formed embryo like structures 50 days after culture in darkness. Subsequent reculture of these embryos in fresh medium resulted in the formation of secondary embryo like structures from the cotyledons. Litz and Conover (1982) reported embryogenesis and plant regeneration from nucellar callus in *Carica papaya*.

### Application

The adventive embryos are of considerable importance to the horticulturists, in that they are genetically uniform and reproduce the characters of the maternal parent without inheriting the variations brought about by gametic fusion. Pseudobulbils of citrus represent tissue banks capable of initiating clones of adventive embryos. In monoembryonic Citrus taxa a serious problem is that of obtaining disease free clones. It is generally believed that most of the viruses of citrus are not transmitted through seedlings, whether of zygotic or nucellar origin. Thus in contrast to clones established from cuttings or by conventional methods of propagation, plants derived from nucellar embryos are free from most viruses. Moreover, many of the desirable plant vigor and fruiting characteristics associated with juvenility are restored in trees established from nucellar seedlings. The formation of nucellar embryos from unpollinated ovules is also of considerable interest in that it can be employed to import seedless citrus varieties without the risk of introducing new viral diseases, thereby avoiding the necessity of subjecting vegetative material to long periods of quarantine.

## ENDOSPERM CULTURE

In flowering plants nutrition of the embryo is an important and vital aspect of the life cycle. In angiosperms it is accomplished by the formation of a new structure, the endosperm produced by a unique event of double fertilization. Endosperm is formed in most cases by fusion of two polar nuclei and one of the male gametes resulting in a triploid number of chromosomes. It is the main source of reserve food for the developing embryo and has an influence on its differentiation. Failure of proper development of endosperm resulted in the abortion of embryo. Generally, the endosperm is a short-lived structure and is consumed during the development of embryo (exalbuminous seed). In some plants, e.g. castor it persists as a massive tissue even in the mature seed (albuminous). The endosperm is an excellent experimental system for morphogenic studies, since it lacks any differentiation and consists mostly of parenchymatous cells.

Attempts to grow endosperm tissue in culture began in the 1930s and immature and mature endosperms of various angiosperm taxa (autotrophic as well as parasitic) have been successfully cultured. Lampe and Mills (1933) were the first to report the proliferation of immature endosperm tissue of maize, grown on a medium containing extract of potato or young kernels. La Rue (1947) observed that in nature, in some maize kernels, the pericarp ruptured and the endosperm exhibited a white tissue mass. Sehgal (1974) cultured the immature endosperm of *Hordeum* and *Triticum* and obtained only proliferation, all efforts to induce organogenesis failed. However, in *Oryza sativa* organogenesis was successfully induced in cultures of immature endosperm by Nakano *et al.* (1975). In apple also organogenesis has been observed (Mu *et al.,* 1977). The reports on immature endosperm culture indicates that the tissue require substances as yeast extract or casein hydrolysate in addition to basal medium. The age of endosperm at culture is critical for its growth *in vitro*. For example, in maize, wheat and barley endosperm tissue younger than 8 or older than 12 days after pollination did not grow in cultures. The endosperm of *Lolium* and *Cucumis* can be grown only when excised 7–10 days after pollination. Nakano *et al.* (1975) pointed out that in rice only cellular endosperm cultured 4–7 days after pollination showed proliferation.

The mature endosperm of several taxa has now been grown successfully. Mostly, the mature endosperm of parasitic taxa shows optimal growth on a medium containing either only a cytokinin or a cytokinin + an auxin. However, in autotrophic members, in addition to an auxin and cytokinin, casein hydrolysate or yeast extract is also essential. The endosperm of *Santalum album*, *Croton bonplandianum* and *Ricinus communis* could be grown on White´s medium containing 2,4-D, Kin and yeast extract. Organ formation has been demonstrated mainly in parasitic species viz. *Dendrophytoe*, *Exocarpus*, *Taxillus* and in some autotrophic species *viz. Croton*, *Citrus*, *Santalum*, *Petroselinum,* etc. In almost all parasitic members differentiation of shoot buds from the mature endosperm occurred without callusing, whereas in autotrophic members the endosperm usually formed a callus mass followed by the differentiation of shoot buds or roots.

Organogenesis from immature or mature endosperm has been observed only in few species like rice, apple, *Citrus*, *Santalum*, *Croton,* etc. In *Ricinus communis*, *Zea mays* and *Cucumis* mature endosperm proliferation resulted in a callus tissue of unlimited growth but organogenesis did not occur, even though endosperm callus showed groups of meristematic cells. The induction of organogenesis in endosperm cultures has always been a challenging problem.

*Application*

Investigations have provided ample evidence that triploid plants can be raised through endosperm culture. Lakshami Sita *et al.* (1980) developed triploid plants of *Santalum album* and Wang and Chang (1978) produced triploid plantlets from *Citrus*. The conventional method for raising triploids is through a cross between a tetraploid and diploid. The crosses may not be successful due to many problems associated in such crosses. In these cases the culture of endosperm may prove to be advantageous. Presently there are a number of crop species viz. banana, apple, beet, tea, mulberry in which triploids are in commercial use and nucellus culture will be helpful.

Endosperm can be used as a nurse tissue for raising hybrid embryos. Kruse (1974) was able to culture 9–12 day old hybrid embryos of *Hordeum* × *Triticum*, *Hordeum* × *Agropyrum* and *Hordeum* × *Secale* on a medium that had endosperm of *Hordeum*. Using *Hordeum* endosperm as a nurse tissue, young embryos of these hybrids can be induced to germinate and form normal hybrid plants. It is desirable that plant breeders apply these techniques to economically important plants where conventional breeding methods prove futile.

## Cell culture

The growing of individual cells which have been obtained from an explant tissue or callus is referred to as cell suspension culture. These are initiated by transferring pieces of explant/callus into flasks with a liquid medium and are then placed on a gyratory shaker to provide aeration and dispersion of cells. It has been discussed in Chapter 7.

## Protoplast culture

Protoplasts are the cells without cell wall. Culture of protoplasts and fusion of protoplasts from different strains has been discussed in detail (see Chapter 9).

# PROTOCOLS

## 5.1: Protocol for seed germination (*Nicotiana*)

*Plant material: Nicotiana tabacum*

1. Wash the seeds by submerging them in water with a few drops of detergent in a beaker and shake by hand, or wrap seeds in two layers of cheese cloth/muslin cloth/nylon pouch and then wash with water.
2. Submerge the seeds in 70% alcohol for 30–60 sec. Decant the alcohol.
3. Transfer the seeds to a flask or beaker containing 20–40% commercial sodium hypochlorite for 15–20 min. Rinse the seeds 5× with sterile distilled water.
4. Place 2–3 seeds per culture vessel on the surface of MS agar medium without growth regulators.
5. Incubate the cultures at 25°C under 16 h photoperiod with ~1000 lux light intensity for 1–2 weeks.
6. Observe regularly for germination of seeds. If need be, transfer the individual plantlets to half MS medium.

## Plant establishment

7. Gently remove well rooted plantlets from the culture vessel, keeping the roots intact. Gently rinse the roots with warm water to remove the agar attached to it.
8. Plant the regenerants in small plastic pots with sterile soil mix. Make sure the soil is moist with water. After transfer of plant to the soil, either cover it with inverted beaker or put a plastic wrap and use a plastic stake to allow the plastic wrap to form a tent over the plant. Plants do not need to be watered for first few days.
9. Place the pots in humid conditions or mist chambers under diffused light at 25°C. Open the tents or remove the inverted container to allow air exchange briefly everyday. After 1 week, let some air in the tent for 1 h each day. After another week, increase gradually to several hours per day. After 2–3 weeks remove the wrap or inverted container from the plants to adjust to natural environmental conditions.

## 5.2: Protocol for embryo culture (cereals—wheat, maize, barley, rice, etc.)

There are numerous examples in which hybrid embryo rescue operations have been carried out. Because hybrid embryos are difficult to obtain for practical classes, the immature embryos of any cereal or legume can be isolated. The same procedure can be applied to hybrid embryo rescue when such crosses are available.

1. Harvest ears of maize/panicle of rice/spikes of wheat or barley after 12–14 days of crossing

or pollination. Developing seeds should be in the milky stage. At this stage embryos are less than 1.5 mm in diameter. Check the size of embryo under binocular microscope. With the help of forceps and/or needle take out the seeds from the ear/ panicle/ spike and remove the lemna and palea wherever it is present.
2. Place few such seeds in a beaker and sterilize them with 20% commercial sodium hypochlorite solution for 10 min. These operations are to be carried out in the laminar airflow hood. Rinse 3–5× with sterile distilled water.
3. Place a sterilized grain on a sterile slide/petri dish and make cuts with needles above the embryo in the endosperm while holding embryo stationary with the second needle.
4. Dissect out immature embryos (1–1.5 mm diameter) from the seeds under a dissection microscope in the hood of laminar flow.
5. Place the embryos on MS or $B_5$ medium without growth regulators for direct plant formation.
6. Incubate the cultures in light at 25°C.
7. Transfer the plantlets after 2 weeks to half strength MS or $B_5$ media with reduced sucrose (1%) concentration.
8. Transfer the plants to plastic pots and follow the conditions of plant establishment as mentioned in Protocol 5.1.

### 5.3: Protocol for Embryo Culture (Legumes—Green Gram, Black Gram, French Bean, Soybean, etc.)

1. Collect developing pods in which the embryos are at the heart stage and the beginning of cotyledon stage. Depending upon the species this stage will approximately be after 9–12 days of pollination.
2. Dip the pods in 70% alcohol for 1 min and then sterilize with 20% commercial sodium hypochlorite solution for 10 min. These operations are to be carried out in the laminar flow hood. Rinse 3–5x with sterile distilled water.

3. Place the pods on a sterile slide/petri dish and slit open the pods and place the developing seeds on a slide or petri dish.
4. Remove the seed coat pieces directly above the embryo and carefully dissect out the embryos with forceps and needle.
5. Inoculate the embryos on the following medium:
MS + casein hydrolysate (200 mg/l) + L-glutamine (100–200 mg/l) + NAA (0.01μM) + BAP (1.0 μM) + sucrose (3–8%).
6. Incubate the cultures in light at 25°C.
7. Transfer the plantlets after 3–4 weeks to half strength MS medium without growth regulators but with reduced sucrose (1%) concentration.
8. Transfer the plants to plastic pots and follow the conditions of plant establishment as mentioned in Protocol 5.1.

### 5.4: Protocol for Callus Induction (*Nicotiana tabacum*)

*Plant material: Nicotiana tabacum*

1. Follow the Steps 1–5 for *in vitro* germination of seeds.
2. Collect the aseptically germinated seedlings when the cotyledons are fully expanded and the epicotyl is beginning to emerge. Usually this will occur when the seedlings are 2 week old. Place each seedling on a sterile slide/ petri dish and prepare the explant.
3. Various explants can be used. Excise the shoot apex with and without cotyledons and insert the stem base into the medium, cotyledons, and hypocotyl section from the decapitated seedling. Place the explants on the following medium:
MS + 1–2 mg/l 2,4-D
4. Incubate the cultures in dark at 25°C. Callus will be produced in 3–4 weeks.
5. Compare the callus induction growth from various explants.
6. Cut small pieces of callus ~0.5 g fresh weight and subculture on the same fresh medium for proliferation.

## 5.5: Protocol for Callus Induction (Cereals—Wheat, Rice, Maize, Barley, etc.)

1. Follow the Steps 1–4 of Protocol 5.2 for isolation of immature embryos.
2. Place 5–6 immature embryos per vessel with scutellum side up on the following medium for callus induction:
   MS + 2 mg/l 2,4-D
3. Incubate the cultures in dark at 25°C. Initially the embryos will show swelling followed by callus proliferation from the scutellum in 3–4 weeks. (Embryos bigger than 2 mm diameter will show precocious plant formation).
4. Scutellar callus can be multiplied at least once on the same callus initiation medium by breaking into 2–3 pieces after 4–6 weeks of initiation of callus.
5. Place hard and compact callus pieces of approximately 0.5cm$^2$ on the same callus induction medium for proliferation.

## REFERENCES

De Fossard, R.A., Baker, P.K. and Bourne, R.A. 1977. The organ culture of nodes of four species of *Eucalyptus*. *Acta Hortic.* **78:** 157–165.

Dieterich, K. 1924. Uber Kultur von Embryonen ausserhalb des Samens. *Flora (Jena)* **117:** 379–417.

Eichholtz, D., Robitaille, H.A. and Hasegawa, P.M. 1979. Adventive embryology in apple. *Hort Sci.* **14:** 699–700.

Hannig, E. 1904. Zur Physiologie pflanzlicher embryonen. I. Uber die cultur von Cruciferen-Embryonen ausserhalb des Embryosacks. *Bot. Ztg.* **62:** 45–80.

Kasha, K.J. and Kao, K.N. 1970. High frequency haploid production in barley (*Hordeum vulgare* L.). *Nature* **225:** 874–876.

Knudson, L. 1946. A new method for the germination of orchid seeds. *Am. Orchid Soc. Bull.* **15:** 214–217.

Kruse, A. 1974. An *in vivo/in vitro* embryo culture technique. *Hereditas* **73:** 157–161.

La Rue, C.D. 1947. Growth and regeneration of the endosperm of maize in culture. *Am. J. Bot.* **34:** 585–586.

Laibach, F. 1925. Das Taubverden von Bastardsmen und die kunsliche Aufzucht fruh absterbender Bastardembryonen. *Z. Bot.* **17:** 417–459.

Laibach, F. 1929. Ectogenesis in plants. Methods and genetic possibilities of propagating embryos otherwise dying in the seed. *J. Hered.* **20:** 201–208.

Lakshami Sita, G., Raghava Ram, N.V. and Vaidyanathan, C.S. 1980. Triploid plants from endosperm cultures of sandalwood by experimental embryogenesis. *Plant Sci. Lett.* **20:** 63–69.

Lampe, L. and Mills, C.O. 1933. Growth and development of isolated endosperm and embryo of maize. *Abstr. Paper Bot Soc.*, Boston.

Litz, R.E. and Conover, R.A. 1982. *In vitro* somatic embryogenesis and plant regeneration from *Carica papaya* L. ovular callus. *Plant Sci Lett.* **26:** 153–158.

Mu, S., Liu, S., Zhou, Y., Qian, N., Zhang, P., Xie, H., Zhang, F and Yan, Z. 1977. Induction of callus from apple endosperm and differentiation of the endosperm plantlets. *Sci. Sinica*, **20:** 370–377.

Mullins, M.G. and Srinivasan, C. 1976. Somatic embryos and plantlets from an ancient clone of the grapevine (cv. Cabernet-Sauvignon) by apomixis *in vitro*. *J. Exp. Bot.* **27:** 1022–1030.

Murashige, T. and Skoog, F. 1962. A revised medium for rapid growth and bioassays with tobacco tissue cultures. *Physiol. Plant.* **15:** 473–497.

Nakano, H., Tashiro, T. and Maeda, E. 1975. Plant differentiation in callus tissue induced from immature endosperm of *Oryza sativa* L. *Z. Pflanzenphysiol.* **76:** 444–449.

Raghavan, V. 1976. *Experimental Embryogenesis in Vascular Plants*. Academic Press, New York.

Rangan, T.S., Murashige, T. and Bitters W.P. *In vitro* initiation of nucellar embryos in monoembryonic *Citrus*. *Hort Sci.*, **3:** 226–227.

Rangaswamy, N.S. 1958. Culture of nucellar tissue of *Citrus in vitro*. *Experientia* **14:** 111–112.

Sehgal, C.B. 1974. Experimental studies on maize endosperm. *Beitr. Biol. Pflanzen.* **46:** 233–238.

Skoog, F. and Miller, C.O. 1957. Chemical regulation of growth and organ formation in plant tissue cultured *In Vitro. Sym. Soc. Exp. Biol.* **11:** 118–131.

Vacin, C and Went. O. 1949. Morphogenesis in aseptic cell culture of *Cymbidium*. *Bot. Gaz.* **110:** 605–610.

Wang, T. and Chang, C. 1978. Triploid Citrus plantlets from endosperm culture. In: *Proc. Symp. Plant Tissue Culture*, Science Press Peking, pp. 463–468.

Went, F. W. 1938. Specific factors other than auxin affecting growth and root formation. *Plant Physiol.* **13:** 55–80.

## FURTHER READING

Johri, B.M. 1982. *Experimental Embryology of Vascular Plants*. Springer-Verlag, Berlin.

Pierik, R.L.M. 1987. *In vitro culture of Higher Plants*. Martinus Nijhoff Publishers, Dordrecht.

Reinert, J. and Bajaj, P.P.S. (eds.) 1977. *Applied and Fundamental Aspects of Plant Cell, Tissue and Organ Culture*. Springer-Verlag, Berlin.

Thorpe, T.A. 1981. *Plant Tissue Culture: Methods and Applications in Agriculture*. Academic Press Inc., New York.

Vasil, I.K. 1984. *Cell Culture and Somatic Cell Genetics of Plants: Laboratory Procedures and Their Applications*. Academic Press, Inc., New York.

# 6

# Micropropagation

Tissue culture techniques are becoming increasingly popular as alternative means of plant vegetative propagation. Plant tissue culture involves asexual methods of propagation and its primary goal is crop improvement. The success of many *in vitro* selection and genetic manipulation techniques in higher plants depends on the success of *in vitro* plant regeneration. Crop breeding and rDNA technology require the widespread use of reliable true to type propagation and better regeneration methods. In this general context, the types of propagation required will vary immensely.

Clonal propagation *in vitro* is called micropropagation. The word 'clone' was first used by Webber for apply to cultivated plants that were propagated vegetatively. The word is derived from Greek (clon = twig, spray or a slip, like those broken off as propagules for multiplication). It signifies that plants grown from such vegetative parts are not individuals in the ordinary sense, but are simply transplanted parts of the same individual and such plants are identical. Thus, clonal propagation is the multiplication of genetically identical individuals by asexual reproduction while clone is a plant population derived from a single individual by asexual reproduction. With the current techniques of single cell and protoplast culture that enable many thousands of plants to be ultimately derived from a single cell in a comparatively short span of time,

the products of this rapid vegetative propagation should by definition be considered a single clone.

The significant advantage offered by the aseptic methods of clonal propagation (micropropagation) over the conventional methods is that in a relatively short span of time and space, a large number of plants can be produced starting from a single individual. It has been estimated that axillary bud proliferation approach typically results in an average 10-fold increase in shoot number per monthly culture passage. In a period of 6 months, it is feasible to obtain as many as 1,000,000 propagules or plants starting from a single explant.

Suitable explants from vascular plants including angiosperms, gymnosperms and pteridophytes can be cultured *in vitro* and induced to form adventitious buds, shoots, embryoids or whole plants. Murashige (1974) outlined three major stages involved in micropropagation (Fig. 6.1).

Stage 1: Selection of suitable explants, their sterilization and transfer to nutrient media for establishment, i.e. initiation of a sterile culture of the explant.

Stage II: Proliferation or multiplication of shoots from the explant on medium.

Stage III: Transfer of shoots to a rooting medium followed later by planting into soil.

Sometimes a Stage IV is also included in cases where establishment of plantlets in soil is

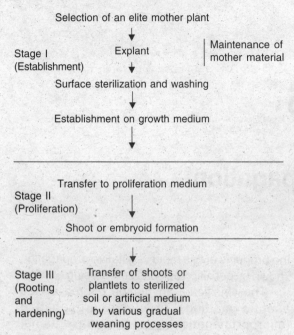

Fig. 6.1: Major stages of micropropagation.

particularly elaborate. Further, not all crop species need to be propagated *in vitro* by means of all three stages. These stages have been designed simply to describe micropropagation processes and to facilitate comparison between two or more systems. Plant regeneration can be achieved by culturing tissue sections either lacking a preformed meristem (adventitious origin) or from callus and cell cultures (*de novo* origin). Axillary bud proliferation approach of plant regeneration is based on preformed meristems. In contrast, adventitious regeneration occurs at unusual sites of a cultured tissue (e.g. leaf blade, cotyledon, internode, petiole, etc.) where meristems do not naturally occur, but it is dependent upon the presence of organized explant tissue. *De novo* plant regeneration occurs from callus and cell cultures in the absence of organized tissue. Thus adventitious or *de novo* plant regeneration can occur by organogenesis and embryogenesis. Micropropagation and plant regeneration can be grouped into the following categories:

(a) Enhanced release of axillary bud proliferation, i.e. by multiplication through growth and proliferation of existing meristems. It can be through apical shoots excised from the parent plant (meristem and shoot tip culture) and by multiplication of existing meristems within axillary shoots, which proliferate on explants after removal from the parent plant (single node and axillary bud culture).

(b) Organogenesis is the formation of individual organs such as shoots and roots either directly on the explant where a preformed meristem is lacking or *de novo* origin from callus and cell culture induced from the explant.

(c) Somatic embryogenesis is the formation of a bipolar structure containing both shoot and root meristems either directly from the explant (adventitive origin) or *de novo* origin from callus and cell culture induced from the explant.

# AXILLARY BUD PROLIFERATION APPROACH

It is the stimulation of axillary buds, which are usually present in the axil of each leaf, to develop in to a shoot. This technique exploits the normal ontogenetic route for branch development by lateral (axillary) meristems. In nature these buds remain dormant for various periods depending upon the growth pattern of the plant. In some species, removal of terminal bud is necessary to break the apical dominance and stimulate the axillary bud to grow into a shoot. Due to continuous application of cytokinin in a culture medium the shoot formed by the bud, which is present on the explant (nodal segment or shoot tip cutting), develops axillary buds. The shoots are then separated and rooted to produce plants or shoots are used as propagules for further propagation. The merit of using axillary bud proliferation from meristem, shoot tip, or bud culture as a means of regeneration is that the incipient shoot has already differentiated *in vivo*. Thus, only elongation and root differentiation are required to establish a complete plant. The induction of axillary bud proliferation seems to be applicable as a means of micropropagation in many cases. This approach can be subdivided into meristem and shoot tip culture, where the explant is taken from the tip of growing shoots while bud cultures are initiated from either terminal or axillary buds, usually with the stem segment attached.

## Meristem and Shoot Tip Culture

Morel and Martin (1952) developed the technique of meristem culture for *in vivo* virus eradication of *Dahlia*. Georges Morel (1965) was the pioneer in applying shoot tip culture for micropropagation of orchid *Cymbidium*. *In vitro* clonal propagation gained momentum in the 1970s when Murashige (1974) gave the concept of developmental stages by defining establishment, proliferation, and rooting and hardening stages. This concept stimulated awareness that a single medium is not sufficient for *in vitro* multiplication, but the propagule needs a series of transfers in specially designed physical and chemical environments to reach regeneration. This method has been more successful in herbaceous plants because of weak apical dominance and strong root regenerating capacities as compared to woody species. Many crops are being currently propagated commercially using these *in vitro* procedures.

Shoots of all angiosperms and gymnosperms grow by virtue of their apical meristems. The apical meristem is usually a dome of tissue located at the extreme tip of a shoot and measures ~0.1 mm in diameter and ~0.25 to 0.3 mm in length. The apical meristem together with one to three young leaf primordia measuring 0.1 to 0.5 mm constitutes the shoot apex. The explant of meristem tip culture may either be the apical meristematic dome or the apical dome with few subjacent leaf primordia. Meristem or shoot tip is cut or isolated from stem by applying a V-shaped cut with a sterilized knife. For meristem tip culture, the cut is applied 0.3 to 0.5 mm below the tip of the dome and the excised tissue (explant) is removed along with

portions of procambial tissue and is immediately planted on the media. Size of the tip explant is the decisive factor which governs the success of culture. When very small explants are used, the presence of leaf primordia appears to determine the capability of an explant to develop. In general, the larger the explant, the better are the chances for survival. It is generally agreed that large explants such as shoot tips measuring up to 2 cm in length should be selected for *in vitro* micropropagation. However, when disease-free plants are the objective, meristem tips should be cultured. Meristem tips between 0.2 to 0.5 mm most frequently produce virus free plants and this method is referred to as meristem tip culture. Like all asexual propagation methods, success of meristem and shoot tip cultures is affected by the season during which the explants are obtained. For plant species with a dormant period, best results may be expected when the explants are dissected at the end of their dormancy period. Further, actively growing shoot tips are recommended for their culture because of their strong growth potential and low virus concentration. A general protocol of shoot tip culture has been given in Fig. 6.2.

MS medium salts have been very satisfactory for such cultures though White's medium and the Gautheret's (1959) were the most widely used media during the early days of meristem culture. The growth additives vary from species to species and in some woody species culture one needs to add certain polyphenol oxidative compounds. Some general conclusions from the published reports have been given.

Murashige reported that there are three stages of culture (Fig. 6.1). Stage I is the culture establishment stage when explants may develop either into single shoots or multiple shoots. It is unlikely that meristem, shoot tip or bud explants have sufficient endogenous cytokinin to support growth and development. Thus at this stage media are supplemented with cytokinin like BA, kinetin, and 2iP. Auxin is another hormone

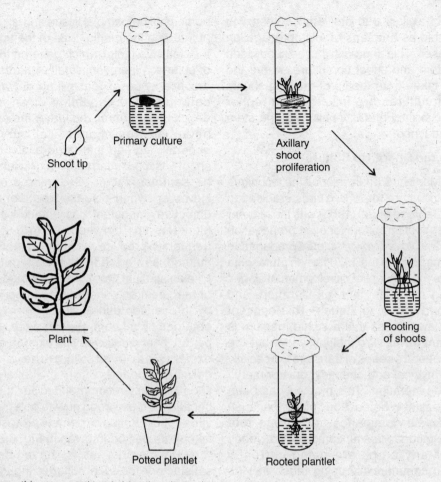

**Fig.6.2:** Schematic representation of shoot tip culture.

required for shoot growth. Since the young shoot apex is an active site for auxin biosynthesis, exogenous auxin is not always needed in Stage I medium, especially when relatively large shoot tip explants from actively growing plants are used. If an auxin is to be added, it should be a weak one like NAA or IBA rather than 2,4-D. The usual range of concentration is between 0.45 and 10.0 µM.

Many plants are rich in polyphenolic compounds. After tissue injury during dissection, such compounds will be oxidized by polyphenol oxidases and the tissue will turn brown or black. The oxidation products are known to inhibit enzyme activity, kill the explants, and darken the tissues and culture media which severely affects the establishment of explant. Some of the procedures used by various workers to combat this problem are: (i) Adding antioxidants to culture medium viz. ascorbic acid, citric acid, polyvinylpyrrolidone (PVP), dithiothreitol, bovine serum albumin, etc.; (ii) Presoaking explants in antioxidant before inoculating into the culture medium; (iii) Incubating the initial period of primary cultures in reduced light or darkness because it is known that phenolic oxidation products are formed under illumination; and (iv) Frequently transferring explants into fresh medium whenever browning of the medium is observed.

In Stage II the objective is to multiply the propagule and for this axillary shoot proliferation is followed as it maintains higher genetic stability and is more easily achievable by most plant species than organogenic or embryogenic path. In axillary shoot proliferation, high levels of cytokinin are utilized to overcome the apical dominance of shoots and to enhance the branching of lateral buds from leaf axils. The cytokinin concentration varies but, 4.5 to 25.0 µM or higher has been used. In general BA is the most effective cytokinin followed by kinetin and 2iP. Exogenous auxin does not promote axillary shoot proliferation, however, if it is used in low concentration it is to nullify the suppressive effect of high cytokinin concentrations on axillary shoot elongation and to restore normal shoot growth.

The purpose of Stage III is *de novo* regeneration of adventitious roots from the shoots obtained at Stage II or in some cases at Stage I itself. Usually *in vitro* produced shoots of sufficient length are cut or separated and transferred to a medium containing auxin. It is well known that *de novo* root initiation depends on a low cytokinin to a high auxin ratio. Since the hormonal requirements for Stage II culture medium are opposite to this particular balance, Stage II medium is rarely used in Stage III. Numerous studies have indicated that NAA followed by IBA, IAA, 2,4-D and other auxins are used for induction of root regeneration. In many species addition of auxin is unnecessary because of its production in young shoots and in such cases it should be avoided because high level of auxin will induce callus formation. Sometimes if roots are unable to initiate in such high salt media irrespective of the hormone present, then the salt concentration in the medium should be lowered to one-half, one-third, or one-fourth of the standard strength. Riboflavin is reported to improve the quality of root system in *Eucalyptus ficifolia*. Similarly, phloroglucinol has been observed to promote rooting in a number of rosaceous trees, but its role has been a subject of controversy.

Optimum inoculation conditions vary but generally the cultures are kept in the range of 20–28°C with the majority in the middle range of 24–26°C. Light intensity and duration tends to vary with species but commonly 16 h day vs. 8 h night is used. The light intensity is usually 1–10 Klux. In tissues where browning is the problem, lower light intensity is better.

### Bud Culture

Buds contain quiescent or active meristems depending upon the physiological state of the plant. Most vascular plants have an indeterminate mode of growth in which the leaf axils contain subsidiary meristems, each of which is capable of growing into a shoot that is identical to the main axis. Depending on the degree of branching displayed by a particular species, only a limited number of axillary meristems develop, the majority being inhibited by apical dominance. The various types used in bud culture are described below.

### Single node culture

Here a bud is isolated along with a piece of stem, with the purpose of forming a shoot by allowing the bud to develop. This method is the most natural method for vegetatively propagating plants *in vitro*, since it is also applicable *in vivo*. The bud found in the axil of a leaf is similar to the one in the stem tip. Bud is isolated and cultured on a nutrient medium for *in vitro* development. The buds in the axil of the newly formed leaves can then also be subcultured and allowed to develop into plantlet. When enough shoots have been obtained, these must be rooted and then transferred to soil. In this technique where buds and shoot tips are isolated and cultured, in principle no cytokinin is added (Fig. 6.3).

The first research into isolation of buds and subsequent rooting of the shoots formed was carried out on asparagus. The single node method has had a good success with many different plants such as potato, *Vitis*, peas, rose, tomato, cucumber, aubergine, etc. Recently, this

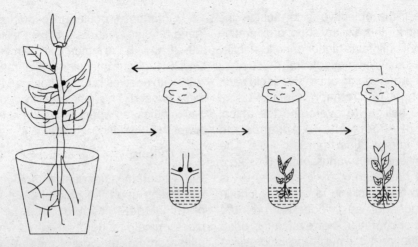

**Fig.6.3:** Schematic representation of vegetative propagation by single node technique.

method has also been applied to *Eucalyptus grandis* and *Araucaria cunninghammii*.

The following points should be kept in mind when applying this method:

i. Isolation by this method is practically impossible when dealing with rosette plants such as gerbera, and when the chances of infection are high.

ii. To reduce the chances of infection it is best to isolate closed buds.

iii. The rate of propagation is dependent on the number of buds available. The more leaves that are initiated with time, the more buds became available, and more rapid is the propagation.

iv. Dormancy, especially of woody plants from temperate climates often presents problems.

v. Starting material, position, age, season of the year at the time of isolation, and physical factors are of great importance as is the composition of nutrient medium.

*Axillary bud method*

In this method a shoot tip is isolated, from which the axillary buds in the axils of leaves develop under the influence of a relatively high cytokinin concentration. This high cytokinin concentration stops the apical dominance and allows axillary buds to develop. If a shoot tip has formed number of axillary or side shoots, then they can be inoculated on to fresh medium containing cytokinin. A schematic diagram is shown in Fig. 6.4. In practice, the single node method is often used in combination with the axillary bud method.

The axillary shoot method was first used for carnation. However, in recent years this method is rarely used for carnation, while it is used on a large scale for strawberry and gerbera. A few of the most important points concerning axillary shoot formation must be kept in mind:

i. The cytokinin requirement is extremely variable and should be adjusted according to the plant species and particular cultivar in use. Most plants react very well to BA and often less to kinetin or 2 iP.

ii. Cytokinin concentration correlates with the developmental stage of the material. Juvenile material requires less cytokinin than adult material.

iii. Cytokinin/auxin ratio is usually 10:1.

iv. It is sometimes recommended that the shoot tip should be allowed to develop before the cytokinin concentration is increased to induce axillary shoot formation.

a.

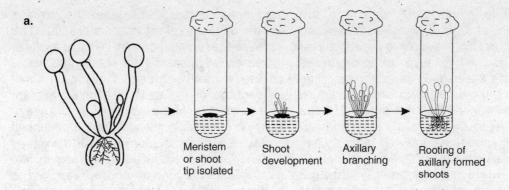

b.

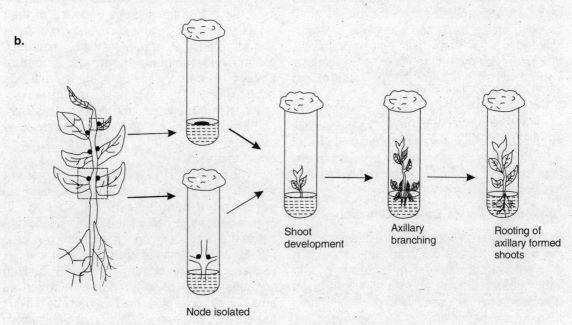

**Fig.6.4:** Schematic representation of axillary bud method of vegetatively propagating plants. (a) rosette plants; (b) elongate plants.

v. If axillary shoot development is not satisfactory then remove or kill the apical meristem.

vi. Liquid medium sometimes promotes axillary shoot formation.

vii. It has been demonstrated in wide range of species that thidiazuron and several substituted pyridylphenylurea compounds rather then classical cytokinins stimulate more axillary branching.

## ORGANOGENESIS

In plant tissue culture, organogenesis is a process of differentiation by which plant organs viz. roots, shoots, bud flowers, stem, etc. are formed while adventitious refers to the development of organs (roots, buds, shoots, flowers, etc.) or embryos (embryo-like structures) from unusual points of origin of an organized explant where a preformed meristem is lacking. Adventitious shoots or roots are induced on tissues that normally do not produce these organs. Adventitious shoots are stem and leaf structures that arise naturally on plant tissues located in sites other than at the normal leaf axil regions. Plant development through organogenesis is the formation of organs either *de novo* or adventitious in origin. Whole plant regeneration via organogenesis is a monopolar structure. It develops procambial strands which establish a connection with the pre-existing vascular tissue dispersed within the callus or cultured explant. This organogenesis process is much more common than somatic embryogenesis and has far more potential for mass clonal propagation of plants.

Plant production through organogenesis can be achieved by two modes: (i) Organogenesis through callus formation with *de novo* origin and (ii) Emergence of adventitious organs directly from the explant. The schematic representation has been shown in Fig. 6.5.

## Organogenesis via Callus Formation

Plant regeneration from cultured explant involves the initiation of basal callus and then shoot bud differentiation. Establishment of callus growth with subsequent organogenesis has been obtained from many species of plants and from numerous explants viz. cotyledons, hypocotyl, stem, leaf, shoot apex, root, young inflorescence, flower petals, petioles, embryos, etc. cultured *in vitro*. For any given species or variety, a particular explant may be necessary for successful plant regeneration. Explants from both mature and immature organs can be induced to form callus and then plant regeneration. However, explants with mitotically active cells are generally good for callus initiation. Immature tissues and organs are invariably more morphogenetically plastic *in vitro*

than mature tissues and organs. The size and shape of the explant is also crucial. The increased cell number present in bigger explants increases the probability of obtaining a viable culture. It has been seen that only a small percentage of the cells in a given explant contribute to the formation of callus. The site of initiation of callus is generally at the peripheral surfaces of the inoculum or at the excised surface. Callus is produced on explants *in vitro* as a result of wounding and in response to hormones, either endogenous or exogenously supplied in the medium. Meristematic tissues or organs like shoot tip, lateral bud, inflorescence, rachis, leaf, petiole, root should be selected in preference to other tissues because of their clonal properties, culture survival, growth rates and totipotency *in vitro*. Meristems, shoot tips, axillary buds, immature leaf, and immature embryos in cereals are particularly suited as good explants. The explants like mature leaves, roots, stems, petioles, and flower parts from herbaceous species can often be successfully cultured to initiate plantlets through organogenesis.

The season of the year, donor conditions of the plant, the age and physiological state of the parent plant contribute to the success of organogenesis in cell cultures. Various culture media are used for organogenesis which include MS, $B_5$, White's medium (White, 1963) and SH (Schenk and Hildebrandt, 1972). MS medium contains a high concentration of nitrogen as ammonium unlike other media. It may be difficult to transfer cells from White's, $B_5$ or SH medium to MS medium. General requirements of vitamins, carbon source etc. have already been discussed in Chapter 3 on nutrition media.

Two modes of cell culture are generally used for organogenic path: (i) the cultivation of cell clusters on a solid medium and (ii) the cultivation of cell suspensions in liquid medium. A suspension cell culture is usually initiated by placing friable callus into liquid culture medium. The suspension usually consists of free cells and aggregates of 2–100 cells. Suspension cultures

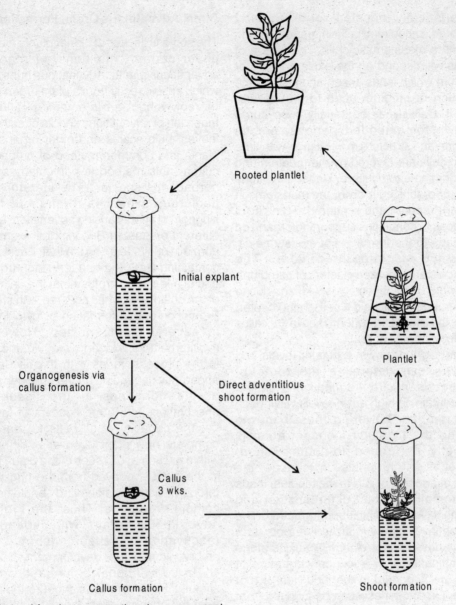

Rooted plantlet

Initial explant

Plantlet

Organogenesis via
callus formation

Direct adventitious
shoot formation

Callus
3 wks.

Callus formation

Shoot formation

**Fig. 6.5:** Protocol for plant regeneration via organogenesis.

should be subcultured at least once a week while callus culture should be subcultured after 3–4 weeks.

Growth regulator concentration in the culture medium is critical for morphogenesis. Skoog and Miller (1957) first indicated its role in morphogenesis. Auxin at a moderate to high concentration is the primary hormone used to produce callus. Often 2,4-D a very potent auxin is used alone to initiate callus. In some species a high concentration of auxin and a low concentration of cytokinin in the medium promotes abundant cell proliferation with the formation of callus. Cytokinins, if supplied are kinetin or benzyladenine. Callus tissue comprises a wide range of cell types and characteristically consists of irregularly differentiated, vacuolated cells interspersed with smaller more meristematic cells. The nature of any callus will depend on the explant tissue or tissues from which it has arisen and also on the composition of medium used to induce and maintain it. Callus may be serially subcultured and grown for extended periods, but its composition and structure may change with time as certain cells are favored by the medium and come to dominate the culture. Under conditions favouring unorganized growth, the meristems in a callus are random and scattered. Transfer of callus pieces to conditions supporting organized growth leads to the formation of meristemoids, also termed as growth centers. The meristemoids are localized clusters of cambium-like cells which may become vascularized due to the appearance of tracheidal cells in the center. These are the sites of organ formation in the callus and can form shoots or roots.

By varying the growth regulator levels and types, one can determine the route of morphogenesis *in vitro*. As mentioned earlier, medium containing high auxin levels will induce callus formation. Lowering the auxin and increasing the cytokinin concentration is traditionally performed to induce shoot organogenesis from callus. In tobacco, the presence of adenine or KIN in the medium leads to the promotion of bud differentiation and development. Auxins inhibit bud formation. The next phase involves the induction of roots from the shoots developed. IAA or IBA auxins alone or in combination with a low concentration of cytokinin are important in the induction of root primordia. Organogenesis can be induced from either cell suspension or callus cultures. One needs to transfer only cells from callus medium to regeneration medium and then continued subculturing on regeneration medium. Thus organ formation is determined by quantitative interaction, i.e. ratios rather than absolute concentrations of substances participating in growth and development.

## Direct Adventitious Organ Formation

The somatic tissues of higher plants are capable, under certain conditions, of regenerating adventitious plants. Adventitious buds are those which arise directly from a plant organ or a piece thereof without an intervening callus phase. Induction of adventitious shoots directly on roots, leaves, bulb scales and other organs of intact plants is a common method of propagation. In culture, this method is particularly suitable to herbaceous species. The literature records numerous examples throughout the plant kingdom of shoots arising adventitiously on many different organs. In *Begonia,* for example, buds normally originate along the leaf. But in a medium containing BAP, the bud formation from the cut end of a leaf is so profuse that the entire surface of the cutting becomes covered with shoot buds. Promotion of bud formation by cytokinin occurs in several plant species. However, the requirement for exogenous auxin and cytokinin in the process varies with the tissue system, apparently depending on the endogenous levels of the hormones in the tissue. These observations led to the concept of totipotency, i.e. the capacity of all cells to regenerate a complete new plant even after differentiation within the somatic tissues of the plant. Every cell in the plant is derived from the original zygote through mitotic divisions and should contain the complete genome. Thus, the formation of adventitious organs will depend on the reactivation of genes concerned with the embryonic phase of development.

In conventional propagation, the main stimulus for adventitious shoot formation arises from the physical separation of cutting from the parent plant, causing changes in the production and distribution of endogenous hormones. The same applies to explants used for *in vitro* procedures, and in some species shoot formation may occur spontaneously on a medium lacking any growth regulators. But in most species the addition of growth regulators to the medium is required to initiate shoot formation. Two main types of growth regulating substances, auxins and cytokinins, are employed at different concentrations with respect to one another and depending upon the explant taken, age of the plant and growing conditions. Adventitious *in vitro* regeneration may give a much higher rate of shoot production than is possible by proliferating axillary shoots. Adventitious shoot proliferation is the most frequently used multiplication technique in micropropagation system.

## EMBRYOGENESIS

In somatic embryogenesis the embryos regenerate from somatic cells, tissues or organs either *de novo* or directly from the tissues (adventive origin), which is the opposite of zygotic or sexual embryogenesis. Various terms for non-zygotic embryos have been reported in literature such as adventive embryos (somatic embryos arising directly from other organs or embryos), parthenogenetic embryos (those formed by the unfertilized egg), androgenetic embryos (formed by the male gametophyte). However, in general context, somatic embryos are those which are formed from the somatic tissue in culture, i.e. *in vitro* conditions. In sexual embryogenesis, the act of fertilization triggers the egg cell to develop into an embryo. However, it is not the monopoly of the egg to form an embryo. Any cell of the gametophytic (embryo-sac) or sporophytic tissue around the embryo-sac may give rise to an embryo. Cells of the nucellus or inner integument of members of Rutaceae family (e.g. *Citrus*) may develop into embryos. There are examples of embryos arising from endospermal cells also. However, occurrence of asexual embryogenesis is generally restricted to intraovular tissues. What is particularly striking about embryogenesis in plant cultures is the development of embryos from somatic cells (epidermis, parenchymatous cells of petioles or secondary root phloem) in addition to their formation from unfertilized gametic cells and tissues typically associated with *in vivo* sexual embryogenesis (e.g. nucellus). Somatic embryogenesis differs from organogenesis in the embryo being a bipolar structure with a closed radicular end rather than a monopolar structure. The embryo arises from a single cell and has no vascular connection with the maternal callus tissue or the cultured explant. Further, induction of somatic embryogenesis requires a single hormonal signal to induce a bipolar structure capable of forming a complete plant, while in organogenesis it requires two different hormonal signals to induce first a shoot organ, then a root organ.

The initiation and development of embryos from somatic tissues in plant culture was first recognized by Steward *et al.* (1958) and Reinert (1958, 1959) in cultures of *Daucus carota*. In addition to the development of somatic embryos from sporophytic cells, embryos have been obtained from generative cells, such as the classic work of Guha and Maheshwari (1964) with *Datura innoxia* microspores. Although the list of species from which somatic embryogenesis has been reported is so long, clear-cut examples are far less. In addition to the members of Umbellifereae and Solanaceae, a range of dicotyledonous families have produced somatic embryos. The Leguminoseae and many monocots of Gramineae family, which are so important agronomically, have proven difficult to grow in culture and regenerate somatic embryos. Though there are reports of successes in these species but manipulation is not so easy as with Solanaceous crops. Somatic embryos should closely resemble their bipolar nature as in the case of zygotic embryos. There should be appropriate root, shoot and cotyledonary development. There should be no vascular connection with the mother tissue. In this chapter embryogenesis will be restricted to sporophytic tissue, the discusssion on androgenesis will be dealt separately (see chapter 8).

Sharp *et al.* (1980) described two routes to somatic embryogenesis.

1. *Direct embryogenesis*: The embryos initiate directly from the explant tissue in the

absence of callus proliferation. This occurs through 'pre-embryogenic determined cells' (PEDC) where the cells are committed to embryonic development and need only to be released. Such cells are found in embryonic tissues (e.g. scutellum of cereals), certain tissues of young *in vitro* grown plantlets (e.g. hypocotyl in *Daucus carota, Ranunculus scleratus, Linum usitatissimum, Brassica napus*), nucellus and embryo-sac (within ovules of mature plants).

2. *Indirect embryogenesis*: Cell proliferation, i.e. callus from explant takes place from which embryos are developed. The cells from which embryos arise are called embryogenically determined cells and forms embryos which are induced to do so, also called as 'induced embryogenic determined cells' (IEDC), e.g. secondary phloem of carrot, leaf tissues of coffee, *Petunia, Asparagus*,. etc. In majority of cases embryogenesis is through indirect method. Here, specific growth regulator concentrations and/or cultural conditions are required for initiation of callus and then redetermination of these cells into the embryogenic pattern of development.

Views expressed favor the idea that the two processes are actually the same and that somatic embryos arise from the continuation of special cells in the original explant. In either case, (embryogenic) callus, once developed, continues to produce embryos. A diagrammatic representation of somatic embryogenesis in carrot has been depicted in Fig. 6.6. Somatic embryos arise from single cells located within clusters of meristematic cells either in the callus mass or in suspension. Such cells develop into proembryos, with polarity following a pattern that tends to mimic the general pattern associated with the development of embryos in the ovule. Proembryo initials may be single cells or multicellular groups. When the conditions are suitable these embryos germinate to produce plantlets.

For some species any part of the plant body serves as an explant for embryogenesis (e.g. carrot) whereas in some species only certain regions of the plant body may respond in culture e.g. cereals. Floral or reproductive tissue in general has proven to be an excellent source of embryogenic material. The physiological state of the plant from which the explant is taken is also extremely important, as is the season during which the material is removed.

Somatic embryogenesis encompasses various stages from callus initiation to embryo development and maturation and subsequently plantlet formation. Equally important is the sequence of media and especially the growth regulators. For many species, one media is used for initial callusing and for the maintenance of callus, a second medium is used for somatic embryo maturation, and a third to allow their growth into plants. An elaborate sequence of media is essential where somatic embryogenesis is lacking or difficult.

The presence of auxin in the medium is generally essential for embryo initiation. Tissue or calluses maintained continuously in an auxin-free medium generally do not form embryos. The callus is initiated and multiplied on a medium rich in auxin which induces differentiation of localized group of meristematic cells called embryogenic clumps. Somatic embryo development generally follows the transfer of cells or callus to media lacking auxin, or with reduced levels of the same auxin, or with similar or reduced levels of a weaker auxin. When transferred to a medium with low auxin or no auxin, the embryogenic clumps develop into mature embryos. In a number of species, somatic embryo initiation and maturation occurs on the primary medium. Transfer to a secondary medium is needed for their growth into plants. Some variables that affect somatic embryogenesis and maturation of embryos in culture are discussed.

Somatic embryos have been grown on a range of media from the dilute White's medium

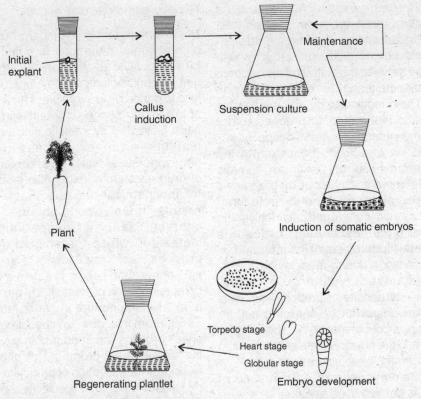

**Fig. 6.6:** Protocol for somatic embryogenesis in carrot.

to very high salt MS medium, but the latter has been extensively employed. The addition of reduced nitrogen in the medium helps in both embryo initiation and maturation. The sources of reduced nitrogen have already been explained. Of all the amino acids, L-glutamine seems to play a special role. Another factor is the chelated form of iron in the media. In the absence of iron, embryo development fails to pass from the globular to the heart-shaped stage.

Growth regulators in the medium, especially auxin or auxin in combination with cytokinin appear essential for the onset of growth and the induction of embryogenesis. Of all the auxins, 2,4-D followed by NAA have proven to be extremely useful. Effective concentration ranges are 0.5–27.6 μM for 2,4-D and 0.5–10.7 μM for NAA. The auxin for the primary (callus initiation) and secondary media (embryo development) may be same or different. One auxin or several

may be used in the same medium. Cytokinins have been important in a number of species. Cytokinins have been used in the primary medium invariably during embryogenesis of crop plants. The effective concentration range for kinetin is 0.5–5.0μM. Cytokinins are important in fostering somatic embryo maturation and especially cotyledon development. Cytokinins are sometimes required for growth of embryos into plantlets. Gibberellins are rarely incorporated in primary culture media. However, they have proven useful in fostering embryo maturation or in stimulation of rooting and subsequent growth of plants in a number of cases. The role of growth inhibitor ABA in somatic embryogenesis has emerged when added at non-inhibitory levels (0.1–1.0 μM). ABA promotes somatic embryo development and maturation and at the same time inhibits abnormal proliferation and initiation of accessory/

adventitious embryos. ABA, antiauxins (5-hydroxynitrobenzyl bromide, 7-azaindole, 2,4,6-trichlorophenoxy acetic acid, p-chlorophenoisobutyric acid) and other growth inhibitors may serve to promote somatic embryo maturation by countering the effects of growth promoters. Their addition to culture media may permit somatic maturation to proceed under conditions when it normally would not occur.

The addition of activated charcoal to the medium has proved to be useful for somatic embryo development. Charcoal media shows lower levels of phenylacetic acid and p-OH benzoic acid compounds, which inhibit somatic embryogenesis. Also, it absorbs 5-hydroxymethyl furfural, an inhibitor formed by sucrose degradation during autoclaving.

Environmental conditions of light, temperature, and density of embryogenic cells in medium are important. Regarding culture vessel, the position of embryos (floating or submerged) and the physical state of the medium (semisolid or liquid) have little effect.

Somatic embryogenesis as a means of propagation is seldom used because:

i. There is a high probability of mutations arising.
ii. The method is usually rather difficult.
iii. The chances of losing regenerative capacity become greater with repeated subcultures.
iv. Induction of embryogenesis is often very difficult or impossible with many plant species.
v. A deep dormancy often occurs with somatic embryogenesis that may be extremely difficult to break.

## Loss of morphogenic potential in embryogenic cultures

Callus or suspension cultures due to aging or prolonged subcultures often show a progressive decline and sometimes complete loss of morphogenic ability. The following hypotheses have been proposed to explain this phenomenon of loss of morphogenesis.

## Genetic and Molecular Aspects

Nuclear changes as polyploidy, aneuploidy and chromosomal mutations in cultured cell may be responsible for the loss of organogenic or embryogenic potential in prolonged cultures. This loss is generally irreversible. The understanding of loss of morphogenic potential in cultures has been facilitated by identification of biochemical markers in the process of somatic embryogenesis. Isolation and characterization of drug-resistant mutants also enable a search for biochemical markers. Some interesting results have been obtained from the characterization of temperature-sensitive mutants in which the embryo development process is impaired. For example, phosphorylation seems to cause a defect in one or more peptides of 'heat shock' proteins induced at higher temperature in carrot mutant cell line Ts 59. From this, one can speculate that (a) heat shock proteins are important in several steps of the development program, (b) phosphorylation is a signal for activation of particular function and (c) kinases have strict specificity. Another mutant line (tsllc) at non-permissive temperatures is unable to acquire polarity because the somatic embryo reaches the globular stage and subsequently enlarges to form secondary embryos or monstrosites.

In orchardgrass and carrot somatic embryogenesis is genetically controlled by a dominant trait. Cytoplasmic factors (Mitochondrial) have also been implicated in control of somatic embryogenesis (Rode et al., 1989). About 21 'embryo-specific' or 'embryo enhanced' genes have been cloned (Zimmerman, 1993, Lin et al., 1996) and it is likely that some of these genes may be useful markers for early embryo development. AGL 15-specific antibodies found to accumulate in microspore embryos in oilseed rape and somatic embryos of alfalfa (Perry et al., 1999) and are found to participate in regulation of programs active during the early stages of embryo development.

Several genes have been isolated which express during somatic embryogenesis. They are classified in three categories: (a) genes involved in cell division (21D7 and CEM1), (b) genes involved in organ formation at globular and torpedo stage embryos), and (c) embryo-specific genes (CHB2 and CEM6). Expression and regulation of the gene action of these genes are described in detail by Komamine (2001).

*Physiological Hypothesis*

Altered hormonal balance within the cells or tissues may also be associated with decline in the embryogenic potential. In such cases it may be possible to restore the potentiality of cells (differentiating organs and embryos) by modifying the growth constituents of the media. The loss of embryogenic potential could also be restored by adding 1–4% activated charcoal in the auxin-free medium (e.g., carrot cultures) or by giving cold treatment to the tissues (Ammirato, 1983). The induction of embryogenic tissue can be aided by various stress factors, such as heat, cadmium and anaerobsis.

*Competitive Hypothesis*

In the complex multicellular explant only a few cells are able to give rise to embryogenic clumps while the remaining cells are non-totipotent. According to the competitive hypothesis, the non-embryogenic cells of the explant will increase under conditions favorable to their growth, resulting in a gradual loss of embryogenic component during repeated subcultures. Restoration of embryogenesis in such cultures is impossible, but if the culture carries few embryogenic cells which are not able to express their totipotency due to the inhibitory effect of the non-embryonic cells, it should be possible to restore the morphogenic potential of these cultures by altering the composition of the medium in such way that selective proliferation of the embryogenic totipotent cells occurs.

# ADVANTAGES OF MICROPROPAGATION

*In vitro* micropropagation techniques are now often preferred to conventional practices of asexual propagation because of the following advantages not only from commercial point of view but also with regard to crop improvement program:

1. A small amount of plant tissue is needed as the initial explant for regeneration of millions of clonal plants in one year. In comparison it would take years to propagate an equal number of plants by conventional methods.
2. Many plant species are highly resistant to conventional bulk propagation practices. Micopropagation techniques provide a possible alternative for these species. Micropropagation helps in bulking up rapidly new cultivars of important trees that would otherwise take many years to bulk up by conventional methods.
3. The *in vitro* techniques provide a method for speedy international exchange of plant materials. If handled properly, the sterile status of the culture eliminates the danger of disease introduction. Thus, the period of quarantine is reduced or unnecessary.
4. The *in vitro* stocks can be quickly proliferated at any time of the year. Also, it provides year round nursery for ornamental, fruit and tree species.
5. Production of disease free plants: Meristem tip culture is generally employed in cases where the aim is to produce disease-free plants. It has been demonstrated that the shoot apices of virus-infected plants are frequently devoid of viral particles or contain very low viral concentration. Though chemotherapeutic and physical agents have been used for production of virus-free plants but with limited success. *In vitro* culture has become the only effective technique to

obtain virus-free plants in potato, *Dianthus*, chrysanthemum, gladiolus, *Pelargonium*, sweet potato (*Ipomea batatus*), yam (*Dioscorea rotundata*), cassava (*Manihot esculenta*), etc. from stocks systemically infected not only with virus but with various other pathogens.

6. Seed production: For seed production in some of the crops, a major limiting factor is the high degree of genetic conservation required. In such cases micropropagation (axillary bud proliferation method) can be used. For example, production of $F_1$ hybrid seed lines in crops like cauliflower where individual parent clones can be bulked for the production of more uniform seed, the production of male sterile lines in crops like onion to provide an alternative to difficult backcrossing methods, and the production of asparagus for producing high quality supermale and female homozygous lines from which desirable all male hybrids can be produced.

7. Germplasm storage: Plant breeding programmes rely heavily on the germplasm. Preservation of germplasm is a means to assure the availability of genetic materials as the need arises. Most seeds and vegetative organs have a limited storage life, research on germplasm preservation has concentrated on the development of procedures to extend usable life spans. Since meristem cells are highly cytoplasmic and non-vacuolated, a high percentage of cells can be expected to survive cryopreservation procedures (for details read Chapter 11 on germplasm storage). Meristems are genetically stable and can be regenerated into pathogen-free plants. Meristems have been identified as excellent material for germplasm preservation of crop species with seed borne viruses also.

8. Artificial seeds: The concept of artificial or synthetic seeds (i.e. encapsulated embryos) produced by somatic embryogenesis has become popular. The aim of somatic embryo

(SE) encapsulation is to produce an analog to true seeds, which is based on the similarity of SE with zygotic embryo with respect to gross morphology, physiology and biochemistry. Two types of synthetic seeds have been developed, namely, *hydrated and desiccated.*

Redenbergh *et al.* (1986) developed hydrated artificial seeds by mixing somatic embryos with an encapsulation matrix so that embryos are well protected in that matrix and it is rigid enough to allow for rough handling. Encapsulation matrix consists of hydrogels such as agar, carrageenan, alginate or plant exudates of arabic (*Acacia senegal*), karaya (*Sterculia* species) or seed gums of guar (*Cyamosis tetragonolobia*) tamarind (*Tamarindus indica*) or microbial products like dextran, xantham gum and divalent salts which forms a complex with ionic bonds of hydrogel matrix. Various divalents salts which can be used are calcium chloride, calcium hydrooxide, ferrous chloride, cobaltous chloride and lanthanum chloride. Sodium alginate has been used successfully with calcium chloride for encapsulation of somatic embryos. Besides somatic embryos, efforts have also been made to use buds, leaves, and meristems. The procedure is mixing somatic embryos with sodium alginate, followed by dropping into a solution of calcium chloride/nitrate to form calcium-alginate beads. About 29–55% embryos encapsulated with this hydrogel germinated and formed seedlings *in vitro*. Calcium alginate capsules tend to stick together and are difficult to handle because they lose water rapidly and dry down to a hard pellet within a few hours of exposure to the atmosphere. These problems can be offset by coating capsules with Elvax 4260 (ethylene vinyl acetate acrylic acid tetrapolymer; Du Pont, USA). From a practical sowing situation, it is necessary to produce high and uniform quality synthetic seeds at large scale. For this an automate encapsulation process has been developed. Processes for use of microcapsules (that release sucrose inside alginate beads), or self-breaking beads, pharmaceutical type capsules, and cellulose acetate mini-plugs, have been recently developed to promote sowing of coated somatic embryos as artificial seeds as well as their germination in non-sterile environment, such as in greenhouse or directly in the field.

Kim and Janic (1989) applied synthetic seed coats to clumps of carrot somatic embryos to develop desiccated artificial seeds. They mixed equal volumes of embryo suspension and 5% solution of polyethylene oxide (polyox WSRN-750), a water-soluble resin, which subsequently dried to form polyembryonic desiccated wafers. The survival of encapsulated embryos was further achieved by embryo 'hardening' treatments with 12% sucrose or $10^{-6}$M ABA, followed by chilling at high inoculum density.

The production of seeds by coating somatic embryos and obtaining plants from these embryos have some problems: (i) artificial seeds that are stable for several months requires the procedures for making the embryos quiescent (ii) artificial seeds need to be protected against desiccation, (iii) recovery of plants from encapsulated embryos is often very low due to incomplete embryo formation or difficulties in creating an artificial endosperm and (iv) the embryo must be protected against micro-organisms.

Disadvantages with artificial seeds:

1. Facilities required are costly and economic considerations may not justify their use in commercial propagation of many kinds of plants.
2. Special skills are required to carry out the work.
3. Errors in maintenance of identity, introduction of an unknown pathogen, or appearance of an unobserved mutant may be multiplied to very high levels in a short time.
4. Specific kinds of genetic and epigenetic modifications of the plant can potentially develop with some cultivars and some systems of culture.

# PROBLEMS ASSOCIATED WITH MICROPROPAGATION

1. Expensive requirements and sophisticated facilities with trained manpower are needed.
2. Though precautions of high order are normally taken during culture, but there are chances of contamination by various pathogens. Contamination could cause very high losses in a short time.
3. The problem of genetic variability may be pronounced in some forms of culture. Shoot tip culture tends to remain stable while those systems that utilize adventitious shoots or multiplication of callus, genetic variability is pronounced. Source of variability may be due to chimeral breakdown, aberrant cell division in callus or cell suspension, epigenetic effects and pre-existing genetic variability.
4. It has been experienced that during repeated cycles of *in vitro* shoot multiplication, cultures show water-soaked, almost translucent leaves. Such shoots exhibit a decline in the rate of growth multiplication, become necrotic and may eventually die. This phenomenon is referred as vitrification. This is also known as hyperhydration. This is due to morphological, physiological and metabolic dearrangements occurring during intensive shoot multiplication in culture. Some preventive measures to reduce hyperhydration includes: (i). increasing concentration of agar; (ii) overlaying medium with paraffin; (iii) using desiccant such as $CuSO_4$; (iv) bottom cooling of the culture vial to improve aeration; (v) lowering cytokinin level or replacing one kind of cytokinin by another; and (vi) manipulating $NH_4^+$ or salt concentrations in the culture medium.

# PROTOCOLS

## 6.1: Meristem and Node Culture of Potato (*Solanum tuberosum*)

*Plant material: Potato tuber*

1. Soak potato tubers in 0.03 M gibberellic acid ($GA_3$) for 1 h or warm potato tubers at room temperature for 2 days before using to break dormancy.
2. Rinse the potato tubers with tap water. Surface sterilize the tubers by soaking in 10–20% commercial sodium hypochlorite solution for 10–20 min.
3. Place whole tubers or cut into 2–3 cm² of 20–30 g sections on the surface of sterile moist vermiculite and keep them in a growth chamber.
4. Harvest sprouts when they reach 2–5 cm in length.

Establishment of *in vitro* plantlets from plant material:

1. Excise nodal sections 1–2 cm from third and fourth nodes from the stem apex with a scalpel and detach their leaves.
2. Surface sterilize the nodal tissue by soaking in 10% commercial sodium hypochlorite for 10 min. Rinse 3x with sterile distilled water. Recut the base of tissue with a scalpel.
3. Place the nodal sections or sprouts on MS medium.
4. Allow growth to 4–6 nodes/stem stage.

Propagation by nodal cutting

1. In the laminar flow hood with the help of large forceps remove a plantlet from the culture vessel and place it on a sterile filter paper or sterile petri dish.
2. Cut the main stem above and below each node into sections of 1–2 cm length. Place one node in each culture vessel containing MS medium with 1–2 mg/l BAP/KIN.

Propagation by meristem

1. Take nodal cuttings from third and fourth nodes of *in vitro* raised plants obtained from plants or sprouts.

2. In the laminar flow hood, peel away protecting leaves on the buds under a dissection microscope at 10–25x.
3. Dissect out meristem domes with one subjacent leaf primordium ~0.5 to 0.75 mm.
4. Culture the meristem domes on the surface of MS agar medium supplemented with KIN (0.1 mg/l = 0.046 µM) and $GA_3$ (0.1 mg/l = 0.29 µM).

## 6.2: Proliferation of Axillary Buds (Strawberry *Fragaria chiloensis*)

*Plant material: Runner shoot tips of strawberry*

1. Collect runners from strawberry plants.
2. Surface sterilize the runners in 10–20% commercial sodium hypochlorite for 10 min and then rinse 4x with sterile distilled water. It is preferable to sterilize further with 0.02% $HgCl_2$ for 3–4 min followed by washing with 1% KCl for 1 min in order to neutralize the mercury ions.
3. Dry the runners in a sterile petri dish.
4. Cut the runners into 1–2 cm or peel the shoot tips and then cut the tip portion of 0.1–0.5 mm size. Inoculate on the establishment medium:
   MS + 0.5–1.0 mg/l BAP + 0.1–0.2 mg/l IBA
5. Incubate the cultures at 25°C under 16 h photoperiod.
6. After 3–4 weeks on this medium, axillary buds will appear.
7. The buds can be proliferated or multiplied if cultured on the following medium:
   MS + 0.8–1 mg/l BAP + 0.2 mg/l IBA + 0.2 mg/l $GA_3$
8. The developed shoots alongwith buds can be further multiplied on the same proliferation medium after 3–4 weeks of culture. Or shoots are separated and transferred to MS medium with auxin (0.5–1 mg/l) IBA. The development of axillary buds ceases and the young plantlets with roots are developed. Profuse rooting should occur within 4–5 weeks.

9. Rooted plantlets can be transferred to soil and follow the procedure of plant establishment and hardening (Protocol 5.1).

## 6.3: Organogenesis–Adventitious Shoot Formation

*Plant material: Leaf explants of tobacco or African violet*

1. Take leaf explants from *in vitro* grown plants or potted plants.
2. If leaves are taken from potted plants, gently wash them with a mild detergent and then rinse under tap water. Dip each leaf into 70% alcohol. Leaves are then surface sterilized with 10% commercial sodium hypochlorite for 5–10 min followed by rinsing 3x in sterile distilled water.
3. Place the leaves in sterile petri dish. Cut leaves into approximately 1 cm$^2$ sections using sterile forceps and scalpel, without vein in the explant.
4. Place the leaf sections on the following media:
   *African violet*: MS + 1.0 mg/l KIN + 0.1 mg/l IAA
   *Nicotiana*: MS + 1–2 mg/l BAP + 0.1–0.2 mg/l NAA
5. Seal the cultures. Incubate at 25°C under 16 h photoperiod at 1000-lux light intensity.
6. Observe the shoot formation after 3 weeks and record the data at regular intervals.
7. Cut tissue masses into small pieces for shoot multiplication, each piece should include at least one tip or rosette, and place them on their respective media.
8. For rooting, transfer the well developed shoots to MS medium without growth regulators or little concentration of auxin (0.5 mg/) to enhance root production.
9. Transfer the rooted plantlets to soil and follow the procedure of plant establishment and hardening (Protocol 5.1).

## 6.4: Organogenesis via Callus Formation (*Nicotiana*)

*Plant material: Nicotiana*

1. Follow Steps 1–4 of Protocol 6.3. Place the leaf sections on the following medium: MS + 1–2 mg/l 2,4-D
2. Incubate the cultures in dark at 25°C. Callus will be produced in 3–4 weeks.
3. For shoot induction, subculture pieces of callus approximately 0.5 cm$^2$ to the medium: MS + 1–2 mg/l BAP + 0.1–0.2 mg/l NAA
4. Follow Steps 5 to 9 of Protocol 6.3.

## 6.5: Organogenesis via Callus Formation (Cereals—Wheat, Barley, Maize, Rice, etc.)

*Plant material: Immature embryos of cereals (wheat, barley, maize, rice)*

1. Collect spikes (for example wheat/barley) 14–15 days after anthesis from plants. Separate each developing seed and remove the lemna and palea.
2. Surface sterilize the seeds in 10–20% commercial sodium hypochlorite containing 1–2 drops of Tween 20 for 10 min. Rinse 3–5x with sterile distilled water.
3. Dissect out immature embryos (1–1.5 mm diameter) from the seeds under dissection microscope in the hood of laminar flow.
4. Place the immature embryos with scutellum side up on the following medium for callus induction: MS + 2 mg/l 2,4-D
5. Incubate the cultures in dark at 25°C. Initially the embryos will show swelling followed by callus proliferation from the scutellum in 3–4 weeks. (Embryos bigger than 2 mm diam. will show precocious plant formation.)
6. Scutellar callus can be multiplied at least once on the same callus initiation medium by breaking into 2–3 pieces after 4–6 weeks of initiation of callus.
7. Place hard and compact calli of approximately 0.5 cm$^2$ on the following shoot induction

medium: MS + 2 mg/l BAP + 0.2 mg/l NAA or MS without growth regulators.

8. Incubate the cultures at 25°C under 16 h photoperiod with ~2 klux light intensity. Shoots would emerge in 3–5 weeks.

9. Transfer shoots to MS medium with 1 mg/l of auxin (IAA/IBA/NAA) for root induction under the same cultural conditions.

10. Transfer rooted plantlets to soil and follow the procedure of plant establishment and hardening (Protocol 5.1).

## 6.6: Embryogenesis (Carrot)

*Plant material: Hypocotyl of carrot seedling*

1. Wash seeds by submerging in water with a few drops of detergent in a beaker and shake by hand, or wrap seeds in two layers of cheese cloth/muslin cloth/nylon pouch and then wash with water.

2. Submerge the seeds in 70% alcohol for 30–60 s. Decant the alcohol.

3. Transfer the seeds to a flask or beaker containing 20–40% commercial sodium hypochlorite for 15–20 min. Rinse 4x with sterile distilled water.

4. Place 2–3 seeds per culture vessel on the surface of MS agar medium.

5. Incubate the cultures at 25°C under 16 h photoperiod with ~1000lux light intensity for 1–2 weeks.

6. Collect the germinated seedlings when the cotyledons are fully expanded. Place each seedling on a sterile petri dish and excise the hypocotyl from each seedling and cut them transversely into two parts.

7. Place the hypocotyl sections on the following medium: MS + 1–2 mg/l 2,4-D

8. Incubate the cultures in dark at 25°C for 4–8 weeks.

9. Maintain the callus by subculturing small pieces on fresh medium every 3–4 weeks. Callus will contain pro-embryo initial cells as well as minute microscopic embryos in the early stages of development.

10. Place 0.5 to 1 cm$^2$ callus pieces on MS agar medium without growth regulators and incubate the cultures at 25°C under 16 h photoperiod with ~1000 lux light intensity. Within 2–3 weeks cultures will exhibit embryos and green plantlets.

11. Tease out individual or group of plantlets from the callus mass and transfer on half strength MS medium under 16 h photoperiod with high light intensity of ~5 klux. Within 4–5 weeks the cultures will resemble seedling carrots.

12. Transfer the plantlets to small pots containing sterile peat moss and vermiculite in a 1:1 ratio. Enclose the plantlets with plastic containers to maintain high humidity.

13. Transfer the plants to soil and follow the procedure of plant establishment and hardening (Protocol 5.1).

## REFERENCES

Ammirato, P.V. 1983. Embryogenesis. In: D.A. Evans *et al.* (Eds.), *Handbook of Plant Cell Culture* Vol. 1. Macmillan, N.Y., pp. 82–123.

Gautheret, R.J. 1959. *La culture des tissus vegetaux: Techniques et realisations.* Masson, Paris.

Guha, S. and Maheshwari, S.C. 1964. *In vitro* production of embryos from anthers of *Datura. Nature* **204**: 497.

Kim, Y.H. and Janick, J. 1989. ABA and polyox encapsulation or high humidity inceases survival of desiccated somatic embryo of celery. *Hort. Science* **24**: 674–676.

Komamine, A. 2001. Mechanisms of somatic embryogenesis in carrot suspension of cultures. *Phytomorphology (Golden Jubilee: Trends in Plant Science)* **51**:277–288.

Lin, X., Hwang, G.J. and Zimmerman, J.L. 1996. Isolation and characterization of a diverse set of genes from carrot somatic embryos. *Plant Physiol.* **112**: 1365–1374.

Morel, G. 1965. Clonal propagation of orchids by meristem culture. *Cymbidium Soc. News* **20**: 3–11.

Morel, G. and Martin, C. 1952. Virus free Dahlia through meristem culture. *C.R. Acad. Sci.*, **235**: 1324–1325.

Murashige, T. 1974. Plant propagation through tissue cultures. *Ann. Rev. Plant Physiol.* **25**: 135–166.

Perry, S.E., Lethi, M.D. and Fernandez, D.E. 1999. The MADS-domain protein AGAMOUS-like 16 accumulates in embryogenic tissues with diverse origins. *Plant Physiol.* **120**: 121–130.

Redenbergh, K., Paasch, B., Nichoi, J.; Kossler, M., Viss, P. and Walker, K. 1986. Somatic seeds: encapsulation of asexual plant embryos. *Biotechnology* **4**: 797–801.

Reinert, J. 1958. Morphogenese und ihre kontrolle an gewebekulturen aus Carotten. *Naturwissenschaften* **45:** 344–345.

Reinert, J. 1959. Uber die kontrolle der morphogenese und die induktion von advientive embryonen an gewebekulturen aus Karotten. *Planta* **58:** 318–333.

Rode, A., Hartmann, C., De Buyser, J. and Henry, Y. 1989. Evidence for a direct relationship between mitochondrial genome organization and regeneration ability in hexaploid wheat somatic tissue cultures. *Curr. Genet.* **14:** 387–394.

Schenk, R.U. and Hildebrandt, A.C. 1972. Medium and techniques for induction and growth of monocotyledonous and dicotyledonous plant cell cultures. *Can. J. Bot.* **50:** 199–204.

Sharp, W.R., Sondahl, M.R., Caldas, L. S. and Maraffa, S. B. 1980. In: *Horticultural Reviews* **Vol. 2.** (Janick, J. ed.), AVI Publishing Co, Westport, Conn, USA, p. 268.

Skoog, F. and Miller, C.O. 1957. Chemical regulation of growth and organ formation in plant tissues cultivated *in vitro.* In: *Biological Action of Growth Substances. Symp. Soc. Exp. Biol.* **11:** 118–131.

Steward, F.C., Mapes, M.O. and Mears, K. 1958. Growth and organized development of cultured cells. II. Organization in cultures grown from freely suspended cells. *Am. J. Bot.* **45:** 705–708.

White, P.R. 1963. *The Cultivation of Animal and Plant Cells.* Ronald Press Co., New York.

Zimmerman, J. L. 1993. Somatic Embryogenesis: A Model for Early Development in Higher Plants. *Plant Cell* **5:** 1411–1423.

## FURTHER READING

Bhojwani, S.S. and Razdan, M.K. 1990. *Plant Tissue Culture: Applications and Limitations.* Elsevier Science Publishers, Amsterdam.

Chawla, H.S. 1998. *Biotechnology in Crop Improvement.* International Book Distributing Co., Lucknow.

Evans, D.A., Sharp, W.R., Ammirato, P.V. and Yamada, Y. 1983. *Handbook of Plant Cell Culture.* Macmillan Publishing Company, New York.

Ignacimuthu, S. 1998. *Plant Biotechnology.* Oxford & IBH Publishing Co. Pvt. Ltd., Delhi.

Mantell, S.H., Mathew, J.A. and McKee, R.A. 1985. *Principles of Plant Biotechnology: An Introduction to Genetic Engineering in Plants.* Blackwell Scientific Publications, Oxford.

Pierik, R.L.M. 1987. *In vitro culture of Higher Plants.* Martinus Nijhoff Publishers, Dordrecht.

Razdan, M.K. 2003 *Introduction to Plant Tissue Culture.* Science Publishers, Inc., Enfield, NH USA.

Street, H.S. (ed.)1973. *Plant Tissue and Cell culture.* Blackwell Scientific Publications, Oxford.

Thorpe, T.A. 1981. *Plant Tissue Culture: Methods and Applications in Agriculture.* Academic Press Inc., New York.

Vasil, I.K. 1984. *Cell Culture and Somatic Cell Genetics of Plants: Laboratory Procedures and Their Applications.* Academic Press, Inc., New York.

# 7

# Cell Suspension and Secondary Metabolites

Haberlandt (1902) made the first attempt to isolate and culture single cells from leaves of flowering plants. Although he failed to achieve the division of free cells, his work stimulated workers to pursue investigation on these lines. Todate, the progress in this field is spectacular that it is possible not only to culture free cells but also to induce divisions in a cell and to raise whole plant from it.

A cell suspension culture consists of cell aggregates dispersed and growing in moving liquid media. It is normally initiated by transferring pieces of undifferentiated and friable calluses to a liquid medium, which is continuously agitated by a suitable device. Suspension cultures have also been started from sterile seedlings or imbibed embryos or leaves by mechanical method. Leaves or the other tissue can be gently grinded or soft tissues can be broken up in a hand operated glass homogenizer. This homogenate, containing intact living cells, dead cells and cell debris is cleared by filtration and centrifugation and then transferred to moving liquid medium. An enzymatic method for isolation of single cells by the use of pectinases, which digest the pectin wall, has also been employed. But the single cell system used in basic and applied research is isolated from cultured tissues. Cultures are initiated by placing freshly cut sections of plant organs on a solidified nutrient medium containing suitable growth regulators. On such a medium the explant exhibits callusing, which is separated from the parent explant and transferred to a fresh medium to build up reasonable amount of callus tissue. This callus is transferred to a liquid medium and agitated to raise fine suspension of cells. Successful establishment of the suspension culture depends upon the initial callus. A wide variety of explants and media compositions have been used with success for callus induction viz. MS, $B_5$, LS (Linsmaier and Skoog, 1965), Blaydes (Blaydes, 1966) etc. Variations and modifications of these media are widely used. To these media are added vitamins, inositol, sucrose and growth regulators, especially auxin (2,4-D at ~1–5 µM concentration), for cells to divide. The addition of 1 µM kinetin may be beneficial. The growth rate and friability of callus produced can vary widely due to explants. The culture medium for suspension culture should be such that it maintains good growth of callus.

Platform (orbital) shakers are widely used for the initiation and serial propagation of plant cell suspension culture. They should have a variable speed control (30–150 rpm) and the stroke should be in the range of 4–8 cm orbital motion. Agitation of medium on a shaker serves two purposes. First it exerts a mild pressure on cell aggregates, breaking them in to smaller clumps and single cells and secondly agitation maintains uniform distribution of cell and cell clumps in the medium. Movement of the medium also provides good gaseous exchange between the culture medium and air. The shaker should be kept in the air-conditioned room with good temperature control. Wide-mouthed Erlenmeyer flasks are widely used as culture vessels. The volume of the culture should be appropriate to the culture vessel e.g. 20 ml per 100 ml flask and 70 ml per 250 ml flask. The flasks are normally sealed with aluminium foil. Flask closure must maintain sterility, allow gas exchange and reduce evaporation. Cotton wool plugs may be used for sealing flasks during autoclaving but not for culturing cells. They are a common source of contamination in flasks that are sitting on a shaker for several weeks. For the first passage, suspensions are selectively subcultured by (i) allowing the cultures to settle briefly and pipetting only suspended material, (ii) gradually reducing the bore size of the pipette, or (iii) sieving before subculture. A reduction in mean aggregate size can be brought by addition of enzymes such as cellulases and pectinases in the medium.

# TYPES OF SUSPENSION CULTURES

### Batch Culture

A batch culture is a cell suspension culture grown in a fixed volume of nutrient culture medium. Cell suspension increases in biomass by cell division and cell growth until a factor in the culture environment (nutrient or oxygen availability) becomes limiting and the growth ceases. The cells in culture exhibit the following five phases of a growth cycle (Fig. 7.1).

i. Lag phase, where cells prepare to divide.
ii. Exponential phase, where the rate of cell division is highest.

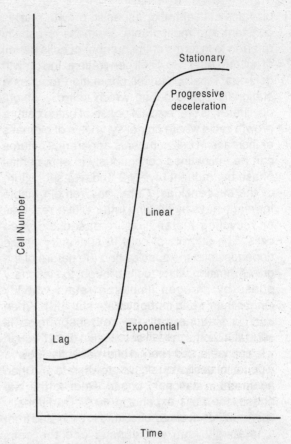

**Fig.7.1:** A model curve showing cell number per unit volume of culture to time in a batch grown cell suspension culture.

iii. Linear phase, where cell division slows but the rate of cells expansion increases.
iv. Deceleration phase, where the rates of cell division and elongation decreases.
v. Stationary phase, where the number and size of cells remain constant.

When cells are subcultured into fresh medium there is a lag phase. It is the initial period of a batch culture when no cell division is apparent. It may also be used with reference to the synthesis of a specific metabolite or the rate of a physiological activity. Then follows a period of cell division (exponential phase). It is a finite period of time early in a batch culture during which the rate of increase of biomass per unit of biomass concentration (specific growth rate) is constant and measurable. Biomass is usually referred to in terms of the number of cells per ml of culture. After 3 to 4 cell generations the growth declines. Finally, the cell population reaches a stationary phase during which cell dry weight declines. It is the terminal phase of batch culture growth cycle where no net synthesis of biomass or increase in cell number is apparent. Cultures can be maintained continuously in exponential phase by frequent (every 2–3 days) subculture of the suspensions. Cells may remain viable for many days by utilizing intracellular reserves or recycling metabolites released by lysed cells. The stability of cells in stationary phase depends upon the species and the nature of the growth limiting factor (cells brought to stationary phase by nitrogen limitation retain viability longer than when carbohydrate starved). When such a stationary phase cell suspension is subcultured, the cells in succession pass through a lag phase, a short period of exponential growth, a period of declining relative growth rate and then again enter stationary phase. Prior to the last phase, the cells expand in size. Traditionally such cultures at each subculture are initiated from a relatively high cell density and the cells therefore accomplish only a limited number of divisions (cell number doublings) before entering stationary phase again. For example, a suspension culture of *Acer pseudoplatanus* initiated at $2 \times 10^5$ cells/ml will reach a final cell density of ca. $3 \times 10^6$ cells/ml corresponding to four successive doublings of the initial population. The cell generation time (doubling time) in suspension cultures varies from 24 to 48 hours in well established cell cultures. Doubling time (td) is the time required for the concentration of biomass of a population of cells to double.

For general purposes, the objective with cell suspension cultures is to achieve rapid growth rates, and uniform cells, with all cells being viable. Frequent subculturing in a suitable medium ensures these qualities. Cells should be subcultured at weekly intervals or less if they are to be used for experimental purposes. The size of the inoculum (dilution rate) will determine the lag period. Optimally suspension cultures passage length is 1–2 weeks, but the exact time and the dilution required must be determined for each cell line. Dilutions of 1:4 after 1 week or 1:10 after 2 weeks are commonly used. There is generally a minimum inoculum size below which a culture will not recover. The pattern of growth induced in multilitre plant cell bioreactors is shown in Fig. 7.2.

## Continuous culture

A culture is continuously supplied with nutrients by the inflow of fresh medium but the culture volume is normally constant. It is of two types:

(i) *Open continuous culture*: A continuous culture is one in which inflow of fresh medium is balanced by outflow of corresponding volumes of culture including harvest of cells. Cells are constantly washed out with the outflowing liquid. The rate of inflow of medium and culture harvest are adjusted so that the cultures are maintained indefinitely at a constant sub-maximal growth rate. In a steady state, the rate of cells washout equals the rate of formation of new cells in the system. A situation of balanced growth is achieved, i.e. majority of cells in the culture are in a similar metabolic state. The growth rate and cell density are held constant by a fixed rate of input of growth limiting nutrients and removal of cells and spent medium. These open continuous systems may be: (a) *chemostats* in which growth rate and cell density are held constant by a fixed rate of input of a growth limiting nutrient medium (nitrogen, phosphorus or glucose). In such a medium, all the constituents other than growth limiting nutrients are present at concentrations higher than that required to maintain the desired rate of cell growth. The growth limiting substance is so adjusted that its increase or decrease is reflected by a corresponding increase or decrease in the growth rate of cells; (b) *turbidostats* in

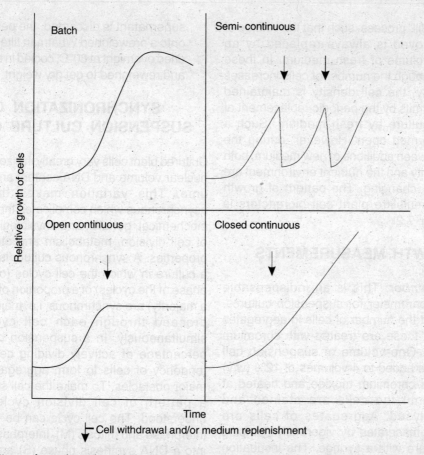

**Fig. 7.2:** Patterns of growth produced in batch, semi-continuous, and continuous plant cell cultures.

which fresh medium flows in response to increase in the turbidity so as to maintain the culture at a fixed optical density of suspension. A pre-selected biomass density is maintained by the washout of cells.

**(ii) _Closed continuous culture_:** A closed culture is one in which cells are retained and inflow of fresh medium is balanced by outflow of corresponding volumes of spent medium only. The cells from the out-flowing medium are separated mechanically and added back to the culture. Thus in closed continuous culture, cell biomass continues to increase as the growth proceeds. It has potential value in studies on cytodifferentiation, where it may be important to grow cells under a particular

regulated environment and then maintain them for a considerable period in a non-dividing but viable state. It can also be used in cases where secondary products e.g. coumarins, lignins, polysaccharides, monoterpene derivatives produced by cell suspension cultures have been shown to be released in significant amounts into their culture medium. In such cases, a maintenance culture in a closed continuous system should enable the chemical product to be continuously harvested from a fixed culture biomass.

## Semi Continuous Culture

In this type of culture the inflow of fresh medium is manually controlled at infrequent intervals by a "drain and refill" process, such that the volume of culture removed is always replaced by an equivalent volume of fresh medium. In these cultures, although the number of cells increases exponentially, the cell density is maintained within fixed limits by the periodic replacement of harvested culture by fresh medium. Such a culture is termed open. However, during the intervals between additions of new medium, both the cell density and the nutrient environment are continuously changing. The pattern of growth induced in multilitre plant cell bioreactors is shown in Fig. 7.2.

## GROWTH MEASUREMENTS

1. *Cell number*: This is an indispensable growth parameter for suspension cultures. To count the number of cells in aggregates of cells, these are treated with chromium trioxide. One volume of suspension cell culture is added to 4 volumes of 12% (w/v) aqueous chromium trioxide and heated at 70°C until the cells are stained and plasmolysed. Aggregates of cells are partially macerated by vigorously pumping the culture with a syringe. The incubation time and maceration required will vary with the culture age. Shrinkage of the protoplasts makes counting of cells in the aggregates easier. The number of cells can be counted under a simple microscope.

2. *Fresh weight*: This can be determined by collecting cells on pre-weighed (in wet condition) nylon fabric filters which can be supported in a funnel. Cells are washed with water to remove the medium, drained under vacuum and weighed. A large sample is always needed to get a reasonably accurate fresh weight estimation.

3. *Dry weight/Packed cell volume (PCV)*: An appropriate volume of suspension culture is centrifuged in a graduated conical centrifuge tube at $2000 \times g$ for 5 min and pellet volume or packed cell volume is noted. PCV is expressed as ml pellet/ml culture. The supernatant is discarded, the pellet washed onto a preweighed Whatman filter paper and dried overnight at 80°C, cooled in a desiccator and reweighed to get dry weight.

## SYNCHRONIZATION OF SUSPENSION CULTURE CELLS

Cultured plant cells vary greatly in size and shape, nuclear volume and DNA content and cell cycle time. This variation makes the culture asynchronous which complicates the studies on biochemical, genetic, and physiological aspects of cell division, metabolism and other cellular properties. A synchronous culture is defined as a culture in which the cell cycles (or a specific phase of the cycles) of a proportion of cells (often a majority) are synchronous, i.e. majority of cells proceed through each cell cycle phase simultaneously. In a suspension culture, low percentage of actively dividing cells and the tendency of cells to form aggregates are the major obstacles. To make the cell synchronous a pattern of cell division cycle must be understood. The cell cycle can be divided into interphase and mitosis (M). Interphase is divided into a DNA synthesis phase (S) and preceded by a pre-synthetic phase (G1) and followed by a post-synthetic phase (G2). Thus, mitosis represents a small proportion of the total cell cycle, which is bordered by G1 and G2. Synchronization would require that all cells proceed simultaneously through specific stages of the cell cycle. The various methods employed to bring this synchronization have been listed below.

1. *Cold treatment*: Temperature shocks at 4°C for few days can be used.

2. *Starvation method*: Deprivation of an essential growth compound from suspension cultures leads to a stationary growth phase. Resupplying the missing compound or subculturing to a fresh complete medium will allow growth to resume and may result in synchronization of cell growth. It has been reported that sycamore cells are arrested in

G1 and G2 by deprivation of phosphorus and carbohydrate respectively. Nitrogen starved cells accumulate in G1.

3. *Use of inhibitors*: Synchronization is achieved by temporarily blocking the progression of events in the cell cycle by addition of DNA synthesis inhibitors and accumulating cells in a specific stage. Upon release of block, cells will synchronously enter the next stage. Generally, 5-fluorodeoxyuridine (FudR), excess thymidine (TdR), and hydroxyurea (HU) are used which accumulate cells at G1/S interface stage. In this method, the cell synchrony is limited to only one cell cycle.

4. *Colchicine method*: Colchicine is one of the most effective spindle fiber inhibitors, which arrests the cells at metaphase stage of cell division. Suspension cultures in exponential growth are supplemented with 0.02% (w/v) colchicine. This synchronization is measured in terms of mitotic index, i.e. the percentage of cells in the mitotic phase of the cell cycle. Procedures effective for synchronization with one species may not be useful for another. Thus, combinations of these methods are often used.

# TECHNIQUE FOR SINGLE CELL CULTURE—BERGMANN CELL PLATING TECHNIQUE

Bergmann (1960) first used the most popular technique of plating out cell suspensions on agar plates. This technique is particularly useful where attempts are being made to obtain single-cell clones. The objective is to establish evenly distributed suspension in a thin layer of culture medium with as low a cellular density as is compatible with the growth of colonies from a high proportion of the cellular units of single cells and cell aggregates (Fig. 7.3). The basic technique of plating is to first count the cell number without maceration stage (see growth measurements). This will enable a known number of cell units to be established per unit volume of plating media. The counted suspension is adjusted by dilution or concentrated by low speed centrifugation. Both the suspension and nutrient medium containing agar are prepared in double the concentration separately. The equal volumes of suspension and the agar medium cooled at 35°C are mixed and then dispensed rapidly into petri dishes in such a manner that cells are evenly distributed in a thin layer (~1 mm thick). The suspension culture, which carries cell aggregates should be filtered through a sieve and only the fine suspension should be plated. The number of cellular units in per square mm are directly counted *in situ* under inverted microscope. The dishes are then sealed and incubated at 25°C for 21 days. After incubation, the number of colonies per plate are counted by making shadowgraph prints of each plate by resting the plate on a sheet of photographic document paper under a photographic enlarger as the light source. This photographic method clearly records colonies not easily observed on the standard type of bacterial colony counter. Plating efficiency (PE) is then calculated as:

$$PE = \frac{\text{No. of colonies/plate at the end of experiment}}{\text{No. of cellular units initially/plate}} \times 100$$

## Applications

1. Embryogenic cell suspension offers the possibility for large-scale clonal propagation.

2. Embryos are natural organs of perennation, many of which typically become dormant. Because of their innate properties, somatic embryos from cell suspensions can prove useful for long-term storage in germplasm banks.

3. Embryogenic suspension cultures present an excellent tool for both theoretical studies and practical applications. Selection schemes with embryogenic cultures give rise to variants for various abiotic and biotic stresses which has been discussed in Chapter 10 on somaclonal variation.

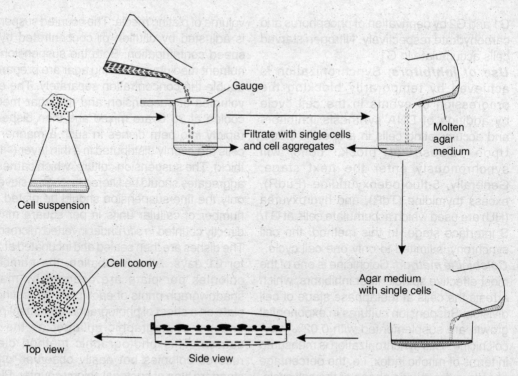

**Fig. 7.3:** Diagrammatic summary of steps involved in Bergmann cell plating technique.

4. Embryogenic cell suspension cultures and population of somatic embryos are a source for the production of important chemicals. Somatic embryos of celery produce the same flavor compounds present in the mature plant but absent in celery callus cultures. Somatic embryos of cacao produce the same lipids, including cacao butter as their zygotic counterparts.

## PRODUCTION OF SECONDARY METABOLITES

Plants are a valuable source of a vast array of chemical compounds including fragrances, flavors, natural sweetners, industrial feedstocks, anti-microbials and pharmaceuticals. In most instances these compounds belong to a rather broad metabolic group, collectively referred to as secondary products or metabolites. If plant cells produce metabolites in vivo or in vitro, which are not directly needed by the plant itself, then these are termed secondary metabolites. These secondary metabolites do not perform vital physiological functions as primary compounds such as amino acids or nucleic acids, but these are produced to ward off potential predators, attract pollinators, or combat infectious diseases.

Plant cell suspension culture is potentially valuable for studying the biosynthesis of secondary metabolites. If plant cells are cultured under conditions that allow the expression of a certain degree of differentiation, they have the potential to produce either by de novo or by biotransformation of specific precursors, a wide range of secondary products.

Despite limitations, there are some major advantages expected from cell culture systems over the conventional cultivation of whole plants for production of secondary metabolites. These are:

(a) Independence from various environmental factors, including climate, pests and microbial diseases, geographical and seasonal constraints.

(b) Useful compounds could be produced under controlled conditions and geared more accurately to market demands.

(c) Any cell of a plant could be multiplied to yield specific metabolites.

(d) A consistent product quality could be assured with the use of characterized cell lines.

(e) There would be greater control over production levels since they would not be so much at the mercy of political interference.

(f) Cell growth could be automatically controlled and metabolic processes could be regulated rationally, all contributing to the improvement of productivity and the reduction of labor and costs.

(g) Production of substances in chemically controlled environment facilitates later processing and product recovery steps.

(h) Culture of cells may prove suitable in cases where plants are difficult or expensive to grow in the field due to their long life cycles, e.g. *Papaver bracteatum*, the source of thebaine takes two to three seasons to reach maturity.

(i) New routes of synthesis can be recovered from mutant cell lines. These routes of synthesis can lead to novel products not previously found in whole plants.

(j) Some cell cultures have the capacity for biotransformation of specific substrates to more valuable products by means of single or multiple step enzyme activity.

As in microorganisms, attempts have been made to obtain substances from cell suspension cultures of higher plants, either through accumulation in the cells (biomass) or sometimes by the release into the nutrient medium. With respect to production of secondary metabolites two main approaches have been followed: (i) The rapid growth of suspension cultures in large volumes, which are subsequently manipulated to produce secondary metabolites. (ii) The growth and subsequent immobilization of cells, which are used for the production of compounds over a prolonged period.

Rapidly growing plant cell suspension cultures initiated and maintained by repeated subculturing on the same medium are the result of an unconscious but systematic selection for those cells in a heterogeneous population with the best growth rate. These cultures often consist mainly of small meristematic type cells. Further, there are evidences to indicate that in many cases, secondary product synthesis typical of intact plant cannot occur in rapidly growing undifferentiated cell cultures, but requires some degree of morphological or biochemical differentiation and slow growth. It is therefore not surprising that the production and yield of metabolite in cell culture will be dependent on a number of factors which have been discussed below.

## Morphological and Chemical Differentiation

Some secondary metabolites accumulate entirely in specific structures, showing so called morphological differentiation. Essential oils are found in glandular scales and secretory ducts (*Mentha piperita*), latex in laticifiers (*Papaver somniferum*), cardiac glycosides of *Digitalis* in leaf cells, quinine and quinidine in the bark of *Cinchona* trees and tobacco alkaloids are primarily synthesized in roots. Thus, accumulation of such products occurs only if specific structures are present in the cultures. The yields in most callus and liquid cell suspension cultures is low because of lack of tissue and organ differentiation. Thus, root differentiating calluses of *Atropa belladona* are capable of producing tropane alkaloids like atropine, while non-differentiating calluses are not. Investigations on cell differentiation versus secondary product synthesis using celery tissue culture revealed that cultures initiated from less differentiated tissues (globular and heart shaped embryos) showed no flavour compounds while cultures initiated from differentiated tissues (torpedo embryos, petiole tissues) possessed the characteristic celery phthalide flavour compounds. Some secondary metabolites accumulate in the vacuoles of cells, while others are formed elsewhere in the cell get accumulated and is then excreted from cells via the vacuole system or by other means. Failure to excrete a metabolite from cells might lead to cessation of biosynthesis, in some cases by feedback regulation mechanisms. As a general routine cell suspensions are established from friable, fast growing types which are suitable for growth and not necessarily for their ability for secondary metabolite production. This fast growth ability is linked in most cases with little tissue and/or biochemical differentiation so that it was believed that cell suspensions capable of good growth in bioreactors were not suitable for product formation. However, lately it has come to notice that under some culture conditions, various degrees of cell differentiation are expressed in cell suspension cultures.

## Medium Composition for Secondary Product Formation

Secondary product formation can be obtained by one-stage or two-stage culture systems. In a two-stage culture system, the first stage involves growing the cells on a standard growth medium and the second stage involves transferring these cells into a production medium suitable for secondary product synthesis. In the one-stage culture system both growth and production steps are combined together.

There are dramatic effects on secondary product synthesis by altering the nutrients in a production medium. In cell cultures of *Morinda citrifolia* a 50% increase in anthraquinone accumulation occurred when phosphate was increased to a concentration of 5mg/l. In suspension cultures of *Catharanthus roseus*, the overall accumulation of secondary metabolites like tryptamine and indole alkaloids have been shown to occur rapidly when cells are shifted to a medium devoid of phosphate.

An insight into the importance of specific media components on production of secondary product shikonin by *Lithospermum erythrorhizon* has been obtained. It had been shown that callus cultures of *L. erythrorhizon* could be induced to produce shikonin on Linsmaier–Skoog (LS) medium supplemented with 1μm IAA and 10μm KIN (Tabata *et al.*, 1974). No shikonin derivatives were produced in cell suspension cultures grown in LS medium, which is suitable for cell growth, while White medium favored the production of shikonin derivatives but it could not support good cell growth. Optimal concentrations of different nutrients were identified, e.g. 6.7 mM for nitrate, 1.2 μM for copper and 13.5 mM for sulfate as well as optimal levels of key organic components. The resultant medium showed 13 times higher shikonin production.

Sucrose is frequently used as a carbon and energy source and leads to optimal growth rate of plant cell cultures. Mannose, galactose, glucose and raffinose have also been used for this purpose. Both the nature and concentration of the sugar generally affects the yield of secondary product.

The type and concentration of growth regulators in cell suspension is probably one of the most important factors influencing their potential for secondary product synthesis. For example 2,4-D can stimulate both cell division and cell expansion, but it can also bring about a dramatic suppression of secondary metabolite synthesis. A range of growth regulator combinations and concentrations should be examined for optimum effect on secondary product formation. For example nicotine production by tobacco cell cultures is best achieved by a two-stage approach. The growth phase is supported on MS medium with $10^{-5}$ M NAA and $10^{-6}$ M KIN, while the production phase is optimal in MS with reduced levels of growth regulators ($10^{-6}$ M NAA and $10^{-7}$ M KIN) and by omitting casein hydrolysate from the growth medium.

The buffering capacity of plant tissue culture media is often very low which causes drift in the pH of medium. Thus, biological buffers such as N-2 hydroxyethyl piperazine-N'-2-ethanesulphonic acid (HEPES) or 2-(N-morpholino) ethane sulphonic acid (MES) are added at a concentration of approximately 100 mM. Undefined medium additives such as casein hydrolysate can also act as buffers.

An exogenous supplement of the biosynthetic precursor added to the culture medium may increase the yield of the end product. However, administration of a direct precursor does not always produce the desired effect.

### Growth Production Patterns

Secondary plant metabolites may be obtained from cell cultures produced by batch or less commonly by continuous processes. Batch culture consists of a heterogeneous mixture of cells at different stages of biochemical and morphological differentiation. Thus during initial and middle phases of culture, varying proportions of cells are in rapid division. As substrates in the medium become more limiting, the proportion of cells undergoing cell divisions decreases and leads to stationary phase, where the majority of cells become quiescent and ageing processes occur in more and more cells. Secondary metabolites synthesis is lowest during the lag phase and during the early exponential phase. Secondary product accumulation begins at the late exponential phase where primary metabolites are converted to secondary metabolites. Product accumulation reaches its peak during the stationary phase when no more cell growth takes place and all the primary metabolites are diverted to the production of secondary metabolites. It is during these later stages of batch culture that most of the secondary metabolites are produced by cells. The accumulation of metabolites at any one particular instant in the culture cycle is the result of a dynamic balance between the rate of its biosynthesis and its biodegradation. Therefore, culture conditions that promote biosynthesis of the desired compound while at the same time limiting those parts of the metabolism which are involved in biodegradation are sought. There is no hard and fast rule as to the best time in the culture cycle to harvest cells in order to obtain maximum yield of the metabolites. Thus, it is best to produce large amounts of biomass under rapid growth conditions and to then transfer the accumulated biomass to a second stage of culture that promotes secondary metabolite production.

Continuous cultures may also be operated in closed and open systems for metabolite production. In closed system the cells are retained and cell density progressively increases as growth continues. Nutrients in excess of requirements are supplied by the continuous inflow of fresh medium, balanced by the continuous harvesting of spent medium. The open system involves the input of new medium and a balancing harvest of an equal volume of culture and cells. This system allows the steady state of growth and metabolism. Regulation involves the use of either chemostats or turbidostats. In the former case, continuous input of fresh medium is set at a fixed rate and determines the nature of the resulting equilibrium. In the latter case, cell density is set at levels monitored by reading the absorbance of the culture.

For products which accumulate most actively when the growth rate is declining, it is possible to maintain cells at a growth rate consistent with the highest cellular content of the compound by chemostat culture. A considerably higher level of the product will be built up in such steady state cells than can be achieved in batch culture. Variation in the medium during this steady state may further enhance biosynthesis of the secondary metabolite. Once the highest level of product is achieved consistent with the maintenance of cell viability, the product can be harvested either through continuous run off of surplus culture or through a single harvest of the whole culture. The latter alternative would probably be used where a relatively slow growth rate is needed for optimal product accumulation. The turbidostat system is used in cases where maximum levels of secondary product develop early during the growth cycle, either during or at the beginning of the decay of exponential growth. This enables studies on the influence of growth regulators, physical factors and precursors.

Semi-continuous culture systems in which large volumes of culture are removed from bioreactors and then replaced with an equivalent volume of fresh medium can be successfully operated as chemostats to either generate biomass or for producing certain metabolites, e.g. different levels of anthocyanins have been produced in carrot cell lines. However, despite clear-cut advantages, continuous cultures are not widely used by plant tissue culture workers, as these cultures probably require a lot of attention.

## Environmental Factors

For a number of plant tissue cultures, it has been shown that environmental conditions, especially light, stimulate the production of secondary compounds. For example, compounds such as carotenoids, flavonoids, polyphenols and plastoquinones are stimulated by light. On the other hand, biosynthesis of certain metabolites is not significantly affected by light and may even be inhibited by it. Thus, irradiation with white or blue light has been seen to result in an almost complete inhibition of alkaloid biosynthesis. The qualitative and quantitative accumulation of $\alpha$ and $\beta$ pinene (monoterpenes) by callus cultures of *Pinus radiata* was markedly influenced by light regime. Lipid compositions of tobacco cell suspensions and alkaloid accumulation in *Papaver* spp. have shown significant fluctuations in response to temperature changes.

## Selection of Cell Lines Producing High Amounts of a Useful Metabolite

It is important that cultured cells should produce high amounts of secondary metabolites from the induced callus. Recently cell strains containing amounts greater than those found in intact plants

have been isolated by clonal selection. The successful selection of these cells producing high amounts of secondary metabolites has been made possible due to the heterogeneity of cultured plant cells.

In selecting specific cell lines producing high amounts of a useful metabolite, two methods have been generally used. (i) Single cell cloning and (ii) cell aggregate cloning.

i. *Single cell cloning*: The simplest procedure is to take single cells from suspension cultures and grow them individually into populations on a suitable medium. Each cell population is then screened for altered phentotype. Only those showing increased yields are finally selected and maintained by subcloning. Alternatively, suspension cultures may be screened for photochemical differences by passing through sieves of definite mesh size followed by density gradient centrifugation. Ozeki and Komamine (1985) obtained two populations of carrot suspension cultures, the heavier of which gave embryos and the lighter accumulated anthocyanin provided auxin was absent in the medium. However, there are difficulties with the isolation and culture of single cells that limit its applicability. Cell aggregate cloning appears to be time consuming, but it is far easier and has been more successful.

ii. *Cell aggregate cloning*: It is simply a tissue culture method that permits the identification of cells with increased biosynthetic capacity for a particular secondary product. It exploits the genetic variation for chemical production of a compound present in the cell population. The variation may be spontaneous, i.e. already present in the original explant material, or that induced during the process of tissue culture initiation and subculture, i.e. somaclonal variation. A typical scheme for the selection of high secondary metabolite producing cell lines by cell aggregate cloning was given by Yamada and Fujita which has been shown in Fig. 7.4 (Yamada and Fujita,

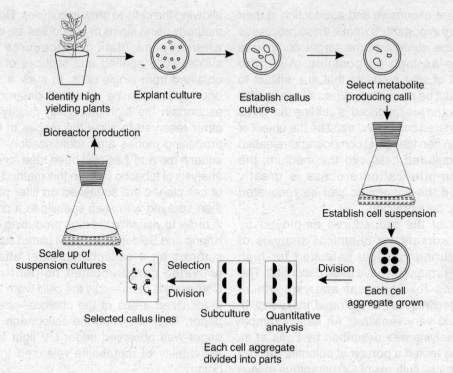

Identify high
yielding plants

Explant culture

Establish callus
cultures

Select metabolite
producing calli

Bioreactor production

Establish cell suspension

Scale up of
suspension cultures

Selection

Division

Division

Each cell
aggregate grown

Selected callus lines

Subculture

Quantitative
analysis

Each cell aggregate
divided into parts

**Fig. 7.4:** Induction and selection of high producing cell lines by cell aggregate cloning method.

1973). The first step in obtaining highly productive cell cultures is to initiate the growth of callus cultures on solid media. Cell suspension cultures are initiated from calli which promote generation of cell aggregates. In this technique, cell aggregates are plated on solid media and cultured. The resulting calluses are then divided; one half for continued growth and the other half for analysis of secondary compounds. In this way, Yamada and Fujita selected high anthocyanin producing cell lines from cultured cells of *Euphorbia millii*. The highest pigmented calluses were continually subcultured in this manner for 24 selections. The amount of pigment found in the highest producing lines after this period of time was seven times greater than in the original cells.

Examples in which it might be possible to enhance the yield using the cell cloning technique are *Daucus carota* (anthocyanin),

*Lithospermum erythrorhizon* (nicotine), *Catharanthus roseus* (serpentine and ajmalicine), *Coptis japonica* (berberine), *Lavandula Vera* (biotin) and *N. tabacum* (ubiquinone-10). Cells used for cloning in most of these cases were plated as units varying from a single cell to a group of small aggregates.

## Product Release and Analysis

One of the problems of the production of herbal drugs is that these compounds are stored intracellularly which renders the product recovery very expensive. DMSO, a permeabilizing agent with concentration of 5–25% has been effectively used. Dornberg and Knorr (1992) showed that the enzymatic permeabilization in *Morinda citrifolia* culture could be used which maintains the cell viability also.

Procedures for the separation and purification of desired metabolic compounds produced by cultured cells are basically the same as those used for plants. Separation-purification processes, however, are expensive and a reduction in their costs is very important. To make these processes cheaper, the content of the target compound should be as high as possible. Also, the production of side products that are difficult to remove must be prevented. When a high level of side product(s) are produced, a cell line that gives only the desired compound must be the object of selection. When the target compound is released from the cultured cells into the medium, the separation-purification process is greatly simplified if the compound can be recovered easily, e.g. by extraction.

Some of the procedures employed by different workers for chemical analysis of product during cell line selection for high metabolite production have been described. The procedures for chemical analysis utilize analytical techniques that are rapid, inexpensive, specific and very sensitive. An early example of such analysis was described by Zenk *et al.* (1977) who tested a portion of colonies derived from single-cell cultures of *Catharanthus roseus* for serpentine and ajmalicine, using a sensitive radioimmunoassay technique. More than 200 samples could be screened each day for selection of high-yielding strains. Hall and Yeoman (1986) used microspectrophotometry to demonstrate that cells accumulating anthocyanin differ from the remaining cells in *Catharanthus* cultures with respect to absorbance spectra. Similarly, differences in UV absorbance spectra by single cells of *Coleus blumei* and *Anchusa officinalis* suspension cultures were correlated to variations in their rosmarinic acid content.

Radioimmunoassay techniques employed in the selection of high-yielding lines of could be made more efficient by using flow cytometric sorting systems. One might also devise staining procedures wherein an added compound would form a complex with the desired compound and enable its detection. Another possibility is to use a fluorescent antibody to detect the desired compound on the cell surface. However, these methods, if applied, must be non-destructive,

allowing the cells to grow in cultures. Destructive methods using stains or antibodies are used only when replica plating procedures can be successfully applied and replicas of colonies obtained from single cells. In such a situation, one replica can be analysed destructively to ascenrtain the high-producing quality and the other reserved to allow the rescue of the high-producing clones after identification. The 'cell squash method' has also been tried for chemical analysis of tobacoo cells. In this method, portions of cell clones are squashed on filter paper and then sprayed with stain specific to a compound in order to visualize the high producing colonies. Knoop and Beiderbeck (1985) plated suspension cultures on a cellophane sheet attached to activated carbon-coated nutrient medium. Compounds released by the cells were absorbed in localized areas of the charcoal-coated filter paper. After removing the cellophane, the filter paper was observed under UV light to assess the quality of metabolite released from each colony.

## Application

Cell cultures have been shown to produce a number of valuable secondary metabolites, which includes alkaloids, glycosides (steroids and phenolics) terpenoids and a variety of flavors, perfumes, agrochemicals, commercial insecticides of plant origin, etc. The most successful example of production of valuable chemical plant cell culture is shikonin, a naphthaquinone used both as dye and a pharmaceutical. According to WHO, nearly 70–80% of the global population depends on herbal drugs. The market value for herbal products is estimated to be over $20 million. *Catharanthus roseus,* a tropical perennial plant, is important as a rich source of terpenoid indole alkaloids. The *Catharanthus* alkaloids represent a market of considerable economic size, viz. vincristine ($3,500,000/kg), vinblastine ($ 1,000,000/kg). The most economically important natural plant insecticides are the pyrethrins. These compounds are of great interest because of their lethal activity against insects, low toxicity to mammals and low persistence after use. Pyrethrins are found in some members of Compositeae, like *Chrysanthemum*, *Tagetus*, etc. The rotenoids are found in *Derris*, *Lonochocarpus* and *Tephrosia*. Some of the antineoplastic compounds and other useful compounds produced by plants are listed in Table 7.1.

Industrial production of secondary metabolites from plant cell and tissue cultures is currently restricted to three products: shikonin, ginseng saponins and berberine (Trigiano and Gray 2000). Shikonin is obtained from reddish purple roots of *Lithospermum erythrorhizon,* but due to uncontrolled mass collection of this plant from nature, it is on threshold of extinction in countries (Japan, China, Korea) where it grows

**Table 7.1:** Secondary metabolites produced by plant species used for increased production through tissue culture techniques

| Compound | Species | Activity |
| --- | --- | --- |
| Pyrethrins | *Chrysanthemum cincerariefolium* | Insecticidal |
| | *Tagetus erecta* | ,, |
| Nicotine | *Nicotiana tabacum* | ,, |
| | *Nicotiana rustica* | ,, |
| Rotenoids | *Derris elliptica* | ,, |
| | *Lonochocarpus utilis* | ,, |
| | *Tephrosia vogaeli* | ,, |
| | *Tephrosia purpurea* | ,, |
| Azadirachtin | *Azadirachta indica* | ,, |
| Phytoecdysones | *Trianthema portulacastrum* | ,, |
| Baccharine | *Baccharis megapotamica* | Antineoplastic |
| Bruceantine | *Brucea antidysenterica* | ,, |
| Cesaline | *Caesalpinia gillisesii* | ,, |
| 3-Deoxycolchicine | *Colchicum speciosum* | ,, |
| Ellipticine, 9-methoxyellipticine | *Ochrosia moorei* | ,, |
| Fagaronine | *Fagara zanthoxyloides* | ,, |
| Harringtonine | *Cephalotaxus harringtonia* | ,, |
| Indicine N-oxide | *Heiotropium indicum* | ,, |
| Maytansine | *Maytenus bucchananii* | ,, |
| Podophyllotoxin | *Podophyllum peltatum* | ,, |
| Taxol | *Taxus brevifolia* | ,, |
| Thalicarpine | *Thalictrum dasycarpum* | ,, |
| Tripdiolide, triptolide | *Tripterygium wilfordii* | ,, |
| Vinblastine, ajmalicine, vincristine | *Catharanthus roseus* | ,, |
| Quinine | *Cinchona officinalis* | Antimalarial |
| Digoxin, reserpine | *Digitalis lanata* | Cardiac tonic |
| Diosgenin | *Dioscorea deltoidea* | Antifertility |
| Morphine | *Papaver somniferum* | Analgesic |
| Thebaine | *Papaver bracteatum* | Source of codeine |
| Scopolamine | *Datura stramonium* | Antihypertension |
| Atropine | *Atropa belladonna* | Muscle relaxant |
| Codeine | *Papaver* spp. | Analgesic |
| Shikonine | *Lithospermum erythrorhizon* | Dye, pharmaceutical |
| Anthraquinones | *Morinda citrifolia* | Dye, laxative |
| Rosamarinic acid | *Coleus blumei* | Spice, antioxidant |
| Jasmine | *Jasmium* spp. | Perfume |
| Stevioside | *Stevia rebaudiana* | Sweetner |
| Crocin, picrocrocin | *Crocus sativus* | Saffron |
| Capsaicin | *Capsicum frutesces* | Chilli |
| Vanillin | *Vanilla* spp. | Vanilla |

wild. Mitsui Petrochemical Co., Japan, however, developed a commercial tissue culture process for production of shikonin using cell cultures. Ginseng saponins are extracted from roots of *Panax ginseng* (ginseng). Plant growth of this species is restricted to 30–40° North latitude. Over the years the demand for this plant has increased considerably and this spurred Meiji Seika Kaisha, Japan, to commercially develop mass scale root cell culture of *Panax ginseng* for extraction of the product. Similarly, berberine is produced from colored cell cultures of *Coptix japonica* for industrial purposes.

## Problems Associated with Secondary Metabolite Production

1. Plant cells have a very long doubling time (16–24 h) when compared to microorganisms (approximately 1 h). This means that plant cells produce relatively low biomass and therefore only small amounts of secondary metabolites.
2. If the production of secondary metabolites in the intact plant is restricted to a particular organ, for instance to a differentiated tissue (e.g. laticifiers or glands), then little production takes place *in vitro* because the cells *in vitro* are often non-differentiated.
3. The production of secondary metabolites *in vitro* is generally very low, certainly when compared to an intact plant.
4. Cell and suspension cultures are genetically unstable; the production of compounds often falls to a level lower than that at the start of the *in vitro* culture through mutation.
5. Plant cells form aggregates *in vitro,* thus vigorous stirring is needed to prevent this clumping, but this may result in damaging the cells.
6. The use of high sugar concentration in the nutrient medium (up to 5% w/v) increases the cost considerably.
7. Infections can considerably limit the process. When a bacterial infection occurs, it quickly spreads over the culture as a result of its shorter doubling time.
8. Much more intensive research for bioreactors and the development of good methods of isolation is needed.
9. Lack of knowledge about the growth of plant cells in bioreactors and fermenters is severely hindering their further development.
10. It is important to consider the cost of the final product per kilogram so that it should not be cost inhibitive.

## Immobilized Cell Systems

There are an increasing number of reports of immobilization of plant cells producing secondary metabolites on a variety of supports. Presently, different processes based on immobilized biocatalysts are in commercial operation, which include production of high fructose syrups, 6-aminopencillanic acid and various amino acids. The technique of immobilization is based on confinement of biocatalyst on or within a matrix by entrapment, adsorption or covalent linkage. Entrapment within a polymeric network is the most appropriate method used for the immobilization of large sensitive plant cells and protoplasts. The cells are physically entrapped in a gel structure and substrate(s)/product(s) can diffuse to/from the immobilized cells. The cells are commonly immobilized by entrapment in calcium alginate, potassium carrageenan or in agarose beads. It is possible to immobilize large clumps of aggregate cells or even calluses, but the most suitable cells for immobilization are fine homogeneous suspensions.

The stages in development of an immobilization system for plant cells are as follows:

(i) Obtain a cell culture with defined and desirable properties, e.g. a cell line with high product yield and a low growth rate.
(ii) Determine a suitable immobilization method which maintains cell viability and product biosynthesis.
(iii) Obtain product release from the cells into the medium while maintaining culture survival.

(iv) Design a suitable vessel for culture of the immobilized cells and operating conditions which maintain cellular characteristics.

During the last 30 years, immobilized enzymes and microbial cells have been extensively studied. Some of the advantages of an immobilized system are as follows:

(i) Reuse of expensive biomass.

(ii) It results in continuous operation.

(iii) Automatic physical separation of product from biocatalyst (cells).

(iv) Stabilization of biocatalyst.

(v) Efficient utilization of expensive fermenters by using high biomass/low volume vessels.

When immobilized whole cells are used they offer additional advantages as: (i) More stable enzyme(s) in natural environment; (ii) Possibility of using whole biosynthetic pathways; (iii) Regeneration of coenzymes and (iv) *In situ* activation when viable cells are used.

The immobilization technique applies only to those plant cell suspensions that show non-growth associated type of products. That means after generation of biomass (cell growth in suspension), the cells can be immobilized and used for various purposes.

## Polymers for immobilization

A number of polymers are available for entrapment of plant cells. These include synthetic—polyacrylamide, epoxy resin, polyurethane; carbohydrates—cellulose, agar, agarose, kappa carrageenan, alginate, chitosan; and proteins—collagen, gelatin and fibrin. The list is rapidly growing and it appears possible to immobilize any plant cell line with preserved viability. Agar and agarose, alginate and kappa-carrageenan have been used quite often to entrap the cells. A different approach is to allow cells to become entrapped within the spaces of a pre-formed matrix such as nylon, polyurethane or metal mesh. When cells in suspension are mixed with these materials (usually in the form of spheres or cubes up to 1 cm diam.), they are rapidly incorporated into the network and subsequently grow into the cavities of the mesh

and become entrapped either by physical restriction or by attachment to the matrix material. Some success with this simple technique has been obtained. Modifications to the methods of using agarose, alginate is to use the mixtures of alginate and gelatin, agarose and gelatin, polyacrylamide and xanthan gum or polyacrylamide and alginate gels.

After the immobilization, tests may be employed to study the viability of immobilized plant cells. The methods used for viability are same as used for cell suspension, e.g. staining with FDA, calcoflour white, phenosafranine, cell growth, mitotic index, etc. (refer to Chapters 8 and 9). In general, all methods employed to study the viability of immobilized plant cells have shown that the cells are not affected to a great extent by the immobilization process.

Cells raised in suspension cultures and immobilized on a suitable matrix are then placed in aerated columns. Most biosynthetic experiments have been carried out in simple batch reactors but there are examples of continuous operation of packed bed reactors. The continuous reactor types are used where the product is extracellular. From the studies made so far, it can be concluded that the immobilized cells can carry out the same biosynthetic reactions as the corresponding freely suspended cells. In fact, in certain cases, a higher productivity (up to 100 times) has been observed for plant cells. Immobilized plant cells may be employed for biosynthetic purposes over an extended period of time. The same preparation of cells has in one case been used for 220 days. The total productivity per biomass unit is in such cases considerably increased.

## Product release

As mentioned earlier, continuous reactors can only be applied to extracellular products. In order to realize the products of intracellular origin, the product is sometimes spontaneously released. It has been observed that products normally stored within the cells leak out into the medium. If it is not so, then induced release is performed.

Secondary products are often stored within the vacuoles of cultivated plant cells. In order to reach such products, two membranes i.e. the plasma membrane and the tonoplast have to be penetrated. For this purpose, a number of permeabilizing agents such as DMSO (dimethyl sulfoxide) have been used. The concentration of DMSO required for quantitative permeabilization varies from 5 to 25% with cell species. For cells of *Catharanthus roseus*, 5% DMSO appears to be sufficient for the release of substances stored in the vacuoles.

The immobilization of plant cells is a relatively new technique. This has not been used on an industrial scale for biosynthesis of plant metabolites. However, its potential is considerable and at least in one instance for the pilot industrial scale biotransformation of β-methyl digitoxin to valuable β-methyl digoxin has been conducted.

## Biotransformation

An area of biotechnology that has gained considerable attention in recent years is the ability of plant cell cultures to catalyze the transformation of a readily available or inexpensive precursor into a more valuable final product. The conversion of a small part of molecule in the structure and composition to industrially important chemicals by means of biological systems is termed as biotransformation. A partial listing of biotransformations with plant cells that can be implemented in a diverse array of enzyme reactions has been shown in Table 7.2.

**Table 7.2:** Biotransformations which can be achieved by plant cell cultures

| Reaction | Substrate | Product |
|---|---|---|
| Reduction | $C = C$ | $CH_2-CH_2$ |
| | $C = C-CO$ | $CH_2-CH_2-CO$ or $CH_2-CH_2-CHOH$ |
| | CO | CHOH |
| | CHO | $CH_2OH$ |
| Oxidation | $CH_3$ | $CH_2OH$ or COOH |
| | $CH_2OH$ | CHO |
| | CHOH | CO |
| | $= S$ | $S = O$ |
| Hydroxylation | CH | C–OH |
| | $CH_2$ | CH–OH |
| | $NH_2$ | NH–OH |
| Glycosylation | OH | O–glucose |
| | COOH | COO–glucose |
| | CH | C–glucose |
| | N | $N^+$–arabinose |
| Esterification | OH | O–palmitate |
| | COOH | COO–malate |
| | NH | N–acetate |
| Methylation and demethylation | OH | $O-CH_3$ |
| | N | $N^+-CH_3$ |
| | $N^+-CH_3$ | $= NH$ |
| Isomerization | *trans* | *cis* |
| | dextra | laevo rotation |
| | β–OH | α–OH |
| Epoxidation | $CH = CH$ | - HC–CH– with O bridge |

The use of plant cell cultures for biotransformation requires the selection of cell types that express the enzymatic capabilities to catalyze the specific reaction of interest. Another factor that weighs heavily in the cell selection process is the specificity of enzyme reaction. Plant cells have been shown to perform more than one biotransformation with a given substrate. Therefore, the selection process needs to focus on the detection of cell lines that specifically catalyze the desired reaction with little or no contamination from other reaction products. Undoubtedly, a complete knowledge of the enzymology of the desired reaction needs to be incorporated into the cell culture selection process.

Two highly desirable features of plant cell biotransformation have been described. First, more than one reaction can be accomplished by a particular cell culture. In other words, multistep transformation can be accomplished using cell cultures that express a series of enzyme activities. The biotransformation of valencene into nootkatone by cell suspensions of *Citrus* is produced by two steps through the 2-hydroxy-derivative nootkatol. Up to 66% conversion rate was achieved with specific cell cultures (Table 7.3). The second unique feature of biotransformation is that in some instances, non-producing cell cultures can be used to synthesize the desired end product. The most celebrated example of this application is the biotransformation of digitoxin into digoxin. To date, plant tissue cultures have been used in achieving a specific biotransformation in the production of a cardiovascular steroid digoxin or $\beta$-methyl digoxin from digitoxin or $\beta$-methyl digitoxin obtained from *Digitalis lanata* (foxglove). Although undifferentiated cell cultures of *D. lanata* do not produce cardiac glycosides; they are nevertheless able to perform special biotransformations on cardenolide glycosides added to the medium. *Datura* cell cultures possess ability to convert hydroquinone into arbutin through glycosylation. Cell cultures of *Stevia rebaudiana* and *Digitalis purpurea* can convert steviol into steviobiocide and stevioside which are 100 times sweeter than cane sugar.

## Secondary Metabolite Production using Genetically-Engineered Plant Cell Cultures

Genetic manipulation of cells cultures has great potential for altering the metabolic profile of plants, thereby making them profitable to the industry. Transgenic cell cultures hold promise for production of flavor components, food additives and colors. Identification of key regulatory enzymes of secondary metabolic pathways is necessary in order to manipulate them in culture cells. Dixon *et al.* (1999) generated cell suspension cultures from transgenic plants over-expressing bean PAL and alfalfa $C_4$ transgenes in tobacco which correspondingly increased the level of phenolic compounds with time in culture. Genetic engineering of metabolic pathways in plants are particularly applicable to such case where elicitation fails to provide the necessary specificity. Attempts at altering phenylpropanoid pathway flux by genetic engineering using plants

**Table 7.3:** Biotransformations performed by plant cells

| Plant cell culture | Substrate | Product |
| --- | --- | --- |
| *Papaver somniferum* | Codeinone | Codeine |
| *Nicotiana tabacum* | Carvoxiome | Cavoxone |
| *Citrus* spp. | Valencene | Nootkatone |
| *Solanum tuberosum* | Solavetivone | Hydroxylated derivatives |
| *Choisya ternate* | Ellipticine | 5-formyl ellipticine |
| *Galium mollugo* | 2-succinyl benzoate | Anthraquinones |
| *Digitalis lanata* | Digitoxin | Digoxin |
| *Datura* spp. | Hydroquinone | Arbutin |
| *Stevia rebaudiana/ Digitalis purpurea* | Steviol | Steviocide, steviobiocide |

cell culture (Dixon and Paiva, 1995) and strategies to improve secondary product synthesis by metabolic engineering of crop plants are reviewed in an edited volume by Fu *et al.* (1999).

# PROTOCOLS

### 7.1: Protocol for Cell Suspension Culture (*Nicotiana tabacum*)

*Plant material: Nicotiana tabacum*

1. Follow Steps 1–5 of Protocol 5.1 for *in vitro* germination of seeds.
2. Collect the aseptically germinated seedlings when the cotyledons are fully expanded and the epicotyl is beginning to emerge. Usually this will occur when the seedlings are 2-week old. Place each seedling on a sterile slide/petri dish and prepare the explant.
3. Various explants can be used. Excise the shoot apex with and without cotyledons and insert the stem base into the medium, cotyledons, and hypocotyl section from the decapitated seedling, etc. Place the explants on the following medium (MSN):
   MS + 1–2 mg/l 2,4-D
4. Incubate the cultures in dark at 25°C. Callus will be produced in 3–4 weeks.
5. Cut small pieces of callus of ~0.5 g fresh weight and subculture on the same fresh medium for proliferation.
6. Prepare MSN liquid medium without agar. For experiment purposes 15 ml medium in a 125 ml Erlenmeyer flask is put. Sterilize the opening of a flask with the flame of burner in the hood.
7. Transfer a piece of callus to a sterile petri dish. Gently break the callus of ~2 cm diameter with forceps into 20–30 small pieces.
8. Transfer the small pieces of callus to the liquid media using forceps.
9. Heat sterilize the opening of a flask and place the sterile cap on the flask. Prepare replicated samples.

10. Incubate on a gyratory shaker set at 125 rpm which should be placed in a temperature controlled room.
11. Subculture every week. For the first few subcultures, remove a portion of the spent medium and replace with fresh medium with the help of a large bore sterilized pipette. When the cell mass has about doubled, carefully split the culture into two flasks with an equal volume of fresh medium. Repeat the incubation cycle.
12. Once the suspension culture becomes established and consists of finely dispersed cell clusters and aggregates, a dilution ratio of 1:4 to 1:10 old culture to fresh medium should be possible on a 7–10 day basis to maintain the cell line. Cell suspension can be seen under a microscope by taking an aliquot from the flask on a glass slide under sterile conditions.

### Cell suspension growth curve

1. Use an established cell suspension line to follow the growth curve over time. Carefully combine several cultures into a larger single batch of suspension, in order to provide a uniform inoculum for each replicate.
2. Pipette 5 ml of inoculum into 25 ml of fresh MSN liquid medium in a 125 ml Erlenmeyer flask. Prepare at least four replicates.
3. Aseptically transfer 10 ml of each culture into separate sterile, conical, calibrated centrifuge tubes. Spin at 2000 g for 5 min. Measure the packed cell volume (PCV) in each tube. This is the zero time value.
4. Resuspend the cells in the centrifuge tube with 10 ml of medium and return to their respective culture flasks. Incubate the suspensions on the gyratory shaker. Excellent sterile techniques and handling is required as this is the most likely source of contamination.
5. Repeat Steps 3–4 every 2–3 days for a total of 3 weeks. Calculate the mean PCV and standard deviation for each sampling time and plot the growth curve.

## 7.2: Protocol for Cell Suspension Culture (Cereals—Wheat, Rice, Maize, Barley, etc.)

1. Follow Steps 1–5 of Protocol 5.4 for induction of callus from embryos of cereals.
2. Follow Steps 6–12 of Protocol 7.1 for initiation of cell suspension culture and then Steps 1–5 for measurement of cell suspension growth. Same liquid medium should be used for cell suspension which was employed for initiation of callus i.e. MS + 2 mg/l 2,4-D.

## REFERNCES

Bergmann, L. 1960. Growth and division of single cells of higher plants *in vitro*. *J. Gen. Physiol*. **43**: 841–851.

Bhojwani, S.S. and Razdan, M.K. 1996. Plant Tissue culture Theory and Practice, a *Revised Edition*.

Blaydes, D.F. 1966. Interaction of kinetin and various inhibitors on the growth of soybean tissue. *Physiol. Plant* **19**: 748–753.

Cell lines of *Catharanthus roseus* with increased tryptophan decarboxylase activity. *Z. Naturforsch. C: Biosci*. **38 C**: 916–922.

Dixon, R.A. and Paiva, N.L.1995. Stress-induced phenylpropanoid metabolism. *Plant Cell* **7**: 1085–1097.

Dornberg, H. and Knorr, D. 1992. Release of intracellularly stored anthraquinones by enzymatic permeabilization of viable plant cells. *Process Biochem* **27**: 161–166.

Dougall, D.K. 1987. Cell cloning and selection of high yielding stains. In: F. Constable and I.K. Vasil (Eds.), Cell Culture and Somatic Cell Genetics of Plants, **Vol 4:** *Cell culture in Phytochemistry*.

Fu, T.J., Singh, G. and Curtis, W.R. (Eds.). 1999. Plants cell and Tissue Culture for the Production of Food Ingredients. *Kluwer, Dordrecht*, 285 pp.

Haberlandt, G. 1902. Culturversuchemit isolierten Pflanzenzellen. *Sitzungsber. Math. Naturwiss. Kl. Kais. Akad. Wiss. Wien*. **111**: 69–92.

Hall, R.D. and Yeoman, M.M. 1986. Temporal and spatial heterogeneity in the accumulations of anthocyanin in cell cultures of *Catharanthus roseus* (L.) *G. Don. Exp. Bot*. **37**: 48–60.

Hunter, C.S. and Kilby, N.J. 1999. Their accumulation and release *in vitro*. In: R.D. Hall (Ed.). Methods in Molecular Biology, Vol. III Plant Cell Culture Protocols Human Press Inc., Totowa, N.J., pp. 403–410.

Kilby, N.J. and Hunter, C.S. 1991. Towards optimization of vacuole- located secondary product from *in vitro* grown plant cells using 1.02 MHz ultrasound. *Appl. Microbiol. Biotechnol*. **33**: 488–451.

Knoop, V. and Beiderbeck, R. 1985. Adsorbent filter—a tool for the selection of plant suspension culture cells producing secondary substance. *Z. Naturforsch. C. Bioci*. **40C**: 297–300

Linsmaier, E.M. and Skoog, F. 1965. Organic growth factor requirements of tobacco tissue cultures. *Physiol. Plant* **18**: 100–127.

Ozeki, V. and Komamine, A. 1985. Induction of anthocyanin synthesis in relation to embryo system for the study of expression and repression of secondary metabois. *In*: K.H. Neumann *et al*. (Eds.), Primary and Secondary Metabolism in Plant Cell Cultures. *Springer- Verlag, Berlin*, pp. 100–106.

Parr, S.J., Robins, R.J. and Rhodes, M.J. C. 1986. Product release form plant cells grown in culture. In: M. Phillips *et al*. (Eds.), Secondary Metabolism in Plant Cell Culture, Cambridge University Press, Cambridge, pp. 173–177.

R.L.1999. The safety assessment of flavor ingredient drived from plant cell and tissue culture. In: T.J. Fu *et al*. (Eds.) Plant cell and Tissue Culture for the Production of Food Ingredients. *Kluwer, Dordrecht, pp*. 251–258.

Sung, Z.R. 1979. Relationship of indole- 3-acetic acid and tryptophan concentrations in normal and 5-methyltryptophan resistant cell linens of wild carrots. *Planta* **145**: 339–345.

Tabata, M., Migukami, H., Hirasoka, N. and Konoshima, M. 1974. Pigment formation in callus cultures of *Lithospermum erythrorhizon. Phytochemistry* **13**: 927–932.

Trigiano, R.N. and D.J. Gray (Eds.). 2000. Plant Tissue Culture Concept and Laboratory Exercise *(Second Edition)*.CRC *Press, Boca Raton,* 454 pp.

Verpoorte, R., Van der Heijden, R., Ten Hoopen, H.J.G. and Memelink, J. 1999. Novel approach to improve secondary metabolite production. In: T.J. Fu, *et al*. (Eds), Plant Cell and Tissue Culture for the production of Food Ingredients. Kluwer, Dordrecht, pp. 85–100.

Widholm, J.M. 1987. Selection of mutants which accumulate desirable secondary compounds. *In*: F. Somatic Cell Genetic of Plants, **Vol 4:** Cell Culture in Phytochemistry. Academic Press, Inc., San Diego, pp. 125–137.

Weiler, E.W. and Dens, B. 1977. Formation of the indole alkaloids serpentine and ajmalicine is suspension cultures *Catharanthus roseus*. In: W. Barze *et al*. (Eds.), Plant Tissue Culture and Its biotechnical Application. Springer- Verlag, Berlin, pp. 27–43.

Yamada, Y. and Fujita, Y. 1983. Production of useful compounds in culture. In: *Handbook of Plant Tissue Culture*. **Vol 1** (Evans, D.A., Sharp, W.R., Ammirato, P.V. and Yamada, Y. eds.), Macmillan, New York, pp. 717–728.

## FURTHER READING

Chawla, H.S. 1998. *Biotechnology in Crop Improvement.* International Book Distributing Co., Lucknow.

Evans, D.A., Sharp, W.R., Ammirato, P.V. and Yamada, Y. 1983. *Handbook of Plant Cell Culture.* Macmillan Publishing Company, New York.

Razdan, M.K. 2003. *Introduction to Plant Tissue Culture.* Science Publishers, Inc., Enfield, NH USA.

Reinert, J. and Bajaj, P.P.S. (eds.) 1977. *Applied and Fundamental Aspects of Plant Cell, Tissue and Organ Culture.* Springer-Verlag, Berlin.

Street, H.S. (ed.) 1973. *Plant Tissue and Cell culture.* Blackwell Scientific Publications, Oxford.

Thorpe, T.A. 1981. *Plant Tissue Culture: Methods and Applications in Agriculture.* Academic Press Inc., New York.

Vasil, I.K. 1984. *Cell Culture and Somatic Cell Genetics of Plants: Laboratory Procedures and Their Applications.* Academic Press, Inc., New York.

Yamada, Y. and Fujita, Y. 1983. Production of useful compounds in culture. In: Handbook of Plant Tissue Culture (eds. Evans, D.A., Sharp, W.R., Ammirato, P.V. and Yamada, Y. 1983), Macmillan Publishing Company, New York, pp. 717–728.

# 8

# *In Vitro* Production of Haploids

The term haploid refers to those plants which possess a gametophytic number of chromosomes (single set) in their sporophytes. The interest in haploids stems largely from their considerable potential in plant breeding, especially for the production of homozygous plants and in their studies on the detection of mutations.

The significance of haploids in the fields of genetics and plant breeding has been realized for a long time. However, their exploitation remained restricted because of low frequency (.001–.01%) with which they occur in nature. Spontaneous production of haploids usually occurs through the process of apomixis or parthenogenesis (embryo development from an unfertilized egg). Artificial production of haploids was attempted through distant hybridization, delayed pollination, application of irradiated pollen, hormone treatment and temperature shock. However, none of these methods were dependable and repeatable. It was in 1964 that Guha and Maheshwari reported the direct development of embryos and plantlets from microspores of *Datura innoxia* by the culture of excised anthers. Later, Bourgin and Nitsch (1967) obtained complete haploid plants of *Nicotiana tabacum*. Since then, anthers containing immature pollen have been successfully cultured for a wide range of economically important species.

Haploids may be grouped into two broad categories: (a) monoploids (monohaploids), which possess half the number of chromosomes from a diploid species, e.g. maize, barley; and (b) polyhaploids, which possess half the number of chromosomes (gametophytic set) from a polyploid species, e.g. potato, wheat. Here, the general term haploid is applied to any plant originating from a sporophyte (2n) and containing (n) number of chromosomes.

Haploid production through anther culture/ microspore has been referred to as androgenesis while gynogenesis is the production of haploid plants from ovary or ovule culture where the female gamete or gametophyte is triggered to sporophytic development.

## ANDROGENIC METHODS

The androgenic method of haploid production is from the male gametophyte of an angiosperm plant, i.e. microspore (immature pollen). The underlying principle is to stop the development of pollen cell whose fate is normally to become a gamete, i.e. a sexual cell, and to force its development directly into a plant. Haploids can be obtained by the culture of excised anthers and culture of isolated pollen (microspore). Here it is worthwhile to explain the structure of anther and development of pollen *in vivo* (Fig. 8.1). An

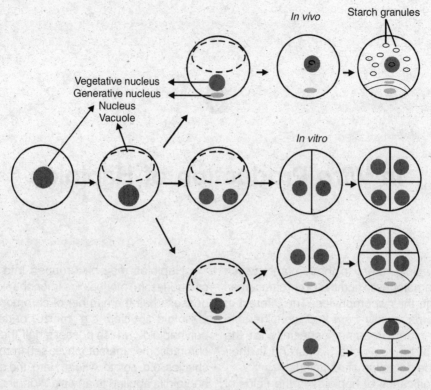

**Fig. 8.1:** Diagrammatic representation showing various modes of division of the microspores under *in vivo* and *in vitro* conditions.

angiosperm stamen consists of a filament, connective tissue and anther. A typical anther shows two anther lobes and each lobe possess two microsporangia or pollen sacs. During microsporogenesis in a young anther, there are four patches of primary sporogenous tissue, which either directly function as pollen mother cells (PMCs) or undergo several divisions. The PMCs form pollen tetrads by meiosis and when the callose wall of tetrad dissolves, the four pollen grains or microspores are liberated. The newly released microspore is uninucleate, densely cytoplasmic with the nucleus centrally located. As vacuolation occurs, the nucleus is pushed towards the periphery. At the first division or first pollen mitosis, the microspore nucleus produces a large vegetative and a small generative nucleus. The second pollen mitosis is restricted to generative nucleus and forms two sperms and takes place in either the pollen or pollen tube.

## Anther Culture

The technique of anther culture is rather simple, quick and efficient. Young flower buds with immature anthers in which the microspores are confined within the anther sac at the appropriate stage of pollen development are surface sterilized and rinsed with sterile water. The calyx from the flower buds is removed by flamed forceps. The corolla is slit open and stamens are removed and placed in a sterile petri dish. One of the anthers is crushed in acetocarmine to test the stage of pollen development. If it is found to be correct stage, each anther is gently separated from the filament and the intact uninjured anthers are inoculated horizontally on nutrient media. Injured anthers may be discarded because wounding often stimulates callusing of the anther wall tissue. When dealing with plants having minute flowers such as *Brassica* and

*Trifolium*, it may be necessary to use a stereo microscope for dissecting the anthers. In case of cereals, spikes are harvested at the uninuclear stage of microspore development and surface sterilized. The inflorescence of most cereals at this stage is covered by the flag leaf. Anthers can be plated on solid agar media in petri dishes. Normally 10–20 anthers are plated in a 6 cm petri dish. Anthers can be cultured on a liquid media and 50 anthers can be cultured in 10 ml of liquid medium (Fig. 8.2). In responsive anthers, the wall tissue gradually turn brown and within 3–8 weeks they burst open due to the pressure exerted by the growing pollen callus or pollen plants. After they have attained a height of about 3–5 cm, the individual plantlets or shoots emerging from the callus are separated and transferred to a medium that would support further development. The rooted plants are transferred to sterile soil mix in the pots.

Anther culture has an advantage over microspore culture in being very quick for practical purposes, and also sometimes the anther wall has an influence on the development of microspores in it, acting as a conditioning factor.

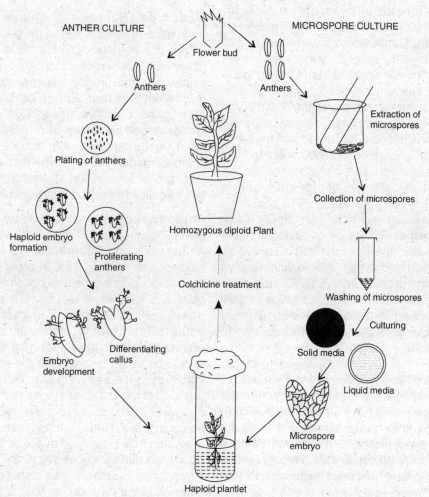

ANTHER CULTURE

MICROSPORE CULTURE

Flower bud

Anthers

Anthers

Plating of anthers

Extraction of microspores

Haploid embryo formation

Collection of microspores

Proliferating anthers

Homozygous diploid Plant

Washing of microspores

Embryo development

Differentiating callus

Colchicine treatment

Culturing

Solid media

Liquid media

Microspore embryo

Haploid plantlet

**Fig. 8.2:** Diagrammatic illustration of anther and microspore culture for production of haploid plants and diploidization.

### Microspore Culture

Haploid plants can be produced through *in vitro* culture of male gametophytic cells, i.e. microspores or immature pollen. In a general procedure for microspore culture, anthers are collected from sterilized flower buds in a small beaker containing basal media (e.g. 50 anthers of *Nicotiana* in 10 ml media). The microspores are then squeezed out of the anthers by pressing them against the side of beaker with a glass rod (Fig. 8.2). Anther tissue debris are removed by filtering the suspension through a nylon sieve having a pore diameter which is slightly wider than the diameter of pollen (e.g. 40 μ for Nicotiana, 100 μ for maize etc.). It has been observed that smaller microspores do not regenerate, thus larger, good and viable microspores can be concentrated by filtering the microspore suspension through nylon sieves. This pollen suspension is then centrifuged at low speed ca. 150x g for 5 min. The supernatant containing fine debris is discarded and the pellet of pollen is resuspended in fresh media and it is washed atleast twice. The microspores obtained are then mixed with an appropriate culture medium at a density of 103 to 104 microspores/ml. The final suspension is then pipetted into small petri dishes. To ensure good aeration, the layer of liquid in the dish should be as thin as possible. Each dish is then sealed with parafilm to avoid dehydration and is incubated. The responsive microspores form embryos or calli and its subsequent development to plant formation can be achieved by transferring to suitable media.

Culture of anthers has proved to be an efficient technique for haploid induction. But it has one main disadvantage that plants not only originate from pollen but also from various other parts of the anther (especially in dicots) with the result that a population of plants with various ploidy levels are obtained. This difficulty can be removed by culture of isolated microspores which offer the following advantages:

1. Uncontrolled effects of the anther wall and other associated tissues are eliminated and various factors governing androgenesis can be better regulated. But disadvantageous where anther wall has a stimulatory effect.
2. The sequence of androgenesis can be observed starting from a single cell.
3. Microspores are ideal for uptake, transformation and mutagenic studies, as microspores may be evenly exposed to chemicals or physical mutagens.
4. Higher yields of plants per anther could be obtained.

**The various factors governing the success of androgenesis have been discussed below.**

*Genotype*: Success in anther culture is predominantly dependent on the genotype of the anther donor material. It has been repeatedly observed that various species and cultivars exhibit different growth responses in culture. It is now known that anther culture ability is genetically controlled. Thus, a general survey for tissue culture response of various cultivars must be undertaken with simple media, as complex media rich in growth regulators tend to favor proliferation of somatic anther tissue giving rise to callus of various ploidy levels.

*Physiological status of the donor plants*: The physiological status of the plants at the time of anther excision influences the sporophytic efficiency of microspores. In order to allow for the in vitro development of pollen into an adult plant it is very important to start with healthy pollen cells. Therefore, it is good to culture anthers from plants grown under the best environmental conditions. The donor plant should be taken care of from the time of flower induction to the sampling of pollen. The use of any kind of pesticide, whether externally applied or systemic should be avoided for 3–4 weeks preceding sampling. The response in culture is predominantly influenced by the different growth conditions during various seasons. The variation in response of anthers from plants grown under different environmental conditions may be due to the differences in the endogenous level of growth regulators. Critical environmental factors are light intensity, photoperiod, temperature, nutrition and concentration of carbon dioxide. It has been generally observed that plants grown outdoors during natural growing seasons are more responsive than greenhouse-grown material. Flowers from relatively young plants at the beginning of the flowering season are more responsive. It is therefore of prime importance that plants be grown under optimal growth conditions, watered with minimal salt solutions periodically, and that relatively young plants be used.

*Stage of pollen*: It has been established that selection of anthers at an appropriate stage of pollen development is most critical. Anthers with microspores ranging from tetrad to the binucleate stage are responsive. But as soon as starch deposition has begun in the microspore, no sporophytic development and subsequently no macroscopic structure formation occurs. Data has established that uninucleate microspores are more prone to experimental treatment for culture just before or during first mitosis. There is an optimum stage for each species. For example, pollen at or just after the pollen mitosis in *Datura innoxia*, *Nicotiana tabacum*, early bicellular stage in *Atropa balladonna* and *Nicotiana sylvestris* and mid- to late uninucleate stage of microspore development in cereals has been found to give the best results.

*Pretreatment of anthers*: The underlying principle of androgenesis is to stop the development of the pollen cell whose fate is normally to become a gamete, and to force its development directly into a plant. This abnormal pathway is possible if the pollen cell is taken away from its normal environment in the living plant and placed in other specific conditions. This induction of androgenesis is enhanced by giving certain treatments to the whole spike, or flower bud or to the anthers.

(i) *Cold pretreatment*: In general, cold treatment between 3 and 6°C for 3 to 15 days gives good response. Maize responds better to a temperature of 14°C. The degree of cold

that should be given is dependent on the species. As a result of cold treatment, weak or non-viable anthers and microspores are killed and the material gets enriched in vigorous anthers. It is possible that cold pretreatment retards aging of the anther wall, allowing a higher proportion of microspores to change their developmental pattern from gametophytic to sporophytic.

(ii) *Hot treatment*: Floral buds or entire plants in some species when subjected to 30°C for 24 h or 40°C for 1 h stimulates embryogenesis (e.g. *Brassica*). The temperature shock appears to cause dissolution of microtubules and dislodging of the spindle which causes abnormal division of the microspore nucleus.

(iii) *Chemical treatment*: Various chemicals are known to induce parthenogenesis. 2-chloro ethylphosphonic acid (Ethrel) has a pronounced effect in increasing the haploid production in various species. Plants are sprayed with an ethrel solution (e.g. 4000 ppm in wheat) just before meiosis in PMCs which results in multinucleated (4–6) pollen with fewer starch grains. It is possible that multinucleated pollen might be induced to form embryos when cultured.

*Culture media*: The composition of medium is one of the most important factors determining not only the success of anther culture but also the mode of development. Normally only two mitotic divisions occur in a microspore, but androgenesis involves repeated divisions. It is difficult to draw a conclusion as to which medium is most suitable as species or even genotypes may demand different nutritional conditions. Basal medium of MS, White (1963), Nitsch and Nitsch (1969) and $N_6$ for solanaceous crops; $B_5$ and its modifications for *Brassica*; and $B_5$, $N_6$, LS (Linsmaier and Skoog, 1965) and Potato2 medium (Chuang *et al.*, 1978) for cereals have been used. Thus, no general recommendations can be given but some general observations have been mentioned.

Sucrose is essential for androgenesis. Sugars are indispensable in the basal medium as they are not only the source of carbon but are also involved in osmo-regulation. The usual level of sucrose is 2–4%; however, higher concentration (6–12%) favors androgenesis in cereals. Chelated iron has been shown to play an important role in the differentiation of globular embryos into heart-shaped embryos and further into complete plants.

Nitrogen metabolism is quite an important feature. The presence of nitrate, ammonium salts as well as amino acids appear to play a very special role at different stages of the developmental process. However, glutamine is probably beneficial for most plant species as an aid to achieving the *in vitro* differentiation of a cell to a complete plant.

Pollen embryogenesis can be induced on a simple mineral-sucrose medium in plants like tobacco, yet for androgenesis to be completed, addition of certain growth regulators is required. For example, cereal anthers require both auxins and cytokinins and optimal growth response depends on the endogenous level of these growth regulators. However to promote direct embryogenesis simple media with low levels of auxins are advisable. When the response of cereal cells and tissues to different growth regulator concentrations and combinations are compared with dicots, generally growth hormones are needed in the former whereas high amount of phytohormone autotropy is present in the dicots.

Activated charcoal in the medium enhances the percentage of androgenic anthers in some species presumably by removing the inhibitors from the medium.

## Process of Androgenesis

The anthers normally start undergoing pollen embryogenesis within 2 weeks and it takes about 3–5 weeks before the embryos are visible bursting out of the anthers. In rice this may take up to 8 weeks. Haploid plantlets are formed in two ways (i) direct androgenesis: embryos originating directly from the microspores of anthers without callusing or (ii) indirect androgenesis i.e. organogenic pathway: microspores undergo proliferation to form callus which can be induced to differentiate into plants.

The process of androgenesis has shown that microspores undergo divisions, which continues until a 40–50 celled proembryo is formed. The embryos, mostly at globular stage, burst out of the exine and are released. The embryos undergo various stages of development, simulating those of normal zygotic embryo formation. However, when the microspore takes organogenic pathway, it looks to be larger than embryonal type of micropspore after 2–3 weeks and contain only a few cells. These cells increase in size, exerting pressure on the exine which bursts open and the contents are released in the form of a callus. These calli then differentiate into plantlets. The plants with well developed shoots and roots are then transferred to pots. The plantlets originating from the callus generally exhibit various levels of ploidy.

The physical environmental conditions in which the cultures are to be placed can enhance the differentiation. The cultures are incubated at 24–28°C. In the initial stages of induction of morphogenesis, darkness is normally more effective or cultures should be kept in low light intensity (500 lux). After induction, macroscopic structures are transferred to a regeneration medium (in cereals with reduced sucrose and auxin concentration) and kept at 14 h day-light regime at 2000–4000 lux.

## The Ploidy Level and Chromosome Doubling

The ploidy of plants derived from anther or microspore culture is highly variable. The wide range of ploidy levels seen in androgenetic plants has been attributed to endomitosis and/or fusion of various nuclei. Variations in ploidy observed in anther cultures appears to be a function of the developmental stage of anthers at the time of excision and culture, with higher ploidy levels more prevalent in anthers cultured following microspore mitosis. Moreover, haploid tissues are quite susceptible to changes in ploidy level during cell proliferation and growth *in vitro*. For obtaining homozygous lines, the plants derived through anther culture must be analysed for their ploidy status. Various approaches to determine ploidy level are:

1. *Counting of plastids in the stomata*: The ploidy level of a plant may be determined by counting the number of plastids in the stomata of a leaf. For example, in potato monohaploids have 5–8, dihaploids 10–15, and tetraploids 18–24 chloroplasts per guard cell.

2. *Chromosome number*: It can be counted from pollen mother cells of buds which can be collected from the regenerated plants and fixed in Carnoy's solution. Acetocarmine or propionocarmine can stain cells. Root tips are also utilized for chromosome counting. Fixed root tips are normally hydrolyzed in 1N HCl at 60°C for 10 min followed by staining with acetocarmine.

3. *Number of nucleoli*: Haploid plants contain one nucleolus while diploids contain 2 nucleoli. Number of nucleoli is directly related to the ploidy status of a plant. Leaves can be incubated overnight with orcein and number of nucleoli can be counted.

4. *Flow cytometric analysis*: Leaves of potential haploid plants are finely chopped and intact interphase nuclei are freed from the cells. At this stage nuclear DNA content reflects the ploidy state of the donor which is determined by flow cytometry. This method is very quick.

**Diploidization**

Haploids can be diploidized to produce homozygous plants by following methods:

1. *Colchicine treatment*: Colchicine has been extensively used as a spindle inhibitor to induce chromosome duplication. It can be applied in the following ways:

(i) The plantlets when still attached to the anther are treated for 24–48 h with 0.5% colchicine solution, washed thoroughly and replanted.

(ii) Anthers can be plated directly on a colchicine-supplemented medium for a week and when the first division has taken place, these are transferred to colchicine free medium for androgenesis process to take place. This method can be followed in maize where male and female flowers are borne separately and diploidization is a problem.

(iii) Colchicine-lanolin paste (0.4%) may be applied to the axils of leaves when the plants are mature. The main axis is decapitated to stimulate the axillary buds to grow into diploid and fertile branches.

(iv) Repeated colchicine treatment to axillary buds with cotton wool plugs over a period of time (e.g. 14 days in potato).

(v) In cereals, vigorous plants at 3–4 tiller stage are collected, soil is washed from the roots and are cut back to 3 cm below the crown. The plants are placed in glass jars or vials containing colchicine solution (2.5 g colchicine dissolved in 20 ml dimethyl sulfoxide and made up to a liter with water). The crowns are covered with colchicine solution. The plants are kept at room temperature in light for 5 h, the roots are washed thoroughly with water and potted into light soil.

Plants should be handled with extra care after colchicine treatment for few days and should be maintained under high humidity.

2. *Endomitosis*: Haploid cells are in general unstable in culture and have a tendency to undergo endomitosis (chromosome duplication without nuclear division) to form diploid cells. This property of cell culture has been exploited in some species for obtaining homozygous plants or for diploidization. A small segment of stem is grown on an auxin-cytokinin medium to induce callus formation. During callus growth and differentiation there is a doubling of chromosomes by endomitosis to form diploid homozygous cells and ultimately plants.

## Significance and Uses of Haploids

1. *Development of pure homozygous lines*: In the breeding context, haploids are most useful as source of homozygous lines. The main advantage is the reduction in time to develop new varieties. A conventional plant breeding programme takes about 6–8 years to develop a pure homozygous line, whereas by the use of anther/microspore culture, the period can be reduced to few months or a year. Thus, homozygosity is achieved in the quickest possible way making genetic and breeding research much easier. Homozygosity is still more important for those plants which have a very long juvenile phase (period from seed to flowering) such as fruit trees, bulbous plants and forestry trees. Even if repeated self-pollination is possible, achievement of homozygosity in this group of plants is an extremely long process.

2. *Developing asexual lines of trees/perennial species*: Rubber tree taller by six meters which could then be multiplied by asexual propagation to raise several clones has been reported by Chinese workers. Another example of pollen haploids has been reported in poplar. In this tree species haploid seedlings selected for desired genotypes are naturally doubled as diploids after 7–8 years which coincides with their flowering. Thus it enables them to be both asexually and sexually propagated. Pollen derived haploid plantlets have been obtained in other woody species also such as *Aesulus hippocastatunum*, *Citrus microcarpa*, *Vitis vinifera*, *Malus prunifolia*, *Litchi chinensis*, *Lycium halinifolium* and *Camellia sinensis*.

3. *Hybrid development*: As a result of complete homozygosity obtained from diploidization of haploids, one can rapidly fix traits in the homozygous condition. Pure homozygous lines can be used for the production of pure $F_1$ hybrids.

4. *Induction of mutations*: Haploid cell cultures are useful material for studying somatic cell genetics, especially for mutation and cell modification. Majority of mutations induced are recessive and therefore is not expressed in the diploid cells because of the presence of dominant allele. Single cells and isolated pollen have the advantage over the entire plant in that they can be plated and screened in large numbers, in a manner similar to microbiological technique. Mutants which are resistant to antibiotics, herbicides, toxins, etc. have been isolated in a number of plant species By subjecting haploid *Nicotiana tabacum* cells to methionine sulfoximine, Carlson (1973) regenerated mutant plants which showed a considerably lower level of infection to *Pseudomonas tabaci*. Wenzel and Uhrig (1981) developed mutants in potato through anther culture which were resistant to potato cyst nematode. Mutants have also been isolated for various temperatures, radiosensitivity to ultraviolet light and gamma radiation, amino acid, e.g. valine, and various antibiotics and drugs.

5. *Induction of genetic variability*: By anther culture not only haploids but also plants of various ploidy levels and mutants are obtained and can be incorporated into the breeding programmes.

6. *Generation of exclusively male plants*: By haploid induction followed by chromosome doubling it is possible to obtain exclusively male plants. For example, in *Asparagus officinalis* male plants have a higher productivity and yield earlier in the season than female plants. If haploids are produced from anthers of male *Asparagus* plants (XY) these are either X or Y, chromosome doubling of Y results in super male plants YY which can subsequently be vegetatively propagated.

7. *Cytogenetic research*: Haploids have been used in the production of aneuploids. Monosomics in wheat, trisomics with 2n = 25 in potato and in tobacco nullisomics were derived from haploids obtained from monosomics which could not produce

nullisomics on selfing. Haploids also give evidence for the origin of basic chromosome number in a species or a genus. For example in pearl millet (*Pennisetum americanum*), occurrence of pairing (upto two bivalents in some cells) suggested that the basic chromosome number may be x = 5.

8. *Significance in the early release of varieties*: Based on anther culture many varieties have been released which are listed in Table 8.1. In Japan, a tobacco variety F 211 resistant to bacterial wilt has been obtained through anther culture. In *Brassica napus*, anther-derived doubled haploid lines had low erucic acid and glucosinolate content. Similarly, in sugarcane, selection among anther culture derived haploids led to the development of superior lines with tall stem and higher sugar content. In bell peppers, dihaploid lines exhibited all shades of color ranging from dark green to light green. These reports have encouraged many plant breeders to incorporate anther culture in breeding methods.

9. *Hybrid sorting in haploid breeding*: One of the essential steps in haploid breeding involves selection of superior plants among haploids derived from $F_1$ hybrids through anther culture. It is properly described as hybrid sorting and virtually means selections of recombinant superior gametes. The haploid method of breeding involving hybrid sorting is considered superior over pedigree and bulk methods. Firstly because of the frequency of superior gametes is higher than the frequency of corresponding superior plants in $F_2$ generation. Secondly haploid breeding reduces significantly the time required for the development of a new variety. For instance if one assumes that the frequency of superior $F_1$ gametes is one in one hundred, then the frequency homozygous $F_2$ plants derived from the fusion of two such superior similar gametes would be one in ten thousand. Therefore, smaller population of double haploids

**Table 8.1:** Varieties released through anther culture

| Crop | Varieties |
| --- | --- |
| Wheat | Lunghua 1, Huapei 1, G. K. Delibab, Ambitus, Jingdan 2288, Florin, Zing Hua-1, Zing Hua-3 |
| Rice | Hua Yu-1, Hua Yu- 2, Zhong Hua 8, Zhong Hua 9, Xin Xiu, Tanfong 1, Nonhua 5, Nonhua 11, Zhe Keng 66, Aya, Hirohikari, Hirohonami |
| Tobacco | Tan Yuh No. 1, Tanyu 2, Tanyu 3, F 211, Hai Hua 19, Hai Hua 30 |

derived from haploids will need to be handled.

10. *Disease resistance*: Haploid production has been used for the introduction of disease resistance genes into cultivars. An established cultivar is crossed with a donor for disease resistance. Either $F_1$ or $F_2$ anthers are plated and haploids are developed. These haploids are screened for resistance and then diploidized. Resistance to barley yellow mosaic virus has been introduced into susceptible breeding lines by haploid breeding (Foroughi and Friedt, 1984). A barley accession Q 21681 was found to be resistant to various diseases. This line was crossed to susceptible breeding lines and anthers of F1 plants were cultured to develop double haploid lines which were resistant to stem rust, leaf rust and powdery mildew (Steffenson *et al.*, 1995). Rice varieties Zhonghua No 8 and No 9 have been developed with blast resistance genes, high yield and good quality using haploids integrated in conventional breeding approaches. Hwasambye, a rice variety bred through anther culture showed resistance to leaf blast, bacterial leaf blight and rice stripe tenui virus (Byeong-Geun *et al.*, 1997). In tobacco, a variant that showed resistance to a highly necrotic strain of potato virus Y (PVY) from a population of doubled haploids was reported by Witherspoon *et al.* (1991).

11. *Insect resistance*: A medium-late maturing rice variety 'Hwacheongbyeo' derived from anther culture showed resistance to brown plant hopper. This variety was also resistant

to blast, bacterial blight and rice tenui virus, and showed cold tolerance (Lee *et al.*, 1989). In rice, promising anther culture lines have been developed which show resistance to rice water weevil (N'Guessan *et al.*, 1994), pests (Zapata *et al.*, 1991).

12. *Salt tolerance*: Salt tolerant breeding lines have been developed in different crop species which have been integrated in conventional breeding. Miah *et al.* (1996) developed doubled haploid salt tolerant rice breeding line that showed tolerance at the level of 8–10 decisiemens/m (ds/m). Sathish *et al.* (1997) established stable NaCl-resistant rice plant line (EC 16–18 ms) by *in vitro* selection of anther derived callus exposed to NaCl. Likewise *in vitro* screening of wheat anther culture derived callus to NaCl resulted in line Hua Bain 124-4 which showed salt tolerance as well as high yield and desiccation tolerance (Zao *et al.*, 1994, 1995).

13. *Doubled haploids in genome mapping*: A rather recent application of DH lines is their use in genome mapping. For molecular screening studies a much smaller sample of doubled haploids is required for desirable recombinants. In a population of DH lines, the identification of markers is much more secure, as most intermediate phenotypic expressions are excluded due to heterozygosity. A gene will segregate in a 1:1 ratio for both molecular marker and the phenotype at the plant level. DH is used for genome mapping for major genes and/or quantitative traits in barley, rice, oil seed rape, etc.

## Problems

1. High level of management and expertise is required to operate the tissue culture production of haploids.
2. Diploids and tetraploids often regenerate at the same rate as the haploids.
3. Selective cell division must take place in the haploid microspores and not in other unwanted diploid tissues. This selective cell division is often impossible.
4. Callus formation whether it has arisen spontaneously or has been induced by regulators is usually detrimental.
5. The relatively high incidence of albinism in some types of anther and pollen culture.
6. The lack of selection of traits during the derivation of haploid material.
7. There is little chance of isolating a haploid from a mixture of haploids and higher ploidy levels since latter ones are easily outgrown.
8. The doubling of a haploid does not always result in the production of a homozygote.

## GYNOGENIC HAPLOIDS

Recent advances in plant tissue culture have resulted in the successful induction of haploid plants from ovary and ovule culture. This means that megaspores or female gametophytes of angiosperms can be triggered *in vitro* to sporophytic development. These plants have been described as gynogenic as compared to androgenic plants derived from microspores. Figure 8.3 shows the various parts of a female

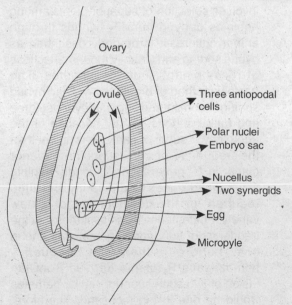

**Fig. 8.3:** A sectional view of an ovary.

reproductive organ. *In vitro* culture of unpollinated ovaries and ovules represents an alternative for the production of haploid plants in species for which anther culture has given unsatisfactory results (e.g. albino plants) or has proven insufficient, e.g. *Gerbera*. The first successful report on the induction of gynogenic haploids was in barley by San Noeum (1976). Subsequently haploid plants were obtained from ovary and ovule cultures of rice, wheat, maize, sunflower, sugar beet, tobacco, poplar, mulberry, etc.

Ovaries can be cultured as pollinated and unpollinated. For haploid production flower buds are excised 24–48 h prior to anthesis for unpollinated ovaries. The calyx, corolla, and stamens are removed and ovaries are then surface sterilized. Before culturing the tip of the distal part of the pedicel is cut off and the ovary is implanted with the cut end inserted in the nutrient medium. The normal Nitsch's (1951), or White's (1954) or MS or $N_6$ inorganic salt media supplemented with growth substances are used. When liquid medium is to be employed, the ovaries can be placed on a filter paper raft or float with the pedicel inserted through the filter paper and dipping into the medium. Sucrose as a carbon source is essential, although maltose and lactose have been shown to be equally favorable. The various species studied so far seem to have few requirements in common for growth *in vitro*. Some species require only the basal medium and sucrose for growth, although addition of an auxin brought about greater stimulation of growth.

The origin of gynogenic haploids differs in species as reported to be from synergids in rice, whereas egg or antipodal cell develops into embryos in barley.

*In vitro* culture of unfertilized ovules has been the most efficient and reliable technique for the production of haploid and doubled haploid plants of sugar beet. In tree species, gynogenic plants were reported in mulberry (*Morus indica*), an important tree for nourishing silkworms for use in the silk industry. Ovaries were cultured on MS medium supplemented with 1 mg/l BA and 1 mg/l kinetin which resulted in direct regeneration of plants without a callus phase.

**Factors affecting gynogenesis**

1. *Genotype*: Genotype of the donor plant is one of the most important factors since each genotype shows a different response. For each genotype a specific protocol must be followed for maximal efficiency.

2. *Growth condition of the donor plant*: It has been found that embryo induction from ovules harvested from lateral branches gives high response than ovules harvested from stem apex. Ovules harvested from first formed lateral branches (at the base of plant) give higher response than from the sixth formed lateral branches in sugar beet.

3. *Stage of harvest of ovule*: Excision of ovules at an early developmental stage where no self pollination can occur or use of highly self incompatible donor plants helps to minimize the problem of fertilization in sugar beet and other crops so that proper stage of unfertilized ovule could be harvested.

4. *Embryo-sac stage*: It has been reported that complete maturation of ovule is not necessary for induction of gynogenesis but an appropriate stage is more important.

5. *Culture conditions*: Although solid media have been used more frequently for gynogenic culture, few investigators have used liquid medium for the induction of gynogenic calli followed by dissection and transfer of embryogenic structures to solid medium for differentiation. MS, B5, Miller's basal medium have been most commonly used. The amount of sucrose has been reported to be important for embryogenesis. In rice 3–6% sucrose was effective, whereas in onion 10% sucrose gives maximum response.

6. *Seasonal effects*: Seasonal variation is an important factor. In sugar beet highest embryo yield was obtained from the summer grown plants. Callus induction from ovules

of Gerbera occurred at a higher frequency during autumn as compared to spring.

7. *Physical factors*: Certain physical factors (treatments) given to explants or plant parts from which explants are taken prior to culture may have a strong influence on embryo induction. For *Beta vulgaris* cold pretreatment of flower buds at 4°C for 4–5 days increases embryo yield from cultured ovules. Similarly cold pretreatment of inflorescence prior to ovule isolation increased haploid callus frequency in *Salvia selarea*. Other physical factors influencing ovary and ovule culture are light and temperature of incubation.

## Advantages

*In vitro* ovary culture has been used to develop haploid plants and has the potential to overcome the problem of albino plant formation in anther culture. The technique has also got superiority over anther culture due to its potential use in the male sterile genotypes.

## Problems

Though haploid plants have been produced by ovary and ovule culture, they have not been used as much as androgenic haploids in crop improvement. Production of gynogenic haploids through female gametophyte still needs more refinement and also there are problems in dissection of unfertilized ovules and ovaries. It has an inherent disadvantage of one ovary per flower to large number of microspores in one anther.

A positive beginning has already been made for development of gynogenic haploids in sugar beet, barley, mulberry and gymnosperms and the future is promising.

## Chromosome Elimination Technique for Production of Haploid in Cereals (Barley and Wheat)

Kasha and Kao (1970) reported a novel technique for the production of haploids in barley. The method involves crossing of *Hordeum*

*vulgare* ($2n = 14$) with *H. bulbosum* ($2n = 14$). In nature, the seeds produced from such a cross develops for about 10 days and then begins to abort. However, if the immature embryos are dissected two weeks after pollination and cultured on $B_5$ nutrient medium without 2,4-D they continue to grow. The plants obtained from such embryos are monoploids with *H. vulgare* chromosomes ($n = 7$) (Fig. 8.4). Evidence indicates that these monoploids are not caused by parthenogenesis, but by the elimination of *H. bulbosum* chromosomes. The evidences were: (i) The occurrence of cells in the embryo with more than seven chromosomes; (ii) The percentage of seed set is higher than observed by induced parthenogenesis; (iii) When diploid *H. vulgare* is pollinated with tetraploid *H. bulbosum*, the percentage of seed set is similar to the diploid cross. Monoploid wheat has also been obtained by crossing *Triticum*

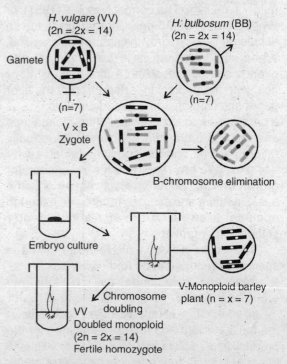

• **Fig. 8.4:** Scheme of monoploid and doubled monoploid barley production via the interspecific hybridization method followed by somatic chromosome elimination.

aestivum (2n = 6x = 42) with diploid *H. bulbosum* (2n = 2x = 14) where the chromosomes of bulbosum are eliminated and haploids of *T. aestivum* are obtained (Barclay, 1975).

*Advantages*

i. The frequency of haploid formation is quite high by this method. In barley varieties namely Mingo, Rodeo, Craig and Gwylan have been released through *bulbosum* technique of chromosome elimination. The varieties were more tolerant to powdery mildew and barley yellow mosaic virus diseases and superior in yield contributing characters as compared to parent.

ii. No aneuploids are obtained by this method.

*Disadvantages*

This technique has a disadvantage of being highly laborious because it involves crossing, embryo excision and then regeneration.

# PROTOCOL

## 8.1: Protocol for Anther Culture of Cereals (Rice, Barley, Wheat, etc.)

1. Panicles or spikes are harvested at the uninucleate stage of microspore development. A stem containing panicle (rice) with pollen at this stage can be identified by the relative position of the flag leaf. Stems are cut with a scalpel below the node from which the panicle arises.

2. Cold pretreatment for a week at 4°C can be given by keeping the cut stems in a beaker containing water. This treatment increases the percentage of cultured anthers that give rise to callus.

3. Remove the flag leaf and surface sterilize the panicles/spikes in a 20% commercial sodium hypochlorite solution for 10 min. Rinse them 4x in sterile distilled water.

4. Place the excised panicle/spike on a sterilized glass slide or petri dish. Anthers are squeezed out with the help of forceps. Each rice floret has 6 anthers.

5. Anthers are placed horizontally on the following $N_6$ medium for callus induction: About 20 anthers in 60 × 15 mm petri dish can be inoculated.

    $N_6$ basic salts + 6% sucrose + 2 mg/l 2,4-D (for rice)

    Constitution of $N_6$ medium is given in Table 3.1

6. Seal the petri dishes and incubate the cultures in dark at 27°C for rice or 20–22°C for barley.

7. After 3–4 weeks, colonies or calluses coming out of the anthers are removed and transferred to the following medium: MS + 0.05 mg/l NAA + 2 mg/l KIN

8. Place the cultures under moderate light intensity with 16 h light: 8 h dark photoperiod.

9. Clumps of shoots that form should be separated in to individual shoots when about 1cm in length and place in individual containers containing the following medium for root induction: MS + 1 mg/l (NAA or IBA)

10. Transfer plants to pots when roots and shoots have developed. Follow the procedure of hardening of plants.

11. Identify the haploid plants.

12. Diploidization of chromosomes can be done by colchicine treatment. Vigorous plants are collected at 3–4 tiller stage, soil is washed from the roots and they are cut back to 3 cm below the crown. The plants are placed in glass jars or vials containing colchicine solution (2.5 g colchicine dissolved in 20 ml dimethyl sulphoxide and made up to a litre with water). Enough of this solution is used so that crowns are covered with colchicine solution. The plants are kept at room temperature in light for 5 h, the roots are washed thoroughly with water and potted into a light soil.

13. Plants should be handled with extra care after colchicines treatment for a few days and must be maintained under high humidity.

## REFERENCES

Barclay, I. R. 1975. High frequencies of haploid production in wheat (*Triticum aestivum*) by chromosome elimination. *Nature* **256:** 410–411.

Bourgin, J.P. and Nitsch, J.P. 1967. Obtentionde Nicotiana haploides a partir d´etamines cultivees *in vitro*. *Ann. Physiol. Veg. (Paris)* **9:** 377–382.

Byeong Geun, O.H., Lim, S.J., Yang, S.J. and Hwang, H.G. 1997. R.D. *J. Crop Sci.* **39:** 94–100.

Carlson, P. 1973. Methionine sulfoximine resistant mutants of tobacco. *Science* **180:** 1366.

Chuang, C.C., Ouang, T.W., Chia, H., Chou, S.M. and Ching, C.K. 1978. A set of potato media for wheat anther culture. In: *Proc. Symp. Plant Tissue Culture*, Science Press, Peking, pp. 51–56.

Foroughi-Wehr, B. and Friedt, W. 1984. Rapid production of recombinant barley yellow mosaic virus resistant *Hordeum vulgare* lines by anther culture. *Theor. Appl. Genet.* **67:** 377–382.

Kasha, K.J. and Kao, K.N. 1970. High frequency haploid production in barley (*Hordeum vulgare* L.). *Nature* **225:** 874–876.

Lee, Y.T., Lim, M.S., Kim, H.S., Shin, H.T., Kim, C.H., Baes, S.H. and Cho, C.I. 1989. An anther derived new high quality variety with disease and insect resistance "Hwacheong byeo". *Res. Rep. of Rural Development Administration, Rice* **31:** 27–34.

Linsmaier, E.M. and Skoog, F. 1965. Organic growth factor requirements of tobacco tissue cultures. *Physiol. Plant* **18:** 100–127.

Miah, M.A.A., Pathan, M.S. and Quayam, H.A. 1996. Production of salt tolerant rice breeding line via doubled haploid. *Euphytica* **91:** 285–288.

N, Guessan, F.K., Quisenberg, S.S. and Croughan, T.P. 1994. Evaluation of rice anther culture line for tolerance to the rice weevil (Coleoptera: Curculionidae). *Environ. Entomol.* **23:** 331–336.

Nitsch, J.P. 1951. Growth and development of *in vitro* excised ovaries. *Am. J. Bot.* **38:** 566–577.

Nitsch, J.P. and Nitsch, C. 1969. Haploid plants from pollen grains. *Science* **163:** 85–87.

San Noeum, L.H. 1976. Haploides d´Hordeum vulgare L. par culture in vitro d´ ovaires non fecondes. *Ann. Amelior Plantes* **26:** 751–754.

Sathish, P., Gamborg, O.L. and Nabors, M.W. 1997. *Theor. Appl. Genet.* **95:** 1203–1209.

Stefenson, B.J., Jin, Y., Lossnagel, B.G., Rasmussen, J.B.K. and Kao, K. 1995. Genetics of multiple disease resistance in a doubled haploid population of barley. *Plant Breed* **114:** 50–54.

White, P.R. 1954. *The Cultivation of Animal and Plant Cells*. Ronald Press Co., New York.

White, P.R. 1963. *The Cultivation of Animal and Plant Cells*. Ronald Press Co, New York.

Witherspoon, W.D., Wornsman, E.A., Gooding, G.V. and Rufby, R.C. 1991. Characterization of gametoclonal variant controlling virus resistance in tobacco. *Theor. Appl. Genet.* **81:** 1–5.

Zao, R.T., Gao, S.G., Qiao, Y.K., Zhu, H.M. and Bi, Y.J. 1995. Studies on the application of anther culture in salt tolerance breeding in wheat. *Acta Agronomica* **21:** 230–234.

Zao, R.T., Gao, S.G., Qiao, Y.K., Zhu, H.M. and Bi, Y.J. 1994. A new approach of screening salt tolerance variants by anther culture to cultivate salt tolerant wheat variety. *Acta Agriculturae Boreali Sinica* **9:** 34–38.

Zapata, F.J., Alejar, M.S., Torrizo, L.B., Novero, A.U., Singh, V.P. and Senadhira, D. 1991. Field performance of anther culture derived lines from F1 crosses of Indica rice under saline and non-saline land. *Theor. Appl. Genet.* **83:** 6–11.

## FURTHER READING

Bhojwani, S.S. and Razdan, M.K. 1990. *Plant Tissue Culture: Applications and Limitations*. Elsevier Science Publishers, Amsterdam.

Chawla, H.S. 1998. *Biotechnology in Crop Improvement*. International Book Distributing Co., Lucknow.

Evans, D.A., Sharp, W.R., Ammirato, P.V. and Yamada, Y. 1983. *Handbook of Plant Cell Culture*. Macmillan Publishing Company, New York.

Mohan, S.J, Sopory, S.K. and Veilleux, R.E. 1996. *In Vitro Haploid Production in Higher Plants*. Kluwer Academic Publishers, Dordrecht.

Pierik, R.L.M. 1987. *In vitro culture of Higher Plants*. Martinus Nijhoff Publishers, Dordrecht.

Reinert, J. and Bajaj, P.P.S. (eds.) 1977. *Applied and Fundamental Aspects of Plant Cell, Tissue and Organ Culture*. Springer-Verlag, Berlin.

Street, H.S. (ed.)1973. *Plant Tissue and Cell culture*. Blackwell Scientific Publications, Oxford.

Thorpe, T.A. 1981. *Plant Tissue Culture: Methods and Applications in Agriculture*. Academic Press Inc., New York.

Vasil, I.K. 1984. *Cell Culture and Somatic Cell Genetics of Plants: Laboratory Procedures and Their Applications*. Academic Press, Inc., New York.

# 9

# Protoplast Isolation and Fusion

The protoplast, also known as naked plant cell refers to all the components of a plant cell excluding the cell wall. Hanstein introduced the term protoplast in 1880 to designate the living matter enclosed by plant cell membrane. The isolated protoplast is unusual because the outer plasma membrane is fully exposed and is the only barrier between the external environment and the interior of living cell. The isolation of protoplasts from plant cells was first achieved by microsurgery on plasmolyzed cells by mechanical method (Klercker, 1892). However, the yields were extremely low and this method is not useful. Cocking (1960) first used enzymes to release protoplasts. He isolated the enzyme cellulase from the culture of fungus *Myrothecium verrucaria* to degrade the cell walls. He applied an extract of cellulase hydrolytic enzyme to isolate protoplasts from tomato root tips. Since that time many enzyme formulations have been used to isolate protoplasts, and the most frequently used enzymes are now commercially available. The use of cell wall degrading enzymes was soon recognized as the preferred method to release large numbers of uniform protoplasts. Under appropriate conditions, in a number of plant species, these protoplasts have been successfully cultured to synthesize cell walls and now genetic manipulations have also been made. Protoplasts are not only useful for cell fusion studies but, these can also take up,

through their naked plasma membrane foreign DNA, cell organelles, bacteria or virus particles.

## PROTOPLAST ISOLATION

Protoplasts are isolated by (i) mechanical and (ii) enzymatic methods.

### Mechanical Method

In this method, large and highly vacuolated cells of storage tissues such as onion bulb scales, radish root and beet root tissue could be used for isolation. The cells are plasmolysed in an iso-osmotic solution resulting in the withdrawal of contents in the center of cell. Subsequently, the tissue is dissected and deplasmolyzed to release the preformed protoplasts. Klercker (1892) isolated protoplasts from *Stratiotes aloides* by this method. Mechanically, protoplasts have been isolated by gently teasing the new callus tissue initiated from expanded leaves of *Saintpaulia ionantha* which was grown in a specific auxin environment to produce thin cell walls. However, this method is generally not followed because of certain disadvantages. (i) It is restricted to certain tissues which have large vacuolated cells. (ii) Yield of protoplasts is generally very low. Protoplasts from less vacuolated and highly meristematic cells do not show good yield. (iii) The method is tedious and laborious. (iv) Viability of protoplasts is low because of the presence of substances released by damaged cells. The mechanical method is useful when there are side effects of cell wall degrading enzymes.

## Enzymatic Method

Cocking in 1960 demonstrated the possibility of enzymatic isolation of protoplasts from higher plants. He used concentrated solution of cellulase to degrade the cell walls. However, there was little work for the next 10 years until the commercial enzyme preparations became available. Takebe *et al.* (1968) for the first time employed commercial enzyme preparation for isolation of protoplasts and subsequently regenerated plants in 1971. Protoplasts are now routinely isolated by treating tissues with a mixture of cell wall degrading enzymes in solution, which contain osmotic stabilizers. A general procedure for protoplast isolation has been shown in Fig. 9.1. The relative ease with which protoplast isolation can be achieved depends upon a variety of factors which have been discussed below.

### Physiological state of tissue and cell material

Protoplasts have been isolated from a variety of tissues and organs including leaves, petioles, shoot apices, roots, fruits, coleoptiles, hypocotyls, stem, embryos, microspores, callus and cell suspension cultures of a large number of plant species. A convenient and most suitable source of protoplasts is mesophyll tissue from fully expanded leaves of young plants or new shoots. Leaf tissue is popular because it allows the isolation of a large number of relatively uniform cells without the necessity of killing the plants. Moreover the mesophyll cells are loosely arranged and enzymes have an easy access to the cell wall. Leaves are taken, sterilized and lower epidermis from the excised leaves is peeled off. These are cut into small pieces and then proceeded for isolation of protoplasts. Since the physiological condition of the source tissue markedly influences the yield and viability of isolated protoplasts, the plants or tissues must

be grown under controlled conditions. Callus tissue and cell suspension cultures used for protoplast isolation should be in the early log phase of growth. Friable tissue with low starch content generally gives better results.

### Enzymes

The release of protoplasts is very much dependent on the nature and composition of enzymes used to digest the cell wall. There are three primary components of the cell wall which have been identified as cellulose, hemicellulose and pectin substances. Cellulose and hemicellulose are the components of primary and secondary structure of cell wall, while pectin is a component of middle lamella that joins the cells. Pectinase mainly degrades the middle lamella, while cellulase and hemicellulase are required to digest the cellulosic and hemicellulosic components of the cell wall respectively. Cellulase (Onozuka) R10 generally used for wall degradation has been partially purified from the molds of *Trichoderma reesei* and *T. viride*. Sometimes additional hemicellulase maybe necessary for recalcitrant tissues and for this Rhozyme HP 150 has been used. The most frequently used pectinase is macerozyme (macerase) which has been derived from the *Rhizopus* fungus. Driselase, another enzyme, has both cellulolytic and pecteolytic activities and has been successfully used alone for isolation of protoplasts from cultured cells. Besides, there are several enzymes, e.g. helicase, colonase, cellulysin, glusulase, zymolyase, meicelase, pectolyase, etc. which are used to treat a tissue that does not release protoplasts easily (Table 9.1). Aleurone cells of barley treated with cellulase results in protoplasts with the cellulase resistant cell wall. These cells called sphaeroplasts are to be treated with glusulase to digest the remaining cell wall.

The activity of the enzyme is pH dependent and is generally indicated by the manufacturer. However, in practice the pH of enzyme solution is adjusted between 4.7 and 6.0. Generally the

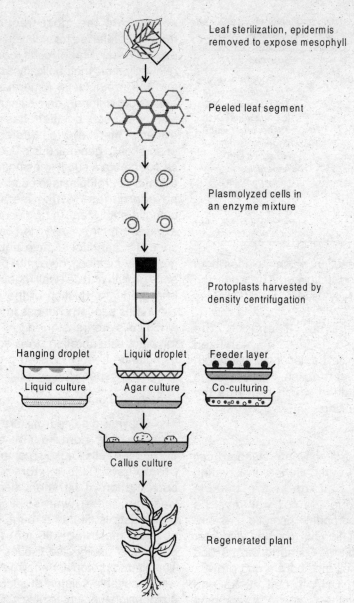

Leaf sterilization, epidermis removed to expose mesophyll

Peeled leaf segment

Plasmolyzed cells in an enzyme mixture

Protoplasts harvested by density centrifugation

Hanging droplet — Liquid droplet — Feeder layer

Liquid culture — Agar culture — Co-culturing

Callus culture

Regenerated plant

**Fig. 9.1:** Schematic illustration of protoplast isolation and culture procedure.

temperature of 25–30°C is adequate for isolation of protoplasts. Duration of enzyme treatment is to be determined after trials. However, it may be short for 30-min duration to long duration for 20 h. The enzymatic isolation of protoplasts can be performed in two different ways.

(a) *Two step or sequential method*: The tissue is first treated with a macerozyme or pectinase enyme which separates the cells by degrading the middle lamella. These free cells are then treated with cellulase which releases the protoplasts. In general, the cells are exposed to different enzymes for shorter periods.

(b) *One step or simultaneous method*: The tissue is subjected to a mixture of enzymes

**Table 9.1:** Some commercially available enzymes used for protoplast isolation (after M.K. Razdan, 2003)

| Enzyme | Source |
|---|---|
| Cellulase Onozuka R-10 | *Trichoderma viride* |
| Cellulase Onozuka RS | *Trichoderma viride* |
| Cellulysin | *Trichoderma viride* |
| Driselase | *Irpex lacteus* |
| Meicelase P-1 | *Trichoderma viride* |
| Hemicellulase | *Aspergillus niger* |
| Hemicellulase H-2125 | *Rhizopus* spp. |
| Rhozyme HP 150 | *Aspergillus niger* |
| Helicase | *Helix pomatia* |
| Macerase | *Rhizopus arrhizus* |
| Macerozyme R-10 | *Rhizopus arrhizus* |
| Pectolyase | *Aspergillus japonicus* |
| Zymolyase | *Arthrobacter luteus* |

in a one step reaction which includes both macerozyme and cellulase. The one step method is generally used because it is less labor intensive.

During the enzyme treatment, the protoplasts obtained need to be stabilized because the mechanical barrier of cell wall which offered support has been broken. For this reason an osmoticum is added which prevents the protoplasts from bursting.

*Osmoticum*

Protoplasts released directly into standard cell culture medium will burst. Hence, in isolating protoplasts, the wall pressure that is mechanically supported by cell wall must be replaced with an appropriate osmotic pressure in the protoplast isolation mixture and also later in the culture medium. Osmotic stress has harmful effects on cell metabolism and growth. A condensation of DNA in cell nuclei and decreased protein synthesis are the two common effects of osmotic stress, although both can be reversed.

Lower (more negative) osmotic potentials are usually generated by the addition of various ionic or non-ionic solutes. Non-ionic substances include soluble carbohydrates such as mannitol, sorbitol, glucose, fructose, galactose or sucrose. Ionic substances are potassium chloride, calcium chloride and magnesium sulphate. Mannitol and sorbitol are the most frequently used with mannitol preferred for isolation of leaf mesophyll protoplasts. Mannitol is considered to be relatively inert metabolically and infuses slowly into the protoplasts. Mineral salts particularly potassium chloride and calcium chloride have also been used, but there has never been good evidence that they are preferable over mannitol or sorbitol. In general, 0.3 to 0.7 M sugar solution can generate a suitable osmotic potential. Upon transfer of protoplasts to culture medium, it may be appropriate to use metabolically active osmotic stabilizers (e.g. glucose and sucrose) along with metabolically inert osmotic stabilizers, such as mannitol. Active substances will be gradually metabolized by the protoplasts during early growth and cell wall regeneration, resulting in a gradual reduction of the osmoticum. This eliminates sudden changes in osmotic potential when cells are transferred to a nutrient medium for plant regeneration. In a solution of proper osmolarity, freshly isolated protoplasts appear completely spherical.

*Protoplast purification*

The enzyme digested mixture obtained at this stage would contain sub-cellular debris, undigested cells, broken protoplasts and healthy protoplasts. This mixture is purified by a combination of filtration, centrifugation and washing. The enzyme solution containing the protoplasts is filtered through a stainless steel or nylon mesh (50–100 μm) to remove larger portions of undigested tissues and cell clumps. The filtered protoplast-enzyme solution is mixed with a suitable volume of sucrose in protoplast suspension medium to give a final concentration of about 20% and then centrifugation is done at about 100x g for 7–10 min. The debris moves down to the bottom of tube and a band of protoplasts appear at the junction of sucrose and protoplast suspension medium. The protoplast band is easily sucked off with a Pasteur pipette into another centrifuge tube. Following the repeated centrifugation and suspension, the protoplasts are washed thrice and finally suspended in the culture medium at an appropriate density. Numerous other gradients have been suggested to aid in protoplast isolation. (i) Two step Ficoll (polysucrose) gradients with 6% on top and 9% below, dissolved in MS medium with 7% sorbitol for purification using centrifugation at 150x g for 5 minutes has been suggested. Following centrifugation, debris settle at the bottom while protoplasts float at the top. (ii) Protoplasts can be separated on to 20% Percoll with 0.25 M mannitol and 0.1 M calcium chloride. After centrifugation at 200x g, protoplasts were recovered above the Percoll. As these high molecular weight substances such as Ficoll and Percoll are osmotically inert, their use may be preferable over sucrose flotation.

*Protoplast viability and density*

The most frequently used staining methods for assessing protoplast viability are:

i. *Fluorescein diacetate (FDA) staining method*: FDA, a dye that accumulates inside the plasma lemma of viable protoplasts can be detected by fluorescence microscopy. As FDA accumulates in the cell membrane, viable intact protoplasts fluoresce yellow green within 5 min. FDA is dissolved in acetone and used at a concentration of 0.01%. Protoplasts treated with FDA must be examined within 5–15 min after staining as FDA disassociated from the membranes after 15 min.

ii. *Phenosafranine staining*: Phenosafranine is also used at a concentration of 0.01% but it is specific for dead protoplasts that turn red. Viable cells remain unstained by phenosafranine.

iii. *Calcofluor White (CFW) staining*: Calcofluor White can ascertain the viability of protoplasts by detecting the onset of cell wall formation. CFW binds to the beta-linked glucosides in the newly synthesized cell wall which is observed as a ring of fluorescence around the plasma membrane. Optimum staining is achieved when 0.1 ml of protoplasts is mixed with 5.0 μl of a 0.1% w/v solution of CFW.

iv. Exclusion of Evans blue dye by intact membranes.

v. Observations on cyclosis or protoplast streaming as a measure of active metabolism.

vi. Variation of protoplast size with osmotic changes.

vii. Oxygen uptake measured by an oxygen electrode which indicates respiratory metabolism.

viii. Photosynthetic studies.

However, the true test of protoplast viability is the ability of protoplasts to undergo continued mitotic divisions and regenerate in to plants.

Protoplasts have both maximum as well as minimum plating densities for growth. Published procedures suggest that protoplasts should be cultured at a density of $5 \times 10^3$ to $10^6$ cells/ml with an optimum of about $5 \times 10^4$ protoplasts/ml. The concentration of protoplasts in a given preparation can be determined by the use of hemocytometer.

*Culture techniques*

Isolated protoplasts are usually cultured in either liquid or semisolid agar media plates. Protoplasts are sometimes allowed to regenerate cell wall in liquid culture before they are transferred to agar media.

*Agar culture*

Agar of different qualities is available but agarose is most frequently used to solidify the culture media. Bergmann cell plating technique as explained earlier (Chapter 7) is followed for plating of protoplasts. Protoplast suspension at double the required final density is gently mixed with an equal volume of double the agar concentration in the medium kept molten at 45°C. The concentration of agar should be chosen to give a soft agar gel when mixed with the protoplast suspension. The petri dishes are sealed with parafilm and incubated upside down. By agar culture method, protoplasts remain in a fixed position so that protoplast clumping is avoided. Protoplasts immobilized in semisolid media give rise to cell clones and allow accurate determination of plating efficiency. Once immobilized, however, hand manipulations are required for transfer to other culture media.

### Liquid culture

It has been generally preferred in earlier stages of culture because (i) it allows easy dilution and transfer; (ii) protoplasts of some species do not divide in agarified media; (iii) osmotic pressure of the medium can be effectively reduced; (iv) density of cells can be reduced after few days of culture. But it has the disadvantage of not permitting isolation of single colonies derived from one parent cell.

Various modifications to these culture methods have been developed.

### Liquid droplet method

It involves suspending protoplasts in culture media and pipetting 100–200 µl droplets into 60 × 15 ml plastic petri dishes. Five to seven drops can be cultured per plate. The plates are then sealed and incubated. This method is convenient for microscopic examination, and fresh media can be added directly to the developing suspensions at 5–7 days interval. A problem encountered in this method is that sometimes the cultured protoplasts clump together at the center of the droplet.

### Hanging droplet method

Small drops (40–100 µl) of protoplast suspension are placed on the inner side of the lid of a petri dish. When the lid is applied to the dish, the culture drops are hanging or suspended from the lid. This method allows culture of fewer protoplasts per droplet than the conventional droplet technique.

### Feeder layer

In many cases it is desirable to reduce the plating density to a minimum for a given protoplast preparation. A feeder layer consisting of X-irradiated non-dividing but living protoplasts are plated in agar medium in petri dishes. Protoplasts are plated on this feeder layer at low density in a thin layer of agar medium. This is especially important when particular mutant or hybrid cells are to be selected on agar plates.

### Co-culturing

It is the culturing of two types of protoplasts viz. slow growing and fast growing. A reliable fast growing protoplast preparation is mixed in varying ratios with protoplasts of a slow growing recalcitrant species. The fast growing protoplasts presumably provide the other species with growth factors and undefined diffusable chemicals, which aid the regeneration of a cell wall and cell division.

### Culture medium

Protoplasts have nutritional requirements similar to those of cultured plant cells. The mineral salt compositions established for plant cell cultures have been modified to meet particular requirements by protoplasts.

Protoplast culture medium should be devoid of ammonium as it is detrimental to its survival and quantity of iron and zinc should be reduced. On the other hand calcium concentration should be increased 2–4 times over the concentration normally used for cell cultures. Increased calcium concentration may be important for membrane stability.

Mannitol and sorbitol are the most frequently used compounds for maintaining the osmolarity, which has already been discussed.

Glucose is perhaps the preferred and most reliable carbon source. Plant cells grow about equally well on a combination of glucose and sucrose, but sucrose alone may not always be satisfactory for plant protoplasts. The vitamins used for protoplast culture include those present in standard tissue culture media. If protoplasts are to be cultured at very low density in defined media there may be a requirement for additional vitamins. Protoplast media frequently contain one or more amino acids. A convenient approach is to add 0.01 to 0.25% vitamin-free casamino acids or casein hydrolysate. Such mixtures appear to meet the needs of protoplasts if inorganic nitrogen is inadequate. The addition of 1–5 mM L-glutamine can improve growth.

The majority of protoplast culture media contain one or more auxins plus one or two cytokinins to stimulate protoplast division and growth. Only a few protoplast systems grow without added growth regulators in the culture media. In general, protoplast culturing starts with a relatively high concentration 1–3 mg/l of NAA or 2,4-D alongwith a lower concentration (0.1 to 1.0 mg/l) of BAP or zeatin. When protoplast division has started, it is often recommended to change the exogenous growth regulator supply. A change in the auxin to cytokinin balance is often used for stimulating morphogenesis.

### Environmental factors

Generally high light intensity inhibits protoplast growth when applied from the beginning of culture. It is better to initiate protoplast culturing in darkness or dim light for few days and later transfer the cultures to a light of about 2000–5000 lux. There are reports of better protoplast growth when the cultures are kept in continuous darkness. In contrast, it has been shown for legume species that light is necessary for initiating protoplast division. Protoplast cultures are generally cultured at temperatures ranging between 20–28°C. A pH in the range of 5.5 to 5.9 is recommended for protoplast culture media and it seems to be satisfactory.

## PROTOPLAST DEVELOPMENT

### Cell Wall Formation

Protoplasts in culture generally start to regenerate a cell wall within a few hours after isolation and may take two to several days to complete the process under suitable conditions. The protoplasts lose their characteristic spherical shape, which is an indication of new wall regeneration. However, newly synthesized cell wall by protoplasts can be demonstrated by staining it with 0.1% calcofluor white fluorescent stain.

In general, protoplasts from actively growing suspension cultures exhibit more rapid deposition of microfibrils than those from differentiated mesophyll cells. A freshly formed cell wall is composed of loosely arranged microfibrils which subsequently become organized to form a typical cell wall.

### Growth, Division and Plant Regeneration

Regeneration of a cell wall is not necessarily a prerequisite for the initiation of nuclear division in protoplast cultures but cell wall formation is required before cytokinesis occurs. Once a cell wall is formed, the reconstituted cells show a considerable increase in cell size. The first cell division generally occurs within 2–7 days. Protoplasts from actively dividing cell suspension enter the first division faster than those from highly differentiated cells of the leaf. The second division occurs within a week, and by the end of the second week in culture, small aggregates of cells are present. After 3 weeks, small cell colonies are visible and colonies of ca. 1 mm diameter are present after approximately 6 weeks in culture. Once small colonies have formed, their further growth is inhibited if they are allowed to remain on the original high osmotic medium. The colonies should, therefore, be transferred to a mannitol or sorbitol-free medium. Macroscopic colonies are transferred to an osmotic free medium to develop callus. The callus may then be induced to undergo organogenic or embryogenic differentiation leading to the formation of plants.

The first report of plant regeneration was in *Nicotiana tabacum* (Takebe *et al.*, 1971). Since then the list of species exhibiting the totipotency has steadily increased. Several reports have suggested that frequency of plant regeneration reported from calluses derived from plant organs differ from the calluses raised from protoplasts. The calluses from intact plant organs often carry preformed buds or organized structures, while such structures are absent in the callus from protoplast origin.

## SOMATIC HYBRIDIZATION

Sexual hybridization between closely related species has been used for years to improve cultivated plants. Unfortunately, sexual hybridization is limited in most cases to cultivars within a species or at best to a few wild species closely related to a cultivated crop. Species barriers thereby limit the usefulness of sexual hybridization for crop improvement. Somatic cell fusion leading to the formation of viable cell hybrids has been suggested as a method to overcome the species barriers to sexual hybridization. Plant protoplasts offer exciting possibilities in the fields of somatic cell genetics and crop improvement. The technique of hybrid production through the fusion of isolated somatic (body) protoplasts under *in vitro* conditions and subsequent development of their product (heterokaryon) to a hybrid plant is known as somatic hybridization. This procedure eliminates sex altogether in hybridization. In somatic hybridization the nucleus and cytoplasm of both parents are fused in the hybrid cell. Sometimes, nuclear genome of only one parent but cytoplasmic genes (plastome) from both the parents are present in the fused hybrid, which is known as cybrid or cytoplasmic hybrid. Thus, protoplast fusion technique can be used to overcome the barriers of incompatibility and acts as a method for the genetic manipulation of plant cells. It provides us with an opportunity to construct hybrids between taxonomically distant plant species beyond the limits of sexual crossability. It also creates cells with new genetic, nuclear as well as cytoplasmic constitutions that otherwise cannot be obtained. Somatic hybridization involves the fusion of protoplasts, selection of hybrid cells and identification of hybrid plants.

### Protoplast fusion

It involves mixing of protoplasts of two different genomes and can be achieved by either spontaneous or induced fusion methods.

### Spontaneous fusion

Cell fusion is a process integral to plant development. The most prominent process is egg fertilization. The breakdown of cell wall during protoplast isolation led people to believe that there would be spontaneous fusion leading to the formation of homokaryons because multinucleate cells were detected as soon as enzymatic protoplast isolation techniques were applied. The argument that cell wall degradation would permit dilation of plasmodesmata, fusion and complete mixing of cells was supported by electron micrographic studies. Spontaneous fusion of protoplasts is observed when protoplasts are isolated from callus cultures. However, spontaneous fusion products do not regenerate in to whole plants except for undergoing a few divisions. Later studies revealed that isolated protoplasts are usually characterized by smooth surfaces and fusion has to be induced by one of a variety of treatments.

### Induced fusion methods

Spontaneous fusion is of no value as fusion of protoplasts of different origins is required in somatic hybridization. To achieve this, a suitable agent (fusogen) is added to fuse the plant protoplasts of different origins. The different fusogens employed are: $NaNO_3$, artificial sea water, lysozyme, high $pH/Ca^{++}$, polyethylene glycol, antibodies, concavalin A, polyvinyl alcohol, electrofusion dextran and dextran sulphate, fatty acids and esters. Some of the methods that have been employed are explained below. Protoplast fusion method has been outlined in Fig. 9.2.

### Treatment with sodium nitrate

Induced fusion by $NaNO_3$ was first reported by Power *et al.* (1970). Isolated protoplasts are suspended in a mixture of 5.5% sodium nitrate in a 10% sucrose solution. The solution containing the protoplasts is incubated in a water bath at 35°C for 5 min and then centrifuged for 5 min at 200x g. Following centrifugation, most of

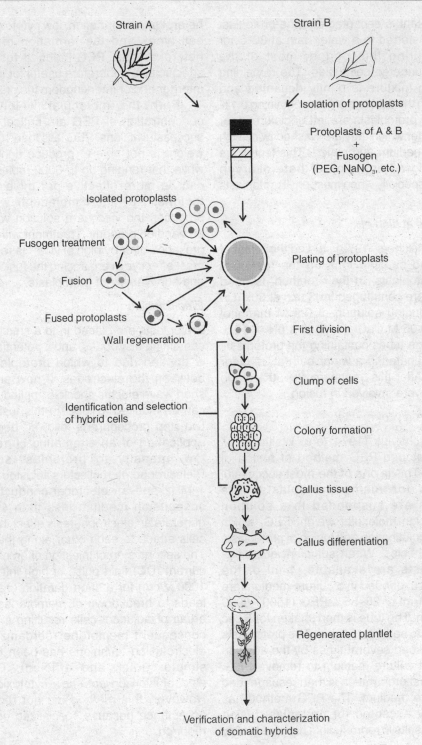

Strain A

Strain B

Isolation of protoplasts

Protoplasts of A & B
+
Fusogen
(PEG, NaNO₃, etc.)

Isolated protoplasts

Plating of protoplasts

Fusogen treatment

Fusion

Fused protoplasts

Wall regeneration

First division

Clump of cells

Identification and selection
of hybrid cells

Colony formation

Callus tissue

Callus differentiation

Regenerated plantlet

Verification and characterization
of somatic hybrids

•**Fig. 9.2:** Schematic illustration of protoplast fusion and regeneration of plants.

the supernatant is decanted and the protoplast pellet is transferred to a water bath at 30°C for 30 min. During this period, most of the protoplasts undergo cell fusion. The remaining aggregation mixture is gently decanted and replaced with the culture medium containing 0.1% $NaNO_3$. The protoplasts are left undisturbed for sometime after which they are washed twice with the culture medium and plated. This technique results in low frequency of heterokaryon formation, especially when mesophyll protoplasts are involved.

### Calcium ions at high pH

Keller and Melchers (1973) studied the effect of high pH and calcium ions on the fusion of tobacco protoplasts. In their method, isolated protoplasts are centrifuged for 3 min at 50x g in a fusion-inducing solution of 0.5 M mannitol containing 0.05 M $CaCl_2 \cdot 2H_2O$ at a pH of 10.5. The centrifuge tubes containing the protoplasts are then incubated in a water bath at 37°C for 40–50 min. After this treatment, 20–50% of the protoplasts were involved in fusion.

### Polyethylene glycol method

Kao and Michayluk (1974) and Wallin et al. (1974) developed PEG method of fusion of protoplasts. This is one of the most successful techniques for fusing protoplasts. The protoplasts are suspended in a solution containing high molecular weight PEG, which enhances agglutination and fusion of protoplasts in several species. When sufficient quantities of protoplasts are available, 1 ml of the protoplasts suspended in a culture medium are mixed with 1 ml of 28–56% PEG (1500 –6000 MW) solution. The tube is then shaken for 5 sec and allowed to settle for 10 min. The protoplasts are then washed several times by the addition of protoplast culture medium to remove PEG. The protoplast preparation is then resuspended in the culture medium. The PEG method has been widely accepted for protoplast fusion because it results in reproducible high-frequency heterokaryon formation, low cytotoxicity to most cell types and the formation of binucleate heterokaryons. PEG-induced fusion is non-specific and is therefore useful for interspecific, intergeneric or interkingdom fusions.

Both the molecular weight and the concentration of PEG are critical in inducing successful fusions. PEG less than 100 molecular weight is not able to produce tight adhesions while that ranging up to 6000 molecular weight can be more effective per mole in inducing fusions. At higher molecular weight PEG produces too viscous a solution which cannot be handled properly. Treatment with PEG in the presence of/or by high pH/$Ca^{++}$ is reported to be most effective in enhancing the fusion frequency and survivability of protoplasts.

### Electrofusion

Protoplasts are placed in to a small culture cell containing electrodes, and a potential difference is applied due to which protoplasts line up between the electrodes. If now an extremely short wave electric shock is applied, protoplasts can be induced to fuse. In this fusion method, two-step procedure is followed beginning with application of an alternating current (AC) of low intensity to protoplast suspension. Dielectrophoretic collectors adjusted to 1.5 V and 1 MHz and an electrical conductivity of the suspension medium less than $10^{-5}$ sec/cm generate an electrophoresis effect that make the cells attach to each other along the field lines. The second step of injection of an electric direct current (DC) field pulse of high intensity (750–1000 V/cm) for a short duration of 20–50 μsec leads to breakdown of membranes in contact areas of adjacent cells resulting in fusion and consequent membrane reorganization. This electrofusion technique has been found to be simple, quick and efficient. Cells after electrofusion do not show cytotoxic response. However, this method did not receive much acceptance because specialized equipment is required.

## Mechanism of fusion

Protoplast fusion consists of three main phases

i. *Agglutination or adhesion*: Two or more protoplasts are brought into close proximity. The adhesion can be induced by a variety of treatments, e.g. concanavalin A, PEG, high pH and high $Ca^{++}$ ions.

ii. *Plasma membrane fusion at localized sites*: Membranes of protoplasts agglutinated by fusogen get fused at the point of adhesion. It results in the formation of cytoplasmic bridges between the protoplasts. Plant protoplasts carry a negative charge from −10 to −30 mV. Due to common charge, the plasma membranes of two agglutinated protoplasts do not come close enough to fuse. Fusion requires that membranes must be first brought close together at a distance of 10Å or less. The high pH–high $Ca^{++}$ ions treatment has shown to neutralize the normal surface charge so that agglutinated protoplasts can come in intimate contact. High temperature promotes membrane fusion due to perturbance of lipid molecules in plasma membrane and fusion occurs due to intermingling of lipid molecules in membranes of agglutinated protoplasts. PEG agglutinates to form clumps of protoplasts. Tight adhesion may occur over a large or small localized area. Localized fusion of closely attached plasma membranes occurs in the regions of tight adhesion and results in the formation of cytoplasmic bridges. It has been further suggested that PEG, which is slightly negative in polarity can form hydrogen bonds with water, protein, carbohydrates, etc. which are positive in polarity. When the PEG molecule chain is large enough it acts as a molecular bridge between the surface of adjacent protoplasts and adhesion occurs.

iii. *Formation of heterokaryon*: Rounding off of the fused protoplasts due to the expansion of cytoplasmic bridges forming spherical heterokaryon or homokaryon.

# IDENTIFICATION AND SELECTION OF HYBRID CELLS

Following fusion treatment, the protoplast population consists of parental types, homokaryotic fusion products of parental cells, heterokaryotic fusion products or hybrids and a variety of other nuclear cytoplasmic combinations. Despite efforts to increase the efficiency of protoplast fusion, usually not more than 1–10% of the protoplasts in a treated population have actually undergone fusion. The proportion of viable heterospecific binucleate A + B type fusion products may be even lower. In the absence of systems that would fuse different protoplasts specifically to produce heterokaryocytes at a very high rate, there is an obvious need to select the products of fusion amongst the unfused and homofused parental cells. Identification and recovery of protoplast fusion products have been based on observation of visual characters, or hybrid cells display genetic complementation for recessive mutations and physiological complementation for *in vitro* growth requirements. In complementation, fusion of two protoplasts each carrying a different recessive marker, will generate a fusion product which is functionally restored since each parent contributes a functional allele that corrects the respective deficiency of the other parent.

## Chlorophyll deficiency complementation

Melchers and Labib (1974) first used genetic complementation to isolate green somatic hybrids following fusion of two distinct homozygous recessive albino mutants of *Nicotiana tabacum*. They used chlorophyll deficient light sensitive varieties "sublethal" and "virescent". After two months of incubation under high light conditions, green colonies develop. This is the most frequently used method to isolate somatic hybrids (Fig. 9.3). Chlorophyll deficient mutants have also been used to raise somatic hybrids of *N. tabacum* + *N. sylvestris*.

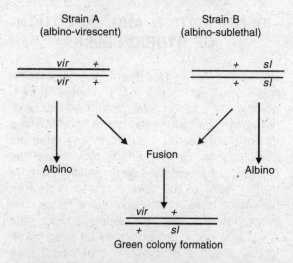

Fig. 9.3: A selection scheme involving complementation of chlorophyll deficient mutations.

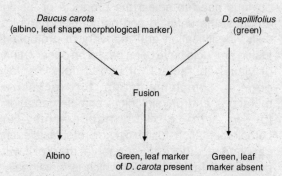

Fig. 9.4: A scheme illustrating selection of interspecific hybrid plants in *Daucus* utilizing morphological marker with albino mutant.

In most cases, it has not been necessary to use two albino mutants to recover somatic hybrid shoots. A single recessive albino mutation could be very useful in combination with a morphological trait or a growth response in the isolation of somatic hybrid plants. Single recessive albino mutation marker in combination with morphological marker(s) to distinguish putative somatic hybrids in *Daucus* has been used (Fig. 9.4). Dudits *et al.* (1977) fused albino *D. carota* protoplasts with wild type *D. capillifolius*. Thus, both hybrid protoplasts and *D. capillifolius* will regenerate into green shoots. However, the morphology of leaves of fused

product resembles those of *D. carota* leaves and can be isolated.

*Auxotroph complementation*

This method of selection has been applied to higher plants on a limited scale. The limitation is due to the paucity of higher plant auxotrophs. There is a report by Glimelius *et al.* (1978) where two parents were nitrate reductase deficient mutants of *N. tabacum* and could not be grown with nitrate as sole nitrogen source, while hybrids could regenerate shoots in the nitrate medium (Fig. 9.5). The lack of nitrate reductase (NR) activity causes an absolute requirement for reduced nitrogen and is caused by a deficiency either in the NR apoenzyme (nia⁻ type mutant) or in the molybdenum cofactor (cnx⁻ type mutant). cnx⁻ and nia⁻ type mutants regularly complement each other upon fusion.

*Complementation of resistance markers*

Dominant characters such as traits conferring resistance to antibiotics, herbicides, amino acid analogues or other toxic compounds have been recognized as potent selectable markers. When protoplasts from two lines are being fused together, the sensitivity trait of each parent will be dominated by the respective resistance trait from the other parent and will grow on a medium containing both the metabolites because of double resistance as compared to single resistance of the parents. This is also referred

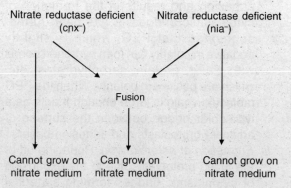

Fig. 9.5: A summary of somatic hybrid selection based on auxotroph complementation.

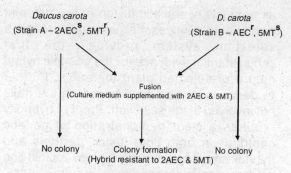

*Daucus carota*
(Strain A – 2AEC$^s$, 5MT$^r$)

*D. carota*
(Strain B – AEC$^r$, 5MT$^s$)

Fusion
(Culture medium supplemented with 2AEC & 5MT)

No colony     Colony formation     No colony
(Hybrid resistant to 2AEC & 5MT)

**Fig. 9.6:** Scheme illustrating selection of somatic hybrids utilizing complementation of resistance markers.

to as biochemical selection. Harms *et al.* (1981) selected double resistant somatic hybrids in *Daucus carota* following fusion of protoplasts from S2-aminoethyl cysteine (AEC) and 5-methyl tryptophan (5MT) resistant parental lines but were sensitive to other compound (Fig. 9.6). Maliga *et al.* (1977) used a kanamycin resistant variant of *N. sylvestris*, KR103, isolated from the cultured cells, as a genetic marker to recover fusion products between *N. sylvestris* and *N. knightiana*. Complementation system involving antibiotics (e.g. methotrexate, streptomycin, cycloheximide, and chloramphenicol) resistance has also been used.

### Use of metabolic inhibitors

In this method, parental cells are treated with an irreversible biochemical inhibitor such as iodoacetate or diethylpyrocarbamate and following treatment only hybrid cells are capable of division. Iodoacetate pretreatment has been used to aid recovery of somatic hybrids between *N. sylvestris* and *N. tabacum* and *N. plumbaginifolia* and *N. tabacum*. In each case the parent protoplasts treated with iodoacetate were unable to reproduce, while the newly formed hybrid protoplasts continued to develop and yield hybrid plants.

### Use of visual characteristics

The most efficient but the most tedious method to select products of protoplast fusion is to visually identify hybrid cells and mechanically isolate individual cells. The approaches used by various workers have been summarized:

i. Use of morphologically distinct cells: Fusion of mesophyll protoplasts containing green chloroplasts with colorless cell culture protoplasts containing distinct starch granules due to growth on sucrose supplemented medium. Fused products can be seen immediately after PEG treatment with one half containing chloroplast and other half with distinct starch granules.

ii. Use of petal protoplasts with leaf protoplasts and petal + cell culture protoplasts can be readily distinguished. The usually vacuolar petal pigment is originally separated within the fused cell but eventually becomes evenly distributed throughout the fused cell.

iii. Fluorescent labeling: Parental protoplasts that are isolated from the same type of cells and not sufficiently different for visual distinction may be loaded with different fluorescent dyes prior to fusion. Fluorescein isothiocyanate (fluorescing in the green) and rhodamine isothiocyanate fluorescing in the red have been used to label two separate batches of mesophyll protoplasts of *Nicotiana* have been used. The labeling is achieved by adding 0.5 mg/l of the compound into the enzyme mixture during incubation period (Galbraith and Mauch, 1980). The double fluorescent label of fusion products can be recognized in a fluorescence microscope and fusion products then cultured after micro-isolation. The double fluorescence provides a possibility to separate fusion products from parental protoplasts when fusion mixtures are run through a fluorescence activated cell sorter (FACS).

iv. The growth pattern of hybrid callus is different from either parental line. Hybrid callus is often more vigorous than parental callus.

v. Differences in the morphology of callus: Parental type protoplasts as well as hybrid cells can develop to form different callus types (Fig. 9.7). No selection pressure is applied at the protoplast or cell level. There are differences in the morphology of callus in hybrid of *Datura innoxia* and *Atropa belladonna* (Krumbiegel and Schieder, 1979).

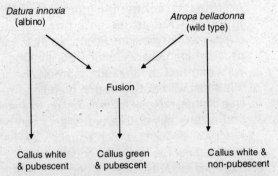

**Fig. 9.7:** Selection system based on morphology of callus employed in intergeneric somatic hybridization.

*Compound selection systems*

Most of the somatic hybrid plants that are known today have been selected using compound selection strategies rather than straightforward marker complementation. Compound selection systems have used differential growth characteristics, regenerability, morphological features and complementation marker systems to distinguish somatic hybrids from parental cells/plants. Whatever the system employed, the utility of a selective system lies in the fact that it provides enrichment for fused cells by reducing or eliminating the non-desired parental type cells. This enrichment thus greatly reduces the number of colonies or plants among which hybrids must be sought.

**Verification and Characterization of Somatic Hybrids**

Protoplasts from any two species can be fused together. However, there are a number of limitations to widespread utilization of somatic hybridization in higher plants, including aneuploidy, species barriers to hybridization, and the inability to regenerate plants from protoplasts. Besides these limitations, hybrids have been developed and these must be verified as products of somatic fusion of two different protoplasts. Successful passage through a selection system provides the first circumstantial evidence for the somatic hybrid nature of selected plant materials. Further evidence must be added based on the traits not involved in the selection. Proof of hybridity requires a clear demonstration of genetic contribution from both fusion partners and hybridity must be sought only from euploid and not from aneuploid hybrids.

1. *Morphology*: When plant regeneration has been accomplished from protoplast fusion, the products show a wide range of morphological features. These can be drawn upon for hybrid verification. In most cases morphological characteristics of either somatic or sexual hybrids are intermediate between the two parents. Both vegetative and floral characters such as leaf shape, leaf area, petiole size, root morphology, trichome (hairs on the leaf surface) length and density, flower shape, colour size and structure, corolla morphology are considered, intensity of flower pigment and seed capsule morphology. The genetic basis for most of these morphological traits has not yet been elucidated, but the intermediate behavior in hybrids suggests control the traits by multiple genes. Sometimes intermediate morphology is not observed in the somatic hybrids because these traits behave as dominant single gene traits as they are present only in one parent, but are also expressed in the somatic hybrids. Such traits include stem anthocyanin pigment, flower pigment, heterochromatic knobs in interphase cells and leaf size. Pollen viability character in general shows a decrease in viability in the hybrids. Whenever possible, additional genetic data should be presented to support hybridity.

2. *Isoenzyme analysis*: Electrophoretic banding patterns of isoenzymes have been extensively used to verify hybridity. Isoenzymes are the different molecular

forms of the enzyme that catalyze the same reaction. For isoenzymic studies, both starch and polyacrylamide gels are employed where electrical properties of proteins (enzymes) are used for separation into different bands. Somatic hybrids may display isoenzyme bands of certain enzymes specific to one or the other parent or to both parents simultaneously. Enzymes that have unique banding patterns versus either parental species and that have been used for identification of somatic hybrids include esterase, aspartate aminotransferase, amylase, isoperoxidase phosphatase, alcohol, malate, and lactate dehydrogenase and phosphodiesterase. If the enzyme is dimeric, the somatic hybrids contain a unique hybrid band intermediate in mobility in addition to the sum of parental bands. This probably represents formation of a hybrid dimeric enzyme unique to the somatic hybrids. Since isoenzymes are extremely variable within plant tissues, it is important to use the same enzyme from each plant and to use plants/tissue at identical developmental ages when comparing parental plants with somatic hybrids. For comparison, zymograms (a diagrammatic representation of isoenzyme banding pattern) should therefore be prepared and interpreted with caution.

3. *Chromosomal constitution*: Counting chromosomes in presumed hybrid cells can be an easy and reliable method to verify hybrid cells and it also provides information on the ploidy state of the cells. Cytologically the chromosome number of the somatic hybrids should be the sum of chromosome number of two parental protoplasts used for fusion. Variation in chromosome number (aneuploidy) is generally observed in hybrids but frequently chromosome number is more than the total number of both the parental protoplasts. Besides numerical number, structural and size differences of their chromosomes should be studied. The presence of marker chromosomes can greatly facilitate the analysis of genetic events in hybrid cells. Extensive use of banding techniques can be made to identify specific chromosomes and investigate rearrangements in somatic hybrid cells.

4. *Molecular techniques*: Recent progress in the development of molecular biological techniques has greatly improved our capabilities to analyze the genetic constitution of somatic plant hybrids. Specific restriction patterns of chloroplast and mitochondrial DNA have been used to great advantage to characterize the nature of plastoms and chondrioms of somatic hybrids and cybrids. Species specific restriction fragments of nuclear DNA coding for ribosomal RNA have been shown to verify somatic hybrids of *Nicotiana glauca* + *N. langsdorffii* (Uchimiya *et al.*, 1983). With the availability of numerous molecular markers such as RFLP, AFLP, RAPD, microsatellites etc., the hybrid identification can be done using these techniques. These are more accurate methods of analysis based on DNA, so no environmental effect is there. RFLP markers have been used for the identification of somatic hybrids between potato and tomato. PCR technology has been utilized for hybrid identification (Baird *et al.*, 1992). RAPD markers established the hybridity between *Solanum ochranthum* and *Lycopersicon esculentum* (Kobayashi *et al.*, 1996).

5. *Genetic characterization*: Somatic hybrid plants that produce fertile flowers can be analyzed by conventional genetic methods. Results obtained from segregation data of selfed $F_2$ generations are generally in good agreement with the expected values and have provided a sound confirmation of the somatic nature of the plants analyzed.

## CHROMOSOME NUMBER IN SOMATIC HYBRIDS

Variation in chromosome number (aneuploidy) is generally observed in hybrids but frequently chromosome number is more than the total number of both the parental protoplasts.

Interspecific and intergeneric somatic hybrids are mostly polyploids. The chromosome number of these hybrids indicates that only few hybrids have the sum total chromosome number of both the parents as expected in an amphiploid. There is an indication that closely related species would yield true amphidiploids through somatic hybridization. Even sexually compatible parents show deviation in the number of chromosomes in their somatic hybrids. The chromosome number of parent species and somatic hybrids obtained has been listed in Table 9.2. The variability in chromosome number of hybrids could be due to following reasons:

i) Multiple fusions give a higher chromosome number.

ii) Fusion of more than two protoplasts with subsequent mitotic irregularities.

iii) In PEG and electro-induced fusions about one-third of fusion products result from fusions among more than two protoplasts.

iv) Asymmetric hybrids result from fusion of protoplasts isolated from actively dividing tissue of one parent and quiescent tissue of the other parent.

v) Unequal rates of DNA replication in two fusing partners may also give asymmetric hybrids.

vi) Somaclonal variation in cultured cells used for protoplast isolation may also lead to variation in chromosome number.

## CYBRIDS

Sexual hybridization is a precise mixture of parental nuclear genes but the cytoplasm is derived from the maternal parent only, while in somatic hybrids the cytoplasm is derived from both the parents. However, somatic hybrids can be obtained where nucleus is derived from one parent and cytoplasm is derived from both, thus producing cytoplasmic hybrids, also called as cybrids (Fig.9.8). Early segregation of nuclei in a fused product can be stimulated and directed so that one protoplast contributes the cytoplasm while the other contributes the nucleus alone or both nucleus and cytoplasm. There are different ways of inactivating, the nucleus of one protoplast can be inactivated. Thus, there will be fusion between protoplasts containing the full complement of nucleus, mitochondria and chloroplasts with functional cytoplasmic component of second parent. The various approaches to achieve this type of fusion are:

1. By application of lethal dosages of $X$ or gamma irradiation to one parental protoplast population. This treatment renders the

**Table 9.2:** Chromosome number in some of the interspecific hybrids produced through protoplast fusion

| Plant species with their chromosome number | | Chromosome number in hybrid |
|---|---|---|
| B. oleracea (2n = 18) | + B. campestris (2n = 18) | Wide variation |
| B. napus (2n = 38) | + B. juncea (2n = 36) | Wide variation |
| Datura innoxia (2n = 24) | + D. stramonium (2n = 24) | 46, 48, 72 |
| N. tabacum (2n = 48) | + N. glutinosa (2n = 24) | 50–58 |
| N. tabacum (2n = 48) | + N. nesophila (2n = 48) | 96 |
| N. tabacum (2n = 48) | + N. sylvestris (2n = 24) | 72 |
| Lycopersicon esculentum (2n = 24) | + L. peruvianum (2n = 24) | 72 |
| Petunia parodii (2n = 48) | + P. hybrida (2n = 14) | 44–48 |
| Solanum tuberosum (2n = 24, 48) | + S. chacoense (2n = 14) | 60 |

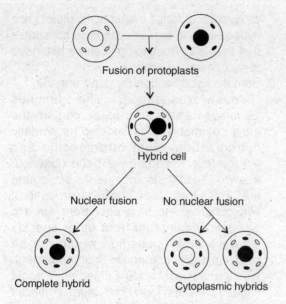

**Fig. 9.8:** Schematic illustration of somatic hybridization which can produce complete hybrids (left) or cytoplasmic hybrids (right).

protoplasts inactivated and non-dividing but they serve as an efficient donor of cytoplasmic genophores when fused with recipient protoplasts. *Nicotiana* protoplasts can be inactivated by 5-kr dose of X-rays. Other protoplasts may require different doses.

2. By treatment with iodoacetate to metabolically inactivate the protoplasts. Pretreatment with iodoacetate will cause the degeneration of non-fused and autofused protoplasts while fusion of iodoacetate pretreated protoplasts with non-treated protoplasts will cause metabolic complementation and result in viable hybrids. In an experiment, iodoacetate-treated *Nicotiana plumbaginifolia* cell suspension was fused with X-irradiated *N. tabacum* mesophyll protoplasts. All regenerated cybrid plants had *N. plumbaginifolia* morphology but most of them contained *N. tabacum* chloroplasts. The iodoacetate treatment does not impair the nucleus of the treated protoplasts. Thus the latter can complement an X-irradiated protoplast.

3. Fusion of normal protoplasts with enucleated protoplasts. Enucleated protoplasts can be obtained by high-speed centrifugation (20,000–40,000x g) for 45–90 minutes in an iso-osmotic density gradient with 5–50% percoll. Additional exposure of isolated protoplasts to cytochalsin B in combination with centrifugation has also been found beneficial for enucleation.

4. Fusion of cytoplasts with protoplasts—Isolated protoplasts can be experimentally induced to fragment into types of subprotoplasts called miniprotoplasts or cytoplasts. The term miniprotoplast was coined by Wallin *et al.* (1978) for subprotoplasts having nuclear material which can divide and may be able to regenerate into plants. Equivalent terms for miniprotoplasts are karyoplast (evacuolated subprotoplast) or nucleoprotoplast. Cytoplasts are nuclear free subprotoplasts which do not divide but are useful in the process of cybridization. Maliga *et al.* (1982) demonstrated that streptomycin resistance encoded by chloroplasts could be transferred by cytoplasts. Similarly, Tan (1987) claimed to have obtained cybrids by fusion of cytoplasts from *Petunia hybrida* and protoplasts of *Lycopersicon peruvianum*.

5. Fusion of a normal protoplast with another in which nuclear division is suppressed.

Cybridization thus opens an exciting avenue to achieve alloplasmic constitution in a single step without the need to perform a series of 8–12 time-consuming backcrosses. Alloplasmic lines contain nucleus of one parent genome with the cytoplasmic constituents from other parent. Application of cybrids would be the directed transfer of cytoplasmic male sterility or herbicide resistance from a donor to a recipient crop plant species. Transfer of cytoplasmic male sterility (CMS) from *N. tabacum* to *N. sylvestris* by protoplast fusion was first reported by Zelcer *et al.* (1978). Resistance of plants to herbicide atrazine has been transferred from *Brassica campestris* to *B. napus* via fractionated protoplast fusions. CMS has been successfully transferred in various crop species of *Oryza, Lycopersicon, Brassica, Nicotiana*, etc. (Table 9.4).

## POTENTIAL OF SOMATIC HYBRIDIZATION

1. Production of novel interspecific and intergeneric crosses between plants that are difficult or impossible to hybridize conventionally. It overcomes sexual incompatibility barriers. For example, fusion between protoplasts of *Lycopersicon esculentum* (tomato) and *Solanum tuberosum* (potato) created the pomato first achieved by Melchers *et al.* (1978). Asymmetric hybrids also develop when there is partial hybridization. These asymmetric hybrids have abnormal or wide variation in chromosome number than the exact total of two species (Table 9.2). Efficient cell fusion can be achieved between sexually compatible and incompatible parents involving interspecific or intergeneric combination (Tables 9.2 and 9.3). Attempts to overcome conventional breeding barriers by interspecific fusion of rice with four different wild species including *Oryza brachyantha*, O. *eichingeri*, O. *officinalis* and O. *perrieri* were more successful. Mature plants with viable pollen could be obtained in all but the first *Oryza* combination. The production of fertile interspecific diploid rice hybrid plants as well as cybrids demonstrates that the fusion technology can be extended to graminaceous crops.

2. Somatic hybridization for gene transfer

i) Disease resistance: The families Solanaceae and Brassiceae contain the most commonly used species for somatic hybridization. Both interspecific and intergeneric hybrids have been obtained. Many disease resistance genes *viz.* potato leaf roll virus, leaf blight, *Verticillium, Phytophthora*, etc. have been transferred to *Solanum tuberosum* from other species where normal crossings would not be possible due to taxonomic or other barriers. Resistance to blackleg disease (*Phoma lingam*) has been found in *B. nigra, B. juncea* and *B. carinata* and after production of symmetric as well as asymmetric somatic hybrids between these gene donors and *B. napus,* resistant hybrids have been developed. Attempts were made to introduce tolerance in *Brassica napus* against *Alternaria brassicae* from *Sinapis alba* (Primard *et al.*, 1988) and beet cyst nematode from *Raphanus sativus* (Lelivelt *et al.*, 1993). Resistance has been introduced in tomato against various diseases like TMV, spotted wilt virus, insect pests and also cold tolerance. Kobayashi *et al.* (1996) reported the incorporation of disease resistance genes from wild species *Solanum ochranthum,* a woody vine like tomato relative, to *L. esculentum*. Hansen

**Table 9.3:** Intergeneric hybrids produced though protoplast fusion

| Plant species with chromosome number | | New genus |
|---|---|---|
| *Raphanus sativus* (2n = 18) | + *B. oleracea* (2n = 18) | *Raphanobrassica* |
| *B. oleracea* (2n = 18) | + *Moricandia arvensis* (2n = 27,28) | *Moricandiobrassica* |
| *Eruca sativa* (2n = 22) | + *B. napus* (2n = 38) | *Erucobrassica* |
| *Diplotaxis muralis* (2n = 42) | + *B. napus* (2n = 38) | *Diplotaxobrassica* |
| *Nicotiana tabacum* (2n = 24) | + *Lycopersicon esculentum* (2n = 24) | *Nicotiopersicon* |
| *Solanum tuberosum* (2n = 24) | + *L. esculentum* (2n = 24) | *Solanopersicon* |
| *Datura innoxia* (2n = 48) | + *Atropa belladona* (2n = 24) | *Daturotropa* |
| *Oryza sativa* (2n = 24) | + *Echinochloa oryzicola* (2n = 24) | *Oryzochloa* |
| *Arabidopsis thaliana* (2n = 10) | + *B. campestris* (2n = 20) | *Arabidobrassica* |

and Earle (1995) reported that black rot disease caused by *Xanthomonas campestris* is a serious disease in cauliflower. The somatic hybrids were produced by protoplast fusion of *B. oleracea* with *B. napus*. Some of the examples of incorporation of resistance genes via protoplast fusion technique have been listed in Table 9.4.

ii) Abiotic stress resistance: Work related to somatic hybridization for abiotic stress has been mainly done on families Fabaceaa, Brassicaceae, Poaceae, Solaneae and relates to cold and frost resistance. Rokka *et al.* (1998) developed somatic hybrids between cultivated potato (*S. tuberosum*) and wild relative (*S. acaule*) possessing several disease and early frost resistance characters.

iii) Quality characters: Somatic hybrids produced between *B. napus* and *Eruca sativa* were fertile and had low concentration of erucic acid content (Fahleson *et al.,* 1993). This hybrid material has been introduced into breeding programme. Likewise, nicotine content character has been transferred to *N. tabacum*.

iv) Cytoplasmic male sterility: Several agriculturally useful traits are cytoplasmically encoded, including some types of male sterility and certain antibiotic and herbicide resistance factors. Pelletier *et al.* (1988) reported *Brassica raphanus* cybrids that contain the nucleus of *B. napus,* chloroplasts of atrazine resistant *B. campestris* and mitochonrida that confer male sterility from *Raphanus sativus*. Cybridization has been successfully used to transfer cytoplasmic male sterility in rice (Kyozuka *et al.*, 1989). Sigareva and Earle (1997) produced cold tolerant cytoplasmically male sterile (cms) cabbage (*Brassica oleracea* ssp. *capitata*) by the fusion of cabbage protoplasts with cold tolerant ogura CMS broccoli lines. Rambaud *et al.* (1997) reported that CMS chicory (*Cichorium intybus*) cybrids have been obtained by fusion between chicory and CMS sunflower protoplasts. Resistance to antibiotics, herbicide as well as CMS has been introduced in so many cultivated species (Table 9.4)

3. Production of autotetraploids: Somatic hybridization can be used as an alternative to obtain tetraploids and, if this is unsuccessful, colchicine treatment can be used.

4. Protoplasts of sexually sterile (haploid, triploid, aneuploid, etc.) plants can be fused to produce fertile diploids and polyploids.

5. Hybridization becomes possible between plants that are still in the juvenile phase.

6. Production of heterozygous lines within a single species that normally could only be propagated by vegetative means, e.g. potato and other tuber and root crops.

7. Somatic cell fusion is useful in the study of cytoplasmic genes and their activities. This information can be applied in plant breeding experiments.

8. The production of unique nuclear-cytoplasmic combinations. Evidence from a number of somatic hybrids suggests that although two types of cytoplasm are initially mixed during protoplast fusion, resulting in heteroplasmons, eventually one parent type cytoplasm predominates, resulting in cytoplasmic segregation. Mitochondrial and chloroplast recombination has also been reported to result in unique nuclear- cytoplasmic combinations. These unique combinations using protoplasts will aid the development of novel germplasm not obtainable by conventional methods.

## PROBLEMS AND LIMITATIONS OF SOMATIC HYBRIDIZATION

There are limitations, however, to the use of these types of somatic hybridization since plants regenerated from some of the combinations are not always fertile and do not produce viable seeds. The successful agricultural application of somatic hybridization is dependent on overcoming several limitations.

1. Application of protoplast methodology requires efficient plant regeneration from protoplasts. Protoplasts from any two species can be fused. However, production of somatic

hybrid plants has been limited to a few species.

2. The lack of an efficient selection method for fused product is sometimes a major problem.

3. The end products after somatic hybridization are often unbalanced (sterile, misformed,

**Table 9.4:** Genetic traits transferred via protoplast fusion

| Somatic Hybrids | | | Traits (resistance) |
|---|---|---|---|
| Nicotiana tabacum | + | N. nesophila | Tobacco mosaic virus |
| Solanum tuberosum | + | S. chacoense | Potato virus X |
| N. tabacum | + | N. nesophila | Tobacco horn worm |
| S. tuberosum | + | S. brevidens | Potato leaf roll virus |
| | | | Late blight and PLRV Potato virus Y |
| S. circalifolium | + | S. tuberosum | Phytophthora |
| S. melongena | + | S. sanitwongsei | Pseudomonas solanacearum |
| S. tuberosum | + | S. commersonii | Frost tolerance |
| B. napus | + | B. nigra | Phoma lingam |
| B. oleracea | + | Sinapis alba | Alternaria brassiceae |
| B. napus | + | Sinapis alba | Alternaria brassiceae |
| Citrus sinensis | + | Poncirus trifoliata | Phytophthora |
| Lycopersicon esculentum | + | L. peruvianum | TMV, spotted wilt virus, cold tolerance |
| S. lycopersicoides | + | L. esculentum | Cold tolerance |
| Solanum ochranthum | + | L. esculentum | Tomato diseases and insect pests |
| Brassica oleracea | + | B. napus | Black rot (Xanthomonas campestris) |
| B. oleracea sp capitata | + | B. oleracea (Ogura CMS line) | Cold tolerance |
| S. tuberosum | + | S. cammersonii | Frost tolerance |
| R. sativa (Japanese radish) | + | B. oleracea var. botrytis | Club rot disease |
| B. oleracea var botrytis | + | S. alba + B. carinata | Alternaria brassicola and Phoma lingam |
| Citrullus lanatus | + | Cucumis melo | Club rot resistance |
| Hordeum vulgare | + | Daucus carota | Frost and salt tolerance |
| Solanum melongena | + | S. sisymbrifolium | Nematode |
| S. tuberosum | + | S. bulbocastanum | Nematode |
| S. tuberosum | + | S. bulbocastanum | Root knot nematode |
| Raphanus sativus | + | Brassica napus | Beet cyst nematode |
| Sinapis alba | + | R. sativus + B. napus | Beet cyst nematode |
| **Quality characters** | | | |
| N. rustica | + | N. tabacum | High nicotine content |
| B. napus | + | Eruca sativa | Low erucic acid |
| **Agronomic characters (transferred via cybrid formation)** | | | |
| N. tabacum | + | N. sylvestris | Streptomycin resistance |
| N. tabacum spp. | | | Triazine resistance |
| S. nigrum | + | S. tuberosum | Triazine resistance |
| B. nigra | + | B. napus | Hygromycin resistance |
| B. napus spp. | | | Triazine resistance |
| N. tabacum | + | N. sylvestris | CMS |
| B. campestris | + | B. napus | CMS |
| N. tabacum | + | N. sylvestris | CMS |
| B. napus+ B. campestris | + | Raphanus sativa | CMS and Triazine resistance |
| Oryza spp. | | | CMS |
| L. esculentum | + | Solanum acaule | CMS |
| B. napus | + | B. tournefortii | CMS |

and unstable) and are therefore not viable, especially if the fusion partners are taxonomically far apart.

4. The development of chimaeric calluses in place of hybrids. This is usually due to the nuclei not fusing after cell fusion and dividing separately. Plants that are regenerated from chimaeras usually lose their chimeric characteristics, since adventitious shoots or embryos usually develop from a single cell.

5. Somatic hybridization of two diploids leads to the formation of an amphidiploid which is generally unfavorable (except when tetraploids are formed intentionally). For this reason in most cases, the hybridization of two haploid protoplasts is normally recommended.

6. Regeneration products after somatic hybridization are often variable due to somaclonal variation, chromosome elimination, translocation, organelle segregation etc.

7. It is never certain that a particular characteristic will be expressed after somatic hybridization.

8. The genetic stability during protoplast culture is poor.

9. To achieve successful integration into a breeding programme, somatic hybrids must be capable of sexual reproduction. In all cases reported, somatic hybrids containing a mixture of genes from two species must be backcrossed to the cultivated crop to develop new varieties. All diverse intergeneric somatic hybrids reported are sterile and therefore have limited value for new variety development. It may be necessary to use back-fusion or embryo culture to produce gene combinations that are sufficiently stable to permit incorporation into a breeding programme.

10. To transfer useful genes from a wild species to a cultivated crop, it is necessary to achieve intergeneric recombination or chromosome substitution between parental genomes.

# PROTOCOL

## 9.1: Protocol for Protoplast Isolation and Fusion

*Plant material: Nicotiana tabacum*

1. Detach fully expanded leaves including petiole. Sterilize in 10% sodium hypochlorite solution containing two drops of Tween 20 for 10 min. Wash 3–4x with sterile distilled water and blot dry on sterile tissue paper.

2. Peel the lower epidermis from sterilized leaves with fine forceps and cut into pieces with a fine scalpel.

3. Transfer the leaf pieces (peeled areas only) with lower surface down into 30ml of 13% mannitol-inorganic salts of cell and protoplast washing media (CPW) solution contained in a petri dish for 30 min–1 h for plasmolysis. Composition of CPW salt solution is as follows:

   | | |
   |---|---|
   | $KH_2PO_4$ | 27.2 mg/l |
   | $KNO_3$ | 101.0 mg/l |
   | $CaCl_2 \cdot 2H_2O$ | 1480.0 mg/l |
   | $MgSO_4 \cdot 7H_2O$ | 246.0 mg/l |
   | KI | 0.16 mg/l |
   | $CuSO_4 \cdot 5H_2O$ | 0.025 mg/l |
   | pH | 5.8 |

4. Remove the mannitol-CPW salt solution with a Pasteur pipette and replace it with filter sterilized enzyme mixture in 13% mannitol solution (~ 20 ml). Enzyme mixture contains macerozyme (0.1–0.5%) and cellulase (0.5–1%).

5. Gently agitate the leaf pieces with sterile Pasteur pipette to facilitate the release of protoplasts, pushing the larger pieces of leaf material to one side and keeping the petri dish at an angle of 15°C or remove the larger debris by filtering through a 60–80 µm mesh.

6. Transfer the filtrate to a screw cap centrifuge tube.

7. Centrifuge at 100x g for 5–10 min to sediment the protoplasts. Remove the supernatant and resuspend the protoplast

pellet in CPW + 21% sucrose (prepared in CPW) and gently disperse the protoplasts in the solution.

8. Centrifuge again at 100x g for 10 min. The viable protoplasts will float to the surface of the sucrose solution while the remaining cells and debris will sink to the bottom of tube.

9. Remove the band of protoplasts from the top with a Pasteur pipette and transfer into another centrifuge tube. Resuspend in CPW + 10% mannitol. Centrifuge again at 100x g for 10 min to separate the contaminating debris.

10. Repeat the washing procedure at least 3 times.

11. After the final washing add enough MS protoplast culture medium with 9% mannitol to achieve a protoplast density of ~ $5 \times 10^4$/ml. Kao and Michayluk (1975) media has been widely used for protoplast culture (Table 9.5). Plate the protoplasts as small droplets (100–150 µl) or a thin layer in small petri dishes for 24h at 25°C.

12. Keep the cultures at low light intensity of 500 lux for the next 2 days and then for rest of the experiment at 2000 lux.

13. Protoplasts form cell walls and begin to divide after 3–5 days.

14. Protoplasts may also be cultured in an agarose medium by mixing 1.5 ml of protoplast suspension with an equal volume of 1.2% agarose medium at 45°C in 35 mm petri dish.

15. Cultures continue to grow in the media and the first colonies become visible after 3–4 weeks.

16. Transfer the colonies to new medium with a reduced mannitol level.

17. Transfer the colonies to a suitable medium for embryo differentiation and plantlet regeneration.

## Protoplast fusion

Fusion treatment: PEG
Material: Green leaf (mesophyll) and cultured cells (albino) of *N. tabacum*

1. Perform the fusion experiment in continuation with protoplast isolation experiment.

2. Take 4–6 ml of freshly isolated protoplast suspension from two sources in CPW salt medium −13 % mannitol (CPW13M) with a density of $5 \times 10^5$/ml.

3. Using a Pasteur pipette place a small drop (~50 µl) of sterile silicone on to the center of a plastic petri dish (35 mm) and gently lower a cover glass on to the drop of silicone fluid.

4. Pipette ~150 µl of protoplast suspension directly on to the middle of cover glass and allow it to settle for 10–15 min. (Protoplasts suspended in CPW13M for convenience are mixed in screw capped centrifuge tube in a 1:1 ratio).

5. Using a Pasteur pipette add 450 µl of PEG fusion solution drop-wise to the edge of the protoplast culture, placing the last drop in the center. PEG fusion solution contains
22.5% (w/v) PEG molecular weight 6000
1.8% (w/v) sucrose
154 mg/l $CaCl_2 \cdot 2H_2O$
9.52 mg/l $KH_2PO_4$
pH 5.8 (adjust with 1M KOH or IN HCl, autoclaved and stored in dark at 4°C)

6. Leave the protoplasts undisturbed for 20–40 min at room temperature and then add one drop of washing solution very gradually whilst withdrawing some of the fusion solution from the coalesced drops. Repeat this procedure at 5 min intervals for the next 20 min. During this procedure it is important that the protoplasts are subjected to minimal disturbance. Washing solution: CPW13M medium to which 0.74g/l $CaCl_2 \cdot 2H_2O$ is added (sterile).

**Table 9.5:** Kao and Michyaluk 8p media for protoplast culture

| Mineral salts (mg/l) | | | |
|---|---|---|---|
| Component | Concentration | Component | Concentration |
| $NH_4NO_3$ | 600 | KI | 0.75 |
| $KNO_3$ | 1900 | $H_3BO_3$ | 3.00 |
| $CaCl_2 \cdot 2H_2O$ | 600 | $MnSO_4 \cdot H_2O$ | 10.00 |
| $MgSO_4 \cdot 7H_2O$ | 300 | $ZnSO_4 \cdot 7H_2O$ | 2.00 |
| $KH_2PO_4$ | 170 | $Na_2MoO_4 \cdot 2H_2O$ | 0.25 |
| KCl | 300 | $CuSO_4 \cdot 5H_2O$ | 0.025 |
| Sesquestrene®330Fe | 28 | $CoCl_2 \cdot 6H_2O$ | 0.025 |
| Vitamins (mg/l) | | | |
| Inositol | 100 | Biotin | 0.01 |
| Nicotinamide | 1 | Choline chloride | 1.00 |
| Pyridoxine HCl | 1 | Riboflavin | 0.20 |
| Thiamine HCl | 1 | Ascorbic acid | 2.00 |
| D-calcium pantothenate | 1 | Vitamin A | 0.01 |
| Folic acid | 0.4 | Vitamin $D_3$ | 0.01 |
| p-Amino benzoic acid | 0.02 | Vitamin $B_{12}$ | 0.02 |
| Sucrose and glucose (g/l) | | | |
| Sucrose | 0.25 | | |
| Glucose | 68.4 | | |
| **Hormones (mg/l)** | | | |
| 2,4-D | 0.2 | | |
| Zeatin | 0.5 | | |
| NAA | 1.0 | | |
| **Organic acids (mg/l) adjusted to pH 5.5 with $NH_4OH$** | | | |
| Sodium pyruvate | 20 | Malic acid | 40 |
| Citric acid | 40 | Fumaric acid | 40 |
| **Other sugars and sugar alcohols (mg/l)** | | | |
| Fructose | 250 | Rhamnose | 250 |
| Ribose | 250 | Cellobiose | 250 |
| Xylose | 250 | Sorbitol | 250 |
| Mannose | 250 | Mannitol | 250 |
| **Amino acids (mg/l)** | | | |
| Glycine | 0.1 | Glutamine | 5.6 |
| Alanine | 0.6 | Glutamic acid | 0.6 |
| Cysteine | 0.2 | Asparagine | 0.1 |
| **Nucleiotides (mg/l)** | | | |
| Adenine | 0.10 | Uracil | 0.03 |
| Guanine | 0.03 | Hypoxanthine | 0.03 |
| Thymine | 0.03 | Xanthine | 0.03 |
| **Others** | | | |
| Casein hydrolysate | 250 mg/l | Coconut water | 20 ml/l) |
| pH | 5.6 | | |

7. Five min after the last wash, carefully suck off the solution with a Pasteur pipette leaving the protoplasts with a thin film of medium and replace it with culture medium (MS protoplast medium with 9% mannitol). Wash at least 3 times with culture medium.

8. For the culture of protoplasts flood the cover glass with culture medium and scatter several drops of medium on the base of petri dish prior to sealing with parafilm.

9. Incubate the cultures initially in the dark for 24 h at 25°C and then transfer to white light (1000 lux). The fusion products can be easily observed using an inverted microscope.

10. Subsequently transfer the colonies for regeneration.

11. Test the hybridity of fusion product and somatic hybrid as per equipment available in the lab, which has been explained in the text.

## REFERENCES

Baird, E., Cooper-Bland, S., Waugh, R., Demaine, M. and Powell, W. 1992. Molecular characterization of inter- and intraspecific somatic hybrids of potato using randomly amplified polymorphic DNA (RAPD) markers. *Mol. Gen. Genet.* **233:** 469–475.

Cocking, E.C. 1960. A method for the isolation of plant protoplasts and vacuoles. *Nature* **187:** 962–963.

Dudits, D., Hadlaczky, G., Levi, E., Fejer, O., Haydu, Z. and Lazar, G. 1977. Somatic hybridization of *Daucus carota* and *D. capillifolius* by protoplast fusion. *Theor. Appl. Genet.* **51:** 127–132.

Fahleson, J., Lagercrantz, U., Eriksson, I. and Glimelius, K. 1993. Genetic and molecular analysis of sexual progenies from somatic hybrids between *Eruca sativa* and *Brassica napus*. Swedish Univ. of Agril. Sci., Dissertation, ISBN 91–567–4684–8.

Glimelius, K., Eriksson, T., Grafe, R. and Muller, A. 1978. Somatic hybridization of nitrate reductase deficient mutants of *Nicotiana tabacum* by protoplast fusion. *Physiol. Plant.* **44:** 273–277.

Hansen, L.N. and Earle, E.D. 1995. Transfer of resistance to *Xanthomonas campestris* pv *campestris* in to *Brassica oleracea* by protoplast fusion. *Theor. Appl. Genet.* **91:** 1293–1300.

Harms, C.T., Potrykus, I. and Widholm, J.M. 1981. Complementation and dominant expression of amino acid analogue resistance markers in somatic hybrid clones from *Daucus carota* after protoplast fusion. *Z. Pflanzenphysiol.* **101:** 377–385.

Kao, K. and Michayluk, C. 1975. Growth of cells and protoplasts at low population density. *Planta* **126:** 105–110.

Keller, W.A. and Melchers, G. 1973. The effect of high pH and calcium on tobacco leaf protoplast fusion. *Z. Naturforsch* **288:** 737–741.

Klercker, J.A. 1892. Eine methods zur isoliering lebender Protoplasten. *Oefvers. Vetenskaps Adad. Stockholm* **9:** 463–471.

Kobayashi, R.S., Stomme, J.R. and Sinden, S.L. 1996. Somatic hybridization between *S. ochranthum* and *L. esculentum*. *Plant Cell Tiss. Org. Cult.* **45:** 73–78.

Krumbiegel, G. and Schieder, O. 1979. Selection of somatic hybrids after fusion of protoplasts from *Datura innoxia* and *Atropa belladona*. *Planta* **145:** 371–375.

Kyozuka, J., Kaneda, T. and Shimamoto, K. 1989. Production of cytoplasmic male sterile rice (*Oryza sativa* L.) by cell fusion. *Bio/Technology* **7:** 1171–1174.

Lelivelt, C.L.C. 1993. Introduction of beet cyst nematode resistance from *Sinapis alba* L. and *Raphanus sativus*

L. into *Brassica napus* L. (oil seed rape) through sexual and somatic hybridization. Ph.D. Thesis Univ. of Wageningen, The Netherlands.

Maliga, P., Lazar, G., Joo, F., Nagy, A.H. and Menczel, L. 1977. Restoration of morphogenic potential in *Nicotiana* by somatic hybridization. *Mol. Gen. Genet.* **157:** 291–296.

Melchers, G. and Labib, G. 1974. Somatic hybridization of plants by fusion of protoplasts. I. Selection of light resistant hybrids of haploid light sensitive varieties of tobacco. *Mol. Gen. Genet.* **135:** 277–294.

Melchers, G., Sacristan, M.D. and Holder, A.A. 1978. Somatic hybrid plants of potato and tomato regenerated from fused protoplasts. *Carlsberg Res. Commun.* **43:** 203–218.

Mglia, P., Lorz, H., Lazr, G. and Nagy, F.1982. Cytoplast–protoplast fusion for interspecific chloroplast transfer in *Nicotiana*. *Mol. Gen. Genet.* **185:** 211–215.

Pelletier, G., Primard, C., Ferault, M, Vedel, F., Cherit, P., Renard, M and Delourme, R. 1988. Use of protoplast in plant breeding: Cytoplasmic aspects. In: *Progress in Protoplast Research* (Puite, K.J., Dons, J.J.M., Huising, H.J., Kool, A.C., Koorneef, M. and Krens, F.A. eds.). Kluwer Academic Publishers, London, pp. 169–176.

Power, J.B., Cummins, S.E. and Cocking, E.C. 1970. Fusion of isolated plant protoplasts. *Nature* **225:** 1016–1018.

Primard, C., Vedel, F., Mathieu, C., Pelletier, G. and Chevre, A.M. 1988. Interspecific somatic hybridization between *Brassica napus* and *Brassica hirta* (*Sinapis alba* L.). *Theor. Appl. Genet.* **75:** 546–552.

Rambaud, G., Bellamy, A. and Dubreuca, A. 1997. Molecular analysis of the fourth progeny plants derived from a Cytoplasmic male sterile chicory hybrid. *Plant Breed.* **116:** 481–486.

Razdan, M.K. 2003. Introduction to Plant Tissue Culture. Science Publishers, Inc. Enfield, USA. pp. 379.

Rokka, V.M., Tauriamen, A., Pietila, L. and Pehu, E. 1998. Interspecific somatic hybrids between wild potato *Solanum acaule* and anther derived dihaploid potato (*Solanum tuberosum* L.). *Plant Cell Rep* **18:** 82–88.

Sigarava, M.A. and Earle, E. D. 1997. Transfer of a cold tolerant ogura male sterile cytoplasm into cabbage. *Theor. Appl. Genet.* **99:** 213–220.

Takebe, I., Otsuki, Y. and Aoki, S. 1968. Isolation of tobacco mesophyll cells in intact and active state. *Plant Cell Physiol.* **9:** 115–124.

Takebe, I., Labib, G. and Melchers, G. 1971. Regeneration of whole plants from isolated mesophyll protoplasts of tobacco. *Naturwissen.* **58:** 318–320.

Tan, M.L. Mc. 1987. Somatic hybridization and cybridization in Solanaceous species. Academisch Proefschrift. Vrije Univ to Amsterdam. (cited by Bajaj, Y.P.S. 1989b, p. 24).

Uchimiya, H., Ohgawara, T., Kato, H., Akiyama, T. and Harada, H. 1983. Detection of two different nuclear genomes in parasexual hybrids by ribosomal RNA gene analysis. *Theor. Appl. Genet.* **64:** 117.

Wallin, A., Glimelius, K. and Eriksson, T. 1974. The induction of aggregation and fusion of *Daucus carota* protoplasts by polyethylene glycol. *Z. Pflanzenphysiol.* **74:** 64–80.

Walllin, A., Glimelius, K. and Eriksson, T. 1978. Enucleation of plant protoplasts by cytochalasin *B.Z. Pflanzenphysiol.* **87:** 333–340.

Zelcer, A., Aviv, D. and Galun, E. 1978. Interspecific transfer of cytoplasmic male sterility by fusion between protoplasts of normal *Nicotiana sylvestris* and X-ray irradiated protoplasts of male sterile *N. tabacum. Z. Pflanzenphysiol.* **90:** 397–407.

## FURTHER READING

Bhojwani, S.S. and Razdan, M.K. 1990. *Plant Tissue Culture: Applications and Limitations.* Elsevier Science Publishers, Amsterdam.

Evans, D.A., Sharp, W.R., Ammirato, P.V. and Yamada, Y. 1983. *Handbook of Plant Cell Culture.* Macmillan Publishing Company, New York.

Fowke, L.C. and Constabel, F. 1985. *Plant Protoplasts.* CRC Press, Inc, USA.

Ignacimuthu, S. 1998. *Plant Biotechnology.* Oxford & IBH Publishing Co. Pvt. Ltd., Delhi.

Pierik, R.L.M. 1987. *In vitro culture of Higher Plants.* Martinus Nijhoff Publishers, Dordrecht.

Razdan, M.K. 1983. *An Introduction to Plant Tissue Culture.* Oxford and IBH Publishing Co. Pvt Ltd., New Delhi.

Reinert, J. and Bajaj, P.P.S. (eds.) 1977. *Applied and Fundamental Aspects of Plant Cell, Tissue and Organ Culture.* Springer-Verlag, Berlin.

Street, H.S. (ed.)1973. *Plant Tissue and Cell culture.* Blackwell Scientific Publications, Oxford.

Vasil, I.K. 1984. *Cell Culture and Somatic Cell Genetics of Plants: Laboratory Procedures and Their Applications.* Academic Press, Inc., New York.

# 10

# Somaclonal Variation

Genetic variability is an essential component of any breeding program designed to improve the characteristics of crop plants. In recent years plant cell culture has been hailed as one of the potential sources of useful genetic variation. The variability generated by the use of a tissue culture cycle has been termed somaclonal variation by Larkin and Scowcroft (1981). They defined a tissue culture cycle as a process that involves the establishment of a dedifferentiated cell or tissue culture under defined conditions, proliferation for a number of generations and the subsequent regeneration of plants. In other words one imposes a period of callus proliferation between an explant and the regeneration of plant. The initiating explant for a tissue culture cycle may come virtually from any plant organ or cell type including embryos, microspores, roots, leaves and protoplasts.

Historically, a culture cycle was essentially seen as a method of cloning a particular genotype. Thus, it became accepted dictum that all plants arising from tissue culture should be exact copies of the parental plant. However, phenotypic variants were frequently observed amongst regenerated plants. These were usually dismissed earlier as tissue culture artifacts due to the recent exposure of exogenous phytohormones, and sometimes they were labeled as epigenetic events. However, evidences have now shown that these variants are not artifacts but variation arising due to culture of cells and this has been termed as somaclonal variation. The cause of variation is attributed to changes in the chromosome number and structure. Genetic heterogeneity in cultures arises mainly due to (i) the expression of chromosomal mosaicism or genetic disorders in cells of explant and (ii) new irregularities brought about by culture conditions through spontaneous mutations. Earlier, terms like 'calliclones' and 'protoclones' were coined to indicate variation arising in regenerated plants from stem and protoplast derived callus respectively. However, a general term somaclones has been given for plants derived from any form of somatic cell culture and somaclonal variation is the variation displayed amongst such plants. Lately, a term gametoclonal variation has been introduced for variation observed among plants regenerated from cultured gametic cells.

## NOMENCLATURE

Though different letters and symbols have been used, two symbols are generally used. Chaleff (1981) has labeled the plants regenerated from tissue culture as R or $R_0$ plants and the self fertilized progeny of $R_0$ plants as $R_1$. Subsequent generations produced by self-fertilization are termed $R_2$, $R_3$, $R_4$, etc. Larkin and Scowcroft (1981) have referred regenerated plants as $SC_1$

$(=R_0)$ and subsequent self fertilized generations as $SC_2$, $SC_3$, $SC_4$, etc.

# SCHEMES FOR OBTAINING SOMACLONAL VARIATION

Two schemes, with and without *in vitro* selection, have been generally followed for getting somaclonal variation in crop plants.

### Without *in vitro* Selection

An explant is cultured on a suitable medium, e.g. small shoot segments (1–2 cm) of sugarcane, cotyledons, hypocotyls, protoplasts, leaves, embryos, etc. The basal medium is supplemented with growth regulators which support dedifferentiation stage, i.e. callus. Normally these cultures are subcultured and then transferred to shoot induction medium for plant regeneration. The plants so regenerated are transferred to pots, grown to maturity and analyzed for variants (Fig. 10.1).

The approach is to find somaclonal variants among the regenerated plants for various characters. Here no directed approach is used and appearance of desirable variant is a chance event. In such a method, both dominant and homozygous recessive traits can be directly selected. If the regenerants are heterozygous, then recessive traits can be selected in the progenies of regenerants. Epigenetic variation will also be avoided when progenies are used. Thus, in case of self-fertilizing crops it is recommended to screen the progenies of the regenerants. Unfortunately, a disadvantage of this approach is that it is time consuming due to the fertilization step and requires screening of many plants.

### With *in vitro* Selection

It is now well known that *in vitro* culture of higher plants can be used for selection of mutants. Protoplast, cell suspension and callus cultures are handled like microorganisms to search for biochemical mutants. Selection for resistance is the most straightforward method for mutant selection, whereby resistant cells in a large population can be selected by their ability to grow in the presence of an inhibitor while the sensitive cells do not. The protocol has been described in Fig. 10.2. Here the dedifferentiated culture (callus) is subjected to selection against inhibitors like antibiotics, amino acid analogs, pathotoxins, etc. These compounds are put in the medium at a concentration such that some cell population survives and can be further grown on a selective medium. Different selection cycles are performed to get tolerant cells/callus cultures that are subsequently regenerated into plants. These plants are then *in vivo* screened against the inhibitor. If the plants are resistant to the inhibitor, then stable transmission of that character is analyzed in subsequent generations. In this approach, variants for a particular character are selected rather than the general variation obtained in first case where selection is done at the plant level.

### *Factors influencing somaclonal variation*

When attempting to produce somaclones for a new crop plant, some factors that are valid for both the schemes must be considered.

***Genotype:*** The genotype of plants used for somaclonal variation is an important variable.

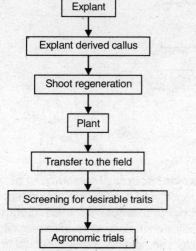

**Fig. 10.1:** A flow diagram for generation of somaclonal variation without *in vitro* selection.

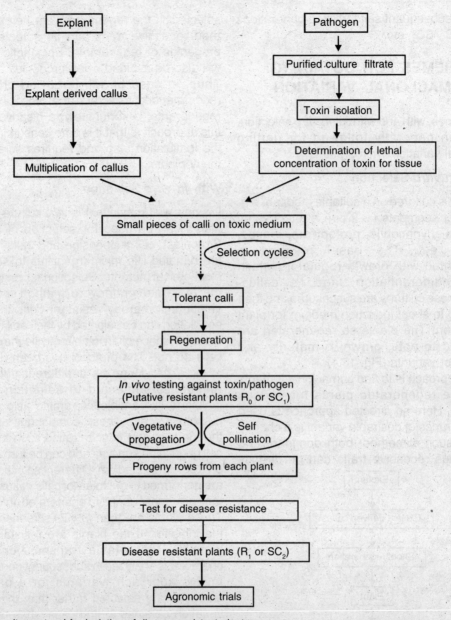

**Fig. 10.2:** *In vitro* protocol for isolation of disease resistant plants.

Genotype can influence both frequency of regeneration and the frequency of somaclones. Sun *et al.* (1983) compared the frequency of polyploidy regenerants in 18 varieties of rice and recovered muliploids in *indica* varieties but not in *japonica* varieties.

***Explant source***: The source of explant has been considered a critical variable for somaclonal variation. For example, work with Geranium has demonstrated that variants were recovered from *in vivo* root and petiole cuttings, but not from stem cuttings. Similarly in sugarcane which is a chromosome mosaic, propagation by stem cuttings can result in plants with variable chromosome number. In addition, since not all explants are equal in terms of regenerability, it is likely that different selective pressures would be exerted against different explants. This could result in different frequencies and spectrums of somaclonal variation among plants from different explants.

***Duration of cell culture***: Chromosome variation has been reported for many established plant cell cultures. It is widely accepted that most long-term cultures are chromosomally variable. Polyploid and aneuploid plants have been frequently recovered in large number of species e.g. Geranium, *Nicotiana*, tomato, alfalfa, *Saccharum,* etc. It has been stated that variation increased with increasing duration of culture.

***Culture conditions***: It has been known that growth regulator composition of the culture medium can influence frequency of karyotypic alterations in cultured cells. The growth regulators 2,4-D, NAA, BAP, etc. have been most frequently considered to be responsible for chromosome variability.

All the above variables are important in the recovery of somaclones among regenerated plants. In many species, these variables cannot be modified and still permit recovery of plants. However, these variables should be carefully examined.

**The following factors are important during *in vitro* selection of somaclonal variants:**

***Selective propagule (cells, protoplasts or calli)***: *In vitro* selection can be applied to any tissue, e.g. cells, protoplasts or calli. There are some advantages as well as disadvantages associated with each type of tissue, but high frequency regeneration must be obtained from that tissue. A main advantage of the use of plant suspension cultures or protoplasts is the capacity to apply nearly standard microbiological techniques. Selection of resistant plant cell lines from protoplasts offer an advantage that these can be prepared not only from suspension cells

and calli, but also directly from whole plant tissue. The clonal origin of cell lines arising from selections using protoplasts is more assured since the digestion of the cell wall frees and separates the individual plant cells. However, the advantages may be offset by the technical requirements necessary to produce viable protoplasts. Generally, protoplast preparation protocols must be developed independently for each plant species or tissue used.

However, selection from calli is probably one of the least complex methods. Callus culture is easy to perform and maintenance and handling can also be done without too much difficulty. The initiation and maintenance of suspension, protoplasts or callus cultures have been explained earlier.

***Selective agent***: Inhibitor used can be an amino acid analog, herbicide, a synthetic toxin(s) isolated from fungal liquid culture. Some of the inhibitors/toxins are commercially available and can be used directly for *in vitro* selection experiments. However, when pathotoxins produced by fungi in liquid culture are to be used, then first these toxins must be isolated. For abiotic stresses of salt and aluminium tolerance, specific compounds like sodium chloride and aluminium chloride can be used respectively. The inhibitor concentration used in the selection experiments will depend on the sensitivity of the plant cell line. Therefore, a dose response curve with respect to growth inhibition must be performed for each selection agent of interest. For amino acid analogs, typical gradients range logarithmically from $\mu$M to 10 mM. For sodium chloride tolerance, range may be from 0.1% to 2% of sodium chloride in the medium. The dilution steps are arbitrarily chosen to facilitate graphic representation of the final results.

Serial dilutions can be made as shown below:

1 mM———→0.1———→0.01———→0.001 mM

The inactivating effect of an inhibitor on cells can be assessed by determining increase in fresh or dry weight or using characters such as plating efficiency, dye uptake, cytoplasmic streaming and incorporation of labeled metabolites into macromolecules. A dose response curve should be prepared to assess the effect of an inhibitor.

For selection of disease resistant mutants, the toxicity of toxin/crude toxin/purified culture filtrate can be tested by leaf or root bioassays. Then, a particular concentration of toxin is used for selection of calli/cells/protoplasts. For this, pieces of calli (3–5) with 50 mg fresh weight can be inoculated on a medium with different concentrations of toxin and incubated for 3–4 weeks. On the basis of fresh weight increase, a particular toxin/inhibitor concentration can be selected for *in vitro* selection experiments.

**Selection technique**: *In vitro* selection of cell lines against an inhibitor is best carried out at a concentration that will kill 70 to 90% of wild type cells. Doses which are too low, should be avoided because cells escaping for purely physiological reasons could outnumber true mutants. On the other hand, inhibitor concentrations that are too high may mean that potentially resistant variants are also unable to grow and are thus lost. Cells should be incubated in an inhibitor-containing medium for a sufficient length of time to ensure expression and growth of the resistant clones. The selection can be carried out at a very high toxin/inhibitor concentration with many or just one cycle of selection, or many cycles with a pause on non-toxic media after 2nd or 3rd selection cycle, or on an increasingly higher inhibitor/toxin level. A selection cycle can be of 3 to 4 week duration.

The selection technique involves placing pieces of cultured tissue on a medium containing a near-lethal concentration of inhibitor and selecting out fast growing visible sectors of tissue from the remainder of the cultured material. Resistant tissue obtained in this way continues to proliferate on a medium containing high level of inhibitor concentration.

**Regeneration of plants**: The next step in the procedure is to regenerate plants from the cultured material. In many instances, the capacity of tissue cultures to regenerate plants is lost over a long period of continuous subculture. Regeneration is generally achieved by reducing the level of auxin in the media and incubating the cultures in light.

**In vivo testing**: It is important that the plants obtained from cultures selected for a particular inhibitor/toxin should be critically examined for expression and inheritance of resistance character. Only in this manner it can be determined whether the resistance observed *in vitro* is due to a mutational event or to the epigenetic adaptation by the cultured cells to the presence of inhibitor/toxin. Normally plants developed against toxin(s) of pathogen are *in vivo* tested against the pathogen. Conditions required for the development of disease symptoms in plants should be provided.

**Agronomic analysis**: The regenerated plants selected against inhibitor/toxin should be analyzed for their agronomic value in the first generation and also in the second generation ($SC_2$) for stability of resistance and other desirable features.

**Resistance stability**: A major difficulty often encountered with plant cell lines selected for resistance to inhibitor/toxin is the instability of the characteristic when removed from selective pressure. There are a variety of reasons for this instability. The genetic lesion responsible may have a high reversion rate or the alteration may simply be a transient genetic or epigenetic adaptation. The most common explanation, however, is the presence of wild type escapes in the cell population. Under non-selective conditions, these cells would rapidly proliferate eventually taking over the culture.

In order to test stability of the resistance trait, establish at least two cultures for each selected liner—one subcultured routinely in the presence and one in the absence of selective conditions. At various intervals over a period of several months perform dose response test. Further, the $R_0$ plants which were found to be resistant should be analyzed for resistance to that character in $R_1$ and subsequent generations.

# APPLICATION OF SOMACLONAL VARIATION

To have an impact on a plant breeding program, an *in vitro* scheme must either produce plants useful for breeding or useful directly.

**1. *Novel variants*:** An implication of somaclonal variation in breeding is that novel variants can arise and these can be agronomically used. A number of breeding lines have been developed by somaclonal variation. An improved scented Geranium variety named 'Velvet Rose' has been developed. An example of heritable somaclonal variation is the development of pure thornless blackberries Lincoln Logan (*Rubus*), Hasuyume, a protoplast derived rice cultivar, and somaclone T-42 has been released. Some of the variants developed as varieties in *Rubus, Torenia, Apium* and others have been shown in Table 10.1, while variation for other useful morphological characters observed in diverse species has been shown in Table 10.2. In India two varieties namely Pusa Jai Kisan in mustard *Brassica* and CIMAP/Bio13 in *Citronella* have been released for cultivation. An improved variety of rice 'DAMA' has been released through pollen haploid somaclone method which combined microspore culture with somatic tissue culture (see Heszky and Simon-Kiss, 1992). Somaclonally derived mutants in tomato with altered color, taste, texture and shelf life are being marketed in USA by Fresh World farms.

**Table 10.1:** A sample of cultivars derived from tissue culture via somaclonal variation

| Crop |
| --- |
| *Eustoma grandiflorum* |
| *Hemerocallis* Yellow Tinkerbell |
| *Paulownia tomentosa* Snowstrom |
| *Pelargonium* Velvet Rose |
| *Torenia fournieri* U Conn White |
| *Rubus laciniatus* Lincoln Logan |
| *Apium* UC-T3 somaclone |
| *Ipomea batatus* Scarlet |
| *Oryza sativa* Hasuyume |
| Mustard *Brassica* Pusa Jai Kisan |
| Citronella CIMAP/Bio13 |

**Table 10.2:** A sample of somaclonal variants obtained in different crop species

| Crop | Characters |
| --- | --- |
| Rice | Tiller no., panicle size, flowering date, plant height, heading date, culm length, leaf shape, leaf color, ear shape, sterility mutants |
| Maize | Glyphosate resistance |
| Wheat | Grain color, height, tiller no., seed storage protein |
| Sorghum | Improved acid soil stress tolerance |
| Oats | Avenin banding pattern variation, increased seed weight, seed number, grain yield |
| Barley | Grain yield, test weight, leaf shape, 1000 grain weight, height, lodging resistance, heading date, β-glucan content, ash |
| Soybean | Varied height, lodging, maturity, seed protein and oil content |
| *Brassica napus* | Growth habit, glucosinolate content |
| Sugarcane | Yield, sugar content, disease resistance |
| Tomato | Violet cotyledon color, dwarf habit, modification of first floral node, early flowering, orange fruit color |
| Carrot | Higher carotene content |
| Potato | Yield, vine quality, maturity and morphology |
| Fennel | Chlorotic foliage, altered stem height, reduced fruit production, varied oil composition |
| Datura | Sulfonylurea, imidazolinone and triazolopyrimidine herbicide resistance |

**2. *Somaclonal variation and disease resistance*:** The greatest contribution of somaclonal variation towards plant improvement is in the development of disease resistant genotypes in various crop species. Resistance was first reported in sugarcane for eye spot disease (*Helminthosporium sacchari*), downy mildew (*Sclerospora sacchari*) and Fiji virus disease by regenerating plants from the callus of susceptible clones and screening the somaclones. Variability from protoplast derived plants was also examined in the tetraploid Russet Burbank variety of potato. Although many of the variants were aberrant, some showed resistance to *Phytophthora infestans*. Table 10.3 lists few successful examples of somaclonal variants obtained without *in vitro* selection at the plant level with an increased disease resistance.

**Table 10.3:** A list of disease resistant crop plants obtained by screening of somaclones at the plant level without *in vitro* selection

| Crop | Pathogen |
| --- | --- |
| **Food crops** | |
| Barley | *Rhynchosporium secalis* |
| Maize | *Helminthosporium maydis* |
| Rice | *Helminthosporium oryzae* |
| Rape | *Phoma lingam, Alternaria brassicicola* |
| Sugar-cane | Fiji virus, *Sclerospora saccharii, Helminthosporium sacchari, Puccinia melanocephala* |
| **Horticultural and other crops** | |
| Potato | *Alternaria solani, Phytophthora infestans,* Potato virus X and Y, *Streptomyces scabie, Verticillium dahlie* |
| Tobacco | *Phytophthora parasitica* |
| Tomato | *Fusarium oxysporum, Pseudomonas solanacearum* |
| Alfalfa | *Verticillium albo-atrum, Fusarium solani* |
| Celery | *F. oxysporum* f.sp. *apii, Septoria apii* |
| Lettuce | Lettuce mosaic virus |
| Apple | *Phytophthora cactorum* |
| Banana | *F. oxysporum* f.sp. *cubense* |
| Peach | *Xanthomonas campestris* pv *pruni, Pseudomonas syringae* pv *syringae* |
| Poplar | *Septoria musiva, Melamspora medusae* |

Somaclonal variation without *in vitro* selection system has the same approach as that for conventional or mutation breeding for screening of disease resistant variants.

The most successful strategy employed so far to select *in vitro* for resistance is that in which either the pathotoxin or partially purified culture filtrate of the pathogen is used as a screening agent. Many facultative fungal and bacterial plant pathogens produce low molecular weight toxic metabolites *in vitro* and *in vivo*. The induction of chlorosis and/or necrosis by a pathogen in a susceptible host suggests that toxins are involved. Carlson (1973) first tested this possibility of *in vitro* selection for disease resistance for wild fire disease of tobacco caused by *Pseudomonas syringae* pv *tabaci*. Haploid cells and protoplasts were first treated with ethylmethane sulfonate mutagen and then cultured in the presence of methyl sulphoximine (MSO) thought to be an analogue of tabtoxin which gave chlorotic halos on tobacco leaves. Most of the calli obtained from the surviving cells soon lost their ability to grow in the presence of MSO. However, three gave resistant plants that were also resistant to tabtoxin. Later, Gengenbach *et al.* (1977) used recurrent selection to select variants carrying resistance to T-toxin of *Helminthosporium maydis*. A very high level of resistance to T-toxin was selected in originally susceptible TmS corn lines by serial subculture of embryo derived callus cultures on a medium containing a just sublethal dose of T-toxin. Most significantly, plants regenerated from resistant TmS cultures proved to be fully toxin resistant. The character was transmitted to the progeny of regenerated plants through the female parent, and gene for this is located on the mitochondrial genome.

The work with T-toxin had a highly significant enriching effect for *in vitro* resistance to a pathotoxin. Nevertheless, several studies have strongly suggested that purified or partially purified toxin preparations may be used to select for novel disease resistance *in vitro*. Behnke (1979) was the first to perform experiments on calli of dihaploid potato clones using a medium containing crude culture filtrate extract of the fungal media employed for *Phytophthora infestans*. However, subsequent workers used purified culture filtrate or crude toxin for *in vitro* selection experiments.

Chawla and Wenzel (1987a, 1987b) used immature embryo derived calli of barley and wheat for screening against crude toxin isolated from culture filtrate of *Helminthosporium sativum* and also synthetic fusaric acid toxin of *Fusarium* species. Two selection methods were employed (a) continuous: 4 selection cycles were performed continuously on toxic media and (b) discontinuous: calluses were transferred to non-toxic media after 2nd or 3rd selection cycle but four cycles were performed on toxic media. Discontinuous approach for *in vitro* selection resulted in a larger number of resistant plants. Table 10.4 lists a few successful examples of *in vitro* selection for disease resistance in crop plants.

Selection of somaclonal variants with disease resistance has resulted in the release of only a few cultivars or genotypes, *viz.* sugarcane cv. with resistance to Fiji disease, tomato variety DNAP-17 a somaclonal variant with monogenic *Fusarium* wilt race 2 resistance, celery line UC-T3 somaclone with resistance to *F. oxysporum* f. sp. *Apii* (Heath-Pagliuso and Rappaport, 1990).

**3. *Somaclonal variation and abiotic stress resistance*:** Somaclonal variation has resulted in several interesting biochemical mutants, which are being successfully used in plant metabolic pathway studies, i.e. amino acid and secondary metabolic pathways. Investigations have shown that the level of free amino acids, especially proline, increases during cold hardening. *In vitro* selection has also been used to obtain plants with increased acid soil, salt, aluminium and herbicide resistance.

***Cold tolerance*:** Lazar *et al.* (1988) developed somaclonal variants for freezing tolerance in Norstar winter wheat. A significant positive correlation between proline level and frost tolerance has been found in a broad spectrum of genotypes. *In vitro* selection and regeneration of hydroxyproline resistant lines of winter wheat with increased frost tolerance and increased proline content has been reported (Dorffling *et al.*, 1997). The results showed

**Table 10.4:** Disease resistant crop plants obtained by *in vitro* selection

| Crop | Pathogen | Selection agent |
|---|---|---|
| **Food crops** | | |
| Rape seed | *Phoma lingam, Alternaria brassicicola* | CF |
| Rice | *Helminthosporium oryzae* | CT |
| | *Xanthomonas oryzae* | Bacterial cells |
| Barley | *Helminthosporium sativum* | CT |
| | *Fusarium* spp. | Fusaric acid |
| Maize | *Helminthosporium maydis* | Hm T toxin |
| Oats | *Helminthosporium victoriae* | Victorin |
| Wheat | *Helminthosporium sativum, Fusarium graminearum* | CT |
| | *Pseudomonas syringae* | Syringomycin |
| Sugar-cane | *Helminthosporium sacchari* | Toxin |
| **Horticultural and other crops** | | |
| Tobacco | *Pseudomonas syringae* pv. *tabaci* | Methionine sulfoximine |
| | *Alternaria alternata; P. syringae* pv *tabaci* | Toxin |
| | Tobacco mosaic virus | Virus |
| | *Fusarium oxysporum* f.sp. *nicotianae* | CF |
| Potato | *Phytophthora infestans, Fusarium oxysporum* | CF |
| | *Erwinia carotovora* | Pathogen |
| Alfalfa | *F .oxysporum* f.sp. *medicagnis* | CF |
| Tomato | Tobacco mosaic virus X | Virus |
| | *Pseudomonas solanacearum* | CF |
| Eggplant | *Verticillium dahlie* | CF |
| | Little leaf disease | Pathogen |
| Peach | *Xanthomonas campestris* pv. *pruni* | CF |
| Hop | *Verticillium albo-atrum* | CF |
| Celery | *Septoria apiicola* | CF |

CF: culture filtrate; CT: crude toxin

strong correlation of increased frost tolerance with increased proline content.

**Salt tolerance:** Plant tissue culture techniques have been successfully used to obtain salt tolerant cell lines or variants in several plant species, *viz.* tobacco, alfalfa, rice, maize, *Brassica juncea, Solanum nigrum,* sorghum, etc. In most cases, the development of cellular salt tolerance has been a barrier for successful plant regeneration, or if plants have been obtained they did not inherit the salt tolerance. Only in a few cases it was possible to regenerate salt tolerant plants. Mandal *et al.* (1999) developed a salt tolerant somaclone BTS24 from indigenous rice cultivar Pokkali. This somaclone yielded 36.3 q/ha under salt stress conditions as compared to 44 q/ha under normal soil. Some of the reports for successful production of healthy, fertile and genetically stable salt tolerant regenerants from various explants, level of selection and other details in various species have been summarized in Table 10.5.

**Aluminium tolerance:** It has long been established that plant species or cultivars greatly differ in their resistance to aluminium stress. In recent years, considerable research has been focused on the understanding of physiological, genetic and molecular processes that lead to aluminium tolerance. Despite the problems encountered in adapting culture media for *in vitro* selection for aluminium resistance, cell lines have been isolated in several species, e.g. alfalafa, carrot, sorghum, tomato, tobacco. Aluminium tolerant somaclonal variants from cell cultures of carrot were selected by exposing the cells to excess ionic aluminium in the form of aluminium chloride (Arihara *et al.*, 1991). Jan *et al.* (1997) elicited aluminium toxicity during *in vitro* selection in rice by making several modifications in the media *viz.* low pH, low phosphate and calcium concentrations, and unchelated iron and aluminium along with aluminium sulphate. After selection on aluminium toxic media, callus lines were maintained on aluminium toxic free media and 9 tolerant plants were obtained. Transmission of aluminium resistance character was identified till the fourth generation.

**Drought tolerance:** Wang *et al.* (1997) reported a sorghum somaclonal variant line (R111) resistant to drought stress. A novel hybrid was developed by crossing R111 with several sterile lines, indicating that selection of somaclonal variants is an effective method for creating new sources of genetic variation. Drought tolerant rice lines were obtained by *in vitro* selection of seed induced callus on a media containing polyethylene glycol as a selective agent which simulated the effect of drought in tissue culture conditions (Adkins *et al.*, 1995)

**4. Herbicide resistance:** Through *in vitro* selection several cell lines resistant to herbicides have been isolated and a few have been regenerated into complete plants. Among the important achievements are tobacco, soybean, wheat, maize, etc. resistant to various herbicides such as glyphosate, sulfonylurea, imidazolinones, etc. Some of the examples have been reported in Table 10.6.

**5. Insect resistance:** Zemetra *et al.* (1993) used *in vitro* selection technique for generation of somaclonal variants for Russian wheat aphid (*Diuraphis noxia*) in wheat. Calli from susceptible wheat cultivar "Stephens" were exposed to an extract from aphid. Resistance to aphid was observed in both $R_2$ and $R_3$ generations. Croughan *et al.* (1994) reported variant of Bermuda grass (*Cynodon dactylon*) called *Brazos R-3* with increased resistance to fall armyworm.

**6. Seed quality:** Recently, a variety Bio L 212 of lathyrus (*Lathyrus sativa*) has been identified for cultivation in central India which has been developed through somaclonal variation and has low ODAP ($\beta$-N-oxalyl -2-$\alpha$, $\beta$ diamino propionic acid), a neurotoxin (Mehta and Santha, 1996), indicating the potential of somaclonal variations for the development of varieties with improved seed quality.

**7. Alien gene introgression:** The increase in genome rearrangement during tissue culture provides a new opportunity for alien gene

**Table 10.5:** *In vitro* selection and regeneration of salt tolerant plants

| Plant | Explant | Salt and Conc. (g/l) | Selection | Inheritance |
|---|---|---|---|---|
| Rice | Embryogenic callus | NaCl (10,20) | One step | $R_0$, $R_1$, $R_2$, $R_3$ and $R_4$ |
| Rice | Mature seeds | NaCl (15) | One cycle | $R_0$, $R_1$ |
| Wheat | Immature embryo | NaCl (2–7) | Two steps | $R_0$, $R_1$ |
| Wheat | Mature embryo | NaCl (5) | Several cycles | $R_0$, $R_1$ |
| Canola | Morphogenic cotyledons | NaCl (7.5 –10) | One step | $R_0$, $R_1$ |
| Canola | Somatic embryos | NaCl (up to 12.5) | Two cycles | $R_0$, $R_1$ |
| *Brassica napus* | Microspore derived embryos | NaCl (6–7)) | One step | $R_0$ |
| *Vigna radiata* | Cotyledons from *in vitro* seedlings | NaCl (up to 150 mM) | Two steps | $R_0$, $R_1$ |
| Alfalfa | Immature ovary derived callus | NaCl (10) | One step | $R_0$, $R_1$, $R_2$ |
| Tobacco | Protoplasts (haploid) | NaCl (11.7) | Short term | $R_0$ × wild type × $F_1$ |
| Flax | Callus | Sulphate salt (26.8) | One step | $R_0$, $R_1$, $R_2$ |
| Coleus | Leaf discs from *in vitro* multiplied shoots | NaCl (5.3) | One step | $R_0$, $R_1$ |

**Table 10.6:** Some examples of somaclonal variants for herbicide resistance

| Crop | Herbicide | Reference |
|---|---|---|
| *Glycine max* | Imazethapyr | Tareghyan *et al.* (1995) |
| *Nicotiana tabacum* | Glyphosate | Stefanov *et al.* (1996) |
| *Gossypium hirsutum* | Sulfonylurea, Imidazolinone | Rajasekaran *et al.* (1996) |
| *Zea mays* | Sethoxydim, cycloxydim | Landes *et al.* (1996) |
| *Zea mays* | Glyphosate | Forlani *et al.* (1992) |
| *Triticum aestivum* | Difenzoquat | Bozorgipour and Snape (1997) |
| *Beta vulgaris* | Imidazolinone | Wright *et al.* (1998) |
| *Datura innoxia* | Chlorosulfuron | Rathinasabpathi and King (1992) |

introgression which can help widen the crop germplasm base, particularly by culturing immature embryos of wide crosses where crop and alien chromosomes cannot replicate through meiosis. Introgression of genes may be better achieved by imposing one or more cell culture cycles on interspecific hybrid material. The enhanced somatic genome exchange is likely to produce regenerants where part of the alien genome has been somatically recombined into the chromosomes of crop species.

# BASIS OF SOMACLONAL VARIATION

*Karyotype changes*: Variant plants with altered chromosome number have been reported by several workers. Polyploidy is the most frequently observed chromosome abnormality e.g. aneuploidy in oats (McCoy, 1982) potato (Sree Ramulu *et al.*, 1985), barley (Kole and

Chawla, 1993). The change in chromosome number in a variant plant is commonly associated with reduced fertility and with altered genetic ratios in the progeny of self-fertilized plants.

*Changes in chromosome structure*: In contrast to gross changes in chromosome number, more cryptic chromosome rearrangements may be responsible for genetic variation. Translocations have been reported in potato, ryegrass, oats, etc. Published work also suggests that chromosome deletions, duplications, inversions and other minor reciprocal and non-reciprocal rearrangements occur among regenerated plants e.g. ryegrass, barley. Chromosome irregularities such as breaks, acentric and centric fragments, ring chromosomes and micronuclei were observed in garlic somaclones. Cytological abnormalities such as multilobed nuclei, multinucleate cells, abnormal anaphase and mixoploidy were observed in $SC_3$ generation of barley.

**Single gene mutations**: Single gene mutations have been detected by several workers (Edallo *et al.*, 1981; Oono, 1978). Recessive single gene mutations are suspected if the variant does not appear in $R_o$ or $SC_1$ generation but the self fertilized $R_1$ or $SC_2$ progeny segregate in 3:1 Mendelian ratio for the trait of interest. This type of analysis has been reported for maize, *Nicotiana sylvestris*, rice, wheat, etc.

**Cytoplasmic genetic changes**: The most detailed work on this has been done on maize. Many studies have implicated mitochondria as the determinants of male sterility and *Drechslera maydis* T toxin sensitivity in maize. Gengenbach *et al.* (1977) and Brettell *et al.* (1979) regenerated plants that were resistant to fungal T toxin. The original lines used had all the Texas male sterile cytoplasm, which results in plants being normally highly susceptible to this disease. However, when selection pressure was applied by subculturing callus cultures in the presence of toxin, some regenerants were found to be not only resistant to the toxin but also fertile. When restriction analysis of the mitochondrial DNA (mtDNA) was evaluated it was evident that significant changes had occurred in the mtDNA of plants derived from cell cultures. Molecular analyses based on restriction endonuclease digest patterns of mitochondrial and chloroplast genomes of regenerated potato plants derived from protoplasts have revealed significant change in mitochondrial genomes but not in chloroplast genomes (Kemble and Shepard, 1994). It was suggested that the variation results from substantial DNA sequence rearrangements (e.g. deletion, addition and intramolecular recombinations) and cannot be explained by simple point mutations. However, point mutations can also occur in the genomes of organelles of regenerated plants. Variations in chloroplast DNA have been detected in tomato and wheat somaclones.

**Mitotic crossing over**: Mitotic crossing over (MCO) could also account for some of the variation detected in regenerated plants. This could include both symmetric and asymmetric recombination. MCO may account for the recovery of homozygous recessive single gene mutations in some regenerated plants. Somatic cell sister chromatid exchange, if it is asymmetric can also lead to deletion and duplication of genetic material and hence variation.

**Gene amplification and nuclear changes**: Studies have shown that nuclear genes are affected by tissue culture stress. De Paepe *et al.* (1982) have shown that heritable quantitative and qualitative changes can be observed in the nuclear DNA content of doubled haploid *Nicotiana sylvestris* obtained from pollen cultured plants. These plants contain generally increased amounts of total DNA and an increasing proportion of highly repeated sequences. Both AT and GC rich fractions are amplified in tissue culture derived plants. Durante *et al.* (1983) showed similar amplification of AT and GC rich fractions in DNA from *Nicotiana glauca* pith explants within hours of culture. These studies suggest the presence of a differential replication process during the early stages of dedifferentiation. It has been seen that plant cells like those of other eukaryotes can increase or decrease the quantity of a specific gene product by differential gene amplification and diminution. For instance, ribosomal RNA gene amplification and diminution are now known to be widespread in wheat, rye, and tobacco, and in flax, ribosomal DNA is known to alter directly in response to environmental and cultural pressures. Reduction in ribosomal RNA genes (rDNA) has been found in potato plants regenerated from protoplasts. Additionally, both structural rearrangement within rDNA and methylation of nucleotide sequences of rDNA have been observed (Brown, 1989; Harding, 1991).

**Transposable elements**: Several authors have speculated that transposable elements may also be responsible for somaclonal variation. Heterozygous light green (Su/su) somaclones with a high frequency of coloured spots on the leaf surface have been detected for a clone of *N. tabacum* (Lorz and Scowcroft, 1983) and for a *N. tabacum* + *N. sylvestris* somatic hybrid. The somatic hybrid has an unstable pattern of inheritance that would be consistent with an unstable gene. A causal relationship between genetic instability possibly related to tissue culture induced transposition and somaclonal variation was speculated.

Recently, genomic changes in a maize line and mutations in tobacco line due to activation of transposable elements under *in vitro* tissue culture conditions have been demonstrated (Peschke and Phillips, 1992). It has also been shown that several types and copies of non-active transposable elements, including retro-transposable elements, are present in the genome of potato (Flavell *et al.*, 1992). Thus, it is likely that a few active transposable elements may also be present in potato and could be responsible for generating some degree of somaclonal variation in regenerated potato plants.

**DNA methylation:** DNA methylation plays an important role in the regulation of gene expression and its implication in somaclonal variation. Many genes show a pattern in which the state of methylation is constant at most sites but varies at others. A majority of sites are methylated in tissues in which the gene is not expressed but non-methylated in tissues where the gene is active. Thus, an active gene may be described as unmethylated. A reduction in the level of methylation (or even the removal of methyl groups from some stretch of DNA) is part of some structural change needed to permit transcription to proceed. Significant changes in the methylation level of genomic DNA during dedifferentiation and somatic embryogenesis have been reported for higher plants. Wherever the gene expression and its regulation is playing a role, DNA methylation becomes an important factor for getting the somaclonal variants. Unstability of gene may arise through the methylation pattern of that particular gene and thus a causal relationship between genetic instability possibly related to tissue culture-induced transposition, and somaclonal variation.

In higher plants many controlling factors act together to achieve the desired gene regulation. One of these levels of control is provided by adding a small "tag" called a methyl group onto "C" one of the bases that make up the DNA code. The methyl group tagged C's can be written as mC. Simpler organisms, such as many types of bacteria and the single celled yeast, usually do not use methyl group tagged C's in regulating their genes, some bacteria but not all use methyl group tagged A's "mA" for this purpose. However, most bacteria have specific patterns of mC and mA for this purpose i.e. for signaling in their DNA called methylated DNA. Thus the process in which the methyl group is added to certain bases as in cytoine (C) and adenine (A) by the help of enzyme DNA methylase at the $C_5$ atom of cytosine and $N_6$ position of adenine is called methylation. When only one strand of DNA is methylated it is called hemimethylated DNA if it is in both strands of DNA this is called fully methylated DNA. In hemimethylated condition gene is able to express while in fully methylated state gene cannot be expressed. In general rule, methylation prevents the gene expression but the de-methylation respond to the expression of a gene. Hence in terms of somaclonal variation (i.e. variation occurs in the *in vitro* culture of plants) the DNA methylation by regulating the gene expression and sometimes by inducing the mutations may cause variation in the culture.

In *Arabidopsis thaliana*, a new variant has been developed from the normal one this is a tiny dwarf plant, shriveled, a mere shadow of its genetically identical neighbour This dwarf plant is named as "bal", because of its shape. It constantly perceives a pathogen attack even thought it has the exact same DNA sequence of parental plant. It was found that the difference between the two plants is not due to changes in the DNA sequence but actually the 'bal dwarf' is caused by increased activation of a single gene. This is something looks like a mutation and behaves like a mutation but it is actually caused by the packaging of DNA (caused basically by DNA methylation) rather than change in the DNA sequence. The gene affected in 'bal variant' is involved in the disease resistance and is called an R-gene. In 'bal plant' the R-gene is more active and consequently the plant defense system becomes hyperactivated, constantly fighting off disease even when no pathogens are present to pose a threat. The resulting dwarfed plant is more resistant to bacterial infection. The precise molecular change leading to the increased R-gene activation is most probably the changes in DNA methylation, chemical modification of cytosine.

The occurrence of somaclonal variants (ca 5%) among populations of somatic embryo derived oil palms (*Elaeis guineensis* Jacq.) has been reported which hampers the scaling-up of clonal plant production. Variant "mantled" palms show a feminisation of flower male parts, which can result in complete sterility in the most severe cases. Given the observed features of the abnormality (spatial heterogeneity, temporal instability, non-Mendelian sexual transmission, absence of any detectable defect in DNA organization as revealed by AFLP/RAPD studies) the hypothesis of an epigenetic alteration of genome expression has been proposed to explain the origin of the variant plants. The relationship between the "mantled" variation and hypomethylation of genomic DNA has been assessed, using two complementary approaches: HPLC quantification of relative amounts of 5-methyl-deoxycytidine and the Methylase Accepting Assay (MAA), based on the enzymatic saturation of CG sites with methyl groups (Rival et al., 2000).

**Disadvantages**: The following are the disadvantages associated with somaclonal variation.

1. Uncontrollable and unpredictable nature of variation and most of the variations are of no apparent use.
2. The variation is cultivar dependent.
3. The variation obtained is not always stable and heritable. The changes occur at variable frequencies.

4. Not all the changes obtained are novel. In majority of cases, improved variants have not been selected for breeding purposes.

# GAMETOCLONAL VARIATION

The variation observed among plants regenerated from cultured gametic cells is termed gametoclonal variation as compared to somaclonal variation that is somatic tissue derived variation. Gametes are products of meiosis, governed by Mendel's laws of segregation and independent assortment. There are three genetic differences which can be pointed as evidence for gametoclonal and somaclonal variations being distinct.

i. Both dominant and recessive mutants induced by gametoclonal variation will be expressed directly in haploid generated plants, since only a single copy of each gene is present. Hence, regenerated gametoclones ($R_0$) can be analyzed directly to identify new variants.

ii. Recombinational events that are recovered in gametoclones would be the result of meiotic crossing over and not mitotic crossing over.

iii. To use gametoclones, chromosome number must be doubled. The most frequently used method to double chromosomes is treatment with colchicine.

Gametoclones can be produced through the culture of either male or female gametic cells, or their derivatives. The most successful and commonly used method for a wide range of species is culture of anthers, or isolated microspores. Anther and microspore culture technology has been well established in many plant species for the production of doubled haploid plants and also for *in vitro* screening of microspores against stresses. Since growth of cells in culture results in single gene mutations, culture of microspores could result in recessive mutations and these can be *in vitro* selected against various stresses. Hence, it is likely that

**Table 10.7:** Variation among plants regenerated from gametophytic tissue culture

| Crop | Characteristics |
|---|---|
| *Brassica napus* | Time to flower, glucosinolate content, leaf shape and color, flower type, pod size and shape |
| *Hordeum vulgare* | Plant height, yield, fertility, days to maturity |
| *Nicotiana sylvestris* | Leaf color, leaf shape, growth rate, DNA content, repeated sequences of DNA |
| *N. tabacum* | TMV resistance, root knot resistance Time of flowering, plant size, leaf shape, alkaloid content, number of leaves Altered temperature sensitivity |
| *Oryza sativa* | Seed size, seed protein level, plant height, level of tillering Waxy mutant |
| *Saintpaulia ionantha* | RuBSase activity |

gametoclonal variants can be detected in the $R_0$ generation, while evaluation of $R_1$ progeny must be completed to detect the full spectrum of variants. It has been observed that mutant cells probably do not regenerate as well as wild type cells. Hence, the mutation spectrum and frequency obtained from regenerated haploids may be strikingly different from that of regenerated diploid tissue.

When referring to gametoclonal variation, one must recognize four distinct sources of variation.

i) New genetic variation induced by cell culture procedures.

ii) Variation resulting from segregation and independent assortment.

iii) New variation at the haploid level induced by the chromosome doubling procedure.

iv) New variation induced at the diploid level resulting in heterozygosity.

In crops such as barley, wheat, rice, *Brassica* spp., tobacco, potato, etc., gametoclonal variants have been generated and used to accelerate and enrich the breeding programmes. Infact, the first report on selection of disease resistant mutants by Carlson (1973) utilized haploid tissue of tobacco. Now, there have been several reports of variation in plants regenerated from gametophytic tissue and some have been documented in Table 10.7.

**REFERENCES**

Adkins, S.W., Kunanuatchaidach, R. and Godwin, I.D. 1995. Somaclonal variation in rice—drought tolerance and other agronomic characters. *Aus. J. Bot.* **43:** 201–209.

Arihara, A., Kumagai, R., Koyama, H and Ojima, K. 1991. Aluminium tolerance of carrot (*Daucua carota* L.) plants regenerated from selected cell cultures. *Soil Sci. Plant Nutr.* **37:** 699–705.

Behenke, M. 1979. Selection of potato callus for resistance to culture filtrates of *Phytopthora infestans* and regeneration of resistant plants. *Theor. Appl. Genet.* **55:** 69–71.

Bozorgipour, R. and Snape, J.W. 1997. An assessment of somaclonal variation as a breeding tool for generating herbicide tolerant genotypes in wheat (*Triticum aestivum* L.). *Euphytica* **94:** 335–340.

Brettell, R.I.S., Goddard, B.V.D. and Ingram, D.S. 1979. Selection of Tms-cytoplasm maize tissue cultures resistant to *Drechslera maydis* T-toxin. *Maydica* **24:** 203–213.

Brown, P.T.H. 1989. DNA methylation in plants and its role in tissue culture. *Genome* **31:** 717–729.

Carlscn, P. 1973. Methionine sulfoximine resistant mutants of tobacco. *Science*, **180:** 1366.

Chaleff, R.S. 1981. *Genetics of Higher Plants.* Cambridge Univ Press, Cambridge.

Chawla, H.S. and Wenzel, G. 1987a. *In vitro* selection of barley and wheat for resistance against *Helminthosporium sativum*. *Theor. Appl. Genet.* **74:** 841–845.

Chawla, H.S. and Wenzel, G. 1987b. *In vitro* selection for fusaric acid resistant barley plants. *Plant Breed.* **99:** 159–163.

Croughan, S.S., Quisenbery, S.S., Eichchorn, M.M., Gloyer, P.D. and Brown, T.F. 1994. Registration of Brazos-R3 Bermudagrass germplasm. *Crop Sci.* **34:** 542.

De Paepe, R., Prat, D. and Huguet, T. 1982. Heritable nuclear DNA changes in doubled haploid plants obtained from pollen cultures of *Nicotiana sylvestris*. *Plant Sci Lett.* **28:** 11–28.

Dorffling, K., Dorffling, H., Lesselich, G., Luck, E., Zimmerman, C., Melz, G. and Jurgens, H.U. 1997. Heritable improvement of frost tolerance in winter wheat by *in vitro* selection of hydroxyproline resistant proline overproducing mutants. *Euphytica* **93:** 1–10.

Durante, M., Geri, C., Grisvad, J., Guille, E., Parenti, R.,and Buaitti, M. 1983.Variation in DNA complexity in *Nicotiana glauca* tissue cultures. I. Pith dedifferentiation *in vitro*. *Protoplasma* **114:** 114–118.

Edallo, S., Zucchinali, C., Perenzin, M. and Salamini, F. 1981. Chromosomal variation and frequency of spontaneous mutation associated with *in vitro* culture and plant regeneration in maize. *Maydica* **26:** 39–56.

Flavell, A.J., Smith, D.B. and Kumar, A. 1992. Extreme heterogeneity of Ty1-Copia group transposons in plants. *Mol Gen Genet.* **231:** 233–242.

Forlani, G., Nielsen, E. and Racchi, M.L. 1992. A glyphosate resistant 5-enol pyruvyl shikimate 3 phosphate synthase confers tolerance to a maize cell line. *Plant Science Limerick* **85:** 9–15.

Gengenbach, B.G., Green, C.E., Donovan, C.D. 1977. Inheritance of selected pathotoxin resistance in maize plants regenerated from cell cultures. *Proc Natl. Acad. Sci. USA* **74:** 5113–5117.

Harding, K. 1991. Molecular stability of the ribosomal RNA genes in *Solanum tuberosum* plants recovered from slow growth and cryopreservation. *Euphytica* **55:** 141–146.

Heath-Pagliuso, S. and Rappaport, L. 1990. Somaclonal variant UC-T3: The expression of *Fusarium* wilt resistance in progeny arrays of celery, *Apium graveolens* L. *Theor. Appl. Genet.* **80:** 390–394.

Heszky, L.E. and Simon-Kiss, I. 1992. 'DAMA', the first plant variety of biotechnology origin in Hungary, registered in 1992. *Hung. Agric. Res.***1:** 30–32.

Jan, V.V.S., Macedo, C.C. de., Kinet, J.M. and Bouharmont, J. 1997. Selection of Al-resistant plants from a sensitive rice cultivar, using somaclonal variation, *in vitro* and hydroponic cultures. *Euphytica* **97:** 303–310.

Kemble, R.J. and Shepard, J.F. 1984. Cytoplasmic DNA variation in a protoclonal population. *Theor. Appl. Genet.* **69:** 211–216.

Kole, P.C. and Chawla, H.S. 1993. Variation of *Helminthosporium* resistance and biochemical and cytological characteristics in somaclonal generations of barley. *Biologia Plantarum* **35:** 81–86.

Landes, M., Walther, H., Gerber, M., Auxier, B., Brown, H. 1996. New possibilities for post emergence grass weed control with cycloxydim and sethoxydim in herbicide tolerant corn hybrids. In: *Proc. Second Int. weed Control Congress,* Denmark pp. 869–874.

Larkin, P.J. and Scowcroft, W.R. 1981. Somaclonal variation—a novel source of variability from cell cultures for plant improvement. *Theor. Appl. Genet.* **60:** 197–214.

Lazar, M.D., Chen, T.H.H., Gusta, L.V. and Kartha, K.K. 1988. Somaclonal variation for freezing tolerance in a population derived from Norstar winter wheat. *Theor. Appl. Genet.* **75:** 480–484.

Lorz, H. and Scowcroft, W.R. 1983. Variability among plants and their progeny regenerated from protoplasts of Su/ su heterozygotes of *Nicotiana tabacum. Theor. Appl. Genet.* **66:** 67–75.

Mandal, A.B., Chowdhury, B. and Sheeja, T.E. 1999. Development and characterization of salt tolerant somaclones in rice cultivar pokkali. *Ind. J. Expt. Biol.* **38:** 74–79.

McCoy, T.J. and Philips, R.L. 1982. Cytogenetic variation in tissue culture regenerated plants of *Avena sativa*: High frequency of chromosome loss. *Can. J. Genet. Cytol.* **24:** 37–50.

Mehta, S.L. and Santha, I.M. 1996. Plant biotechnology for development of non-toxin strains of *Lathyrus sativus.* In: *Summer short course on "Enhancement of Genetic Yield Potential of Pulse Crops.* Kanpur, India, pp. 65–73.

Oono, K. 1978. Test tube breeding of rice by tissue culture. Proc Symp Methods in Crop Breeding. *Tropical Agriculture Res Series No. 11.* Ministry of Agriculture and Forestry, Ibaraki, Japan.

Peschke, V.M. and Phillips, R.L. 1992. Genetic variation of somaclonal variation in plants. *Advances in Genetics* **30:** 41–75.

Rajasekaran, K., Grula, J.W. and Anderson, D.M. 1996. Selection and characterization of mutant cotton (*Gossypium hirsutum* L.) cell lines resistant to sulfonylurea and Imidazolinone herbicides. *Plant Science Limerick* **119:** 115–124.

Rathinasabapathi, B. and King, J. 1992. Physiological response of herbicide resistant cell variants of *Datura innoxia* to branched chain amino acids. *Plant Science Limerick* **81:** 191–198.

Rival, A., Jaligot, E., Laroche, A., Beule, T., Oakeley, E.J., Verdeil, J.L. and Tregear, J. 2000. DNA methylation v/s somaclonal variation in oil palm. In: Int. Symp. Tropical and subtropical fruits. p. 575.

Skoog, F. 1944. Growth and organ formation in tobacco tissue cultures.

Sree Ramulu, K., Dijkhuis, P., Hanisch ten Cate,Ch.H. and Groot de,B. 1985. Patterns of DNA and chromosome variation during *in vitro* growth in various genotypes of potato. *Plant Sci.* **41:** 69–78.

Stefanov, K.L., Djilianov, D.L., Vassileva, Z.Y., Batchvarova, R.B., Atanassov, A.I., Kuleva, L.V. and Popov, S.S. 1996. Changes in lipid composition after *in vitro* selection for glyphosate tolerance in tobacco. *Pesticide Sci.* **46:** 369–374.

Sun, Z.X., Zhao, C.Z., Zheng, K.L., Qi, X.F. and Fu, Y.P. 1983. Somaclonal genetics of rice, *Oryza sativa* L. *Theor. Appl. Genet.* **67:** 67–73.

Tareghyan, M.R., Collin, H.A., Putwain, P.D. and Mortimer, A.M. 1995. Characterization of somaclones of soybean resistant to imazethapyr. In: *Brighton crop protection conference: weeds. Proc. Int. Conf,* Brighton, U.K., pp. 375–380.

Wang, C.X., Wang, L.Q., Bai, Z., Wang, F., Wang, F., Zheng, L.P., Li, A.J. and Wang, C. 1997. Tissue culture selection of somaclonal variation line R111 of sorghum. *Acta Agriculturae Borealli-Sinica* **12:** 49–53.

Wright, T.R., Bascomb, N.F., Sturner, S.F. and Penner, D. 1998. Biochemical mechanism and molecular basis for ALS-inhibiting herbicide resistance in sugar beet (*Beta vulgaris*) somatic cell selections. *Weed Sci.* **46:** 13–23.

Zemetra, R.S., Schotzko, D.J., Smith, C.M. and Lauver, M. 1993. *In vitro* selection for Russian wheat aphid (*Diuraphis noxia*) resistance in wheat (*Triticum aestivum*). *Plant Cell Rep.* **12:** 312–315.

## FURTHER READING

Chawla, H.S. 1998. *Biotechnology in Crop Improvement*. International Book Distributing Co., Lucknow.

Evans, D.A., Sharp, W.R., Ammirato, P.V. and Yamada, Y. 1983. *Handbook of Plant Cell Culture*. Macmillan Publishing Company, New York.

Larkin, P.J. and Scowcroft, W.R. 1981. Somaclonal vatiaion—a novel source of variability from cell cultures for plant improvement. *Theor, Appl. Genet.*, **60**: 197–214.

Pierik, R.L.M. 1987. *In vitro culture of Higher Plants*. Martinus Nijhoff Publishers, Dordrecht.

Reinert, J. and Bajaj, P.P.S. (eds.) 1977. *Applied and Fundamental Aspects of Plant Cell, Tissue and Organ Culture*. Springer-Verlag, Berlin.

Vasil, I.K. 1984. *Cell Culture and Somatic Cell Genetics of Plants: Laboratory Procedures and Their Applications*. Academic Press, Inc., New York.

# 11

# Germplasm Storage and Cryopreservation

The aim of germplasm conservation is to ensure the availability of useful germplasm at any time. Germplasm is the sum total all genes and their alleles present in a crop and its related species. This is represented by a collection of various strains and related species of the concerned crop species. Breeding programs rely heavily on locally adapted ancient plant varieties and their wild relatives as sources of germplasm. Further, the continuing search for high yielding varieties of crop plants with resistance to pathogens and pests warrants the availability and maintenance of large collections of germplasm. Also, the modern agricultural practices and other developmental activities have caused a rapid decline in genetic variability, termed as genetic erosion. Due to genetic erosion, it is necessary to conserve germplasm, which could be done either *in situ* or *ex situ*. *In situ* germplasm conservation is done by protecting areas of diversity, which is generally ideal for wild relatives of crops. *Ex situ* germplasm conservation is done either in seed banks or field collection. In species which are propagated through seeds, it is economical to preserve the seeds. These seeds are dried to water content of 5–8% and then stored at a temperature of –18°C or lower and low humidity. Sometimes it is not feasible to store seeds because:

i) some plants do not produce fertile seeds;
ii) some seeds remain viable only for a limited duration;

iii) some seeds are very heterozygous and therefore, not suitable for maintaining true to type genotypes; and
iv) seeds of certain species deteriorate rapidly due to seed borne pathogens.

In case of vegetatively propagated species, preservation of germplasm heavily taxes manpower and land resources. As *in vitro* techniques are becoming more important in crop improvement through the use of somatic cell genetics, the material produced *in vitro* will have to be conserved *in vitro*. The *in vitro* system is extremely suitable for storage of plant material, since in principle it can be stored on a small scale, disease free and under conditions that limit growth. Germplasm storage *in vitro* is crucial for the future development and safety of agriculture.

The conservation of plants *in vitro* has a number of advantages over *in vivo* conservation. These are as follows:

i) *In vitro* culture enables plant species that are in danger of being extinct to be conserved.
ii) *In vitro* storage of vegetatively propagated plants can result in great savings in storage space and time.
iii) Sterile plants that cannot be reproduced generatively can be maintained *in vitro*.
iv) It is possible *in vitro* to efficiently reduce growth, which decreases the number of subcultures necessary.

v) If a sterile culture is obtained, often with great difficulty, then subculture *in vitro* is the only safe way of ensuring that it remains sterile. There are two main approaches for *in vitro* storage of germplasm.

# CRYOPRESERVATION

Cryopreservation (Gr., Kryos = frost) means "preservation in the frozen state". In practice, this is generally meant to be storage at very low temperatures, e.g. over solid carbon dioxide (–79°C), in low temperature deep freezers (–80°C or above), in vapor phase of nitrogen (–150°C) or in liquid nitrogen (–196°C). Generally the plant material is frozen and maintained at the temperature of liquid nitrogen (–196°C). At this temperature, the cells stay in a completely inactive state. The theoretical basis of freeze preservation is the transfer of water present in the cells to the solid state. While pure water becomes ice at 0°C,

the cell water needs a much lower temperature because of freezing point depression by salts or organic molecules. The lowest temperature at which water has been demonstrated to be liquid is –68°C. Cryopreservation of plant cells, organs and tissues received widespread importance mainly due to the emphasis placed in recent years on *in vitro* manipulations with cultured cells. Storage at very low temperature considerably slows down or halts metabolic processes and biological deterioration. Two major considerations have to be taken into account for freezing of specimens. (i) Susceptibility or degree of freeze tolerance displayed by a given genotype to reduced temperatures. (ii) The formation of ice crystals within the cells. The cryopreservation of plant cell culture and seeds and eventual regeneration of plants from them has been shown in Fig. 11.1, and it involves the following steps:

1. Raising sterile tissue cultures
2. Addition of cryoprotectants and pretreatment

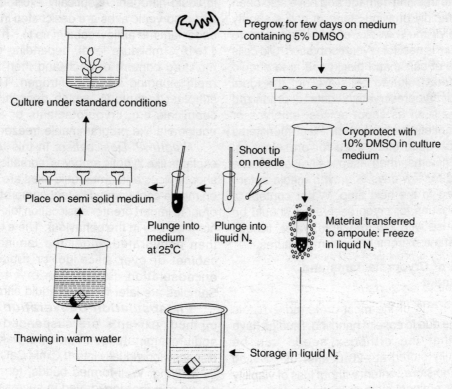

Culture under standard conditions

Pregrow for few days on medium containing 5% DMSO

Cryoprotect with 10% DMSO in culture medium

Shoot tip on needle

Place on semi solid medium

Plunge into medium at 25°C

Plunge into liquid N₂

Material transferred to ampoule: Freeze in liquid N₂

Thawing in warm water

Storage in liquid N₂

**Fig. 11.1:** Standard protocol for cryopreservation of shoot tips.

3. Freezing
4. Storage
5. Thawing
6. Determination of survival/viability
7. Plant growth and regeneration

## Raising Sterile Tissue Cultures

The morphological and physiological conditions of the plant material influence the ability of an explant to survive freezing at −196°C. Different types of tissues can be used for freezing such as the apical and lateral meristems, plant organs (embryos, endosperm, ovules, anther/pollen), seeds, cultured plant cells, somatic embryos, protoplasts, calluses, etc. In general, small, richly cytoplasmic, meristematic cells survive better than the larger, highly vacuolated cells. Cell suspensions have been successfully frozen in a number of species. These should be in the late lag phase or exponential phase. Callus tissues, particularly among tropical species, are more resistant to freezing damage and have also been used. A rapidly growing stage of callus shortly after one or more weeks of transplantation to the medium is best for cryopreservation. Old cells at the top of callus and blackened area should be avoided. Cultured cells are not the ideal system for cryopreservation. Instead, organized structures such as shoot apices, embryos or young plantlets are preferred. Plant meristems have been quite often used for preservation. Plant meristems when aseptically isolated are incubated for few days in growth media before proceeding to the next step. Water content of cells or tissues for cryopreservation should be low because with low freezable water, tissues can withstand extremely low temperatures.

## Addition of Cryoprotectants and Pretreatment

Seeds are one of the most preferred explants for storage due to ease of handling. Studies have shown that the orthodox seeds can be successfully stored at −20°C after desiccating to 5–7% moisture content without loss of viability using conventional storage methods. However, many crops and species, which are indigenous to the tropics and belong mainly to the category of recalcitrant seeds, do not produce orthodox seeds and hence are not amenable to the conventional storage methods e.g. Arecanut (*Areca catechu* L.), Black Pepper (*Piper nigrum* L), Cardamom (*Elettaria cardamomum* Maton), Coffee (*Coffea arabica* L. *C. canephora* Pierre *C. liberica* Bull), Coconut (*Cocos nucifera* L.), Oil palm (*Elaeis guineensis* Jacq), Rubber (*Hevea brasiliensis Muell*-Arg), Tea (*Camellia sinensis* L.O. Kuntze). These cannot be desiccated to low moisture content without substantial loss of viability and hence are difficult to conserve even for short periods at low temperatures. Thus the following pretreatment for desiccation can be used before freezing.

*Air Desiccation-Freezing*: In this technique, the seeds are desiccated to different target moisture contents by maintaining over charged silica gel in desiccators for 4–48 h before freezing in liquid nitrogen. Aseptically excised embryos and embryonic axes are desiccated in sterile air under laminar airflow cabinet up to 5 h to around 11–16% moisture level, depending on critical moisture content of each and then frozen by rapid plunging in liquid nitrogen. This is the simplest technique, which does not require chemicals e.g. cryoprotectants or expensive equipment like programmable freezer, etc.

*Pregrowth-Desiccation*: In this technique, explants like zygotic embryos, somatic embryos, shoot apices and embryonic axes are pregrown on media containing different cryoprotectants in order to impart greater desiccation tolerance and homogeneity in the behaviour. The explants are then dehydrated under the laminar airflow cabinet or over silica gel or processed by encapsulation-dehydration or vitrification. Samples are later frozen in liquid nitrogen.

***Encapsulation-Dehydration***: In this method, explants are suspended in 3–5% sodium alginate solution and picked up to dispense individually into 100 mM $CaCl_2$ followed by shaking. Well-formed beads, formed within 10–20 min are dehydrated in sucrose-enriched media for a minimum period of 17 h. Beads are later desiccated in laminar airflow cabinet before freezing in liquid nitrogen.

*Vitrification*: It involves the treatment of tissues with cryoprotectants in vitrification solution followed by fast freezing. Most commonly used vitrification solution is PVS2 (Plant Vitrification Solution 2) which contains glycerol 30%, dimethyl sulphoxide (DMSO) 15% and ethylene glycol 15%. Initially this procedure was developed for shoot apices, cell suspensions and somatic embryos but recently this technique has been applied to zygotic embryos and embryonic axes also.

**Addition of cryoprotectants**: The tissue is prepared for freezing by treating with various cryoprotectants to withstand the freezing conditions of −196°C. The effect of temperature on plants depends equally on genotype and environment, as well as physiological conditions. While genotype for practical purposes is an unchangeable factor, the other two factors can be altered according to circumstances. The freeze tolerance of a culture can be improved by timing the harvest of cells (in exponential phase) or by hardening process (growing the plants for shorter duration of one week at a lower temperature). Culturing of shoot apices for a brief period under optimum conditions before freezing has proved beneficial. Growing plants at 4°C for 3–7 days before taking explant also increases the survival of specimens. There are two potential sources of cell damage during freeze preservation.

I. Formation of large ice crystals inside the cells, leading to the rupture of organelles and the cell itself.

II. Intracellular concentration of solutes increases to toxic levels before or during freezing as a result of dehydration.

Addition of cryoprotectants controls the appearance of ice crystals in cells and protects the cells from the toxic solution effect. A large number of heterogeneous groups of compounds have been shown to possess cryoprotective properties with different efficiencies. These are glycerol, dimethyl sulfoxide (DMSO), glycols (ethylene, diethylene, propylene), acetamides, sucrose, mannose, ribose, glucose, polyvinylpyrrolidone, proline, etc. Most frequently used are DMSO, sucrose, glycerol and proline. Cryoprotectants depress both the freezing point and the supercooling point of water i.e. the temperature at which the homogeneous nucleation of ice occurs thus retarding the growth of ice crystal formation. DMSO has proved to be an excellent cryoprotectant because of its good properties. It has a low molecular weight, is easily miscible with the solvent, non-toxic at low concentrations, easily permeable into the cells and easily washable from the cells. Generally a concentration of 5–10% for DMSO and 10–20% for glycerol is adequate for most materials. In instances where application of a single cryoprotectant does not result in higher survival, a mixture of cryoprotectants has proven beneficial. For example, sorbitol does not enter the cells which would reduce cellular water content and hence decrease the rate of initial ice crystal formation. However, DMSO enters the cells and it would reduce cellular dehydration during freezing. Thus, initial ice crystal formation and subsequent dehydration during freezing are reduced when two different cryoprotectants are used.

In practice, the material suspended in the culture medium and treated with a suitable cryoprotectant is transferred to sterile polypropylene cryovials or ampoules with a screw cap. A dilute solution of the cryoprotectant (e.g. 5–10% DMSO) is added gradually or at intervals to the ampoules containing the tissue material to prevent plasmolysis and protects cells against osmotic shocks. After the last addition of cryoprotectant, but before cooling, there should be an interval of 20–30 min. The ampoules containing the cryoprotectants are then frozen.

**Freezing**

The type of crystal water within stored cells is very important for survival of the tissue. For this reason, three different types of freezing procedures have been developed. Various types of cryostats and freezing units are available by which different cooling rates can be easily regulated.

**(a)** *Rapid freezing*: The plant material is placed in vials and plunged into liquid nitrogen, which has a cooling rate of $-300$ to $-1000°C/$min or more. This method is technically simple and easy to handle. The quicker the freezing is done, the smaller the intracellular ice crystals are. A somewhat slower temperature decrease is achieved when the vial containing plant material is put in the atmosphere over liquid nitrogen ($-10$ to $-70°C/min$). Dry ice ($CO_2$) instead of nitrogen can also be used in a similar manner which has a temperature of $-79°C$. It is also possible that ultra-rapid cooling may prevent the growth of intra-cellular ice crystals by rapidly passing the cells through the temperature zone in which lethal ice crystal growth occurs. Rapid freezing has been employed to cryopreserve shoot tips of carnation, potato, strawberry and *Brassica napus* and somatic embryos.

**(b)** *Slow freezing*: In this method, the tissue is slowly frozen with a temperature decrease of $0.1–10°C/min$ from $0°C$ to $-100°C$ and then transferring to liquid nitrogen. Survival of cells frozen at slow freezing rates of $-0.1$ to $-10°C/min$ may have some beneficial effects of dehydration, which minimizes the amount of water that freezes intracellularly. Slow cooling permits the flow of water from the cells to the outside, thereby promoting extracellular ice formation instead of lethal intracellular freezing. It is generally oberved that upon extracellular freezing the cytoplasm will be effectively concentrated and plant cells will survive better when adequately dehydrated. Slow freezing is generally employed to freeze the tissues for which computer-programmed freezers have been developed. Successful examples of cryopreservation of meristems include diverse species as peas, strawberry, potato, cassava, etc.

**(c)** *Stepwise freezing*: This method combines the advantages of both rapid and slow methods. A slow freezing procedure down to $-20$ to $-40°C$, a stop for a period of time (approximately 30 min) and then additional rapid freezing to $-196°C$ is done by plunging in liquid nitrogen. A slow freezing procedure initially to $-20$ to $-40°C$ permits protective dehydration of the cells. An additional rapid freezing in liquid nitrogen prevents the growing of big ice crystals in the biochemically important structures. The stepwise freezing method gives excellent results in strawberry and with suspension cultures.

## Storage

Storage of frozen material at the correct temperature is as important as freezing. The frozen cells/tissues are immediately kept for storage at temperature ranging from $-70°C$ to $-196°C$. A liquid nitrogen refrigerator running at $-150°C$ in the vapor phase or $-196°C$ in the liquid phase is ideal for this purpose. The temperature should be sufficiently low for long-term storage of cells to stop all metabolic activities and prevent biochemical injury. Long-term storage is best done at $-196°C$. Thus as long as regular supply of liquid nitrogen is ensured in liquid nitrogen refrigerator, it is possible to maintain the frozen material with little further care. The injury caused to cells could be due to freezing or storage. The latter occurs primarily when the cells are not stored at sufficiently low temperatures.

## Thawing

Thawing is usually carried out by putting the ampoule—containing the sample in a warm water bath ($35–45°C$) for thawing. Frozen tips of the sample in tubes or ampoules are plunged into the warm water with a vigorous swirling wrist action just to the point of ice disappearance. It is important for survival of the tissue that the tubes should not be left in the warm bath after the ice melts. Just at the point of thawing, quickly transfer the tubes to a water bath maintained at room temperature ($20–25°C$) and continue the swirling action for 15 sec to cool the warm walls of the tubes. Then continue with washing and culture transfer procedures. It is necessary to avoid excessive damage to the fragile thawed tissues or cells by minimizing the amount of handling at pre-growth stage.

## Determination of Survival/Viability

Regrowth of plants from stored tissues or cells is the only realistic test of survival of plant materials. However, at any stage, the viability of frozen cells can be determined. Cell viability tests *viz*. FDA staining, growth measurements by employing parameters like cell number, packed cell volume, dry and fresh weight, mitotic index have already been discussed in Chapters 7 and 9 on cell suspension and protoplast isolation. Other staining methods of triphenyl tetrazolium chloride (TTC) and Evan's blue have been described.

### TTC method of staining

In this method, cell survival is estimated by the amount of formazan produced as a result of reduction of TTC. This reaction results in a pink colour. The procedure involves the following steps.

1. Buffer solution: 78% $Na_2HPO_4 \cdot 2H_2O$ solution 0.05 M (8.9 g/l):22% $KH_2PO_4$ solution 0.05 M (6.8 g/l).
2. TTC solution: 0.18M TTC dissolved in buffer solution.
3. About 150 mg of cell sample is put into 3 ml of TTC solution and incubated for 15 h at 30°C.
4. The TTC solution is drained off and the cells washed with distilled water.
5. Cells are centrifuged and extracted with 7 ml of ethanol (95%) in a water bath at 80°C for 5 min.
6. The extract is cooled and made to 10 ml volume with 95% ethanol.
7. The absorbance (pink colour) is then recorded with a spectrophotometer at 530 nm.

The amount of formazan produced by the frozen cells is expressed as a percentage (survival) of formazan produced by the control cell suspension.

### Evan's blue staining

One drop of 0.1% solution of solid Evan's blue in culture medium is added to one drop of cell suspension on a microscopic slide and observed under light microscope. Living cells remain unstained while dead cells are stained blue.

It should be noted that viability of cells based on staining reaction alone would be insufficient and misleading and cannot replace the plant regrowth experiments.

## Plant Growth and Regeneration

Major biophysical changes occur in the cell during freezing and thawing. The freshly thawed cell is strongly prone to further damage and requires appropriate nursing. Regrowth of cryopreserved materials would depend upon the manner in which specimens are handled, such as plating the cells without washing away the cryoprotectants, incorporation of special additives in regrowth medium and the physical environmental conditions the cells are exposed to during the early stages of regrowth. The cryoprotectants may be removed either by washing with the culture medium or by plating the thawed cultures on to a filter paper kept on the medium. Washing of cell suspensions and embryogenic callus cultures prevented their growth (e.g. *Gossypium hirsutum*), while without washing of culture is deleterious in *Saccharum*. Addition of certain compounds in the media during regrowth has shown increased surviving ability e.g. $GA_3$ in the medium for freeze preserved shoot tips of tomato, activated charcoal for carrot.

## SLOW GROWTH METHOD

Cells or tissues can be stored at non-freezing conditions in a slow growth state rather than at the optimum level of growth. This technique is comparable to the traditional technique of Japanese gardeners known as Bonsai, in which trees are kept alive in small pots over hundreds of years. Growing processes are reduced to a minimum by limitation of a combination of factors like temperature, nutrition medium, hormones, high osmoticum etc. By this method, water in the tissue is maintained in the liquid condition, but all biochemical processes are delayed. All cultures in slow growth are static and therefore require only shelf space in suitable environments. This approach can be applied to tissue cultures mainly by following methods:

(a) *Temperature*: The tissues (shoot tips) can be cultured at low temperatures of 1–4°C. The explants such as callus, shoot tip cultures, nodal segments and somatic embryos are established and maintained for 3–4 weeks under standard growth conditions. The explants are then transferred to lower temperatures of 1–4°C for cultures normally grown at 20–25°C and to 4–10°C for cultures normally grown at 30°C. A 16 hr photoperiod with reduced lighting (~50 lux) may be beneficial. Storage of cultured cells under low temperature has shown limited success. Cultures have been stored at low temperature for prolonged period of 9 months to 17 months in *Musa* and *Lolium*.

(b) *Low oxygen pressure*: The oxygen pressure can be regulated by atmospheric or partial pressure. Generally tissue growth reduces when oxygen pressure declines below 50 mm Hg and viability of callus culture increases when stored under 2–4 ml of mineral oil at 22°C. This is because low oxygen intake by cultured materials results in a decrease in $CO_2$ production during low oxygen storage. Consequently, the photosynthetic process is reduced in these tissues, thereby inhibiting their further growth. The idea that low partial pressures of oxygen may be advantageous for the storage of plant tissue culture was propounded by Caplin (1959). His experiment with carrot tissue cultures demonstrated that the amount of growth under mineral oil (liquid petroleum widely used for conservation of cultures of various microorganisms) is controlled by the supply of oxygen to the tissue. Since mineral oil prevented the diffusion of gases, therefore the level of oxygen below the oil reduced gradually due to intake by cultured tissues. This ultimately resulted in suspension of tissue growth.

The major limitation in using low oxygen storage for germplasm conservation is the long-term effects of low partial pressures of oxygen on plants. Studies on tobacco and chrysanthemums revealed that after the initial 6-week period there was considerable difference in the growth of stored materials. The medium desiccation observed in these treatments would further limit the storage time of cultures unless there is a control mechanism for monitoring the relative humidity within the storage chamber.

(c) *Nutrient restriction*: Lowering the nutrient concentration in the medium to 1/2 or 1/4th is useful in retarding the growth of cultures in some species. It is better if individual nutrients are restricted or altogether eliminated. For example, grape shoot apices grown on MS medium with 6% of ammonium nitrate showed 70–80% survival and greatly reduced the growth rate.

(d) *Growth regulators*: It is possible to use plant growth regulators like succinic acid, 2,2-dimethyl hydrazide (B-9), cycocel (CCC), abscisic acid (ABA), paclobutrazole, daminozide, tri-iodobenzoic acid (TIBA), etc. to delay culture growth. These compounds are generally used in the range of 10–20 mg/l. These chemicals cause reduced growth, mainly shorter internodes of the plants.

(e) *Osmotics*: Cultures can be grown on a medium supplemented with high levels of different osmoticums such as mannitol, sorbitol or sucrose with concentrations ranging from 3–6% (w/v). The high level of compounds acts as osmotic inhibitors, which slows down the growth.

By this method, virus free strawberry plants were maintained for 6 years at 4°C. Grape plants have been stored for 15 years at 9°C by yearly transfer to fresh medium. This method holds great promise in the industry employing micropropagation techniques. During the period of low demand for a particular genotype the cultures may be simply shelved in refrigerator to avoid frequent sub-culturing.

### Achievements

In India at National Bureau of Plant Genetic Resources the slow growth approach is being used for the maintenance of large number of cultures belonging to asexually propagated and woody crops (banana, potato, plantain, cassava, apple, pear). Besides, the material of some species is also conserved by the technique of cryopreservation for which protocols have been standardized.

## Applications

1. Conservation of genetic material: Large number of plant species have been successfully maintained by means of cryopreservation of cultured embryos, tissues, cells or protoplasts. A list of cryopreserved tropical plants has been shown in Table 11.1.
2. Freeze storage of cell cultures: A cell line to be maintained has to be subcultured and transferred periodically and repeatedly over an extended period of time. It also requires much space and manpower. Freeze preservation is an ideal approach to suppress cell division and to avoid the need for periodical sub-culturing.
3. Maintenance of disease free stocks: Pathogen-free stocks of rare plant materials could be frozen, revived and propagated when needed. This method would be ideal for international exchange of such materials.
4. Cold acclimation and frost resistance: Tissue cultures would provide a suitable material for selection of cold resistant mutant cell lines, which could later differentiate into frost resistant plants.

## Limitations

1. Greater skill in handling and maintenance of cultures is required.
2. Sophisticated facilities are required.
3. Plants may show genetic instability.
4. Cell/tissues get damaged during cryopreservation.
5. Cryopreservation procedures are genotype dependent.
6. Cost of maintenance of large collection is very high.
7. Slow growth cultures are vulnerable to contamination.

**Table 11.1:** A list of tropical plants cryopreserved in various forms

| | |
|---|---|
| **Cell suspensions** | **Somatic Embryos** |
| *Berberis dictyophilla* | *Citrus sinensis* |
| *Capsicum annum* | *Coffea arabica* |
| *Glycine max* | *Daucus carota* |
| *Nicotiana plumbaginifolia* | |
| *Nicotiana tabacum* | |
| *Oryza sativa* | |
| *Zea mays* | |
| **Protoplasts** | **Pollen Embryos** |
| *Glycine max* | *Arachis hypogea* |
| *Nicotiana tabacum* | *Citrus* spp. |
| *Zea mays* | *Nicotiana tabacum* |
| | *Atropa belladona* |
| **Callus** | **Zygotic Embryos** |
| *Gossypium arboreum* | *Camellia sinensis* |
| *Oryza sativa* | *Manihot esculenta* |
| *Saccharum* spp. | *Zea mays* |
| *Capsicum annum* | *Hordeum vulgare* |
| | *Triticum aestivum* |
| **Meristems** | |
| *Arachis hypogea* | |
| *Cicer arietinum* | |
| *Solanum tuberosum* | |

8. Large storage space is required especially for slow growth cultures.

## REFERENCE

Caplin, S.M. 1959. Mineral-oil overlay for conservation of plant tissue cultures. *Am. J. Bot.* **46:** 324–329.

## FURTHER READING

Bhojwani, S.S. and Razdan, M.K. 1990. *Plant Tissue Culture: Applications and Limitations.* Elsevier Science Publishers, Amsterdam.

Pierik, R.L.M. 1987. *In vitro culture of Higher Plants.* Martinus Nijhoff Publishers, Dordrecht.

Reinert, J. and Bajaj, P.P.S. (eds.) 1977. *Applied and Fundamental Aspects of Plant Cell, Tissue and Organ Culture.* Springer-Verlag, Berlin.

Singh, B.D. 2006. *Plant Biotechnology,* Kalyani Publishers, India.

Thorpe, T.A. 1981. *Plant Tissue Culture: Methods and Applications in Agriculture.* Academic Press Inc., New York.

Vasil, I.K. 1984. *Cell Culture and Somatic Cell Genetics of Plants: Laboratory Procedures and Their Applications.* Academic Press, Inc., New York.

# PART II

# *Genetic Material and Its Organization*

# 12

# Genetic Material

Cell is the basic unit of life. All living organisms including bacteria, plants, and animals are built from cells. Generally cells are very small, with diameters much less than 1 mm, so they are invisible to the naked eye. In the simplest cells, bacteria, a cell wall surrounds a very thin fatty acid containing outer plasma membrane, which in turn surrounds a superficially unstructured inner region. Within this inner region is located the bacterial DNA that carries the genetic information. These nuclei-free bacteria and their close relatives are known as prokaryotes. Thus, in prokaryotes, nuclear material is not separated from the cytoplasm by a discrete membrane. These also lack an extensive cellular architecture: membranous organelles are absent and the genetic material is not enclosed in a distinct structure (Table 12.1). In virtually all cells other than bacteria, the inner cellular mass is partitioned into a membrane-bound spherical body called nucleus and an outer surrounding cytoplasm. In the nucleus is located the cellular DNA in the form of coiled rods known as chromosomes. Cells that contain a nucleus and with other distinctive membranous organelles such as mitochondria, vesicles, and Golgi bodies are referred to as eukaryotes. In the cells of eukaryotes, a nuclear membrane separates the genetic material from the cytoplasm, which is then further subdivided by other distinct membranous structures.

Eukaryotes can be unicellular or multicellular and comprise all macroscopic forms of life such as plants, animals, and fungi. Plant cells differ from animal cells in the presence of outer cell wall surrounding the cell membrane and the chloroplasts, whereas animal cell contain centrosomes, which are involved in cell division and lysosomes.

The essence of a cell is its ability to grow and divide to produce progeny cells, which are likewise capable of generating new cellular molecules and replicating themselves. Thus, cells are in effect tiny factories that grow by taking up simple molecular building blocks. Considering the complexity of a living organism, one is struck by the relatively small number of organic components used in the formation of substances. Cells contain molecules of great variety, but a living cell is composed of restricted set of elements, six of which (C, H, N, O, P, S) make up more than 99% of its weight. Certain simple combinations of atoms such as methyl ($-CH_3$), hydroxyl ($-OH$), carboxyl ($-COOH$), and amino ($-NH_2$) groups occur repeatedly in biological molecules. Each such group has distinct chemical and physical properties. Molecules can be broadly grouped into two categories.

1. *Small organic molecules*: These are compounds with molecular weight in the 100–1000 range containing up to 30 or so carbon atoms. Molecules of this kind are usually found

**Table 12.1:** Characteristics of pro- and eukaryotic organisms

| Feature | Prokaryote | Eukaryote |
| --- | --- | --- |
| Structural unit | Bacteria, blue green algae | Unicellular organisms, Multicellular plants and animals |
| Size | <1–2 × 1–4 μm | >5 μm in width or diameter |
| Location of genetic material | Nucleoid, chromatin body, or nuclear material | Nucleus, mitochondria and chloroplasts |
| Genetic material | Naked DNA | DNA complexed with proteins |
| Chromosomes | 1 | >1 |
| Nuclear membrane | Absent | Present |
| Histone proteins | Absent | Present |
| Nucleolus | Absent | Present |
| Cell wall | Mainly consist of carbohydrates and amino acid complex | Mainly consist of cellulose, hemicellulose, pectin, etc. |
| Membrane-bound organelles: mitochondria, endoplasmic reticulum, etc. | Absent | Present |
| Ribosome | 70 S structure | 80 S structure |
| Site of electron transport | Cell membrane | Organelles |
| Genetic exchange mechanisms | Conjugation, transformation, transduction | Gamete fusion |
| Cytoplasmic movement | Absent | Present |
| Cell division | Binary fission | Mitosis |
| Life-cycle | Short | Long |

free in solution in the cytoplasm where they form a pool of intermediates from which large molecules called macromolecules are made. Some of the small organic molecules are sugars, amino acids and fatty acids. At least 750 different types of small molecules are found.

2. *Macromolecules*: These are invariably polymeric molecules that are formed by joining together specific types of small molecules A large polymeric molecule, or more than one polymer associated into a large molecule can be referred to as macromolecule. Some of the macromolecules such as carbohydrates, proteins, and nucleic acids are most important. Three simple molecules—sugars, amino acids and nucleotides—which form macromolecules have been briefly described.

## SUGARS

The simple monosaccharide sugars are molecules composed of carbon, hydrogen, and oxygen, which participate in the proportion $(CH_2O)_n$ and for this reason they are called carbohydrates (hydrates of carbon). All sugars contain hydroxyl groups (–OH) and either an aldehyde or ketone group. The hydroxyl group of one sugar can combine with aldehyde or ketone group of a second sugar with the elimination of water to form a disaccharide. The addition of more monosaccharides in the same way results in oligosaccharides and polysaccharides.

Glucose, a hexose monosaccharide, is the principal food component of many cells. Units of glucose are capable of polymerizing to form large polysaccharide molecules such as cellulose and starch in plants. Starch is the major source of food for humans. Cellulose is the most abundant molecule on earth, as it is the main constituent of plant fiber and wood. Besides, glucose, ribose and other simple sugars, derivatives of sugars such as uronic acids, amino sugars, or acetylated sugars may also form polymers. These polysaccharides play different role in organisms.

## AMINO ACIDS

Amino acids are chemically varied but they all contain a carboxyl acid group and amino group both linked to a single carbon atom. Proteins are long linear polymers of amino acids joined head to tail by a peptide bond between the carboxyl acid group of one amino acid and amino group of the next amino acid. There are twenty amino acids in proteins, each with a different side chain attached to $\alpha$-carbon atom. All amino acids contain nitrogen atoms; therefore, the proteins represent a class of nitrogenous compounds. The properties of amino acid side chains in aggregates determine the properties of proteins they constitute. This results in diverse structure and sophisticated functions of proteins. Specific terms commonly used to refer to different levels of protein structure are given below:

*Primary structure*: The sequence of amino acids linked by peptide bonds along the polypeptide chain is the primary structure.

*Secondary structure*: An ordered structure of a protein in which the polypeptide chain is folded in a regular manner with a characteristic helical or pleated structure. Much of the folding is the result of linking carbonyl and amide groups of the peptide chain backbone by hydrogen bonds. In many proteins the hydrogen bonding produces a regular coiled arrangement called the $\alpha$ helix.

*Tertiary structure*: A final structure of a protein in three-dimensional form from a single polypeptide chain. The final conformation of a protein from a single polypeptide, which is bent or tightly folded and compact is determined by a variety of interactions including hydrogen bonds, salt bonds, hydrophobic or nonpolar bonds and van der Walls forces. It is by virtue of their tertiary structure that proteins adopt a globular shape.

*Quaternary structure*: Used for proteins that contain more than a single polypeptide chain to indicate the spatial relationships among the separate chains or subunits. It means how the two or more chains are arranged in relation to each other. Most larger proteins contain two or more polypeptide chains, between which there are usually no covalent linkages. Proteins with two or more polypeptide chains are known as oligomeric proteins. A well-known example of an oligomeric protein is hemoglobin.

## NUCLEOTIDES

Nucleotides are subunits of DNA (deoxyribonucleic acid) and RNA (ribonucleic acid). The basic control and functioning of the organism is exercised by DNA and RNA. There are no molecules as important as nucleic acids, as they carry within their structure the hereditary information that determines the structure of proteins. The instructions that direct cells to grow and divide are encoded by them. In prokaryotes, DNA is present as a naked molecule, but in eukaryotes, DNA is complexed with proteins to form chromosomes. In some viruses, RNA is the genetic material. These DNA and RNA consist of basic monomer units called nucleotides, which are linked by phosphodiester bridges to form the linear polymers. The nucleotides in turn consist of three molecular compounds: a phosphate radical, a ribose sugar and a nitrogenous base (Fig. 12.1).

Five-carbon sugars are called pentoses. RNA contains the pentose D-ribose, whereas 2-deoxyribose is found in DNA. In both cases it is a five-membered ring known as *furanose*: D-ribofuranose for RNA and 2-deoxy-D-ribofuranose for DNA. When these ribofuranoses are found in nucleotides, their atoms are numbered as 1′, 2′, 3′ and so on to distinguish them from the ring atoms of nitrogenous bases.

Nitrogenous bases are ring compounds, which can combine with $H^+$. They are of two types: (1) Pyrimidines: These are single derivatives of a six-membered pyrimidine ring (C: cytosine, T: thymine, and U: uracil), and various pyrimidine derivatives such as dihydrouracil are present as minor constituents in certain RNA molecules, and (2) Purines: They have a second five-membered ring fused to a six-membered ring (A: adenine, G: guanine).

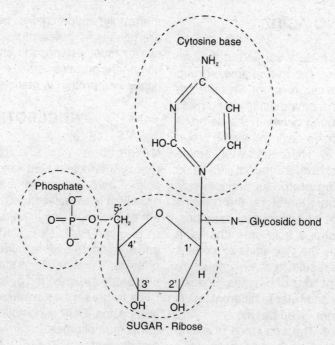

Fig. 12.1: A nucleotide.

Adenine (6-amino purine) and guanine (2-amino-6-oxy purine) are the common purines found in both DNA and RNA. Other naturally occurring purine derivatives include hypoxanthine, xanthine, and uric acid. Hypoxanthine and xanthine are found rarely in nucleic acids whereas uric acid, the most oxidized state for a purine derivative, is never found in nucleic acids. The nitrogenous bases are numbered in a clockwise fashion in the usual system for organic compounds (Fig. 12.2).

## Structural formulae of nucleotides

One may find that the structural formulae of bases are written in different forms. This is a result of their so-called tautomeric behavior. For instance, guanine can be written in two forms, as shown in Fig. 12.2.

## Nomenclature of nucleoside and nucleotide compounds

When the ribose and nitrogenous base complex is separated from the phosphate group, the remaining structure is called a nucleoside. So a phosphonucleoside is a nucleotide

Base + Sugar = Nucleoside

Base + Sugar + Phosphate = Nucleotide

So nucleosides are compounds formed when a base is linked to a sugar by a glycosidic bond (Fig. 12.1). Glycosidic bonds involve the carbonyl carbon atom of the sugar, which in cyclic structures is joined to the ring O atom. Such carbon atoms are called anomeric. In nucleosides, the bond is an N-glycoside because it connects the anomeric C-1′ to the N-1 of a pyrimidine or to the N-9 of a purine.

Nucleotides may contain several phosphate residues, but are present in DNA and RNA only as monophosphates. The letter M in the compound name signifies monophosphate, D means di-, and T is triphosphate. The conventional nomenclature is shown in Table 12.2. Nucleosides are named by adding the ending –idine to the root name of a primidine or –osine to the root name of a purine. The common nucleosides are thus cytidine, uridine, thymidine, adenosine, and guanosine. Because thymine was originally thought to occur only in DNA, the use of the term deoxythymine is considered to be redundant. Both TMP and dTMP are currently

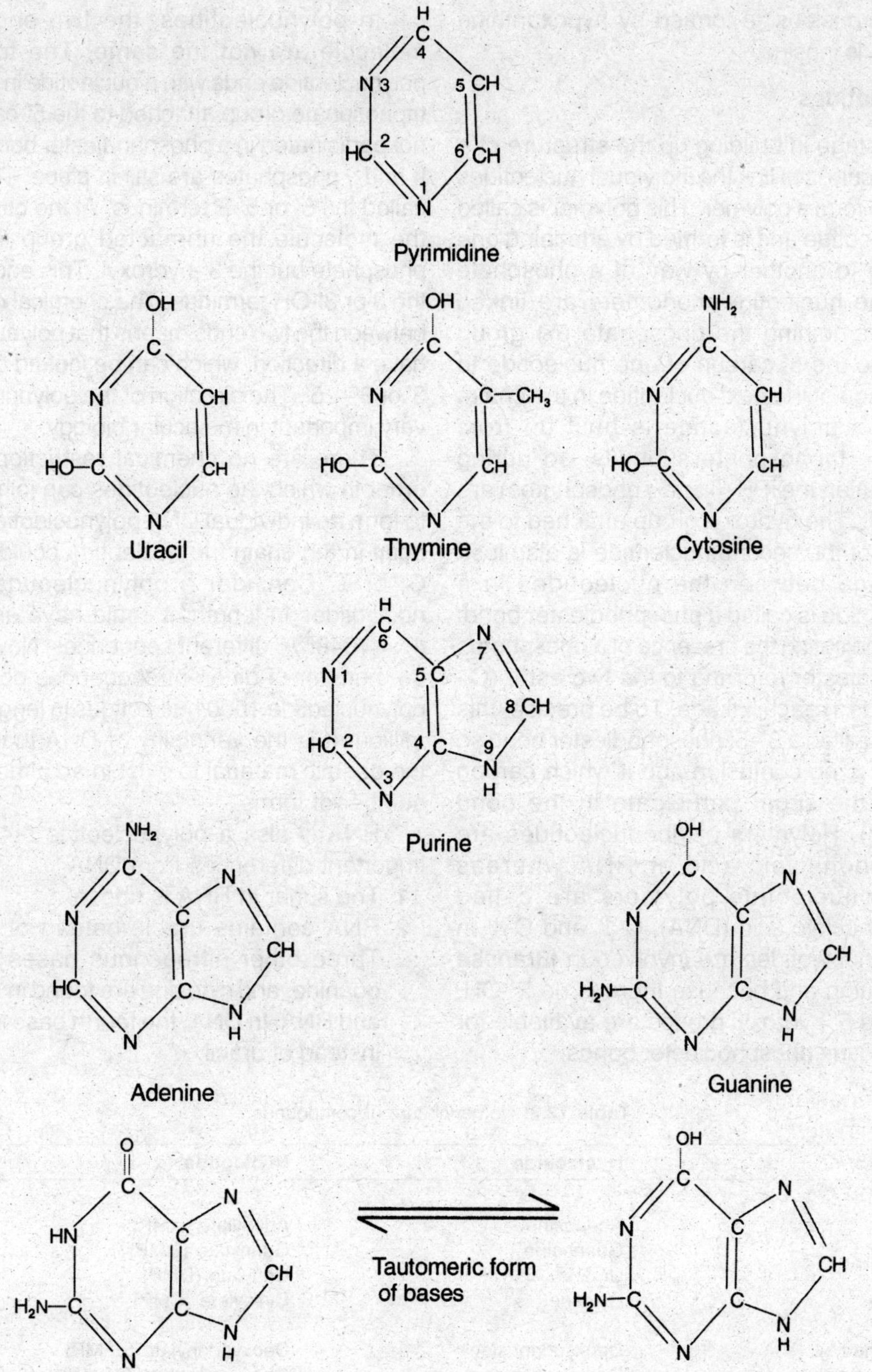

Fig. 12.2: The nitrogenous bases involved in DNA and RNA formation.

used. The nucleoside formed by hypoxanthine and ribose is inosine.

## Polynucleotides

The next stage in building up the structure of a DNA molecule is to link the individual nucleotides together to form a polymer. This polymer is called a polynucleotide and is formed by attaching one nucleotide to another by way of a phosphate group. The nucleotide monomers are linked together by joining the phosphate ($\alpha$) group attached to the 5' carbon of one nucleotide to the 3' carbon of the next nucleotide in the chain. Normally a polynucleotide is built up from nucleoside triphosphate subunits, so during polymerization the iPP ($\beta$ and $\gamma$ phosphates) are cleaved off. The hydroxyl group attached to the 3' carbon of the second nucleotide is also lost. The linkage between the nucleotides in a polynucleotide is called a phosphodiester bond: phospho indicating the presence of a phosphorus atom and diester referring to the two ester (C–O–P) bonds in each linkage. To be precise, this should be called a 3'–5' phosphodiester bond so that there is no confusion about which carbon atoms in the sugar participate in the bond (Fig. 12.3). Polymers of ribonucleotides are named ribonucleic acid or RNA whereas deoxyribonucleotide polymers are called deoxyribonucleic acid (DNA). C-1' and C-4' in deoxyribonucleotides are involved in furanose ring formation and because there is no 2'-OH, only 3' and 5' hydroxyl groups are available for internucleotide phosphodiester bonds.

In polynucleotides, the two ends of the molecule are not the same. The top of the polynucleotide ends with a nucleotide in which the triphosphate group attached to the 5'-carbon has not participated in a phosphodiester bond and the $\beta$ and $\gamma$ phosphates are still in place. This end is called the 5' or 5'-P terminus. At the other end of the molecule the unreacted group is not the phosphate but the 3'-hydroxyl. This end is called the 3 or 3'-OH terminus. The chemical distinction between the two ends means that polynucleotides have a direction, which can be looked on as 5'$\rightarrow$ 3' or 3'$\rightarrow$ 5'. The direction of the polynucleotide is very important in molecular biology.

There are no chemical restrictions on the order in which the nucleotides can join together to form an individual DNA polynucleotide. At any point in the chain the nucleotide could be A, G, C, or T. Consider a polynucleotide just 10 nucleotides in length: it could have any one of $4^{10} = 1048576$ different sequences. Now imagine the number of different sequences possible for polynucleotide 1000 nucleotides in length, or one million. It is the variability of DNA that enables the genetic material to exist in an almost infinite number of forms.

RNA is also a polynucleotide but with two important differences from DNA

1. The sugar in RNA is ribose.
2. RNA contains uracil instead of thymine. Three other nitrogenous bases adenine, guanine, and cytosine are found in both DNA and RNA. In DNA, the fourth base is thymine instead of uracil.

**Table 12.2:** Nomenclature of compounds

| Base | Nucleoside | Nucleotide |
| --- | --- | --- |
| **RNA** | | |
| Adenine (A) | Adenosine | Adenylate (AMP) |
| Guanine (G) | Guanosine | Guanylate (GMP) |
| Uracil (U) | Uridine | Uridylate (UMP) |
| Cytosine (C) | Cytidine | Cytidylate (CMP) |
| **DNA** | | |
| Deoxyadenine (A) | Deoxyadenosine | Deoxyadenylate (dAMP) |
| Deoxyguanine (G) | Deoxyguanosine | Deoxyguanylate (dGMP) |
| Deoxythymidine or Thymine (T) | Deoxythymidine or Thymidine | Deoxythymidylate or Thymidylate (dTMP) |
| Deoxycytosine (C) | Deoxycytidine | Deoxycytidylate (dCMP) |

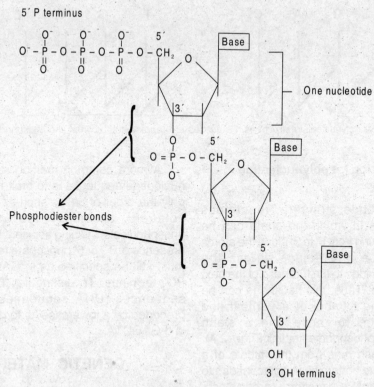

**Fig. 12.3:** Structure of a short polynucleotide showing phosphodiester bonds.

RNA in the cell usually exists as a single polynucleotide. The structure of RNA polynucleotide is exactly similar to DNA. DNA is invariably in the form of two polynucleotides wrapped around one another to form the famous double helix.

*Significance of differences between DNA and RNA*

Due to the chemical differences between DNA and RNA, the greater stability of DNA has been postulated.

DNA contains thymine instead of uracil. The key observation is that cytosine deaminates to form uracil at a finite rate in vivo. Because C in one DNA strand pairs with G in the other strand, whereas U would pair with A in the DNA. Conversion of C to U would result in a heritable change of a CG pair to a UA pair. Such a change in nucleotide sequence would constitute a mutation in the DNA. To prevent this, a cellular proof reading mechanism exists, and U arising from deamination of C is treated as inappropriate and is replaced by C. Now, if DNA normally contained U rather than T, the repair system of the host could not really distinguish U formed by C deamination from U correctly paired with A. However, the U in DNA is 5-methyl U, or as it is commonly known, thymine. That is, the 5-methyl group on T labels it as if to say this U belongs; do not replace it.

The ribose 2'-OH group of RNA is absent in DNA. Consequently, the ubiquitous 3'-O of polynucleotide backbones lacks a vicinal hydroxyl neighbor in the DNA. This difference leads to a greater resistance of DNA to alkaline hydrolysis. In other words, RNA is less stable than DNA because its vicinal 2'-OH group makes the 3'-phosphodiester bond susceptible to nucleophilic cleavage.

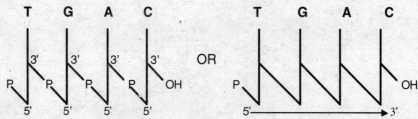

**Fig. 12.4:** Notation for polynucleotide structures. Furanoses are represented by lines and phosphodiesters by diagonal slashes.

## Shorthand notation for polynucleotide structures

In a polynucleotide structure, repetitious uniformity exists in which the chain can be visualized as running from 5′ to 3′ along the atoms of one furanose and thence across the phosphodiester bridge to the furanose of the next nucleotide in line. Thus, this backbone can be portrayed by the symbol of a vertical line representing the furanose and a slash representing the phosphodiester link (Fig. 12.4). The diagonal slash runs from the middle of a furanose line to the bottom of an adjacent one to indicate 3′– (middle) to 5′–(bottom) carbon atoms of neighboring furanoses joined by the phosphodiester bridge. The base attached to each furanose is indicated above it by one-letter designations: A, C, G, or T (U). The convention in all notations of nucleic acids structure is to read the polynucleotide chain from the 5′ end to the 3′ end.

It is known that bases are not part of the sugar-phosphate backbone, but instead serve as distinctive side chains, as the R groups of amino acids along a polypeptide backbone. A simple notation of these structures is to list the order of bases in the polynucleotide using single capital letters—A, G, C, and T (or U). Occasionally, a lowercase 'p' is written between each successive base to indicate the phosphodiester-bridge as in GpApGpCpTpA. A 'p' preceding the sequence indicates that the nucleic acid carries a $PO_4$ on its 5′ end as in pGpApGpCpTpA; a 'p' terminating the sequence denotes the presence of a phosphate in the 3′-OH end as in GpApGpCpTpAp.

A more common method representation of nucleotide sequence is to omit the 'p' and write only the order of bases such as GAGCTA. This notation assumes the presence of phosphodiesters joining adjacent nucleotides. The presence of 3′- or 5′ phosphate termini, however, must still be specified as in GAGCTAp for a 3′ $PO_4$ terminus. To distinguish DNA from RNA sequences, DNA sequences are typically preceded by a lowercase 'd' to denote deoxy as in d-GAGCTA.

## GENETIC MATERIAL

Mendel's experiments in 1865–66 on peas led him to conclude that traits of the pea were under the control of two distinct factors (later named genes), one coming from the male parent and the other from the female parent. About the same time, it became known that heredity is transmitted through the egg and sperm, and since a sperm contains little cytoplasm in comparison to the amount of material in its nucleus, the obvious conclusion was that nucleus carries the hereditary determinants of a cell. Soon afterwards, the chromosomes within the nucleus were made visible through the means of special dyes. The cells of members of a given species were seen to contain a constant number of chromosomes. A few years later, the number of chromosomes in the sex cells of male and female gametes were shown to be exactly $n$ (the haploid number), half the $2n$ (diploid number). The chromosome number is reduced to $n$ during gamete formation and fertilization restores the $2n$ chromosome number. Chromosomes thus behave exactly as they should if they were the bearers of Mendel's genes.

Several lines of indirect evidence have long suggested that DNA contains the genetic information of the living organisms. Results obtained from several different experiments showed that most of the DNA is located in chromosomes, whereas RNA and proteins are abundant in the cytoplasm. Moreover, a correlation exists between the amount of DNA per cell and the number of sets of chromosomes per cell.

## Discovery of DNA

The discovery of the double helical structure of DNA by James Watson and Francis Crick at Cambridge in 1953 was one of the greatest triumphs in biology as a whole. But the structure of polynucleotide was already known when Watson and Crick came to the scene. Besides, certain other observations made at that time helped Watson and Crick to elucidate its structure. These have been briefly mentioned.

Friedrich Miescher (1868) isolated cells that contained an unusual phosphorus containing compound, which he named "nuclein" (today's nucleoprotein).

Miescher (1872) reported that sperm heads, isolated from salmon sperm, contained an acidic compound (today's nucleic acid) and a base which he named "protamine."

Altman (1889) described a method for preparing protein-free nucleic acids from animal tissue and yeast. At this stage it was recognized that nucleic acids in the organisms are enmeshed with proteins.

Chargaff revealed a simple mathematical relationship between the proportion of bases in any one sample of DNA. The data of Chargaff showed that the four bases commonly found in DNA (A, C, G, and T) do not occur in equimolar amounts and that the relative amounts of each vary from species to species. However, certain pairs of bases, namely, adenine residues, equal the number of thymine, and the number of guanine equals the number of cytosine, i.e.

$A = T$ and $G = C$ (Chargaff, 1951). It means total purines (A+G) will equal the total pyrimidine (T+C), i.e. there is a 1:1 ratio. These findings are known as *Chargaff's rules*: [A] = [T]; [C] = [G]; [pyrimidines] = [purines].

By the early 1950s, X-ray studies on DNA by Wilkins, Franklin, and others indicated a well-organized, multiple-stranded fiber about 22 Å in diameter that was also characterized by the presence of groups spaced 3.4 Å apart along the fiber and a repeating unit every 34 Å.

Taking into account the facts known at that time, Watson and Crick in 1953 proposed a double helix structure of DNA as shown in Fig. 12.5 (Watson and Crick, 1953). According to them,

1. DNA double helix comprises two polynucleotide chains.
2. The nitrogenous bases are stacked on the inside of the helix, while the sugar-phosphate forming the backbone of the molecule is on the outside.
3. The two polynucleotide chains interact by hydrogen bonding. There is a very strict combining rule for hydrogen bonding.

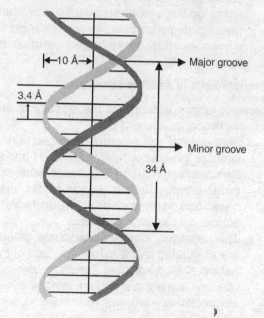

**Fig. 12.5:** The Watson–Crick model of DNA double helix.

4. Ten base pairs occur per turn of the helix. The pitch of the helix is 34 Å, meaning that the spacing between the adjacent base pairs is 3.4 Å. The helix diameter is 20 Å.

5. The two strands of the double helix are antiparallel, with one polynucleotide chain running in 5'→3' direction and the other chain in 3'→5' direction. Only antiparallel polynucleotides form a stable double helix.

6. The double helix has two different grooves: major and minor. Thus, the helix is not absolutely regular. This feature is important in the interaction between the double helix and proteins involved in DNA replication and expression.

7. The double helix is right handed.

8. The two polynucleotide chains base pairs are complementary. The sequence of one determines the sequence of the other. This complementary base pairing is of fundamental importance in molecular biology. The rule is that A base pairs with T and forms two hydrogen bonds, and G base pairs with C and forms three hydrogen bonds. Other base pairs are not normally permitted in the double helix as they will be either too large (purine–purine) or too small (pyrimidine–pyrimidine), or will not align in a manner allowing hydrogen bonding to occur (as with A–C and G–T).

**Double helix is a stable structure**

Several factors account for the stability of the double helical structure of DNA.

i. Both internal and external hydrogen bonds stabilize the double helix. In DNA, H-bonds form between A:T and C:G bases, whereas polar atoms in the sugar-phosphate backbone form external H bonds with surrounding water molecules.

ii. The negatively charged phosphate groups are all situated on the exterior surface of the helix in such a way that they have minimal effect on one another and are free to interact electrostatically with cations in solution, such as $Mg^{++}$.

iii. The core of the helix, which consists of base pairs, is bonded together with hydrophobic interactions and van der Walls forces besides H bonding.

## DNA replication

If genetic information is to be faithfully inherited from generation to generation, the nucleotide sequence of DNA must be faithfully copied from the original parental DNA molecule to the two daughter molecules. The clue as to how this duplication is achieved lies in the double helical structure of the DNA molecules. Since complementary DNA strands are held together by hydrogen bonding, their separation can be readily achieved and when separated, each strand serves as a template for the synthesis of a complementary copy. After replication, the original (parental) DNA duplex will give rise to two identical daughter duplex molecules.

Following DNA replication, one strand of a daughter molecule has been conserved from the parental duplex, while the other strand has been newly synthesized. This mode of replication is called semiconservative. Although semiconservative replication requires strand separation, the disruption of the double helical structure occurs only transiently and only a small part of a chromosome is single stranded at any given time. DNA replication is extremely complex and involves the coordinated participation of many proteins with many different enzymatic activities.

When a DNA molecule is being replicated, only a limited region is in a non base-paired form. The breakage in base pairing starts at a distinct position called the replicating origin and gradually progresses along the molecule in both the directions. The important region at which the base pairs of the parent molecule are broken and the new polynucleotides are synthesized is called the replication fork.

An enzyme able to synthesize a new DNA polynucleotide is called a DNA polymerase. In 1959, Arthur Kornberg and his colleagues isolated from E. coli an enzyme capable of DNA

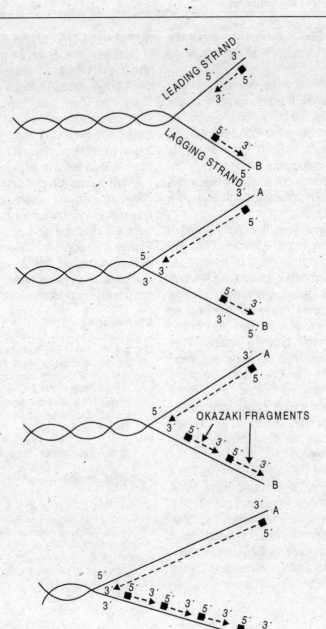

**Fig.12.6:** DNA replication showing sequence of unwinding and replication at the fork of a DNA double helix. ■ RNA primer.

synthesis on a polynucleotide template and named it as polymerase I. But later studies revealed that DNA polymerase III and not DNA polymerase I is the primary enzyme involved in DNA replication. There are two main properties of DNA polymerase III.

1. It can synthesize DNA only in the 5′→3′ direction.

2. It cannot initiate DNA synthesis unless there is already a short double-stranded region to act as a primer.

During DNA replication there is a breakage of base pairing between the two strands of the parent DNA molecule, which is carried out by helicase enzyme in conjunction with single-strand binding proteins (SSBs). This prevents the two strands from immediate reannealing. The result is the replication fork that provides the template on which DNA polymerase III can work (Fig. 12.6).

Since DNA polymerase III can synthesize only in the 5′→3′ direction, it means that the template strands must be read in the 3′→5′ direction. Thus, one strand is called the leading strand, where new polynucleotides can be synthesized continuously. However, the second strand where DNA strand cannot be copied continuously is called a lagging strand. This strand has to be replicated in sections: a portion of the parent helix is dissociated and a short stretch of lagging strand replicated, a bit more of the helix dissociated and another segment of the lagging strand replicated and so on. These are called Okazaki fragments, which are 100 to 1000 nucleotides in length.

DNA polymerase III cannot initiate DNA replication. To initiate DNA replication, RNA polymerase first synthesizes a short primer of 6 to 30 ribonucleotides on the template DNA strand. Once replication has been initiated, then DNA polymerase III continues DNA replication. On the lagging strand, DNA polymerase III can synthesize DNA only for a short distance before it reaches the RNA primer at the 5′ end of the next Okazaki fragment. At this point, DNA polymerase III stops and DNA polymerase I carries out the functions of removing the ribonucleotides of the primer of the adjacent Okazaki fragment and replacing them with deoxyribonucleotides. When all the ribonucleotides are replaced, DNA polymerase I either stops or possibly carries on a short distance into the short region of the Okazaki fragment, continuing to replace nucleotides before it dissociates from the new double helix. Finally the adjacent Okazaki fragments, which are separated by a single gap between neighbouring nucleotides, are sealed by ligase enzyme, which forms a phosphodiester bond.

## REFERENCES

Chargaff, E. 1951. Structure and function of nucleic acids as cell constituents. *Fed. Proc.* **10:** 654–659.

Miescher, F. 1871. On the chemical composition of pus cells. *Hoppe-Seyler's Med.-Chem. Untersuch.* **4:** 441–460. (Reprinted in the collection of Gabriel and Fogel. 1955. *Great Experiments in Biology.* Prentice-Hall, Englewood Cliffs, N.J.)

Watson, J.D. and Crick, F.H.C. 1953. The structure of DNA. *Cold Spring Harbor Symp. Quant. Biol.* **18:** 123–131.

## FURTHER READING

Garret, R.H. and Grisham, C.M. 1999. *Biochemistry.* Saunders College Publishing, Orlando, Florida.

Lewin, B. 1995. *Genes V.* Oxford University Press, Oxford, U.K.

Snustad, D.P., Simmons, M.J. and Jenkins, J.B. 1997. *Principles of Genetics.* John Wiley & Sons, New York.

# 13

# Organization of DNA and Gene Expression

When the double helix of DNA was found, it was believed that many distinct molecules are present in a chromosome. But this picture has suddenly changed due to the realization that long DNA molecules are fragile and easily breakable into smaller fragments. By the late fifties when great care was taken to prevent shearing of DNA, it was found that DNA molecules contained as many as 200,000 base pairs. But now it is known that a single DNA molecule of *E. coli* contains more than 4 million base pairs. Likewise, a single DNA molecule is thought to exist within each of the chromosome of higher plants. These chromosomes contain an average some 20 times more DNA than present in the *E. coli* chromosome. A chromosome is thus properly defined as a single genetically specific DNA molecule to which are attached large number of proteins that are involved in maintaining chromosome structure and regulating gene expression.

## DIFFERENT FORMS OF DNA

DNA molecules can have more than one conformation in solution depending upon the pH and ionic environment. Under physiological conditions the dominant form of DNA is the B-form. The rare **A-DNA** exists only in dehydrated state. The A-form differs from the B-form in relatively minor ways. A-DNA is still a double helix but more compact than the B-form of DNA. A-form has 11 base pairs per turn of the helix and a diameter of 23Å. The pitch or distance required to complete one helical turn in B-DNA is 3.4 nm while in A-DNA it is 2.46 nm. The base pairs in A-DNA are no longer nearly perpendicular to the helix axis, but instead are tilted 19 degree with respect to this axis. Successive base pairs occur every 0.23 nm along the axis, as opposed to 0.332 nm in B-DNA. The B-DNA is thus longer and thinner than the short, squat A-form, which has its base pairs displaced around, rather then centered on the helix axis (Table 13.1). It has been shown that relatively dehydrated DNA fibers can adopt the A-conformation under physiological conditions, but it is still unclear whether DNA ever assumes this form *in vivo*. However, double helical DNA:RNA hybrids probably have an A-like conformation. The 2′-OH in RNA sterically prevents double helical regions of RNA chains from adopting the B-form helical arrangement. Importantly, double stranded region in RNA chains assumes an A-like conformation, with their bases strongly tilted with respect to the helix axis. Since 1953 other forms of DNA have also been found: C, D, E, F and Z, each with a slightly

**Table 13.1:** A comparison of the structural properties of A, B and Z-DNA (Adapted from Dickerson, R.L. *et al.*, 1982)

| Character | A-DNA | B-DNA | Z-DNA |
|---|---|---|---|
| Overall proportions | Short and broad | Longer and thinner | Elongated and slim |
| Rise per base pair | 2.3 Å | 3.32 Å ± 0.19 Å | 3.8 Å |
| Helix packing diameter | 25.5 Å | 23.7 Å | 18.4 Å |
| Helix rotation sense | Right handed | Right handed | Left handed |
| Base pairs per helix repeat | 1 | 1 | 2 |
| Base pairs per turn of helix | ~11 | ~10 | 12 |
| Mean rotation per base pair | 33.6 Å | 35.9 ± 4.2° | – 600/2 |
| Pitch per turn of helix | 24.6 Å | 33.2 Å | 45.6 Å |

Table adapted from Dickerson, R.L. et al. (1982). Cold spring Harbor Symposium on Quantitative Biology **47**:14.

different conformation to the double helix. There is no strong evidence that C,D,E and F conformations exist *in vivo*.

**Z-DNA** exists *in vivo* and is the only left-handed form of the helix and takes its name from the zigzag path that the sugar-phosphate backbone follows. It was first recognized by Alexander Rich and his associates in X-ray analysis of synthetic deoxynucleotide dCpGpCpGpCpG, which crystallized into an antiparallel double helix of unsuspected conformation (Fig. 13.1a). The alternating

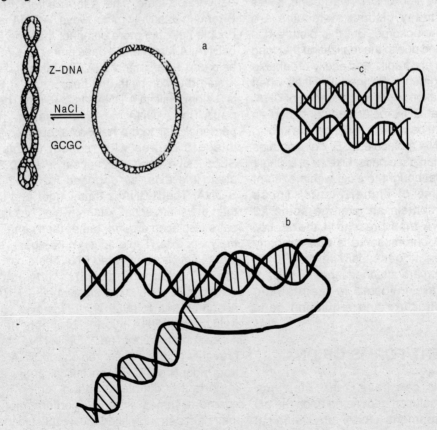

**Fig. 13.1:** DNA conformations. **a:** Z-DNA; **b:** the triple helix conformation of DNA; **c:** cruciform DNA.

pyrimidine-purine (Py-Pu) sequence of this oligonucleotide is the key to its unusual properties. Because alternate nucleotide assumes different conformations, the repeating unit on a given strand in the Z-helix is the dinucleotide. That is for any number of bases, n along one strand, $n-1$ dinucleotides must be considered. For example, a GpCpGpC subset of sequence along one strand is comprised of the successive dinucleotide units GpC, CpG, and GpC. In B-DNA, the nucleotide conformations are essentially uniform and the repeating unit is the mononucleotide. It follows that the CpG sequence is distinct conformationally from the GpC sequence along the alternating copolymer chains in the Z-double helix. The conformational alterations going from B to Z realign the sugar-phosphate backbone along a zigzag course that has a left handed orientation, thus the designation Z-DNA. In any GpCpGp subset, the sugar-phosphate of GpC form the horizontal 'zig' while the CpG backbone forms the vertical 'zag'. The mean rotation angle circumscribed around the helix axis is $-15°$ for a CpG step and $-45°$ for a GpC step (giving $-60°$ for the dinucleotide repeat). The minus sign denotes a left handed or counterclockwise rotation about the helix axis. Z-DNA is more elongated and slimmer than B-DNA.

Segments of DNA in which purine and pyrimidine residues alternate are able to form DNA under certain conditions, and Z-DNA may be stabilized in vivo by binding Z-specific proteins. One possible biological role for left handed DNA might be based on the fact that small regions of Z-DNA relieve the superhelical strain in a negatively supercoiled DNA molecule, in the same way that small regions of single stranded DNA can. These local regions of Z-DNA may bind to regulatory proteins that are specific for Z-DNA.

The Z form can arise in sequences that are not strictly alternating Py-Pu. For example, the hexanucleotide $^{m5}CGAT^{m5}CG$, a Py-Pu-Pu-Py-Py-Pu sequence containing two —methyl

cytosine ($^{m5}C$), crystallizes as Z-DNA. Indeed, the in vivo methylation of C at 5 position is believed to favor a B to Z switch because, in B-DNA, these hydrophobic methyl groups would protrude into the aqueous environment of the major groove and destabilize its structure. In Z-DNA, the same methyl group can form a stable hydrophobic patch. It is likely that Z-conformation naturally occur in specific regions of cellular DNA, which otherwise is predominantly in B-form. Furthermore because methylation is implicated in gene regulation, the occurrence of Z-DNA may affect the expression of genetic information.

## DNA SUPERCOILING—TERTIARY STRUCTURE OF DNA

In vivo, the majority of DNA molecules do not have free ends but are present as closed circular structures. When a closed circular duplex is twisted around its axis, supercoils appear in the DNA, rather like the coiling seen when a rubber band is twisted. When B-DNA is exposed to forces that twist it in a direction opposite to the twist of the double helix, it forms negative supercoils. This allows the DNA molecules to relieve the twisting force by loosening the winding of two strands around each other. Supercoiling can only occur in a closed circular structure and the greater the number of supercoils, the greater the torsion in the closed duplex. It has become clear that the degree of supercoiling in cells is strongly controlled by specific enzymes (topoisomerases and gyrases) that are capable of either adding or subtracting supercoiled twist in the DNA. A DNA molecule cannot form a supercoil if one or both of the DNA strands are broken. If this occurs, the DNA supercoil unwinds to a relaxed form. Supercoils are classified in two ways. Negative supercoils twist the DNA around its axis opposite to the direction of the turns of right-handed helix. Positive supercoils twist the DNA in the same direction as the right-handed helix. It is thought that only negatively supercoiled DNA exists *in vivo*, although positively supercoiled DNA may be generated *in vitro*.

*Linking number*: The basic parameter characterizing supercoiled DNA is the linking number ($L$). This is the number of times the two strands are intertwined, and, provided both strands remain covalently intact, $L$ cannot change. In a relaxed circular DNA duplex of 400 bp, $L$ is 40 (assuming 10 bp per turn in B-DNA). The linking number for relaxed DNA is usually taken as the reference parameter and is written as $L_0$. $L$ can be equated to twist ($T$) and write ($W$) of the duplex, where twist is the number of helical turns and write is the number of supercoils:

$$L = T + W$$

In any closed circular DNA duplex that is relaxed, $W = 0$. A relaxed circular DNA of 400 bp has 40 helical turns, $T = L = 40$. This linking number can only be changed by breaking one or both strands of the DNA, winding them tighter or looser, and rejoining the ends.

Values of $T$ and $W$ for various positively and negatively supercoiled circular DNAs have been given below:

(a) Positively supercoiling

$T = 0$, $W = 0$ then $L = 0$
$T = +3$, $W = 0$ then $L = +3$
$T = 2$, $W = +1$ then $L = +3$
$T = +2$, $W = +1$ then $L = +3$

(b) Negative supercoiling

$T = 0$, $W = 0$ then $L = 0$
$T = -3$, $W = 0$ then $L = -3$
$T = -2$, $W = -1$ then $L = -3$
$T = -2$, $W = -1$ then $L = -3$

Negative supercoils favor local unwinding of the DNA, allowing processes such as transcription, DNA replication and recombination. Negative supercoiling is also thought to favor the transition between B-DNA and Z-DNA, and moderate the interactions of DNA binding proteins involved in gene regulation.

# CRUCIFORMS—TERTIARY STRUCTURE OF DNA

Besides the conformations, DNA may also form unusual structures that appear to be important in regulating gene expression. These include a triple helix conformation and a cross-shaped structure called cruciform DNA (Fig. 13.1 b and c). Triple helices are characterized experimentally by their susceptibility to digestion *in vitro* with S1 nuclease (an enzyme that specifically degrades regions of single stranded DNA). Cruciform DNA takes its name from a locally formed cross-shaped segment. A DNA segment is only able to fold into a cruciform if the DNA sequence contains an inverted repeat or palindromic sequence. Palindromes are words, phrases or sentences that are the same when read backward or forward, such as "radar", "Madam, I'am Adam", Malayalam. Inverted repeat sequences (palindromes) are symmetrically repeated DNA sequences that are same in both chains of a double helix when read in the 5′ to 3′ direction. The example shows a double stranded 16 base pair inverted repeat with the center of symmetry (dyad axis) indicated by a dash.

    5′  TATACGAT-ATCGTATA  3′
    3′  ATATGCTA-TAGCATAT  5′

In naturally occurring DNA sequences the two halves of the inverted repeat are often separated by a non-inverted repeat sequence. Inverted repeat sequences are present in both prokaryotic and eukaryotic genomes, and the cruciform structures have been postulated to be the sites recognized or induced by regulatory DNA binding proteins.

# EUKARYOTIC DNA ORGANIZATION INTO NUCLEOSOMES

A typical human cell is 20 μm in diameter. Its genetic material consists of 23 pairs of ds DNA molecules in the form of chromosomes. Thus the average length of DNA would be $3 \times 10^9$ bp /23 or $1.3 \times 10^8$ bp. At 0.34 nm/bp in B-DNA this represents a DNA molecule 5 cm long. If we consider all the 46 dsDNA molecules together it would be approximately 2 m of DNA that must be packaged into a nucleus of perhaps 5 μm in diameter. It makes it very clear that DNA must be condensed by a factor of $10^5$.

In eukaryotes the DNA molecules are tightly complexed with proteins to form chromatin, which is a nucleoprotein fiber with a beaded appearance. The key to the bead-like organization of chromatin is its histone proteins, which are divided into five main classes: H1, H2A, H2B, H3 and H4. They come together to form nucleosome units containing an octomeric core composed of two molecules each of the H2A, H2B, H3 and H4 components. Around each octomeric histone core 146 bp of B-DNA in a flat, left handed superhelical conformation makes 1.65 turns. Histone 1, a three domain protein serves to seal the ends of the DNA turns to the nucleosome core and to organize the additional 40–60 bp of DNA that link consecutive nucleosomes.

A higher order of chromatin structure is created when the nucleosomes in their characteristic beads on a string motif, are wound in the fashion of a solenoid having six nucleosomes per turn. The resulting 30 nm filament contains about 1200 bp in each of its solenoid turns. Interactions between the respective H1 components of successive nucleosomes stabilize the 30 nm filament. This filament then forms long DNA loops of variable length, each containing an average between 60,000 and 150,000 bp. Electron microscopic analysis of human chromosome 4 suggests that 18 such loops are then arranged radially about the circumference of a single turn to form a miniband unit of the chromosome. According to this model, approximately 106 of these minibands are arranged along a central axis in each of the chromatid of human chromosome 4 that form at mitosis. Despite studies much of the higher order structure of chromosomes remains a mystery. Also found within chromatin are a large number of different DNA binding proteins. These proteins are likely to play a variety of specific roles in regulating gene expression.

## DNA CONTENT

The nuclear genome is the largest in the plant cell, both in terms of picograms of DNA and in the number of genes encoded. Nuclear DNA is packaged into chromosomes along with histones and non–histone proteins, all of which play an important role in gene expression. These various components are held together to form chromatin. While the DNA encodes the genetic information, the proteins are involved in controlling the packaging of DNA and regulating the availability of DNA for transcription. One process, which is basic to all development, is the replication of nuclear genetic material, and its equal distribution between two daughter cells during mitosis. The DNA contained in each of the daughter cells must be identical to that contained in the original cell.

The amount of DNA in the haploid genome of an eukaryotic cell is known as its C-value where the 1C amount is an unreplicated haploid DNA content. The nuclear DNA content as C-value varies widely in higher plants from 0.15 to over 200 pg. *Arabidopsis thaliana* has 0.15 pg/haploid genome to some members of the Loranthaceae (Mistletoes) which have over 100 pg/haploid genome. Even the smallest plant genome is larger than that found in *Drosophila melanogaster* (165,000 kb) and contains much more DNA than is required to specify all the proteins synthesized during the course of development. The nuclear content and number of base pairs of some crop species have been given in Table 13.2 adapted from Arumuganathan and Earle (1991).

All higher eukaryotes contain significantly more DNA than what appears to be necessary to encode their structures and functions. For example, the human haploid genome has a

**Table 13.2:** Nuclear DNA content of plant species

| Species | Common name | pg DNA/ 2C | Mbp/1C |
|---|---|---|---|
| *Arabidopsis thaliana* | Arabidopsis | 0.30 | 145 |
| *Arachis hypogaea* | Peanut/ groundnut | 5.83 | 2813 |
| *Avena sativa* | Oats | 23.45 | 11315 |
| *Brassica campestris* ssp *oleifera* | Turnip rape | 0.97–1.07 | 468–516 |
| *Brassica juncea* | Brown mustard | 2.29 | 1105 |
| *Carica papaya* | Papaya | 0.77 | 372 |
| *Cicer arietinum* | Chick pea | 1.53 | 738 |
| *Citrus sinensis* | Orange | 0.76 | 367 |
| *Daucus carota* | Carrot | 0.98 | 473 |
| *Glycine max* | Soybean | 2.31 | 1115 |
| *Gossypium hirsutum* | Cotton | 4.39, 4.92 | 2118, 2374 |
| *Helianthus annus* | Sunflower | 5.95–6.61 | 2871–3189 |
| *Hordeum vulgare* | Barley | 10.10 | 4873 |
| *Lens culinaris* | Lentil | 8.42 | 4063 |
| *Lycopersicon esculentum* | Tomato | 1.88–2.07 | 907–1000 |
| *Oryza sativa* | Rice | 0.87–0.96 | 419–463 |
| *Saccharum officinarum* | Sugarcane | 5.28–7.47 | 2547–3605 |
| *Solanum tuberosum* | Potato | 3.31–3.86 | 1597–1862 |
| *Sorgum bicolor* | Sorghum | 1.55 | 748 |
| *Triticum aestivum* | Wheat | 33.09 | 15966 |
| *Tulipa* sp. | Garden tulip | 51.2, 63.6 | 24704, 30687 |
| *Vigna mungo* | Black gram/urd | 1.19 | 574 |
| *Zea mays* | Maize | 4.75–5.63 | 2292–2716 |

1C: unreplicated haploid genome; 2C: Diploid genome;
1 picogram (pg) = 965 million or mega base pairs (Mbp)

C-value of about $3 \times 10^6$ kb, yet estimates of the number of genes in the human genome, calculated by a variety of means, suggest that there are only about 30,000 genes. If one takes 3 kb as a generous estimate for the coding sequence of an average gene, then there exists at least a 10-fold excess of DNA in the genome over that required to produce a human being. A comparison of values from other organisms show that pea, for example, contains 1000 times as the DNA of *E. coli*, which probably contains about 4000 genes. Thus, if the entire pea DNA were expressed in the conventional sense, it would represent four million genes. While it is difficult to estimate the number of genes necessary to account for the differences in biological complexity between pea and *E. coli*, a value as high as this seems extraordinary.

Estimates of the haploid DNA content of higher plant nuclei have also been made by studying the kinetics of DNA renaturation. The kinetic analysis shows that a large proportion of the DNA is composed of highly repetitive sequences and is unlikely to represent genes coding for proteins. Most of the protein coding regions are represented only once per haploid genome and are therefore often referred to as unique or single copy sequences. The proportion of higher plant DNA present as single copy sequences ranges from 20 to 80%, with a general tendency for plants with large genomes to have higher proportion of repetitive DNA.

Much of the genome comprises sequences that are repeated many times. The number of repeats can vary from a few to millions. It has been convenient to classify repeated sequences broadly according to their repetition frequency. Sequences that are repeated at about $10^4$ to $10^6$ copies per haploid genome are highly repetitive and other repeated sequences are called moderately repetitive. Much of the highly repetitive DNA is composed of simple sequences of 1 to 200 bp repeated often in large tandem arrays, and these are known as satellite DNA.

Satellites usually consist of long tandem arrays of thousands of related sequences. In many cases the repeats are short tandem repeats of oligonucleotides (1–6 nucleotides tandem repeats as microsatellites or tandem repeats of 6–200 bp as minisatellites), but there are examples of repeats of several thousand base pairs also. As observed by microscopy many satellites are located around the centromeres and telomeres where they appear as constrictions or densely staining regions called heterochromatin in the metaphase chromosome. They are usually not transcribed. Most eukaryotic genomes also contain discrete segments of DNA called transposons that are capable of moving from one part of a genome to another.

Plants with a small genome and a low proportion of repetitive DNA have short somatic and meiotic cycles and a short generation time. A good example is *Arabidopsis thaliana* which is popular in plant molecular biology because of low DNA content and five-week generation time. Amongst cereals, rice has a small genome with 415 Mbp compared to the wheat genome, which has 16 billion base pairs. In general, horticultural plants have a small genome size and this should facilitate molecular studies of these crops as is being done on rice.

## Denaturation

When duplex DNA molecules are subjected to conditions of pH, temperature or ionic strength that disrupt hydrogen bonds and results in the separation of two strands of DNA is referred to as denaturation. If temperature is the denaturing agent, the double helix is said to melt. The course of dissociation of two strands of DNA can be followed spectrophotometrically because the relative absorbance of DNA solution at 260 nm increases as much as 40% as the bases unstuck. This absorbance increase or hyperchromic shift is due to the fact that the aromatic bases in DNA interact via their $\pi$ electron clouds when stacked together in the double helix. Because the UV absorbance of the bases is a consequence of $\pi$ electron transitions, and because the potential for these transitions is diminished when the bases stack, the bases in duplex DNA absorb less 260 nm radiation than expected for their numbers. Unstacking alleviates the suppression of UV absorbance. The rise in absorbance coincides with strand separation and midpoint of the absorbance increase is termed the melting temperature, $T_m$. DNAs differ in their $T_m$ values because they differ in relative G+C content of a DNA. The higher the G+C content of a DNA, the higher its melting temperature because G:C pairs are held by three H bonds whereas A:T pairs have only two H bonds.

$T_m$ is also dependent on the ionic strength of solution; the lower the ionic strength, the lower the melting temperature. At 0.2M Na⁺, $T_m = 69.3 + 0.41$ (% G + C). Ions suppress the electrostatic repulsion between the negatively charged phosphate groups in the complementary strands of the helix, thereby stabilizing it. DNA in pure water melts even at room temperature. At high concentration of ions, $T_m$ is raised and the transition between helix and coil is sharp.

At pH values greater than 10, extensive deprotonation of the bases occurs that destroy the hydrogen bonding potential and denatures the DNA double helix. Similarly at pH below 2.3 extensive protonation of bases take place and disrupts the pairing of two strands of DNA. Alkali is generally preferred because it does not hydrolyze the glycosidic linkages in the sugar-phosphate backbone. Small solutes like formamide and urea also act as denaturants for DNA as they readily form H bonds at temperatures below $T_m$, if present in sufficiently high concentration to compete effectively with the H-bonding between the base pairs.

## DNA Renaturation

Denatured DNA will renature to reform the duplex structure if the denaturing conditions are removed (i.e. solution is cooled, pH is returned to neutrality or the denaturants are diluted out). Renaturation requires reassociation of the DNA strands into a double helix, a process known as reannealing. This occurs when complementary bases in the two strands come together and helix can be zippered up. The general nature of the eukaryotic genome can be assessed by the kinetics with which denatured DNA reassociates. Reassociation between complementary sequences of DNA occurs by base pairing, which is a reversal of denaturation by which they were separated.

## Renaturation rate and DNA sequence complexity—*Cot* curves

The renaturation rate of DNA is an excellent indicator of the sequence complexity of DNA. For example T4 DNA contains about $2 \times 10^4$ base pairs, *E. coli* possesses $4.64 \times 10^6$ bp while human $3.3 \times 10^9$ bp. *E. coli* DNA is considerably more complex as compared to T4 phage in that it encodes more information. Expressed in another way, for any given amount of DNA (in grams), the sequences represented in an *E. coli* sample are more heterogeneous, i.e. more dissimilar from one another, than those in an equal weight of phage T4 DNA. Therefore it will take the *E. coli* DNA longer to find out their complementary partners and reanneal. This renaturation of DNA can be analysed quantitatively. The rate of reaction is governed by the equation

$$\frac{dC}{dt} = -kC^2$$

where, *C* is the concentration of single stranded DNA at time *t*, and *k* is a second order reassociation rate constant. Starting with an initial concentration of DNA, $C_0$ at time $t = 0$, the amount of single stranded DNA remaining after time *t* is

$$\frac{C}{C_0 t} = \frac{1}{1 + k.C_0 t}$$

Where the unit of *C* is moles of nucleotides per liter and that for *t* is seconds.

This equation shows that the parameter controlling the reassociation reaction is the product of DNA concentration ($C_0$) and time of incubation (*t*) usually described as the **Cot**.

A useful parameter is derived by considering the conditions when the reaction is half complete at time $t_{1/2}$. Then

$$\frac{C}{C_0} = \frac{1}{2} = \frac{1}{1 + k.C_0 t_{1/2}}$$

Yielding

$$C_0 t_{1/2} = 1/k$$

The value required for half reassociation ($C_0 t_{1/2}$) is called **Cot$_{1/2}$**. Since the Cot$_{1/2}$ is the product of the concentration and time required to proceed halfway, a greater Cot implies a slower reaction. Cot$_{1/2}$ is given as moles nucleotides × seconds/liter as Cot value.

The reassociation of DNA usually is followed in the form of a cot curve which plots the fraction of DNA that has reassociated ($1 - C/C_0$) against the log of the Cot. Fig. 13.2 gives Cot curves for several genomes. The rate of reassociation can be followed spectrophotometrically by the UV absorbance decrease as duplex DNA is formed. Note that relatively more complex DNAs take longer to renature as reflected by their greater Cot$_{1/2}$. The form of each curve is similar, but the Cot required in each case is different. It is described by the Cot$_{1/2}$.

The **Cot$_{1/2}$** is directly related to the amount of DNA in the genome. It means as the genome becomes more complex, there are fewer copies of any particular sequence within a given mass of DNA. For example, if the $C_0$ of DNA is 12 pg, then the bacterial genome whose size is .004 pg will contain 3000 copies of each sequence, whereas eukaryotic genome of size 3 pg will contain only 4 copies of each sequence in an initial concentration of 12 pg of DNA. Thus, the same absolute concentration of DNA measured in moles of nucleotides per liter (the $C_0$) will provide a concentration of each eukaryotic sequence.

Since the rate of reassociation depends on the concentration of complementary sequences, for the eukaryotic sequences to be present at the same relative concentration as the bacterial sequences, it is necessary to have more DNA or to incubate the same amount of DNA for longer

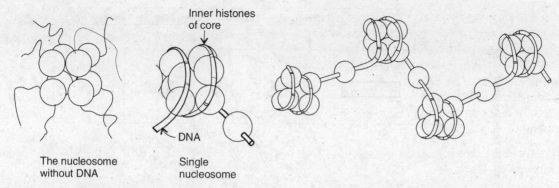

**Fig. 13.2:** The core of nucleosome formed by eight histone molecules, made up of two molecules each of H2A, H2B, H3 and H4 joined at their globular portions with the ends of the proteins free (left). A single nucleosome with its inner core of histones, the DNA wrapped around the core and bound by the free ends of the histones, and a stretch of linker DNA to which is bound a single molecule of H1. Extreme right—a chain of four nucleosomes.

time. Thus, the $Cot_{1/2}$ of the eukaryotic reaction is much greater than the $Cot_{1/2}$ of the bacterial reaction.

The $Cot_{1/2}$ of a reaction therefore indicates the total length of different sequences that are present. This is described as the complexity, usually given in base pairs.

The renaturation of the DNA of any genome should display a $Cot_{1/2}$ that is proportional to its complexity. Thus the complexity of any DNA can be determined by comparing its $Cot_{1/2}$ with that of a standard DNA of known complexity. Usually *E. coli* DNA is used as a standard. Its complexity is taken to be identical with the length of the genome (implying that every sequence in the *E. coli* genome of $4.2 \times 10^6$ bp is unique. So it can be written as:

$$\frac{Cot_{1/2} \text{ (DNA of any genome)}}{Cot_{1/2} \text{ (}E.\text{ }coli\text{ DNA)}} =$$

$$\frac{\text{Complexity of any genome}}{4.2 \times 10^6 \text{ bp}}$$

# FLOW OF GENETIC INFORMATION: CENTRAL DOGMA

How is the genetic information held within DNA translated into protein products? In 1958, Francis Crick postulated that the biological information contained in the DNA of the gene is transferred first to RNA and then to protein. Crick also stated that the information is unidirectional, proteins cannot themselves direct the synthesis of RNA and RNA cannot direct the synthesis of DNA (Crick, 1958). However, the latter part of central dogma was shaken in 1970 when Temin and Baltimore (Baltimore, 1970; Temin and Mizutani, 1970) independently discovered that certain viruses do transfer biological information from RNA to DNA by an enzyme reverse transcriptase that synthesizes DNA from an RNA template. This central dogma representing the flow of information from DNA to RNA is not unidirectional but it can be from RNA to DNA also (Fig. 13.3).

# ORGANIZATION OF GENES IN DNA MOLECULES

In the cells of higher organisms all the genes are carried by a small number of chromosomes, each of which contains a single DNA molecule. Each of these DNA molecules must carry several hundreds of genes. Gene is a unit of inheritance. Gene can be defined as a segment of genome with a specific sequence of nucleotides, which has a specific biological function. Genes are transcribed in to mRNA and then translated in to polypeptides. Ribosomal RNA (rRNA) and

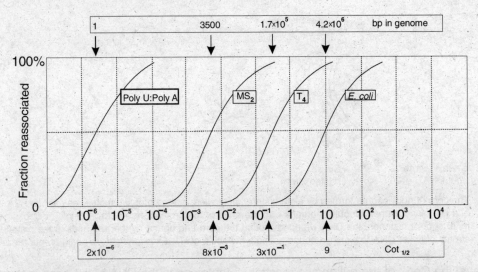

**Fig. 13.3:** *Cot* curves for different genomes. The form of each curve is similar but the *Cot* required in each case is different.

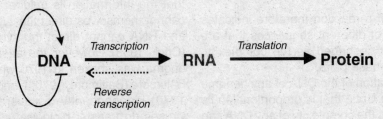

**Fig. 13.4:** Central dogma: .... line indicates that reverse transcription was added later to the central dogma proposed by Crick in 1958.

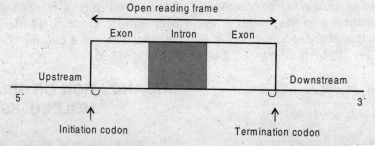

**Fig. 13.5:** Key features of an eukaryotic gene.

transfer RNA (tRNA) genes are transcribed but not translated in to polypeptides. These are non coding RNAs, which have various functions in the cell.

The part of a protein coding gene that is translated in to protein is called open reading frame (ORF). The ORF is read in the 5′ to 3′ direction along the mRNA. The ORF starts with an initiation codon and ends with a termination codon (Fig. 13.5). The majority of genes are spaced more or less randomly along the length of a DNA molecule. In some cases genes are grouped into distinct clusters. The clusters are made up of genes that contain related units of biological information. Examples of such cluster of genes are operons and multigene families.

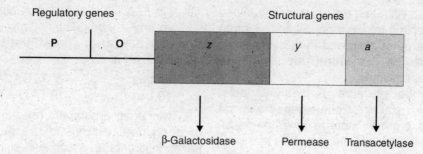

• **Fig. 13.6:** Lactose operon of *E. coli* showing different genes and products.

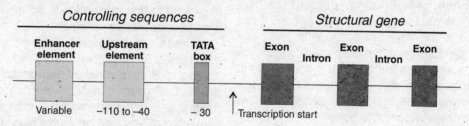

• **Fig. 13.7:** Diagrammatic representation of a typical nuclear gene.

## Operon

Operons are fairly common in the organization of genes in bacteria. An operon is a cluster of genes coding for a series of enzymes that work in concert in an integrated biochemical pathway. For example lactose operon of *E. coli* is a cluster of three genes each coding for one of three enzymes involved in conversion of disaccharide lactose into monosaccharide units of glucose and galactose (Fig. 13.6).

## Multigene family

These are found in many organisms. A multigene family is also a cluster of related genes, but here the individual genes have identical or similar nucleotide sequences and therefore contains identical or similar pieces of biological information. There are simple multigene families in which all of the genes are exactly or virtually the same, e.g. all the 5S rRNA genes are clustered into a single multigene family. Complex multigene families are made up of similar but non-identical genes, e.g. globin polypeptides of vertebrates. In humans, four globin polypeptides 2$\alpha$ and 2$\beta$ combines with a heme cofactor to make a single molecule of hemoglobin.

## PLANT GENE STRUCTURE AS DISCONTINUOUS GENE

A number of plant structural genes have been analyzed in a variety of species. Fig 13.7 represents a typical nuclear gene with some of its sequences. Several investigators in 1971 discovered that the biological information carried by the genes is split into distinct units separated by intervening segments of DNA. The sections containing biological information or coding sequences are called exons and the intervening sequences are referred to as introns. These discontinuous genes also called split or mosaic genes, are now known to be common in higher organisms. These introns are transcribed but not represented in mature mRNA and are hence not translated. No introns have been found in rRNA genes but they have been demonstrated in a number of plant structural genes. In higher organisms a gene may contain no introns or may have as many as hundred. Often introns are much longer than exons.

Introns can be detected by comparing the nucleotide sequences of DNA to the mRNA using genomic clones. Cloned genomic DNA is annealed with RNA and the products examined under an electron microscope. Since the intron sequences are not contained in the RNA, they will form single stranded loops in the heteroduplex molecule.

The number of introns is highly variable. There are a number of plant genes with no introns at all, e.g. zein storage genes of maize, soybean protein genes, one rare class leaf gene and most *cab* genes (nuclear genes encoding chlorophyll *a/b* binding protein of the photosystem). Plant genes with two or three introns are the soybean actin gene, soybean leghemoglobin gene, and soybean glycinine genes, a rare class leaf gene and genes for the small subunit of RuBP carboxylase. There are nine introns in the maize *Adh* genes and five intron in the phytochrome gene.

When intervening sequences are present, all plant genes are identical in the first and last two bases of the introns GT and AG respectively. Conservation of the two dinucleotides at the intron/exon junctions in all eukaryotic genes suggests that similar RNA splicing mechanisms are involved.

## CONTROL SEQUENCES

In prokaryotes, leader sequence of bacterial transcript contains a short sequence called Shine-Dalgarno sequence (SDS). It is complementary to a polypyrimidine sequence at 3' end of 16s rRNA. Base pairing between these sequences is the mean by which bacterial ribosome select the correct AUG codon for initiation (Fig. 13.8). Such type of sequences are not found in eukaryotes.

Initiation of transcription occurs when RNA polymerase binds to a specific DNA sequence called the promoter at the beginning of an operon that occurs just upstream of genes. The maximum rate of transcription from a particular segment of DNA depends on the sequence of bases in its promoter. The actual DNA sequence at a promoter site varies from gene to gene, but all promoters have a consensus sequence,

which in *E. coli* has two distinct components, a –35 box and –10 box. The –10 box is also called Pribnow box after the scientist who first characterized it.

| | |
|---|---|
| **–35 box** | 5' TTGACA 3' |
| **–10 box** | 5' TATAAT 3' |

The negative numbers refer to the position of each box relative to the site at which transcription starts, which is designated as +1. The +1 position therefore corresponds to the site at which the first nucleotide is polymerized into the encoded RNA molecule.

These consensus sequences were determined by sequencing the DNA of a large number of *E. coli* promoters and identifying those bases that predominate at certain positions relative to known transcriptional start sites. The sequence of any one promoter may differ from the consensus sequence and certain variations give rise to promoters with reduced efficiency of RNA polymerase binding. This in turn leads to lower levels of transcription and is an important factor in the control of gene expression. The sequences of a few of the promoters recognized by the *E. coli* RNA polymerase have been shown in Table 13.3.

In eukaryotes, the RNA polymerase attachment site is most important, which is usually just upstream of the gene to be

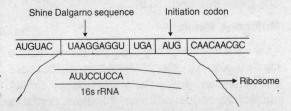

**Fig. 13.8:** Shine-Dalgarno sequences in bacterial transcript.

**Table 13.3:** The sequences of a few of the promoters recognized by the *E. coli* RNA polymerase

| Gene | –35 sequence | –10 sequence |
|---|---|---|
| *E. coli* consensus | TTGACA | TATAAT |
| *Lac* operon | TTTACA | TATGTT |
| *Trp* operon | TTGACA | TTAACT |
| TRNA gene | TTTACA | TATGAT |

transcribed. This attachment site is equivalent to the *E. coli* promoter but is not the only nucleotide sequence that has a role in transcription.

### TATA box

Upstream from the cap site are one or two short sequence elements that are common to many eukaryotic genes which are involved in normal transcription. Of the sequences, TATA-box or Goldberg Hogness box is the best characterized. This sequence normally occurs −25 to −30 nucleotides upstream of the cap site in almost all plant and animal genes. This sequence is required for correct expression of eukaryotic genes *in vitro* and *in vivo*.

### AGGA box

Another sequence that may be involved in regulation of transcription of some eukaryotic genes is the consensus sequence "CAAT" box or GG (CT) CAATCT. This sequence is often found −40 and −180 nucleotides upstream of the cap site. The sequence of these elements is much more variable but often includes the sequence CAAT, termed the CAAT box. In plants it is referred to as the AGGA box.

### Other regulatory elements

Enhancer sequences are another class of elements that show high level of transcription. These sequences were originally discovered in viral genes. Now these have also been found in plants. Enhancers are located within a few hundred nucleotides upstream of the start of transcription. Although most enhancer elements are found in flanking sequences on the 5′ ends of genes, they are also found to occur elsewhere. An enhancer core sequence such as GGTGTGGAAAG, or more generally GTGGT/AT/AT/AG written as GTGGWWWG (T/A means that either T or A is now referred as W is found in this position—See appendix), can be identified by sequence comparisons, although it is clear that sequences other than the core also are important. Enhancer-like sequences in plants stimulate the tissue specific expression of genes.

# TYPES OF RNA MOLECULES

The RNA molecules produced by transcription can be grouped into three major classes according to function. These are ribosomal or rRNA, transfer or tRNA and messenger or mRNA. Besides, heterogeneous nuclear RNA and small nuclear RNA are also present. Ribosomal and transfer RNAs are the end products of gene expression and perform their roles in the cell as RNA molecules. Messenger RNA undergoes the second stage of gene expression, acting as the intermediate between a gene and its final expression product, a polypeptide.

Ribosomal and transfer RNAs are often referred to as stable RNAs, indicating these molecules are long lived in the cell, in contrast to mRNA, which has a relatively rapid turnover rate. Messenger RNA is involved in transcription and both types of stable RNA—rRNA and tRNA—are involved in translation of mRNA information to polypeptide synthesis. Details of these RNAs have been described before the actual process of protein synthesis. The aim is to describe various sequences involved in gene regulation.

### Messenger RNA and processing

Messenger RNA acts as the intermediate between the gene and the polypeptide translation product. mRNA molecules are not generally long lived in the cell. Most bacterial mRNAs have a half-life of only a few minutes, while eukaryotic cells mRNA molecules have longer half lives (e.g. 6 h for mammalian mRNA). In prokaryotes, the mRNA molecules that are translated are direct copies of the gene. In eukaryotes, most mRNAs undergo a fairly complicated series of modifications and processing events before translation occurs. These are:

1. Chemical modifications to the two ends of the mRNA molecules.
2. Removal of introns.
3. RNA editing.

*Chemical modifications to the two ends of the mRNA molecules*

That the eukaryotic mRNA undergoes modification became clear when fractions present in the nucleus and cytoplasm were compared. The nucleus can also be divided into two regions: the nucleolus, in which rRNA genes are transcribed, and the nucleoplasm, where other genes including those for mRNA, are transcribed. The nucleoplasmic RNA fraction is called heterogeneous nuclear RNA or hnRNA, the name indicating that it is made up of a complex mixture of RNA molecules, some over 20 kb in length. The mRNA in the cytoplasm is also heterogeneous, but its average length is only 2 kb. If the mRNA in the cytoplasm is derived from hnRNA, then modification and processing events, including a reduction in the length of the primary transcripts, must occur before mRNA leaves the nucleus.

An RNA molecule synthesized by transcription and subjected to no additional modification will have at its 5′ end the chemical structure **pppNpN...** where N is the sugar-base component of the nucleotide and p represents a phosphate group. Note that a 5′ terminus will be a triphosphate.

It was found that the 5′ end of eukaryotic mRNAs are blocked by the addition of 7-methyl-Gppp caps (7 methyl guanosine residues joined to the mRNAs by triphosphate linkages) that are added during the synthesis of the primary transcript (Fig. 13.9).

Some eukaryotic mRNAs may undergo further modification of the 5′ end with the addition of methyl groups to one or both of the next two nucleotides of the molecule. The 5′ caps may guide ribosomes to begin translation at the correct AUG codon.

A second modification of most eukaryotic RNAs is the addition of a long stretch of up to 250 A nucleotides to the 3′ end of the molecule, producing a poly(A) tail:

...pNpNpA(pA)$_n$pA

Polyadenylation does not occur at the 3′ end of the primary transcript. Instead the final few nucleotides are removed by a cleavage event that occurs between 10 and 30 nucleotides downstream of a specific polyadenylation signal (consensus sequences 5′–AAUAAA–3′) to produce an intermediate 3′ end to which the poly(A) tail is subsequently added by the enzyme poly(A) polymerase.

The reason for polyadenylation is not known, although several hypotheses have pointed that the length of poly(A) tail determines the time in which the mRNA survives in the cytoplasm before being degraded.

*Removal of introns*

Eukaryotic genes contains introns, and transcription produces a faithful copy of the template strand. However, the introns must be removed and exon regions of the transcript attached to one another before translation occurs. This process is called splicing. Different types of introns have different splicing mechanisms. All the introns found in nuclear genes that are transcribed into mRNA fall into GT-AG class of introns. This designation GT-AG refers to a particular class in which introns possess the same dinucleotide sequences at their 5′ and 3′ ends. These sequences are GT for the first two nucleotides of the intron (GT is the sequence in the DNA and GU is the transcript) and AG for the last two nucleotides. In fact, the GT-AG rule represents only a component of slightly longer consensus sequences found at the 5′ and 3′ splice junctions of GT-AG introns. The consensus sequences are

5′ splice site    5′–AGG↓TAAGT–3′
3′ splice site    5′–PyPyPyPyPyPyNCA↓G–3′/

5′–YYYYYYNCA↓G–3′

These are slightly more complicated consensus sequences. 'Py' indicates a position where either of the two-pyrimidine nucleotides (C or T referred as Y) can occur, and 'N'

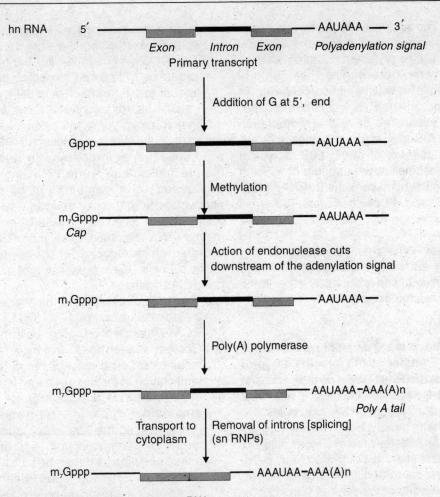

• **Fig. 13.9:** Processing of eukaryotic messenger RNA.

represents any nucleotide. The consensus sequence for the 3′ splice site is therefore six pyrimidines in a row, followed by any nucleotide, followed by a C and then the AG. The actual sequences in an intron may be slightly different, but the GT and AG components are always present.

In outline, the splicing reaction involves the following steps:

1. Cleavage occurs at the 5′ splice site.
2. The resulting free 5′ end is attached to an internal site within the intron to form a lariat structure.
3. The 3′ splice site is cleaved and the two exons joined together.

## RNA editing

In RNA editing the nucleotide sequence of a mRNA is altered by inserting new nucleotides or by deleting or changing existing ones.

## Ribosomal RNA

Ribosomal RNA molecules are components of ribosomes that act as factories for protein synthesis. During translation, ribosomes attach to mRNA molecules and migrate along them, synthesizing polypeptides. Ribosomes are made up of rRNA molecules and proteins, and are numerous in most cells. They comprise about 80% of the total cell RNA and about 10% of the total protein. The eukaryotic ribosomes are large, having a total molecular mass of 4,220,000 daltons and a size of 32nm × 22nm. Sizes of ribosomes are determined by velocity sedimentation. The sedimentation coefficient is expressed as S-value (S = Svedberg unit). Prokaryotic ribosomes have a sedimentation coefficient of 70S while eukaryotic ribosomes are variable but average of about 80S. A typical eukaryotic ribosome has two subunits of 40S and 60S. The large subunit contains 3 rRNAs (28S, 5.8S and 5S) and 49 polypeptides. The small subunit has a single 18S rRNA and 33 polypeptides.

Ribosomal RNAs characteristically contain a number of specially modified nucleotides including pseudouridine residues, ribothymidylic acid and methylated bases.

## Transfer RNA

Each cell contains many different types of tRNA molecules. Transfer RNAs are the adapter molecules that read the nucleotide sequences of mRNA transcript and convert it into a sequence of amino acids. Transfer RNA molecules are relatively small, mostly between 74 and 95 nucleotides, for different molecules in different species. Almost every tRNA molecule in every organism can be folded into a base paired structure referred to as clover-leaf (Fig. 13.10). This structure is made up of the following components

1. The acceptor arm formed by a series of usually seven base pairs between nucleotides at the 5′ and 3′ ends of the molecule. During protein synthesis an amino acid is attached to the acceptor arm of the tRNA.
2. The D or DHU arm, so named because the loop at its end almost invariably contains the unusual pyrimidine dihydrouracil.
3. The anticodon arm plays the central role in decoding the biological information carried by the mRNA.
4. The extra, optional or variable form that may be a loop of just 3–5 nucleotides, or a much larger stem loop of 13–21 nucleotides with up to 5 base pairs in the stem.
5. The TΨC arm, named after the nucleotide sequence T, Ψ, C (Ψ = nucleotide containing pseudouracil, another unusual pyrimidine base), which its loop virtually always contain.

In both prokaryotes and eukaryotes, tRNAs are transcribed initially as precursor tRNA which are subsequently processed to release the mature molecules. A pre-tRNA molecule is processed by a combination of different ribonucleases that make specific cleavages at the 5′ (RNase P) and 3′ (RNase D) ends of the mature tRNA sequence. Eukaryotic tRNA genes are also clustered and in addition occur in multiple copies. The processing of eukaryotic pre-tRNAs includes

(a) The three nucleotides at the 3′ end may be added after transcription. All tRNAs have 5′–CCA–3′ trinucleotide sequence at the 3′ end.
(b) Certain nucleotides undergo chemical modification. The most common types are methylation of the base or sugar component of the nucleoside, e.g. guanosine to 7-methyl guanosine; base rearrangements (interchanging the position of atoms in the purine or pyrimidine ring, e.g. uridine to pseudouridine; double bond saturation (uridine to dihydrouridine); deamination (removal of an amino group, e.g. guanosine to inosine); sulphur substitution (uridine to 4-thiouridine); addition of complex groups (guanosine to queosine).

Over 50 types of chemical modifications have been discovered, each catalyzed by a different tRNA modifying enzyme. Each tRNA molecule is able to form a covalent linkage with its specific amino acid at the end of the acceptor arm of tRNA clover-leaf by a group of enzymes called aminoacyl tRNA synthetases and the process is called aminoacylation or charging.

## Small Nuclear RNA

Small nuclear RNAs or snRNAs are a class of RNA molecules found in the nucleus of eukaryotic cells. They are neither tRNA nor small

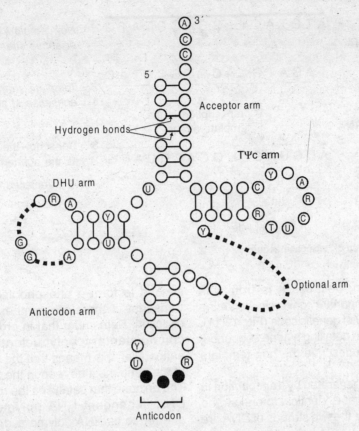

**Fig.13.10:** Cloverleaf model of tRNA.

rRNA molecules, although they are similar in size to these species. They contain from 100 to 200 nucleotides, and some of the bases are methylated and modified as in tRNA and rRNA. The snRNA does not exist as naked RNA but is found as stable complexes with specific proteins forming small nuclear ribonucleoprotein particles or snRNPs, which are about 10S in size. It is present in 1 to 10% of the total amount of ribosomes. The snRNPs are important in the processing of eukaryotic gene transcripts (hnRNA) into mature mRNA for export from the nucleus to the cytoplasm.

# TRANSCRIPTION

The first stage of gene expression, where the template strand of the gene directs the synthesis of a complementary RNA molecule, is known as transcription. The RNA molecule that is synthesized is called a transcript. The notation used to describe the nucleotide sequence of a gene and the actual process of transcription in prokaryotes and eukaryotes has been explained.

## Nucleotide sequences

To illustrate aspects of gene structure and expression, a standard procedure is followed for nucleotide sequences. Usually the sequence presented will be that of the non-template strand of the gene in the 5′→3′ direction. This may seem illogical as we know that biological information is carried by the template strand and not by the complement. However, it is the sequence of the RNA transcript (which is the same as that of non-template strand) that the molecular biologist is

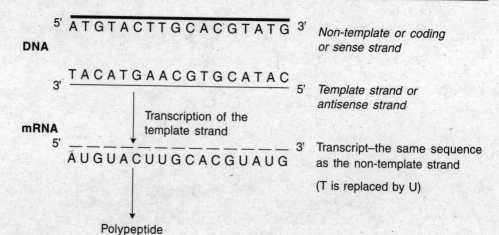

Fig 13.11: Representation of nucleotide sequence.

really interested in, because it is from this sequence that the important conversion to amino acids is made by way of genetic code (Fig. 13.11). For this reason, when detailing a gene sequence, the convention is to describe the non-template strand.

The template is generated by transcription in the form of mRNA that is identical to one strand of DNA duplex. Thus, the two strands of DNA are as follows:

The DNA strand that bears a sequence identical to the mRNA (except for possessing T instead of U) is called the **coding strand** or **sense strand** or **non-template strand**.

The other strand of DNA, which directs the synthesis of mRNA via complementary base-pairing is called the **template strand** or **antisense strand**. The antisense strand has a 3'→ 5' direction which directs the synthesis of mRNA in the 5'→ 3' direction.

The underlying chemical reaction in transcription is the synthesis of an RNA molecule. This occurs by polymerization of ribonucleotide subunits in the presence of a DNA template which directs the order in which the individual ribonucleotides are polymerized in to RNA. During polymerization the 3' OH group of one ribonucleotide reacts with the 5' P group of the

second to form a phosphodiester bond. The transcript is therefore synthesized in 5'→ 3' direction. Remember that in order to base pair, complementary polynucleotides must be antiparallel: this means that the template strand of the gene must be read in the 3'→5' direction. The enzyme that catalyzes the reaction is called DNA dependent RNA polymerase (usually referred to as RNA polymerase).

# TRANSCRIPTION IN PROKARYOTES

### RNA polymerase in *E. coli*

A single enzyme species catalyzes the synthesis of all types of *E. coli* RNA molecules (messenger, transfer and ribosomal). This is in contrast to the finding in eukaryotic cells where several different RNA polymerases have been identified.

The complete *E. coli* RNA polymerase complex is composed of five protein subunits: two alpha ($\alpha$) subunits, one beta ($\beta$) and one beta prime ($\beta'$) subunit, and a sigma ($\sigma$) subunit. Collectively, the complex is known as the holoenzyme and is represented as $\alpha_2\beta\beta'$ $\sigma$. The $\alpha_2\beta\beta'$ core of the enzyme is required for the polymerase activity. The $\sigma$ subunit is required solely for DNA sequence recognition at a promoter site.

The actual process of transcription can be divided into three phases: initiation, elongation and termination.

## Initiation

Initiation of transcription occurs with the binding of RNA polymerase enzyme to a DNA molecule at a specific position, just in front (upstream) of the gene. These attachment points are called promoters. A promoter is a short nucleotide sequence that is recognized by an RNA polymerase as a point at which to bind to DNA in order to begin transcription. The actual DNA sequence at a promoter site varies from gene to gene, but all promoters have a consensus sequence which for *E. coli* has two distinct components a −35 box and −10 box. This has been explained earlier.

The σ subunit attaches to the promoter sequence (following recognition by σ), while the other subunits cooperate in catalyzing RNA synthesis during transcription. Once transcription has been initiated, the σ subunit dissociates from the rest of the enzyme after approximately 8 ribonucleotides have been polymerized into RNA, leaving the core enzyme to continue transcription. At initiation the RNA polymerase complex has an elongated shape and covers up to 80bp of DNA; during movement along the DNA the core enzyme covers approximately 30bp of DNA. The optimal rate of polymerization is estimated to be approximately 40 nucleotides per second at 37°C.

## *Elongation*

During the elongation stage of transcription the RNA polymerase moves along the DNA, reading the template information in the 5′→3′ direction. An RNA-DNA hybrid is thus formed in the unwound regions. It must be noted that requirement of all double stranded nucleic acid structures is that antiparallel nature be maintained. As the RNA polymerase moves further along the DNA template, the newly synthesized RNA strand is displaced and the DNA duplex reforms behind the transcriptional complex. Three points should be noted:

1. The transcript is not longer than the gene: Position +1 is the point where synthesis of the RNA transcript actually begins, but it is not the start of the gene itself. Almost invariably the RNA polymerase enzyme transcribes a leader segment before reaching the gene. The length of the leader sequence varies from gene to gene. Similarly, when the end of the gene is reached, the enzyme continues to transcribe a trailer segment before termination occurs.

2. Only a small region of the double helix is unwound at any one time. Once elongation has started, the polymerized portion of the RNA molecule can gradually dissociate from the template strand, allowing the double helix to return to its original state. The open region or transcription bubble contains between 12 and 17 RNA-DNA base pairs. This is especially important because unwinding a portion of the double helix necessitates overwinding the adjacent areas and there is a limited molecular freedom for this to occur.

3. The rate of elongation is not constant. It has been observed that occasionally the enzyme will slow down, pause and then reaccelerate, or even go so far as to reverse over a short stretch, removing ribonucleotides from the end of the newly synthesized transcript before resuming its forward course.

## *Termination*

Termination signals for *E. coli* genes vary immensely but all possess a common feature: they are complementary palindromes. The complementary palindromes exert their influence on transcription by enabling a stem loop to form in the growing RNA molecule. Some stem loop terminators are followed by a run of 5 to 10 As in the DNA template which after transcription into Us will leave the transcript attached to the template by a series of weak A–U base pairs. It is thought that RNA polymerase pauses just after the stem loop structure and that the A–U base pairs then break, so that the transcript detaches from the template.

Other terminators do not have the run of As. These are called Rho-dependent terminators. Rho is a large protein and is not a true subunit of the RNA polymerase. Rho attaches to the growing RNA molecule and when the polymerase pauses at the stem loop structure, it disrupts the base pairing between the DNA template and RNA transcript.

Termination of transcription results in dissociation of the RNA polymerase enzyme from the DNA and release of the RNA transcript. The core enzyme is now able to reassociate with a subunit and begin a new round of transcription on the same or a different gene.

# TRANSCRIPTION IN EUKARYOTES

## Eukaryotic RNA polymerases

Eukaryotic cells contain three different RNA polymerases, each with a distinct functional role. Ribosomal RNA (rRNA) is transcribed off rDNA gene by RNA polymerase I. The synthesis of messenger RNA is catalyzed by RNA polymerase II, with the subsequent addition of poly(A) tails carried out by the poly(A) polymerase. Transfer RNA and a variety of smaller nuclear and cytoplasmic RNAs are made by RNA polymerase III. The existence of three different types of eukaryotic RNA polymerases suggests the existence of three different types of promoters that can be regulated independently. Eukaryotic RNA polymerases are composed of many proteins that interact both with each other and with multiple sites on the DNA.

*Initiation*: Initiation of transcription in eukaryotes depends on recognition and binding of RNA polymerase II to a promoter sequence. This attachment site is not the only nucleotide

sequence that has a role in transcript initiation. Other possible sequences spread out over several hundred base pairs will also be involved in the binding. Besides this, certain transcription factors (TFs) are also involved. The eukaryotic core promoter element is called a 'TATA box' because it contains the attachment site for RNA polymerase II. The attachment site has the following consensus sequence

5′-TATAAAT-3′

### The TATA box

The TATA box is usually located about 25 nucleotides upstream of the position at which transcription starts, and is therefore also called –25 box. RNA polymerase attaches to TATA box, which is mediated by a set of proteins called transcription factors (Fig. 13.12). These TFs have been designated the names TFIIA, TFIIB and so on, distinguishing them from TFI and TFIII transcription factors that work with RNA polymerase I and III respectively. The first step in this attachment is the binding of TFIID to the TATA box, which is probably mediated by TFIIA. Once TFIID has attached to the DNA template the transcription complex starts to form. TFIIB binds next, followed by RNA polymerase II itself, and then the two final transcription factors TFIIE and TFIIF. Transcription complex is now ready to begin synthesizing RNA.

## Termination

Identification of termination signals for RNA polymerase has proven difficult because the 3′ ends of the transcripts are removed immediately after synthesis. But it is clear that the trailer regions may be very long, extending 1000 to 2000 bp downstream of the gene. This led to the suggestion that there are in fact no set termination points downstream of genes transcribed by RNA polymerase II. Possibly one of the transcription factors detaches from the complex at the end of the gene, destabilizing the complex and leading it to fall off the template at some later point.

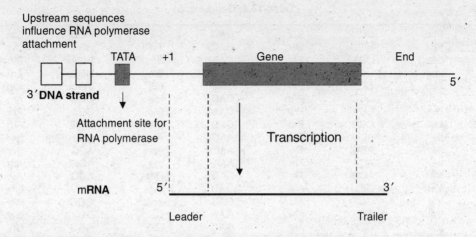

**Fig 13.12** A diagrammatic representation of different sequences involved in transcription.

## GENETIC CODE AND TRANSLATION

Translation is the process by which the information in the mRNA molecule is decoded into a polypeptide chain. This process is carried out by ribosomes and aminoacyl transfer RNAs. The central fact about translation is that the sequence of amino acids in the polypeptide being synthesized is specified by the sequence of nucleotides in the mRNA molecule being translated. The rule that determines which sequence of nucleotides specifies which amino acids is embodied in the genetic code.

### Genetic code

The problem of exactly how the genetic code works was the major preoccupation of biologists when double helix of DNA was postulated. Since proteins are constructed from a population of 20 different amino acids, the code that specifies them must have at least 20 different possible words. In order to decipher this coded information, two basic assumptions were made, which were subsequently confirmed.

1. Colinearity between gene and protein: It means the order of nucleotides in the gene directly correlates with the order of amino acids in the corresponding polypeptide.

2. Each code word or codon is a triplet of nucleotides that code for a single amino acid. Since there are only four different nucleotide bases in DNA (A, G, C, T) and correspondingly in mRNA (A, G, C, U), a one base code could only specify four different amino acids. A two base code could specify 16 amino acids (4 x 4), but this is still not sufficient. A three base code is capable of specifying all 20 amino acids since there are 64 possible combinations (4 x 4 x 4). Information in mRNA is indeed translated in units of three nucleotides, and these units are called triplet codons. The assignment of particular codons to particular amino acids is called the genetic code, and is shown in Table 13.4.

Of the 64 possible codons, 61 specify a particular amino acid, with the remaining 3 codons (UAA, UAG, UGA) being used as stop signals which indicate to the translation machinery that protein synthesis is complete. The codon AUG (which specifies methionine) is also known as a start codon since in almost all cases the start site for translation is determined by an AUG in the mRNA. Note that only two amino acids (methionine and tryptophan) have a single codon; all other amino acids have more than one.

**Table 13.4:** Genetic code

| UUU | Phe | UCU | Ser | UAU | Tyr | UGU | Cys |
|-----|-----|-----|-----|-----|-----|-----|-----|
| UUC | Phe | UCC | Ser | UAC | Tyr | UGC | Cys |
| UUA | Leu | UCA | Ser | UAA | Stop | UGA | Stop |
| UUG | Leu | UCG | Ser | UAG | Stop | UGG | Trp |
| CUU | Leu | CCU | Pro | CAU | His | CGU | Arg |
| CUC | Leu | CCC | Pro | CAC | His | CGC | Arg |
| CUA | Leu | CCA | Pro | CAA | Gln | CGA | Arg |
| CUG | Leu | CCG | Pro | CAG | Gln | CGG | Arg |
| AUU | Ile | ACU | Thr | AAU | Asn | AGU | Ser |
| AUC | Ile | ACC | Thr | AAC | Asn | AGC | Ser |
| AUA | Ile | ACA | Thr | AAA | Lys | AGA | Arg |
| AUG | Met | ACG | Thr | AAG | Lys | AGG | Arg |
| GUU | Val | GCU | Ala | GAU | Asp | GGU | Gly |
| GUC | Val | GCC | Ala | GAC | Asp | GGC | Gly |
| GUA | Val | GCA | Ala | GAA | Glu | GGA | Gly |
| GUG | Val | GCG | Ala | GAG | Glu | GGG | Gly |

Key to amino acid abbreviations:
Ala: alanine; Arg: arginine; Asn: asparagine; Asp: aspartic acid; Cys: cysteine; Gln: glutamine; Glu: glutamic acid; Gly: glycine; His: histidine; Ile: isoleucine; Leu: leucine; Lys: lysine; Met: methionine; Phe: phenylalanine; Pro: proline; Ser: serine; Thr: threonine; Trp: tryptophan; Tyr: tyrosine; Val: valine.

The genetic code has three important features:

1. The code is degenerate: All amino acids except methionine and tryptophan have more than one codon, so that all possible triplets have a meaning. Most synonymous codons are grouped into families (GGA, GGU, GGG and GGC, for example all code for glycine.

2. The code contains punctuation codons: Three codons UAA, UAG and UGA do not code for amino acid and if present in the middle of a heteropolymer cause protein synthesis to stop. These are called termination codons and one of the three always occurs at the end of a gene at the point where translation must stop. Similarly, AUG virtually always occurs at the start of a gene and marks the position at which translation should begin. AUG is called initiation codon and it codes for methionine. Thus, most newly synthesized polypeptides will have this amino acid at the amino terminus, though it may be subsequently removed by post-translational processing of the protein.

3. Universality of genetic code: When the genetic code was completed in 1966 it was assumed to be universal, i.e. it was used by all organisms and there is considerable evidence that genetic code is essentially universal. However, there is some variation in the preference for different codons in different organisms. For example, AUA codon is for isoleucine in eukaryotes but *E. coli* does not use AUA instead it uses AUU and AUC. Mitochondrial genes also use a slightly different code in which UGA (normally a stop codon) codes for tryptophan. AGG and AGA are stop codons in mitochondria while usually it codes for arginine and AUA codes for methionine in mitochondrial genome while isoleucine is usually coded. Other rare differences have also come to light, but it is true to say that the code shown in Table 13.4 is followed by virtually all nuclear genes.

# TRANSLATION

There are three major steps in the synthesis of polypeptide
  1. Chain initiation
  2. Chain elongation
  3. Chain termination

## Chain initiation

The first step in translation is the attachment of the small 40S subunit of ribosome (30S in prokaryotes) to the mRNA molecule. The ribosome binding site ensures that translation starts at the correct position. The correct attachment point for ribosomes in *E. coli* has the following consensus sequence

  5′ A G G A G G U 3′

An eukaryotic mRNA does not have an internal ribosome binding site equivalent to the above consensus sequence. Instead, the small subunit of the ribosome recognizes the cap structure as its binding site and therefore attaches to the extreme 5′ end of the mRNA molecule. The translation process itself starts when an aminoacylated tRNA base pairs with an initiation codon that has been located by the small subunit of ribosome. The initiator tRNA is charged with methionine, because methionine is the amino acid coded by AUG. The initiation process involves the ordered formation of a complex which constitute an initiator factor (eIF2), GTP, the initiating methionine tRNA (met-tRNA) and the small 40S subunit ribosomal unit. This complex along with other initiation factors binds to the 5′ end of mRNA and then moves along the mRNA until it reaches the first AUG codon (Fig. 13.13).

### Chain elongation

Once the initiation complex has formed, the large subunit joins the complex, which requires the hydrolysis of GTP and the release of initiation factors. Two separate and distinct sites are now available at the ribosome to which tRNAs can bind. The first of these called the peptidyl or P-site is at that moment occupied by the aminoacylated t-RNA-met which is still base paired with the initiation codon. The aminoacyl or A-site is positioned over the second codon of the gene and at first it is empty. Elongation begins when the correct aminoacylated tRNA enters the A-site and base pairs with the second codon. This requires elongation factor eEF1 and a second molecule of GTP is hydrolyzed to provide the required energy. Now both sites of the ribosome are occupied by aminoacylated tRNAs and the two amino acids are in close contact. A peptide bond is formed between the carboxyl group of methionine and the amino group of the second amino acid. The reaction is catalyzed by peptidyl transferase. Peptidyl transferase acts in conjunction with tRNA deacylase that breaks the link of met-tRNA. This results in a dipeptide attached to the tRNA present in the A-site.

Translocation occurs when the ribosome slips along the mRNA by three nucleotides so that aa-aa-tRNA enters the P-site, expelling the now uncharged tRNA$^{met}$ and making the A-site vacant again. The third aminoacylated tRNA enters the A-site and the elongation cycle is repeated. Each cycle requires hydrolysis of a molecule of GTP and is controlled by an elongation factor eEF2.

After several cycles of elongation, the start of the mRNA molecule is no longer associated with the ribosome, and a second round of translation can begin by binding of 40S subunit and forming a new initiation complex. The end result is a polysome, a mRNA being translated by several ribosomes at once.

### Chain termination

Termination of translation occurs when a termination codon (UAA, UAG or UGA) enters the A-site. There are no tRNA molecules with anticodons able to base pair with any of the termination codons. A termination factor eRF recognizes all the termination codons in eukaryotes and it requires hydrolysis of a GTP molecule. The ribosome releases the polypeptide and mRNA and subsequently

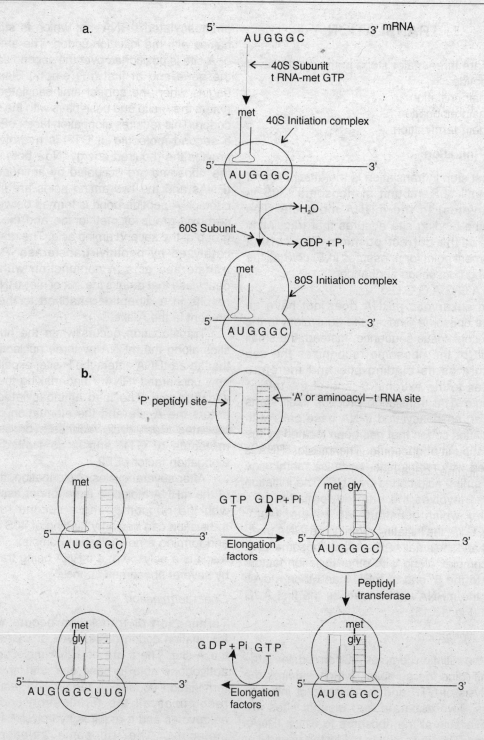

**Fig. 13.13:** Initiation (a) and elongation (b) process of polypeptide chain synthesis.

dissociates into subunits, which enter the cellular pool before becoming involved in a new round of translation. The mechanism of protein synthesis is essentially the same for both plants and animals. Ribosomes from one species will faithfully translate mRNA from the other.

## Post-translation modifications

Primary translation products undergo a variety of modifications vital to the production of a fully functional protein. These post-translational processing steps are varied and more important in eukaryotes. One reason for this is the existence of more membrane-bound compartments into which proteins must be directed.

The membrane-bound polysomes synthesize proteins which are either secreted from the cell, or form part of a membrane, or are packaged into specialized organelles (e.g. seed storage protein bodies). It has been postulated that translocated proteins carry an N-terminal extension of about 20 amino acids, termed as signal peptide. It binds to a receptor in the membrane as soon as it is synthesized and emerges from the ribosome. Signal peptides have been found in the primary translation products of a number of mRNAs, e.g. zein from maize. These signal peptides are to be removed. Besides, glycosylation occurs during passage of proteins through a membrane. Many proteins remain inactive until their polypeptide chains are cleaved by proteases. In some cases,

the original single polypeptide chain is not a functional product, fragments produced by proteolytic cleavage are needed. Other types of processing involve the addition of moieties (carbohydrates, lipids, cofactors and phosphate groups) to the amino acid residues.

In conclusion it can be seen that there are many steps in eukaryotes, each of which must function properly for the successful expression of a gene. Thus, there are many potential points at which gene expression may be controlled.

## REFERENCES

Arumuganathan, K. and Earle, E.D. 1991. Nuclear DNA content of some important plant species. *Plant Mol. Biol. Rep.* **9:** 208–221.

Baltimore, D. 1970. RNA dependent DNA polymerase in virions of RNA tumor virus. *Nature* **226:** 1209–1211.

Crick, F.H.C. 1958. On protein synthesis. *Symp. Soc. Expt. Biol.* **12:** 138–163.

## FURTHER READING

Adams, R.L.P., Knowler, J.T. and Leader, D.P. 1992. *The Biochemistry of the Nucleic acids*, 11th ed, Chapman & Hall, London.

Garret, R.H. and Grisham, C.M. 1999. *Biochemistry*, Saunders College Publishing, Orlando, Florida.

Lewin, B. 1995. *Genes V*. Oxford University Press, Oxford, U.K.

Snustad, D.P., Simmons, M.J. and Jenkins, J.B. 1997. *Principles of Genetics*. John Wiley & Sons, Inc., New York.

# PART III

# Recombinant DNA Technology

# 14

# Basic Techniques

The last few years have seen the rapid development of new methodologies in the field of molecular biology. The initial impetus for rDNA technology came about in the early 1970s with the simultaneous development of techniques for the transformation of *E. coli*, cutting and joining of DNA molecules, and monitoring of the products of cutting and joining reaction. New techniques have been regularly introduced and the sensitivity of older techniques greatly improved upon. To explain the processes of gene cloning and gene manipulations, understanding of certain basic techniques is essential. These techniques are discussed in this chapter.

**Agarose gel electrophoresis**

Electrophoresis through agarose gels is the standard method for the separation, identification, and purification of DNA and RNA fragments ranging in size from a few hundred to 20 kb. Besides agarose, polyacrylamide gel electrophoresis is also used for the same purpose, but is preferred for small DNA fragments. The technique of agarose gel electrophoresis is simple, rapid to perform, and capable of resolving DNA fragments that cannot be separated adequately by other procedures such as density gradient centrifugation. Furthermore, the location of DNA within the gel can be determined directly by staining with low concentrations of intercalating fluorescent

ethidium bromide dye under ultraviolet light. If necessary, these bands of DNA can be recovered from the gel and used for a variety of cloning purposes.

Agarose is a linear polymer extracted from seaweed. Agarose gels are cast by melting the agarose in the presence of desired buffer until a clear transparent solution is obtained. The melted solution is poured into a mould and can be made into a variety of shapes, sizes, and porosities (4–10%) and the gel is allowed to harden. On hardening, the agarose forms a matrix, the density of which is determined by the concentration of agarose. When an electric field is applied across the gel, DNA, which is negatively charged at neutral pH, migrates towards the anode. The rate of migration depends on a number of parameters: (1) molecular size of DNA, (2) agarose concentration, (3) conformation of DNA, and (4) composition of electrophoresis buffer. Larger molecules migrate more slowly than the smaller molecules do because they have to find their way through the pores of the gel. By using gels of different concentrations it is possible to resolve a wide range of DNA molecules. The electrophoretic mobility of DNA is also affected by the composition and ionic strength of electrophoresis buffer. In the absence of ions, electrical conductivity is minimal and DNA migrates very slowly. In buffers of high ionic strength, electrical conductance is very efficient. Several different buffers are available, viz. TAE (Tris-acetate), TBE (Tris-borate), TPE (Tris-phosphate) and alkaline buffer, at a pH range of 7.5–7.8. Electrophoresis is normally carried out at room temperature. Gels are stained with intercalating fluorescent ethidium bromide dye and as little as 0.05 µg of DNA in one band can be detected as visible fluorescence under ultraviolet light. Ethidium bromide is a powerful mutagen and toxic, and therefore proper care should always be taken while handling the solution, and the staining solution should always be decontaminated after its use.

## Pulsed field gel electrophoresis

In recent years the means to analyze entire genomes have increased enormously, culminating in the construction of physical maps of the genomes of *E. coli*, *Bacillus subtilis*, *Mycobacterium tuberculosis*, *Caenorhabditis elegans,* and the sequencing of the yeast and human genomes. In plants the progress has not been as fast, though the situation is improving rapidly. *Arabidopsis* and rice are being used as model systems and have been completely sequenced.

The pulsed field gel electrophoresis (PFGE) technique was conceived by Schwartz and Cantor (1984) to allow the electrophoretic separation of DNA molecules several megabases in size. This was necessary for the development of megabase mapping and cloning techniques because all linear double-stranded DNA molecules larger than a certain size migrate through agarose gels at the same rate. Agarose gel electrophoresis is the standard technique for separating nucleic acids; however, conventional agarose gels cannot resolve fragments >20 kb. This is because large DNA fragments are unable to migrate through the gel matrix by the same sieving mechanism that small fragments do. Ordinary agarose gel electrophoresis fails to separate extremely long DNA molecules because the steady electric field stretches them out so that they travel end-first through the gel

matrix in a snake-like fashion. This mode of migration is known as reptation. This DNA can no longer be sieved according to its size. The greater the pore size of the gel, the larger the DNA that can be sieved. Thus, agarose gels cast with low concentrations of agarose (0.1–0.2%) are capable of resolving extremely large DNA molecules. However, such gels are quite fragile and must be run very slowly. Even then, they are incapable of resolving linear DNA molecules larger than 750 kb in length. Pulsed field gel electrophoresis is a variation of agarose gel electrophoresis, which takes advantage of the electrophoretic property of large DNA fragments. PFGE resolves DNA molecules of 100–1000 kb by intermittently changing the direction of electric field in a way that causes large DNA fragments to realign more slowly with the new field direction than do smaller molecules. In the PFGE method, pulsed alternating, orthogonal electric fields are applied to a gel, which forces the molecules to reorient before continuing to move snake-like through the gel. This reorientation takes much more time for larger DNA molecules, so that progressively longer DNA molecules move more and more slowly. Large DNA molecules become trapped in their reptation tubes every time the direction of electric field is altered and can make no further progress through the gel until they have reoriented themselves along the new axis of the electric field. The larger the DNA molecule, the longer the time required for this alignment (Fig. 14.1a). DNA molecules whose reorientation times are less than the period of electric pulse will therefore be fractionated according to size. In conventional electrophoresis, a single constant electric field is employed, which limits the separation range to 50 kb. Separation in PFGE is achieved by applying two perpendicularly orientated electric fields in an alternating mode, which enabled Schwartz and Cantor initially to separate electrophoretically individual yeast chromosomes ranging in size from 230 kb to 2000 kb. Thus, the gap in resolution between standard techniques in molecular biology and the 100–5000 kb

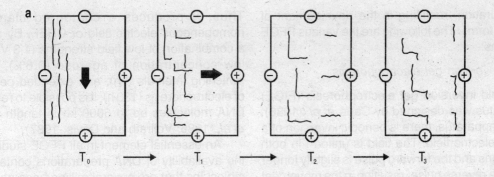

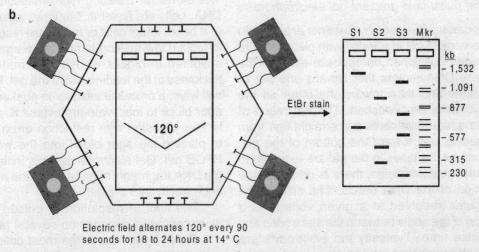

Electric field alternates 120° every 90
seconds for 18 to 24 hours at 14° C

**Fig.14.1:** Pulsed field gel electrophoresis. (a) very large DNA molecules that are migrating parallel to the electric field in one direction takes more time (T) to reorient in an alternating electric field than do smaller molecules. (b) CHEF pulsed field gel systems use a hexagonal gel box that alters the angle of the fields relative to the agarose gel. Concept of the figure taken from Miesfeld: *Applied Molecular Genetics.* Copyright © 1999 John Wiley and Sons. This material is used by permission of John Wiley and Sons,Inc.

size range covered by genetic analysis was closed.

The understanding of PFGE is still incomplete, and several models to explain the behavior of DNA molecules during electrophoresis have been proposed. The separation of molecules is basically achieved by forcing them to reorient periodically to changing field conditions. This change of direction is quicker by smaller molecules, which obtain a larger net mobility and are thus separated from larger molecules. Since its introduction, PFGE has been rapidly accommodated to the

analysis of chromosomal DNA from a wide variety of organisms. Currently, techniques have advanced to allow the separation and analysis of DNA molecules up to 12 megabases.

In the original Schwartz and Cantor PFGE apparatus, the speed, migration and direction of DNA molecules was dependent on the position at which the DNA is loaded on the gel. Consequently, DNA migrates in a complex path and skews towards the edges of the gel, making size estimates difficult. These problems were solved by several modifications of the electrode configuration, resulting in the development of variant forms. The following are the various PFGE systems:

### Field inversion gel electrophoresis

The field inversion gel electrophoresis (FIGE) apparatus was designed by Carle *et al.* (1986). In this apparatus, there is periodic inversion of a single electric field. The field is uniform in both directions and the forward pulse is slightly longer than the reverse pulse, resulting in the movement of the DNA along a perfectly straight track. FIGE resolves DNA fragments up to 2000 kb in length.

### Vertical pulse field gradient gel electrophoresis

The apparatus designed by Gardiner *et al.* (1986) uses a vertical gel apparatus with platinum wire electrodes positioned on opposite sides of the gel. The DNA moves first toward one set of electrodes and then toward the other as the electric fields are switched. The net result of these zigzag movements is a straight line from the loading well toward the bottom of the gel. Because all the lanes in the gel are exposed to equivalent electric fields, there is no horizontal distortion of the DNA bands. The size of the molecules resolved at a given voltage is a function of the angle between the electrodes and the pulse time. Typically the electrodes are arranged at a 90° angle and 10 sec pulses are used to separate molecules between 50 and 450 kb in length and 50–60 second pulses are used to fractionate molecules larger than 100 kb in length.

### Contour-clamped homogeneous electric field gel electrophoresis

In contour-clamped homogeneous electric field gel electrophoresis (CHEF), electric field is generated from multiple electrodes that are arranged in a square or hexagonal contour around the horizontal gel and are clamped to predetermined potentials (Fig. 14.1b). A square contour generates electric fields that are oriented at right angles of 120° or 60° depending on the placement of the gel and polarity of electrodes.

Thus, it produces a contoured clamped homogeneous electric field or CHEF. By using a combination of low field strengths (1.3 V/cm), low concentration of agarose (0.6%), long switching intervals (1h), and extended periods of electrophoresis (130 h), it is possible to resolve DNA molecules up to 5000 kb in length (Chu *et al.*, 1986; Vollrath and Davis, 1987).

An essential element in all PFGE studies is the availability of DNA preparations containing molecules that are several million base pairs in size. These large DNA molecules are extremely sensitive to shearing and nuclease activity. Therefore, for isolating high molecular weight DNA, cells are lysed *in situ* in an agarose plug or a bead. Intact cells or nuclei are resuspended in molten low-melting-temperature agarose and solidified in blocks whose size matches the thickness of the loading slot of the gel. Following cell lysis, a protease inhibitor is soaked into the agar block to inactivate proteinase K, and DNA is then digested with restriction enzymes prior to placing the agar block into the well of the PFGE gel. Gel electrophoresis is initiated, and the DNA fragments migrate out of the agar block and directly into the gel.

The limit of resolution of pulsed field gel electrophoresis depends on several factors:

**Pulse time:** This is the most critical factor in all PFGE systems, and is the time the electric field is applied in each direction. The longer the pulse time, the more the separation is extended to higher molecular weight DNAs.

**Temperature:** Mobility of DNA molecules increases with temperature. In PFGE, the temperature is usually maintained at 10–15°C. When higher temperatures are used, separation over a somewhat larger range can be obtained with the separation window moving to larger molecules at the expense of more diffuse bands.

**Field strength:** Electric field strength also has a strong influence on the separation range. Depending on the desired separation range, field strengths vary from 0.5 to 10 V/cm. Above a critical value, which depends on DNA size, molecules becomes trapped within the agarose matrix and no longer migrate. Trapping can occur after various times of electrophoresis and leads to a strong base smearing, especially for molecules in the megabase size range. Very high field strength, that above 10 V/cm, should be avoided, as bands become too diffuse. A good compromise is usually a field strength of 5–7 V/cm.

*Field angle*: The angles of the two electric fields to the gel is an important factor. Smaller angles, between 95 and 120 increases the speed at which separation is attained by a factor of 2 to 3. Especially for separating very large molecules in the magabase size range this may reduce the run times by several days.

*Agarose*: The agarose concentration is kept at 1% (w/v). When higher agarose concentration is used separation occurs over a smaller range but with sharper bands and a decrease of the overall mobility.

*DNA concentration*: The mobility of DNA molecules in PFGE is sensitive to the concentration of DNA, with high concentrations leading to lower mobilities. This effect may be substantial and results in size estimates that differ by as much as 50 kb.

*DNA topology*: The behavior of nonlinear DNA molecules strongly deviates from linear molecules. Typically, circular DNA molecules (supercoiled and relaxed) show a much lower mobility in PFGE than do linear molecules.

One has to keep in mind that changing one parameter will influence the effect of other parameters.

**Detection**: The standard technique of ethidium bromide is used for staining the gels. For detection of minor species of DNA, the gels are usually destained in water for extended periods of time (24–48 h) before photography.

In plant molecular biology, PFGE has already found widespread applications in the following aspects:

1. Determination of size of genomic DNA;
2. Characterization of the large-scale organization of telomeric and sub-telomeric regions of rice, tomato, and barley;
3. Construction of long-range restriction maps of centromeric and non-centromeric regions;
4. Organization of gene clusters such as the pea leghemoglobin genes, the wheat α-amylase genes, the patatin genes of potato and tomato, and the tandemly repeated tomato 5S RNA genes,
5. Study of the higher order chromatin structure in maize;
6. Assessment of *in vivo* damage and repair of DNA following exposure to radiation; and
7. Map-based cloning to isolate genes for which only a variant phenotype and genetic linkage map position are known.

**Polyacrylamide gel electrophoresis**

In native polyacrylamide gel electrophoresis (PAGE), proteins are applied to a porous polyacrylamide gel and separated in an electric field. When proteins are placed in an electric field, molecules with a net charge, such as proteins, will move toward one electrode or the other, a phenomenon known as electrophoresis. There are different types of PAGE. In native polyacrylamide gel electrophoresis, the molecules separate in an electric field on the basis of their net charge and size of the protein, since the electrophoresis separation is carried out in a gel, which serves as a molecular sieve. Small molecules move faster through the pores as compared to large molecules.

The gels are made of polyacrylamide, which is chemically inert. It is readily formed by the polymerization of acrylamide. Choosing an appropriate concentration of acrylamide and the crosslinking agent, methylene bisacrylamide, can control the pore sizes in the gel. The higher the concentration of acrylamide, the smaller is the pore size of the gel. The gel is usually cast between the two glass plates of 7–20 cm² separated by a distance of 0.5–1.0 mm. The protein sample is added to the wells in the top of the gel, which are formed by placing a plastic or Teflon comb in the gel solution before it sets (Fig. 14.2). A bromophenol blue dye is mixed with the protein sample to aid its loading on to the gel.

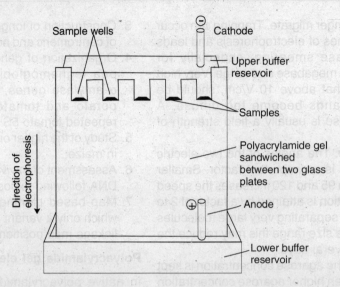

**Fig. 14.2:** Native polyacrylamide gel electrophoresis.

Because the bromophenol blue dye is a small molecule, it also migrates quickly through the gel during electrophoresis, thus indicating the progress of electrophoresis. In the simple native PAGE, buffer is same in upper and lower reservoirs and in the gel with a pH of approximately 9, such that most proteins have net negative charges and migrate toward the anode in the lower reservoir. An electric current (~300 V) is applied across the gel from top to bottom for a period to move the protein through the gel. When blue indicator of bromophenol dye reaches at the bottom of the gel, electric current is switched off and the gel is removed from the electrophoresis apparatus. The gel is stained and the protein bands within it are visualized.

A modification of simple PAGE is SDS-PAGE (sodium dodecyl sulfate-PAGE). In SDS-PAGE, the protein sample is treated with a reducing agent such as 2-mercaptoethanol or dithiothreitol to break all disulfide bonds. It is then denatured with SDS, a strong anionic detergent, which disrupts nearly all the noncovalent interactions and covers them with an overall negative charge. Approximately one molecule of SDS binds via its hydrophobic alkyl chain to the polypeptide backbone for every two amino acid residues, which gives the denatured protein a large net negative charge that is proportional to its mass. The SDS-protein mixture is then mixed with the bromophenol blue dye and applied in to the sample wells, and electrophoresis is performed as for native PAGE. As all the proteins now have an identical charge to mass ratio, they are separated on the basis of their mass. The smallest proteins move farthest. Thus, if proteins of known molecular mass are electrophoresed alongside the samples, the mass of unknown proteins can be determined. SDS-PAGE is a rapid, sensitive, and widely used technique from which one can determine the degree of purity of a protein sample, the molecular mass of unknown sample, and the number of polypeptide subunits with a protein.

## Isoelectric focusing

In isoelectric focusing (IEF), proteins are separated by electrophoresis in a pH gradient in a gel. They separate on the basis of their relative content of positively and negatively charged groups. Each protein migrates through the gel until it reaches the point where it has no net charge—this is its isoelectric point (pI); here, the protein's net charge is zero and hence it does not

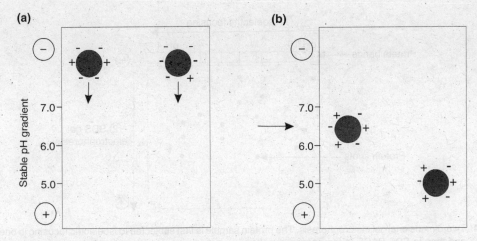

**Fig. 14.3:** Isoelectric focusing. (a) Before applying an electric current, (b) after applying an electric current the proteins migrate to a position at which their net charge is zero (isoelectric point, pI).

move in an electric field. In IEF, a polyacrylamide gel is used that has large pores so as not to impede protein migration and contains a mixture of polyampholytes (small multicharged polymers that have many pI values). If an electric field is applied to the gel, the polyampholytes migrate and produce a pH gradient. To separate proteins by IEF, they are electrophoresed through such a gel. Each protein migrates through the gel until it reaches a position at which the pH is equal to its pI (Fig. 14.3). If a protein diffuses away from this position, its net charge will change as it moves into a region of different pH and the resulting electrophoretic forces will move it back to its isoelectric position. In this way each protein is focused into a narrow band (as thin as 0.01 pH unit) about its pI.

## Two-dimensional gel electrophoresis

Isoelectric focusing can be combined with SDS-PAGE to obtain very high-resolution separations by a procedure known as two-dimensional gel electrophoresis. In this technique, the protein sample is first subjected to isoelectric focusing in a narrow strip of gel containing polyampholytes. This gel strip is then placed on top of an SDS polyacrylamide gel and electrophoresed to produce a two-dimensional (2D) pattern of spots in which the proteins have

been separated in the horizontal direction on the basis of their pI, and in the vertical direction on the basis of their mass. The overall result is that proteins are separated on the basis of their size and charge. Thus two proteins that have very similar or identical pIs and produce a single band by isoelectric focusing will produce two spots by 2D gel electrophoresis (Fig.14.4). Similarly, proteins with similar or identical molecular masses, which would produce a single band by SDS-PAGE, also produces two spots because of the initial separation by isoelectric focusing. This 2D gel electrophoresis has enormous use in proteomics study.

### Staining of proteins

The most commonly used protein stain is the dye Coomassie brilliant blue. After electrophoresis, the gel containing separated proteins is immersed in an acidic alcoholic solution of the dye. This denatures the proteins, fixes them in the gel so that they do not wash out, and allows the dye to bind to them. After washing away the excess dye, the protein bands are visible as discreet blue bands. A more sensitive stain is soaking the gel in a silver salt solution. However, this technique is rather difficult to apply.

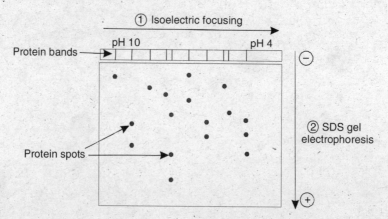

**Fig. 14.4:** Two-dimensional gel electrophoresis. The protein sample is first subjected to isoelectric focusing in one direction and then to SDS-PAGE in the second direction.

## Nucleic acid blotting

Blotting describes the immobilization of sample (nucleic acids, proteins) on to a solid support, generally nylon or nitrocellulose membranes. In nucleic acid blotting, the blotted samples are nucleic acids (DNA or RNA) on nylon or nitrocellulose membranes, which are then used as targets for hybridization with labeled probes. Labeling of probe nucleic acids and hybridization has formed the basis for a range of experimental techniques that help understand the organization and expression of genetic material.

The blotting method was originally developed by Southern (1975), who demonstrated that DNA restriction fragments that had been electrophoretically fractionated on agarose gels could be transferred to a solid support (nitrocellulose) and detected as discreet bands following hybridization to a complementary DNA probe. When the blotting method is applied to RNA, it is termed *Northern blotting*. Western blot analysis refers to the transfer of proteins to membranes and their detection with antibody probes.

### Southern blot analysis

Southern (1975) gave the method for detecting DNA fragments in an agarose gel by blotting on a nylon or nitrocellulose membrane followed by detection with a probe of complementary DNA or RNA sequence. The blotting technique has not changed over the years since its first description, except for the availability of increasingly sophisticated blotting membranes, kits for labeling, and apparatus for electrophoresis transfer. Although membranes originally used for DNA transfer were made of nitrocellulose, nylon-based membranes used nowadays perform much better due to their higher nucleic acid binding capacity and physical strength. In Southern blotting, agarose gel electrophoresis is performed. (The procedure of agarose gel electrophoresis has already been explained.)

**Procedure:** An outline of a procedure for Southern hybridization is shown in Fig. 14.5. Genomic DNA is digested with one or more restriction enzymes, and the resulting fragments are separated according to size by electrophoresis on an agarose gel. The agarose gel is mounted on a filter paper wick. A piece of nylon membrane slightly larger than the size of gel is placed over it, followed by 2–4 pieces of prewet filter papers. Thus the membrane is sandwiched between the gel and a stack of filter papers and paper towels (Fig. 14.6). Air bubbles between each layer are removed. A 10-cm stack of dry filter paper towels cut to size is placed on top of the pyramid and a glass plate and weight of around 0.5 kg placed on top. Strips of parafilm or plastic wrap are placed around the gel to

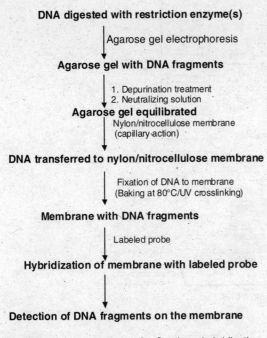

DNA digested with restriction enzyme(s)

↓ Agarose gel electrophoresis

Agarose gel with DNA fragments

↓ 1. Depurination treatment
2. Neutralizing solution

Agarose gel equilibrated
Nylon/nitrocellulose membrane
(capillary action)

DNA transferred to nylon/nitrocellulose membrane

↓ Fixation of DNA to membrane
(Baking at 80°C/UV crosslinking)

Membrane with DNA fragments

↓ Labeled probe

Hybridization of membrane with labeled probe

↓

Detection of DNA fragments on the membrane

**Fig. 14.5:** A general scheme for Southern hybridization.

prevent direct contact between the stack of paper towels and the wick, which can cause a short circuit in the flow of the transfer solution.

DNA in the agarose gel is denatured *in situ*. For efficient blotting, after electrophoresis the gel is generally pretreated. The electrophoresed DNA is exposed to a short depurination treatment (0.25 N HCl), followed by alkali treatment. It shortens the DNA fragments at depurinated sites and denatures the fragments so that the DNA is single stranded and accessible for probing. The gel is then equilibrated in neutralizing solution prior to blotting. The use of positively charged nylon membranes removes the need for extended gel pretreatment.

DNA fragments from the gel get transferred to a membrane. The fragments are transferred from the gel to membrane by the capillary transfer method. DNA fragments are carried from the gel in a flow of liquid and deposited on the surface of the solid support of membrane. The liquid is drawn through the gel by capillary action, which is established and maintained by stack of

dry and absorbent paper towels. This transfer is allowed to proceed for 12 h or longer. The paper towels and the 3MM filter papers above the gel and membrane are removed and the membrane laid on a dry sheet of 3MM filter papers. The membrane is dried on filter paper, followed by baking in an oven at 80°C for 2 h. DNA can also be fixed to the membranes by the UV cross-linking method. It is based on the formation of crosslinks between a small fraction of thymine residues in the DNA and positively charged amino groups on the nylon membrane.

Following the fixation step, the membrane can then be used for the hybridization labeling reaction. Probes can be labeled single-stranded DNA, oligodeoxynucleotides, or an RNA that is a complementary sequence to the blotted DNA. Labeling of probe can be done by radioactive or nonradioactive methods, which have been discussed in Chapter 17. For maximum hybridization of probe nucleic acid sequence to the blotted DNA the perfect conditions are chosen solution hybridizations, membrane hybridizations tend not to proceed to completion. In membrane hybridization some of the bound nucleic acid is embedded and is inaccessible to the probe. Prolonged incubations may not generate any significant increase in detection sensitivity. The composition of hybridization buffer can greatly affect the speed of reaction and the sensitivity. Dextran sulphate and other polymers act as volume extruders to increase the rate and extent of hybridization. Dried milk, heparin and detergents such as SDS have been used to depress non-specific binding of the probe. Formamide can be used to depress the melting temperature of the hybrid so that reduced temperatures of hybridization can be used.

After hybridization, the membrane is washed to remove unbound probe and regions of hybridization are detected as per the protocol for a particular probe used. If a radioactive probe sequence has been employed, the regions of hybridization are detected by the autoradiographic method.

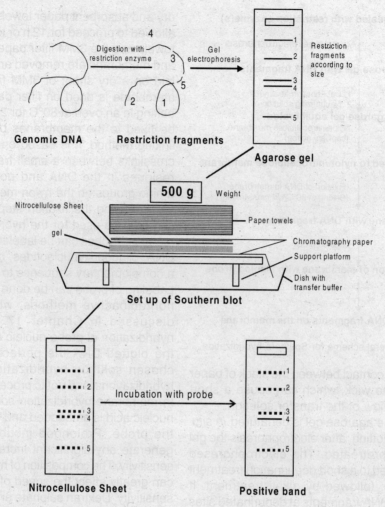

**Fig. 14.6:** Agarose gel electrophoresis and Southern blotting.

## Northern blot analysis

Northern hybridization is the blotting of RNA separated by gel electrophoresis on nylon or nitrocellulose membranes followed by hybridization with nucleic acid (RNA or DNA) sequences. The procedure of northern blotting is similar to Southern blotting except that RNA rather than DNA is blotted on the membrane. Further, in Southern hybridization, DNA has to be denatured before blotting, which is not needed in northern hybridization. Since RNA was not able to bind nitrocellulose, initially northern blotting was carried out on chemically reactive paper, where RNA sequences are bound covalently. The RNA was blotted or immoblized on diazotized cellulose [diazobenzyloxymethyl (DBM) cellulose)] paper (Alwine et al., 1977). Subsequently o-aminophenylthioether (APT) cellulose was developed, which was easier to prepare and more stable than DBM cellulose. Subsequently, various other methods were developed where it was shown that RNA denatured by various chemical treatments binds tightly to nitrocellulose and hybridizes to radioactive probes with high efficiency. Thus, presently both nylon based membranes like Amersham Biosciences Hybond N+ and nitrocellulose membranes are used for blotting of RNA under appropriate conditions because of the convenience of using these membranes. Attachment of denatured RNA to nitrocellulose is presumed to be noncovalent but is essentially irreversible. It is therefore possible to hybridize sequentially RNA immobilized on nitrocellulose to a series of radioactive and non-radioactive

probes without significant loss of the bound nucleic acid.

*Proteins blotting*

**Western blotting**: This refers to blotting of electrophoresed protein bands from a SDS-polyacrylamide gel on to a membrane (nylon or nitrocellulose) and their detection with antibody probes (Fig. 14.7). The sample is electrophoresed on SDS-PAGE, which separates the proteins on

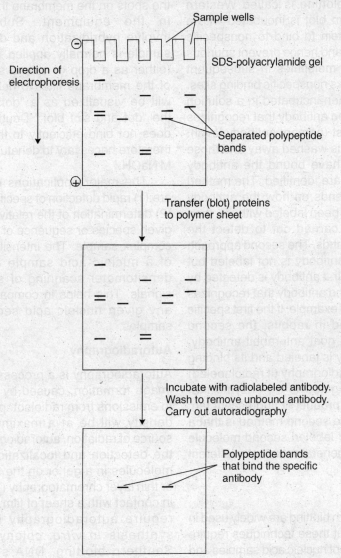

Sample wells

SDS-polyacrylamide gel

Direction of electrophoresis

Separated polypeptide bands

Transfer (blot) proteins to polymer sheet

Incubate with radiolabeled antibody.
Wash to remove unbound antibody.
Carry out autoradiography

Polypeptide bands that bind the specific antibody

**Fig. 14.7:** Western blotting using a radiolabeled antibody. Protein bands are detected by autoradiography.

the basis of size, resulting in a series of bands down the gel. Because the gel matrix does not allow large antibodies to enter rapidly, the sample proteins must be transferred to a solid support. The gel is overlaid with a nitrocellulose or nylon sheet, and an electric field is applied so that proteins migrate from the gel to the sheet where they become bound. The nitrocellulose membrane thus has an exact image of the pattern of proteins that was in the gel. This particular form of blotting is called Western blotting. The western blot is incubated with a protein such as casein to bind to nonspecific protein binding sites and hence prevent spurious binding of antibody molecules in subsequent steps. This step blocks nonspecific binding sites. The membrane is then incubated in a solution containing the labeled antibody that recognizes the protein of interest (primary antibody level). Unreacted antibody is washed away and those protein bands that have bound the antibody become visible and are identified. The method of visualization depends on how the antibody was labeled. If it has been labeled with $^{125}I$, then autoradiography is carried out to detect the radioactive protein bands. The second approach is that the primary antibody is not labeled but the membrane with first antibody is detected by incubating with second antibody that recognizes the first antibody. For example, if the first specific antibody was raised in rabbits, the second antibody could be a goat anti-rabbit antibody. The second antibody is labeled and its binding is detected by autoradiography (if radiolabeled) or it could be conjugated to an enzyme that generates a colored product or fluorescent tag. The advantage of the second method is that a single preparation of labeled second molecule can be employed as general detector for different probes.

## Dot blot technique

Southern and Northern blotting are widely used in molecular biology, but these techniques require extensive purification of nucleic acid samples and the entire process is time consuming and expensive, although indispensable for structural analysis of nucleic acids. However, if only the detection and quantitation of any given sequence is required, it is possible to shorten the procedure to avoid purification steps, electrophoresis, and blotting of the gel.

The purified nucleic acid sample (DNA or RNA) may be applied directly to nitrocellulose filter paper either by "dotting" it on with a pipette or an apparatus is used for placing the spots on the membrane through slots made in the equipment. Subsequent steps involve hybridization and detection. As the sample is normally applied in a circular form (either as a drop or oblong slot) the exposure of the membrane filter to detection procedure will be visualized as a "dot" or "slot" hence the "dot and slot blot." Double-stranded DNA does not bind efficiently to the filters and it is therefore necessary to denature the DNA by 0.4 M NaOH.

The major applications of this technique are:(1) rapid detection of specific sequences; and (2) determination of the relative amounts of any given species or sequence of RNA or DNA in a complex sample. The intensity of hybridization of a nucleic acid sample is quantified by densitometer scanning of autoradiographic signals. This helps in comparing quantities of any given nucleic acid sequence in many samples.

## Autoradiography

Autoradiography is a process of photographic image formation, caused by exposure of film to emissions from radioisotopes. As the image density will be at a maximum close to the source of radiation, autoradiography allows both the detection and localization of radioactive molecules in a gel or on the surface of a filter or thin layer chromatography (TLC) plate placed in contact with a sheet of film. Techniques that require autoradiography include protein synthesis *in vitro,* colony hybridization, Southern blotting, DNA sequencing, dot blotting, etc.

The power of autoradiography lies not only in its very high sensitivity but also in its ability to discriminate between labeled and unlabeled molecules. For example, high sensitivity is needed for the detection of radioactively labeled "probes," hybridized to complementary sequences of nucleic acids which are immobilized on sheets of nitrocellulose by blotting processes or colony hybridization. One of the most dramatic examples of autoradiography is the sequencing of DNA, in which the sequence is actually read from an autoradiograph of the sequencing gel.

The procedure used for autoradiography depends on the radioisotope being employed, the levels of radioactivity being incorporated, the medium in which the radioactive compounds are found, and on the degree of resolution and speed of analysis required. In every case, photographic film is used to detect the radiation emitted by isotopes, and it is because of the nature of photographic process that there is no single, universally acceptable method of autoradiography. A basic understanding of this process can be useful in deciding which method of autoradiography will produce the optimum results.

Photographic film consists of a support, such as plastic sheet, coated with a thin layer of an emulsion that contains microscopic crystals of a silver halide suspended in gelatin. If these crystals are treated with a reducing agent or developer, they will be converted to metallic silver. However, the activation energy for this reduction is greatly decreased if there are already some silver atoms in the crystals, and in such cases the crystal is converted to silver far more rapidly. Once conversion of $Ag^+$ to $Ag$ atom begins in a crystal, the process is autocatalytic, accelerating as more Ag atoms are formed. Clearly with limited development, only those crystals that already contain some silver atoms will be further reduced to metallic silver. This provides the basis of the photographic process since exposure of the emulsion to radiation will result in the formation of a few silver atoms in some of the crystals,

thus producing a latent image, which can be revealed by limited development. After development, and before further exposure to light, any undeveloped silver halide is removed by being dissolved in a fixer solution such as sodium thiosulphate, thus preventing any further formation of silver.

The exact mechanism is not known, but it appears that radiation absorbed by a crystal of silver halide can promote an electron into a "conduction band," allowing it to move through the crystal until trapped in a sensitivity speck. Once trapped, the electron can combine with a mobile interstitial $Ag^+$ ion to form $Ag$, and this then act as an efficient trap for the next electron produced in the crystal, which will also combine with an $Ag^+$ ion. Formation of the first Ag is a reversible process, but two or more silver atoms forms a stable speck in the crystal. Hence the latent image is stable only if it is composed of at least two Ag atoms per crystal. Absorption of either $\beta$ or $\gamma$ emissions by a silver halide will result in the formation of several Ag atoms, and so a latent image is formed.

In direct autoradiography, the sample present in a gel or filter is placed in contact with a sheet of X-ray film, and left to form a latent image. However, this process will work if emitted radiations can reach and be absorbed by the film. This method is ideal for use with $^{14}C$ or $^{35}S$, particularly when the compounds are on a surface (e.g. nitrocellulose filter or TLC), since $\beta$ emissions can penetrate into the X-ray film, where they are be absorbed by the photographic emulsion. However, this cannot be used for detecting tritiated compounds since they produce very low-energy $\beta$-emissions and can rarely penetrate the protective coating of X-ray film. When direct autoradiography is used with $^{32}P$ or $^{125}I$, unfortunately they are so highly penetrating that majority of the emissions pass completely through the film without being absorbed and hence without forming a latent image.

The procedure for direct autoradiography is very simple. The sample (a dried gel, membrane) is placed in firm contact with a sheet of "direct" type X-ray film in the dark. This is done using a commercial cassette, where the sample and film can be sandwiched between two sheets of perplex, glass, or aluminium taped or clipped together. The assembly is then left at room temperature for several hours or days (depending on the level of radioactivity in the sample) to allow adequate exposure of the film. Development and fixing of the film are then carried out according to the manufacturer's instructions. If the sample is a non-dried gel, e.g. sequencing gel, it is necessary to carry out exposure at −70°C to stop any diffusion of molecules within the gel.

Indirect autoradiography (fluorography) is employed when either low-energy radioactive compounds (e.g. $^{3}$H) or high-energy compounds ($^{32}$P or $^{125}$I) are used. The principle of fluorography is that the energy of emissions is converted by a scintillant into light, which then causes the formation of a latent image in the film. For low-energy emission compounds ($^{3}$H) , if the sample is gel or membrane, it is impregnated with a solution of scintillant. The most commonly used scintillant is 2,5-diphenyloxazole (PPO). After drying under vacuum, the impregnated gel is covered with a sheet of thin plastic film (Saran wrap or Cling film) and placed in contact with a sheet of preexposed screen or medical X-ray film, in the dark. The gel and film are sealed in a cassette and left at −70°C for exposure of the film. Since the latent image is formed by light rather than by β-emissions, the process is known as fluorography.

When radioactive compounds that emit high-energy emissions are used, the gels or membranes are not impregnated with scintillant (PPO). In such cases, "intensifying screens," which contain a dense solid scintillant (usually calcium), tungstate are used. The intensifying screens are placed against the film, on the far side from radioactive sample. Any emissions passing through the photographic emulsion are absorbed by the screen and converted to light, effectively superimposing a photographic image upon the direct autoradiographic image.

It has been reported that small amounts of radioactivity are under-represented with the use of fluorography and intensifying screens. This problem can be overcome by a combination of preexposing a film to an instantaneous flash of light (preflashing) and exposing the autoradiograph at −70°C. Preflashing provides many of the silver halide crystals of the film with a stable pair of silver atoms and −70°C temperature increases the stability of a single silver atom increasing the time available to capture a second photon.

### E. coli transformation

The process of transferring exogenous DNA into cells is called "transformation." Thus it refers to any application in molecular biology in which purified DNA is actively or passively imported into host cells using nonviral methods. There are basically two general methods for transforming bacteria.

1. *Chemical transformation*: The chemical method utilizes $CaCl_2$ and heat shock to stimulate the entry of DNA into cells (Fig. 14.8). Cohen *et al.* (1972) showed that $CaCl_2$-treated *E. coli* cells are effective recipients for plasmid DNA. Almost any strain of *E. coli* can be transformed with plasmid DNA with varying efficiency. But it is known that many bacteria contain restriction systems, which can influence the efficiency of transformation. Thus it is essential to use a restriction-deficient strain of *E. coli* as a transformable host. Bacterial cells in the log phase which are competent (cells that are in a state in which they can be transformed by DNA in the environment) are used for transformation. Cells are incubated with DNA on ice in a solution containing $CaCl_2$, followed by a brief heat shock at 42°C. These cells are allowed to recover for a short time in the growth media and then plated on agar plates containing appropriate antibiotics. In most cases it is possible to obtain ~$10^{6}$–$10^{8}$ antibiotic-resistant colonies per microgram of circular

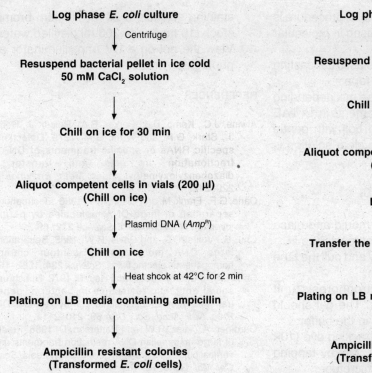

**Fig. 14.8:** Calcium chloride method of *E. coli* transformation.

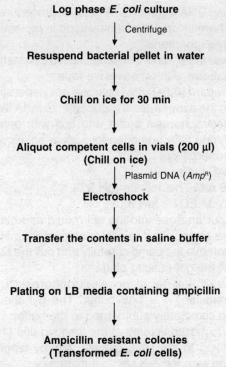

**Fig. 14.9:** Electroporation method of *E. coli* transformation.

plasmid DNA using this method. The mechanism of bacterial transformation is not known, but some studies have suggested that the DNA enters the cell through protein pores that are activated by both calcium and heat. Numerous modifications of the original CaCl$_2$ procedure have been developed in an attempt to increase the transformation efficiency of various *E. coli* strains. For example, addition of hexamine cobalt chloride or dimethyl sulfoxide to the CaCl$_2$ solution was found to increase transformation efficiencies by 10 to 100-fold.

2. *Electroporation method*: It uses a short pulse of electric charge to facilitate DNA uptake (Fig. 14.9). Bacterial cells are suspended in a low ionic strength solution in a special cuvette and electric shock is given. Electroporation was first applied to animal cells, but lately by increasing the field strength using high voltage power sources,

electroporation parameters were found that worked well for bacteria. The mechanism of DNA transformation is poorly understood but it is thought that short pulse (~3 msec) of electric charge (2.5 kV, ~200 ohms, ~30 microfaradays) may cause a transitory opening in the cell wall or perhaps activate membrane channels. The major advantages of electroporation are that high transformation frequencies of $10^8$–$10^{10}$ transformants/$\mu$g plasmid DNA are routinely obtained, and once optimized, electroporation works well with most bacterial strains.

# PROTOCOLS

## Agarose gel electrophoresis

Agarose gel electrophoresis is a simple and highly effective method for separating, identifying, and purifying DNA fragments. A general procedure is given here since it is routinely used in molecular biology experiments.

1. Seal the ends of a suitable plastic gel-casting platform with an adhesive tape.
2. Prepare (0.6–0.8%) agarose mix depending on the experimental requirements in 1× TAE electrophoresis buffer and boil with gentle stirring till a homogeneous clear solution is formed. Cool to 55°C.
   TAE buffer (1×)
   40 mM Tris acetate (pH 8.0)
   1 mM EDTA (pH 8.0)
3. Pour agarose into the gel mould and place the comb; allow to set for at least 30 min.
4. Remove the comb carefully and pull the tape off the gel-casting platform.
5. Keep the gel in an electrophoresis unit containing 1 × TAE buffer. The gel should be completely submerged in the buffer.
6. Add 1/10th volume of the tracking dye (10× stock) to the DNA sample. Mix by tapping and spin for 2–3 sec in a microfuge.
   Tracking dye (10× stock)
   0.1M EDTA pH 8.0
   0.01M Tris.Cl pH 8.0
   Bromophenol blue (0.25%)
   Glycerol (50%)
7. Load the DNA sample carefully. Load a suitable molecular weight marker in the first and as well as the last lane along with the DNA samples. Attach the electrodes. (Make sure that the negative black terminal is at the same end of the unit as the wells.)
8. Turn on the power supply and run at 20 mA power current. Run for an appropriate time by looking at the tracking dye.
9. Disconnect the current after an appropriate time. Remove the gel from electrophoresis unit and immerse it in an ethidium bromide stain (10 mg/ml) for 20–30 min. (Gloves should be worn while handling ethidium bromide.)
   Ethidium bromide (10 mg/ml)
   Dissolve 100 mg ethidium bromide in 10 ml distilled water and store at 4°C. Working staining solution is 20 µl ethidium bromide stock (10 mg/ml) to 200 ml distilled water.
10. View the gel on a UV transilluminator and photograph the gel.

## REFERENCES

Alwine, J.C., Kemp, D.J., Parker, B.A., Reiser, J., Renart, J., Stark, G.R. and Wahl, G.M. 1979. Detection of specific RNAs or specific fragments of DNA by fractionation in gels and transfer to diazobenzyloxymethyl paper. *Meth. Enzymol.* **68:** 220–242.

Carle, G.F., Frank, M. and Olson, M.V. 1986. Electrophoretic separation of large DNA molecules by periodic inversion of electric field. *Science* **232:** 65.

Chu, G., Vollrath, D. and Davis, R.W. 1986. Separation of large DNA molecules by contour clamped homogeneous electric fields. *Science* **234:** 1582–1585.

Cohen, S.N. Chang, A.C.Y. and Hsu, L. 1972. Nonchromosomal antibiotic resistance in bacteria: genetic transformation of *Escheichia coli* by R-factor DNA. *Proc. Natl. Acad. Sci., USA* **69:** 210–2114.

Gardiner, A., Lass, R.W. and Patterson, D. 1986. Fraction of large mammalian DNA restriction fragments using vertical pulse field gradient gel electrophoresis. *Somat. Cell Mol. Gene.* **12:** 185.

Schwartz, D.C. and Cantor, C.R. 1984. Separation of yeast chromosome sized DNA by pulsed field gradient gel electrophoresis. *Cell* **37:** 67–75.

Southern, E.M. 1975. Detection of specific sequences among DNA fragments separated by gel electrophoresis. *J. Mol. Biol.* **98:** 503–517.

Vollrath, D. and Davis, R.W. 1987. Resolution of DNA molecules greater than 5 Mb by contour-clamped homogeneous electric fields. *Nucl.c Acids Res.* **15:** 7865.

## FURTHER READING

Hames, B.D., Hooper, N.M. and Houghton, J.D. 1998. *Instant Notes in Biochemistry*, Bios Scientific Publishers Ltd., Oxford.

Miesfeld, R.L.1999. *Applied Molecular Genetics*. Wiley-Liss, New York.

Old, R.W. and Primrose, S.B. 2001. *Principles of Gene Manipulation: An Introduction to Genetic Engineering*. Blackwell Scientific Publications, Oxford.

Sambrook, J., Fritsch, E.F. and Maniatis, T. 1989. *Molecular Cloning: A Laboratory Manual*. Cold Spring Harbor Laboratory Press, USA.

Walker, J.M. and Gaastra, W. (Eds.). 1987. *Techniques in Molecular Biology*. Croom Helm, Australia.

# 15

# Gene Cloning:
# Cutting and Joining DNA Molecules

The term *gene cloning* can be defined as the isolation and amplification of an individual gene sequence by insertion of that sequence into a bacterium where it can be replicated. Cloning a fragment of DNA allows indefinite amounts to be produced from even a single original molecule. Cloning technology involves the construction of novel DNA molecules by joining DNA sequences from different sources. The product is often described as recombinant DNA. The construction of such composite or artificial DNA molecules has also been termed *genetic engineering* or *gene manipulation* because of the potential for creating novel genetic combinations by biochemical means. A precise legal definition of gene manipulation has been given as follows: The formation of new combinations of heritable material by the insertion of nucleic acid molecules, produced by whatever means outside the cell, into any virus, bacterial plasmid or any other vector system so as to allow their incorporation into a host genome in which they do not naturally occur but in which they are capable of continued propagation.

The basic events in gene cloning have been described in the following steps and shown in Fig. 15.1.

1. Isolation of the gene of interest.

2. A fragment of DNA (gene of interest) to be cloned is incorporated into a small replicating (usually circular) DNA molecule called a *vector*. This can be an *E. coli* plasmid, a virus, a cosmid, etc. The vector with an incorporated gene is called a *recombinant vector or molecule*.

3. The recombinant vector is introduced into a host cell by transformation.

4. Cells that have acquired recombinant DNA molecule are selected.

5. Recombinant DNA molecule is multiplied within the host cell to produce a number of identical copies of the cloned gene.

In gene cloning, a fragment of DNA that codes for a required gene product has to be removed from the host organism and moved to a vector molecule (e.g. plasmid, phage, cosmid) to construct recombinant DNA molecule. This involves cutting DNA molecules at specific sites and joining them together again in a controlled manner. In order to understand the process of gene cloning, i.e. to insert a piece of foreign DNA into a vector for construction of recombinant DNA molecule, one must understand the various vectors, mechanism of cutting and joining DNA molecules at specific sites, the use of agarose gel electrophoresis as a method for monitoring

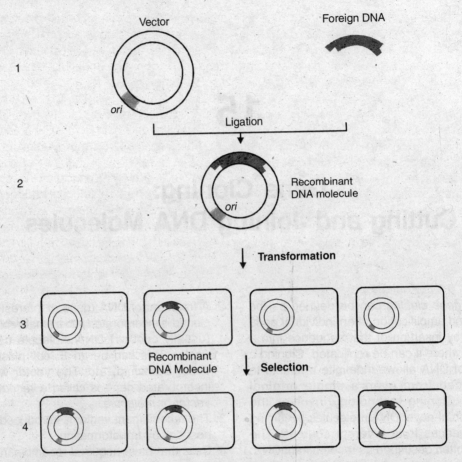

**• Fig. 15.1:** Fundamental steps of a cloning experiment. The cloning experiment involves (1) the production and preparation of vector and foreign (passenger) DNA, (2) the ligation of vector and passenger DNA so as to construct a recombinant DNA molecule, (3) the transformation of suitable host cells, and (4) the selection of those cells which have acquired recombinant DNA molecules.

the cutting and joining reaction, and a means of monitoring transformation of *E. coli*. Thus, a gene cloning experiment has four essential components, which have been dealt in details to explain the process of gene cloning.

1. Enzymes for cutting and joining the DNA fragments into vector molecules.
2. Cloning vehicles or vectors.
3. DNA fragments, i.e. gene libraries.
4. Selection of a clone of transformed cells that has acquired the recombinant chimeric DNA molecule.

# ENZYMES FOR CUTTING— RESTRICTION ENDONUCLEASES

Present-day DNA technology is totally dependent on the ability to cut DNA molecules at specific sites with restriction endonucleases. Before 1970, there was no method available for cutting a duplex DNA molecule. It became apparent that related phenomenon of host-controlled restriction and modification might lead towards a solution to this problem when it was discovered that restriction involves specific endonucleases. What is host-controlled restriction and modification? Restriction means the identification of incoming DNA to the cell (e.g. bacteria) and its destruction by cleaving into pieces if it is recognized as foreign DNA. Modification is the protection of the cell's own DNA by methylation of certain bases so that host DNA is not cleaved. The phenomenon of restriction and modification has been explained by the behavior of λ phage on two *E. coli* host strains C and K (Fig. 15.2). In the bacteria/bacteriophage system, when the incoming DNA is a bacteriophage genome to the bacteria, the effect is to reduce the efficiency of plating (EOP), i.e. to reduce the number of plaques formed in plating test.

If a stock preparation of phage lambda is grown on *E. coli* strain C and is then titred on C and K strains of *E. coli*. The titres observed on these two strains differed with lower titer on K strain of *E. coli*. The phage is said to be restricted by the second host strain *E. coli* K. If those phages that result from infection on *E. coli* K are collected and replated on *E. coli* K, then they are no longer restricted. But if they are again first grown on *E. coli* C and then on *E. coli* K, they are again restricted when plated on *E. coli* K. Thus the efficiency with which the phage plates on a particular host strain depends on the strain on which it was last propagated. This nonheritable change conferred upon the phage by the second host strain (*E. coli* K) that allows it to be replated on that strain without any restriction is called modification. It was

observed by radioactive labeling experiments that when restricted phages adsorb on the surface of restrictive hosts and inject their DNA, it is degraded soon. The endonuclease responsible for this degradation is called restriction endonuclease or restriction enzyme. The restrictive host must protect its own DNA from degradation, and its DNA must be modified. The modification involves methylation of certain bases at a very limited number of sequences within the DNA, which constitutes the recognition sequences for the restriction endonucleases. Thus, the restriction enzyme consists of two enzymatic elements: nuclease and methylase. The genes that specify host-controlled restriction and modification systems may reside on the host chromosome itself or may be located on a plasmid or prophage such as P1. Different enzymes recognize different but specific sequences, which range from 4 to 8 base pairs.

During construction of a recombinant DNA molecule, the circular vector must be cut at a single point into which the fragment of DNA to be cloned can be inserted. Not only must each vector be cut just once, but also all the vector molecules must be cut at precisely the same position. Restriction endonucleases are enzymes that make site-specific cuts in the DNA. The first enzyme was isolated from *Haemophilus influenzae* in 1970. Now, many types of restriction endonucleases have been isolated from a wide variety of bacteria, which calls for a uniform nomenclature and classification. According to the recommendation of Smith and

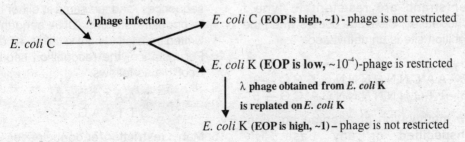

**Fig. 15.2:** Host controlled restriction and modification of l phage on *E. coli* strain C and K analyzed by efficiency of plating.

Nathans (1973), each enzyme is represented by a three-letter code in italics. For example, *Hae* for *Haemophilus aegypticus* and *Hin* for *Haemophilus influenzae*. Sometimes a four-letter code is used when the enzymes are isolated from different serotypes of a species. The letter code denotes the origin of enzyme. For example, *Hinf* for *Haemophilus influenzae* serotype f. If more than one enzyme is isolated from a single origin, then these are denoted by Roman numerals (I, II, III, etc.). For example, *Eco*RI for the first enzyme of *Escherichia coli* serotype R.

There are three types of restriction endonucleases (Yuan, 1981).

## Type I endonucleases

1. They are complex nucleases, e.g. *Eco*K and *Eco*B.
2. They function simultaneously as an endonuclease and a methylase and require ATP, Mg$^{++}$, and S-adenosyl methionine as cofactors.
3. They are single multifunctional enzymes with three different subunits: a restrictive subunit, a modification subunit, and a specificity subunit, which determines the recognition specificity.
4. The recognition site of class I enzymes is 15 bp in length, which can be methylated at the adenine position in the recognition site, and the cleavage site is approximately 1000 bp away from the 5′ end of the sequence TCA located in the 15-bp recognition site. DNA molecules with recognition sequences methylated at both adenine residues on either strand are resistant to type I endonucleases. Cleavage occurs when the recognition site is unmethylated.
   e.g.

*Eco*K   5′ - A A*C N N N N N N G T G C — 3′
         3′ - T T G N′N′N′N′N′N′C A*C G — 5′

* Show methylation at adenine position; N–unspecified or any base; N′–corresponding complementary base to N. Heteroduplex molecules consisting of one

modified and one unmodified strand will always be methylated by the enzyme. Unmodified DNAs are restricted, i.e. cleaved.
5. These enzymes exhibit high ATPase activity. For each phosphodiester cleavage, more than 10,000 molecules of ATP are hydrolyzed.
6. These enzymes show specificity for the recognition site but not for their cleavage site. This produces a heterogeneous population of DNA fragments. In view of this, it is very undesirable to use in the gene cloning techniques.

## Type II endonucleases

1. They are simple enzymes that consist of a single polypeptide. They have separate endonuclease and methylase activities.
2. They recognize a specific nucleotide sequence and cut a DNA molecule at this sequence and nowhere else. Most of these enzymes recognize a hexanucleotide sequence, but others recognize tetra-, penta-, or octanucleotide sequences on the DNA. For example, the restriction endonuclease *Pvu*I isolated from *Proteus vulgaris* cuts DNA at the hexanucleotide CGATCG, whereas a second enzyme *Pvu*II isolated from the same bacterium cuts at a different hexanucleotide CAGCTG.
3. They are very stable and only require Mg$^{++}$ as cofactors.
4. The majority of the known recognition sequences have an axis of rotational symmetry (palindromic sequences), i.e. the sequences read the same in either direction in opposite strands. DNA sequences are written from left to right in the 5′→3′ direction. For instance, the recognition sequence of *Eco*RI is as follows.

5′ — G A A T T C — 3′
3′ — C T T A A G — 5′

5. Many restriction endonucleases make a simple double-stranded cut in the middle of the recognition sequence, resulting a blunt

| Recognition sequence | Cleavage pattern |
|---|---|

**(Staggered cut)**

*Eco*RI      ↓

5′ ---- G A A T T C --- 3'         5′ ---G $_{OH}$ 3'

3′ ---- C T T A A G --- 5'         3′ ---C T T A A $_P$ 5'

         ↑

*Pst*I      ↓

5′ ----- C T G C A G --- 3'         5′ --- C T G C A $_{OH}$ 3'

3′ ----- G A C G T C --- 5'         3′ --- G $_P$ 5'

      ↑

**(Blunt cut)**

*Hae*III      ↓

5′ ---- G G C C --- 3'         5′ --- G G $_{OH}$ 3'

3′ ---- C C G G --- 5'         3′ --- C C $_P$ 5'

      ↑

**Fig. 15.3:** Type II restriction endonuclease cleavage patterns.

or flush end. *Pvu*II, *Alu*I, *Hae*III, etc. are examples of blunt-end cutters (Fig. 15.3).

6. Other restriction endonucleases do not cut the two DNA strands at exactly the same position. Instead, the cleavage is staggered usually by 2 or 4 nucleotides, so that the resulting DNA fragments have short single-stranded overhangs at each end. These are called sticky or cohesive ends, as base pairing between them can stick the DNA molecules back together again. The staggered cut may result in overhanging ends either at 5′ (*Eco*RI) or 3′ (*Pst*I). The 3′ end always carries a hydroxyl group whereas the 5′ end always carries a phosphate group (Fig. 15.3). The cleavage patterns of some restriction enzymes are shown in Table 15.1.

**Table 15.1:** Catalogue of some restriction endonucleases with their recognition sequences

| Enzyme | Producing organism | Recognition sequence[a] |
|---|---|---|
| **Staggered cut (sticky ends)** | | |
| *Bam*HI | *Bacillus amyloliquefaciens* H | G↓GATCC |
| *Bgl*II | *Bacillus globigi* | A↓GATCT |
| *Eco*RI | *Escherichia coli* R | G↓AATTC |
| *Hind*III | *Haemophilus infuenzae* Rd | A↓AGCTT |
| *Pst*I | *Providencia stuartii* | CTGCA↓G |
| *Sal*I | *Streptomyces albusG* | G↓TCGAC |
| *Taq*I | *Thermus aquaticus* | T↓CGA |
| *Pst*I | *Providencia stuartii* | CTGCA↓G |
| **Blunt ends** | | |
| *Alu*I | *Arthrobacter luteus* | AG↓CT |
| *Hae*III | *Haemophilus aegypticus* | GG↓CC |
| *Hpa*I | *Haemophilus parainfuenzae* | GTT↓AAC |
| *Pvu*II | *Proteus vulgaris* | CAG↓CTG |
| *Sau*3A | *Staphylococcus aureus* 3A | ↓GATC |
| *Sma*I | *Serratia marcescens* | CCC↓GGG |

↓ indicates the site of enzyme cut within the specified nucleotide sequence.

[a] The sequence shown is of one strand in the 5′→ 3′ direction.

```
5' --- G 3'              +              A A T T C --- 3'
3' --- C T T A A 5'                             G --- 5'

                         ↓

              5' --- G A A T T C --- 3'
              3' --- C T T A A G --- 5'
```

**Fig. 15.4:** Cleavage pattern of *Eco*RI.

7. The DNA ends of one molecule generated by a given enzyme with a given recognition sequence can form complementary base pairs with any other DNA molecule, provided they possess the same ends. The cleavage products of *Eco*RI is shown in Fig. 15.4. Hence, these enzymes are widely used in gene technology.

8. Some restriction endonucleases possess identical recognition and cleavage sites that are called isoschizomers, e.g. *Hin*dIII and *Hsu*I

```
5' — A A G C T T — 3'
3' — T T C G A A — 5'
```

9. Some enzymes possess the same recognition site but differ in the cleavage site. *Sma*I and *Xma*I have recognition site as 5' C C C G G G 3', but their cleavage sites are different (Fig. 15.5).

10. Some restriction endonucleases with different recognition sequences may produce the same sticky ends, that is, generate identical overlapping termini. Examples are *Bam*HI (recognition sequence GGATCC), *Bgl*II (recognition sequence AGATCT), and *Sau*3A (recognition sequence GATC). All these produce GATC sticky ends. Fragments of DNA produced by cleavage with either of these enzymes can be joined to each other, as each fragment will have a complementary sticky end.

The number of type II enzyme cleavage sites in a DNA molecule depends on the size of DNA, its base composition, and the GC content of the recognition sequences. The probability $P$ for a particular tetranucleotide consisting of all four bases to occur in a DNA molecules containing 50% AT and 50% GC bp is $(1/4)^4$. Tetranucleotide cleavage occurs at every 256 bp distance and hexanucleotide at every 4096 bp.

Star activity is another property shown by type II restriction endonucleases, i.e. the enzyme changes its specificity. It arises mainly due to the change in reaction conditions from the optimum, e.g. increasing the pH, lowering the concentration of NaCl, or replacing magnesium by manganese, and also due to the methylation of DNA. For example, *Eco*RI normally recognizes and cleaves the sequence 5' - G A A T T C - 3'. When the reaction condition changes, it recognizes a tetranucleotide sequence 5'-A A T T –3'.

```
          ↓
Smal   5' ---- C C C G G G 3' --------> 5' C C C_OH 3'
       3' ---- G G G C C C 5' --------> 3' G G G_P 5'
                      ↑

          ↓
Xmal   5' ---- C C C G G G 3' --------> 5' C_OH 3'
       3' ---- G G G C C C 5' --------> 3' G G G C C_P 5'
                    ↑
```

**Fig. 15.5:** Cleavage pattern of type II enzymes with same recognition site but differs in cleavage site.

## Type III endonucleases

1. They have two different subunits. One unit is responsible for site recognition and modification while the other is responsible for nuclease action.
2. They require ATP and $Mg^{++}$ as cofactors.
3. The recognition sites are asymmetric. These nucleases cleave DNA at specific sequences, which are nonpalindromic. Cleavage takes place by nicking one strand in the immediate vicinity of the recognition site, usually 25–27 bp downstream of the recognition site. Two sites in opposite orientations are necessary to break the DNA duplex. Recognition sequences of some of the type III enzymes are given below.

$$5'- G \ A \ C \ G \ C \ (N)_5 \ - 3'$$
$$3'- C \ T \ G \ C \ G \ (N)_{10} \ - 5'$$
*Hga*I

$$5'- G \ A \ A \ G \ A \ (N)_8 \ - 3'$$
$$3'- C \ T \ T \ C \ T \ (N)_7 \ - 5'$$
*Mbo*II

$$5'- G \ G \ A \ T \ G \ (N)_9 \ - 3'$$
$$3'- C \ C \ T \ A \ C \ (N)_{13} \ - 5'$$
*Fok*I

4. In contrast to the staggered ends generated by type II endonucleases, the single-strand ends produced by type III endonucleases always differ from each other and thus cannot be recombined at random.
5. They lack ATPase activity, and S-adenosyl methionine is not required.
6. They produce a relatively homogenous population of DNA fragments, unlike fragments obtained by cleavage of type I enzymes.
7. They are not used for gene cloning because cleaved products are not uniform.

### Other restriction enzymes

Some restriction enzymes have characteristics different from those of the above-mentioned three types. For example, the *Eco*571 consists of a single polypeptide that has both modification and nuclease activities, with the nuclease activity stimulated by S-adenosyl methionine. It has other properties of type II. Thus, this system should be reclassified from type II to type IV.

Another system *Mcr* (modified cytosine reaction) also falls outside the type I–IV classification. It was found that DNA from some bacterial and eukaryotic sources was restricted in some of the commonly used host strains of *E. coli*. The phenomenon is caused by methyl cytosine in the DNA, and was called modified cytosine reaction. There are two *Mcr* systems: (*Mcr*A) encoded by a prophage-like element and (*Mcr*BC) encoded by two genes, *mcr*B and *mcr*C, which are located very close to the genes encoding the *E. coli* K type I system. The recognition site for *Mcr*BC is

$$R^mC \ (N_{40-80}) \ R^mC, \text{ where } R = A \text{ or } G$$

Cleavage occurs at multiple sites in both strands between the methyl cytosines and it requires GTP. The *Mcr*BC system is analogous to the *Rgl*B (restricts glucoseless phage) system discovered in T-even phages. It was observed that wild type T-even phages have glucosylated 5-hydroxymethyl cytosine instead of cytosine in their DNA. If glucosylation of 5-hydroxymethyl cytosine is prevented by mutation of the phage or host genes, the phage is restricted. It was not originally clear why *E. coli* should possess a system for restricting DNA containing 5-hydroxymethyl cytosine, since this is an uncommon form of DNA. But it is now apparent that the *Rgl*B (= *Mcr*BC) system that restricts DNA containing 5-hydroxymethyl cytosine, methyl cytosine or 4-methyl cytosine.

## JOINING DNA MOLECULES

The enzymes used to join DNA fragments are called DNA ligases. When the vector DNA and foreign DNA are both cut with the same restriction enzyme, the overlapping ends of the vector and foreign DNA are compatible and complementary. These DNA fragments and vector molecules are mixed together, which forms complementary base pairs between overlapping terminal single-stranded DNA sequences. Ligases act on DNA substrates with 5′ terminal phosphate groups and form the phosphodiester bond between the two DNA sequences (vector molecule and the DNA to be cloned) to join them together. This is the final step in construction of a recombinant DNA molecule. This process is called *ligation*.

### DNA ligase

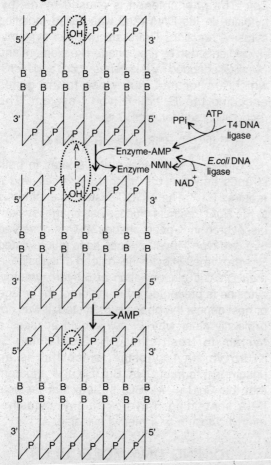

**Fig. 15.6:** Action of DNA ligase. An enzyme–AMP complex binds to a nick bearing 3′OH and 5′ P groups. The AMP reacts with the phosphate group. Attack by the 3′OH group on this moiety generates a new phosphodiester bond, which seals the nick. Reproduced from *Principles of Gene Manipulation,* R.W. Old and S.B. Primrose, by kind permission of Blackwell Science, Osney Mead, Oxford, U.K.

*E. coli* and bacteriophage T4 encode the enzyme DNA ligase, which seals single-stranded nicks between adjacent nucleotides in a duplex DNA chain. The reaction catalyzed by the enzymes of *E. coli* and T4 phage infected *E. coli* is very similar, but they differ in their cofactor requirements. The T4 enzyme requires ATP, whereas the *E. coli* enzyme requires NAD⁺. Both of these catalyze the formation of a phosphodiester bond between adjacent 3′ hydroxyl and 5′ phosphate termini in DNA (Fig. 15.6). In each case, the cofactor is split and forms an enzyme–AMP complex. The complex binds to the nick and makes a covalent bond in the phosphodiester chain.

Bacteriophage T4 DNA ligase is often used in gene cloning because it can join DNA molecules of annealed cohesive DNA termini or nicks as well as join blunt-ended double-stranded DNA molecules, whereas *E. coli* DNA ligase is ineffective in ligating blunt-ended DNA fragments.

### DNA-modifying enzymes

#### Kinase

Bacteriophage T4 polynucleotide kinase catalyzes the transfer of $\gamma$-$^{32}$P phosphate of ATP to a 5′ terminus of DNA or RNA.

$$5′ - G_{OH}\ 3′ \qquad\qquad 5′ - G_{OH}\ 3′$$
$$3′ - CpTpTpApA_{OH} - 5′ \xrightarrow[\text{Polynucleotide kinase}]{\gamma - ^{32}P\text{-ATP}} 3′ - CpTpTpApAp^* - 5′$$

Two types of reactions are commonly used. In the forward reaction, the gamma-phosphate is transferred to the 5′ terminus of dephosphorylated DNA (i.e. the DNAs that lack the required phosphate residue). In the exchange reaction, an excess of ADP causes bacteriophage T4 polynucleotide kinase to transfer the terminal 5′ phosphate from phosphorylated DNA to ADP; the DNA is then rephosphorylated by transfer of a radiolabeled gamma phosphate from $\gamma^{32}$P-ATP. Thus, this enzyme of T4 kinase has phosphorylation activity as well as

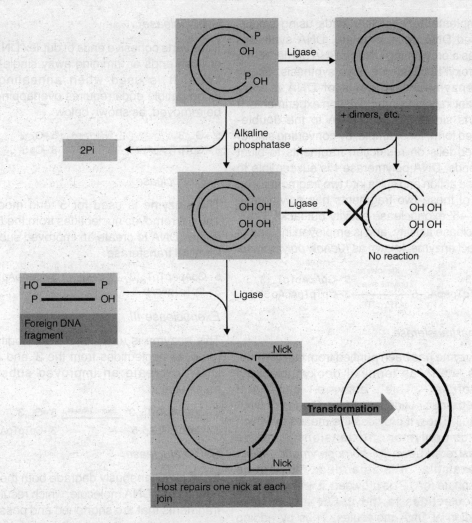

• **Fig. 15.7** Action of alkaline phosphatase to prevent recircularization of vector plasmid without insertion of foreign DNA. Reproduced from *Principles of Gene Manipulation*, R.W. Old and S.B. Primrose, by kind permission of Blackwell Science, Osney Mead, Oxford, U.K.

phosphatase activity. This enzyme is used in radiolabeling the 5′ termini for sequencing by the Maxam-Gilbert technique (see DNA sequencing methods) and for other uses requiring terminally labeled DNA. It is also used in phosphorylating synthetic linkers and other fragments of DNA that lack terminal 5′ phosphates in preparation for ligation.

### Alkaline phosphatase

It catalyzes the removal of 5′ phosphate groups from the DNA and thus modify the termini of DNA.

$$5′-G_{OH}\ 3′ \qquad\qquad\qquad\qquad 5′-G_{OH}\ 3′$$
$$3′-CpTpTpApAp-5′ \xrightarrow[\text{Phosphatase}]{\text{Alkaline}} 3′-CpTpTpApA_{OH}-5′$$

By treatment with alkaline phosphatase, both recircularization and plasmid dimer formation are prevented because ligase cannot join the ends (Fig. 15.7).

### DNA polymerase

DNA polymerase I of *E. coli,* also known as DNA-dependent DNA polymerase, directs the synthesis of complementary nucleic acids using single-stranded DNA as a template. DNA synthesis requires a preexisting DNA or RNA primer with a 3′ hydroxyl to initiate de novo synthesis. The *in vitro* enzymatic synthesis of DNA is very important in genetic manipulation experiments. It converts single-stranded DNA to the double-stranded form, for example, by converting single-stranded tails on restriction fragments to blunt flush ends. DNA polymerase I is susceptible to protease action and splits into two fragments. The larger of these two fragments has polymerase and 3′→5′ exonuclease activity but lacks 5′→3′ exonuclease activity, and is employed instead of the intact enzyme, known as *Klenow polymerase.*

$$5'-G_{OH}\,3' \qquad \xrightarrow[\text{dATP, dTTP}]{\substack{\text{Klenow} \\ \text{polymerase}}} \qquad 5'-GpApApTpT_{OH}\,3'$$
$$3'-CpTpTpApAp-5' \qquad\qquad\qquad\qquad 3'-CpTpTpApAp\,5'$$

### Terminal transferase

This enzyme has been purified from calf thymus, and is known as terminal deoxynucleotidyl transferase. This enzyme can add oligodeoxynucleotide tails to the 3 ends of DNA duplexes. Thus, it provides the means by which the homopolymeric extensions can be synthesized, known as *homopolymeric tailing.* For example, if single deoxynucleotide triphosphate (dATP) is provided, it will repeatedly add nucleotides to the 3′-OH termini of a population of DNA molecules. Thus by adding oligo d(A) sequences to 3′ ends of one population and oligo d(T) to the 3′ ends of another population of DNA molecules, the two types of DNA molecules can anneal to form mixed dimeric circles. Jackson *et al.* (1972) applied for the first time the homopolymer tailing method when they constructed a recombinant in which a fragment of phage λ DNA was inserted into SV40 DNA using dA.dT homopolymers (Fig. 15.8). This has been extensively used in the construction of recombinant plasmids. An example of using dC.dG tailing is shown in Fig. 17.5.

### S 1-Nuclease

It converts cohesive ends of duplex DNA to blunt or flush ends or trimming away single-stranded ends. It is used when annealing of two incompatible ends requires overlapping ends to be removed, as shown below:

$$5'-G_{OH}\,3' \qquad \xrightarrow{\text{SI-Nuclease}} \qquad 5'-G_{OH}\,3'$$
$$3'-CpTpTpApAp-5' \qquad\qquad\qquad 3'-Cp\,5'$$

### λ-Exonuclease

This enzyme is used for 5′ end modification. Thus, it removes nucleotides from the 5′ ends of duplex DNA to create an improved substrate for terminal transferase.

$$5'-GpApApTpT_{OH}\,3' \xrightarrow{\text{λ-Exonuclease}} 5'-GpApApTpT_{OH}\,3'$$
$$3'-CpTpTpApAp-5' \qquad\qquad 3'-Cp-5' \quad +Np5'$$

### Exonuclease III

This enzyme is used for 3′ end modification. It removes nucleotides from the 3′ end of duplex DNA to create an improved substrate for manipulation.

$$5'-GpApApTpT_{OH}\,3' \xrightarrow{\text{ExonucleaseIII}} 5'-G_{OH}\,3' \quad +Np5'$$
$$3'-CpTpTpApAp-5' \qquad\qquad\quad 3'-CpTpTpApAp-5'$$

### Bal 31 Nuclease

It can simultaneously degrade both the 3′ and 5′ strands of a DNA molecule, which results in DNA fragments that are shortened and possess blunt ends at both the termini. It is a highly specific single-stranded endodeoxyribonuclease

$$5'-GpApApTpT_{OH}\,3' \xrightarrow{\text{Bal 31}} 5'pApA_{OH}\,3'$$
$$3'-CpTpTpApAp-5' \qquad\qquad 3'TpTpA-5'$$
$$\text{+ Oligo and mono nucleotides}$$

### Linkers and adaptors

Recombinant DNA and gene cloning experiments take advantage of the ability of restriction enzymes to cut DNA at particular positions. If suitable enzyme sites are not available for manipulation on the DNA, then these cleavage sites can be added as linker or

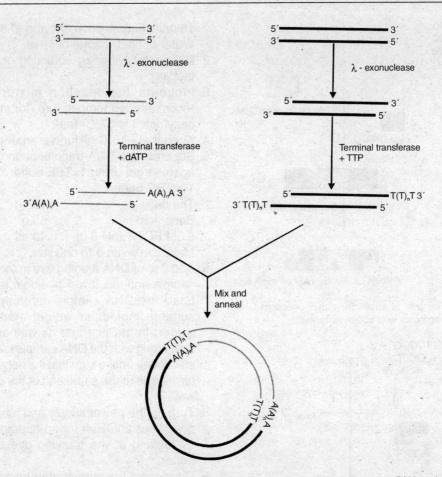

**Fig. 15.8:** Use of terminal transferase enzyme to add complementary homopolymer tails to two DNA molecules.

adaptor molecules. *Linkers* are synthetic single-stranded oligonucleotides, which self-associate to form symmetrical double-stranded molecules containing the recognition sequence for a restriction enzyme. For example, the oligonucleotide dCCGAATTCGC will self-associate to give a duplex structure containing the *Eco*RI recognition sequence (indicated by asterisks).

5′ C C G*A*A*T*T*C*G G 3′
3′ G G C T T A A G C C 5′

These nucleotides can be ligated to the blunt end of any DNA fragment and then cut with the specific restriction enzymes to generate DNA fragments with cohesive termini (Fig. 15.9).

*Adaptor* molecules are chemically synthesized DNA molecules within preformed cohesive ends. They are required when the target site for the restriction enzyme used is present in within the DNA sequence to be cloned. Consider a blunt-ended foreign DNA. A synthetic adaptor molecule is ligated to the foreign DNA. The adaptor is used in the 5′ hydroxyl form to prevent self-polymerization (Fig.15.10). The foreign DNA plus ligated adaptors is phosphorylated at the 5′ terminus and ligated to the vector previously cut with a particular restriction enzyme.

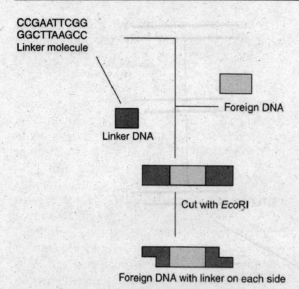

CCGAATTCGG
GGCTTAAGCC
Linker molecule

Foreign DNA

Linker DNA

Cut with *Eco*RI

Foreign DNA with linker on each side

• **Fig. 15.9:** The use of linker for construction of DNA fragments with cohesive termini.

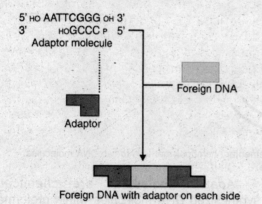

5' HO AATTCGGG OH 3'
3'      HOGCCC P  5'
Adaptor molecule

Foreign DNA

Adaptor

Foreign DNA with adaptor on each side

• **Fig. 15.10:** The use of adaptors for construction of DNA fragments with cohesive termini.

## PROTOCOL

### 15.1: Restrictive digestion of plasmid DNA

1. Pipette the following components in a clean micro centrifuge tube kept on ice. Add the restriction enzyme at the end. (Always make a control in which DNA is not added).
Appropriate 10x buffer
DNA sample (~ 0.5 μg)

10 units of restriction enzyme (0.5 to 1 μl)
Water to final volume of 10 μl.

2. Mix gently either by vortexing or by finger tapping.

3. Incubate the reaction mixture at the recommended temperature (for most of the enzymes it is 37°C) for 2 h.

4. Keep it at –40°C till further analysis.

5. Separate the DNA fragments on 0.8% mini agarose gel using 1×TBE buffer
10×TBE buffer

Tris base                108 g
Boric acid               55 g
0.5M EDTA (pH 8.0)       40 ml
Make the volume to one liter.

6. Add 2 μl of DNA loading dye in the digested sample and mix it well by finger tapping.

7. Load the DNA sample carefully. Load a suitable molecular weight marker (1-kb ladder) in the first and as well as the last lane along with the DNA samples. Attach the electrodes (make sure that the negative black terminal is at the same end of the unit as the wells).

8. Turn on the power supply and run at 20 mA power current. Run for an appropriate time by looking at the tracking dye of the gel buffer.

9. Disconnect the current after an appropriate time. Remove the gel from electrophoresis unit and immerse it in an ethidium bromide stain. Working staining solution is 20 μl ethidium bromide stock (10 mg/ ml) to 200 ml distilled water. Keep the gel in the staining solution for 20–30 min. (Gloves should be worn while handling ethidium bromide.)

10. View the gel on UV transilluminator and photograph the gel.

### REFERENCES

Jackson, D.A., Symons, R.H. and Berg, P. 1972. Biochemical method for inserting new genetic information into DNA of Simian virus 40: circular SV 40 DNA molecules containing lambda phage genes and the galactose operon of *Escherichia coli*. *Proc. Natl. Acad. Sci. USA* **69:** 2904–2909.

Smith, H.O. and Nathans, D. 1973. A suggested nomenclature for bacterial host modification and restriction systems and their enzymes. *J. Mol. Biol.* **81:** 419–423.

Temin, H.M. and Mizutani, S. 1970. RNA dependent DNA polymerase in virions of Rous Sarcoma virus. *Nature* **226:** 1211–1213.

Yuan, R. 1981. Structural and mechanism of multifunctional restriction endonucleases. *Annu. Rev. Biochem.* **50:** 285–315.

## FURTHER READING

Chawla, H.S. 1998. *Biotechnology in Crop Improvement.* International Book Distributing Co., Lucknow, India.

Grierson, D. and Covey, S.N. 1988. *Plant Molecular Biology.* Blackie, Glasgow.

Maniatis, T., Fritsch, E.F. and Sambrock, J. 1982. *Molecular Cloning: A Laboratory Manual.* Cold Spring Harbor Laboratory Press, USA.

Mantell, S.H., Mathew, J.A. and McKee, R.A. 1985. *Principles of Plant Biotechnology: An Introduction to Genetic Engineering in Plants.* Blackwell Scientific Publications, Oxford.

Old, R.W. and Primrose, S.B. 2001. *Principles of Gene Manipulation: An Introduction to Genetic Engineering.* Blackwell Scientific Publications, Oxford.

Sambrook, J., Fritsch, E.F. and Maniatis, T. 1989. *Molecular Cloning: A Laboratory Manual.* Cold Spring Harbor Laboratory Press, USA.

Winnacker, E.L. 1987. *From Genes to Clones: Introduction to Gene Technology*, VCH, Weinheim, Germany.

# 16

# Gene Cloning: Vectors

One of the most important elements in recombinant DNA (rDNA) technology or in gene cloning is the vector. A random DNA or cDNA segment or specific gene is linked into a vector to form rDNA molecule, which can be propagated in suitable host cells to a large number, is a cloning vector. There are different types of cloning vectors for use with different types of host cells. The largest number exists for *E. coli* and the best known of these is the plasmid vector. Besides, there are cosmids, phages, phagemids, yeast artificial chromosomes, transposons, bacterial artificial chromosomes, etc. A few of the vectors have been described.

A vector should have the following features:
1. It must contain a replicon that enables it to replicate in host cells.
2. It should have several marker genes, which help to differentiate the transformed cells from the nontransformed cells and also from the transformed cells, which contain recombinant DNA molecules, e.g. genes for ampicillin and tetracycline resistance.
3. It should have a unique cleavage site within one of the marker genes so that insertion of foreign DNA into the marker gene leads to its inactivation and identification of recombinant DNA molecule.
4. For the expression of cloned DNA, the vector DNA should contain suitable control elements, such as promoters, terminators, and ribosome binding sites.

## Biology of *E. coli* K-12

Stanley Cohen and Herb Boyer chose *Escherichia coli* strain K-12 for their molecular biology experiments in the early 1970s because it was easy to grow and amenable to metabolic studies. During the last 30 years, *E. coli* K-12 has proved to be an innocuous biological host for the propagation of recombinant DNA molecules. The attenuated *E. coli* K-12 strain does not thrive outside of the laboratory environment and it is unable to compete against the more genetically robust *E. coli* serotypes normally found in the human intestine.

*E. coli* is a gram negative rod-shaped bacterium about 2.5 μm long, which contains flagella and a genome of 4,639,221 bp encoding at least 4000 genes. The natural habitat of *E. coli* is the intestine, and it is able to grow in simple media containing glucose, ammonia (or peptide digests), phosphate and salts ($Na^+$, $K^+$, $Ca^{2+}$, and $Mg^{2+}$). *E. coli* contains both an outer and inner membrane separated by a protein-rich periplasmic space. The outer membrane contains proteins that function in the import and export of metabolites and also provide binding sites for infectious bacteriophage. Like all gram-negative bacteria, *E. coli* have no nuclear membrane and the chromosome is a large circular duplex molecule with a membrane attachment site and a single origin of replication. *E. coli* cells undergo cell division every 20 minutes when grown under optimal aerobic conditions, and they are facultative anaerobes, which allow them to be grown to high density using fermentation devices. Single *E. coli* colonies are isolated on standard bacterial agar plates and individual strains can be stored anaerobically in agar stabs or frozen as glycerol stocks.

*E. coli* K-12 is a safe, nonpathogenic bacterium, but specific laboratory practices are always warranted and specific biosafety guidelines must be followed for recombinant DNA research. *E. coli* K-12 as a standard host used in gene cloning experiments has several added properties, which has greatly increased the efficiency.

(i) Bacterial restriction modification system poses the biggest problem for gene cloning. The *E. coli* K-12 restriction system (*Eco*K) requires three proteins encoded by *hsd*R, *hsd*M, and *hsd*S genes that function together as a protein complex for both cleaving and methylation of DNA. If the recognition system 5′–AAC(N)$_5$GTGC–3′ is unmethylated, it cleaves the DNA at a variable distance from this recognition site. *E. coli* K-12 strains with mutations in the endogenous restriction modification system (*hsd*R⁻) are used.

(ii) *E. coli* K-12 strains used for gene cloning contain mutations in the *rec*A gene, which encodes a recombination protein. *rec*A⁻ enhances the biosafety of *E. coli* K-12 strains because *rec*A⁻ strains are sensitive to ultraviolet light.

(iii) Most *E. coli* K-12 strains contain mutations in *end*A gene, which encodes DNA-specific endonuclease I. Loss of this endonuclease greatly increases the plasmid DNA yields and improves the quality of DNA that is isolated using standard biochemical preparations.

# PLASMIDS

Plasmids are double-stranded, closed circular DNA molecules, which exist in the cell as extrachromosomal units. They are self-replicating, and found in a variety of bacterial species, where they behave as accessory genetic units. Plasmids replicate and are inherited independently. There are three general classes of plasmids: (i) virulence plasmids, which encode toxin genes, (ii) drug resistance plasmids, which confer resistance to antibiotics, and (iii) plasmids, which encode genes required for bacterial conjugation. Basically all molecular cloning methods involve the manipulation of specific DNA fragments that utilize some aspects of different plasmids. Plasmids range in size from 1 to 200 kb and depend on host proteins for maintenance and replication functions. Most of the plasmid-encoded genes that have been characterized impart some growth advantage to the bacterial host. These are present in a characteristic number of copies in a bacterial cell, yeast cell or even in organelles (mitochondria) of eukaryotic cells. These plasmids can be single copy plasmids that are maintained as one plasmid DNA per cell or multicopy plasmids that are maintained as 10–20 genomes per cell (Table 16.1). There are also plasmids that are under relaxed replication control, thus permitting their accumulation in very large numbers (up to 1000 copies per cell). Most plasmids used in cloning have a relaxed mode of replication. The best-studied plasmid replication system is that of *ColE1*, which is a small *E. coli* plasmid encoding antibacterial proteins called colicins.

**Table 16.1:** Replicons carried by the currently used plasmid vectors.

| Plasmid | Replicon | Copy number |
|---|---|---|
| pBR322 and its derivatives | pMB1 | 15–20 |
| pUC | pMB1 | 500–700 |
| pACYC and its derivatives | p15A | 10–12 |
| pSC101 and its derivatives | pSC101 | ~15 |
| ColE1 | ColE1 | 15–20 |

*ColE1* replication has been shown to involve transcription of a DNA sequence near the origin, which creates the RNA primer required for initiation of DNA synthesis. These plasmids are used as cloning vectors due to their increased yield potential.

Most plasmid vectors in current use carry a replicon derived from the plasmid pMB1, which was originally isolated from a clinical specimen (its close relative is *ColE1* replicon). Under normal conditions of growth, 15–20 copies of plasmids carrying this replicon are maintained in each bacterial cell. The pMB1 (or *ColE1*) replicon does not require plasmid-encoded functions for replication; instead, it relies entirely on long-lived enzymes supplied by the host. Replication occurs unidirectionally from a specific origin and is primed by an RNA primer.

Replication of plasmid DNA is carried out by a subset of enzymes used to replicate the bacterial chromosome. Usually, a plasmid contains only one origin of replication together with its associated *cis*-acting control elements. The control of plasmid copy number resides in a region of the plasmid DNA that includes the origin of DNA replication.

The F plasmids encode the gene required for bacterial conjugation. This extrachromosomal genetic element is ~95 kb, and was named because of its role in bacterial fertility. The F plasmid is present as one copy per cell and encodes ~100 genes, some of which are required for F plasmid transfer during conjugation (*tra* genes). As a result of integration into *E. coli* chromosome and then an imprecise excision resulting in the reformation of a circular plasmid, the F plasmid is able to transfer portions of the *E. coli* genome into another bacterium via conjugation. F plasmid containing *E. coli* strains are required in molecular cloning involving M13 filamentous bacteriophage. The F plasmid carries the sex pili gene, which encodes the cell-surface-binding site for M13 phages.

Frequently, plasmids contain some genes advantageous to the bacterial host, for example, resistance to antibiotics, production of antibiotics, degradation of complex organic compounds, and production of colicins, and enterotoxins. Antibiotics are antimicrobial agents that can be classified based on their mode of action. Bacterial antibiotic-resistance genes encode proteins that inactivate these agents either by preventing their accumulation in the cell or by inactivating them once they have been imported. By utilizing these antibiotic-resistance genes as dominant genetic markers in plasmid cloning vectors, it is possible to select for *E. coli* cells that have maintained high copy replication of plasmid DNA molecules. Table 16.2 lists examples of commonly used antibiotics and the corresponding bacterial antibiotic resistance genes that have been exploited for specific cloning experiments.

Discussions about the danger involved in cloning foreign DNA into bacteria have led to requirements for cloning vectors regarding containment. As genetically engineered plasmids should not be able to spread in nature by means of conjugation, cloning vectors should not be self-transmissible plasmids. Plasmids are self-transmissible when they contain transfer (*tra*) genes, which allow the production of pili involved in conjugal transfer of plasmid DNA. If a second plasmid, which does not contain these *tra* genes, is also present in the same cell, it will also be transferred as long as it contains the so-called mobilization (*mob*) genes. This plasmid is then said to have been mobilized. Laboratory-used plasmids, therefore, have their *tra* and *mob* genes deleted to prevent the possibility of self-transmission or mobilization of recombinant plasmids should they escape from the laboratory environment.

A plasmid vector used for cloning is specifically developed by adding certain features:

i. Reduction in size of vector to a minimum to expand the capacity of vector to clone large fragments. Since the efficiency of transformation of bacterial cells drops drastically when plasmids larger than 15 kb are used, the size of cloning vector should

**Table 16.2:** Antibiotics and antibiotic resistance genes used in gene cloning

| Antibiotic | Mode of action | Resistance gene | Application |
|---|---|---|---|
| Ampicillin | Inhibits cell wall synthesis by disrupting peptido-glycan cross-linking | β-lactamase (amp$^R$) gene product is secreted and hydrolyzes ampicillin | Amp$^R$ gene is included on plasmid vector as positive selection markers |
| Tetracycline | Inhibits binding of aminoacyl tRNA to the 30S ribosomal subunit | tet$^R$ gene product is membrane bound and prevents tetracycline accumulation by an efflux mechanism | tet$^R$ gene product is a positive selection marker on some plasmids |
| Kanamycin | Inactivates translation by interfering with ribosome function | Neomycin or amino glycoside phospho transferase (neo$^R$) gene product inactivates kanamycin by phosphorylation | Neo$^R$ gene is a positive selection marker on plasmids |
| Bleomycin | Inhibits DNA and RNA synthesis by binding to DNA | The bla$^R$ gene product binds to bleomycin and prevents it from binding to DNA | Bla$^R$ gene is a positive selection marker used in plasmids |
| Hygromycin B | Inhibits translation in prokaryotes and eukaryotes by interfering with ribosome translocation | Hygromycin-B-phospho transferase (hph or hyg$^R$) gene product inactivates hygromycin B by phosphorylation | Hyg$^R$ gene is used as a positive selection marker in eukaryotic cells that are sensitive to hygromycin B |
| Chloramphenicol | Binds to the 50S ribosomal subunit and inhibits translation | Chloramphenicol acetyl transferase (CAT or CM$^R$) gene product metabolizes chloramphenicol in the presence of acetyl CoA | CAT/CM$^R$ gene is used as a selectable marker, and as transcriptional reporter gene of promoter activity in eukaryotic cells. |

be small, preferably 3–4 kb. In this way foreign DNA fragments of 10–12 kb can be accommodated.

ii. It should contain an origin of replication that operates in the organism into which the cloned DNA is to be introduced.

iii. Introduction of selectable markers.

iv. Introduction of synthetic cloning sites termed *polylinker*, restriction site bank, or polycloning sites that are recognized by restriction enzymes. This polycloning site is usually present inside a marker gene so that with the insertion of foreign DNA it will inactivate that marker gene and the recombinants can be selected.

v. Incorporation of axillary sequences such as visual identification of recombinant clones by histochemical tests, generation of single-stranded DNA templates for DNA sequencing, transcription of foreign DNA sequences *in vitro*, direct selection of recombinant clones and expression of large amounts of foreign proteins.

Some of the plasmid vectors have been described below.

### pBR322

This was one of the first artificial cloning vectors to be constructed, and undoubtedly the most widely used cloning vector up to now (Bolivar and Rodriguez, 1977a; 1977b). It is a 4.36-kb double-stranded cloning vector. This plasmid vector has been put together from fragments originating from three different naturally occurring plasmids. Figure 16.1 indicates all the salient features of this vector. It contains *ColE1 ori* of replication with relaxed replication control. Generally there are 15–20 molecules present in a transformed *E. coli* cell, but this number can be amplified by incubating a log phase culture of *ColE1* carrying cells in the presence of chloramphenicol. Like *ColE1*, pBR322 and all its derivatives are amplifiable, which makes it very simple to isolate large amounts of these plasmids. pBR322 contains two antibiotic—resistance genes, one is the ampicillinresistance

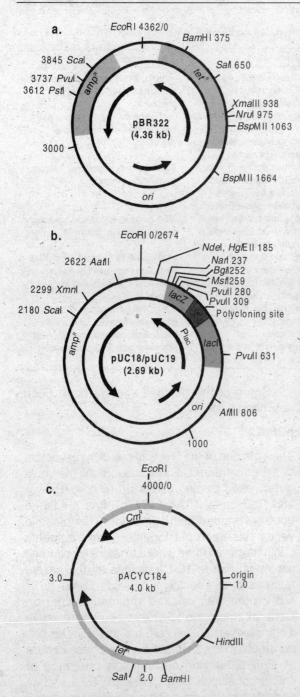

• **Fig. 16.1:** Plasmid vectors. a–pBR322; b–pUC18/19; c–pACYC184.

($amp^R$) gene coding for a β-lactamase (which modifies ampicillin into a form that is nontoxic to *E. coli*) and the other is the tetracycline-resistance ($tet^R$) gene (a set of genes coding for enzymes that detoxify tetracycline). The β-lactamase gene originates from the transposon *Tn*1, originally located on the R plasmid R1 drd 19. Due to the way in which pBR322 was constructed, this gene is no longer transposable. The gene determining tetracycline resistance was derived from the naturally occurring plasmid pSC101, a low-copy-number plasmid which in itself has also been used as a cloning vector. The presence of these two resistance genes indicates that a bacterium transformed with pBR322 will be able to grow on a medium containing ampicillin and tetracycline, whereas the cells lacking the plasmid will not be able to grow.

The complete nucleotide sequence of pBR322 has been determined. The plasmid contains 20 unique recognition sites for restriction enzymes. Six of these sites, i.e. *Eco*RV, *Bam*HI, *Sph*I, *Sal*I, *Xma*III and *Nru*I, are located within the gene coding for tetracycline resistance; two sites for *Hind*III and *Cla*I lie within the promoter of the tetracycline resistance gene; and three sites for *Pst*I, *Pvu*I and *Sca*I lie within the β-lactamase gene. Cloning of a DNA fragment into any of these 11 sites results in the insertional inactivation of either one of the antibiotic resistance markers. The gene of interest is spliced into $tet^R$ gene cluster, and then the *E. coli* cells are transformed. Thus, three types of cells are obtained:

1. Cells that have not been transformed and so contain no plasmid molecules and will be $amp^S$ $tet^S$.
2. Cells that have been transformed with pBR322 but without the inserted DNA fragment or gene will be $amp^R$ $tet^R$. These are transformed cells.
3. Cells that contain a recombinant DNA molecule, that is, the DNA fragment has been inserted into the pBR322 at $tet^R$ gene cluster. These cells will lose tetracycline

resistance because the fragment has inserted in the middle of tetracycline resistance gene cluster. These are recombinants and will be *amp*[R] *tet*[S].

The bacteria are spread on an agar medium and colonies grown. By checking for growth on agar media modified with ampicillin, one can exclude untransformed cells, but both transformants and recombinants will produce colonies. Replica plates are made and then grown on agar with ampicillin and tetracycline. Colonies that do not grow on tetracycline agar are recombinants whereas both transformants and recombinants will grow on ampicillin agar. By comparing the replica plates, recombinants can be picked up from agar ampicillin plates. Plasmids can take up an insert of up to 10 kb.

Like *ColE1*, pBR322 is not self-transmissible, but it can be mobilized by other self-transmissible plasmids such as the F-factor.

## pACYC184

This is a small 4.0 kb cloning vector developed from the naturally occurring plasmid p15A (Fig. 16.1). It contains the genes for resistance to chloramphenicol and tetracycline. The chloramphenicol gene has a unique *Eco*RI restriction site and the tetracycline gene has a number of restriction sites as mentioned in pBR322. Insertional inactivation of one of the antibiotic-resistance genes facilitates the recognition of strains carrying plasmids with cloned DNA fragments. Like pBR322, it is not self-transmissible, and can be amplified. It has been a very popular cloning vector since it is compatible with the *ColE1*-derived plasmids of the pBR322 family and can therefore be used in complementation studies.

## pUC vectors

The name pUC is derived because it was developed in the University of California (UC) by Messings and his colleagues (Norrander *et al.*, 1983). These plasmids are of 2.7 kb and possess the *ColE1 ori* of replication. These vectors contain *amp*[R] gene and a new gene called *lac*Z, which was derived from the *lac* operon of *E. coli* that codes for β-galactosidase enzyme (Fig.16.1).

**The lac operon**: A number of bacterial operons in *E. coli* serve to illustrate various principles of prokaryotic gene regulation. But the *lac* operon has application in molecular biology experiments also. The *lac* operon encodes three enzymes: *lac*Z gene encodes β galactosidase, which cleaves lactose into glucose and galactose; *lac*Y encodes permease; and *lac*A encodes transacetylase. When glucose levels are low and lactose levels are high, transcription of the *lac* operon is maximally induced through a mechanism of transcriptional derepression and activation, which requires *lac* repressor and catabolite activator protein (CAP). The lac repressor protein is encoded by *lac*I gene. Binding of the lac repressor to the operator sequence is extremely tight and specific, which cause transcriptional repression through a mechanism involving repressor interference with RNA polymerase binding to the *lac* promoter. When an effector molecule such as isopropyl thiogalactoside (IPTG), a nonmetabolizable allolactose analog, binds to the lac repressor protein, its DNA binding affinity for the *lac* operator decreases and transcription of the *lac* operon is derepressed. The β galactosidase activity of the *lac* operon has been used as a tool for identification of clones. The β galactosidase activity can be measured in live cells by using the chromogenic substrate 5-bromo-4-chloro-3-indolyl-β galactoside (X-gal). In wild type *E. coli* cells, cleavage of X-gal produces a blue color that can be visualized as a blue colony on agar plates. The X-gal assay can be used to identify *lac* operon mutants, which appear as white colonies against a blue background. Another way the X-gal assay has been exploited is by taking advantage of β galactosidase mutants, which function through a mechanism called α complementation. Characterization of these mutants showed that β galactosidase activity could be reconstituted in *E. coli* cells that expressed both nonfunctional

N-terminal (α fragment) and C-terminal (ω fragment) β galactosidase polypeptides. α complementation has been used to monitor insertional cloning into plasmids containing restriction sites within the ~150 amino acid coding sequence of the α fragment. When these plasmids are grown in *E. coli* strains expressing the ω fragment on an F′ plasmid, it is possible to score antibiotic-resistant cells that contain the plasmid cloning vector and are either blue (no gene insertion) or white (coding sequences are disrupted). The advantage of using α complementation for DNA cloning is that plasmid vectors are limited by the size of insert DNA that can be accommodated, and the α fragment (~450 bp) is small compared to the entire *lacZ* gene (~3500 bp).

This enzyme splits lactose into glucose and galactose. A recombinant pUC molecule is constructed by inserting new DNA into one of the restriction sites that are clustered near the start of the *lacZ*. This means that

1. Cells harboring normal pUC plasmids are ampicillin resistant and able to synthesize β-galactosidase (transformant but nonrecombinant). When an inducer of *lac* operon IPTG (isopropyl thio-galactoside) and X-gal (5-bromo-4-chloro-3-indolyl-β-D-galactopyranoside), a substrate for β-galactosidase, is put in the media, it will be broken down by the enzyme into a product. It will give deep-blue colonies.

2. Cells harboring recombinant pUC plasmids are ampicillin resistant but unable to synthesize β-galactosidase. When IPTG and X-gal are put in the media, it will give white colonies because the substrate cannot be broken down. Recombinants can therefore be selected by the color of colonies and there will be no need of replica plating as done in the case of pBR322.

Cells plated + agar + ampicillin + IPTG + X-gal = Nonrecombinant—**Blue colonies**
Cells plated + agar + ampicillin + IPTG + X-gal = Recombinant—**White colonies**

## pUN121

This is a 4.4-kb vector derived from pBR322 (Nilsson *et al.*, 1983). This vector has been developed for rapid selection of bacteria containing a recombinant plasmid. This is required when genomic libraries are constructed, which involves the construction of several thousands of clones and handpicking of individual clones with inserted DNA would be extremely tedious. Direct-selection vectors have therefore been constructed. In these vectors (Fig. 16.2), the original promoter region of the tetracycline-resistance gene is exchanged for the $P^R$ promoter of bacteriophage (). Transcription from this

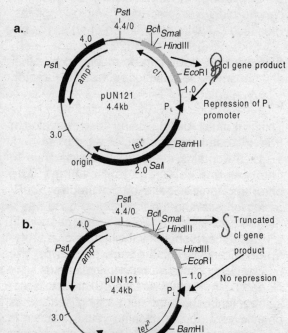

**Fig. 16.2:** pUN121 plasmid vector (a positive selection vector). a–When no foreign DNA is inserted in one of the restriction sites in the *cl* gene, the *cl* gene product is synthesized and represses the $P_L$ promoter. The cells harboring such plasmids are sensitive to tetracycline. b–Plasmids with foreign DNA inserted in the *cl* gene produce a truncated repressor. The $P_L$ promoter is derepressed and the cells become tetracycline resistant.

promoter is suppressed by the protein encoded by the lambda *cl* gene, which was also cloned on pUN121. Cells harboring pUN121 will therefore have the phenotype *amp*[R] *tet*[S] as the *cl* gene product suppresses expression of the tetracycline-resistance gene. The *cl* gene contains unique sites for the restriction enzymes *Eco*RI, *Hind*III, *Bcl*I, and *Sma*I. Cloning of foreign DNA into these restriction sites will cause insertional inactivation of the *cl* gene, abolishing the suppression of P$_R$ promoter. Cells harboring such a plasmid will become *tet*[R]. Plating of transformed cells on plates containing tetracycline will therefore only result in growth of cells harboring the pUN121 vector with inserted DNA fragments.

### Yeast plasmid vectors

Yeast (*Saccharomyces cerevisiae*) is a single-celled eukaryotic microbe, which can be cultured and manipulated using the standard techniques applied to bacteria. The genetic material of yeast is packaged into chromosomes within a membrane-enclosed nucleus and partitioned at cell division by mitosis and meiosis. In 1976, Struhl *et al.* reported that a fragment of yeast DNA when cloned into *E. coli* restored histidine-independent growth to strains carrying the *his*B mutation. Similarly, Ratzkin and Carbon (1977) ligated fragments of yeast genome into plasmid *ColE1* and obtained a small number of leucine-independent transformants from a *leu*B *E. coli* strain. In each case, the yeast chromosomal segment carried the gene for the yeast enzyme equivalent to that absent in the bacterial strain. Thus, the yeast *HIS3* gene was able to be expressed in a bacterial cell and produce the yeast version of the enzyme imidazole glycerol phosphate dehydratase and thereby complement the mutation in the bacterial *his*B strain (With yeast genes it is customary to indicate a wild-type allele in capital letters and the mutant allele in lower case letters). Other yeast genes isolated in this way include *LEU2*, *URA3*, *TRP*1, and *ARG*4. In general, genes from eukaryotic organisms rarely function in bacterial cells, but

the success with yeast genes arises from the lack of introns in most such genes coupled with the ability of some yeast promoters to function in bacteria.

The wild-type yeast genes thus isolated provided essential markers in subsequent attempts to introduce exogenous DNA into yeast cells. Hinnen *et al.* (1978) demonstrated that the yeast *LEU2* gene carried by the plasmid *ColE1* could be taken up by *leu2* yeast sphaeroplasts and transform them to a leucine-independent phenotype. In this case the transformation resulted from the integration of the added DNA into a yeast chromosome. Thus, a variety of approaches have resulted in a cloning system that makes use of many of the powerful techniques developed for *E. coli.*

*Vector systems*

Most yeast vectors developed share the following common features:

1. Most of the vectors are derived from bacterial plasmids and retain both the ability to replicate in bacteria and selection markers suitable for use in bacterial systems. As plasmid amplification is not high in yeast as compared to that in *E. coli*, this is a most useful feature. Thus, initial plasmid construction is done in *E. coli* before transfer to yeast. Such vectors permitting cloning in two different species are referred to as *shuttle vectors.*

2. In most cases, the selection markers employed are nutritional markers such as *LEU2* or *HIS3*. In some cases, such markers are not suitable. For example, Brewing yeasts are polyploid and it is hence not possible to obtain auxotrophic mutants. In such cases, resistance to copper or to the drug chloramphenicol have been used.

There are different classes of yeast vectors (Fig. 16.3) and the main distinguishing factors are the replication and transmission of the vector once inside the yeast cells.

***Yeast integrating plasmids (YIps).*** These are not actually yeast plasmid vectors, because

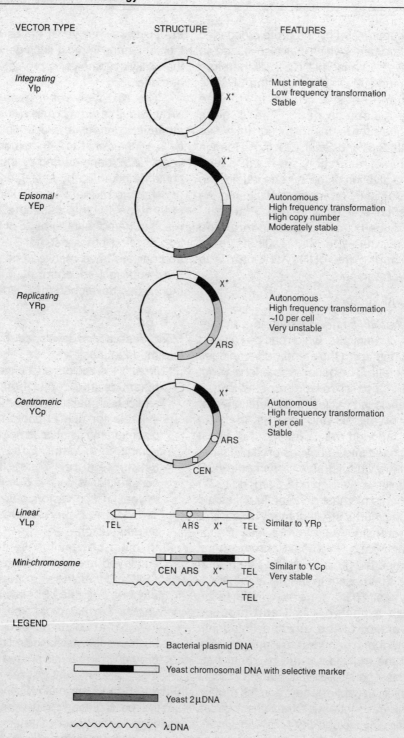

**Fig. 16.3:** Features of yeast vectors.

they are unable to replicate within a yeast cell without integration into a chromosome. Such vectors contain a yeast marker as the only addition to the bacterial plasmid. As integration is essential for a successful transformation, the frequencies are very low, with a typical yield of one transformant per µg of DNA. Here the plasmid vector integrates into the host chromosomal DNA. Integration is due to homologous recombination between the yeast gene carried by the YIp as the selectable marker and the inactive version of the gene present in one of the yeast chromosomes. Thus, the LEU2-carrying plasmid would integrate at the chromosomal leu2 region as a result of crossover either within the gene itself or the flanking region carried by the plasmid. The product is a yeast chromosome with an integrated bacterial plasmid flanked by a duplication of the gene. Low frequency excision events involving an intrachromosomal crossover between the homologous regions will eliminate either the incoming LEU2 allele or the original leu2 allele as well as the bacterial segment. This latter case represents a replacement of a chromosomal allele with an added allele from outside the cell.

An important development in the use of these vectors has been the observation that opening the plasmid within the yeast segment before adding it to the yeast cells greatly increased the transformation frequency. For that purpose, the plasmid is treated with a restriction enzyme that has a unique recognition site within the yeast segment. If a plasmid carried two yeast segments (e.g., URA3 and LEU2), cutting the plasmid within one of these regions will target the integration to that chromosomal locus.

A YIp will not replicate as a plasmid because it lacks autonomously replication sequences (ARS); therefore, it cannot be multiplied in large numbers inside the yeast host.

**Episomal plasmids–Yeast episomal plasmid (YEp)**. It is a yeast vector with the ability to replicate autonomously without integration into a yeast chromosome. Yeast vectors with this property have been built around a naturally

occurring yeast plasmid, the so-called 2µ circle. This molecule has no function and hence its name is based on contour length. It is present in most yeast strains with a copy number of 40–50 per haploid genome. The 2-µm plasmid, as it is called, is 6 kb in size. These are autonomous and show high frequency transformation.

The first 2-µm-based vectors carried the whole of this plasmid and gave high frequency transformation (up to $2 \times 10^4$ transformants per µg DNA) combined with reasonable stability levels. Such plasmids are, however, large and carry only a limited number of unique restriction sites. Smaller vectors carrying only a limited region of the 2-µm circle have been developed. Such plasmids are more easily handled in cloning experiments, but the loss of 2-µm sequences has cost in terms of stability. Hence, stability levels acceptable for most purposes can be obtained in $cir^+$ cells, and the high copy numbers obtained have made these plasmids the main workhorse for yeast cloning.

**Yeast replicating plasmids (YRp)**. This is a vector based on chromosomal elements. Chromosomal sequences with the ability to support autonomous replication of plasmids in yeast were first discovered when it was observed that a region closely linked to TRP1 gene would allow high frequency transformation and propagation without the need for integration. Such so-called ARS have been isolated from many regions of the yeast genome and from a range of other eukaryotic organisms. Vectors based on ARS are highly unstable once in the yeast cell, but they show high frequency transformation. The copy number in cells retaining the plasmid is high, in the order of 10 per cell.

**Yeast centromeric plasmid (YCp)**. This vector is also based on chromosomal elements. It is an ARS-based plasmid, which carries in addition a yeast chromosomal centromeric DNA sequence. This greatly increased the stability of an ARS-based plasmid. Such plasmids have a copy number of one per cell and are thus useful in situations in which a low copy number is essential, such as cloning genes expressing a product lethal to the cell in excess. *ARS* centromeric plasmids have also been referred to as mini chromosomes, which are circular.

**Yeast linear plasmid (YLp)**. These are ARS centromeric plasmids, but are linear rather than circular. Linear plasmids have been developed by adding telomeric (chromosome end) sequences to the ends of a cut ARS-based plasmid. The original construction made use of telomeres from ciliate protozoan *Tetrahymena*. These linear plasmids are present like a genuine chromosome, in a single copy, and are the most stable yeast plasmid vectors. Linear plasmids cannot replicate in *E. coli* and are hence technically difficult to work with. They have nonetheless attracted much interest as they raise the exciting possibility of putting together complete sets of genes in artificial eukaryotic chromosomes.

## Ti plasmids

Most cloning vectors for plants are based on the Ti plasmid, which is not a natural plant plasmid, but belongs to a soil bacterium *Agrobacterium tumefaciens*. This bacterium invades plant tissues, causing a cancerous growth called a crown gall. During infection, a part of the Ti plasmid called the T-DNA is integrated into the plant chromosomal DNA. Cloning vectors based on Ti plasmid have been explained in greater details in Chapter 21. Likewise, there are Ri plasmids present in *Agrobacterium rhizogenes*, which cause hairy root disease in plants.

# COSMIDS

Cosmids are plasmid vectors that contain a bacteriophage lamda *cos* site, which directs insertion of DNA into phage particles (Fig. 16.4). The development of cosmid vectors was based on the observation that a ~200-bp DNA sequence in the λ genome called *cos* is required for DNA packaging into the phage particle during lytic infection. Cleavage at *cos* by the λ terminase protein during phage packaging produces a 12-nucleotide cohesive end at the termini of the linear λ genome. Recircularization of the λ genome after bacterial infection is facilitated by base pairing between the complementary cohesive ends. Cosmids were developed to overcome the technical problem of introducing large pieces of DNA into *E. coli*. Thus, bacteriophage *cos* DNA sequence was introduced that is required for packaging DNA into preformed phage heads. For successful packaging of DNA, there must be two *cos* sequences separated by a distance of 38–52 kb. Cosmid cloning vectors with DNA inserts of 30–45 kb can be packaged in vitro into λ phage particles, provided that the ligated double-stranded DNA contains λ *cos* sequences on either side of the insert DNA. Because the λ phage head can hold up to 45 kb of DNA, the optimal ligation reaction in cosmid cloning produces recombinant molecules with *cos* sequences flanking DNA segments of ~40 kb. Following adsorption of these phage particles on to suitable *E. coli* host cells, the cosmid vector circularizes via the cohesive ends and replicates as a plasmid. Cosmid vectors possess an origin of replication, a selectable genetic marker (antibiotic resistance), and suitable cloning sites. For this reason, cosmids are ideal vectors for genome mapping. Like plasmids, cosmids can multiply in large copy number using bacterial *ori* of replication (*ColE1*) inside the bacterial cell and do not carry the genes for lytic development. The advantages of cosmids are that (a) relatively large size of insert DNA (up to 45 kb) can be cloned; and (b) DNA can be introduced into the host using bacteriophages derived by in vitro packaging. The disadvantages are that (a) it is difficult to store bacterial host as glycerol stock; and (b) in vitro packaging is needed to maintain cosmids inside the viral heads.

# BACTERIOPHAGE VECTORS

Bacteriophages are viruses that infect bacteria. These are usually called phages. The phages are constructed from two basic components:

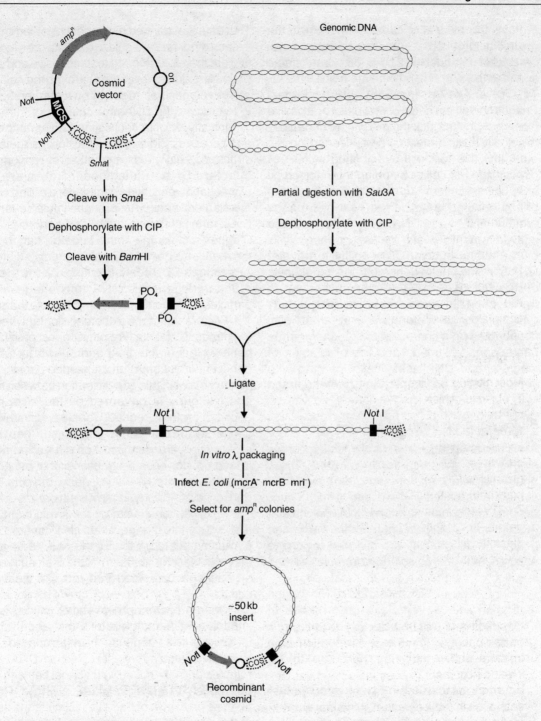

**Fig. 16.4:** Schematic diagram of cosmid vector and cloning of genomic DNA. Concept of the figure taken from Miesfeld: *Applied Molecular Genetics*. Copyright © 1999 John Wiley and Sons. This material is used by permission of John Wiley and Sons, Inc.

proteins form a coat or capsid within which the nucleic acid genome is enclosed. Phages can be double stranded (T2, T4, T6, λ) or single stranded (φX174, M13). There are both DNA (T2, T4, T6, λ, φX174) and RNA (MS2) phages. Phages contain an origin of replication, but unlike plasmids, phages lack the machinery necessary to actually make proteins. Phages generally gain entry into the host cell by injecting their DNA directly into the cell or by being internalized by host cell processes. In most cases, once the phage genome is inside the host cell, phage proteins and phage DNA are synthesized and infectious particles are released through lysis or membrane budding. The ability to transfer DNA from the phage genome to bacterial hosts during the bacterial infection is known as transfection. In gene cloning, two specially designed phage vectors are used, which are constructed from components of the temperate bacteriophage lambda and the production of single-strand DNA using cells infected with vectors based on filamentous bacteriophage M13.

### Biology of bacteriophage λ

λ is a bacteriophage capable of infecting certain strains of E. coli, such as E. coli K-12. The λ phage particle contains two major structural components called a head and a tail. The λ genome is a double-stranded DNA molecule of 48,503 bp that is packaged within the phage head. The tail proteins are required for phage attachment to the E. coli outer membrane and injection of phage DNA into the host cell.

λ infection of E. coli can result in phage replication and cell lysis leading to release of ~100 infective phage particles (lytic phase) or λ DNA can integrate into the E. coli genome as a prophage and remain dormant, creating a lysogenic host cell.

Having established the basic sequence of events, we can now look at the molecular level. λ DNA exists as a double-stranded linear DNA molecule about 48.5 kb long. At each 5′ end there is a single-stranded region 12 bases in length.

These single-stranded regions are complementary to each other and are able to pair with one another and are therefore referred to as cohesive end sites (cos). After injection of the DNA into the cell, the cohesive ends pair together by hydrogen bonds to form a circular DNA molecule.

Initially, λ DNA replication begins at the ori and proceeds bidirectionally to form a theta (θ) structure. This yields a number of monomeric circular DNA molecules. However, as bidirectional replication continues, a rolling circle mechanism begins to produce multiple-length linear forms of double-stranded λ genomic DNA. Staggered cleavage of this large concatamer at the regularly spaced cos sites is mediated by the λ terminase protein. Packing of the newly replicated genomic DNA into phage head particles is coincident with cos site cleavage.

Factors that control the decision of whether the phage will take a lytic or lysogenic path have been well characterized and shown to reflect transcriptional control of λ encoded genes. The primary event in this genetic switch between lysis and lysogeny is governed by the level of λ encoded proteins required for the lytic pathway. If one set of the λ immediate early genes is preferentially transcribed, then the lytic pathway is favored; however, if a transcriptional repressor protein encoded by λ cI gene inhibits this transcription, then lysogeny is favored.

Physical constraints on the amount of DNA that will fit into phage head also control DNA packaging. Although the λ genome is 48 kb, DNA that is at least 38 kb but no more than 52 kb can be packaged provided that cos site cleavage occurs.

Often a bacteriophage vector is introduced into the host E. coli cells by transfection, which is exactly the same as transformation, but involves phage rather than plasmid DNA. The transfectants are mixed with normal bacteria and poured on to an agar plate, giving a set of plaques on a lawn of bacteria, each plaque being a clone consisting of bacteriophages produced by a single infected cell.

The portions of phage genome that determine lysogenic growth are removed from the DNA, and the phage is made into a DNA which always enters lytic lifecycle and ensures that transfected cells immediately give rise to plaques. With most phage vectors, nonessential region of the genome can be disrupted and deleted without harming the genes for the important replicative and structural proteins of the phage, and the DNA fragments to be cloned are ligated into it.

The utility of λ vectors for gene cloning was significantly enhanced following the development of two major improvements in the general strategy.

First, to overcome low efficiency of $CaCl_2^-$ mediated DNA transformation (0.1% of the cells), as compared to phage infection (10% of the cells), protein extracts were developed that permitted ligated DNA to be packaged in vitro into infectious phage particles. Following the ligation of foreign DNA to the purified λ vector arms, the ligation reaction is then carefully mixed with an enriched protein fraction containing phage-packaging proteins. Through an energy-independent association process involving highly ordered protein-protein interactions, the λ DNA is cleaved at appropriately spaced cos sites by λ terminase and packaged into infectious particles. Commercially available packaging extracts are commonly used to achieve phage titers of ~$10^7$ plaque forming units/microgram vector DNA.

A second improvement in λ cloning systems led to an increase in the ratio of recombinant to nonrecombinant phage in the pool of infectious particles. In the first generation of EcoRI-based λ vectors because of uncleaved λ vector or religation of vector without an insert, resulted in high numbers of nonrecombinant phages. Therefore, several genetic strategies were devised to select genetically against nonrecombinant λ genomes. Two of these selection schemes are illustrated in two different λ-phage-based cloning vectors. One involves small DNA insertions (usually cDNA) into the λ

cI gene, which controls the lysogenic pathway (λgt10, λgt11), and the other relies on replacement of vector "stuffer" DNA encoding two λ genes that promote lysogeny (red, gam), with large fragments of foreign DNA (EMBL3, EMBL4).

Bacteriophage cloning vectors based on λ are λgt10, λgt11, EMBL3 (European Molecular Biology Lab), EMBL4, Charon, etc.

### λ phage cloning vectors

The bacteriophage λ contains double-stranded DNA of length 48.5 kb, which is a linear molecule with short complementary single-stranded projections of 12 nucleotides at its 5′ ends. These cohesive termini are referred to as cos sites, which allow the DNA to be circularized after infection of the host cell. Derivatives of bacteriophage have been developed as cloning vectors (Fig. 16.5). The main advantage of a λ-based vector is that large fragments of up to 25 kb can be inserted as compared to plasmid vectors, which can take up 10 kb fragments. When a bacteriophage DNA is digested with restriction enzyme, it generates three fragments: a 19.6-kb left arm carrying the genes for the lambda head and tail proteins, a 12- to 14-kb central fragment carrying the red and gamma genes under pL control; and a 9- to 11-kb right arm carrying the lambda replication and lysis genes. Genomic DNA is partially digested with restriction endonucleases to produce a population of DNA fragments, having sizes from 15 to 20 kb. The size-selected fragments are ligated to vector arms. Viable phages containing recombinant DNA are obtained by in vitro packaging. In the replacement vector, "stuffer" DNA of λ vector encoding two λ genes red and gam that promotes the lysogeny, is replaced with large fragments of foreign DNA. Lack of red/gam genes permits lytic growth. A permanent collection of recombinant phages is then established after amplifying the phages through several generations of growth on a strain. A number of strategies are used to identify recombinant plaques. Some vectors carry the

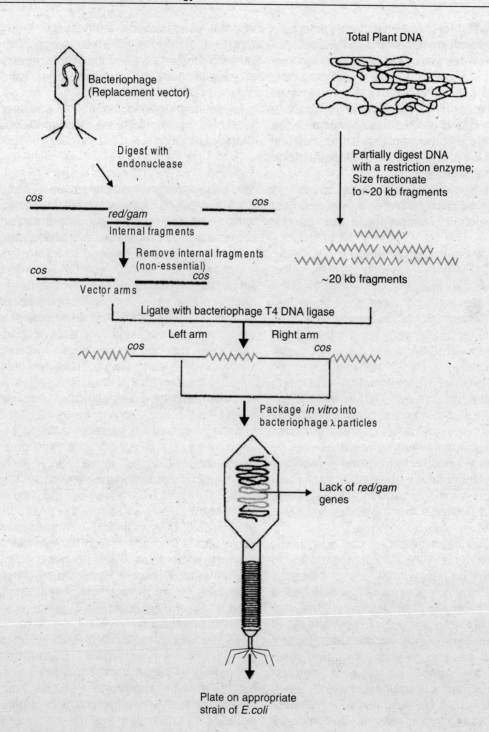

Total Plant DNA

Bacteriophage
(Replacement vector)

Digest with
endonuclease

Partially digest DNA
with a restriction enzyme;
Size fractionate
to ~20 kb fragments

*cos*                                    *cos*

*red/gam*
Internal fragments

Remove internal fragments
(non-essential)

~20 kb fragments

*cos*                              *cos*
Vector arms

Ligate with bacteriophage T4 DNA ligase

Left arm          Right arm

*cos*                    *cos*

Package *in vitro* into
bacteriophage λ particles

Lack of *red/gam*
genes

Plate on appropriate
strain of *E.coli*

•**Fig. 16.5:** Genomic DNA cloning in bacteriophage λ.

*lacZ* gene and are therefore plated on agar, IPTG, X-gal plates (as explained earlier in pUC plasmid), whereas others rely on differences in plaque morphology.

Advantages of this vector are as follows: (i) The size of cloned DNA is relatively large (9–25 kb); (ii) The multiplication of cloned DNA is very easy by infecting a suitable host with bacteriophages containing recombinant DNA; (iii) There is direct selection of clones containing recombinant DNA during *in vitro* packaging; and (iv) Storage of clones is easy.

**λgt10 and λgt11.** These are insertion vectors. These are modified lambda phages designed to clone cDNA fragments. λgt10 is a 43-kb double-stranded DNA for cloning fragments that are only 7 kb in length. The insertion of DNA inactivates the *cl* (repressor) gene, generating a *cl⁻* bacteriophage (Fig. 16.6). Nonrecombinant λgt10 is *cl⁺* and forms cloudy plaques on appropriate *E. coli* host, whereas recombinant *cl⁻* λgt10 forms clear plaques. Further, in an *E. coli* strain carrying an *hfl*A150 mutation, only *cl⁻* phage will form plaques because *cl⁺* will form lysogens and will not undergo lysis to form any plaques. Thus, recombinant λgt10 plaques can be easily selected. λgt11 is a 43.7-kb double-stranded phage for cloning DNA segments that are less than 6 kb in length. Foreign DNA can be

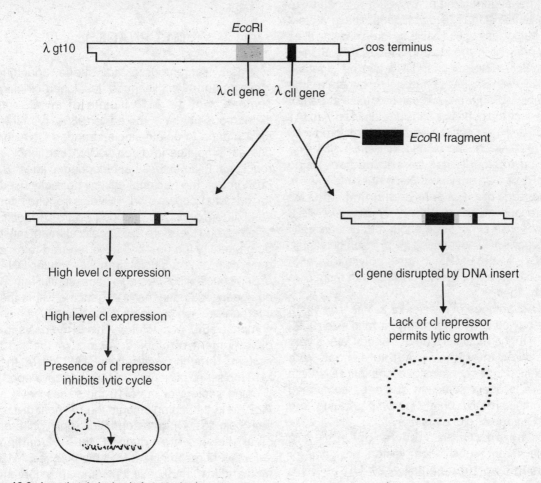

**Fig. 16.6:** Insertional cloning in λgt 10 phage vector.

expressed as a β-galactosidase fusion protein. Recombinant λgt11 can be screened using either nucleic acid or antibody probes. The recombinant λgt11 becomes gal⁻ whereas nonrecombinant λgt11 remains gal⁺, so that an appropriate E. coli host with recombinant phage will form white or clear colonies. Nonrecombinant phage (gal⁺) forms blue colonies, thus permitting screening in the presence of IPTG and X- GAL.

**EMBL 3 and EMBL 4**. The EMBL (European Molecular Biology Lab) vectors are replacement vectors used for cloning large (up to 20 kb) fragments of genomic DNA to develop genomic libraries. λ DNA containing the red and gam genes in λ EMBL3 is replaced with DNA fragments generated by BamHI. Fragments are substituted at the site of trpE gene. These vectors are particularly useful because three enzyme sites are present in the cloning site (EcoRI, BamHI, and SalI). The genomic DNA can be digested partially with Sau3A1 (tetracutter) to generate large number of fragments and can be cloned at BamHI site since the cohesive ends produced by Sau3A1 and BamHI are same. These fragments produced by Sau3A1 can be cloned at BamHI site. The cloned fragments can be excised back using EcoRI or Sal1 enzymes. Recombinants can be selected on E. coli containing the P2 bacteriophage lysogen. Nonrecombinant λEMBL3 vector DNA is not large enough to be efficiently packaged. It is never used as an insertion vector. EMBL 3 and EMBL 4 have polylinkers with reverse orders of restriction sites with respect to each other.

Advantages of this vector are as follows: (i) The cloned DNA is relatively large in size (9–25 kb); (ii) The multiplication of cloned DNA is very easy by infecting a suitable host with bacteriophages containing recombinant DNA; (iii) There is direct selection of clones containing recombinant DNA during in vitro packaging; and (iv) Storage of clones is easy.

**Charon**: Charon bacteriophages are replacement (substitution) vectors designed for cloning large fragments (9–20 kb) of DNA. Charon 40 is unique in that its stuffer consists of small DNA fragments polymerized head-to-tail, which can be digested by a single restriction enzyme Nae. The charon bacteriophage can be used to screen recombinant genomic DNA libraries by in vivo recombination. Charon 4A vector is used chiefly as a substitution vector to construct genomic libraries of eukaryotic DNA. Typically, fragments of genomic DNA of size 7–20 kb obtained by partial digestion with EcoRI can be cloned. The EcoRI sites used for cloning are located in the lacZ gene and the bio gene. This can be used as insertion vector if digested with XbaI, which will create single cut and up to 6 kb fragments can be cloned. There are other charon series vectors, such as charon 34 and 35.

# M13 PHAGES

M13 is a filamentous E. coli phage. Joachim Messing and his colleagues developed cloning vectors from the M13 isolate of an E. coli filamentous phage in the early 1980s, primarily as a means to obtain single-stranded DNA for use as templates in dideoxy DNA sequencing reactions. Filamentous bacteriophages infect E. coli through the bacterial sex pili normally used for bacterial conjugation. Proteins required for sex pili formation are encoded on the E. coli F plasmid, and therefore this genetic element must be present in the host cell (also referred to as a male cell). It is a single-stranded circular DNA of 6407 nucleotides. A single nonessential region has been located in the M13 genome within the intergenic region between genes II and IV, which comprises 507 nucleotides. This is the basis for development of cloning vectors in the M13 system. Through modifications of M13 DNA, the M13 mp series of cloning vectors was developed. Cloning vectors of the M13 mp series have a lacZ gene that complements the gal host giving blue colonies. On transformation, only white or clear plaques are obtained, thus permitting selection of recombinant M13 mp plaques. M13 mp8 and M13 mp9 have the same arrangement of genes, but differ in the direction of polycloning sites. This permits sequencing from both ends of the double-stranded DNA molecule using Sanger's method of dideoxy chain termination reaction. The advantages associated with this vector are that about 100–200 copies of RF (replicative form) accumulate in the cell, the DNA is obtained from the virus as single strands, and this avoids the technical difficulties involved in the separation of single-stranded DNA from double-stranded DNA using alkaline CsCl gradient. M13-based cloning vectors have disadvantages: (i) It is difficult to obtain sufficient amounts of the double-stranded DNA (RF type) required for recombinant DNA reactions; (ii) the rolling circle mechanism of M13 replication can be affected by insert sequence and size, causing under representation of certain recombinant molecules following infection and (iii) since the intergenic region is very small, the cloning range is very less.

# PHAGEMIDS

These are plasmid vectors having both bacterial as well as phage *ori* of replication, for example, pBluescript II KS. M13-based cloning vectors have disadvantages as mentioned above. To overcome these problems, hybrid M13 phage/plasmid cloning vectors called phagemids were constructed in the early 1990s. The original phagemid vectors were derived from the high copy pUC18/pUC19 plasmids first developed by Messing.

pBluescript II KS is a pUC-based plasmid and is 2958 bp long with *lacZ* gene precedes from polycloning site (Fig. 16.7). The main features of this vector are as follows:

1. Multiple cloning site bordered by T3 and T7 promoters to read in opposite directions on the two strands to direct the in vitro synthesis of RNA.

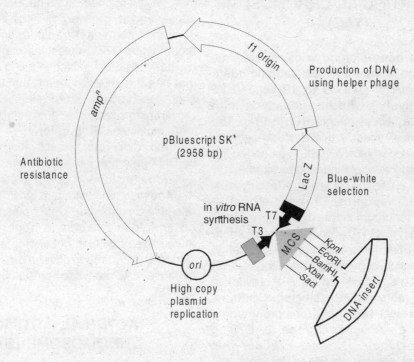

**Fig. 16.7** pBuescript SK+ phagemid cloning vector. Concept of the figure taken from Miesfeld: *Applied Molecular Genetics.* Copyright © 1999 John Wiley and Sons. This material is used by permission of John Wiley and Sons, Inc.

2. Selection of the recombinant DNA molecules through blue-white colonies on X-gal plates through a *lac*I gene in upstream portion of *lacZ* gene.
3. The fl (+) and fl (−) origins of replication from filamentous phage yields sense (+) and antisense (−) strands. The f1 origin sequence allows the production of single-stranded circular DNA following infection of the host bacterial cell with M13 helper phage.
4. Modified *ColEI ori* of replication, which results in high copy plasmid replication.
5. The ampicillin resistant gene (*amp*^R) helps in selection of cells containing phagemids. These vectors can thus be used to generate single-stranded DNA *in vivo* or RNA *in vitro* that is complementary to either of the two strands of foreign DNA inserted in to the polycloning site.

## YEAST ARTIFICIAL CHROMOSOME (YAC)

Very simple gene cloning vectors, such as plasmids and λ vectors, can accommodate an insert DNA of size up to 10 kb and 25 kb respectively, whereas cosmids can accommodate DNA segments of up to 40 kb. The development of yeast artificial chromosomes (YAC) by David Burke *et al.* (1987) extended the cloning range to several thousand base pairs (i.e. about 1000–2000 kb) by using a totally new approach. YAC vectors use DNA sequences that are required for chromosome maintenance in *S. cerevisiae*. A typical YAC consists of centromere elements (*CEN*4) for chromosome segregation during cell division, two telomere sequences (*TEL*) for chromosome stability, an origin of replication and one or more growth selectable markers (*URA*3) to select positively for chromosome maintenance. In essence, the YAC is a minichromosome. Normal eukaryotic chromosomes are always linear. In yeast, the sequence elements required for circular plasmid vectors to be replicated and stably maintained in an extrachromosomal state have been identified as ARS. CEN (centromere) sequences ensure segregation and telomeres seal the chromosome ends. Linear plasmids must carry telomeres, which are a tandem array of simple satellite-like repetitive sequences. For cloning purposes, circular YAC is digested with restriction enzymes. From the resulting fragments, the fragments that contain *his* gene are deleted and the other fragments are utilized for cloning insert DNA in the YAC. The most commonly used YAC for genome mapping of higher eukaryotes is pYAC4 (Fig. 16.8). The left and right arms have all the essential yeast chromosomal elements to enable replication in a yeast host, as well as markers *TRP* and *URA* on the left and right arm respectively, which permits selection for transformants. The two arms can be ligated to *Eco*RI ends of a large fragment of genomic DNA insert to create a recombinant chromosome. This molecule can then be maintained as a linear chromosome in yeast, and is referred to as a yeast artificial chromosome. Yeast transformants containing recombinant YAC molecules can be identified by red/white color selection using a yeast strain that has the *ADE*2–1 ochre mutation, which is suppressed by the *SUP*4 gene product.

Transformed yeast containing recombinant YAC molecules–**Red colony**

(red color is due to inactivation of *SUP*4 gene by DNA insertion into *Eco*R1 site)

Non-transformed yeast–**white colony**

The plasmid vector can be grown in bacteria because it contains a *ColE1* plasmid origin and an antibiotic resistance gene (*amp*^R). The advantage of this kind of vector is that only few clones are required to cover a particular genome because YAC can accommodate large inserts of DNA.

## BACTERIAL ARTIFICIAL CHROMOSOME (BAC)

BAC is another cloning vector system in *E. coli* developed by Mel Simon and his colleagues as an alternative to YAC vectors. BAC vectors are

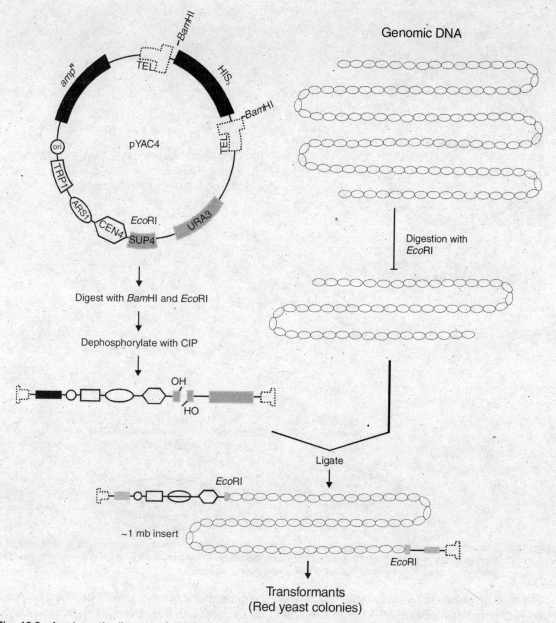

**Fig. 16.8:** A schematic diagram of pYAC4 vector and cloning of genomic DNA. Concept of the figure taken from Miesfeld: *Applied Molecular Genetics.* Copyright © 1999 John Wiley and Sons. This material is used by permission of John Wiley and Sons, Inc.

maintained in *E. coli* as large single copy plasmids that contain inserts of 50–300kb. BAC vectors contain F-plasmid origin of replication (*ori*S), F plasmid genes that control plasmid replication (*rep*E) and plasmid copy number (*par*A, B, C), and the bacterial chloramphenicol acetyltransferase (*CM*^R) gene for plasmid selection. The methods used to construct a BAC library are essentially the same as those for a standard plasmid library, except that the insert

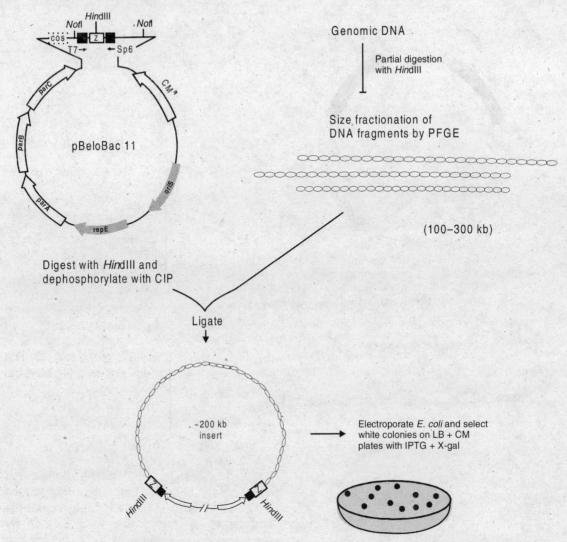

**Fig. 16.9:** Diagrammatic representation of pBeloBAC11 vector and cloning of high molecular weight DNA. Concept of the figure taken from Miesfeld: *Applied Molecular Genetics.* Copyright © 1999 John Wiley and Sons. This material is used by permission of John Wiley and Sons, Inc.

DNA must be prepared by preparative pulse field gel electrophoresis. pBeloBAC 11 is an example of a BAC vector developed in the Simon lab (Fig. 16.9). High molecular weight DNA is partially digested with the enzyme *Hin*dIII, size fractionated by PFGE, and ligated to the *Hin*dIII digested and phophatase-treated pBeloBAC 11 vector. *E. coli* cells are electroporated with the ligated material and white colonies (DNA insertion into *lacZ* α subunit gene) are isolated on the basis of chloramphenicol resistance. The insert DNA can be excised with the rare cutting enzyme *Not*I (8 bp recognition site). The plasmid can also be linearized at the *cos*N site by lambda terminase to facilitate end labeling for restriction enzyme mapping. The bacteriophage T7 and SP6 transcriptional promoter sequences flanking the *Hin*dIII cloning site provide a means to generate end specific RNA probes for library screening. DNA segments from the bithorax gene of

*Drosophila* have already been cloned using this vector.

Advantages of this vector are as follows: (1) The clones are stable over many generations and (2) BAC can be transformed into *E. coli* very efficiently through electroporation, thus avoiding the packaging extract. The disadvantages are that it lacks positive selection for clones containing inserts and very low yield of DNA.

## P1 PHAGE VECTORS

The P1 cloning system was developed as an alternative to YAC and cosmid vectors for cloning of high molecular weight DNA fragments. The vector, which is 30.3 kb, can accommodate an insert DNA of size up to 95 kb. With 1–2 μg of vector arms and 2–4 μg of size-fractionated DNA fragments, which have *Sau*3A1 sticky ends, greater than $10^5$ clones can be generated using the P1 system. In spite of *cos* site in vector, the packaging of P1 phage DNA into the viral head is determined by a 162-bp DNA sequence known as *pac* site. P1 phage head can accommodate DNA of length up to 110 kb.

pAD10SacBII, and pAD10 are the most common P1 phage vectors. The former pAD10 Sac BII vector and its elements are shown in Fig. 16.10. It contains a *ColE1* replicon, P1 *pac* site, an 11-kb *Bam*HI-*Sca*I stuffer fragment from adenovirus DNA, kanamycin resistant (*kan*<sup>R</sup>) gene from a transposon *Tn*903, and an ampicillin resistant gene (*amp*<sup>R</sup>). The P1 vector also contains P1 *lox*P recombination site. The cells containing this vector DNA can be induced by IPTG, but the cells containing this vector will die if there is sucrose in the media, which helps in the selection of transformed cells. There are T7 and Sp6 promoters and rare cutting sites for *Sal*I, *Sfi*I, and *Not*I endonucleases, which permits recovery of the insert. The host bacterial cell should have P1 recombinase gene for multiplying the cloned DNA as a plasmid. It offers certain advantages: (1) Multiple copies of genome are produced from the representative library; (2) Particular DNA cloned can be

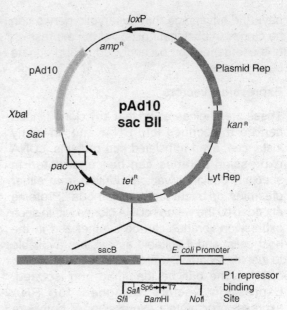

**Fig. 16.10:** pAD10sacBII vector and its elements.

produced in large quantities very easily; (3) The cloning process is very easy; and (4) Physical dissection of cloned DNA is easy.

## P1-DERIVED ARTIFICIAL CHROMOSOME (PAC)

This cloning system was developed by Ioannou *et al.* (1994). The constructed vector incorporates features of both P1 and F factor systems and can be transformed into *E. coli* by electroporation. In a PAC vector, inserts of size 100–300 kb can be cloned. It is devoid of problems such as chimerism and instability of cloned DNA.

### Shuttle vectors

These are plasmids capable of propagating and transferring (shuttling) genes between two different organisms, one of which is typically a prokaryote (*E. coli*) and the other a eukaryote (e.g. yeast). Shuttle vectors must have unique origins of replication for each cell type as well as different markers for selection of transformed host cells harboring the vector. Shuttle vectors have the advantage that eukaryotic genes can be cloned in bacterial hosts, yet the expression of these genes can be analyzed in appropriate backgrounds.

### Expression vectors

These are engineered so that any cloned insert can be transcribed into RNA, and, in many instances, even translated into proteins. cDNA expression libraries can be constructed in specially designed vectors derived from either plasmids or bacteriophage lambda. Proteins encoded by the various cDNA clones within such expression libraries can be synthesized in the host cells, and if suitable assays are available to identify a particular protein, its corresponding cDNA clone can be identified and isolated. Expression vectors designed for RNA expression or protein expression, or both, are available.

# PROTOCOLS

## 16.1: Isolation of plasmid DNA: mini preparation

1. Preparation of bacterial culture: Plasmids can be isolated from their bacterial cultures. The Luria-Bertinin (LB) agar medium with antibiotics can be used.
2. The medium has the following constituents:
   Tryptone          10 g/l
   NaCl              10 g/l
   Yeast extract     5 g/l
   Adjust pH to 7.0 with 1N NaOH; Add agar 20 g/l for solid media.
3. Dissolve all the constituents in water and adjust pH. For solid medium add 20 g/l agar. The medium is aliquoted 50–100 ml in each conical flask. The medium is autoclaved at 15 p.s.i. pressure, 121°C for 20 min. The flasks are cooled to 50°C and appropriate antibiotics are added (50 µg/100 ml) depending upon the plasmid selectable marker genes.

4. Streak glycerol stock of bacterial culture on solid media plates containing appropriate antibiotics and incubate at 37°C.
5. After 16 h of incubation, pick a single colony from the petri dish and inoculate in 3 ml volume of LB broth containing appropriate antibiotics. Incubate the cultures for 16 h at 37°C on a shaker at a speed of 120–150 rpm.

## Mini preparation of plasmid DNA

1. Distribute the bacterial culture in two sterile eppendorf tubes (1.5 ml). Spin for 30 sec and pour off the supernatant, completely leaving the bacterial pellet as dry as possible.
2. Suspend the pellet in 100 µl of Solution 1 by vortexing.
   Solution 1
   Glucose          (50 mM)
   EDTA             (10 mM)
   Tris-Cl          (25 mM)
   pH 8.0, stored at 4°C
3. Leave on ice for 5–10 min.
4. Add 200 µl freshly prepared Solution 2 and gently mix by inverting
   Solution 2: Alkali lysis buffer
   10N NaOH        20 µl
   10% SDS         100 µl
   Distilled water   880 µl
5. Leave on ice for 10 min.
6. Add 150 µl of Solution 3 (ice cold) to the tube and gently mix by inverting. Incubate on ice for 15 min.
   Solution 3: Potassium acetate buffer (pH 5.5) for 100 ml
   Potassium acetate (5M)   60 ml
   Glacial acetic acid       11.5 ml
   Water                    28.5 ml
7. After incubation spin at 15,000 rpm for 10 min at 4°C. Collect the supernatant carefully in a fresh tube.
8. Add equal volume of phenol:chloroform in a 1:1 ratio and spin to remove the aqueous phase.
9. Precipitate the DNA with two volumes of ethanol at room temperature. Invert the tubes

10–15 times and incubate at –20°C for 20 min.

10. Centrifuge at 15,000 rpm for 5 min at 4°C. Collect the pellet and dry.

11. Rinse the pellet with 70% ethanol, decant ethanol and dry the pellet.

12. Dissolve the pellet in appropriate volume (50 μl) of Tris-EDTA (TE) buffer.
    Tris EDTA buffer pH 8.0
    Tris (10 mM) pH 8.0
    EDTA (1 mM) pH 8.0

13. Estimate the concentration of DNA: DNA has a maximum absorbance at about 260 nm. Based on extinction coefficient an optical density (OD) of 1.0 at 260 nm corresponds approximately to 50 μg/ml of double stranded DNA. The ratio of $OD_{260}/OD_{280}$ provides the estimate of purity. A typically pure preparation of DNA has a ratio of approximately 1.8. If there is a contamination with protein or phenol, the ratio will be less than this value. The steps for UV quantification of DNA are as follows:

 i. To measure the concentration of DNA, dilute it appropriately with TE buffer (either 1:50 or 1:100).

 ii. Standardize the spectrophotometer using TE buffer as blank.

 iii. Measure the absorbance of the sample at 260–280 nm.

 iv. Calculate the DNA concentration of the sample as follows:
    Concentration of DNA (μg/ml) = $OD_{260}$ × 50 × dilution factor
    Concentration of DNA (μg/μl) = $OD_{260}$ × 50 × dilution factor/1000

## 16.2: Isolation of genomic DNA by SDS–Proteinase K method

1. Any plant tissue can be used for DNA extraction. While collecting material it is advisable to avoid brown plant tissue material.

2. Grind tissue with liquid nitrogen in a precooled mortar and pestle.

3. Take 3.8 ml extraction buffer (600 μl of proteinase K (1mg/ml) and 600 μl of 5% SDS) in a centrifuge tube. Add the ground tissue in the above buffer solution and homogenize it.
   Extraction buffer (pH 8.0)
   Tris base                0.05 M 6 g/l
   EDTA (pH 8.0)            0.03 M 87.7 g/l
   Total volume 1000 ml

4. Transfer the homogenized mixture to a tube and incubate at 37°C for 1 h.

5. Add 1 g sodium perchlorate to the mixture and slightly roll the tubes for homogenization. Centrifuge at 8,000 rpm for 5 min.

6. Transfer the supernatant into another tube and add 9 ml ethanol perchlorate. A precipitate of DNA material will be observed. Ethanol perchlorate solution: Dissolve 40 g $NaClO_4 \cdot H_2O$ in 320 ml absolute ethanol, then add 120 g $NaClO_4 \cdot H_2O$ and 80 ml distilled water.

7. Scoop out DNA into a tube.

8. Add 4 ml of phenol saturated with TE buffer (pH 8.0) and 4 ml of chloroform.

9. Spin at 5000 rpm for 3 min and transfer the aqueous phase to a fresh tube.

10. Repeat the phenol chloroform procedure as in steps 8 and 9.

11. Precipitate the DNA with two volumes of ethanol at room temperature. Invert the tubes 10–15 times and incubate at –20°C for 20 min.

12. Centrifuge at 15,000 rpm for 5 min at 4°C. Collect the pellet and dry.

13. Rinse the pellet with 70% ethanol, decant ethanol and dry the pellet.

14. Dissolve the pellet in appropriate volume (250 μl) of Tris-EDTA (TE) buffer.
    Tris EDTA buffer pH 8.0
    Tris (10 mM) pH 8.0
    EDTA (1 mM) pH 8.0

15. Estimate the concentration of DNA: Follow the Step 13 of Protocol 16.1.

## REFERENCES

Bolivar, F., Rodriguez, R.L., Betlach, M.C., and Boyer, H.W. 1977. Construction and characterization of new cloning vehicles I. Ampicillin resistant derivatives of the plasmid pMBq. *Gene* **2**: 75–93.

Bolivar, F., Rodriguez, R.L., Greene, P.J., Betlach, M.C., Heynecker, H.L., Boyer, H.W, Crosa, J.H. and Falkow, S. 1977. Construction and characterization of new cloning vehicles II. A multipurpose cloning system. *Gene* **2**: 95–113.

Burkee, D.T., Carle, G.F. and Olson, M.V. 1987. Cloning of large segments of exogenous DNA in to yeast by means of artificial chromosome vectors. *Science* **236**: 806–812.

Hinnen, A., Hicks, J.B. and Fink, G.R. 1978. Transformation of yeast. *Proc. Natl. Acad. Sci. USA* **75**: 1929–1933.

Ioannou, P.A., Amemiya, C.T., Garnes, J., Kroisel, P.M., Shizuya, H., Chen, C., Batzer, M.A. and de Jong, P.J. 1994. A new bacteriophage P1 derived vector for propagation of large human DNA fragments. *Nature Genet.* **6**: 84–89.

McClintock, B. 1951. Chromosome organization and genic expression. *Cold Spring Harbor Symp. Quant. Biol.* **16**: 13–47.

Nilsson, B., Uhlen, M., Josephson, S., Gatenbeck, S. and Philipson, L. 1983. An improved positive selection plasmid vector constructed by oligonucleotide mediated mutagenesis. *Nucl. Acids Res.* **11**: 8019–8039.

Norrander, J., Kempe, T. and Messings, J. 1983. Construction of improved M13 vectors using oligo-deoxynucleotide directed mutagenesis. *Gene* **26**: 101–106.

Ratzkin, B. and Carbon, J. 1977. Functional expression of cloned yeast DNA in *Escherichia coli. Proc. Natl. Acad. Sci. USA* **74**: 487–491.

Struhl, K., Cameron, J.R. and Davis, R.W. 1976. Functional genetic expression of eukaryotic DNA in *Escherichia coli. Proc. Natl. Acad. Sci. USA* **73**: 1471–1475.

## FURTHER READING

Chawla, H.S. 1998. *Biotechnology in Crop Improvement*. International Book Distributing Co., Lucknow.

Gupta, P.K. 1996. *Elements of Biotechnology*. Rastogi Publishers, Meerut.

Kingsman, S.M. and Kingsman, A.J. 1988. *Genetic Engineering: An Introduction to Gene Analysis and Exploitation in Eukaryotes*. Blackwell Scientific Publication, Oxford.

Maniatis, T., Fritsch, E.F. and Sambrock, J. 1982. *Molecular Cloning: A Laboratory Manual*. Cold Spring Harbor Laboratory Press, USA.

Mantell, S.H., Mathew, J.A. and McKee, R.A. 1985. *Principles of Plant Biotechnology: An Introduction to Genetic Engineering in Plants*. Blackwell Scientific Publications, Oxford.

Miesfeld, R.L. 1999. *Applied Molecular Genetics*. Wiley-Liss, New York.

Old, R.W. and Primrose, S.B. 2001. *Principles of Gene Manipulation: An Introduction to Genetic Engineering*. Blackwell Scientific Publications, Oxford.

Sambrook, J., Fritsch, E.F. and Maniatis, T. 1989. *Molecular Cloning: A Laboratory Manual*. Cold Spring Harbor Laboratory Press, USA.

Walker, J.M. and Gaastra, W. (Eds.). 1987. *Techniques in Molecular Biology*. Croom Helm, Australia.

# 17

# Gene Cloning: cDNA and Genomic Cloning and Analysis of Cloned DNA Sequences

In order to clone a DNA sequence that codes for a required gene product, the gene has to be removed from the organism and cloned in a vector molecule. Libraries of such cloned fragments are made. A *gene library* is a random collection of cloned fragments in a suitable vector that ideally includes all the genetic information of that species. This is sometimes also called the shotgun collection because for obvious reasons an appropriate selection has to be made. There are two ways by which gene libraries are made, referred to as complementary (cDNA) and genomic DNA libraries. Cloning of DNA fragments obtained from cDNA and genomic DNA have been discussed in terms of cDNA and genomic DNA cloning.

## COMPLEMENTARY DNA (cDNA) LIBRARIES AND CLONING

For many plants and animals, a complete genome library contains such a vast number of clones that the identification of the desired gene is a mammoth task. cDNA libraries are used when it is known that the gene of interest is expressed in a particular tissue or cell type. Since all the genes are not expressed at the same time, relatively few clones will be needed to make a cDNA library that contains copies of all the genes that are active in the tissue or cells from which the mRNA is obtained. It will be a relatively easy task to screen the cDNA library for a clone carrying a particular expressed gene. mRNA cannot itself be ligated into a cloning vector, but it can be converted into DNA by cDNA synthesis and then ligated into a cloning vector.

The strategy of cDNA cloning involves the copying of mRNA transcripts into DNA, which are then inserted into bacterial plasmids and then placed into bacteria by transformation. At this stage it should be known that mRNA used for cDNA preparation is a processed transcript and not the original one transcribed from DNA. The processing of mRNA has been explained in Chapter 13. Almost all RNA transcripts are processed to form the mature RNA.

The primary transcript formed from the DNA in eukaryotes is known as heterogeneous nuclear (hn) RNA. This hnRNA is often modified in the nucleus, which involves a number of steps before it reaches cytoplasm for translation. A cap structure is added to the 5′ end, which consists of a 7-methyl guanosine residue, and a tail of adenine residues is added to the 3′ end poly (A) tail. The final processing step is the removal (or splicing out) of intervening sequences (introns), which are transcribed along the rest of the gene.

Endonucleases specifically cleave the mRNA molecule, and the coding regions (exons) are joined to form a continuous coding sequence. After completion of these processing steps, the mature mRNA molecule is transported into the cytoplasm where it can participate in protein synthesis.

The principal steps involved in cDNA cloning are summarized in Fig. 17.1.

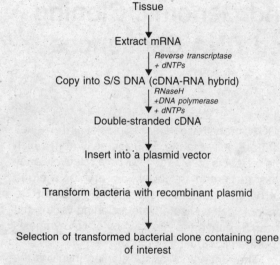

Fig. 17.1: General scheme for cDNA cloning.

## Isolation or extraction of mRNA

When it is known that the gene of interest is expressed in a particular tissue or cell type, a crude extract of the tissue is prepared and then freed from proteins, polysaccharides, and other contaminants. Many eukaryotic mRNAs can be further purified from the total or polysomal fraction using the technique of oligo-dT cellulose chromatography. It is known that mRNAs have poly(A) (adenosine residues) tail at their 3′ end. Under appropriate conditions this tail will bind to a string of thymidine residues immobilized on cellulose and then poly (A)⁺ fraction can be eluted. Two or three passages of the poly (A)⁺ fraction through such a column produces a fraction highly enriched for mRNA.

The mRNA fraction will contain many different mRNA sequences, but certain techniques can be employed to enrich for a particular mRNA species. At the beginning itself, careful selection of a tissue may be made so that a particular mRNA is abundantly produced for its isolation, for example, the use of bean cotyledons harvested during their phase of storage of protein synthesis. Having prepared a mRNA fraction, it is necessary to check that the sequence of interest is present. This is done by translation of mRNA in vitro and identification of appropriate polypeptides in the products obtained.

## Synthesis of first strand of cDNA

The mRNA fraction is copied into the first strand of DNA as single-stranded, using RNA-dependent DNA polymerase, also known as reverse transcriptase (Fig. 17.2). This enzyme, like other DNA polymerases, can only add residues at the 3′-OH group of an existing primer, which is base paired with the template. For cloning of cDNAs, the most frequently used primer is oligo-dT, 12–18 nucleotides in length, which binds to the poly(A) tract at the 3′ end of mRNA molecules. The RNA strand of the resulting RNA–DNA hybrid is destroyed by alkaline hydrolysis prior to second-strand synthesis.

## Synthesis of second strand of cDNA

The second strand of cDNA can be synthesized by two methods.

1. *Self priming*: cDNA: The mRNA hybrid so obtained is denatured so that a second strand can be synthesized on the single strand of cDNA by the Klenow fragment of DNA polymerase I, which is self-priming (Fig. 17.3). Due to unknown reasons, the single-stranded DNA has a transitory hairpin structure at the 3′ end, which can be used to prime the synthesis of the second strand of cDNA by the Klenow fragment of *E. coli* DNA polymerase I. The hairpin loop and any single-stranded overhang at the other end are then digested away with the single-strand-specific S1 nuclease. The final

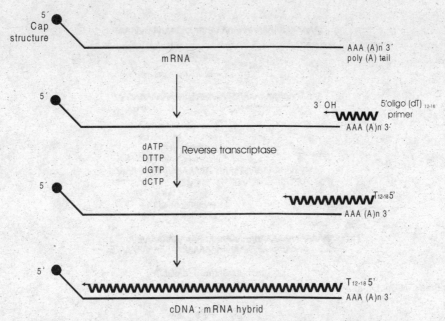

**Fig. 17.2:** Synthesis of the first strand of cDNA using an oligo (dT) primer and reverse transcriptase.

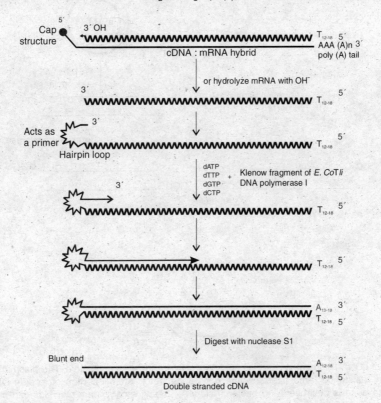

**Fig. 17.3:** Synthesis of double-stranded cDNA by the self-priming method.

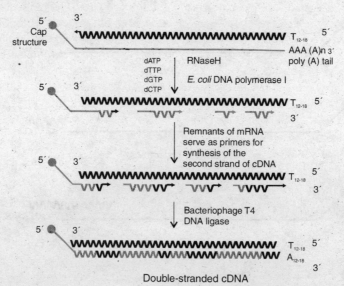

**Fig. 17.4:** Synthesis of double-stranded cDNA by replacement synthesis method.

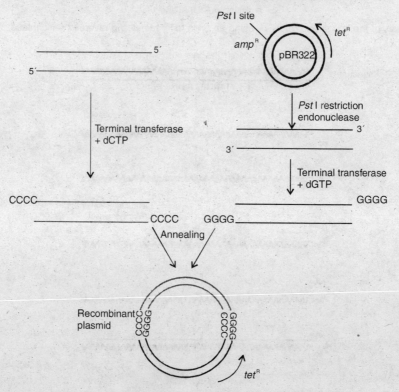

• **Fig. 17.5:** Cloning of cDNA fragment into a plasmid (.....) cDNA and (—) the plasmid. The plasmid, pBR 322 carries ampicillin resistance ($amp^R$) tetracycline resistance ($tet^R$). After insertion of foreign DNA the ampicillin resistance gene is destroyed and bacterial clones carrying recombinant plasmids can be selected on the basis of drug resistance.

product is a population of double-stranded, blunt-ended DNA molecules complementary to the original mRNA fraction.

2. *Replacement synthesis*: The cDNA:mRNA hybrid is used as a template for a nick translation reaction (Fig. 17.4). RNase H produces nicks and gaps in the mRNA strand of the hybrid, creating a series of RNA primers that are used by *E. coli* DNA polymerase I during the synthesis of the second strand of cDNA. The reaction has three advantages: (1) It is very efficient, (2) It can be carried out directly using the products of the first strand reaction, and (3) It eliminates the need to use nuclease S1 to cleave the single-stranded hairpin loop in the double stranded cDNA. Most cDNA libraries are now being constructed using a replacement reaction to synthesize the second strand of cDNA.

## Cloning of cDNA

The most commonly used procedure for cloning cDNAs involves the addition of complementary homopolymeric tracts to double-stranded cDNA and to the plasmid vector (Fig. 17.5). A string of cytosine residues are added to the cDNA using the enzyme terminal transferase, to form oligo-dC tails on the 3' ends. Similarly, a plasmid is cut open at a unique restriction endonuclease site and tailed with oligo-dG. The vector and the double-stranded cDNA are then joined by hydrogen bonding between the complementary homopolymers to form open circular hybrid molecules capable of transforming *E. coli*.

## Introduction to host cells

The recombinant plasmids are used to transform bacteria, usually the *E. coli* K-12 strain. *E. coli* cells treated with calcium chloride will take up plasmid molecules from the surrounding medium and the host cell will repair any gaps in the recombinant plasmid. If the plasmid has been chosen carefully, it is possible to select transformed from nontransformed bacteria on the basis of antibiotic resistance. Many cloning plasmids contain two antibiotic resistance genes, one of which is destroyed during cloning. In the case of pBR322, cloning into unique *Pst*I site destroys ampicillin resistance but leaves tetracycline resistance intact (Fig. 17.5). Bacteria transformed with a recombinant plasmid will be sensitive to ampicillin but resistant to tetracycline. This simple selection tells the investigator which colonies carry a cDNA copy of some sort.

## Clone selection

The antibiotic resistance selection already carried out has identified which clones carry a recombinant plasmid, but there will be thousands of different inserts. The cloning procedure usually begins with a whole population of mRNA sequences. The most difficult part of cDNA is selecting which of these clones carries the sequence of interest. If the gene is expressed, then the simplest selection is to screen for the presence of the protein, either by bacterial phenotype it produces or by the protein detection methods usually based on immunological or enzymological techniques. More often than not the protein is not expressed, and other methods such as nucleic acid hybridization are used. Identification of the gene is discussed after the genomic DNA cloning. The procedure is same for both the cDNA and genomic cloning.

# GENOMIC CLONING

Genomic libraries and clones are necessary in addition to cDNA clones because cDNA clones are generated from processed mRNA, which lack introns and other sequences surrounding the gene. From the biotechnological point of view, if genes are to be modified and returned to plants, it is likely that genomic sequences will be more useful.

The genome of plants is remarkably complex, and a particular fragment of interest comprises only a small fraction of the whole genome. Therefore, construction of a useful and representative recombinant genomic library is dependent on the generation of a large population of clones. This is necessary to ensure that the library contains at least one copy of every sequence of DNA in the genome. A general strategy to construct genomic library from random fragments of DNA in phage is shown in Fig. 17.6.

## Isolation of DNA

For genomic cloning, nuclear DNA must be isolated from the tissue in a high molecular weight form, largely free from organelle DNA and RNA. The first step for construction of genomic library is isolation of genomic DNA. Different methods are available for the isolation of DNA, but proteinase K, SDS method, and the CTAB method are most commonly employed. Preparation of high molecular weight DNA with minimum amount of shearing and nonspecific damage and whose size range is 100–150 kb is adequate for production of genomic DNA libraries. The size of large molecular weight DNA can be checked by pulsed field gel electrophoresis. DNA preparation can be made RNA free by digestion with ribonucleases.

## Partial digestion

The high molecular weight genomic DNA is partially digested with a restriction enzyme, such as Sau3A, to obtain fragments with desired range of 9–25 kb, with about 30% of the genomic DNA remaining undigested.

## Vectors for cloning

Plasmids are usually used for cDNA cloning, but they accommodate a relatively small piece of DNA due to problems with transformation using very large plasmids. It is generally desirable to clone larger pieces of genomic DNA as a large clone is more likely to contain an intact copy of a gene plus some of the flanking sequences. Cloning large pieces also brings the number of different clones needed to cover the entire genome down to a manageable number. Thus, genomic cloning is often carried out by using bacteriophage λ and cosmids as vectors, which can accommodate relatively large DNA molecules. Bacteriophage λ has high cloning and packaging efficiency and is easy to handle and store as compared to plasmid vectors. The phage genome is double stranded and 50 kb in length. It is amenable to genetic manipulation and the structure of the genome is well understood. Since its first use as a cloning vector in 1974, many different vectors have been constructed that enable replacement of part of the λ genome with foreign DNA. Part of the central genome is nonessential, so it can be replaced without any detriment to phage reproduction. This type of vector is termed as a replacement vector. Here, the vector is cleaved at two different sites and a portion of the vector is removed and foreign DNA can be inserted into this. The replacement vectors such as EMBL3 (European molecular biology lab) and the lambda DASH and lambda FIX vectors contain a stuffer fragment that is replaced by the genomic DNA and will hold 9–23 kb of inserted DNA. There are insertion vectors also, which are derivatives of wild type phage with a single target site at which DNA can be inserted. These are also used in which the foreign DNA is directly inserted into the digested arms of the vector which generally accommodate up to 12-kb fragments depending on the vector, for example, λgt10, λgt11 and lambda ZAP (see Fig. 16.6). However, these insertion vectors are not commonly used for genomic cloning because the insert size is small.

Cosmids can accept approximately 45 kb of foreign DNA, almost twice as much as bacteriophage vectors (see Fig. 16.4). However, genomic DNA libraries are considerably more difficult to construct and maintain in cosmids than in bacteriophages. In the past few years, it has become obvious that certain genes are too large to be accommodated as a single fragment in cosmids also. For this, yeast artificial chromosomes (YACs) and bacterial artificial chromosome (BAC) as vectors are used for cloning (see Figs. 16.8 and 16.9).

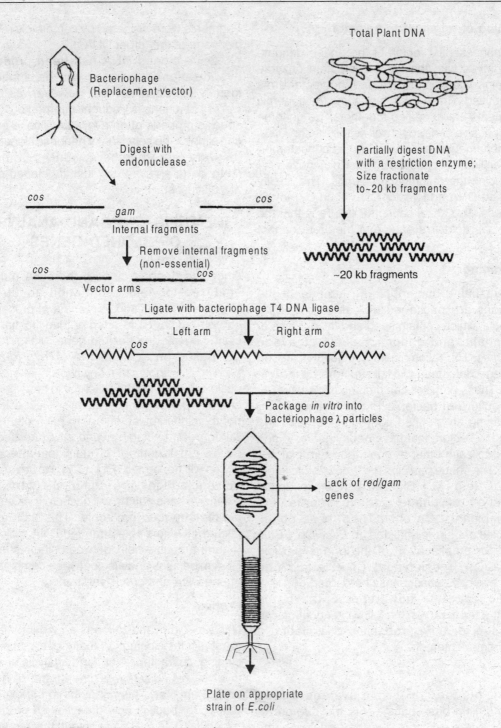

**Fig. 17.6:** Genomic DNA cloning in bacteriophage λ.

## Ligation of fragment to a vector

The nonessential region of the phage genome is removed with the same restriction enzyme used for digestion of DNA and the vector arms are purified. These vector arms and the genomic fragments are annealed together. The genome has cohesive ends so as well as the foreign DNA, the genome will anneal to form a concatamer. Proper ligation conditions must be chosen for producing the recombinant clones. The quality of insert DNA and vector DNA is crucial to the success. The usual cause for failure is that the DNA has defective ends and therefore cannot ligate properly.

## Packaging

The resulting bacteriophage lambda of the ligations are packaged by using packaging extracts, which contain head and tail proteins of the phage and endonucleolytic cleavage enzymes. The recombinant λ molecules can be packaged into head and tail structures in the test tube. These particles can then be added directly to a culture of bacteria. Proteins required for packaging are purified from infected cells and added to vectors that have been linearized and joined to each other through their terminal *cos* sites. The phage particles are then used to infect *E. coli* [(e.g. XL1 Blue MRF (P2)] for initial selection of phage particles containing recombinant DNA only. Then these phage particles are further infected on *E. coli* XL1 Blue MRF for amplification of the phage particle containing recombinant DNA, and clone selection can subsequently be carried out.

To calculate how representative the constructed library is, the following equation can be used for 99% probability of isolating a fragment of interest:

$$I \times N = 4.6 \times G$$

where *I* is the size of the average cloned fragment (in base pairs), *N* is the number of clones screened, and *G* is the target genome size (in base pairs). This assumes that the cloned fragments were prepared by completely random cleavage of the target DNA.

Once cloned DNA has been produced, hybridization techniques can be applied to identify individual clones. However, as cDNA cloning of a gene is frequently carried out first, a specific probe is often available. These probes can simplify the selection of genomic clones from the very large number produced. Different methods to analyze and identify cloned genes are discussed.

# IDENTIFICATION AND ANALYSIS OF CLONED GENES

Once the gene is cloned, the DNA must be analyzed to yield information about gene structure and function. A number of procedures can be employed to attempt identification of a clone carrying the desired gene. Most of these are based on a technique of *hybridization probing*. It is known that any two single-stranded nucleic acids have the potential to base pair with one another. If two strands of nucleic acids are not exactly complementary to each other, they will form an unstable hybrid structure because only a small number of individual interstrand bonds will be formed (Fig. 17.7). However, if the polynucleotides are complementary, then extensive base pairing will occur to form a stable double-stranded molecule. This pairing can occur between single-stranded DNA molecules to form a double helix between single-stranded RNA molecules and also between combinations of one DNA and one RNA strand.

## Probes

Nucleic acid hybridization is a fundamental tool in molecular cloning. Virtually every aspect of cloning, characterization, and analysis of genes involves hybridization of one strand of nucleic acid to another. This brings us to the aspect of probes for nucleic acid detection. A probe is a molecule (DNA or RNA segment) that recognizes the complementary sequence in DNA or RNA

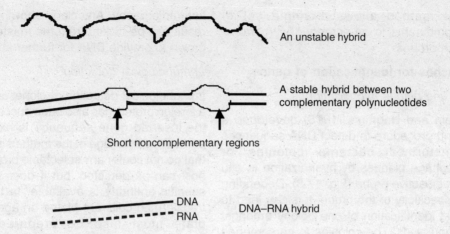

**Fig. 17.7:** Nucleic acid hybridization.

molecules by molecular hybridization and thus allows identification and isolation of these specific DNA sequences from an organism. The success of colony or plaque hybridization as a means of identifying a particular recombinant clone depends on the availability of a DNA molecule that can be used as a probe. In practice, the nature of the probe is determined by the information available about the desired gene.

The following probes that can be used for hybridization:

i. *Homologous probes*: These contain at least part of the exact nucleic acid sequence of the desired gene.

ii. *Partially homologous probes*: A similarity in the nucleotide sequence is seen when two genes for the same protein, but from different organisms are compared. These similarities mean that a single-stranded probe prepared from the gene of one organism will form a hybrid with the gene from the second organism. Although the two molecules will not be entirely complementary, enough base pairs will be formed to produce a stable structure. This strategy is called *heterologous probing*, and has proved very useful in screening genomic libraries for genes that have already been identified in other organisms.

iii. *Synthetic oligonucleotide probes*: The gene that is being sought will be one that codes for a protein. The amino acid sequence of this protein is determined. From this amino acid sequence it is possible to use the genetic code to predict the nucleotide sequence of the relevant gene. This prediction will always be an approximation. Once the sequence of the gene has been predicted, a mixture of oligonucleotides that will hybridize to the gene can be synthesized. Usually a region of five or six codons where the prediction is fairly unambiguous is chosen, and a set of synthetic oligonucleotides including all the possible sequences is used as a probe. One of these oligonucleotides will have the correct sequence and so will hybridize to the gene being sought.

The advantages of hybridization method are as follows:

i. It does not require expression of the inserted sequence.

ii. It can be applied to any sequence, provided a suitable radioactive probe is available.

iii. Several identical prints can be easily made from a single phage plate, allowing the screening to be performed in duplicate and hence with increased reliability.

iv. This method allows a single set of recombinants to be screened with two or more probes.

## Approaches for identification of genes

### Colony and plaque hybridization

Grunstein and Hogness (1975) developed a screening procedure to detect DNA sequences in transformed bacterial colonies or bacteriophage plaques by hybridization in situ with a radioactive probe (Fig. 17.8). Depending on the specificity of the probe, this can lead to the rapid identification of one colony amongst many thousands. The colonies to be screened are first replica plated onto a nitrocellulose or nylon membrane by placing on the surface of an agar plate prior to inoculation. A replica agar plate is retained as a reference set. The colonies are grown and the nitrocellulose membrane disc carrying the colonies removed. The membrane bearing the colonies is removed and treated with an alkali so that the bacterial colonies are lysed and their DNAs are denatured. The membrane disc is then treated with proteinase K to remove proteins and leave the denatured DNA bound to the nitrocellulose, for which it has a high affinity, in the form of a DNA print of the colonies. The single-stranded DNA molecules can then be bound tightly to the membrane by a short heat treatment at 80°C or by UV irradiation. The molecules will be attached to the membrane through their sugar-phosphate backbones, so the bases will be free to pair with complementary nucleic acid molecules. The probe must now be labeled, denatured by heating, and applied to the membrane in a solution of chemicals that promote nucleic acid hybridization. After a period to allow hybridization to take place, the membrane is washed to remove unbound probe molecules, dried, and the positions of bound probe detected. The nature of probes that can be used has been explained earlier. During hybridization, any colony carrying a sequence complementary to the probe will become radiolabeled and can be identified by autoradiography. Any clone showing a positive result can be picked from the master plate and grown to provide DNA for further analysis.

### Immunological detection

Immunological detection of clones synthesizing a foreign protein has also been successful where the inserted gene sequence is expressed. A particular advantage of the method is that genes that do not confer any selectable property on the host can be detected, but it does require that specific antibody is available. In this method transformed cells are plated on agar. A replica plate must also be prepared because subsequent procedures kill these colonies. The bacterial colonies are then lysed to release the antigen from positive colonies. Nitrocellulose filters imprinted with detritus of bacterial lysis are soaked in a solution containing the antibody (unlabeled). The antigen complexes with the bound IgG antibody. The filter is removed and exposed to $^{125}$I labeled IgG. The $^{125}$I-IgG can react with the bound antigen via antigenic determinants at sites other than those involved in the initial binding of antigen to the IgG coated filter (Fig. 17.9). Positively reacting colonies are detected by washing the filter and making an autoradiographic image. The required clones can then be recovered from the replica plate. Nowadays, the secondary ligand $^{125}$I-IgG is available, which is covalently linked to an enzyme whose activity can be detected histochemically (e.g. alkaline phosphatase). The key to the success lies in the quality of the antibody and it is essential that the antibody efficiently recognize the denatured protein.

### Southern blot analysis of cloned genes

The method was originally developed by Southern (1975), who demonstrated that DNA restriction fragments that had been electrophoretically fractionated on agarose gels could be transferred to a solid support (nitrocellulose) and detected as discreet bands following hybridization to a complementary

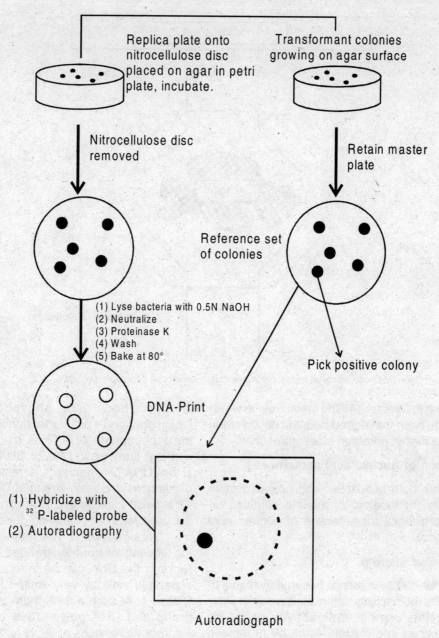

Replica plate onto nitrocellulose disc placed on agar in petri plate, incubate.

Transformant colonies growing on agar surface

Nitrocellulose disc removed

Retain master plate

Reference set of colonies

(1) Lyse bacteria with 0.5N NaOH
(2) Neutralize
(3) Proteinase K
(4) Wash
(5) Bake at 80°

Pick positive colony

DNA-Print

(1) Hybridize with $^{32}$P-labeled probe
(2) Autoradiography

Autoradiograph

**Fig. 17.8:** Grunstein-Hogness method for detection of recombinant clones by colony hybridization.

DNA probe. The procedure of agarose gel electrophoresis has been explained in Chapter 14.

*Blotting procedure.* Genomic DNA is digested with one or more restriction enzymes and the resulting fragments are separated according to size by electrophoresis on an agarose gel. The DNA is denatured in situ and transferred from the gel to a membrane. The procedure of Southern blotting has already been

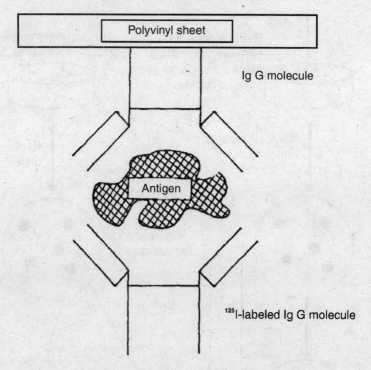

**Fig.17.9:** Antigen–antibody complex formation in the immunochemical detection method.

explained in Chapter 14. The membrane to which DNA has been transferred from the gel can then be used for hybridization labeling reaction.

## Detection of nucleic acid sequences

There are both radioactive and nonradioactive methods for labeling of nucleic acids to be used as probes for detection of nucleic acid sequences.

### Radioactive labeling

During the 1970s, methods became available to introduce radioactivity into nucleic acids *in vitro*. Before this period, radioactivity could be incorporated into nucleic acids by metabolic labeling, in which radioactive precursors usually $^{32}P$ were introduced into the cells that were synthesizing the DNA of interest. This procedure required large quantities of radioactivity and involved laborious methods for purification of nucleic acids. Further, this could be applied to small number of DNAs, e.g. viral genomes. But

later methods use phosphorylation by bacteriophage T4 polynucleotide kinase in which the γ phosphate of ATP is transferred to 5′ hydroxyl terminus in DNA or RNA or the use of *E. coli* DNA polymerase 1 to replace unlabeled nucleotides in double-stranded DNA with alpha $^{32}P$ labeled nucleotides. The following methods are used for *in vitro* labeling of nucleic acids.

i. *Nick translation*: This method is employed for labeling of double-stranded DNA probes. Nicks in the DNA can be introduced at widely separated sites by very limited treatment with DNase 1. At such a nick, there is a free 3′ OH group and DNA polymerase of *E. coli* will incorporate residues at 3′OH terminus that has been created (Fig. 17.10). DNA polymerase 1 has a 5′→3′ exonuclease activity and removes or hydrolyzes nucleotides from the 5′ side of the nick. The simultaneous elimination of nucleotides from the 5′ side and the addition of nucleotides to the 3′ side results in movement of the nick along the DNA, i.e. nick translation. If the four

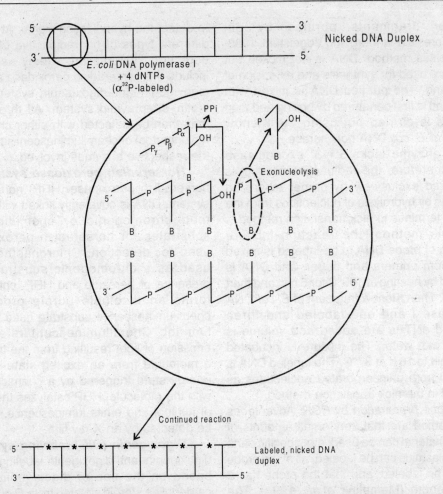

**Fig. 17.10:** $^{32}$P-labeling of duplex DNA by nick-translation. Asterisks indicate radiolabeled phosphate groups.

deoxynucleoside triphosphates are radiolabeled (e.g. $^{32}$P- dNTPs), the reaction progressively incorporates the label into a duplex that is unchanged except for translation of the nick along the molecule. Because the original nicking occurs virtually at random, a DNA preparation is effectively labeled uniformly to a degree depending on the extent of nucleotide replacement and specific radioactivity of the labeled precursors. Generally, the reaction is performed with only one of the four deoxynucleoside triphosphates (e.g. alpha $^{32}$P-dCTP) in labeled form and three unlabeled dNTPs.

A typical reaction mixture contains plasmid DNA (probe DNA to be labeled), nick translation buffer, one labeled and other unlabeled dNTPs, DNase 1, *E. coli* DNA polymerase I, and water to make up the volume. The reaction mixture is incubated for 2–4 h at 14°C. Reaction is stopped by adding EDTA. Normally, the radiolabeled DNA is separated from the unincorporated dNTPs on a Sephadex column and the probe is ready for detection and hybridization studies.

ii. *Random primed radiolabeling of probes*: This method is used to generate probes from denatured, closed circular DNA or denatured, linear, double stranded DNA especially the restriction fragments purified by gel electrophoresis (Feinberg and Vogelstein, 1983; 1984). In this method, DNA is not nicked but primers are used for synthesis and extension of DNA strand. The purified DNA is mixed with primers and is first denatured by boiling and then synthesis is carried out using the Klenow fragment of *E. coli* DNA polymerase.

This enzyme lacks 5′→3′ exonuclease activity; therefore, the radioactive product is synthesized exclusively by primer extension rather than by hydrolysis of nucleotides from the nick at 5′ terminus in nick translation method.

In this method, the reaction mixture consisting of probe DNA to be labeled is mixed with random primers and buffer, and DNA is denatured by heating at 100°C for 5 min and then put on ice. Then Klenow fragment of *E. coli* DNA polymerase I and one labeled and three unlabelled dNTPs are added and volume is made up with water. The mixture is incubated from 30 min to 16 h at 37°C. The labeled DNA is separated from unincorporated nucleotides as described in the nick translation method.

iii. *Probe preparation by PCR*: Advantages of this method are that very small amounts of template material can be used, e.g. specific parts of whole plasmids can be labeled, all of the probe DNA can be labeled, and that the probe has a defined length (Memelink *et al.*, 1994). The procedure is same as described in PCR reactions. A reaction mixture contains template DNA (linearized), amplification buffer, *Taq* DNA polymerase, primers, one labeled and other unlabelled dNTPs, and the volume is made up with water. Forty cycles are run in the PCR thermal cycling apparatus. (For more details refer to Chapter 18 on PCR.) The probe is separated from the unincorporated nucleotides as described for nick translation.

### Nonradioactive labeling

Nonradioactive methods to label nucleic acids are used more often. Nonradioactive detection methods typically require shorter exposure times to detect the hybridization signal. At least three different types of nonradioactive labeling and detection systems are widely used. These include the horseradish peroxidase system, the digoxigenin–anti-digoxigenin system, and the biotin–streptavidin system. All three systems may then be detected with either chromogenic (colorimetric) or chemiluminescent substrates for the respective enzymes involved.

*Horseradish peroxidase system.* In the horseradish peroxidase (HRP) nonradioactive system, DNA is covalently linked with HRP and then chromogenic or chemiluminescent substrates for horseradish peroxidase are used for detection. Chloronapthol is often used as a chromogenic substrate. In the presence of peroxide and HRP, chloronapthol forms an insoluble purple product. The chemiluminescence substrate used for HRP is luminol. Chemiluminescent refers to the emission of light resulting from the transition of a molecule from an excited state to a lower energy state triggered by a chemical reaction with the molecule. HRP catalyzes the oxidation of luminol and emits luminescence, which can be detected by an X-ray film.

*Digoxigenin (DIG) labeling system.* The digoxigenin–anti-digoxigenin labeling system is based on the use of digoxigenin (DIG), a cardenolide steroid isolated from *Digitalis* plants, as a hapten. A nucleotide triphosphate analog (digoxigenin [11]-dUTP) containing the digoxigenin moiety can be incorporated into DNA to be used as probes by nick translation or random prime labeling. The DIG-labeled probe is subsequently detected by enzyme-linked immunoassay using an antibody to digoxigenin (anti-DIG) to which alkaline phosphatase has been conjugated. A chromogenic or chemiluminescent substrate for alkaline phosphatase can then be used to detect the DIG-labeled probe. Chromogenic substrates such as 5-bromo-4-chloro-3-indolyl phosphate (BCIP) and nitroblue tetrazolium chloride (NBT) are added to give purple/blue color (Fig. 17.11).

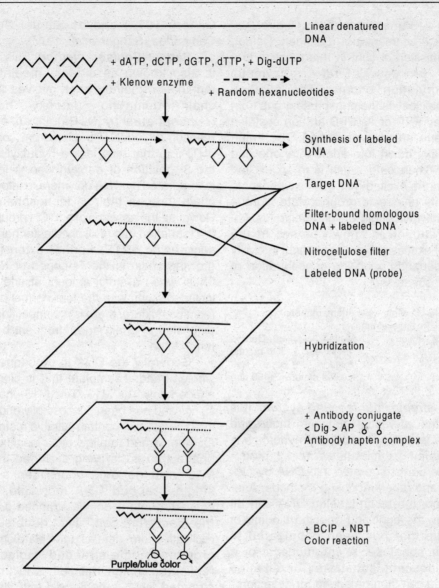

Linear denatured DNA

+ dATP, dCTP, dGTP, dTTP, + Dig-dUTP

+ Klenow enzyme

+ Random hexanucleotides

Synthesis of labeled DNA

Target DNA

Filter-bound homologous DNA + labeled DNA

Nitrocellulose filter

Labeled DNA (probe)

Hybridization

+ Antibody conjugate
< Dig > AP
Antibody hapten complex

+ BCIP + NBT
Color reaction

Purple/blue color

**Fig. 17.11:** Non-radioactive DIG-DNA labeling.

The use of dioxetane chemiluminescent probes has been developed in the last few years. Dioxetane substrates can be used with the digoxigenin or the biotin–streptavidin systems. There are a number of stable 1,2-dioxetane derivatives that emit light when activated by enzymes. The 1,2-dioxetane that is commonly used has a phosphate group attached and can be enzymatically activated by alkaline phosphatase. The alkaline phosphatase can be directly linked to the DNA used as a probe or it can be covalently linked to streptavidin and then bound to a biotin-labeled probe. Alternatively, alkaline phosphatase can be covalently linked to the antibody directed against a hapten such as digoxigenin. Chemiluminescent substrates for alkaline phosphatase include PPD and CSPD. PPD, also called AMPPD, is

4-methoxy-4-(3-phosphatephenyl) spiro (l,2-dioxetane-3,2' adamantane) (Lumigen, Detriot, MI). The emission of light in this reaction is a two-step process. First, enzymatic dephosphorylation occurs. The alkaline phosphatase cleaves the phosphate group from the protected PPD or AMPPD and an unstable 1,2 dioxetane anion is produced. The anion rapidly breaks down into adamantanone and methyl *m*-oxybenzoate which is in an excited state. When the excited methyl *m*-oxybenzoate decays to its electronic ground state, light is emitted. The wavelength of light emitted is 480 nm. The reaction is shown below. In the presence of excess substrate, the light intensity produced depends upon the concentration of alkaline phosphatase.

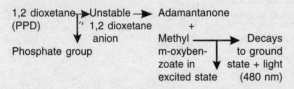

### Biotin–streptavidin labeling system.

*Biotin–streptavidin labeling system.* It is based on the interaction between biotin and avidin, and is widely used in immunology. A biotin-containing nucleotide analog (biotin-[14]dATP) can be incorporated into DNA by nick translation and random priming methods. After hybridization, the biotin-labeled DNA can be detected by the tight and specific binding of streptavidin that has been conjugated to alkaline phosphatase. A chromogenic or a chemiluminescent substrate for alkaline phosphatase can then be added for detection.

## DNA Sequencing

DNA sequencing is an important technique by which the precise order of nucleotides in a segment of DNA can be determined. Rapid and efficient DNA sequencing methods became available in late 1970s and that revolutionized the work on structure of gene. Two different techniques were developed simultaneously by Maxam and Gilbert at Harvard (Maxam and

Gilbert, 1977) and by Sanger-Coulson at Cambridge (Sanger *et al.,* 1977).

*Sanger-Coulson method.* This technique is also known as dideoxy-mediated chain termination method as it involves the use of chain-terminating dideoxy nucleoside triphosphates (ddNTPs) in an enzymatic reaction. 2',3'ddNTPs differ from conventional dNTPs in that they lack a hydroxyl residue at the 3' position of deoxyribose. They can be incorporated by DNA polymerase into a growing chain through their 5' triphosphate groups. However, the absence of a 3'-OH group prevents the formation of a phosphodiester bond with the succeeding dNTP. Further extension of the growing chain is then stopped or terminated. Thus, when a small amount of one ddNTP is included along with the four normal dNTPs in a reaction mixture, there is a competition between extension of chain and infrequent but specific termination.

Generally the DNA to be sequenced is a single-stranded template that is cloned into a phage known as M13. This M13 phage infects *E. coli* and has been successfully engineered in the laboratory to contain suitable restriction sites for cloning and to allow easy identification of a phage that is carrying a cloned insert. The M13 viral genome comprises a circular, single-stranded DNA molecule, which is converted into a double-stranded circle inside the bacterial cell. The double-stranded replicative form acts as a template on which the progeny single-stranded molecules are replicated inside the cell, and this single-stranded DNA is packaged into viral protein coats and extruded from the cell to complete the life cycle of the phage. However, it is also possible to use denatured double-stranded DNA template, although the best results are obtained from single-stranded DNA templates.

A short, synthetic oligonucleotide primer is annealed onto the single-stranded template of M13 DNA just upstream of the restriction sites used for cloning (Fig. 17.12). A universal primer that anneals to the vector sequences flanking

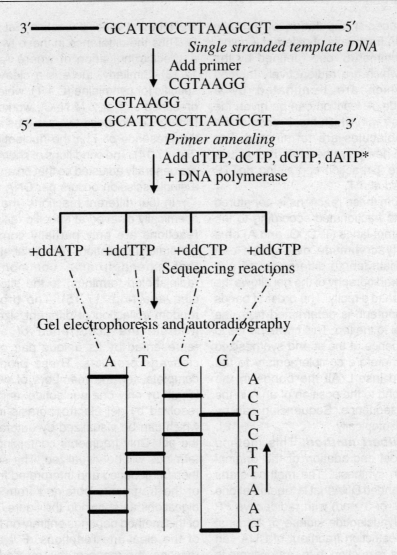

**Fig. 17.12:** DNA sequencing by the dideoxy chain termination method. * indicates the presence of radioactively labeled P or S.

the target DNA can also be used. When this complex is incubated *in vitro* with purified DNA polymerase and the four dNTPs, the primed site will act as a starting point for DNA synthesis in the 5′ to 3′ direction through the cloned sequence. If a ddNTP is present, synthesis stops when this molecule is incorporated into the growing chain. Four separate reactions are

therefore performed. In each reaction a different ddNTP is present in addition to all four dNTPs (dTTP, dCTP, dGTP, dATP). One of the four dNTPs (usually dATP) is radiolabeled. The ratio of dNTP to ddNTP in the reaction mixture is carefully adjusted so that the dideoxy moiety is incorporated only occasionally in place of its deoxy homolog. Hence, each of the four reactions produces a population of partially synthesized DNA fragments of different lengths, all sharing a common 5′ end (defined by the primer), all of which are radioactively labeled and all of which are terminated by a dideoxynucleotide. A reaction can be given the name of the base with which the 3′ ends of the synthesized molecules are terminated. For example, all the newly synthesized, radioactive molecules in the T-reaction containing ddTTP will be terminated at a T.

The DNA from these reactions is denatured and the products fractionated according to the size in four different lanes (T, C, G, and A) on a denaturing polyacrylamide gel, which can resolve DNA chain length differences of one nucleotide. Autoradiography of the gel allows the sequence to be read directly. The order of bands on the autoradiograph is determined from the bottom of the gel to the top. This corresponds to the 5′ to 3′ sequence of the strand synthesized *in vitro* and is therefore complementary to the template-cloned insert. All the bands in the T-track correspond to the position of an A in the template DNA sequence. Sequencing can be also done non-isotopically.

***Maxam–Gilbert method.*** This method involves chemical degradation of the original DNA rather than synthesis. The method starts with a single-stranded DNA that is labeled at one end (either its 5′ or 3′ end) with radioactive $^{32}P$ using either polynucleotide kinase or terminal transferase. A restriction fragment of DNA can be used. From a restriction map, an enzyme is chosen that removes a small piece from one end of the molecule, leaving just one end labeled. Double-stranded DNA can be used if only one strand is labeled at only one of its ends.

The method involves modification of a base chemically and then the DNA strand is cleaved by reactions that specifically fragment its sugar-phosphate backbone only where certain bases have been chemically modified. There is no unique reaction for each of the four bases. However, there is a reaction specific to G only

and a purine specific reaction that removes A or G. Thus the difference in these two reactions is a specific indication of where A occurs (Fig. 17.13). Similarly, there is a cleavage reaction specific for pyrimidine (C + T), which if run in the presence of 1 or 2 M NaCl, works only with C. Differences in these two are thus attributable to the presence of T in the nucleotide sequence (Fig. 17.14). The conditions of chemical cleavage are generally adjusted so that on an average only a single scission occurs per DNA molecule.

In four different reactions, the DNA is then chemically cleaved at specific residues, but the reactions are only partially completed. This generates populations of radiolabeled molecules that extend from a common point (the radiolabeled terminus) to the site of chemical cleavage (Fig. 17.15). The products are a random collection of different sized fragments wherein the occurrence of any base is represented by its unique pair of 5′- and 3′-cleavage products. These products form a complete set, the members of which differ in length by only one nucleotide, and they can be resolved by gel electrophoresis into a ladder, which can be visualized by autoradiography of the gel. Only fragments containing the labeled terminus will be visualized. The sequence can then be deduced and interpreted from the order of the fragments obtained from the different digestions as shown in the figure. The success of this method depends entirely on the specificity of the cleavage reactions. For a number of reasons, the range of this method is less than that of the Sanger method.

In principle, this method can provide the total sequence of a dsDNA molecule just by determining the purine positions on one strand and then the purines on the complementary strand. Complementary base pairing rules then reveal the pyrimidines along each strand, T complementary to where A is, C complementary to where G occurs. With current knowledge, it is possible to read the order of 400 bases from the autoradiogram of a sequencing gel.

## Cleavage of G

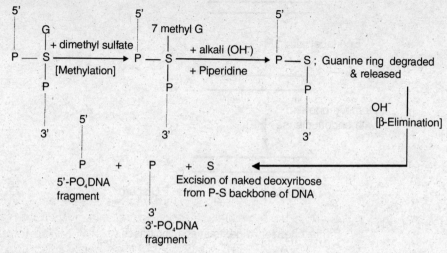

## Cleavage of A or G

*DNA is treated with acid followed by treatment with dimethyl sulfate which results in methylation of adenine at 3-position as well as guanine at 7 position. Subsequent reaction with OH⁻ and piperidine results in scision of DNA strand and displacement of purines.*

**Fig. 17. 13:** Cleavage of purines using dimethyl sulfate in the Maxam-Gilbert method.

## Cleavage of pyrimidines (C or T) *reaction for T is shown*

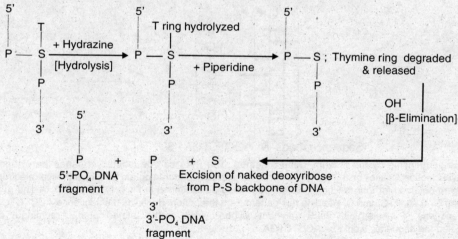

## Cleavage of C

*If cleavage reaction for pyrimidine is performed in the presence of 1 or 2M NaCl, it works only with C.*

**Fig. 17.14:** Cleavage of pyrimidines using hydrazine hydrolysis in the Maxam-Gilbert method.

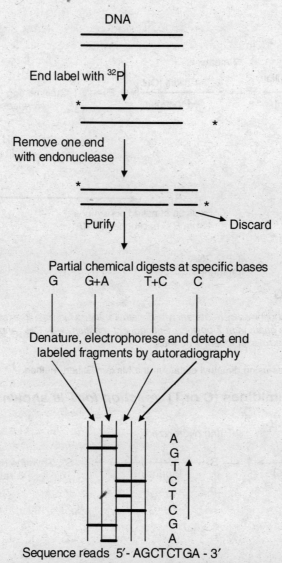

DNA

End label with $^{32}$P

Remove one end
with endonuclease

Purify ————→ Discard

Partial chemical digests at specific bases
G      G+A      T+C      C

Denature, electrophorese and detect end
labeled fragments by autoradiography

A
G
T
C
T
C
G
A

Sequence reads 5′- AGCTCTGA - 3′

**Fig. 17.15:** DNA sequencing by the Maxam–Gilbert method. The DNA can be labeled at either the 5′ end (using polynucleotide kinase) or 3′ end (by terminal transferase). Autoradiogram of a hypothetical electrophoretic pattern obtained for four reaction mixtures performed as described in Figs. 17.13 and 17.14 and run in four lanes G, G+A, T+C, and C. Reading this pattern from the bottom up yields the sequence AGCTCTGA. If the 5′ end was $^{32}$P labeled, only the 5′ fragments will be evident on the autoradiogram. Assuming 5′ end was labeled the sequence would be AGCTCTGA.

The Sanger method required S/S templates, specific primers, and Klenow fragment of *E. coli* DNA polymerase I whereas the Maxam-Gilbert method uses simple chemical reagents. However, with the development of bacteriophage M13 and phagemid vectors, the availability of synthetic primers and improvements to the sequencing reactions, the dideoxy-mediated chain termination method is now used much more extensively than the Maxam-Gilbert method. Nevertheless, the chemical degradation method has one advantage in that the sequence is obtained from the original DNA molecule and not from an enzymatic copy. Therefore, one can sequence synthetic oligonucleotides with the chemical degradation method, analyze DNA modifications such as methylation, and study both DNA secondary structure and the interaction of proteins with DNA by either chemical protection or modification interference experiments. However, because of its ease and rapidity, the Sanger technique is the best choice for DNA sequencing.

**High throughput DNA sequencing:** The Maxam-Gilbert and Sanger techniques, which are highly conceptual, elegant and efficacious, are in practice time consuming and partly labor intensive, partly because a single radioisotopic reporter is used for detection. Using one reporter to analyze each of the four bases requires four separate reactions and four gel lanes. The resulting autoradiographic patterns, obtained after a delay for exposure and development, are complex and require skilled interpretation and data transcription.

Prober *et al.* (1987) developed a rapid DNA sequencing system based on fluorescent chain terminating dideoxynucleotides. DNA sequence determination using the dideoxy chain termination method results in the synthesis of reaction products that have the same 5′ terminus, but random 3′ ends due to the incorporation of dideoxynucleotides (ddNTPs). This sequencing system is based on the use of a novel set of four chain-terminating dideoxynucleotides, each carrying a different chemically tuned succinyl fluorescein dye tag distinguished by its fluorescent emission. Dideoxy sequencing involves template-directed enzymatic extension of a short oligonucleotide primer in the presence of chain-terminating dideoxyribonucleotide triphosphates (ddNTPs). The fluorescent dyes used are 9-(carboxyethyl)-3-hydroxy-6-oxo-6H-xanthanes or succinyl fluoresceins (SF-xxx, where xxx refers to the emission maximum in nanometers) (Fig. 17.16). These fluorescent dyes give different colors (e.g. A—red; T—blue; G—green; C—yellow). The wavelengths of the

absorption and emission maxima in the succinyl fluorescein system are tuned by changing the substituents $-R_1$ and $-R_2$ as shown in Fig. 17.16. The set of four fluorescence tagged chain-terminating reagents have been designed. These are ddNTPs to which succinyl fluorescein has been attached via a novel acetylenic linker to the heterocyclic base. (These reagents are designated N-SF-*xxx* where N refers to the ddNTP, SF - succinyl fluoresceins and *xxx* refers to the emission maximum in nanometers). The linker is attached to the 5 position in the pyrimidines and to the 7 position in the 7-deazapurines. Avian myeblastosis virus reverse transcriptase is used in a modified dideoxy sequencing protocol to produce a complete set of fluorescence-tagged fragments in one reaction mixture. Labeled DNA fragments produced by enzymatic chain extension reactions are separated by polyacrylamide gel electrophoresis, detected and identified as they migrate past the fluorescence detection system specially matched to the emission characteristics of this dye set. A scanning system allows multiple samples to be run simultaneously and computer-based automatic base sequence identifications are made. The DNA sequencing system has been commercialized by Applied Biosystems (AB), and it soon became the workhorse of the Human Genome Project. The AB system uses four spectrally distinct fluorescent ddNTPs. This strategy makes it possible to perform the entire sequencing reaction in a single tube and to resolve the chain-terminated products in one lane of a sequencing gel. There are now 96-capillary sequencers. The AB DNA detection system permits direct real time data acquisition by laser activated dye excitation at a point in the electrophoresis run that maximizes fragment resolution. Using this method, reliable sequence information can be obtained for DNA segments >500 nucleotides long in less than 4 hours. Based on the number of lanes per gel, one ABI instrument can generate ~50,000 nucleotides of sequence information in an 8-hr day. An initial drawback to automated DNA sequencing using

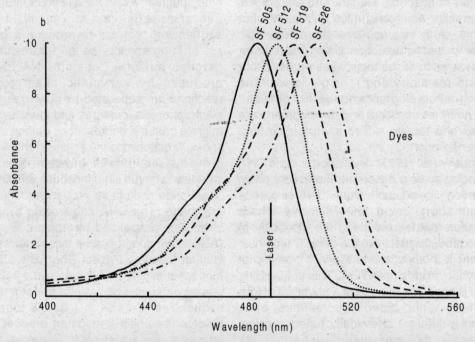

**Fig. 17.16:** Succinyl fluorescein dyes. (A) Chemical structure of the four dyes used to label dideoxynucleotide triphosphates for use as chain-terminators in modified dideoxy DNA sequencing protocols. (B) Normalized absorption spectra of the dyes shown in (A). Absorption coefficient at the maximum for SF-505 is 72,600$M^{-1}cm^{-1}$. Spectra were measured in pH 8.2, 50 mM aqueous Tris-HCl buffer. The other dyes have coefficients within 10 percent of this value. Vertical bar (laser) indicates the position of the argon ion laser line at 488 nm used for fluorescence excitation.

a fluorescence-based detection system was the reduced signal-to-noise ratios that occur when small amounts of template are used in the reaction. The solution to this problem was to develop a one-sided PCR amplification strategy, called cycle sequencing, that utilizes a modified *Taq* DNA polymerase and temperature cycling to generate high levels of chain terminated products from a small amount of template. The use of cycle sequencing has greatly extended the automation capabilities of DNA sequencing because it requires much less starting material for each round of sequencing.

## REFERENCES

Feinberg, A.P. and Vogelstein, B. 1983. A technique for radiolabelling DNA restriction endonuclease fragments to high specific activity. *Anal Biochem.* **132:** 6.

Feinberg, A.P. and Vogelstein, B. 1984. Addendum: A technique for radiolabelling DNA restriction endonuclease fragments to high specific activity. *Anal Biochem.* **137:** 266.

Grunstein, M. and Hogness, D.S. 1975. Colony hybridization-A method for the isolation of cloned DNAs that contain a specific gene. *Proc. Natl. Acad. Sci. USA* **72:** 3961–3965.

Maxam, A.M. and Gilbert, W. 1977. A new method for sequencing DNA. *Proc. Natl. Acad. Sci. USA* **74:** 560–564.

Memelink, J., Swords, K.M.M., Staehlin, L.A. and Hoge, J.H.C. 1994. Southern, Northern and Western blot analysis. *Plant Mol. Biol. ManualN* **F1:** 1–21.

Prober, J.M., Trainor, G.L., Dam, R.J., Hobbs, F.W., Robertson, C.W., Zagursky, R.J., Cocuzza, A.J., Jensen, M.A. and Baumeister, K. 1987. A system for rapid DNA sequencing with fluorescent chain terminating didexoynucleotides. *Science* **238:** 336–341.

Sanger, F., Nicklen, S. and Coulson, A.R. 1977. DNA sequencing with chain termination inhibitors. *Proc. Natl. Acad. Sci. USA* **74:** 5463–5467.

Southern, E.M. 1975. Detection of specific sequences among DNA fragments separated by gel electrophoresis. *J. Mol. Biol.* **98:** 503–517.

## FURTHER READING

Chawla, H.S. 1998. *Biotechnology in Crop Improvement.* International Book Distributing Co., Lucknow.

Kingsman, S.M. and Kingsman, A.J. 1988. *Genetic Engineering: An Introduction to Gene Analysis and Exploitation in Eukaryotes.* Blackwell Scientific Publication, Oxford.

Maniatis, T., Fritsch, E.F. and Sambrock, J. 1982. *Molecular Cloning: A Laboratory Manual.* Cold Spring Harbor Laboratory Press, USA.

Mantell, S.H., Mathew, J.A. and McKee, R.A. 1985. *Principles of Plant Biotechnology: An Introduction to Genetic Engineering in Plants.* Blackwell Scientific Publications, Oxford.

Miesfeld, R.L.1999. *Applied Molecular Genetics.* Wiley-Liss, New York.

Old, R.W. and Primrose, S.B. 2001. *Principles of Gene Manipulation: An Introduction to Genetic Engineering.* Blackwell Scientific Publications, Oxford.

Sambrook, J., Fritsch, E.F. and Maniatis, T. 1989. *Molecular Cloning: A Laboratory Manual.* Cold Spring Harbor Laboratory Press, USA.

Walker, J.M. and Gaastra, W. (eds.). 1987. *Techniques in Molecular Biology.* Croom Helm, Australia.

# 18

# Polymerase Chain Reaction

The investigation of an organism's genome was greatly enhanced during the early 1970s with the development of recombinant DNA technology. This technology allows for *in vivo* replication (amplification) of genomic DNA regions that are covalently linked with bacterial plasmid or virus clones. In 1985–86, a second major development occurred at Cetus Corporation, USA, where researchers developed an *in vitro* method for the amplification of DNA fragments, referred to as the polymerase chain reaction (PCR). The idea for PCR is credited to Kary Mullis, who along with five other researchers demonstrated that oligonucleotide primers could be used specifically to amplify specific segments of genomic DNA or cDNA. Mullis was awarded the Nobel Prize in chemistry in 1993 for his contribution to the development of PCR. The key feature of the PCR technique is the exponential nature of amplification. This PCR is based on the features of semiconservative DNA replication carried out by DNA polymerases in prokaryotic and eukaryotic cells. PCR results in the selective amplification of a chosen region of a DNA molecule. In this technique, DNA molecule is not cloned in a bacterial plasmid or virus. The only requirement is that the sequence at the borders of the selected DNA region must be known so that two short oligonucleotides can anneal to the target DNA molecule for amplification. These oligonucleotides delimit the region that will be amplified. In normal DNA replication within a cell, DNA duplex opens up and a small single-stranded RNA primer is synthesized to which DNA polymerase adds further nucleotides. In PCR, a specific region of DNA, which is flanked by two oligonucleotide (deoxy) primers that share identity to the opposite DNA strands, is enzymatically amplified. Amplification of selected region from a complex DNA mixture is carried out *in vitro* by the DNA polymerase I from *Thermus aquaticus*, a bacterium that lives in hot springs.

This amplification is achieved by a repetitive series of cycles involving three steps:

1. Denaturation of a template DNA duplex by heating at 94°C.
2. Annealing of oligonucleotide primers to the target sequences of separated DNA strands at 55–65°C.
3. DNA synthesis from the 3′-OH end of each primer by DNA polymerase at 72°C.

By using repetitive cycles, where the primer extension products of the previous cycle serve as new templates for the following cycle, the number of target DNA copies has the potential to double each cycle (Fig. 18.1).

This technique became important with the discovery of *Taq* DNA polymerase by Kary Mullis. *Taq* DNA polymerase is a thermostable enzyme, which retains activity even after denaturation by heat treatment at 94°C, whereas the DNA

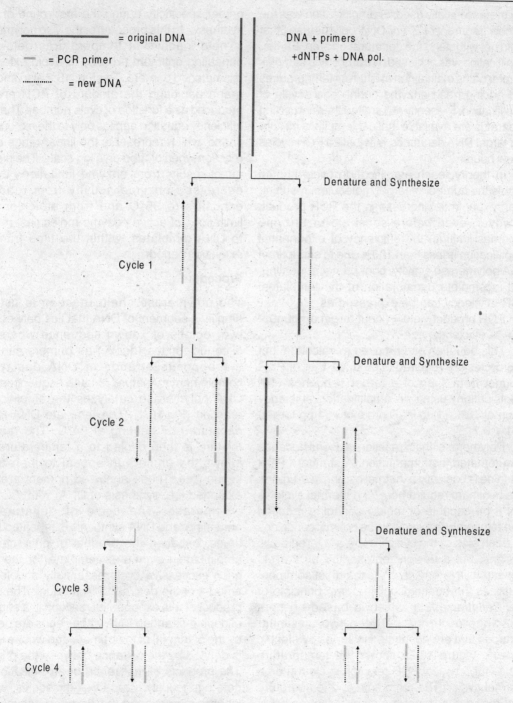

= original DNA

= PCR primer

= new DNA

DNA + primers
+dNTPs + DNA pol.

Denature and Synthesize

Cycle 1

Denature and Synthesize

Cycle 2

Denature and Synthesize

Cycle 3

Cycle 4

**Fig. 18.1:** Schematic diagram of the polymerase chain reaction. Only the first two cycles are shown completely. Beginning with the third cycle, the diagram does not show the fate of original DNA and the extension products made from it.

polymerase initially used for amplification was the Klenow fragment of *E. coli* DNA polymerase I that could not withstand the denaturation step. Earlier, amplification was achieved by transferring samples in water baths maintained at different temperatures and adding fresh enzyme during each cycle of amplification. Presently automated thermal cycling apparatus are available that can amplify a million-fold target DNA sequence with just 20 PCR cycles in few hours.

In theory, each amplification cycle should double the number of target molecules, resulting in an exponential increase in the PCR product. However, even before substrate or enzyme becomes limiting, the efficiency of exponential amplification is less than 100% due to suboptimal DNA polymerase activity, poor primer annealing, and incomplete denaturation of the templates. PCR efficiency can be expressed as

PCR product yield = (input target amount) × $(1 + \% \text{ efficiency})^{\text{cycle number}}$

This equation can be used to calculate that ~26 cycles are required to produce 1 µg of PCR product from 1 pg of a target sequence ($10^6$ amplification) using an amplification efficiency value of 70%. [1 µg PCR product = (1 pg target) × $(1 + 0.7)^{26}$].

Polymerase chain reaction is useful because of automated instrumentation. A standard PCR machine is basically a thermal cycling instrument. The minimum requirement for a thermal cycler is that it be capable of rapidly changing reaction tube temperature to provide the optimal conditions for each step of the cycle. There are three basic designs, but a design represented by the MJ Research DNA engines™, is the most common. It utilizes the heating/cooling pump principle of the *Pelletier effect*, which is based on heat exchange occurring between two dissimilar surfaces that are connected in series by electric current. Figure 18.2a shows the temperature profile for a typical PCR reaction cycle run on a thermal cycler. The ramp time and temperature accuracy are the most important parameters that must be controlled by the thermal cycler. A hot start is a method that can be used to increase

primer specificity during the first round of DNA synthesis. In this procedure, the *Taq* polymerase is held inactive at temperatures below the annealing optimum (<55°C) to avoid extending mismatched primers. The Fig. 18.2 also shows a graph depicting the amount of PCR product produced as a function of cycle number. The PCR efficiency equation applies only to the exponential phase, which represents the linear range of the reaction when plotted on a log scale. The plateau effect results from enzyme limitations due to decreased enzyme activity from repeated exposure to 95°C and from stoichiometric limitations of active enzyme molecules, relative to DNA templates, within the time frame of extension period.

## Procedure

The polymerase chain reaction is used to amplify a segment of DNA that lies between the two regions of known sequence where two oligonucleotides (deoxy) as primers can bind the opposite strands of DNA due to the complementary nature of base sequences. *Taq* DNA polymerase catalyzes the amplification reaction (Fig.18.1). The template DNA is first denatured by heating at 94°C. The reaction mixture is then cooled to a temperature that allows the primers to anneal to their target sequences. These annealed primers are then extended (i.e. synthesis of DNA) with *Taq* DNA polymerase. The cycle of denaturation, annealing, and DNA synthesis is repeated many times, because the product of one round of amplification serves as template for the next, each successive cycle essentially doubles the amount of the desired DNA product. The major product of this exponential reaction is a segment of double-stranded DNA whose ends are defined by the 5′ termini of the primers and whose length is defined by the distance between the primers. The products of first round of amplification are heterogeneously sized DNA molecules, whose lengths may exceed the distance between the binding sites of the two primers. In cycle 2, the original strands and the new strands from cycle

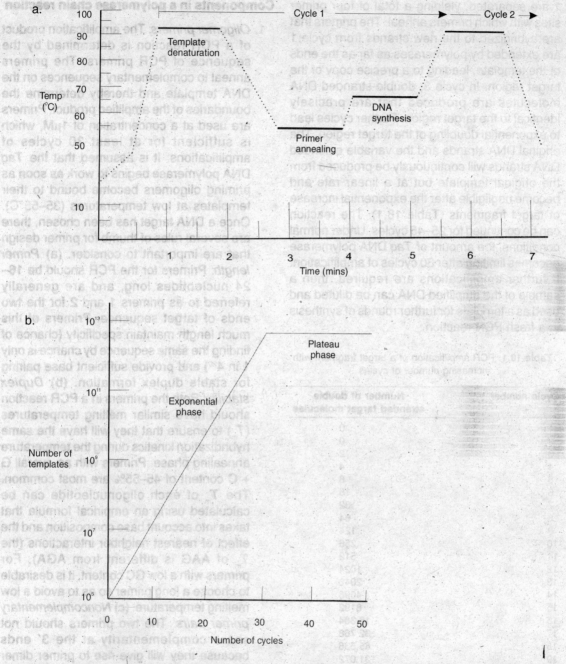

**Fig. 18.2:** a—Temperature profile of a typical PCR cycle. The dotted lines represent the ramping period which is usually Δ1°C per second with most thermal cyclers. b—Accumulation of PCR target molecules as a function of cycle number. The exponential phase lasts about 30 cycles under standard reaction conditions. The plateau phase results from limiting amounts of enzyme and reduced enzyme activity. Concept of the figure taken from Miesfeld: *Applied Molecular Genetics.* Copyright © 1999 John Wiley and Sons. This material is used by permission of John Wiley and Sons, Inc.

1 are separated, yielding a total of four primer sites with which primers anneal. The primers that are hybridized to the new strands from cycle 1 are extended by polymerases as far as the ends of the template, leading to a precise copy of the target region. In cycle 3, double-stranded DNA molecules are produced that are precisely identical to the target region. Further cycles lead to exponential doubling of the target region. The original DNA strands and the variable extended DNA strands will continuously be produced from the original template but at a linear rate and become negligible after the exponential increase of target fragments (Table 18.1). The reaction can be continued for 25–45 cycles. Under normal conditions, the amount of *Taq* DNA polymerase becomes limiting after 30 cycles of amplification. If further amplifications are required, then a sample of the amplified DNA can be diluted and used as a template for further rounds of synthesis in a fresh PCR reaction.

**Table 18.1:** PCR amplification of a target fragment with increasing number of cycles

| Cycle number | Number of double stranded target molecules |
| --- | --- |
| 1 | 0 |
| 2 | 0 |
| 3 | 2 |
| 4 | 4 |
| 5 | 8 |
| 6 | 16 |
| 7 | 32 |
| 8 | 64 |
| 9 | 124 |
| 10 | 256 |
| 11 | 512 |
| 12 | 1024 |
| 13 | 2048 |
| 14 | 4096 |
| 15 | 8192 |
| 16 | 16,384 |
| 17 | 32,768 |
| 18 | 65,536 |
| 19 | 131,072 |
| 20 | 262,144 |
| 21 | 524,288 |
| 22 | 1,048,576 |
| 23 | 2,097,152 |
| 24 | 4,194,304 |
| 25 | 8,388,608 |

**Components in a polymerase chain reaction**

1. *Oligomer primers*: The amplification product of a PCR reaction is determined by the sequence of PCR primers. The primers anneal to complementary sequences on the DNA template and thereby determine the boundaries of the amplified product. Primers are used at a concentration of 1µM, which is sufficient for at least 30 cycles of amplifications. It is assumed that the *Taq* DNA polymerase begins to work as soon as priming oligomers become bound to their templates at low temperatures (35–55°C). Once a DNA target has been chosen, there are several rules of thumb for primer design that are important to consider. (a) *Primer length*: Primers for the PCR should be 16–24 nucleotides long, and are generally referred to as primers 1 and 2 for the two ends of target sequence. Primers of this much length maintain specificity (chance of finding the same sequence by chance is only 1 in $4^{20}$) and provide sufficient base pairing for stable duplex formation. (b) *Duplex stability*: Both the primers in a PCR reaction should have similar melting temperatures ($T_m$) to ensure that they will have the same hybridization kinetics during the temperature annealing phase. Primers with an overall G + C content of 45–55% are most common. The $T_m$ of each oligonucleotide can be calculated using an empirical formula that takes into account base composition and the effect of nearest neighbor interactions (the $T_m$ of AAG is different from AGA). For primers with a low GC content, it is desirable to choose a long primer so as to avoid a low melting temperature. (c) *Noncomplementary primer pairs*: The two primers should not share complementarity at the 3′ ends because they will give rise to primer dimer products. For example, if one primer has the sequence 5′–...GGCG–3′ and the other primer has a terminal sequence 5–...CCGC–3′, then they can form a short hybrid that will become a substrate for DNA synthesis. Once

the primer dimer product is formed, it is a competing target for amplification. (d) *No hairpin loops*: Each primer must be devoid of palindromic sequences that can give rise to stable intrastrand structures that limit primer annealing to the template DNA. (e) *Optimal distance between primers*: This rule is specific for application, but for most diagnostic PCR assays, it is best when the opposing primers are spaced 150–500 bp apart. (f) Sequences with long runs (i.e. more than three or four) of a single nucleotide should be avoided. Presently, a variety of computer algorithms has been developed, which takes all these parameters into account, and the researcher only needs to provide a DNA target sequence.

2. *Amplification buffer*. The standard buffer for PCR reaction contains KCl, Tris.Cl, and 1.5 mM magnesium chloride. When incubated at 72°C, the pH of the reaction drops by more than a unit. To maintain the pH, presence of divalent magnesium cations is critical.

3. *Deoxyribonucleoside triphosphates*: dNTPs are used at a saturating concentration of 200 $\mu$M for each dNTP.

4. *Target sequence*: Template DNA containing the target sequences can be added in a single or double stranded form. Linear target sequences are amplified better as compared to closed circular DNAs. Purity of template is flexible and even crude DNA extracts (e.g. pulp of a fossil tooth, a hair follicle) can be used as long as the addition of template does not inhibit the activity of polymerase. The concentration of the target sequence in the PCR reaction is generally in nanograms (5–100 ng).

5. *Taq DNA polymerase*: Two forms of *Taq* DNA polymerase are available: the native enzyme purified from *Thermus aquaticus* and a genetically engineered form synthesized in *E. coli* (Ampli Taq-TM). *Taq* DNA polymerase is the most frequently used and preferred enzyme. The original PCR protocol used the Klenow fragment of *E. coli* DNA polymerase I to perform the primer extension reaction. However, this meant that fresh enzyme had to be added after each round of denaturation because this enzyme is easily heat inactivated. Another problem with *E. coli* DNA polymerase is that the optimal activity level of the enzyme is 37°C, which greatly limits the specificity of the reaction due to degenerate primer annealing at this low temperature. Both these problems were solved by the use of *Taq* DNA polymerase. This enzyme retains activity even after repeated exposure to a temperature of 95°C, and is also fully active at ~75°C, which essentially eliminates degenerate heteroduplex formation. More recently, several other thermostable DNA polymerases have been isolated and characterized that offer advantages for specialized PCR assays (Table 18.2). The biggest difference between these enzymes is their inherent $3' \rightarrow 5'$ exonuclease proof reading activity, which is important if the sequence of PCR product must be error free.

**Table 18.2:** Characteristic features of different thermostable DNA polymerases

| Enzyme | Relative efficiency | Error rate | Processivity | Extension rate | $3' \rightarrow 5'$ exonuclease | $5' \rightarrow 3'$ exonuclease |
|---|---|---|---|---|---|---|
| Taq Pol | 88 | $2 \times 10^{-4}$ | 55 | 75 | No | Yes |
| Tli Pol (Vent) | 70 | $4 \times 10^{-5}$ | 7 | 67 | Yes | No |
| Pfu Pol | 60 | $7 \times 10^{-7}$ | n.d. | n.d. | Yes | No |
| RTth | n.d. | n.d. | 30 | 60 | No | Yes |

*Note*. Relative efficiency–Percent conversion of template to product per cycle; Error rate–Frequency of errors per base pairs incorporated; Processivity–Average number of nucleotides added before dissociation; Extension rate–Average number of nucleotides added per second; n.d.–not determined.

DNA polymerases with 3'→5' exonuclease activity are *Tli* from *Thermococcus litoralis* and *Pfu* from *Pyrococcus furiosus*, which have a 10–100 times lower error rate than DNA polymerases lacking this activity. *Taq* DNA polymerase and *E. coli* DNA polymerase I encode a 5'→3' exonuclease activity. Two other parameters are processivity (number of nucleotides polymerized before template dissociation) and extension rate (nucleotides polymerized/second). Some thermostable DNA polymerases can use RNA templates as a substrate, which can be useful for PCR applications that require a separate cDNA synthesis reaction using viral reverse transcriptase. An enzyme of this is the recombinant form of *Tth* polymerase (*rTth*) from *Thermus thermophilus*, which can catalyze high temperature reverse transcription of RNA in the presence of $MnCl_2$. Approximately 2 units of either of the enzymes are required to catalyze a typical PCR. Addition of excess enzyme may lead to amplification of nontarget sequences. Both forms of the polymerase (*Taq* and Ampli *Taq*™) carry a 5'→3' polymerase-dependent exonuclease activity, but they lack a 3'→5' exonuclease activity. It means that *Taq* polymerase does not correct the incorporation of mismatched bases.

The reaction is carried out in a sterile 0.5-ml microfuge tube by mixing the components in the order of sterile water, amplification buffer, dNTPs, primers 1 and 2, template DNA and finally the *Taq* DNA polymerase. Total volume can be 50 or 100 µl. The reaction mixture is overlayered with 2–3 drops of mineral oil. This prevents evaporation of the sample during repeated cycles of heating and cooling. But now PCR thermal cyclers are available with heated lid; thus, mineral oil is not required in those reactions. Also, reactions can be carried out in a much smaller volume of 10–25 µl.

The amplified DNA product is detected by agarose gel electrophoresis, followed by staining with ethidium bromide.

### Inverse PCR

Conventional PCR is used frequently to amplify DNA segments lying between the two priming oligonucleotides. But PCR can also be used to amplify unknown DNA sequences that lie outside the boundary of known sequences for which primer pairs can be designed. This is referred to as inverse PCR (iPCR). The technique for amplifying unknown flanking DNAs was developed by different research groups (Ochmann *et al.*, 1988; Triglia *et al.*, 1988). In this technique, genomic DNA is digested with restriction enzyme(s) to produce DNA fragments of appropriate size (1 to 4 kb). It must be noted that there should be no sites of cleavage within the core target sequence. The DNA fragments are then ligated to form monomeric circular molecules, followed by PCR using oligomer primers that are designed to extend outwardly from the known core target sequence, i.e. in the direction opposite to those used for normal PCR (Fig.18.3). The major product of the amplification reaction is a linear double-stranded DNA molecule that consists of a head-to-tail arrangement of sequences flanking the original target DNA. Inverse PCR is therefore useful as a method of chromosome crawling to explore the chromosomal sequences that are contiguous to a known segment of DNA. But the length of DNA that can be amplified by PCR is limited to relatively short steps along the chromosome.

## REVERSE TRANSCRIPTASE-MEDIATED PCR (RT-PCR)

Combining the process of cDNA synthesis by reverse transcriptase and PCR amplification into a single application leads to reverse transcriptase PCR (RT-PCR). Thus, amplification of double-stranded DNA by PCR

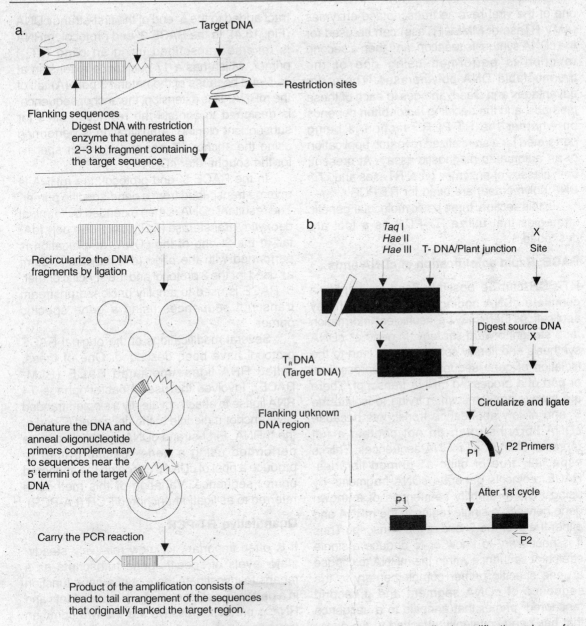

a.

Target DNA

Restriction sites

Flanking sequences

Digest DNA with restriction enzyme that generates a 2–3 kb fragment containing the target sequence.

Recircularize the DNA fragments by ligation

Denature the DNA and anneal oligonucleotide primers complementary to sequences near the 5' termini of the target DNA

Carry the PCR reaction

Product of the amplification consists of head to tail arrangement of the sequences that originally flanked the target region.

b.

*Taq* I
*Hae* II
*Hae* III   T- DNA/Plant junction

X Site

$T_R$ DNA (Target DNA)

Digest source DNA

Flanking unknown DNA region

Circularize and ligate

P2 Primers

P1

After 1st cycle

P1

P2

**Fig. 18.3:**  a—Major steps involved in inverse polymerase chain reaction. b—Generalized amplification scheme for unknown DNA flanking the $T_R$ DNA border of pTiT37 by iPCR.

is not limited to genomic DNA targets, but can also be applied to double-stranded cDNA that has been synthesized from RNA using a reverse transcription reaction. It is possible to detect, and even isolate, very rare mRNA transcripts from cell samples. RT-PCR acts as a means to obtain

material for sequence determination and a step in cloning a cDNA copy of the RNA.

As described earlier, there are thermostable DNA polymerases such as *rTth*, which can utilize RNA templates for cDNA synthesis and thus permit single enzyme RT-PCR. Alternatively, one of the viral reverse transcriptase enzymes (AMV RTase or MMLV RTase) can be used for the cDNA synthesis reaction, and then a second reaction is performed using one of the thermostable DNA polymerases. There are advantages and disadvantages to each of these methods, and the deciding factor often depends on whether the RT-PCR reaction is being performed for a specialized research application or an automated diagnostic assay. At present, both classes of enzymes (viral RTases and *rTth* DNA polymerase) are used for RT-PCR.

In this section, three basic molecular genetic strategies that utilize RT-PCR as a tool are described.

## RACE: Rapid amplification of cDNA ends

RT-PCR can be possibly used to amplify a complete cDNA coding region. But in many applications there may be insufficient information for a straightforward strategy. In general, cDNA synthesis and library screening can lead to the isolation of gene sequences that represent all or part of a processed mRNA transcript. There are always problems when trying to isolate the 5′ end of low abundance transcripts because cDNA libraries often do not contain a full representation of all mRNA sequences. This is especially true of oligo-dT-primed libraries. RACE protocols generate cDNA fragments by using PCR to amplify sequences of a known gene between a single region in the mRNA and either the 3′ or the 5′ end of the transcript. Thus it is necessary to know or to deduce a single stretch of sequence within the mRNA to choose a gene specific primer complementary to the sequence of cDNA segment and a second "anchored" primer that anneals to a sequence that has been covalently attached to the newly created cDNA terminus. This strategy was first described by Frohman *et al.* (1988).

In the two RACE protocols, extension of the cDNAs from the ends of the transcript to the specific primers is accomplished by using primers that hybridize either at the natural 3′ poly (A) tail of the mRNA, or at a synthetic poly (dA)

tract added to the 5′ end of the first-strand cDNA (Fig. 18.4). In the RACE 3′ end protocol, mRNA is reverse transcribed using an oligo (dT$_{17}$) primer, which has a 17-nucleotide extension at its 5′end because of the natural 3 poly (A) tail of the mRNA. This extension, the anchor sequence, is designed to contain the restriction sites for subsequent cloning. Amplification is performed using the anchor 17-mer and a primer specific for the sought-after cDNA.

In the RACE 5′ end protocol, the mRNA is reverse transcribed from a gene-specific primer. The resultant cDNA is then extended by terminal deoxytidyl transferase (TdT) to create a poly (dA) tail at the 3′ end of the cDNA. Amplification is performed with the oligo (dT$_{17}$)/anchor system as used for the 3 protocol and the specific primer. 5′ RACE is used to amplify uncloned upstream transcript sequences using a gene specific primer.

Several modifications of the original RACE protocol have been described. One of these, called RNA ligase-mediated RACE (RLM-RACE), involves the use of bacteriophage T4 RNA ligase to attach covalently a single-stranded RNA anchor molecule to the decapped 5′ end of the mRNA. First-strand cDNA synthesis can be performed using a gene-specific primer to produce a pool of cDNAs encoding the anchored primer sequence. Variation of this method is referred to as ligation-anchored PCR (LA-PCR).

## Quantitative RT-PCR

It is often important to know relatively steady-state levels of specific gene transcripts as a means to investigate the role of a gene function in a particular cell phenotype. Northern blots and RNase protection assays are the two common ways to measure steady-state mRNA levels. However, both of these assays are labor intensive and may require up to 25 µg of total RNA for each assay. Therefore, most gene expression studies cannot be performed using conventional RNA-based assays. It is possible to develop a quantitative RT-PCR assay that takes into account variation in the number of

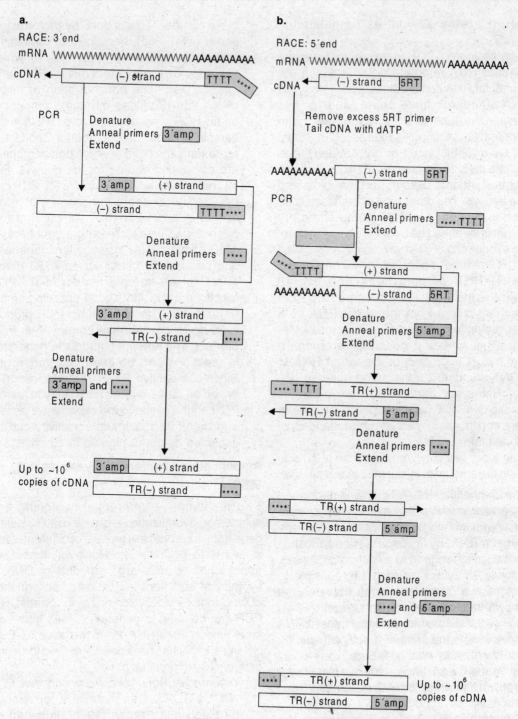

**a.**

RACE: 3′end

mRNA WWWWWWWWWWWWWWWW AAAAAAAAAA

cDNA ← (−) strand  TTTT ****

PCR

Denature
Anneal primers  3′amp
Extend

3′amp  (+) strand
(−) strand  TTTT****

Denature
Anneal primers  ****
Extend

3′amp  (+) strand
TR(−) strand  ****

Denature
Anneal primers
3′amp  and  ****
Extend

Up to ~10⁶
copies of cDNA  3′amp  (+) strand
TR(−) strand  ****

**b.**

RACE: 5′end

mRNA WWWWWWWWWWWWWWWW AAAAAAAAAA

cDNA ← (−) strand  5RT

Remove excess 5RT primer
Tail cDNA with dATP

AAAAAAAAA  (−) strand  5RT

PCR

Denature
Anneal primers  **** TTTT
Extend

**** TTTT  (+) strand
AAAAAAAAAA  (−) strand  5RT

Denature
Anneal primers  5′amp
Extend

**** TTTT  TR(+) strand
TR(−) strand  5′amp

Denature
Anneal primers  ****
Extend

****  TR(+) strand
TR(−) strand  5′amp

Denature
Anneal primers
****  and  5′amp
Extend

****  TR(+) strand
TR(−) strand  5′amp

Up to ~10⁶
copies of cDNA

• **Fig. 18.4:** Rapid amplification of cDNA ends (RACE). **a**—3′ protocol; **b**—5′ protocol. Reproduced from *Principles of Gene Manipulation*, R.W. Old and S.B. Primrose, by kind permission of Blackwell Science, Osney Mead, Oxford, U.K.

target molecules as well as amplification efficiency.

Quantitative RT-PCR measurements require a standard RNA template against which the experimental RNA can be measured. Two types of RNA standards have been used for this purpose. One type is an endogenous gene product that is ubiquitously expressed in all cell types, for example, actin or glyceraldehyde-3-phosphate dehydrogenase (GAPDH), which are the same standards used in northern blots and RNase assays. The other type of standard is an exogenously added cRNA (complementary RNA) that is synthesized by in vitro transcription. These exogenous cRNAs are sometimes called RT-PCR "mimics" because they contain the same RT-PCR priming sites and overall sequence as the target RNA, but produce a PCR product that differs from the target RNA by a unique restriction site or a shift in molecular weight. This makes it possible to compare directly the PCR product of the target RNA to the cRNA standard on the same agarose gel. Mimic cRNAs are transcribed from plasmids containing SP6 or T 7 bacteriophage promoters and are constructed by in vitro mutagenesis of the cloned target cDNA.

There are two ways to use mimic cRNAs to measure the amount of target mRNA in a sample.

(1) *Noncompetitive RT-PCR*: This is performed by adding mimic cRNA to the RNA sample at approximately the concentration of the target mRNA. By limiting PCR amplification to the linear range (mid-exponential phase), usually 25 cycles or less, it is possible to establish a direct correlation between the known amount of cRNA and target mRNA. However, the absolute number of target mRNA molecules in the sample greatly affects the number of cycles needed to detect the product within the linear range. This is a significant problem if this method is used for diagnostic purposes.

(2) *Competitive RT-PCR*: It circumvents the problem of variable target copy number by using product ratios that are independent of cycle number. This is done by performing a series of RT-PCR reactions that contain the same amount of sample RNA to which various known amounts of mimic cRNA have been added. The concentration of mimic cRNA that produces the same amount of mimic PCR product as the target RNA in the sample (a mimic: target product ratio of 1.0) represents the point at which both templates are competing equally for the primers. Fig. 18.5 illustrates how competitive RT-PCR can be used to quantify the amount of AMG mRNA in a cellular RNA sample that has been doped with increasing amounts of an AMG mimic cRNA. Insertional mutagenesis can be used to construct an AMG mimic cRNA for an internal marker in RT-PCR reactions using AMG gene-specific primers (P1 and P2). The AMG mimic PCR product is 30 bp longer than the normal AMG PCR product. The amount of products generated in reactions can be seen in agarose gel. Since competitive RT-PCR quantification is based on ratios, it is not necessary to restrict PCR amplification to the exponential phase, and therefore maximum sensitivity can be achieved by performing up to 35 cycles.

### Amplification of differentially expressed genes

Two techniques—differential screening and subtraction hybridization—can be used to isolate gene transcripts that are present at different levels in two RNA populations. However, these two approaches require large amounts of RNA to synthesize sufficient quantities of an enriched cDNA probe for library screening. In contrast, RT-PCR can be used to generate cDNAs from very small amounts of mRNA. Here, two basic RT-PCR cloning approaches have been described to identify differentially expressed genes.

(a) *Differential display reverse transcriptase PCR (DDRT-PCR)*: This technique, developed by Liang and Pardee (1992), is based on the principle of two-dimensional gel electrophoresis as a rapid screening

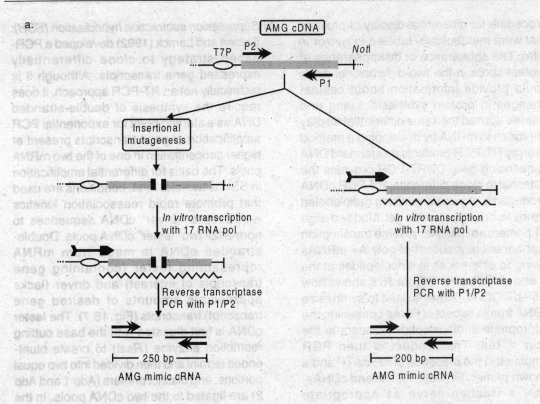

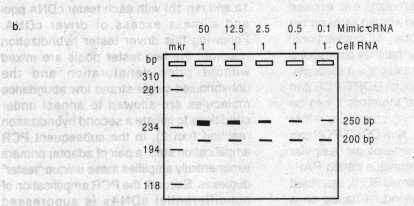

**Fig. 18.5:** Competitive RT-PCR uses a mimic cRNA to quantify the amount of target mRNA in an experimental sample based on a ratio of PCR products. a—Flow scheme showing how insertional mutagenesis can be used to construct an AMG mimic cRNA for use as an internal marker in RT-PCR reactions using AMG gene specific primers (P1 and P2). b—Schematic representation of an agarose gel showing the relative amounts of RT-PCR products generated in reactions using a constant amount of cellular RNA and decreasing amounts of known *in vitro* synthesized AMG mimic cRNA. Concept of the figure taken from Miesfeld: *Applied Molecular Genetics.* Copyright © 1999 John Wiley and Sons. This material is used by permission of John Wiley and Sons, Inc.

procedure for differential display of proteins that were metabolically labeled *in vivo* or *in vitro*. The appearance or disappearance of protein spots in the two-dimensional array could provide information about cellular changes in protein synthesis. Liang and Pardee applied the same differential display approach to mRNA by developing a method to array RT-PCR products on standard DNA sequencing gels. DDRT-PCR provides the potential to clone differential RT-PCR cDNA products by physically excising radiolabeled bands from an acrylamide gel. Modified oligo dT primers are used for reverse transcription that anneal to a subset of poly A+ mRNAs owing to differences in dinucleotides at the 3′ end of the primer. Figure 18.6 shows how a 5′–dT-GC–3′ primer is used to synthesize cDNA from a subset of RNAs containing the appropriate 3′ dinucleotide adjacent to the poly A tail. The product is then PCR amplified in the presence of $^{33}$P-dATP and a known primer. Of these first-strand cDNAs, only a fraction serve as appropriate templates for separate PCR reactions containing a known arbitrary primer. Candidate RT-PCR products are excised from the gel, reamplified with the same primer pair, and used as probes on conventional Northern blots to verify that they correspond to differentially expressed gene transcripts. The false positive rate with DDRT-PCR can be highly variable and therefore it can be best used as a screening procedure rather than a cloning strategy. Now DDRT-PCR can be also performed as non-isotopic. A similar RT-PCR screening technique called RAP-PCR (RNA arbitrarily primed PCR), described by Welsh and McClellend, is based on a genomic DNA fingerprinting strategy. The primary difference between these two methods is that RAP-PCR uses the same random primer for both reverse transcription and PCR amplification steps, which eliminates bias toward the noncoding 3′ poly A+ sequences.

*(b) Suppression subtraction hybridization (SSH)*: Siebert and Larrick (1992) developed a PCR-based strategy to clone differentially expressed gene transcripts. Although it is technically not an RT-PCR approach, it does require the synthesis of double-stranded DNA as a starting point for exponential PCR amplification of gene transcripts present at higher concentration in one of the two mRNA pools. The basis for differential amplification in SSH is twofold. First, conditions are used that promote rapid reassociation kinetics with excess "driver" cDNA sequences to normalize two "tester" cDNA pools. Double-stranded cDNA is made from mRNA representing tester (containing gene transcripts of interest) and driver (lacks appreciable amounts of desired gene transcript) transcripts (Fig. 18.7). The tester cDNA is first digested with the base cutting restriction enzyme (*Rsa*I) to create blunt-ended termini and then divided into two equal portions, and distinct primers (Adp 1 and Adp 2) are ligated to the two cDNA pools. In the next step, separate rapid reassociation hybridization reactions are performed (rxn 1a and rxn 1b) with each tester cDNA pool and a mass excess of driver cDNA. Following this driver tester hybridization reaction, the two tester pools are mixed without prior denaturation and the unhybridized single-strand low abundance molecules are allowed to anneal under conditions to initiate a second hybridization reaction (rxn 2). In the subsequent PCR amplification step, a pair of adapter primers exponentially amplifies these unique "tester" duplexes. Second, the PCR amplification of nondifferential cDNAs is suppressed because undesirable sequences contain inappropriate priming sites or are capable of forming intrastrand "panhandle" structures that are poor substrates for PCR. The combined PCR effect of exponential amplification of selective tester duplex cDNAs, with the resulting suppression of

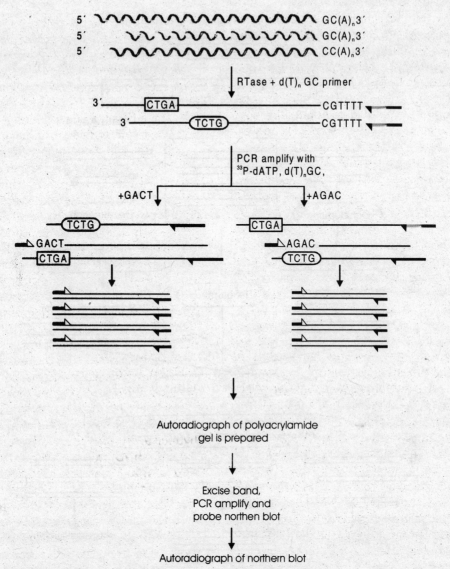

**Fig. 18.6:** A schematic diagram of DDRT-PCR showing how a 5'—dT-GC—3' primer is used to synthesize cDNA from a subset of RNAs containing the appropriate 3' dinucleotide adjacent to the poly A tail. Of these first strand cDNAs, only a fraction of will serve as appropriate templates for separate PCR reactions containing a known arbitrary primer. Concept of the figure taken from Miesfeld: *Applied Molecular Genetics.* Copyright © 1999 John Wiley and Sons. This material is used by permission of John Wiley and Sons, Inc.

common sequence cDNAs, can potentially result in a $10^2$- to $10^3$-fold enrichment of differentially expressed tester cDNA sequences. The amplified cDNAs are used to make an enriched plasmid library, which can be randomly sequenced or screened with a differential cDNA probe.

### Cloning of PCR products

There are numerous situations in which a specific DNA segment must be cloned into a plasmid vector. Although PCR generates sufficient material for analysis, when extensive manipulations of the sequence is needed or when

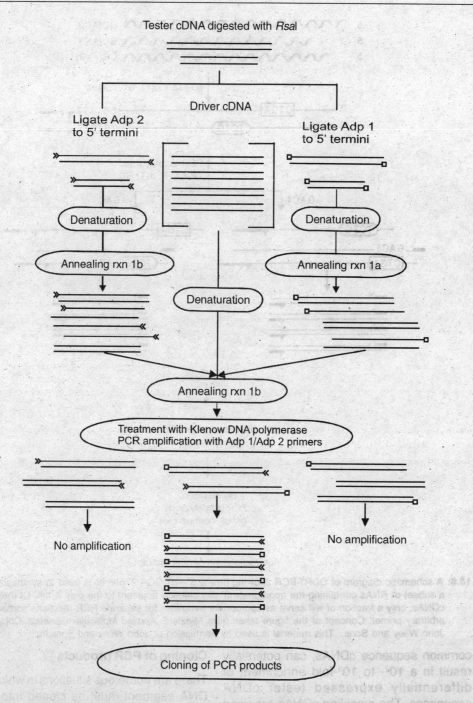

**Fig. 18.7:** A flow diagram of suppression subtraction hybridization technique. It combines cDNA normalization with selective PCR amplification to achieve enrichment of differentially expressed gene transcripts. Concept of the figure taken from Miesfeld: *Applied Molecular Genetics.* Copyright © 1999 John Wiley and Sons. This material is used by permission of John Wiley and Sons, Inc.

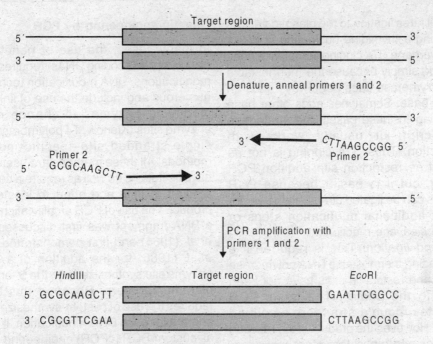

**Fig. 18.8:** Incorporation of extra sequences at the 5' end of a primer. Primer 1 has an extra sequence near its 5'-end which forms a *Hin*dIII site (AAGCTT) and Primer 2 has an extra sequence near its 5'-end which forms an *Eco*RI (GAATTC) site. Each primer has an additional 5'-terminal sequence of four nucleotides so that hexanucleotide restriction sites are placed within the extreme ends of amplified DNA. Reproduced from *Principles of Gene Manipulation,* R.W. Old and S.B. Primrose, by kind permission of Blackwell Science, Osney Mead, Oxford, U.K.

large amounts of DNA are required, it is necessary to clone the PCR product. But amplified DNA fragments do not clone efficiently because there are no convenient restriction enzyme sites to facilitate the design of subcloning strategy. Three basic strategies are commonly used to subclone PCR products.

1. *Restriction site addition*: This is done by including the nucleotide sequence of an appropriate restriction enzyme into the 5' ends of the primer. A PCR primer may be designed which, in addition to the sequence required for hybridization with the input DNA, includes an extra sequence at its 5' end. Figure 18.8 illustrates the addition of a *Hin*dIII and an *Eco*RI sites to the ends of an amplified DNA fragment. Four nucleotides are placed between the hexanucleotide restriction sites and the extreme ends of the DNA for ensuring that the restriction sites are good substrates for the restriction

endonucleases. The extra sequence does not participate in the first hybridization step; only the 3' portion of the primer hybridizes, but it subsequently becomes incorporated into the amplified DNA fragment. The restriction enzyme recognition site does not interfere with template annealing during the first PCR cycle. However, all subsequent cycles contain PCR products that have the restriction enzyme site incorporated into the DNA termini. By digesting the PCR products with the corresponding enzymes, the amplified DNA fragment can be cloned by a standard ligation reaction.

2. *T/A cloning*: This relies on the fact that a proportion of amplified DNA fragments are not truly blunt ended because thermostable DNA polymerases (*Taq* and *Tth*) lack 3' to 5' exonuclease activity and have a tendency to add an extra A residue at the 3' end of each strand. This one base overhang at each

end facilitates ligation to the plasmid vectors containing a thymidine nucleotide (TMP or TTP) overhang. T/A cloning sometimes does not work, simply because the thermostable DNA polymerase does not always adds an extra A base. Sometimes adds other base or nothing In these cases, the stratagene PCR-Script kit usually works with satisfaction. Thus, T/A cloning is not as efficient as restriction site addition PCR cloning, but it is easier because PCR products can be ligated directly to the vector without additional purification steps or enzyme cleavage reactions.

3. *Blunt-end ligation*: This is required if a proofreading thermostable DNA polymerase is used and restriction sites have not been added to the primer. The basic PCR generates double-stranded amplified product. However, the product of the reaction does not contain a large proportion of perfectly blunt-ended molecules. This can be carried out in a combined exonuclease/repair reaction with Klenow polymerase in the presence of all four dNTPs. *Taq* DNA polymerase in general remains tightly bound to the 3′ ends of the DNA, so proteinase K treatment is recommended. Excess primer is removed by preparative gel electrophoresis as a means of purifying the amplified DNA. The DNA fragment can then be blunt ligated into a suitable vector (e.g. pBluescript) linearized with *Sma*I. The chemically synthesized primers, which form the 5′ end of each strand of the DNA do not bear 5′-terminal phosphates. Therefore, if the linearized vector has been treated with phosphatase to prevent simple reclosure, it is essential that 5′ phosphate groups be added to the PCR product by using a polynucleotide kinase. Blunt-ligated recombinants can be identified by loss of β-galactosidase activity using a blue-white screening protocol.

## Genetic engineering by PCR

In many cases, the use of genetic regions (regulatory and coding DNAs) requires nucleotide modifications. DNA modification techniques are numerous and include the use of linkers, DNA-modifying enzymes to change restriction enzyme sites (Klenow, T4 polymerase, etc.) and single-stranded site-specific modification methods. All these methods are useful, but they have limitations as well as require a considerable amount of time and effort to locate modified product. The use of PCR amplifications to modify a DNA fragment was first discussed by Mullis *et al.* (1986) and first demonstrated by Scharf *et al.* (1986) for the addition of a restriction enzyme site synthesized into the 5′-ends of PCR oligomer primers. Because this technique requires the use of custom-synthesized oligomer primers for each DNA modification, it is referred as custom-PCR (cPCR) engineering (Slightom, 1991). The major strength of cPCR engineering is that most of the complex DNA modifications can be incorporated into the sequence of the synthesized oligomer primers, allowing the engineering process to proceed as a series of straightforward ligation events.

## Applications

The usefulness of PCR was quickly recognized by scientists in many different disciplines and it, either directly or after some modifications, has been applied to resolve many problems that were unapproachable by other techniques.

1. It has extensive applications in the diagnosis of genetic disorders such as phenylketonuria, hemophilia, sickle cell anemia, thalassemia, etc. and the detection of nucleic acid sequences of pathogenic organisms in clinical samples. Sometimes, the RFLP pattern of PCR products in healthy and defective foeti can be studied or the PCR product may also be sequenced to reveal the differences. For detection of pathogens, immunochemistry and histocytochemistry

have long been used by pathologists to analyze cell phenotypes in histological specimens. However, a major limitation of these procedures is the inability to detect infectious agents such as viral pathogens and other microorganisms that persist at a very low level in infected cells. When it became clear that PCR could offer increased specificity and sensitivity, researchers began to develop PCR strategies to detect as few as one viral genome in tissue specimens. Two basic PCR approaches have been followed for pathogen detection. The first approach is competitive quantitative PCR, with an added emphasis on low-level detection and strain identification. The second approach is to develop in situ PCR (ISPCR) procedures that allow the same level of sensitivity and specificity as solution PCR assays, with the advantage of cell-specific resolution. Similar to solution PCR, the ISPCR amplification parameters must be optimized with regard to temperature, time, and reaction constituents. ISPCR products are detected by using labeled dNTPs (radioactive or biotinylated) in the reaction or by subsequently processing the ISPCR specimen using standard in situ hybridization conditions to detect the amplified hybridization target.

2. The genetic identification of forensic samples, even DNA extracted from a hair or a single sperm are sufficient for PCR amplification. Some well-characterized sequences of microsatellites are being used for designing primers, so that DNA finger printing may be achieved through PCR.

3. For the analysis of homologous genes in evolutionary biology.

4. PCR amplification has several plant molecular biology related uses. These include

i. Generation of specific sequences of cloned double-stranded DNA for use as probes.

ii. Generation of probes for genes or cDNAs that have not yet been cloned.

iii. Generation of libraries of cDNA from small amounts of mRNA. The main advantage of this method is that it allows large cDNA libraries to be established from as few as one or two cells.

iv. Generation of large amounts of template DNA for sequencing to be carried out either by the Maxam-Gilbert chemical degradation method or Sanger dideoxy-mediated chain termination method.

v. Identification of genetic mutations: Deletions and insertions at defined loci can be detected by a change in the size of the amplified product. Alternatively, deletions can be recognized by their failure to amplify when one of the priming oligonucleotides maps within the deleted sequence. It is also useful for the identification of point mutations. The PCR technique of *single-strand conformational polymorphism (SSCP)* is the widely used method for detecting single base pair changes in genomic DNA. It is based on the principle that two single-strand DNA molecules of identical length but unlike sequences migrate differentially in a nondenaturing acrylamide gel. This is due to sequence-dependent intramolecular folding reactions that contribute to the overall structure of the molecule. Single-strand conformational polymorphism is done by amplifying a specific DNA segment using a radiolabeled (or fluorescently labeled) primer, or by including trace amounts of $^{32}P$-$\alpha$-dNTP in the PCR reaction. The PCR products are then heat denatured and cooled on ice to promote rapid formation of intrastrand duplexes. Figure 18.9 shows how SSCP can be used to identify mutant alleles in known target DNA sequences. Two other PCR-based approaches for detecting single nucleotide differences are *base excision sequence scanning (BESS)*, which uses an enzymatic sequencing method to detect thymidine residues in PCR products, and *artificial mismatch hybridization (AMH)*, a method that exploits heteroduplex

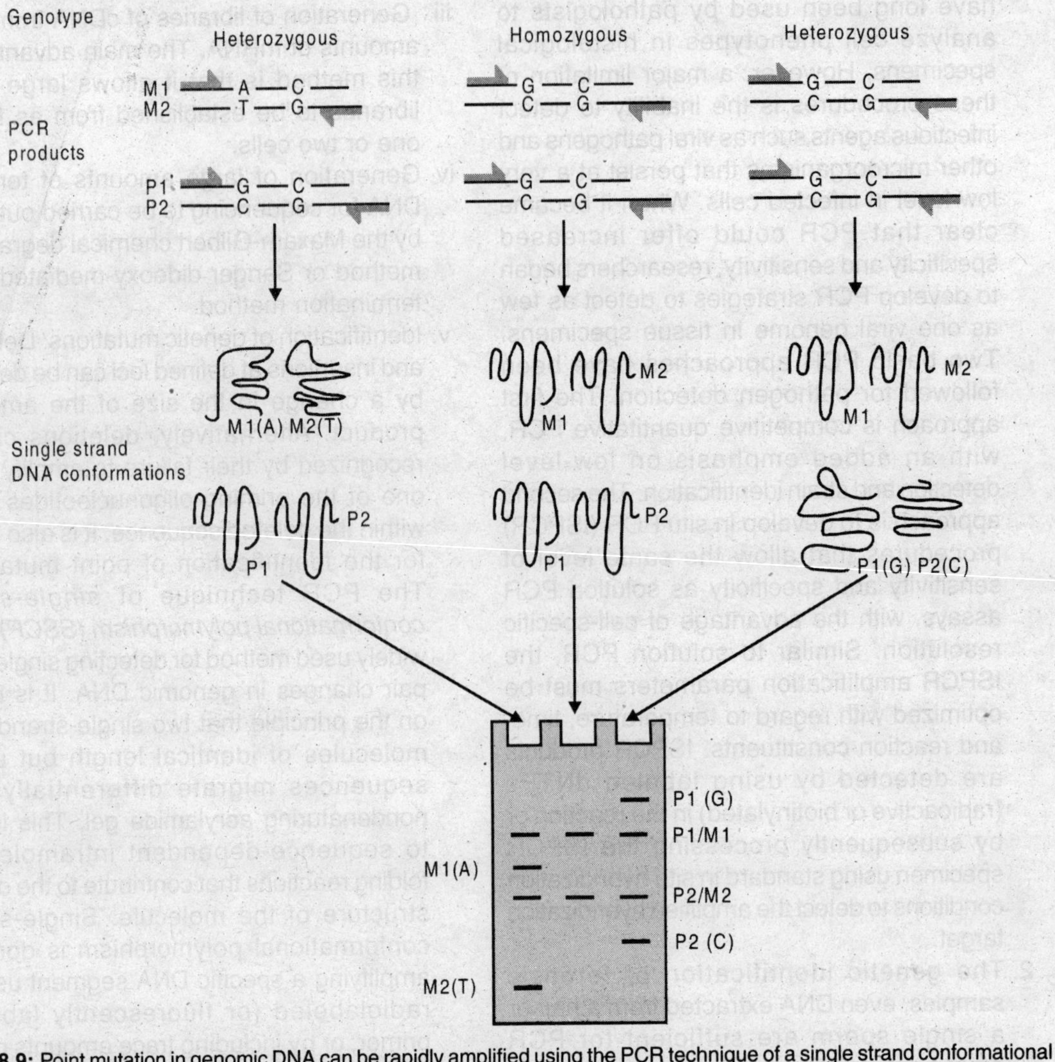

**Fig. 18.9:** Point mutation in genomic DNA can be rapidly amplified using the PCR technique of a single strand conformational polymorphism (SSCP). In this example, AMG primer pairs were used to detect nucleotide alterations in a specific region of the AMG gene known to cosegregate as a dominant mutation. With DNA from three individuals (two disease heterozygotes and a homozygote normal), PCR amplification was performed with AMG specific primers that flank a region where multiple dominant alleles have previously been identified. Because the primary nucleotide sequences of each of the DNA strands in a heterozygote individual are different, four separate bands are resolved in the corresponding lanes. Maternal (M1, M2) and paternal (P1, P2) strands are shown. Concept of the figure taken from Miesfeld: *Applied Molecular Genetics*. Copyright © 1999 John Wiley and Sons. This material is used by permission of John Wiley and Sons, Inc.

stability assays to detect single base pair mismatches.

vi. PCR detection of a gene transferred into a plant genome: The ability to transfer genes into a plant genome is well established. The determination of those plant tissues that are transformed from those that are not can be accomplished using selectable marker or reporter genes in most cases. However, there are some situations in which the use of selectable or reporter gene is not effective or where a reporter gene is not informative, either because of its lack of expression or due to its incomplete transfer. In the PCR method, genomic DNA is isolated and the PCR target gene; that is, the gene that was used for transformation experiments is selected for amplification. With designing of appropriate primers the gene is amplified and seen on a gel.

vii. PCR technique is employed in various molecular markers like RAPD; microsatellites or simple sequence repeats, amplified fragment length polymorphism (AFLP), etc. Variations observed in the amplification products are used as markers (examples and details have been dealt in Chapter 22).

viii. Identification of unknown plant DNA regions that flank an *Agrobacterium* T-DNA boundary. Although PCR is frequently used to amplify DNA segments lying between the two priming oligonucleotides, it can also be used to amplify unknown DNA sequences that lie outside the boundary of known sequences. This is referred to as inverse PCR (iPCR). Inverse PCR is also useful if the border sequences of a DNA segment are not known and those of a vector are known, then the sequence to be amplified may be cloned in the vector and border sequences of a vector may be used as primers in such a way that the polymerization proceeds in inverse direction, that is away from the vector sequence flanked by the primers and towards the DNA sequence of inserted segment. Similarly, if the gene sequence is known, one can use its border sequences as primers, for an iPCR to amplify the sequences flanking this gene.

## Advantages

1. It can produce large amounts of identical DNA molecules from a minute amount of starting material. In fact, the DNA isolated from a single cell (even plant protoplast) is sufficient for gene detection.
2. The technique is quick and simple.

3. The technique is extremely sensitive.
4. The DNA need not be pure for amplification provided the sample does not contain contaminants that inhibit *Taq* polymerase.

## Problems

1. Nucleotide sequence of at least the boundary regions must be known so that oligonucleotide primers can bind and synthesize the DNA. This limits PCR to the study of genes that have already been characterized in part by cloning methods.
2. PCR is a very sensitive technique and is prone to generating false signals, which in many cases are the result of contaminants from previous PCR amplifications being present in the amplification reaction components. False positives can also result from minute contamination by any plasmid or lambda clone that may contain the target gene sequence.

## PROTOCOL

### 18.1: PCR for detection of transgenes

The protocol given below can be taken as a standard "template" but depending on the length (in base pairs) and percentage of G+C, etc. of the primers used to detect the particular transgene, it may be necessary to change the amplifying temperature.

1. Isolate the genomic DNA
2. Add the reagents in 0.2 ml microcentrifuge tube placed on ice
3. Primers: In many experiments nptII gene is used as a selectable marker along with gene of interest for transformation. Therefore by using the nptII primers, the nptII gene is amplified to confirm its integration and also of gene of interest into the nuclear genome.

### Primer sequences

Forward (Upper) primer (23 mer)
5′ ACGTTGTCACTGAAGCGGGAAGG–3′
Reverse (Lower) primer (24 mer)
5′ GTAAAGCACGAGGAAGCGGTCAGC–3′

| Stock | Final | 25 µl |
|---|---|---|
| *Taq* DNA polymerase buffer (10x) 2.5 µl | 1x | |
| dNTPs (10 mM each) | 0.2 mM each | 0.5 µl each |
| Primer 1 forward (10 µM) | 0.2 µM | 0.5 µl |
| Primer 2 reverse (10 µM) | 0.2 µM | 0.5 µl |
| DNA (100 ng/µl) | 100 ng | 1 µl |
| *Taq* DNA polymerase (5 units/µl) | 1U | 0.2 µl |

Sterile water to make up the final volume to 25 µl.

4. Prepare cocktail of components. Always keep one tube with only cocktail and water (without DNA) as control.
5. Mix well and spin briefly.
6. Overlay the reaction with a 25 µl of mineral oil (in those PCR models where hot lid is not provided) to prevent evaporation of reagents at high temperatures.
7. Place the tubes in thermal cycler block and set the machine for following cycling conditions.

| | |
|---|---|
| Step 1 | 94°C 2 min |
| Step 2 | (25–30 cycles) |
| | 94°C 1 min |
| | 55°C 1 min |
| | 72°C 1 min |
| Step 3 | 72°C 5 min |
| Step 4 | 4°C end |

After amplification, add 2.5 µl of tracking dye in the sample and store at 4°C.

8. Prepare agarose gel of 1.2%.
9. Load the samples carefully in the wells.
10. Add suitable molecular weight (1kb ladder) in one of the lanes.
11. Carry out electrophoresis, stain the gel with ethidium bromide and photograph it.

## REFERENCES

Frohman, M.A., Dush, M.K. and Martin, G.R. 1988. Rapid production of full length cDNAs from rare transcripts: Amplification using a single gene specific oligonucleotide primer. *Proc. Natl. Acad. Sci. USA* **85:** 8998–9002.

Liang, P. and Pardee, A.B. 1992. Differential display of eukaryotic messenger RNA by means of the polymerase chain reaction. *Science* **257:** 967–971.

Mullis, K.B, Faloona, F., Scharf, S.J., Saiki, R.K. Horn, G.T. and Erlich, H.A. 1986. Specific enzymatic amplification of DNA *in vitro*: the polymerase chain reaction. *Cold Spring Harbor Symp. Quant. Biol.* **51:** 263–273.

Ochmann, H., Gerber, A.S. and Hartl, D.L. 1988. Genetic application of an inverse polymerase chain reaction. *Genetics* **120:** 621–623.

Scharf, S.J., Horn, G.T. and Erlich, H.A. 1986. Direct cloning and sequence analysis of enzymatically amplified genomic sequences. *Science* **233:** 1076–1078.

Siebert, P. and Larrick, J. 1992. Competitive PCR. *Nature* **359:** 557–558.

Slightom, J.L. 1991. Custom polymerase chain reaction engineering of a plant expression vector. *Gene* **100:** 251–255.

Triglia, T., Peterson, M.G. and Kemp, K.J. 1988. A procedure for *in vitro* amplification of DNA segments that lie outside the boundaries of known sequences. *Nucl. Acids Res.* **16:** 8186.

## FURTHER READING

Maniatis, T., Fritsch, E.F. and Sambrock, J. 1982. *Molecular Cloning: A Laboratory Manual.* Cold Spring Harbor Laboratory Press, USA.

Miesfeld, R.L.1999. *Applied Molecular Genetics.* Wiley-Liss, New York.

Old, R.W. and Primrose, S.B. 2001. *Principles of Gene Manipulation.* Blackwell Science, UK.

Saiki, R.K., Gyllenstein, U.B. and Erlich, H.A. 1995. The polymerase chain reaction. In: Davis, K.E. (ed.), *Genome Analysis–A Practical Approach* IRL Press, Oxford.

Slightom, J.L., Drong, R.F. and Chee, P.P. 1995. Polymerase chain reaction: gene detection, inverse PCR, and genetic engineering. *Plant Molecular Biology Manual,* **F4:**1–24.

# 19

# *In vitro* Mutagenesis

Present-day knowledge of biological macromolecules originates from the study of spontaneous or induced mutations. Mutants are essential prerequisites for any genetic study also. In the past, isolation and characterization of DNA, RNA, or protein variants had to await the identification of phenotypically deviant host organisms. Because of this and moreover because most mutations are phenotypically silent, there was only a very limited subset of genes or gene products amenable for the study of structure-function relationships. Classically, mutants are generated by treating the test organism with chemical or physical agents that modify DNA. This method though extremely successful suffers from a number of disadvantages. (1) Any gene in the organism can be mutated, and the frequency with which mutants occur in the gene of interest can be very low; (2) Mutants with the desired phenotype when isolated do not guarantee that the mutation has occurred in the gene of interest; (3) Prior to the development of gene cloning and sequencing techniques, there was no way of knowing where in the gene the mutation had occurred and whether it arose by a single base change, an insertion of DNA, or a deletion.

The advent of recombinant DNA methodology offered a change from a passive to a more active mode of study. The basis was laid by the discovery of restriction enzymes, ligases, polymerases, and the enzymes which modify DNA molecules (nucleases, methylases), and by the development of techniques of gene cloning. This enabled one to isolate any DNA in pure form and in large quantities, introduce changes in the DNA, selecting them in the absence of phenotypic expression and studying the effects of these deliberate alterations in the DNA either *in vitro* or after reintroduction *in vivo*.

Sometimes the objective of isolating gene sequences from DNA libraries is simply to obtain inserts that can be analyzed by DNA sequencing, for example, when performing comparative studies of the same gene between different species. However, in many cases, the isolation of novel gene sequences requires the use of functional assays to characterize the cloned DNA. In conventional genetics, a biological comparison between wild type and mutant phenotypes often reveals the normal gene function. This same approach can be used in molecular biology, except that the researcher is able to choose precisely what type of mutations will likely to be most informative.

## Site-directed mutagenesis

With the development of techniques in molecular biology, isolation and study of a single gene is now possible and mutagenesis has been refined. Instead of mutagenizing many cells of an organism and then analyzing many thousands of offsprings to isolate a desired mutant, it is now possible to specifically change any given base in a cloned DNA sequence. This technique is known as site-directed mutagenesis. The strategies recently developed for *in vitro* site-directed mutagenesis allow one to create virtually any change at will and this—in principle— enables us to correlate any piece of gene sequence information with gene regulation and product formation. This methodology invariably yields gene mutations, but it is only applicable to fully characterized cloned DNA. The importance of site-directed mutagenesis goes beyond gene structure-function relationships because the technique enables mutant proteins to be generated with very specific changes in particular amino acids. Targeted mutagenesis is therefore virtually impossible without prior sequence determination of the DNA.

The type of mutagenesis protocol chosen largely depends on the desired alterations in target DNA. It may involve the creation of single base transitions and/or transversions, deletions, and insertions or precise replacement of whole sequence blocks of variable size. This chapter describes some basics *in vitro* mutagenesis strategies, each of which is used for creating specific types of mutations in cloned DNA sequences.

### Deletion mutagenesis

In deletion mutagenesis, nucleotides are removed from the termini or internally. Information from restriction maps can be used to design specific deletion that allows convenient recloning of the modified DNA segment. Alternatively, exonucleases can be used to degrade DNA nonspecifically from the termini of a linearized plasmid. The extent of exonuclease treatment, usually as a factor of reaction time, determines the amount of deleted DNA. Fig. 19.1 illustrates how a plasmid insert can be mutated by restriction enzyme mediated deletion mutagenesis using compatible restriction enzyme mediated termini, or by

digesting linearized plasmid DNA with exonucleases followed by treatment with Klenow DNA polymerase. Based on the restriction map of a plasmid DNA insert *Bgl*II and *Bam*HI digestion results in compatible termini (GATC). Exonuclease III (*Exo*III) degrades one strand of double-stranded DNA from the 3′ end, but only if the DNA termini contain either a 5′ overhang or a blunt end, but not an extended single strand 3′ overhang. This property can be exploited to generate unidirectional deletions by digesting target DNA with two restriction enzymes, one that leaves a 5′ overhang (*Eco*RI or *Bam*HI) and the other a 3′ overhang (*Kpn*I or *Apa*I). *Exo*III deleted DNA must be treated with the single strand nuclease S1 to remove the undigested 5′ strand. *Bal*31 exonuclease degrades double-stranded DNA to yield bidirectional deletions that can be blunt ligated without prior treatment with S1 nuclease.

*a. Unidirectional deletions.* Yanisch-Perron *et al.* (1985) and Barcak and Wolf (1986) described a method for making a series of deletions, as shown in Fig.19.2. This method is based on the principle that α-thiophosphate-containing phosphodiester bonds are resistant to hydrolysis by the 3′→5′ exonucleolytic activity of T4 phage DNA polymerase. Linear duplex DNA molecules blocked at one 3′ terminus with a thiophosphate are then degraded from the other end with the exonuclease. Digestion for different lengths of time followed by treatment with nuclease S1 and ligation allows the preparation and recovery of a set of deletion mutants.

*b. Gap-sealing mutagenesis.* This method involves the use of pancreatic deoxyribonuclease or restriction endonucleases in the presence of ethidium bromide to introduce one single-strand scission per closed circular molecule. The nick is then extended into a small gap by the exonucleolytic activity of *Micrococcus luteus* DNA polymerase I or a more extended gap in the presence of exonuclease III. Gapped DNA can be linearized by nuclease S1, and the

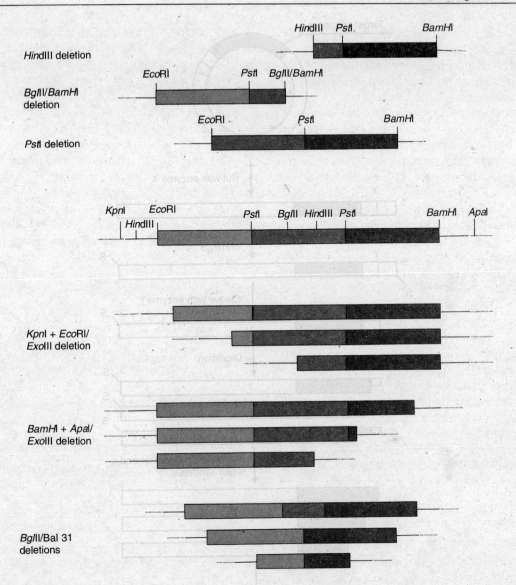

**Fig. 19.1:** Deletion mutagenesis, using restriction enzymes and exonucleases. Concept of the figure taken from Miesfeld: *Applied Molecular Genetics. In vitro* Mutagenesis This material is used by permission of John Wiley and Sons, Inc.

shortened molecules can be cyclized by T4 ligase directly or after addition of a synthetic linker molecule (Fig.19.3).

    ***c. Linker scanning mutagenesis.*** McKnight and Kingsbury (1982) reported an elegant method combining deletion and substitution mutagenesis in their study of promoter elements of the thymidine kinase gene

of HSV. Two opposing sets of 5′ and 3′ deletion mutants were obtained by sequential exonuclease III and nuclease S1 action and were flanked by synthetic restriction enzyme *Bam*HIlinker sequences. Opposing 5′ and 3′ deletion mutants were sorted out and recombined via the synthetic restriction site. This was done in such a way that the residues in the

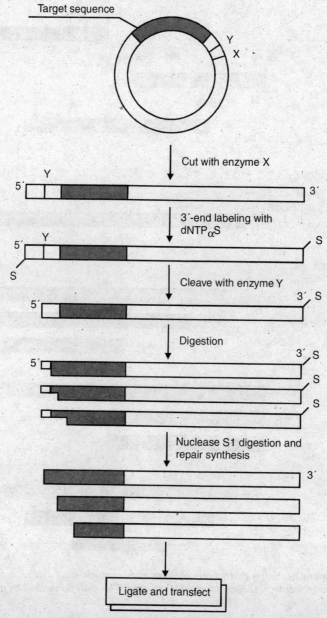

**Fig. 19.2:** A procedure for making unidirectional deletions in a DA molecule. X and Y represent recognition sequences for two different restriction endonucleases. Reproduced from *Principles of Gene Manipulation,* R.W. Old and S.B. Primrose, by kind permission of Blackwell Science, Osney Mead, Oxford, U.K.

linker exactly replaced a stretch of nucleotides in the wild-type sequences (Fig.19.4). For the thymidine kinase gene, an almost ideal set of mutants was obtained, which exhibited no net increase or decrease in length, but now the linker scanned across nearly every individual residue within the region −120 to +20 nucleotides of the transcriptional start. A similar series of linker scanning mutants of the 5′ flankage region of mouse β-globin gene was reported by Charnay

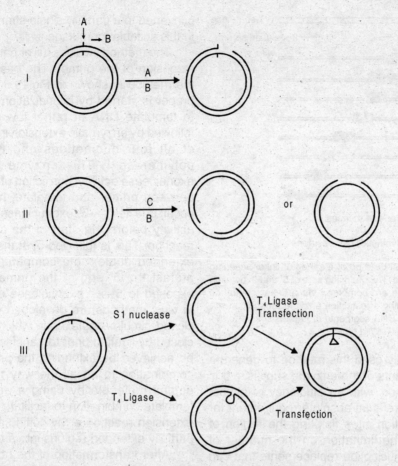

**Fig. 19.3:** Construction of gapped single-stranded regions and internal deletions in duplex circular DNA. A—Restriction endonuclease in presence of ethidium bromide; B—*Micrococcus luteus* DNA pol I or exonuclease III; C—DNase I in presence of ethidium bromide. Scheme I–Use of restriction enzyme in presence of ethidium bromide to produce a nick at a defined position. Scheme II–Use of DNase I and ethidium bromide to create nick. In both the schemes I and II, nicks are converted to limited gaps by exonucleolytic action of *Micrococcus luteus* DNA pol I or extended gaps by exonuclease III. Scheme III–Two pathways for construction of internal deletions either by sequential action of S1 nuclease and T$_4$ ligase or by direct gap misligation and *in vitro* repair.

*et al.* (1985). The success of this method goes with the initial choice for matching partner molecules from the opposing libraries of unidirectional mutants (5′ and 3′). If the choice is limited, one gets constructs that either have additional residues or small deletions flanking the linker sequence.

Deletion mutagenesis is sometimes used to create novel proteins lacking one or more functional domains, for example, deleting the hormone-binding domain of a steroid receptor to produce a constitutively active transcription factor. Internal deletions of protein coding sequences are more complicated because two-thirds of all deletions result in frame-shift mutations.

*Cassette mutagenesis*

In cassette mutagenesis, a synthetic DNA fragment containing the desired mutant sequence is used to replace the corresponding sequence in the wild type gene. Wells *et al.*

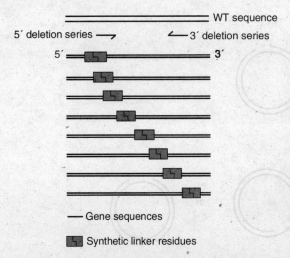

**Fig. 19.4:** A typical build up of a series of linker-scanning mutants. The opposing sets of 5′ and 3′ deletion mutants are combined via a linker sequence containing a restriction enzyme recognition site to give DNA segments of lengths similar to WT DNA. WT–Wild type.

(1985) originally used this method to generate improved variants of the enzyme subtilisin. It is a simple method, with an efficiency of almost 100%. But there is an absolute requirement for unique restriction sites flanking the region of interest and the limitation on the number of different oligonucleotide replacements that can be synthesized.

### Oligonucleotide-directed mutagenesis

This method is based on the use of a relatively short oligonucleotide (7–20 nucleotides long) containing a base mismatch for defined mutation. The short oligonucleotide is used as a primer, which is strand specific to synthesize one complete mutant DNA strand in an *in vitro* DNA synthesis reaction.

To use this strategy, the nucleotide sequence of the region to be mutated must be known to permit the synthesis of a complementary mutagenic oligonucleotide. An oligonucleotide is designed such that it is complementary to a single-stranded template DNA and cloning the gene in an M13-based vector. However, DNA cloned in a plasmid and obtained in duplex form can also be converted to a partially single-stranded molecule that is suitable for mutagenesis.

The method involves full or partial enzymatic extension of the primer. The basic principle of the method is shown in Fig.19.5. The reaction proper is started by hybridization of the primer to template DNA at rather low temperatures followed by enzymatic extension in the presence of all four nucleotides and Klenow DNA polymerase I. This enzyme lacks 5′→3′ exonuclease activity; correction of the mismatch between primer and template, however, may occur due to 3′→5′ exonuclease proofreading activity before the start of the actual priming reaction. This is followed by transfection of the extended duplex into competent *E. coli*. To protect the 5′ end of the primer from being exposed to 5′→3′ exonuclease editing activity *in vivo*, methods are developed to create fully closed circular duplexes in which all nicks are closed by ligation prior to transfection. This can be achieved by extending the primer/template combination to an all the way round circular duplex molecule by using a double primer/template combination to facilitate the full length extension reaction or by cutting and ligation of partially extended regions into a closed duplex.

After transformation of the host *E. coli*, the heteroduplex gives rise to homoduplexes whose sequences are either that of original wild-type DNA or that containing the mutated base. Generally the frequency of mutated clones is rather low. To pick out mutants, the clones can be screened by nucleic acid hybridization with a [32]P labeled oligonucleotide as a probe. The mutant can then be sequenced to check that the procedure has not introduced any adventitious changes. This was necessary with early versions of the technique where *E. coli* DNA polymerase was used. The more recent use of high-fidelity DNA polymerases from T 4 and T 7 phages has minimized the problem of extraneous mutations as well as shortened the time for copying the second strand. Also, these polymerases do not strand displace the oligomer, a process which would eliminate the original mutant oligonucleotide.

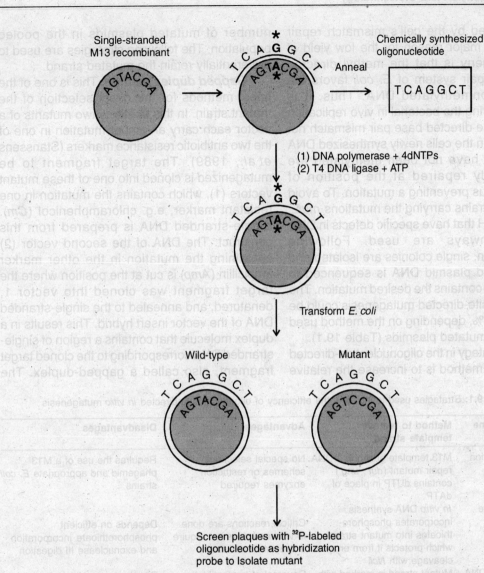

**Fig. 19.5:** Oligonucleotide-directed mutagenesis. Asterisks indicate mismatched bases. Reproduced from *Principles of Gene Manipulation*, R.W. Old and S.B. Primrose, by kind permission of Blackwell Science, Osney Mead, Oxford, U.K.

The efficiency with which the single primer method yields mutants depends on many factors. The double-stranded heteroduplex molecules that are generated will be contaminated both by any single-stranded nonmutant template DNA that has remained uncopied and by partially double-stranded molecules. The presence of these species considerably reduces the proportion of mutant progeny. They can be removed by sucrose gradient centrifugation or by agarose gel electrophoresis, but this is time consuming and inconvenient.

Following transformation and *in vivo* DNA synthesis, segregation of the two strands of the heteroduplex molecule can occur, yielding a mixed population of mutant and nonmutant progeny. Mutant progeny have to be purified away from parental molecules, and this process is complicated by the cell's mismatch repair system. The major reason for the low yield of mutant progeny is that the methyl directed mismatch repair system of *E. coli* favors the repair of nonmethylated DNA. Thus, it is important during the bacterial in vivo replication phase that the directed base pair mismatch not be repaired. In the cells newly synthesized DNA strands that have not been methylated are preferentially repaired at the position of mismatch, thus preventing a mutation. To avoid this, *E. coli* strains carrying the mutations *mut*L, *mut*S, or *mut*H that have specific defects in DNA repair pathways are used. Following transformation, single colonies are isolated and the recovered plasmid DNA is sequenced to determine if it contains the desired mutation. The efficiency of site-directed mutagenesis could be as high as 80%, depending on the method used to enrich for mutated plasmids (Table 19.1).

A key strategy in the oligonucleotide-directed mutagenesis method is to increase the relative number of mutated plasmids in the pooled population. The following strategies are used to preferentially retain the mutated strand.

***Gapped duplex method***. This is one of the finest methods for the direct selection of the mutant strain. In this strategy, two mutants of a vector each carry an amber mutation in one of the two antibiotic resistance markers (Stanssens *et al.*, 1989). The target fragment to be mutagenized is cloned into one of these mutant vectors (1), which contains the mutation in one resistant marker, e.g. chloramphenicol (*Cm*). Single-stranded DNA is prepared from this construct. The DNA of the second vector (2) containing the mutation in the other marker ampicillin (*Amp*) is cut at the position where the target fragment was cloned into vector 1, denatured, and annealed to the single-stranded DNA of the vector insert hybrid. This results in a duplex molecule that contains a region of single-stranded DNA corresponding to the cloned target fragment, also called a gapped-duplex. The

**Table 19.1:** Strategies used to increase the efficiency of oligonucleotide-directed *in vitro* mutagenesis

| Descriptive name | Method to remove template strand | Advantages | Disadvantages |
|---|---|---|---|
| dUTP incorporation | M13 template grown in a DNA repair mutant (*dut*⁻, *ung*⁻) contains dUTP in place of dATP | No special selections schemes or restriction enzymes required | Requires the use of a M13 phagemid and appropriate *E. coli* strains |
| Phosphorothioate incorporation | *In vitro* DNA synthesis incorporates phosphoro-thioates into mutant strand which protects it from enzyme cleavage with *Nci*I | Critical reactions are done *in vitro* and does not require a special *E. coli* strain | Depends on efficient phosphorothioate incorporation and exonuclease III digestion |
| Gapped duplex DNA | Mutant strand is marked with both gain of chloramphenicol resistance and loss of *amp*ᴿ phenotypes | Reversal of two antibiotic resistant phenotypes allows for calculation of mutation frequency | Requires a special vector and depends on efficient in vitro heteroduplex formation |
| Site elimination | Mutant strand is marked with a second oligo-directed mutation in a non-essential enzyme site | Nonessential enzyme site can be anywhere on vector | Appropriate enzyme site is required and differences in oligo annealing temperature may affect efficiency |
| *amp*ˢ site reversion | Mutant strand is marked with a second oligo-directed reversion of an *amp*ˢ to *amp*ᴿ phenotype | Both the *amp*ᴿ and *tet*ᴿ genes can be used sequentially to obtain mutations | Requires a special *amp*ᴿ cloning vector that can be used for reversion |
| PCR amplification | Methylated template strand is eliminated after PCR using the restriction enzyme *Dpn*I | Requires only one transformation and can be done quickly | PCR reaction can introduce mutations at non target sites |

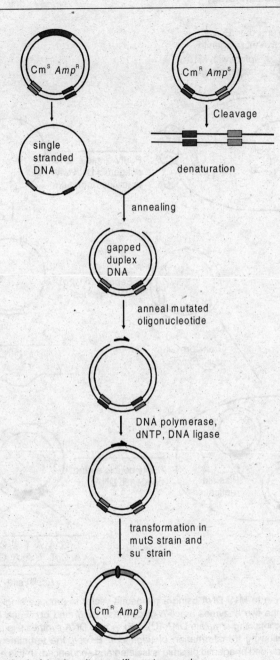

**Fig. 19.6:** Gapped duplex method of *in vitro* site-specific mutagenesis.

synthetic oligonucleotide containing the desired base changes is annealed to the single-stranded region. Thereafter, the remaining single-stranded regions are filled in and the DNA is transformed into hosts that are deficient in mismatch repair, e.g. a *mut*S strain unable to suppress amber mutations. Selection for clones expressing the resistant marker ensures that only molecules containing the mutated strands are present in the obtained clones (Fig. 19.6).

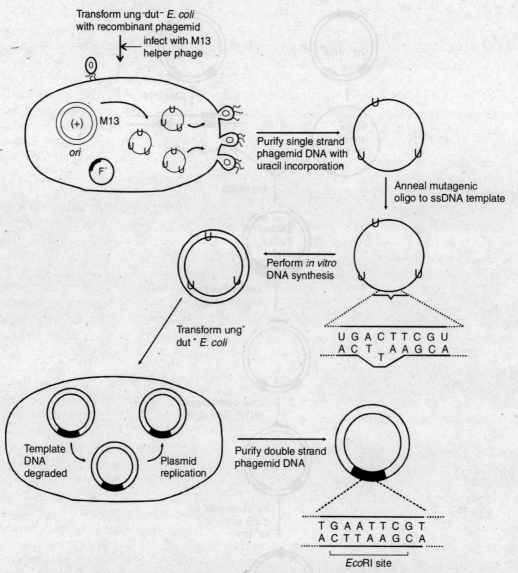

**•Fig. 19.7:**   dUTP incorporation in to M13 DNA using a phagemid vector to produce single stranded DNA in an *E. coli* strain defective in the two enzymes uracil-*N*-glycosylase (*ung*) and dUTPase (*dut*). The mutagenic oligo is annealed to uracil containing template DNA (U), and *in vitro* DNA synthesis is performed to generate double stranded DNA. Following transformation of *ung⁺*, *dut⁺ E. coli*, the template strand is degraded and the surviving double stranded phagemid plasmid is isolated and sequenced. In this example, the mutant oligo was used to engineer a novel *Eco*RI site into the DNA insert. Concept of the figure taken from Miesfeld: *Applied Molecular Genetics.* Copyright © 1999 John Wiley and Sons. This material is used by permission of John Wiley and Sons,Inc.

***The dUTP incorporation strategy***. This strategy (shown in Fig. 19.7) relies on the degradation and repair of template DNA that has been marked with deoxyuridine prior to the in vitro DNA synthesis step. Thomas Kunkel developed the dUTP incorporation strategy by taking advantage of an *E. coli* strain that contains defects in genes responsible for preventing dUTP incorporation into DNA (Kunkel *et al.,* 1987). In this method, M13 phages are grown on a specialized host before mutagenesis, which carries two mutations. One of the mutations is in the *ung* gene, which encodes uracil-N-glycosylase, an enzyme that normally removes uracil bases from DNA that result from cytosine deamination. The second defect is in the *dut* gene (dUTPase), which prevents the buildup of dUTP pools in the cell. Growing single- stranded M13 phage in an *ung⁻/dut⁻ E. coli* strain results in the phages containing 20–30 residues of uracil in the template DNA. These phages are unable to grow in an *ung⁺* host. The mutagenic oligo is then annealed to uracil containing template DNA (U) and *in vitro* DNA synthesis is performed to generate double-stranded DNA. Following primer annealing and *in vitro* DNA synthesis, a heteroduplex composed of a uracil-containing strand and a mutant strand is synthesized *in vitro* in the presence of dTTP. This DNA is transformed into a *ung⁺ dut⁺ E. coli* strain. Because the template DNA contains a large number of dU residues in place of dA, it is degraded by the activity of uracil-N-glycosylase and results only in the mutant progeny.

*Phosphorothioate method*. This strategy is based on the use of nucleotide thioanalogue phosphorothioate (e.g. dCTPαS) in the *in vivo* reaction (Taylor and Eckstein, 1985). Since some endonucleases (e.g. *Ava*I, *Ava*II, *Ban*II, *Hind*II, *Nci*I, *Pst*I and *Pvu*I) cannot cut phosphorothioate containing DNA strand, only the opposite strand is cut. The mutant oligonucleotide is annealed to the single-stranded DNA template and *in vivo* DNA synthesis takes place in the presence of a thionucleotide by DNA polymerase (Fig. 19.8). A heteroduplex is generated in which the mutant strand is phosphorothioated. After sealing the gap in the mutant strand with DNA ligase, the heteroduplex is treated with endonuclease (*Nci*I), which cleaves only the parental strand. Exonuclease III is used to partially degrade the template strand starting from the generated nicks. This strand is subsequently resynthesized

with DNA polymerase I, resulting in a homoduplex molecule containing only the mutant sequences on both strands. After transformation, a high yield of mutant clones is obtained. Originally nitrocellulose filters were used with phosphorothioate method to remove single-stranded template DNA, but these have been replaced with phage T5 exonuclease, which has single- and double-stranded exonuclease activities with a copurifying single-stranded endonuclease activity. The advantages of this method are as follows:

1. This method permits strand selection *in vitro* and gives very high efficiencies of point mutations, deletions, and insertions.
2. It does not require specialized hosts or phage vectors.

A disadvantage of all primer extension methods is that they require a single stranded-template DNA. The phosphorothioate method has been used with a double stranded template, but the procedure is very laborious.

### Chemical mutagenesis

Chemical modification of DNA *in vitro* is a technique of increasing importance for the simple generation of nucleotide substitutions. Intrinsically mutagenic attack on DNA by chemical treatment is a random process and may therefore at first sight appear nonapplicable. Yet the present state of the art is such that essentially any mutation can be generated by a combination of optimized procedures for mutagenic treatment and highly effective screening. For chemical mutagenesis a variety of chemical reagents have been advocated, including nitrous acid or hydroxylamine (convert C and A to T or G respectively), formic acid (protonates purine ring nitrogen of A and G, leading to depurination and ultimately to transversion and/or transition mutations), hydrazine (splits thymine and cytosine rings, causing all possible base changes), dimethylsulphate at neutral or acid pH (methylates G or A respectively and creates all possible base changes), sodium bisulphite (deaminates C residues, leading to the formation

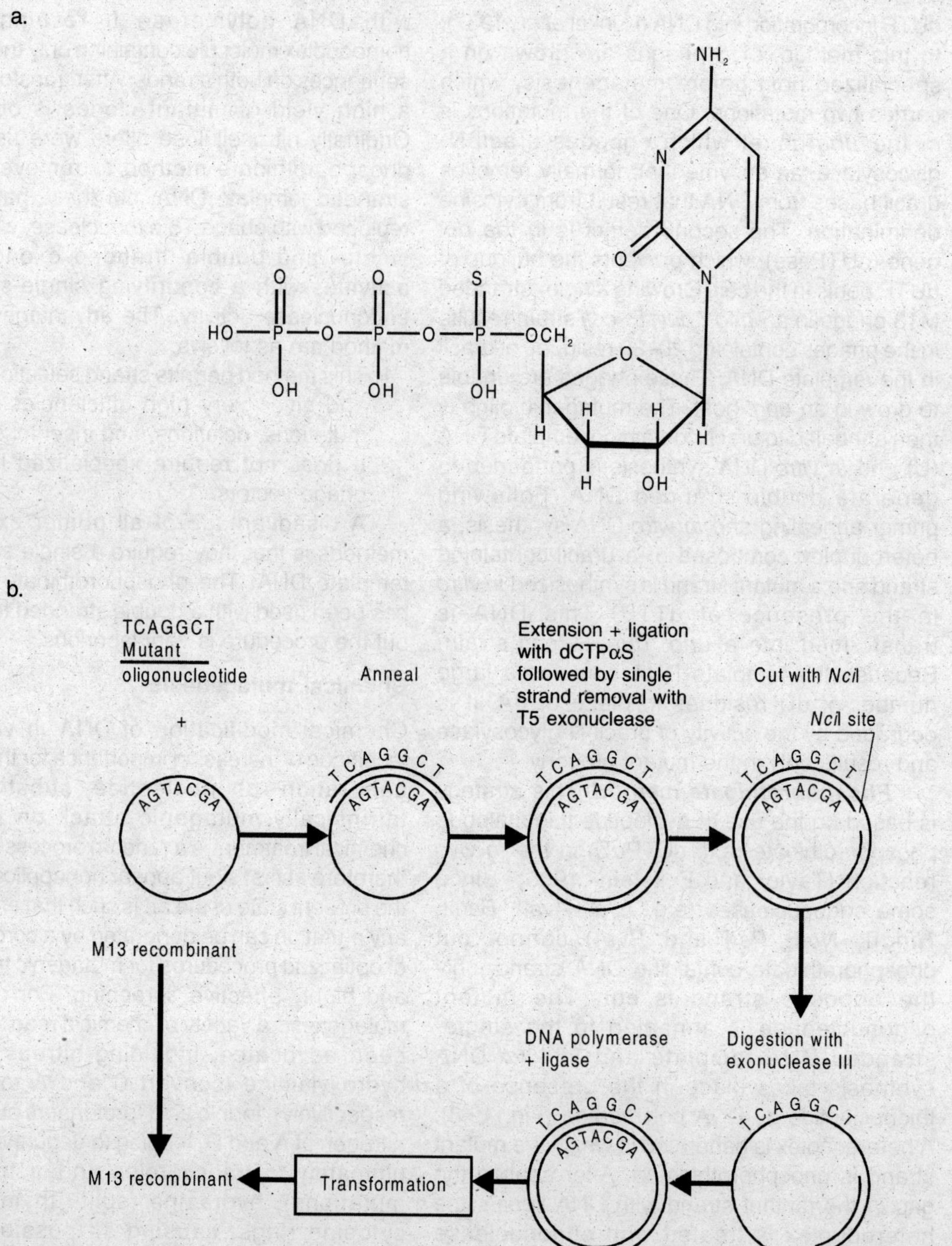

a.

b.

Fig. 19.8: (a) Structure of a thionucleotide (dCTPαS) (b) Phosphorothioate method for *in vitro* strand selection of mutants. Reproduced from *Principles of Gene Manipulation,* R.W. Old and S.B. Primrose, by kind permission of Blackwell Science, Osney Mead, Oxford, U.K.

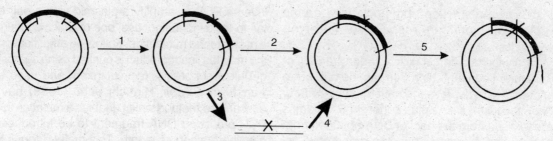

**Fig. 19.9:** A commonly used strategy for chemical mutagenesis using single stranded specific reagents. The different steps are:1–Treatment with single stranded specific mutagen; 2–Reverse transcriptase + dNTPs + T$_4$ ligase; 3–Reverse transcriptase + dNTPs followed by fragment isolation; 4–Recloning in nonmutagenized acceptor plasmid; 5–Selection for mutants.

of U residues which after DNA replication causes a C to T change), and methoxylamine (targets C, giving N$^4$-methoxycytosine and causes change to T). Unlike most chemicals the last two reagents are strictly specific for single-stranded DNA.

Sodium bisulphite in particular has found wide use in protocols developed a few years ago, where duplexes of DNA are formed that leave a short but defined single-stranded region for interaction with the mutagen (Fig. 19.9). In the strategy the region of interest is cloned into a vector designed for generating single-stranded DNA. Such target DNAs can be formed from two nonequal complementary strands by hybridization yielding circular heteroduplexes with a gapped single-stranded region. Other groups have employed molecules that have a defined nick extended to a small gap by 5′→3′ exonucleolytic activity of *Micrococcus luteus* DNA polymerase or 3′→5′ exonuclease III activity. The circular duplex DNA containing the defined single-stranded region is dissolved in sodium acetate or citrate buffer and a high concentration (0.5–4 M) of reagent is added. After a few hours of incubation at slightly elevated temperature (25–50°C) the bisulphite is removed by exhaustive dialysis. Bisulphite treated single-stranded DNA is then reconstituted in vitro with double-stranded DNA by using a DNA polymerase. For *in vitro* synthesis, DNA is incubated in the presence of all four deoxynucleotide triphosphates to repair

the gapped region by DNA polymerase I or Klenow DNA polymerase action, prior to ligation by T4 DNA ligase (some prefer the use of reverse transcriptase instead of DNA polymerase I for gap repair). A fragment comprising the region of interest is then recloned into the original construct, replacing the nonmutagenized homologue. Mutant progeny plasmids or mutant RF or recombinant phage DNA is finally obtained on transfection of the DNA into bacterial host cells.

Meyers *et al.* (1985) developed a protocol that utilizes formic acid or hydrazine, which cause damage to specific bases. The target DNA of rather short length is cloned into a single-stranded vector M13 adjacent to a so-called G–C clamp. This clamp is a short stretch of DNA rich in GC residues introduced in the mutagenesis vector by molecular cloning. It functions to keep the strands of the target DNA segment from complete denaturation during certain steps in the procedure. Single-stranded DNA is generated and treated with chemical mutagens. The mutagenic reagents are removed by ethanol precipitation, and subsequently an oligonucleotide primer is annealed to the insert DNA in the CG clamp region. Next, strand extension with reverse transcriptase in the presence of dNTPs ensures that the region of insert DNA, including the GC clamp of the vector, is double stranded. This enzyme incorporates random nucleotides at sites of damage with a high frequency. The region of interest is then excised by restriction endonucleases and is recloned, replacing respective wild-type sequences.

Polyacrylamide gel with a linear gradient of denaturant concentration (formamide and urea) is used which depress the melting temperature. First, the insert DNAs plus adherent GC clamps are excised from the pool of DNA obtained from bacterial transformants run on a preparative gel to separate mutant DNAs from the parental (nonmutant) wild type. The gradient gel causes DNAs that contain nucleotide substitutions to run faster or slower than originally, due to an altered melting behavior. Partially denatured DNAs are eluted from the acrylamide matrix after visualization by ethidium bromide fluorescence and religated again in the vector above containing the M13 origin now to yield only mutant genotypes. The chimeric DNAs are transfected into E. coli and individual colonies are prescreened again by restriction analysis of DNA minipreps and analytical gradient gel electrophoresis. Genuine mutants having altered mobility from wild-type DNA are identified and single-stranded DNA from the minipreps of mutant candidates are prepared for dideoxy sequencing to characterize the nucleotide changes.

This approach is attractive for two reasons. First, although it requires a fair number of steps, it yields no nonmutageized background colonies and distinct genotypes can already be identified prior to characterization by sequencing. Second, depending on the chemical mutagen and the regime of treatment applied, a given DNA segment can really be saturated with transversions and/or transitions.

## PCR-mediated *in vitro* mutagenesis

The development of optimal reaction conditions for long-distance PCR made it possible to design site-specific mutagenesis strategies that exploit unique primer designs to direct the amplification of entire plasmids. One example of this is to use inverse PCR to amplify a plasmid with a pair of tail-to-tail oligonucleotides, one of which contains a mismatched nucleotide. Following amplification, the mutated linear product is purified using agarose gel electrophoresis, recircularized, and used to transform E. coli. Higuchi et al. (1988) have described a method which enables a mutation in a PCR-produced DNA fragment to be introduced anywhere along its length. This method requires four primers and three PCRs. Two primary PCR reactions produce two overlapping DNA fragments, both bearing the same mutation in the overlap region (Fig. 19.10). One PCR uses primer A and primer B whereas the second PCR utilizes primer A' and primer C. Primers A and A' are complementary. These two primary reactions generate two overlapping DNA fragments bearing the mutation. The overlap in the sequence allows the fragments to hybridize. These are mixed, denatured, and annealed. One of the two possible hybrids is extended by DNA polymerase to produce a duplex fragment. The other hybrid has recessed 5' ends and as it is not a substrate for polymerase is lost from the reaction mixture. The third PCR reaction is conducted with primers B and C.

Strategies have also been developed that permit the construction of unique gene fusions using consecutive PCR reactions that use the PCR products of one reaction to prime DNA synthesis during the second amplification (Fig. 19.11). This technique is sometimes called PCR-mediated gene SOEing (splicing by overlap extension). In addition to creating gene fusions, this method also provides a PCR-based approach for inserting nucleotide mutations into the middle of a gene coding sequence. Sarkar and Sommer (1990) described a method that utilizes three oligonucleotide primers to perform two round of PCR. The product of the first PCR is used as a megaprimer for the second PCR.

The advantage of the PCR method of mutagenesis is that it is 100% efficient in generating desired mutations.

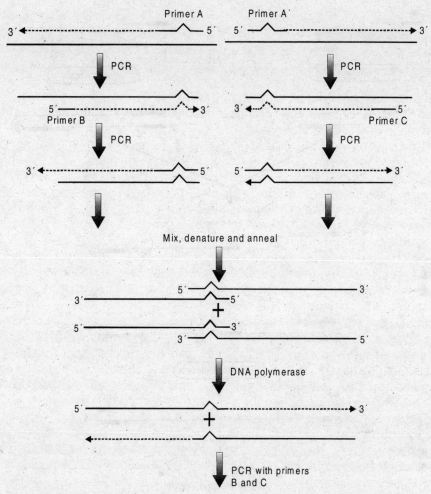

• **Fig. 19.10:** PCR method of site directed mutagenesis. Primers A and A′ are complementary and shown as dark lines. Reproduced from *Principles of Gene Manipulation,* R.W. Old and S.B. Primrose, by kind permission of Blackwell Science, Osney Mead, Oxford, U.K.

The disadvantages of the PCR method are as follows:

1. The PCR product usually needs to be ligated into a vector.

2. *Taq* polymerase copies DNA with low fidelity. Thus the sequence of the entire amplified segment generated by PCR mutagenesis must be determined to ensure that mutation has taken place.

### Advantages of site-directed mutagenesis

1. It serves as a reliable tool for gene manipulations and has simplified earlier techniques, which were extremely laborious, tedious, and time consuming.

2. It helps to study gene structure and function relationship as it enables mutant proteins to be generated with specific changes in particular amino acids. Such mutations facilitate the study of mechanism catalysis, substrate specificity, stability, etc.

3. The ability to manipulate enzymes has resulted in dramatic advances in understanding biological catalysis and has led to the novel field of protein engineering.

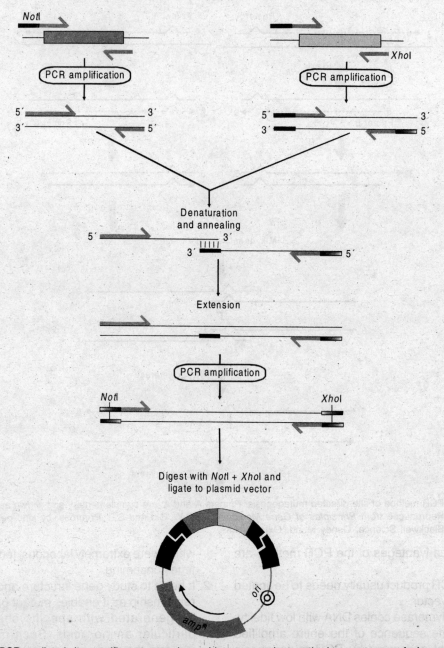

**Fig. 19.11:** PCR mediated site specific mutagenesis provides a convenient method to create gene fusion and nucleotide substitutions rapidly. Novel gene fusions can be created by performing an initial pair of PCR reactions that introduce complementary DNA sequences at the ends of two sequences that will be spliced together. The overlapping segment functions as a bridge to generate a new template for a second amplification reaction that uses the two most outside primers from the initial reactions. The net result is a spliced PCR product that can be directly inserted into a cloning vector through the addition of restriction enzyme recognition sequences to the primers. Concept of the figure taken from Miesfeld: *Applied Molecular Genetics.* Copyright © 1999 John Wiley and Sons. This material is used by permission of John Wiley and Sons, Inc.

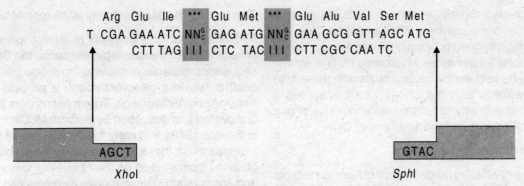

**Fig. 19.12:** Mutagenesis by means of doped oligonucleotides. During synthesis of the upper strand of oligonucleotide, a mixture of all four nucleotides is used at the position indicated by the letter N. When the lower strand is synthesized inosine (I) is inserted at the position shown. The double-stranded oligonucleotide is inserted into the relevant position of the vector. Reproduced from *Principles of Gene Manipulation*, R.W. Old and S.B. Primrose, by kind permission of Blackwell Science, Osney Mead, Oxford, U.K.

4. Through *in vitro* mutagenesis techniques it is now possible to assess in detail the identity and function of an amino acid by replacing it with an alternative residue and determining the effect this has on the enzymatic activity.

## Random mutagenesis

Random mutagenesis of a gene or gene segment is best carried out using doped oligonucleotides. These are synthesized using oligonucleotide precursors that are contaminated by a small amount of other three nucleotides. Reidhaar-Olson and Sauer (1988) have used this method to probe functionally and structurally important amino acids in a bacteriophage. The principle of this method is that the first strand of the oligonucleotide cassette is synthesized with equal mixtures of all four bases in the first two codon positions (indicated by the letter N) and an equal mixture of G and C in the third position (Fig. 19.12). The resulting population of the base combinations will include codons for each of the 20 naturally occurring amino acids at each of the mutagenized residue positions. When the complementary strand is synthesized, inosine is inserted at each randomized position because it is able to pair with each of the four conventional bases. The two strands are annealed and the mutagenic cassette ligated into the gene of interest.

The advantage of this approach is that it reduces the degeneracy.

The limitations of this approach are:
1. It is limited by the length of oligonucleotide that can be synthesized.
2. In a random mutagenesis for even a short oligonucleotide sequence, the resultant mutant library becomes too large to be screened. For example, mutagenesis of a 9 amino acid peptide using NNN triplets results in a DNA complexity of $4^{27}$ and a protein complexity of $20^9$. The use of NN (G, C) as described above, represents a slightly improved dope. It should be remembered that since different codons may translate to the same amino acid, each protein sequence does not appear equiprobably in the mutant library. Arkin and Youvan (1992) described a set of mixtures e.g. NTG/C encodes the hydrophobic amino acids phenylalanine, isoleucine, leucine, methionine and valine. Other mixtures can be designed that encode primarily hydrophilic or aromatic amino acids.

Another method of generating mutants is by PCR. Mutagenic PCR is a random mutagenesis technique that exploits the elevated error rate of *Taq* polymerase in the presence of $MnCl_2$ and high $MgCl_2$. *Taq* DNA polymerase lacks a $3' \rightarrow 5'$ exonuclease activity and is therefore unable to proof-read copied sequences. This results in a high error frequency in PCR-amplified DNA, the average frequency being one mistake per 10,000 nucleotides per cycle. Mutagenic PCR is similar to random chemical mutagenesis with the added advantage that the chosen pair provides a convenient way to target the random base pair mutation to a defined segment of DNA.

## Insertional mutagenesis

A highly efficient procedure for obtaining mutants in genes identified in sequencing programs takes advantage of the availability of large collections of plants mutagenized by an insertion element. In the plant system, the insertional element is either a T-DNA or a transposon. Insertional mutagenesis is known to occur naturally in many species through the excision and reintegration of endogenous transposable elements, which insert into a gene. The disrupted gene is not functionally redundant; that is, if other genes are unable to complement partially or completely the function of that gene, the progeny homozygous for the mutation exhibits a detectable altered phenotype. *In vitro* insertional mutagenesis involves using known DNA sequences as the insertion elements. In addition to providing a powerful approach to gene identification through their loss of or change in function, it has the added advantage of facilitating gene cloning since the insertion element acts as a tag of the mutant gene (Walbot, 1992). Random or directed insertion of DNA sequences into a gene mutates and tags the gene. Both T-DNA and transposon-mediated gene tagging have been used successfully in the identification and isolation of several plant genes.

The use of insertional mutagenesis in principle provides a more rapid way to clone a mutated gene. Because the sequence of inserted element is known, the gene in which it is inserted can be easily recovered using various cloning or PCR-based strategies.

## Transposon-mediated insertional mutagenesis

All higher organisms possess multiple copies of several kinds of transposable elements, the DNA sequences capable of moving from one site to another within a genome through a process of excision and reintegration. Transposons were first discovered and described by Barbara McClintock in the late 1940s and early 1950s. Majority of the information on these mobile elements has been obtained from studies on their behavior in maize and snapdragon. These mobile elements can be transposed from one site to another either within the same replicon or to another replicon present in the cell. When inserted at a given locus, transposons disrupt the function of the gene at the site of their integration and cause large-scale rearrangements of adjacent DNA sequences. If the site of transposon integration is in, or in the vicinity of, a particular gene, the expression of that gene can be altered in many ways depending on the proximity of the element to the structural or regulatory region of the gene. This lead to transposon-mediated insertion mutagenesis and is one of the best means to induce a high rate, frequency, and types of *in vitro* mutagenesis. Transposable elements cause mutations in two principal ways.

1. *Transposition in a coding region of DNA*: In a natural state, transposable elements are present in the noncoding regions of the genome and are therefore silent and harmless. But they can insert into a coding region of a gene and disrupt its function. Because most transposable elements contain their coding regions, their transcription interferes with the transcription of the gene and gets terminated within the transposable element. Even if the transcription proceeds through the element, the phenotype is altered and a mutant results. Its coding region when compared to the nonmutated organism contains altered and incorrect sequences.

2. *Recombination*: Transposable elements induce in vitro mutagenesis by introducing in vitro recombinations. Transposable elements are present in multiple copies, often with two or more copies in the same chromosome. Whenever the two copies are either together or near to each other, they pair during synapsis, undergo crossing over, resulting in the deletion of the DNA region between these two elements. Alternatively, the downstream element can pair with the upstream element in the homologous chromosome, producing a chromatid that lacks the element in the region between the elements and one chromatid in which this region is duplicated.

Details on transposon structure and function and transposon as a gene tag are discussed in Chapter 20.

Advantages of transposon based approaches are as follows:

1. They generate large populations of insertions.
2. They have the ability to use the propensity of many transposable elements to transpose to linked sites, which makes it possible to remobilize the element for insertion in the vicinity of the starting insertion site.

Limitations of transposon-based approaches are as follows:

1. Not many isolated and fully characterized transposons are available except those from maize and snapdragon. This limits the approach of transposon tagging to these plants. To overcome this restriction, transposon elements in heterologous systems are being used.
2. Transposons characterized so far all have a tendency to tag a gene located at some distance from the transposon-associated region in the parent. To overcome such a problem, a heterologous system is used. For this, nonautonomous elements such as *Ds* or *dSpm* are introduced at various locations of the plant genome. When this plant is crossed with a plant containing an active element such as *Ac* or *Spm*, this element activates the defective elements and induces new mutations. The selfed progeny yields new transposon-tagged mutations. Once the mutation of interest is identified, the autonomous element is outcrossed, thereby stabilizing the mutation. Through backcrossing, the defective element cosegregating with the phenotype can be identified and the tagged gene isolated. Another advantage of using heterologous systems is that the species receiving a transposable element does not contain any element of the same family in its genome. This makes identification of a transposable element cosegregating with the mutant phenotype easy.

### T-DNA mediated insertional mutagenensis

The soil-borne plant pathogen *Agrobacterium tumefaciens* is a natural genetic engineer of plants. Under the inductive effects of phenolics released by wounded dicotyledonous plant tissues to which the bacteria attach, the virulence region of the Ti plasmid is activated and endonucleases encoded by the *vir*D operon cleave the borders of the T-DNA region. The procedure of T-DNA transfer has been explained in Chapter 21. Transgene integration in plants is by illegitimate recombination, which occurs at a low frequency. T-DNA is transferred as a single-stranded molecule associated with the VirD2 and VirE proteins comprising the so-called T-complex. Breaks in chromosomal DNA are required for insertion of the T-DNA strand, and analyses of insertion sites indicate sequence similarities of only a few nucleotides between T-DNA and plant DNA. There are some rearrangements or deletions induced in the host DNA during this process. In some cases, the T-DNA molecules may insert in tandem arrays or at a single or multiple loci as inverted or direct repeats. The characteristic of *Agrobacterium*-mediated plant transformation is that the T-DNA can be used as an insertional mutagen and by extension, a gene tag. A large number of mutants have been obtained by T-DNA insertion. Versatility of this experimental system can be further exploited to tag DNA sequences that control gene expression. A number of plant species have been used to identify and clone of genes by T-DNA insertional mutagenesis, the principal requirements being ease of transformation and a diploid chromosome complement. But T-DNA tagging is rather inefficient in most plant species because of the high percentage of noncoding and highly repetitive DNA. For dicots, the main tool is T-DNA insertion mutants with the most effort in *Arabidopsis*. This weedy crucifer offers several advantages as its genome being too small facilitates the screening of genomic libraries with cloned T-DNA flanking probes. *Arabidopsis* has only 10–14% highly repetitive DNA; the detailed integrated genetic/RFLP linkage map is available and genetic transformation is routine in several laboratories.

T-DNA tags provide dominant markers for genetic mapping and for mutations. To identify a gene by T-DNA tagging, a large population of transgenic plants is generated that contains T-DNA harboring selectable and screenable reporter genes. This population is screened for a phenotype of choice and its stability over a number of generations is ascertained. The segregation of the mutant phenotype is due to a single mutation or the mutation is consequence of a T-DNA insertion event. If the phenotype and T-DNA cosegregate 100% in a sample of atleast 100–200 progeny following both selfing and backcross, cloning can be done using inverse PCR, plasmid rescue of the screening of genomic libraries made from individual transgenic lines of interest. The proof that the cloned gene is allelic to the mutant gene of interest can be obtained by genetic complementation in which the wild type is used to transform the mutant and the particular mutant is rescued. Introduction of the wild-type allele is done by *Agrobacterium*-mediated transformation of a line homozygous for the mutation; self-fertilization of the complemented line results in segregation of the mutant and wild type alleles in a Mendelian fashion.

Advantages of T-DNA mediated mutagenesis:
1. Results in fewer (1–2) insertions per line.
2. Insertions are stable.
3. Easy to maintain and does not show strong insertional biases.

Limitations of T-DNA mediated mutagenesis:
1. It is more difficult to achieve saturation with T-DNA insertional mutagenesis.
2. Most of the transformation systems developed for *Arabidopsis* involve tissue culture, which often induces high frequencies of somaclonal variation. To overcome this problem, a non-tissue culture transformation system based on infecting germinating seeds with *Agrobacterium* has been developed. In *Arabidopsis*, highly efficient plant transformation techniques have been instrumental in generating large populations of T-DNA insertion lines while minimizing the effect of somaclonal variation (Azpiroz-Leehan and Feldman, 1997).

The first published examples of targeted mutations in *Arabidopsis* show that careful analysis of mutants is often required to detect the deleterious effects of a given mutation (Gilliard *et al.,* 1998; Hirsch *et al.,* 1998).

A further extension of this strategy involves the use of an EST-like approach to systematically sequence the flanking regions of insertions. Using PCR-based techniques, it is possible to isolate and sequence a large number of insertion sites from pooled or individual lines. The insertion sites can be mapped by comparing the sequences with the genomic sequences. Databases of such flanking sequences will be established and can be searched for hits in any particular gene.

**REFERENCES**

Arkin, A.P. and Youvan, D.C. 1992. Optimizing nucleotide mixtures to encode specific subsets of amino acids for semi random mutagenesis. *Biotechnology* **10:** 297–300.

Azpiroz-Leehan, R. and Feldman, K.A. 1997. T-DNA insertion mutagenesis in *Arabidopsis* going back and forth. *Trends Genet.* **13:** 152–156.

Barcak, G.J. and Wolf, R.E. Jr. 1986. A method for unidirectional deletion mutagenesis with application to nucleotide sequencing and preparation of gene fusions. *Gene* **49:** 119–128.

Charnay, P., Mellon, P. and Maniatis, T. 1985. Linker scanning mutagenesis of the 5′ flanking region of the mouse β- major globin gene: sequence requirements for transcription in erythroid and nonerythroid cells. *Mol. Cell Biol.* **5:** 1498–1511.

Gilliard, L.U., Mckinney, E.C., Asmussen, M.A. and Meagher, R.B. 1998. Detection of deleterious genotypes in multigenerational studies. I. Disruptions in individual *Arabidopsis* actin genes. *Genetics* **149:** 717–725.

Higuchi, R., Krummel, B. and Saiki, R.K. 1988. A general method of *in vitro* preparation and specific mutagenesis of DNA fragments: study of protein and DNA interactions. *Nucl. Acids Res.* **16:** 7351–7367.

Hirsch, R.E., Lewis, B.D., Spalding, E.P. and Sussman, M.R. 1998. A role of AKT1 potassium channel in plant nutrition. *Science* **280:** 918–921.

Kunkel, T.A., Roberts, J.D. and Zakour, R.A. 1987. Rapid and efficient site specific mutageneis without phenotypic selection. *Meth. Enzymol.* **154:** 367–382.

McKnight, S.L. and Kingsbury, R. 1982. Transcriptional control signals of a eukaryotic protein coding gene. *Science* **217:** 316–324.

Meyers, R.M, Lerman, L.S. and Maniatis, T. 1985. A general method for saturation mutagenesis of cloned DNA fragments. *Science* **229:** 242–246.

Reidhaar-Olson, J.F. and Sauer, R.T. 1988. Combinatorial cassette mutagenesis as a probe of the informational content of protein sequences. *Science* **241:** 53–57.

Sarkar, G. and Sommer, S.S. 1990. The megaprimer method of site directed mutagenesis. *Biotechniques* **8:** 404–407.

Stanssens, P., Opsomer, C., Mc Keown, Y.M., Kramer, W., Zabeau, M. and Fritz, H.J. 1989. Efficient oligonucleotide directed construction of mutations in expression vectors by the gapped duplex DNA using alternating selectable markers. *Nucl. Acids Res* **17:** 4441–4454.

Taylor, J.W. and Eckstein, F. 1985. The rapid generation of oligonucleotide directed mutations at high frequency using phosphorothioate modified DNA. *Nucl. Acids Res.* **13:** 8765–8785.

Walbot, V. 1992. Strategies for mutageneis and gene cloning using transposon tagging and T-DNA insertional mutagenesis. *Annu. Rev. Plant Physiol. Plant Mol. Biol.* **43:** 49–82.

Wells, J.A., Vasser, M. and Powers, D.B. 1985. Cassette mutagenesis: an efficient method for generation of multiple mutations at defined sites. *Gene* **34:** 315–323.

Yanisch-Perron, C., Vieira, J. and Messing, J. 1985. Improved M13 phage cloning vectors and host strains: nucleotide sequences of the M13 mp18 and pUC19 vectors. *Gene* **33:** 103–119.

## FURTHER READING

Kaul, M.L.H. and Nirmala, C. 1999. Biotechnology: Miracle or mirage. *In vivo* and *In vitro* mutagenesis. In: Siddiqui, B.A. and Khan, S. (Eds.), *Breeding in Crop Plants: Mutations and In Vitro Breeding.* Kalyani Publishers, India.

Old, R.W. and Primrose, S.B. 2001. *Principles of Gene Manipulation.* Blackwell Science, UK.

Miesfeld, R.L. 1999. *Applied Molecular Genetics.* Wiley-Liss, New York.

Walker, J.M. and Gaastra, W. (Eds.) 1987. *Techniques in Molecular Biology.* Croom Helm, London and Sydney.

# 20

# Transposon Genetic Elements and Gene Tagging

Transposons are mobile genetic elements. They are segments of DNA that are not autonomous and can insert at random into the genome, plasmids, or bacterial chromosomes independent of host cell recombination system. They have been described as introgenic parasites and have the ability to integrate at different sites within the genome and to move around the genome i.e. transpose. The process of integration and excision of transposable elements is called transposition. All transposable elements share two basic properties:

1. The ability to move from place to place in the genome—hence their designation as mobile DNAs or transposable elements.
2. The ability to amplify their copy number within the genome via this transposition, thereby providing a selectable function that can make them selfish or parasitic DNAs.

Barbara McClintock discovered transposons almost 50 years ago (McClintock, 1951). By insertion, transposons can abolish gene function, resulting in mutant phenotypes. Transposon-induced mutations are often unstable because on excision of the transposon, gene function is restored, and as a consequence, reversion of the phenotype can be recognized by variegated pattern. Most of the activities of a transposable element give rise to changes in gene and/or genome structure, often with accompanying alterations in gene activity. It was for this reason that McClintock originally named these entities as "controlling elements."

Transposons contain two functional regions: (1) DNA repeat sequences at the termini of the transposon, which are required for genomic integration and excision and (2) an internal segment encoding a transposase enzyme, which is required for transposition.

Transposons have been identified in many organisms both prokaryotes and eukaryotes. Bacterial transposons were the first to be characterized at the molecular level. In the last decade, plant transposons were isolated and characterized. In maize, a variety of transposons have been identified, which are classified into families with several members. Transposons have also been identified in other plant species such as snapdragon, petunia, *Arabidopsis*, etc. In many plants with large and complex genomes, transposable elements make up over 50% of the nuclear DNA.

## Transposable elements in bacteria

Genetic instabilities have been found in bacteria, and in many cases led to the identification of transposable elements. Bacterial transposons were the first to be studied at the molecular level, and have been categorized as follows.

## IS elements

The simplest bacterial transposons are the insertion sequences or IS elements. Each type is given the prefix IS followed by a number that identifies the type reflecting the history of their isolation. A standard strain of *E. coli* may contain several (<10) copies of any one of the more common IS elements. The IS elements are autonomous units, each coding for the proteins needed to sponsor its own transposition. Each IS element is different in sequence, but there are some common features in organization. An IS element is compactly organized, less than 1500 nucleotide pairs long. Typically, it contains a single coding sequence with short, identical or nearly identical sequences at both the ends. The gene codes for a protein called transposase, which

is involved in promoting or regulating transposition. The terminal sequences at both the ends are always in inverted orientation with respect to each other, and therefore they are called terminal inverted repeats (TIRs), their length ranging from 9 to 40 nucleotide pairs (Fig. 20.1). The structure of a generic transposon before and after insertion at a target site is shown in Fig. 20.2. When IS elements insert into a chromosome or plasmid, they create a duplication of the DNA sequence at the site of the insertion. One copy of the duplication is located at each side of the element. These short (3–12 bp) direct repeat sequences are therefore called target site duplications and are thought to arise from staggered breaks in double-stranded DNA (Table 20.1). Thus an IS element displays a

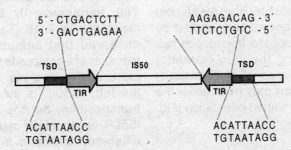

**Fig 20.1:** Structure of an IS50 element. TIR-Terminal inverted repeat; TSD- Target site duplication. Terminal inverted repeats are imperfect because the fourth nucleotide pair from each end is different.

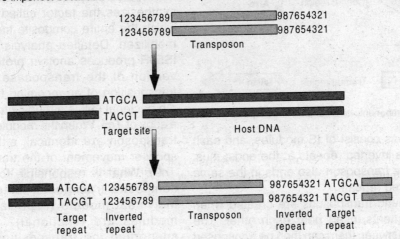

**Fig. 20.2:** Transposons with terminal inverted repeats generate direct repeats of flanking DNA at the target site. The ends of the transposon consist of inverted repeats of 9 base pairs (1 to 9).

**Table 20.1:** Features of some IS elements

| Name | Element size (bp) | ITR Left/Right (bp) | TSDs (bp) | Target site preference |
|------|-------------------|---------------------|-----------|------------------------|
| IS1 | 768 | 20/23 | 9 | At rich terminal G or C |
| IS2 | 1327 | 32/41 | 5 | Hot spots in $P_2$ genome |
| IS3 | 1258 | 29/40 | 3 | ? |
| IS4 | 1426 | 16/18 | 11–13 | $AAAN_{20}TTT$ |
| IS5 | 1195 | 16 | 4 | Hot spots |
| IS10 | 1329 | 17/22 | 9 | NGCTNAGCN |
| IS50R | 1531 | 9 | 9 | Hot spots |
| IS91 | ~1800 | 8/9 | 0 | ? |

*Note*: ITRs–Inverted terminal repeats; TSD–Target site duplication.

characteristic structure wherein its ends are identified by inverted terminal repeats whereas the adjacent ends of flanking host DNA are identified by short direct repeats.

*Composite transposons*

Two homologous IS elements combine with other genes (generally those for drug resistance) to form a composite transposon, denoted by symbol Tn. Thus, a composite transposon has a central region carrying the drug marker(s) flanked on either side by "arms" that consist of the IS elements. The arms may be in either the same (direct repeat) or inverted orientation (Fig. 20.3).

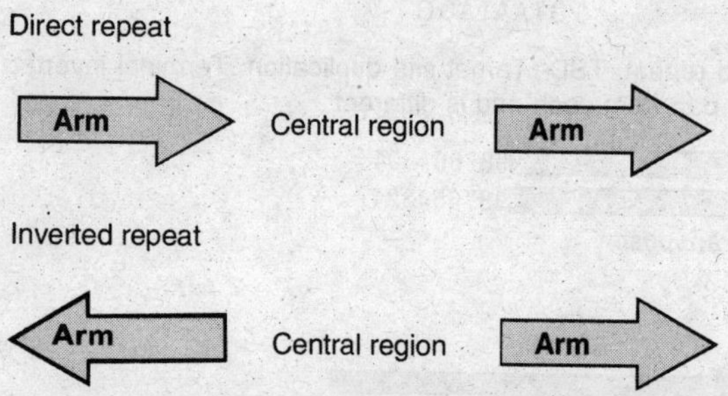

Direct repeat

Inverted repeat

**Fig. 20.3:** Components of a composite transposon.

The arms consist of IS modules, and each module has inverted repeats at the ends; thus, a composite transposon also ends in the same short inverted repeats.

Compound transposons are created when two IS elements insert near each other. The sequence between them can then be transposed by the joint action of the flanking elements. The gene present in the composite transposon is not necessary for transposition. Fig. 20.4 shows three examples of composite transposons. In Tn9, the flanking IS elements are in direct orientation with respect to each other, whereas in Tn5 and Tn10, the orientation is inverted. Each of these composite transposons carries a gene for antibiotic resistance—chloramphenicol in Tn9, kanamycin in Tn5, and tetracycline resistance in Tn10 (Table 20.2). It has been observed that sometimes the flanking IS elements in a composite transposon are not quite identical. For example, in Tn5, the element on the left, IS50L, is incapable of stimulating transposition, but the element on the right, IS50R, is capable of causing transposition. This difference is due to a single nucleotide pair change that prevents IS50L from synthesizing a necessary transposition factor. However, IS50R synthesizes the factor called transposase by which the entire composite transposon can be mobilized. Detailed analysis has shown that IS50R produces another protein, a shortened version of the transposase, which inhibits transposition of an incoming transposon when the bacterial cell is already infected and carries a copy of Tn5. When the modules of a composite transposon are identical, either module can sponsor movement of the transposon (e.g. in Tn9). What is responsible for transposing a composite transposon instead of just the individual module (especially where both modules are functional)? A major force supporting the transposition of composite transposons is the selection for the marker(s) carried in the central region.

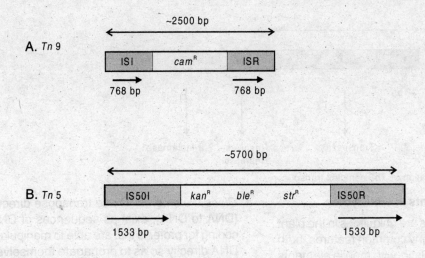

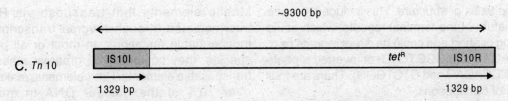

**Fig. 20.4:** Genetic organization of composite transposons. *cam*–Chloramphenicol; *kan*–kanamycin; *ble*–bleomycin; *str*–streptomycin; *tet*–tetracycline.

**Table 20.2:** Characteristic features of some composite transposons

| Element | Length (bp) | Marker genes | Flanking IS elements | IS elements orientation |
|---------|-------------|--------------|----------------------|-------------------------|
| Tn903 | 3100 | kan$^R$ | IS903 | Inverted |
| Tn5 | 5700 | kan$^R$ | IS50R IS50L | Inverted |
| Tn9 | 2500 | cam$^R$ | IS1 | Direct |
| Tn10 | 9300 | tet$^R$ | IS10R IS10L | Inverted |

## Complex transposons: the Tn3 family

Complex transposons do not contain IS elements. Elements in this group of transposons have inverted terminal repeats that are 38–40 nucleotide pairs long and produce target site duplications of 5 nucleotide pairs on insertion. They are larger than IS elements (typically 5000 nucleotide pairs long) and usually contain accessory genes as well as genes needed for

transposition. The genetic organization of Tn3, the most thoroughly studied is shown in Fig. 20.5. Three genes, *tnp*A, *tnp*R and *bla*, encode for transposase, a resolvase/repressor, and beta lactamase, respectively. The beta lactamase confers resistance to ampicillin while the other two proteins play important role in transposition. The transposition stage involves the ends of the elements as in IS type elements.

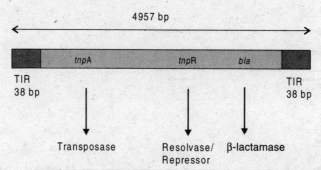

**Fig. 20.5:** Genetic organization of *Tn3* complex transposon.

## Transposable elements in eukaryotes

Transposable elements are ubiquitous in the plant kingdom and share many common features, both structural and mechanistic, with mobile elements from other eukaryotes.

*Classification*

*On the basis of structure.* This includes the size and nature of the terminal repeats, which could be long or short and could be direct repeats (e.g. CCCAGAC and GGGTCTG) or inverted repeats (e.g. CCCAGAC and GTCTGGG). There are four types of transposons:

i. With long terminal direct repeats (*Copia* in *Drosophila*, *Ty* in yeast, *IAP* in mice)
ii. With long terminal inverted repeats (*FB, TE* in *Drososphila*)
iii. With short terminal inverted repeats (*P* and *I* in *Drosophila*, *Ac/Ds* in maize, *Tam*1 in snapdragon)
iv. Without terminal repeats (*Alu* in mammals)
*On the basis of mechanism of transposition.*
According to this classification, transposable elements have been grouped into two major superfamilies.
i. *Class I elements (RNA transposable elements)*: These transpose by reverse transcription of an RNA intermediate (DNA–RNA–DNA). They are mobile because of their ability to make DNA copies of their RNA transcripts, and these DNA copies then integrate at new sites in the genome. Hence, these elements do not excise when they transpose; instead they make a copy that inserts elsewhere in the genome.

ii. *Class II elements*: These transpose directly (DNA to DNA), exist as sequences of DNA coding for proteins that are able to manipulate DNA directly so as to propagate themselves within the genome.

## Class I elements

Mobile elements that transpose via RNA intermediates through reverse transcription induce stable mutations. In most or all plant species, they comprise the greatest mass of transposable elements. Retroelements make up over 70% of the nuclear DNA in maize (SanMiguel and Bennetzen, 1998) and are equally or even more numerous in other plant species with large complex genomes. Three main transposable element families belonging to class I: (1) Retroviruses' (2) long terminal repeat (LTR) retrotransposons, and (3) non-LTR retrotransposons. Only the last two families have been found in plants.

*Retroviruses*: Retroviruses have genomes of single-stranded RNA that are replicated through a double-stranded DNA intermediate. Retroposons/retrotransposons are genomic (duplex DNA) sequences that occasionally transpose within a genome but do not migrate between cells. Reverse transcription is the unifying mechanism for reproduction of retroviruses and perpetuation of retrotransposons. The cycle of each type of element is in principle similar, although retroviruses are usually regarded from the perspective of the free viral RNA form, wheras retroposons are regarded from the stance of the

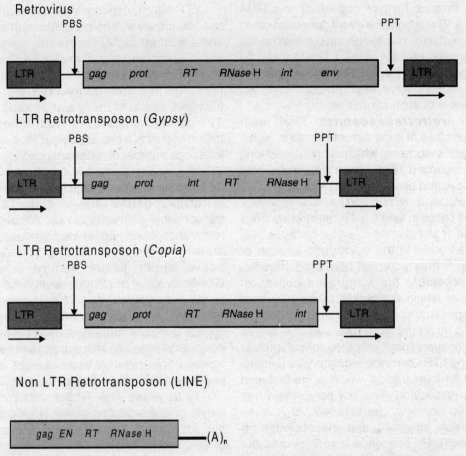

**Fig. 20.6:** Class I mobile elements. PBS is the prime binding site and PPT is polypurine tract, key sequence regions needed for replication/transposition of the LTR containing elements.

genomic (duplex DNA). Retroposons differ from retroviruses in not passing through an independent infectious form and resemble them in the mechanism used for transposition.

A typical retroviral sequence contains *gag*, *pol*, and *env* genes, which are translated into polyproteins, each of which is cleaved into smaller functional proteins (Fig. 20.6). The *gag* gene gives rise to the protein components of the nucleoprotein core of the virion and the *env* gene codes for components of the envelope of the particle, thus packing and generating virion. The *pol* gene codes for proteins concerned with nucleic acid synthesis and recombination. The *pol* gene codes for reverse transcriptase, which

converts the genome into a complementary DNA strand. *Pol* gene also has a DNA polymerase activity, which enables it to synthesize a duplex DNA from the single-stranded reverse transcript of the RNA. The enzyme has an RNase H activity, which can degrade the RNA part of the RNA–DNA hybrid. The DNA copy synthesized is longer than the RNA template; by switching template strands, reverse transcriptase copies the 3′ sequence of the RNA to the 5′ end of the DNA, and copies the 5′ sequence of the RNA to 3′ end of DNA. This generates the characteristic LTRs. Linear duplex DNA is inserted into the host genome by the enzyme integrase. Transcription of the integrated DNA from a promoter in the left

LTR generates further copies of the RNA sequence. The integration event generates direct target repeats. An inserted provirus therefore has direct terminal repeats of the LTRs, flanked by short repeats of the target DNA. Mammalian and avian genomes have endogenous (inactive) proviruses with such structures.

***LTR retrotransposons***: The most numerous class of large retroelements in plants is the retrotransposons, which contain direct long terminal repeats (LTRs). These vary in size from several hundred bases to over 10 kb, with LTRs that are usually a few hundred bases to to several thousand bases in length. LTR retrotransposons are found in all eukaryotes, either because they originated early in the eukaryotic lineage or because of their potential horizontal transfer. These resemble the integrated copies of retroviruses which are flanked by long terminal direct repeats and contain an internal domain encoding proteins analogous to the group specific antigen (*gag*) and the *pol* polyprotein (Fig. 20.6). LTRs carry the regulatory sequences required for transcription, which is the first step in the transposition. The *pol* polyprotein has conserved domains characteristic of reverse transcriptase, integrase, and endonuclease (rt-int-RNaseH). By sequence homology and the order of these domains, retrotransposons have been divided into two major subclasses, the *Ty 1-copia* group and the *gypsy* group (named after their first representatives observed in *Drosophila*). They differ in the position of integrase within the encoded polyprotein. Both these groups are found in plants, but the *Ty 1-copia* group has been better characterized. Recent studies have shown that retrotransposons are a significant fraction of large plant genomes. Due to this character, most of the plant retrotransposons were identified as repetitive sequences or insertion elements in cloned DNA. Several *Ty 1-copia* group elements have been fully sequenced in the plant kingdom. Even algae contain LTR retrotranspsons, indicating that such sequences are from a very ancient lineage.

LTR retrotransposons within individual plant species are always heterogeneous in sequence and are often highly repetitious, varying from about 20 copies per genome for the *Ta* elements of *Arabidopsis thaliana* to 13,000 for *Del 1* elements in *Lilium henryi*. The *Bis 1* family accounts for 5% of the wheat genome and one *Ty 1-copia* subgroup is present in more than one million copies in a bean species (*Vicia faba*). The total copy number of retrotransposons in the rice genome was estimated to be about 1000, and 32 families were isolated showing that retrotransposons are a major class of transposable elements in rice. Although these retrotransposons appear inactive during normal growth conditions, 5 out of 32 families were active under tissue culture conditions. Grandbastein *et al.* (1989) reported the isolation of the first tobacco (*N. tabacum*) transposable element *Tnt 1* which seems to be the most complete mobile retrotransposon characterized in higher plants. The retrotransposon have been isolated from different plant species and their characteristics are shown in Table 20.3.

***Ty in yeast***: The *Ty* elements comprise a family of dispersed repetitive DNA sequences that are found at different sites in different strains of yeast (*Saccharomyces cerevisiae*). *Ty* is an abbreviation for transposon yeast. The yeast carries about 35 copies of a *Ty* transposable element in its haploid genome. These transposons are about 5900 nucleotide pairs and are bounded at each end by a DNA segment called the δ sequence, which is approximately 340 base pairs. Each δ sequence is oriented in the same direction, known as long terminal repeats or LTRs (Fig. 20.7). *Ty* elements are flanked by five base pair direct repeats created by the duplication of DNA at the site of the *Ty* insertion. These target site duplications do not have a standard sequence, but they tend to contain AT base pairs. This may indicate that *Ty* elements preferentially insert into AT rich regions of the genome. The genetic organization of the *Ty* elements resembles that of the eukaryotic retroviruses. These single-stranded

**Table 20.3:** Characteristic features of some retrotransposons

| Organism | Retrotransposon | Length of internal domain | LTR Left/Right (bp) | TSD (bp) |
|---|---|---|---|---|
| Yeast | Ty 912 | 5250 bp | 334 | 5 |
| Drosophila | Copia | 4594 bp | 276 | 5 |
| Tobacco | Tnt 1 | 3984 bp | 610 | 5 |
| Arabidopsis | Ta 1–3 | 4190 bp | 514 | 5 |
| Rice | Tos 3–1 | 5200 bp | 115 | 5 |
| Barley | Bare-1 | 1255 bp | 1829 | 4 |
| Maize | BS–1 | 3203 bp | 302 | 5 |
| Wheat | WIS 2–1A | 5114 bp | 1755 | 5 |

*Note*: LTR–Long terminal repeat; TSD–Target site duplication; bp–base pair.

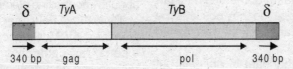

**Fig. 20.7:** Genetic organization of yeast *Ty* element.

RNA viruses synthesize DNA from their RNA after entering a cell. The DNA then inserts itself into a site in the genome, creating a target site duplication. The simplest proviruses possess three genes, *gag*, *pol* and *env*, which encode structural, catalytic, and membrane proteins, respectively. *Ty* elements have only two genes, *TyA* and *TyB*, which are analogous to the *gag* and *pol* genes of the retroviruses. Biochemical studies have shown that the product of these two genes can form virus-like particles inside yeast cells. However, it is not known whether these particles are genuinely infectious. But it has been shown that transposition of *Ty* elements involves anRNA intermediate and hence *Ty* elements are called retrotransposons.

***Non-LTR retrotransposons***: These are also called nonviral retrotransposons. This class of transposons is identified by the absence of long terminal repeats, besides other features. They are targets for a transposition event by an enzyme system coded elsewhere. They do not code for proteins that have transposition functions. A significant part of the moderately repetitive DNA of genomes consists of retrotransposons. Two families account for most of this material: (1) long interspersed nuclear elements (LINEs) and (2) short interspersed

nuclear elements (SINEs). LINEs are derived from transcripts of RNA polymerase II, whereas SINEs are derived from transcripts of RNA polymerase III.

Non-LTR transposonshave been best characterized in mammals and *Drosophila*, but their list is growing steadily in plants. LINEs have the structure of an integrated DNA version of an mRNA (Figure 6). A fully intact LINE encodes for gag proteins (involved in intracellular packaging of RNA transcript) and a polymerase (*pol*) function that includes the enzyme reverse transcriptase (RT). The *pol* functions have the ability to reverse-transcribe a LINE RNA into DNA, while an endonuclease (EN) also encoded by the element is probably associated with integration into the genome. LINE insertions are flanked by short direct duplications of target DNA, like those seen for all other mobile DNAs, usually created by the action of a transposase or integrase. LINEs terminate in an A-rich tract.

The mammalian genome contains 20–50,000 copies of LINEs, called L1. The typical member is ~6,500 bp long, terminating in an A-rich tract. The first plant non-LTR retro-transposons to be discovered were the *Cin4* elements in maize. *Cin4* elements range in size from 1 kb to 6.5 kb. The copy number of plant non-LTR retrotransposons can be very high: *Del2* in *Lilium speciosum* is present in 250,000 copies, occupying 4% of the genome. Members of this family have been discovered in tobacco also. Sequencing of LINEs suggests that there is homology between an open reading frame in L1 and reverse transcriptase. This suggests that L1 could have originated as a mobile gene coding for its own transposition. Since LINEs originate from RNA polymerase II transcripts, the genomic sequences are inactive because they lack the promoter that was upstream of the original start point for transcription. Because they possess the features of the mature transcript, they are called processed *pseudogenes.*

SINEs are highly abundant in animals but rare in most plant genomes. SINEs are usually only 100 bp to 300 bp in size and appear to be derived from reverse transcription of RNA polymerase III products. They encode no known peptides and must use *trans*-acting polymerase and integrase functions in order to transpose. SINEs are usually derivatives of tRNA or snRNA genes that have mutated to a structure that can be reverse transcribed and integrated. Because these RNA polymerase III transcribed genes carry a promoter specified within the RNA itself, a newly inserted element can usually be transcribed in any active part of the genome, thus creating a high potential for amplification. Both LINEs and SINEs lack introns because they are derived from mature mRNA, often have an integrated poly(A) tail at their 3′ end. These integrated RNA copies are seen as intronless pseudogenes.

The most prominent SINEs are the Alu family in humans and B1 family in mouse. In the human genome, a large part of the moderately repetitive DNA exists as sequences of ~ 300 bp interspersed with nonrepetitive DNA. The duplex DNA is cleaved by the restriction enzyme *Alu*I at a single site located 170 bp along the sequence. Cleaved sequences are members of a single family and are named as the Alu family after their origin of identification. There are ~ 300,000 members in the haploid genome.

## Class II elements

These mobile elements transpose directly, i.e. DNA to DNA. These transposable elements are a mixed bunch. The three superstars of this class are the maize transposons *Ac-Ds*, *Spm/dspm* (*En/I*), and *Mu*. *Ac-Ds* elements were the first transposable elements of any organism to be studied systematically by Barbara McClintock (1951). Genetic studies on maize initiated in the 1940s identified changes in the genome during somatic cell division. These changes are brought about by controlling elements, recognized by their ability to move from one site to another. Originally identified by McClintock using purely genetic means, the controlling elements can now be recognized as transposable elements directly comparable to those in bacteria. The basic structure of the three major plant class II transposable element families is simple if one ignores the intron/exon structure (Fig. 20.8). All of them contain terminal inverted repeats and probably only one or two genes essential for transposition. A family of DNA transposable elements is characterized by their sharing the same terminal inverted repeat (TIR) sequences. Hence, all *Ac* and all *Ds* transposable elements have approximately the same 11-bp TIR, while the 13-bp inverted repeat termini shared by all *Spm/dspm* elements are completely different from the *Ac/Ds* TIR. Within a family, one or more members will encode an enzyme called transposase, which recognizes the family's TIRs. All families come in two varieties, i.e. two-element system. The elements can be classified into two classes: (1) Autonomous elements, which have the ability to excise and transpose autonomously, rendering sites of their insertion mutable or *m* alleles. This element is analogous to bacterial transposons in its ability to mobilize (2) Nonautonomous elements, which are deleted defective elements that do not transpose autonomously but require helper functions from an autonomous element. These are stable and can be derived from autonomous element (often internal deletion derivatives) of the same family. They become unstable only when an autonomous member of the same family is

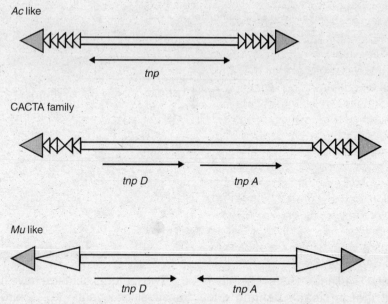

**Fig. 20.8:** Schematic representation of the structural features of class II plant transposons. Filled triangles are short terminal repeats and unfilled triangles are short repeats, with the polarity shown by the direction of triangle. The exact number and polarity of the short repeats is constant for particular element types in the same family.

present elsewhere in the genome. When complemented by an autonomous member of the same family, the nonautonomous element has the ability to transpose. Families of controlling elements are defined by the interactions between autonomous and nonautonomous elements. A family consists of a single type of autonomous element accompanied by many varieties of nonautonomous elements. A nonautonomous element is included in the family by its ability to be activated in *trans* by the autonomous elements. A nonautonomous element of one family cannot be cross-inactivated by the autonomous elements of any other family; therefore, the relationship between autonomous and nonautonomous elements of a family seems to be quite specific.

**Ac-Ds elements**: The activator (*Ac*) and dissociation (*Ds*) elements of *Zea mays* represent a relatively simple transposable element system (Fig. 20.9). The autonomous *Ac* element is 4565 bp long, which encodes a major transcript of 3.5 kb that contains a long GC-rich untranslated leader (652 nucleotides). Following splicing out of four introns from the primary transcript, a mature mRNA is obtained with 3312 nucleotides that is translated into a protein of 807 amino acids, and the protein is of 92 kDa. The *Ac* element is bounded by a direct repeat of 8 nucleotide pairs. This direct repeat is created when the element inserts into a site on a chromosome. Other repeat sequences are found within the element itself. The most conspicuous is 11-bp sequence at one end is repeated at the other end in the opposite orientation. These terminal inverted repeats (TIRs) are thought to play an important role in transposition. The product of *Ac*, termed as transposase, is sufficient for transposition of *Ac* in a growing number of transgenic heterologous plant species. The *cis* determinants for transposition of *Ac* are represented by about 200 bp at each end. Mutational alterations within these regions reduce or abolish excision of *Ac* elements. DNA-binding tests showed that the *Ac* transposase binds to the hexamer motif AAACGG that occurs in clusters within these

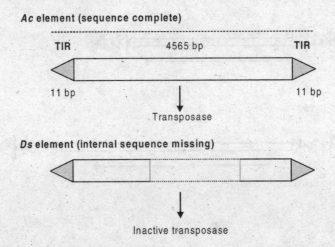

**Fig. 20.9:** Structural organization of the members of *Ac/Ds* family of transposable elements in maize.

regions (Kunze and Starlinger, 1989). The other *cis* determinant that is clearly defined and absolutely required for excision is 11-bp terminal inverted repeats, which seem to be recognized by *Ac* transposase. The asymmetry of the two *Ac* ends is important with respect to excision. *Ac* elements with identical ends are defective in excision.

Defective element of the *Ac* family, termed dissociation (*Ds*), form a rather heterogeneous group. Common to all *Ds* elements is the 11-bp TIR. Nonautosomal elements are derived from autonomous elements by deletions that inactivate the *trans*-acting transposase, but leave the sites intact (termini) on which the transposase acts. Because of deletions, *Ds* elements cannot produce transposase. However, transposase produced by *Ac* element is diffusable and *trans* acting and causes transposition of *Ds* elements.

Two features of maize have helped to follow transposition events: (1) Controlling elements often insert near genes that have visible but nonlethal effects on the phenotype; and (2) maize displays clonal development and due to this occurrence and timing of a transposition event can be visualized by the phenotype. Mitotic descendents of a given cell remain in the same location and give rise to a sector of tissue with changed phenotype, which is called variegation.

*Ac* seems to be the prototype of a family of elements whose members are widely distributed in plants (Gierl *et al.,* 1992). Characteristics of the *Ac* transposable element family are shown in Table 20.4.

***En/Spm transposable element family***: Another maize transposon family discovered by McClintock is the suppressor mutator (*Spm*) and the enhancer (*En*) family. In this family the autonomous element is called the *Spm* and the

**Table 20.4:** Ac transposable element family

| Element | TIR | TSD | Size | Species |
|---|---|---|---|---|
| *Ac* | CAGGGATGAAAT | 8 | 4565 bp | *Zea mays* |
| *Bg* | CAGGG | 8 | 4869 bp | *Zea mays* |
| *Tam 3* | TAAAGATGTGAA | 8 | 3629 bp | *A. majus* |
| *Tpc 1* | TAGGGTGTAAA | 8 | 927 bp | *Pisum crispum* |
| *Tps-r* | TAGGGGTGGCAA | 8 | 8000 bp | *Pisum sativum* |
| *Tst 1* | CAGGGGCGTAT | 8 | 736 bp | *Solanum tuberosum* |

*Note*: TIR–Terminal inverted repeat; TSD–Target site duplication; kb–kilo base; bp–base pair.

**Table 20.5:** The *En/Spm* or CACTA transposable element family

| Elements | TIR | TSD (bp) | Size | Species |
|---|---|---|---|---|
| En/Spm | CACTACAAGAAAA | 3 | 8.287 kb | Zea mays |
| MB 11 | CACTACCGGAATT | 3 | 9 kb | Zea mays |
| Tam 1 | CACTACAACAAAA | 3 | 15.164 kb | Zea mays |
| Tam 2 | CACTACAACAAAA | 3 | 5.187 kb | Zea mays |
| Tam 4 | CACTACAAAAAAA | 3 | 4.329 kb | Zea mays |
| Tam 7 | CACTACAACAAAA | 3 | 7 kb | Zea mays |
| Tam 8 | CACTACAACAAAA | 3 | 3 kb | Zea mays |
| Tam 9 | CACTACAACAAAA | 3 | 5.5 kb | Zea mays |
| Tgm | CACTATAAGAAAA | 3 | 1.6–1.2 kb | Glycine max |
| Pis 1 | CACTACGCCAAAA | 3 | 2.5 kb | Pisum sativum |

*Note*: TIR–Terminal inverted repeat; TSD–Target site duplication.

nonautonomous elements are called *dSpm* (d stands for deleted or defective while *En* is autonomous and inhibitor (*I*) is nonautonomous element. The *Spm* and *En* autonomous elements are virtually identical. The 8287-bp long autonomous enhancer (*En*) or suppressor mutator element (*Spm*) encodes at least two functional products derived by alternative splicing from a precursor transcript. *En/Spm* has 13-bp terminal inverted repeats. These TIRs are *cis*-acting sequences that are absolutely required for excision. When they insert into a chromosome, they create a nucleotide pair target site duplication. The *dSpm* elements are smaller than the *Spm* elements because part of their DNA sequence has been deleted. The deleted *dSpm* elements are unable to stimulate their own transposition. Two functional products–TNPA, a 67 kDa protein, and TNPD, a 131 kDa protein– are encoded by the two genes. TNPA is about 100 times more abundant than TNPD, but both proteins are required for transposition of *En/Spm*. Similar to *Ac* transposon, the asymmetry at the two ends is also required for *En/Spm*. A possible autoregulatory role for TNPA has been suggested, because the innermost TNPA binding motif at the 5′ end overlaps with the "TATA" box of the single *En/Spm* promoter (Gierl *et al.*, 1988). It has been speculated from a comparison with *En/Spm* related elements from other plant species that the TNPD interacts with 13-bp TIRs and may accomplish endonucleolytic cleavage

at the elements ends. *En/Spm* belongs to the so-called CACTA family of elements. These elements produce a3-bp target site duplication on insertion and share nearly identical 13-bp TIRs (Table 20.5).

**Unclassified elements:** In addition to the elements described previously, a number of less well-characterized mobile elements are known. Small transposable elements in plants, the miniature inverted repeat transposable elements (MITEs), have a structure indicating that they are likely to be DNA transposable elements. The MITEs are unusual in having no identified autonomous elements. The existence of undiscovered autonomous elements that encode MITE-specific transposases is likely, although it is also possible that these tiny elements utilize a *trans*-acting transposition that is itself not encoded on a mobile DNA. Such an activity might be specified by a standard host gene involved in some other cellular process (e.g. DNA replication, recombination or repair), although there are no obvious candidates at this time. For example, Tourist–Stowaway family elements are found in high copy number in maize. These Tourist–Stowaway sequences are present in a wide variety of flowering plants. These elements are small (about 100–350 bp) and heterogeneous in sequence. The structures of these elements are similar to several different types of transposable elements, but identical to none.

## Transposon Gene Tagging

Isolation of plant genes by conventional techniques requires knowledge of a gene product. In routine procedures for gene cloning, genomic or expression libraries are screened with probes made of mRNA or protein respectively. Also, based on the cDNA or protein sequence, oligonucleotide primers can be designed to apply a PCR approach. However, for genes with unknown products, alternative strategies that have a genetic basis are required. Recently, gene isolation techniques have been successfully applied, starting with the genomic map position of the gene. These map-based cloning approaches, however, do not involve mutant phenotypes that may hamper the identification of genes. To obtain mutants that allow simultaneous isolation of the corresponding genes, strategies have been developed using transposons. By insertional mutagenesis, mutant phenotypes are obtained and the genes responsible for this are molecularly tagged and hence can be cloned.

It has been demonstrated that *Ac* and *Ds*, like other transposons, can be used to isolate plant genes. The procedure of gene tagging by transposons is known as transposon tagging. This has been shown to be generally applicable and is one of the most useful methods to identify and isolate specific gene loci of an organism. The advantage of transposon cloning over most other gene isolation methods is that except for a mutant phenotype, no other information on the gene product or its function is required. In principle, any gene can be mutated by a transposon and if its inactivation can be recognized on the basis of a phenotype, it can be cloned using the transposon as a probe. Only in plant species with a well-defined endogenous transposon system, such as maize and snapdragon, was transposon tagging previously possible; in other plant species lacking characterized transposons, transposon tagging could not be applied. However, lately it has been demonstrated that the *Ac* element from maize

can carry out transposition successfully in tobacco. Since then, *Ac* and *Ds* have been introduced in other heterologous hosts. There are two major features to the method of transposon tagging:

1. Integration of a transposon into a genetic locus often leads to a loss of function whereas excision from that locus may result in partial or full restoration of its original tasks.

2. Excision and integration of a transposon result in a change in the physical size of the donor site as well as that of the new insertion site. As a consequence, the result of transposition becomes detectable as restriction fragment length polymorphisms (RFLPs) when the transposon is used as a molecular probe.

Isolation of genes through transposon tagging involves insertion of a transposon within a gene (Fig. 20.10). This can induce a mutation that can be detected phenotypically. A DNA fragment containing the transposon and gene sequences can then be isolated using the transposon sequences as a probe. Then sequences flanking the transposon can be used to screen a library made from nonmutant plants

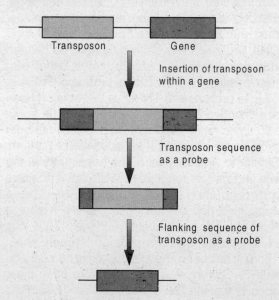

Fig. 20.10: The major steps in isolation of gene by transposon tagging.

for isolation of an intact gene. The major steps are:

1. Isolation of the transposable element
2. Gene tagging
3. Characterization of the transposon-induced mutation
4. Isolation of the mutated gene by using transposon as a probe
5. Isolation of the intact gene by using sequences flanking the transposon as a probe.

### 1. *Isolation of the transposable element*

Many transposons are isolated by trapping them in genes that have been previously isolated (rev. Balcells *et al.*, 1991). For example, the waxy (*Wx*) locus encodes a glycosyl transferase, which is required for amylose production in the kernel. Kernels that harbor insertions of the maize transposon activator (*Ac*) at the *Wx* locus were identified because they contained sectors of tissue with and without amylose. The first step in obtaining *Ac* from this material was the isolation of a cDNA clone for the *Wx* gene made from mRNA expressed in nonmutant plants. Once the nonmutant copy of the gene was cloned, it was straightforward to use this as a hybridization probe to isolate the mutant copy, and therefore the *Ac* element (Fedoroff *et al.*, 1983). Other transposable elements were isolated in maize by similar methods. Once the transposable elements were isolated, they could be used to clone genes previously tagged by insertion of one of these elements.

### 2. *Gene tagging*

The insertion of a transposon causes genetic instability, which has to be identified in a genetic screening procedure. Subsequently, the gene can be isolated by cloning the DNA sequences flanking the transposable insertion. This technique, called transposon tagging, has been widely used to isolate genes. The phenomenon of genetic instability caused by insertion of transposons can be illustrated by an example of coloration of flowers. There is a wild type gene which gives colored flowers. When a transposon gets inserted into the colored gene, it inactivates the gene and cells are not able to synthesize the pigment during petal development resulting in white flowers. Due to any excision events in mitotically active cells, the cell autonomous function of the color gene may be restored. Clonally derived daughter cells of such somatic reversions form pigmented sectors will thus display a variegated color system of the petals. Excisions occurring at an early stage result in large sectors, whereas later-stage excisions result in the development of small sectors of color in the petal. When excisions occur in cells or their ancestors that contribute to the germ cells (germinal excision), selfing may lead to the appearance of wild-type colored plants in the progeny. In fact, an excision event early in the development of a plant may result in a fully revertant and wild-type colored flower. Assuming that no further transposition occurs, RFLP analysis using the colored gene as a probe will reveal a 1:1 ratio between the wild type (revertant) allele and the transposon carrying the mutant allele. From selfing of such flowers, a Mendelian 3:1 segregation of wild type to variegated colored plants in the progeny is expected.

However, the same situation occurs with respect to the transposon carrying the allele, and upon excision, the new transposition insertion site. If the new insertion has caused a recessive mutation, a quarter of the progeny of the selfed revertant flower will also reveal the new mutant phenotype. This is the basis of the nontargeted gene tagging approach. The large number of revertants needed for this approach can be obtained from the progeny of selfed variegated flowers.

Two different approaches to isolate genetic loci with the help of transposons are based on the following observations: (1) *Targeted*, with the goal to recover a specific gene and (2) *Nontargeted,* not aiming for a specific gene but any phenotype that is correlated with transposon insertion. These two approaches require different experimental set-ups involving different crosses (Lonnig and Huijser, 1994).

**Nontargeted or random gene tagging:** This is designed to identify the reinsertion of an active transposon at any new position in the genome where it causes a detectable mutation. The basis for a nontargeted gene tagging approach is the relatively rare germinal excision events. As a first step, a plant population is generated that contains active transposable elements. Many individuals of this population will contain a transposable element at a new chromosomal location. If this new position is within a gene, the transposon will probably inactivate it. However, as the insertion will almost certainly be heterozygous in this generation, no mutant phenotype will be detected. These plants are therefore allowed to self-fertilize and the subsequent generation screened for mutants.

This approach is used where it is supposed that the newly tagged mutant allele is fully recessive and, furthermore, that the transposon donor site and the insertion site are unlinked. In such a case, the revertant is self-fertilized so that $F_2$ progenies inherit the transposition events. It means the progeny will segregate for plants with a wild-type and the (new) mutant phenotype in an approximate ratio of 3:1.

The frequency of insertion events is generally correlated with that of excision events. Thus one can increase the probability of finding such new mutations by selecting for germinal revertants from lines carrying an active transposon in an already characterized locus. In such a program one never knows beforehand which mutants will appear in the next generation. Therefore, this approach is preferred when the aim is to mark any new locus in a genetically controlled process involving many genes.

This method is advantageous over targeted tagging in that although one may be screening for a mutation in a particular gene, other interesting mutations can be detected. Strategies of this type are used if instead of a particular gene, a group of genes that belong, for example, to a particular physiological pathway or to a developmental program are to be isolated. Such strategies have been used successfully in snapdragon, maize, and petunia.

**Targeted gene tagging:** This strategy is directed toward tagging a specific gene of interest, and for this stable mutant alleles are available. Wild-type plants containing active transposable elements as female are crossed with plants that are homozygous recessive for the gene of interest (stable mutation). Majority of the $F_1$ progeny will exhibit the wild-type phenotype, since these plants are heterozygous for this gene. In exceptional cases new (recessive) mutant alleles might arise in the active transposon carrying WT line, where a transposable element in the wild-type parent will have been inserted into the gene and be transmitted to the gametes. This rare situation should alter the WT appearance of plants. These rare events are then directly uncovered in the $F_1$ generation by individuals that exhibit the mutant phenotype. An $F_1$ population of at least 10,000 plants is needed, but to improve the probability of getting alleles with transposon insertions, this number should be raised to 30,000. The new mutants are selfed and revertants looked for in the next generation. Transposons are probably involved if revertants appear. The frequency of mutations caused by transposable elements may vary because of the nature of the active element system used and the position of the element in chromosome. This method has been used successfully to isolate mutations of the *deficiens* locus of snapdragon (Sommer et al., 1990) as well as transposon-induced mutations in maize (Table 20.5).

### Problems of targeted gene tagging

1. Insertion within the gene of interest is a rare event. In the above-cited example of the *deficiens* mutant, out of 45,000 plants only 17 mutants could be isolated. If a mutation in the desired gene does not arise, it is unlikely that any other transposon-induced mutation will be apparent.

2. There is difficulty in cross-fertilizing some plant species and success depends on obtaining large number of hybrid progeny, which is difficult in plants such as *Arabidopsis*.

3. *Characterization of transposon insertion mutation*

Characterization of a transposon insertion mutation involves testing which type of transposon and which copy of that particular type has caused the mutation.

First, it is necessary to prove that the mutation is due to the insertion of a transposon. The instability of transposon insertion mutation can be used to achieve this. Most mutations caused by the insertion of a transposon should produce sectors of revertant tissue due to somatic transposition events and should produce nonmutant progeny as a consequence of germinal transposition.

Second, species such as *Zea mays* and *Antirrhinum majus* contain several different active transposable element systems and these elements occur in more than one copy per genome. In any case, cosegregation of transposon copy with the mutant phenotype has to be demonstrated. This is done by analyzing DNA from the progenitor, mutant, and revertant plants in genomic Southern experiments. The success of gene tagging therefore increases with the extent to which the different potentially mobile transposable elements of a given organism are molecularly characterized.

4. *Isolation of mutant gene*

Once the transposable elements have been isolated, they can be used to clone genes that have been tagged by the insertion of one of these elements. For example, the Bronze (*Bz*) locus encodes an enzyme required for anthocyanin synthesis in kernels and *Ac*-induced mutations of the *Bz* locus were identified genetically. This locus could then be isolated relatively easy by using the *Ac* element sequences to screen a library made from mutant plant DNA.

5. *Isolation of intact gene*

A nonmutant copy of the gene is later identified by using DNA flanking *Ac* to screen a library made from nonmutant plants.

**Transposon tagging in heterologous species**

A major limitation in transposon tagging is the low frequency with which a specific gene may be targeted. Conceivably, there are two aspects of the process, which could be optimized to increase the frequency of transposon inserting within a gene of interest: (1) the transposon frequency could be increased; (2) the proportion of transposition events which lead to the transposon inserting within the gene of interest could be raised.

For the use in heterologous species, *Ac* and *Ds* components are designed separately and are then used to transform heterologous species (Fig. 20.11). These are also known as transposon traps. It has two components. The first component carryies an immobile transposase source, i.e. which expresses transposase (*Ac* protein) but terminal sequences essential for transposition have been deleted. This type of element is referred to as *Ac* stable and can be modified further. For example, the low level of expression of *Ac* mRNA in maize and tobacco suggests that this is a major factor limiting transposition frequency. To overcome this, the *Ac* promoter is removed and replaced with plant promoters, which are capable of much higher transcription rates. The second component is similar to the *Ds* element of maize; it comprises an element that possesses both termini but does not encode the Ac protein and can, therefore, only move in the presence of *Ac* stable elements. To follow transposition of this internally deleted element, it is inserted within genes whose activity can be followed phenotypically. For example, *Ds* insertion within genes required for resistance to antibiotics results in individuals in which transposition can be followed with the appropriate

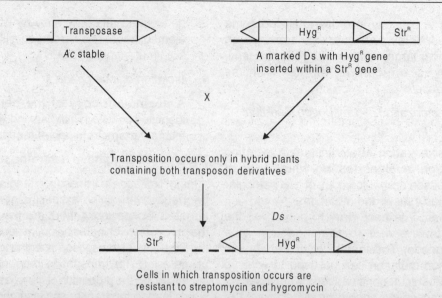

**Fig. 20.11:** A diagram illustrating two component transposon systems based on *in vitro* engineered derivatives of the maize transposon *Ac*. *Ac* stable component comprises the *Ac* transposase gene expresed from a strong plant promoter but cannot transposase as one terminus has been removed. The second derivative consists of *Ds* element marked by the insertion of a hygromycin (Hyg) resistance gene. The marked *Ds* is inserted between the promoter and coding sequence of streptomycin (str) resistance gene, which inactivates the streptomycin resistance gene. Plants containing these constructs are crossed. Transposition of *Ds* occurs in the hybrid plants. In the progeny this can be detected as green clones of green streptomycin-resistant cells on a white background. In the subsequent generation fully streptomycin-resistant individuals occur which are derived from pollen or egg cells in which *Ds* has been excised from the streptomycin resistance gene. If these seedlings are also resistant to hygromycin then the *Ds* must be present at a new location in the plant genomes.

drug. For many plant species, the most appropriate antibiotic for this purpose is streptomycin. Thus internal deletion derivative of *Ac* is inserted within a streptomycin resistance gene in vitro. The insertion inactivates the gene, which is then introduced into plants.

Progeny of these plants are grown and crossed to plants containing *Ac* stable elements. The hybrid seeds, which have inherited both *Ac* derivatives, are then germinated on a medium containing streptomycin. If no transposition occurs, the streptomycin gene remains inactive and the seedlings appear white because streptomycin prevents normal chloroplast development. However, if the transposon excises from streptomycin-resistance gene, the activity of the gene is restored, enabling those cells harboring an excision event to develop normal chloroplasts. This results in clones of

green cells on the cotyledons of germinating seedlings. Excision of the transposon also occurs in cell lineages destined to form gametes, and in this case fully streptomycin-resistant seedlings might appear in subsequent generations. These seedlings are fully green when grown on a streptomycin-containing medium, and might contain the transposon at a new location in the plant genome. However, seedlings in which excision has occurred do not necessarily still contain the transposon. This difficulty can be overcome in an engineered transposon tagging system by inserting a second antibiotic-resistance gene (e.g. hygromycin) within the transposon. Thus the second component, a *Ds* element, is marked by the insertion of a hygromycin-resistance gene (*Hyg$^R$*). The marked *Ds* is inserted between the promoter and coding sequence of a streptomycin-resistance gene. This inactivates the gene. The *Ds* can only transpose when the same cell has *Ac* stable element. A plant containing one of these components is cross-pollinated with a plant containing the second derivative and transposition of *Ds* occurs in the progeny. In this way excision and reinsertion is selected for with a combination of streptomycin and hygromycin antibiotics. This can be detected as clones of green streptomycin-resistant cells on a white background. In the subsequent generation, fully streptomycin-resistant individuals are obtained, which are derived from pollen or egg cells in which *Ds* has excised from the streptomycin-resistance gene. If the seedlings are resistant to hygromycin and streptomycin then the *Ds* must be present or got inserted at a new location in the plant genome.

Transposon trap elements (Fig. 20.12) can be designed to isolate either the gene itself (gene trap) or its control elements (enhancer trap). Gene traps are based on the proper integration of a mobile element carrying a reporter gene that lacks its own promoter sequences into an actively expressed genetic locus. When the trap element integrates into an exon, in the proper reading frame and orientation, the reporter gene is expressed. To ensure that the trap works, appropriate splicing signals are provided immediately upstream of the reporter gene sequence if gene trap element get inserted into an intron. The enhancer trap construct carries a minimal weak promoter upstream of the reporter gene. When this element gets inserted in the vicinity of an actively transcribed region, the reporter gene is activated by the neighboring regulatory sequences. The technique of enhancer trapping allows the expression pattern of a large number of genes to be visualized and screened rapidly.

*Improving targeting of transposition events*: How can one increase the frequency with which transposons insert within the gene of interest? In maize, majority of transposition events result in the transposon moving relatively short genetic distances. A study found that new positions in the genome of *Ac* elements which excised from

the *P* gene of anthocyanin were approximately 65% and were shown to be genetically linked to *P* gene itself (Balcells *et al.*, 1991). This suggests that in maize *Ac* transposes preferentially to genetically linked sites. It was later shown that plants in which *Ac* had transposed from the *P* gene produced a relatively high proportion of progeny in which *Ac* reinserted within the gene. This was interpreted as subsequent transposition from the new locations resulting in reinsertion of transposon within the *P* gene. These experiments suggest that the most effective way to tag a specific gene with a transposon would be to start with a transposon closely linked genetically to the gene. This assumption is supported by recent experiments showing that *Ac* also transposes to linked sites in tobacco and this approach has now been incorporated into most gene tagging experiments with *Ac*.

## Isolation of genes

Transposon tagging is now being adopted in a wide range of plant species to isolate genes involved in developmental and metabolic processes, which have been very difficult to approach by traditional biochemical methods. Genes controlling resistance to several plant diseases, floral initiation, and development as well as synthesis and detection of hormonal signals are among the various types of genes tagged (Tables 20.6 and 7). The *deficiens* gene, which is necessary for normal development of petals and stamens and *floricaula* gene which seems to govern the developmental switch from shoot to floral meristems have been isolated (Sommer *et al.*, 1990). Transposon tagging has been used to isolate genes necessary for resistance to fungal pathogen (Jones *et al.*, 1994; Izawa *et al.*, 1997).

## Advantages of transposon tagging

1. Generation of large number of transposon insertions by simply crossing a small number of transgenic parental lines.
2. Visual selection based on reporter gene expression.

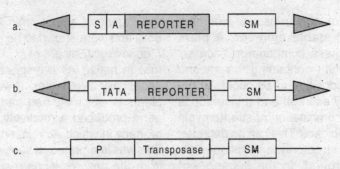

**Fig. 20.12:** Transposable element constructs for trapping genetic regions. a. Gene trap construct; b. enhancer trap construct; c. immobolized transposase construct without the terminal inverted repeats. The arrows represent the conserved terminal repeat sequences required for transposition. SM–selectable marker; S–conserved splicing signal; A–splice acceptor; TATA–minimal promoter sequence; P–strong promoter.

## Significance of transposon tagging

### Transposable elements and mutations

Transposable elements are responsible for mutations in a wide variety of organisms. Transposons have a number of effects on the genomes. DNA sequence alterations caused by transposons can range from one or a few base pairs to large chromosomal rearrangements via deletion, inversion, and duplication. Transposons affect genes directly by inserting into them, leading to subtle or major effects on gene expression. Many insertions even create a null phenotype because insertion into exons results in altered proteins. Transposable element insertions into introns can also have a variety of phenotypic effects, depending on the creation or destruction of splice sites, damage to transcriptional control regions, or transcriptional interference.

Insertion of class II elements into 5′ transcriptional control regions can also lead to mutations that depend upon the presence' or absence of active elements in the plants. Insertion of transposable element into 3′ untranslated regions can create hybrid 3′ sequences and new polyadenylation sites; for instance, *Mu* and LTR transposons have contributed polyadenylation signals to some genes (Ortiz and Strommer, 1990).

Loss of mobile elements by excision often results in mutations. Excision is virtually always imprecise at the nucleotide level, leaving behind a few extra base pairs (a footprint) to mark the elements passage through the region.

The natural ability of transposable elements to cause mutations has been exploited to stimulate the transposition of a particular family of elements, thereby increasing the mutation rate.

### Transposons and gene expression

The sequences introduced by insertion or altered by excision have the ability to impose new temporal and spatial patterns of gene expression. This can be mediated by alterations to transcriptional regulatory control regions or post-transcriptional RNA processing. Transposons are therefore not only capable of producing mutations but can also generate new regulatory units. This suggests that transposable elements have played and are playing a major role in plant gene evolution.

### Transposons in vector development

Transposable elements are useful in the genetic transformation of higher organisms. Perhaps the most sophisticated system has been developed by using the P elements of *Drosophila* for construction of cloning vectors.

Members of P elements vary in size. The largest elements are 2907 bp long. These transposons terminate in 31-bp long terminal inverted repeats and are flanked by 8-bp target

**Table 20.6:** Genes isolated using transposon tagging in plants

| Gene | Gene function | Species | Transposon |
|------|---------------|---------|------------|
| A1 | Enzyme required for anthocyanin biosynthesis | Maize | Mu; En/Spm |
| A2 | Enzyme required for anthocyanin biosynthesis | Maize | En |
| Bz2 | Anthocyanin pathway | Maize | Ds; Mu |
| C1 | Regulatory gene | Maize | En/Spm |
| C2 | Chalcone synthase | Maize | En/Spm |
| hcf–106 | Chloroplast development | Maize | Mu |
| Kn1 (knotted) | Regulatory gene | Maize | Ds |
| Opaque 2 | Regulatory gene | Maize | Ac; En/Spm |
| R | Regulatory gene | Maize | Ac |
| P | Regulatory gene | Maize | Ac |
| d1 and d5 | GA biosynthesis | Maize | Ac-Ds |
| Rp1 | Resistance gene | Maize | Mu; Ac-Ds |
| Deficiens | Regulatory gene | A. majus | Tam7 |
| Delila | Regulatory gene | A. majus | Tam2 |
| Floricaula | Regulatory gene | A. majus | Tam3 |
| Globosa | Regulatory gene | A. majus | Tam3; Tam9 |
| Olive | Chloroplast development | A. majus | Tam3 |
| An9 | Anthocyanin biosynthesis | Petunia | dTph-4 |
| An2 | MyB domain protein | Petunia | dTph–1 |

**Table 20.7:** Genes isolated using transposon tagging in heterologous species

| Gene | Gene function | Species | Transposon |
|------|---------------|---------|------------|
| ALB-3 | Accumulation of anthocyanin | Arabidopsis | En/I |
| Ms-2 | Chloroplast differentiation | Arabidopsis | Ac-Ds |
| N | Male sterility | Arabidopsis | En/I |
| cf-4A | TMV resistance | Tobacco | Ac |
| Dem | Avr-4 response | Tomato | Ac-Ds |
| Cf-9 | Organization of shoot apical tissues of developing embryos | Tomato | Ds |
| Pi-z$^t$ | Resistance gene | Tomato | Ds |
| Ei-Ef-1 | Blast resistance | Rice | Ds |
| Ei-Ef-1 | Flowering time gene | Rice | Ds |

site duplication. These complete elements are autonomously mobile because they carry a gene that encodes a transposase protein. Other P elements are structurally incomplete because they lack the ability to produce transposase, but they possess terminal and subterminal sequences needed for transposition. Cells of higher organisms can be transformed if DNA fragments are inserted into the transposable elements. A system has been developed by using the P elements of Drosophila wherein a nonautonomous element serves as the transformation vector and a complete element serves as the source of transposase that is needed to insert the vector into the chromosome of a Drosophila cell.

The nonautonomous element carried a fragment of DNA with a rosy gene and a marker gene to identify transgenes within a P element and was inserted in a plasmid. The second plasmid carried the P element for the expression of transposase, but without the capability of its own integration. The two plasmids were coinjected into the embryos of a strain of Drosophila melanogaster homozygous deficient for the rosy gene. It was found that 50% of the progeny did carry rosy+ progeny, indicating successful transfer. Ac and Ds elements can also be used in a manner similar that for Drosophila.

## REFERENCES

Balcells, L., Swinburne, J. and Coupland, G. 1991. Transposons as tools for the isolation of gene. *Trends Biotech.* **9:** 31–37.

Fedoroff, N., Wessler, S. and Shure, M. 1983. Transposable elements. *Cell* **35:** 235–242.

Gierl, A., Lutticke, S. and Saedler, H. 1988. TnpA product encoded by the transposable element En-1 of *Zea mays* is a DNA binding protein. *EMBO J.* **7:** 4045–4053.

Gierl, A. and Saedler, H. 1992. Plant transposable elements and gene tagging. *Plant Mol. Biol.* **19:** 39–49.

Grandbastein, M.A., Spielman, A. and Caboche, M. 1989. Tnt 1, a mobile retroviral like transposable element of tobacco isolated by plant cell genetics. *Nature* **337:** 376–380.

Izawa, T., Tohru, O. and Nakano, T. 1997. Transposon tagging in rice. *Plant Mol. Biol.* **35:** 219–229.

Jones, D.A., Thomas, C.M., Kim, E. and Jones, D.J. 1994. Isolation of the tomato cf-9 gene for resistance to *Cladosporium fulvum* by transposon tagging. *Science* **266:** 789–793.

Kunze, R. and Starlinger, P. 1989. The putative transposase of transposable element Ac from *Zea mays* interacts with subterminal sequences of *Ac. EMBO J.* **8:** 3177–3185.

Lonnig, W.E. and Huijser, P. 1994. Gene tagging by endogenous transposons. *Plant Mol. Biol. Manual* **K1:** 1–15.

McClintock, B. 1951. Chromosome organization and genic expression. *Cold Spring Harbor Symp. Quant. Bio.* **16:** 13–47.

Ortiz, D.F. and Stromner, J.M. 1990. The *Mu 1* maize transposable element induces tissue specific aberrant splicing and polyadenylation in two *Adh 1* mutants. *Mol. Cell Biol.* **10:** 2090–2095.

SanMiguel, P. and Bennetzen, J.L. 1998. Evidence that a recent increase in maize genome size was caused by the massive amplification of intergene retrotransposons. *Ann. Bot.* **82:** 37–44.

Sommer, H., Beltran, J.P., Huijser, P., Pape, H., Lonnig, W.E., Saedler, H. and Schwartz-Sommer, Z.S. 1990. Deficiens, a homoeotic gene involved in the control of flower morphogenesis in *Antirrhinum majus*: The protein shows homology to transcription factors. *EMBO J* **9:** 605–613.

## FURTHER READING

Balcells, L., Swinburne, J., and Coupland, G. 1991. Transposons as tools for the isolation of gene. *Trends Biotech.* **9:** 31–37.

Bennetzen, J.L. 2000. Transposable element contributions to plant gene and genome evolution. *Plant Mol. Biol.* **42:** 251–269.

Gierl, A., Saedler, H., and Peterson, P.A. 1989. Maize transposable elements. *Annu. Rev. Genet.* **23:** 71–85.

Flavell, A.J., Pearce, S.R., and Kumar, A. 1994. Plant transposable elements and the genome. *Curr. Opin. Genet. Develop.* **4:** 838–844.

Lewin, B. 1995. *Genes V.* Oxford University Press, Oxford, UK.

Lonnig, W.E. and Huijser, P. 1994. Gene tagging by endogenous transposons. *Plant Mol. Biol. Manual* **K1:** 1–15.

Snustad, D.P., Simmons, M.J., and Jenkins, J.B. 1997. *Principles of Genetics.* John Wiley & Sons, New York.

# 21

# Gene Isolation

Molecular manipulation of specific genes is a very popular area of research in biotechnology. Since no crop is perfectly suited to human needs, plant breeders are routinely concerned with improving traits such as tolerance to disease or environmental stress, yield, quality, and plant architecture. Several recent technological developments promise to facilitate greatly in the identification and isolation of genes.

## Chromosome maps

Genome mapping also known as chromosome mapping establishes the road map of a genome. Mapping of chromosome helps to locate important genes and manipulate them. Chromosome mapping helps to identify the molecular environment of both coding and non-coding DNA sequences. The map of a chromosome can be of two types: genetic map and physical map.

### Genetic map

A genetic map identifies the linear arrangement of genes on a chromosome and is assembled from meiotic recombination data. This is based on the work of Thomas Hunt Morgan on *Drosophila*. Morgan concluded that there must be exchange of materials between homologous chromosomes, in accordance with an earlier hypothesis of chromosome exchange formulated by Hugo De Vries in 1903. Morgan and Cattell

(1912) proposed the term 'crossing over' to describe this physical exchange, which we now accept as manifested cytologically by a chiasma, a cross shaped structure commonly observed between non-sister chromatids during meiosis. Morgan's student, Alfred H. Sturtevant, realized the potential of this observation by pointing out that the proportion of crossovers could be used as an index of the distance between any two genes. He named this proportion as centiMorgan (cM) after his mentor Thomas Hunt Morgan. One cM is defined as the genetic distance between two loci with a recombination frequency of 1% and is also often called a 'map unit'. These are theoretical maps which give the chromosomal order of genes and the distance of separation expressed as the percent of recombination between them. They cannot pinpoint the physical whereabouts of genes or determine how far apart they are in base pairs. Distances on this map are not directly equivalent to physical distances.

### Physical Map

Physical maps identify the actual physical position of genes on a chromosome. Distances are measured in base pairs (bp), kilobase (kb=1,000 bases) or megabases (mb =1 million bases).

Both genetic and physical maps can be global or local, indicating coverage of the entire genome or a genome section, respectively, depending on the objectives of mapping and resources available. Complete genetic maps are now available in almost al the plant species of academic and economic importance. Complete physical maps are available in only a few plant species (e.g., *Arabidopsis*, rice poplar, peach). Information from genetic mapping is now used for crop improvement purposes.

## General strategies for cloning genes from plants

All techniques for gene isolation exploit one or more of the following characteristics that define genes:

  i. They have a defined primary structure (sequence).
 ii. They encode an RNA with a particular expression pattern.
iii. They occupy a particular location within the genome.
 iv. Most genes have a function.

Some techniques may permit the isolation of genes from any plant, whereas others are only applicable to one or a few plant species. Consequently, genes isolated from one flowering plant can generally be used to isolate the corresponding genes from other plants by heterologous hybridization. Therefore, genes isolated from non-crop plants, such as *Arabidopsis*, may be useful in manipulating crop plants.

## Isolation of genes coding for specific proteins

The most widely used method for isolation of genes based on their function involved protein purification and complementation of mutant phenotypes. Isolation of genes for specific proteins became possible only after the discovery of reverse transcriptase enzyme because this enzyme can be used for the synthesis of cDNA from mRNA. This cDNA can then be used for isolating the corresponding gene from genomic DNA. Thus, techniques should be available for the isolation of specific mRNA. For this purpose, antibodies are produced against a specific protein for which the gene is to be isolated. The different steps involved are shown in Fig. 21.1. cDNA may be cloned in expression vectors such as λgt11, which can accept gene insertions into the β galactosidase gene. The chimeric vector will then produce a hybrid protein, which can be identified using antibodies. Plaques belonging to the chimeric vector carrying the desired gene can also be identified by the reaction of an antibody attached to a radioactive protein (see immunological detection procedure in Chapter 17).

## Isolation of genes that are tissue specific in function

Genes that are expressed in specific tissues are much easier to isolate. For example, genes for storage proteins are expressed only in developing seeds. Such genes can be isolated because mRNA extracted from these specific tissues will either exclusively belong to the gene of interest or be rich in this specific mRNA. Other mRNA molecules in minor quantities can be eliminated since these can be identified through their isolation from tissues where this gene is silent. The strategy is outlined in Fig. 21.2.

Specific molecular probes (DNA or RNA), if available, can be used for isolation of specific genes. These probes may be available from another species for the same gene or may be artificially synthesized using a part of the amino acid sequence of the protein (see probes in Chapter 17). The probes obtained from one species and used for another species are called heterologous probes. These are very effective in identifying gene clones during colony hybridization/ plaque hybridization/ Southern blots. Heterologous probes are generally used with cDNA library and not with genomic library.

Synthetic oligonucleotide probes can also be used if protein sequencing is known for 5–15 consecutive amino acids, and this information can be used for the synthesis of a synthetic oligonucleotide probe. These oligo probes can be used for screening of cDNA or genomic libraries for isolation of gene.

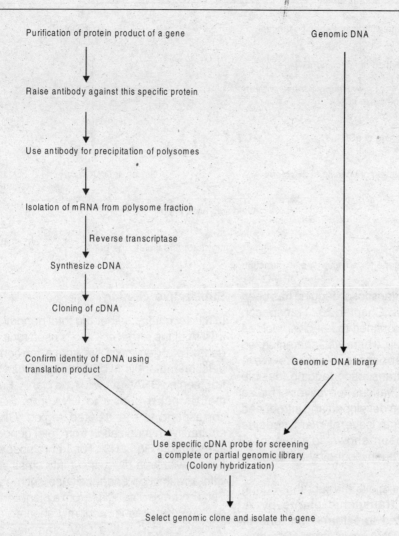

Purification of protein product of a gene

Raise antibody against this specific protein

Use antibody for precipitation of polysomes

Isolation of mRNA from polysome fraction

Reverse transcriptase

Synthesize cDNA

Cloning of cDNA

Confirm identity of cDNA using
translation product

Genomic DNA

Genomic DNA library

Use specific cDNA probe for screening
a complete or partial genomic library
(Colony hybridization)

Select genomic clone and isolate the gene

**Fig. 21.1:** Isolation of a gene for a specific protein.

## Cloning methods based on DNA insertions

Several techniques exploit the fact that certain kinds of mutations lead to relatively large alterations of chromosomal structure, which can be used to isolate the corresponding genes. In some cases when the phenotypic effect is known but the gene product is not known or not identified, genes can be isolated. This phenomenon is known as reverse genetics, when one starts from the phenotype and then traces it back to the gene without knowing the gene product. The methods used for isolating these genes are different from those used for genes coding for known proteins. Two approaches involve either transposon or T-DNA from the Ti plasmid of *Agrobacterium tumefaciens* are frequently employed. In both these cases, transposons or T-DNA insert into or near the gene of interest, thereby causing a mutation that marks the gene. The chromosomal region surrounding this molecular tag of transposon or T-DNA can then be isolated by probing a genomic library of the tagged line with the tag (transposon or T-DNA).

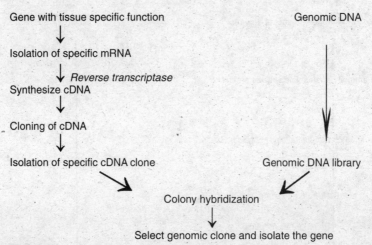

**Fig. 21.2:** Isolation of a gene for a tissue specific function.

Transposon and transposon tagging has been discussed in detail in Chapter 20. Transposon tagging was earlier confined to species such as snapdragon and maize, where well-characterized endogenous transposable elements were available. To make transposon tagging feasible in other plant species, various derivatives of maize *Ac* element have been developed and introduced by transformation into several plant species where they have been shown to transpose. Several genes have been cloned by transposon tagging.

T-DNA tagging exploits the fact that during the production of transgenic plants by *A. tumefaciens* mediated transformation, one or more copies of the T-DNA are inserted at random locations into the host genome. The procedure of T-DNA transfer and integration of DNA is explained in Chapter 23. For *Arabidopsis*, a method for large-scale transformation of intact plants has been developed. When this transformation procedure is used, 35–40% of the mutations generated are tagged by a T-DNA insert. The major limitation of this approach is the relatively large amount of effort required to produce and propagate considerable numbers of intact transformed plants.

## Subtractive cloning

In this technique, either the total genomic DNA or mRNA of the target cell line is used for isolating a gene. Total genomic DNA from a line containing a deletion mutant is hybridized in excess to highly fragmented DNA from wild type plants. Fragments that do not hybridize to DNA from the mutant correspond to the deleted region. Cloning by subtractive hybridization from total genomic DNA is shown in Fig. 21.3. Total chromosomal DNA from a wild type plant is cut into small segments with a restriction endonuclease such as *Sau*3A. Total chromosomal DNA from a plant containing a deletion mutation is randomly sheared and then biotinylated. The two DNA samples are then denatured and the nonbiotinylated DNA (wild) is hybridized to a large excess of the biotinylated DNA (mutant) in solution. The DNA is then applied to a column containing avidin-coated beads that bind biotinylated DNA molecules. Most of the DNA from the wild type plant hybridizes to the excess biotinylated DNA from the mutant plant and therefore is bound to the column. However, DNA from the wild type plant (DNA fragment B) that corresponds to the DNA fragment missing from the mutant plant cannot hybridize to a biotinylated strand and so will be enriched in the column eluate.

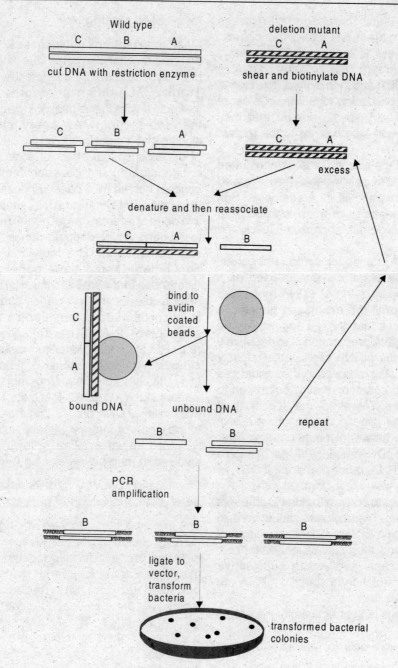

**Fig. 21.3:** A strategy for cloning by subtractive hybridization for genomic DNA.

The eluated DNA is then rehybridized to an excess of biotinylated mutant DNA and the column enrichment process is repeated. After several cycles of enrichment, the resulting DNA is amplified using PCR and then cloned. The clones are tested by hybridization to Southern blots of the mutant and wild type to ensure that they hybridize to the wild type but not to the mutant DNA.

The utility of the method was demonstrated by Sun *et al.* (1992) by isolating gene (*ga*1) involved in gibberellin biosynthesis in *Arabidopsis*. Salt stress induced genes have been isolated from wheatgrass by using this approach (Gulick and Dvorak, 1990).

When mRNA is used for isolation of genes, the mRNA of the target cell line is used as a substrate to prepare a set of cDNA molecules corresponding to all the expressed genes. To remove sequences that are not specific for the target cell, the cDNA preparation is exhaustively hybridized with the mRNA of the other closely related cell. This step removes all the sequences from the cDNA preparation that are common to the two cell types. Therefore, the specificity of the technique will depend on the closeness of the relationship between the two cells. After discarding all the cDNA sequences that hybridize with the other mRNA, those that are left (<5%, if technique works well) are hybridized with mRNA from the target cell to confirm that they represent coding sequences. These clones should contain sequences specific to the mRNA population of the target cell and they can be characterized. The different steps involved in subtractive hybridization for isolation of specific mRNA are outlined in Fig. 21.4.

The limitations of subtractive cloning are as follows:

1. It is practicable only for plants with small genomes.
2. Methods have not been developed for reliably producing mutagenized plant populations with small deletions.

## Map-based cloning

Most important eukaryotic genes, for example, disease resistance genes, are known by their phenotype and not by their biochemical effects. Traditional cloning strategies have been ineffective with these genes. Researchers have therefore sought new approaches to gene cloning that are not based on the prior knowledge of the biochemical effect of a gene, and map-based cloning is one such technique. The concept behind map based or positional cloning is to find a DNA marker (DNA sequence) linked to a gene of interest (target gene) and then "walk" to the gene via overlapping clones. It has certain requirements: (i) The individuals within the population should have genetically based differences in the trait of interest; (ii) The genes responsible for this difference can be mapped to a chromosomal position(s) adjacent to segments of DNA that have already been cloned (e.g. RFLP or microsatellite markers); (iii) The availability of comprehensive genomic libraries of relatively large DNA fragments, typically in vectors like YAC, BAC, PAC, etc., and (iv) The availability of DNA probes that are closely linked to the gene of interest (ideally less than a few hundred kb apart. The following steps are involved in map-based cloning (Fig. 21.5).

*Identification of DNA markers linked to target gene and construction of genetic maps*

The first step in map-based cloning is that large number of cloned DNA sequences are mapped in terms of their genetic location (e.g. RFLP mapping). Earlier, only RFLP marker was used, but now various other molecular marker techniques are also used for genetic location of a DNA sequence. Those DNA sequences that are tightly linked to the gene of interest are identified. The procedure is outlined in Fig. 21.6 for identification of DNA markers linked to target gene. For example, when RFLP analysis is conducted on three regions of chromosomes from two strains of the same species, which is indicated by straight lines in Fig. 21.6. Restriction

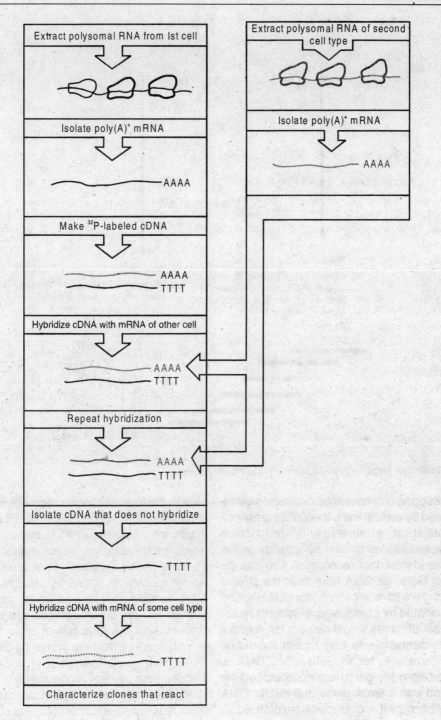

**Fig. 21.4:** A scheme for isolation of mRNA specific to one cell type by subtractive hybridization.

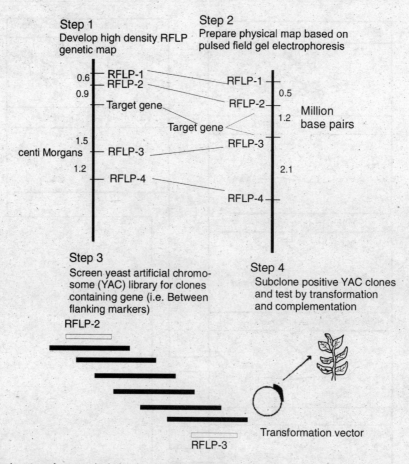

**Fig. 21.5:** The major steps for gene isolation in map-based cloning.

enzyme recognition sequences on these regions are indicated by vertical lines. In strain 2, presence of a mutation eliminates a restriction endonuclease site as shown by asterisk in the figure. The loss of this recognition site can be visualized. Genomic DNA from both the strains is isolated, cut with restriction endonuclease, and size fractionated by agarose gel electrophoresis. The number of bands would be very large and it would be impossible to distinguish individual bands. Therefore, for visualization, DNA is transferred from the gel to a nitrocellulose filter and probed with a small piece of genomic DNA in a Southern blot experiment (explained in Chapter 14). The probe DNA will hybridize to two DNA restriction fragments from strain 1 but,

due to the loss of a recognition site from strain 2, it will hybridize to only one larger DNA restriction fragment. Likewise, RAPDs can also be used to mark the target gene. Assuming the same three regions of chromosomes from two strains of the same species indicated by straight lines in Fig. 21.6. Binding sites for a particular oligonucleotide primer on the chromosomes are indicated by arrows, with the direction of arrow showing the orientation in which the primer binds at that site. When PCR is performed, DNA will be amplified between the pairs of oligonucleotide primers that bind close together and with opposite orientation with respect to each other. The chromosomal regions that will be amplified are indicated by the numbers showing the distance in kilobase pairs

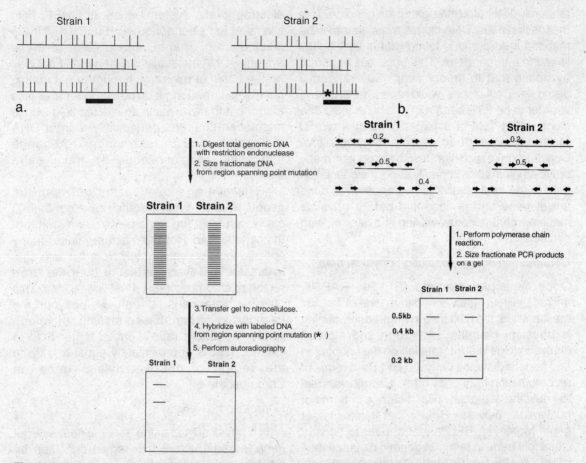

**Fig. 21.6:** A procedure for identification of molecular markers linked to target gene using: a.–RFLP; b.–RAPD.

between the relevant pairs of binding sites. The asterisk indicates the mutation that has eliminated an oligonucleotide primer binding site from the strain 2. The loss of this binding site leads to the production of one less PCR product from strain 2 as compared to strain 1. The loss of this PCR product can be visualized by size fractionating the PCR products on a gel.

Rice is the most extensively analyzed species by RFLPs. The polymorphism detected by RFLP markers is very reliable as it involves the recognition by specific restriction enzymes and hybridization with a specific probe. Genetic maps for rice with more than 2300 DNA markers, tomato with more than 1400 markers distributed over 12 chromosomes, *Arabidopsis* with 500

markers distributed over five chromosomes, and likewise for barley, wheat, maize, etc. have been developed. In many cases, disease resistance genes have also been mapped relative to RFLP markers. One of the major problems with RFLPs as starting points for map-based cloning is the difficulty in specifically saturating chromosomal regions near gene of interest with sufficient numbers of RFLP markers. This is due to the large amount of DNA in the genome of most plants. Plant species have typically more than 1 billion base pairs ($10^9$) of nuclear DNA: rice— 400 mb; tomato—1 billion bases; maize—3 billion bases; and wheat genome—more than 16 billion base pairs. DNA clones used for generating an RFLP map are, by their very nature, randomly distributed throughout the genome. This means that many DNA clones must generally be mapped before one is found that is very tightly linked to a target gene. This becomes apparent by noting that in theory over 1500 randomly distributed DNA clones would have to be analyzed in order to be 95% certain of finding at least one clone within 1cM of a target gene in a typical plant genome of total size of 1000 cM. Construction of such high-resolution genetic maps provides information primarily about the orientation of DNA markers relative to one another, and the target gene and is essential before physical mapping, chromosome walking and gene cloning can begin.

### Construction of high-resolution physical maps

Once molecular markers (RFLPs, RAPDs, microsatellites) tightly linked to a gene of interest are identified, the next step in map-based cloning is physical mapping. Although segregation analysis yields vital information on the orientation and genetic distance between RFLPs in terms of recombination frequency (cM), it is still essential to know the actual physical distance in terms of nucleotides between RFLPs that flank a target gene. Moreover, RFLPs that appear to be very close to a gene in terms of genetic distance may still be very far away in terms of physical distance. It is clearly essential to know the magnitude of the physical distance between RFLPs before initiating the next step of chromosome walking and map-based cloning.

Until recently, detailed characterization of the large DNA molecules necessary for map-based cloning was impossible. Routine agarose gel electrophoresis only separates DNA molecules up to 50 kb, whereas map-based cloning generally involves distance up to 1 megabase (Mb) or more. Now, DNA molecules up to 10 Mb can be separated by pulsed field gel electrophoresis (PFGE). Recent developments in PFGE, which is capable of separating DNA molecules upto 10 million bp in length, has made long-range physical mapping practical. PFGE includes several types of related electrophoretic systems such as CHEF, field inversion gel electrophoresis (FIGE), which are discussed in Chapter 14. In each of these systems, DNA molecules are separated not only on the basis of migration through a gel matrix, but also on how long it takes for DNA molecules to reorient themselves in an electric field whose orientation changes periodically. Larger DNA molecules tend to take longer to change directions, and, consequently, travel more slowly.

The average relationship between the genetic and physical distance of a genome is calculated from the genome size and length of the genetic map. However, actual values for any specific location vary widely from the average. In tomato, 1 cM averages 900 kb, but in the *Tm2a* region of chromosome 9, 1 cM is 4–16 Mb. This physical distance is currently well beyond the capabilities of map-based cloning. In wheat, where the kb:cM ratio is much larger than in tomato, 1cM is approximately equal to 1Mb in the region of the $\alpha$-amylase gene on Chromosome 6.

### Chromosome walking

After a physical map of the genome near a target gene is constructed, this region can then be cloned by chromosome walking. Walking along the chromosome is a term used to describe an approach that allows the isolation of gene sequences whose function is quite unknown. It is a technique that has been in use for more than a decade to identify genomic clones contiguous to a cloned DNA starting point. The principle is that a cloned genomic fragment is used as a starting point for the walk. This should be as close as possible to the target point. The starting point may be RFLP sequence. Two examples are given to illustrate chromosome walking.

Fig. 21.7 shows that gene A is to be cloned, which has been identified genetically but no probe is available. Sequence of a nearby gene B is available in cloned fragment 1. The genomic library is then screened with the chosen clone as probe to identify other clones containing DNA

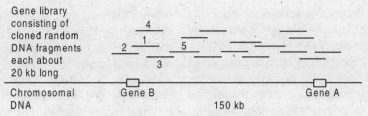

**Fig. 21.7:** Chromosome walking to clone DNA sequences.

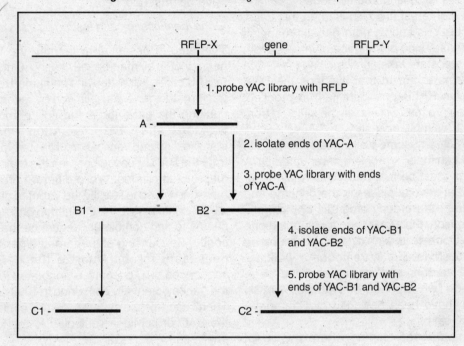

**Fig. 21.8:** Gene isolation by map-based cloning using chromosome walking.

with which it reacts and which represent clones overlapping with it. The overlap can be to the right or left. This step will identify overlapping clones 2, 3, and 4. Clone 3 can in turn be used as a probe to identify clone 5. Repetition of this walking step along the chromosome is done. A large-scale map of the walk can be made by assembling contiguous clones.

In the second example, molecular markers such as RFLPs have been identified and are used as a starting point for chromosome walking. Two RFLPs lying closest to the gene of interest, yet flanking the gene on either side, are chosen. Chromosome walking begins by identifying

genomic clones that overlap the initial RFLPs, which is accomplished by probing a complete genomic library with radiolabeled nucleic acid probes synthesized from the initial RFLPs. The newly identified genomic clones are then isolated and the ends of these clones become starting points for the next step in the chromosome walk. In Fig. 21.8, a mutation in a gene has been determined by genetic mapping to lie between two particular RFLP markers. DNA from RFLP-X is used to probe a genomic DNA library cloned in YAC vector in order to identify clones that contains an overlapping DNA insert. DNA from the end of this clone (YAC-A) is then isolated by inverse PCR and is used to screen the genomic DNA library to identify clones containing additional overlapping DNA fragments (YAC-B1 and YAC-B2). Each cycle of isolating the ends of the clone and then using this end to identify clones containing overlapping DNA fragments represents one step in chromosome walking. The process is continued until clones spanning the gene of interest are identified. In this case, since RFLP-Y is known from genetic mapping to lie on the far side of the gene of interest from RFLP-X, steps are taken until a clone (YAC-C2) that contains an insert that overlaps the DNA contained in RFLP-Y is identified. In practice, the walking process will usually be initiated from both RFLPs simultaneously.

When chromosome walking is performed in plants and animals, a problem of repeated DNA sequences must be realized. These may occur dispersed at several places in the genome and could disrupt the orderly progress of the walk. For this reason, the probe used for stepping one genomic clone to the next must be a unique sequence clone. For chromosome walking, improved vectors and strategies have been developed. Two λ phage derivatives, λDASH and λFIX, have been designed by Stratagene Cloning Systems.

For map-based cloning where positional location is done and the distances to be traversed may be as great as 1 Mb or more, cloning vectors in the form of yeast artificial chromosomes (YACs) and bacterial artificial chromosomes (BACs) are used. YACs and BACs can have fragments of 400–600 kb and 100–150 kb, respectively. YAC libraries have been constructed for Arabidopsis, tomato, maize, rice, etc. Although a fewer number of clones are needed to make a contig for a specific genomic region by using YACs, chimaerism in the inserted DNA fragment can be a problem. BAC carries rather small DNA fragments compared to that by YACs, but is less prone to chimaerism. Using these libraries and DNA markers for overlapping contigs, physical maps are being developed. The

most advanced map developed in plants is for Arabidopsis. In rice, about 1200 RFLP and STS markers were used for screening a YAC library containing about 7000 clones with the average insert size of 350 kb. As a result, about 2200 independent YACs have been identified and ordered. Both Arabidopsis and rice genomes have been sequenced.

*Gene identification and testing by transformation*

This step is essential to identify the targeted gene or QTL by map-based cloning. Once a BAC or BAC contig is isolated and confirmed to contain the targeted gene, as delimited by recombinants flanking the gene locus through chromosome walking and high-resolution mapping, identification and verification of the targeted gene from the BAC or contig become the major task of map-based cloning. While different procedures have been used to identify the targeted gene from a BAC containing it, the straightforward approach seems to first completely sequence the BAC or contig. Sequencing of the targeted gene is also necessary to characterize the gene. The sequenced BAC or contig is annotated to identify the genes potentially contained in BAC or contig. Then, each of the putative genes contained in the BAC or contig is subjected to the following analyses.

**BLAST search**: This is the most economical and fastest method. Comparative analysis of each candidate gene contained in the BAC or contig against those structurally and functionally characterized in GenBank will provide useful information about the gene of the targeted trait. For instance approximately 3/4 of the plant disease resistance genes cloned to date encode proteins characterized by the conserved nucleotide-binding site (NBS) and leucine-rich repeat (LRR) domains. If the targeted gene of map based cloning is responsible for resistance to a pathogen and the predicted protein of a candidate gene also contains NBS and LRR domains, the candidate gene is more likely the targeted gene.

***Comparative sequence analysis***: Since the mutation of the targeted gene is responsible for the phenotypic variation of the targeted trait, the wild and mutated alleles of the gene must somehow differ in coding sequences, regulatory sequences or both. Comparative sequence analysis of the targeted genes isolated from phenotypically contrasting lines of the species will reveal the nature of mutation, thus providing useful information about the targeted genes (Li *et al.,* 2006). According to the sequences of the annotated genes contained in the BAC or contig, primer pairs are designed for each gene and used for PCR to amplify the corresponding genes from the phenotypically contrasting lines such as resistance versus susceptible line for resistant gene cloning. The PCR products are then used as templates to sequence the corresponding gene form the lines. If difference is consistently detected for a gene at the nucleotide sequence and its predicted amino sequence between the wild and mutated lines studied, the gene is likely to be the targeted gene.

***Gene expression analysis***: Many of the genes express tissue-specificity and/or are environment inducible. These features provide another characteristic of assisting at identification of the targeted gene from the candidate genes contained in the BAC or contig. For instance, the wheat vernalization gene *VRN2* tends to express at the apices, the critical point for transition from vegetative growth to reproductive growth and responses to low temperature (Yan *et al.,* 2004). The rice grain shattering gene *sh4* prefers to express at the junction between flower and pedicel (Li *et al.,* 2006). Quantitative real-time reverse transcription PCR (RT- PCR) and northern blot analysis have been widely used to analyze the expression of the candidate genes in particular tissue and/or with particular treatments. The RT-PCR is a relatively new technique to quantify gene expression, in which mRNA is first reverse transcribed into cDNA and the cDNA is then used as templates of conventional PCR.

BLAST search, comparative sequencing and gene expression analysis provide useful information on which of the candidate genes contained in the BAC or contig is more likely to be the targeted gene. However, a genetic complementation experiment is necessary to verify the prediction of the above results on the targeted gene. Several methods have been used for the genetic complementation experiment.

***Genomic DNA Transformation***: Genomic DNA constructs are made for each candidate gene that is likely the targeted gene in a plant-transformation-competent binary vector (e.g. p CLD04541) from the BAC or contig containing the targeted gene according to the sequence of the BACs. To facilitate the expression of the transgene in the host plant, the construct must include both the coding region and the regulatory sequences of the genes. The DNA constructs are then transformed into a mutant genotype of the targeted trait, which is often the mutant parent of the mapping population by *Agrobacterium*-mediated transformation or bombardment method. The conversion of the host plant mutant type into the wild type will be indicative of that the DNA construct contains the targeted gene.

***cDNA transformation***: An alternative method is to transform cDNA clones into plant for genetic complementation experiment. This method is especially used when no sequence information is available for the BAC or contig. The BAC or contig containing the targeted gene is employed as a probe to screen a cDNA library potentially containing the targeted gene. The positive cDNA clone are isolated, sequenced and subjected to fine- mapping, BLAST, comparative sequence analysis and expression study as described above. The most likely candidate cDNA clone(s) is/are subcloned into a plant-transformation-competent binary vector (e.g. pBI121) with a promoter to drive the inserted gene to express. There are two possible orientations of cloning a cDNA insert into a binary vector. One orientation is the start codon of the cDNA gene is just in the downstream of or proximal to the promoter, which is called sense construct. The other orientation is that the start codon of the cDNA gene distal to the driving promoter, which is called anti-sense construct. This could be determined by PCR analysis using is just in downstream of or proximal to the promoter, which is called anti-sense construct. This could be determined by PCR analysis using the cDNA constructs as templates and the primer pair, one from the vector and the other from insert cDNA. The sense construct is transformed into the mutant genotype plants either by the *Agrobacterium*-mediated and/or by the bombardment method. As the transformation of genomic DNA construct above, if the cDNA construct converts the mutant phenotype into the wild phenotype, the cDNA must contain the targeted gene. The anti-sense construct is transformed into the wild genotype of the targeted trait. If the expression of the wild type gene is repressed significantly, the cDNA construct should contain the targeted gene (also see RNAi).

*RNA interference (RNAi)*: RNA interference or RNA-mediated interference (RNAi) was discovered by Fire *et al.*, (1998). It is a mechanism for RNA-guided regulation of gene expression that is commonly observed in plants and animals. RNAi involves a process of generation of double-stranded RNA (dsRNA.) and small inhibitory RNA (siRNA) that interferes with the expression of the gene complementary to the dsRNA. dsRNA is trimmed in cells by RNA ribonuclease named as dicer into short dsRNA called siRNA of 20–25 nucleotides long. An siRNA can be processed into single-stranded sense and anti-sense RNAs. The anti-sense RNA interacts with several proteins to form the ribonuclease containing multi-protein complex (RISC) that binds to and degrade the sense mRNA that are complementary to the anti-sense RNA, thus the translation of the targeted mRNA being inhibited (See Chapter 23). This phenomenon has been recently used in the genetic complementation experiments for gene identification in map based cloning of genes and

QTLs (Yan *et al.*, 2004; Uauy *et al.*, 2006). A segment unique to each candidate gene contained in the BAC or contig is cloned in forward and reverse orientations into a plant-transformation competent, RNAi- specific binary plasmid vector such as pMDC161 to facilitate dsRNA formation and transformed into the wild genotype of the targeted traits. If the performance of the targeted trait of the transgenic plants is significantly repressed, the candidate gene must be the targeted gene.

For all three transformation experiments described above, it is necessary to verify the phenotypes of the primary transgenic plants ($T_0$) by analyzing their $T_1$ progeny. The $T_0$ plant, if the transgene is dominant, will immediately complement the mutant type and convert it into wild type in phenotype. However, it is hemizygous and thus the phenotyping result is dependent on a single primary transformant plant. Therefore, it is desirable to further genetically analyze its $T_1$ progeny to verify the genotype and phenotype of the $T_0$ plant. This is especially true for the targeted gene controlling quantitative trait because the progeny test is essential to accurately phenotype a quantitative trait.

## Difficulties in map-based cloning

Although simple in concept, map-based cloning is generally very difficult in practice due to the following reasons:

1. *The enormous size of eukaryotic genome*: The genomes of higher plants are typically more than 1 billion base pairs, and in crops such as wheat and oats, they are more than 16 billion base pairs in length. This means that even when 1000 RFLP markers have been placed on a linkage map, the average physical distance between RFLPs will be 1 Mb in most plants and over 16 Mb in wheat and oats. Such distances are formidable even with the best chromosome walking technique.

2. Relationship between genetic distance, as measured by recombination frequency, and

physical distance, as measured in number of nucleotides, is not uniform throughout eukaryotic genomes. Thus, even RFLP markers that appear to be close to a gene of interest in terms of recombination frequency may still, in-fact, be far away in terms of actual number of nucleotides.

3. Most eukaryotic genomes contain significant amounts of repetitive DNA, which makes it extremely difficult to perform chromosome walking. Dispersed, repetitive DNA sequences can cross-hybridize with the overlapping stretch of DNA required to identify the next step of a chromosome walk and when this happens, further progress in chromosome walking along the genome may be almost impossible.

## Chromosome jumping

Chromosome walking is simple in principle, but walking on a large-scale is too demanding. For large distances, walking is usually combined with chromosome jumping, which overcomes the difficulties caused by repeated, unclonable, or unstable genomic segments. Chromosome jumping has been described by Collins and Weissman (1984) and Poustka and Lehrach (1986). This technique depends on circularization of very large genomic DNA fragments, followed by cloning DNA from the region covering the closure site of these circles, thus bringing together DNA sequences that were originally located far apart in the genome. These cloned DNAs from the closure sites make up a jumping library, which greatly speeds up the process of long-range chromosome walking. Chromosome jumping utilizes type II restriction endonucleases that cut DNA sequences infrequently. Such enzymes have octanucleotide or longer recognition sequences, for example, NotI (GCGGCCGC), PacI (TTAATTAA) and SfiI (GGCCNNNNNGGCC). Some restriction endonucleases have shorter recognition sequences, but have the property of being uncommon sequences in mammalian DNA, for example, NruI (TCGCGA) and BssHII

(GCGCGC). Their target sequences include two CG dinucleotides, which is rare in mammalian DNA. Additional specificities in target sites may be created by combining the specificities of certain methylases with that of restriction endonucleases. For example, DpnI cuts DNA at the sequence G-$^m$A-T-C to produce flush ends. This enzyme requires the methylation of adenines in both strands of DNA for cleavage to occur.

A strategy for creating a jumping library is shown in Fig. 21.9. Poustka et al. (1987) used NotI restriction enzyme to digest genomic DNA, which cuts rarely. Partial digestion of genomic DNA would results fragments of 50–200 kb. As shown in the figure, X and Y genes are far apart. The vector is also digested with NotI enzyme, and the fragments are ligated in the vector molecules, which favors chimaeric circle formation. This chimaeric vector is then cleaved with a restriction endonuclease that cuts frequently but has no target sites in the vector. Thus, it will bring together two marker genes (X and Y) near, which will be separated only by small portion of vector DNA. This technique has been applied to a jump of 100 kb in the cystic fibrosis locus in humans.

## Chromosome landing approach

In map-based cloning, the central step is chromosome walking, which is a very time consuming process. When map-based cloning was first proposed for use in species with large genomes, molecular linkage mapping was in its infancy and molecular maps were not available for most plant species. It could take months, or even years, to find a single molecular marker linked to a target gene. Recent developments both in technology and analytical procedures led Tanksley et al. (1995) to propose the chromosome landing approach. In this technique, efforts are directed toward selectively identifying molecular markers that are so close physically to the target gene that little or no chromosome walking is required. These tightly linked markers are used to screen a genomic

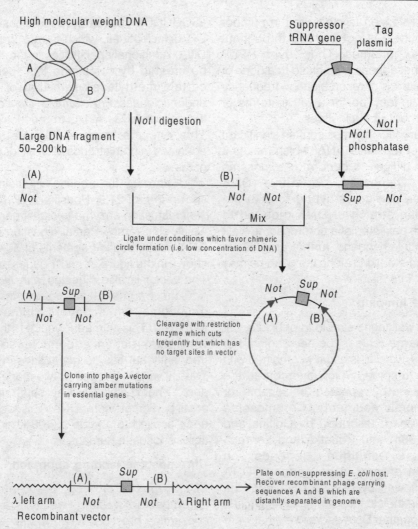

**• Fig. 21.9:** Gene isolation by the chromosome jumping technique. Reproduced from *Principles of Gene Manipulation,* R.W. Old and S.B. Primrose, by kind permission of Blackwell Science, Osney Mead, Oxford, U.K.

library to directly isolate a single clone containing the gene of interest. Molecular linkage maps have now been constructed for more than 30 species, and work is underway to construct maps for many more species by using all the new molecular techniques that have been already described and also by many new techniques. The availability of high-density molecular linkage maps provides a ready starting point to search for molecular markers tightly linked to a target gene. Additional help can be sought from comparative genetic linkage maps of species

that have evolved from a common ancestral genome and thus share portions of their linkage maps that are conserved in both gene content and order. Extensive comparative maps have been constructed for a number of plant taxa, including species in the grass family (rice, maize, sorghum, barley, and wheat), the night-shade family (tomato, potato and pepper), and the mustard family (*Arabidopsis*, cabbage, cauliflower and rape).

Although high-density maps and comparative maps can provide some of the

**Table 21.1:** Example of genes and QTLs coned by map based cloning from crop plant

| Name | Gene or QTLs | Functions | Species | Reference |
|---|---|---|---|---|
| Hdl | QTL | Photoperiod response/heading | Rice | Yano et al. (2000) |
| Hd6 | QTL | Photoperiod response/heading | Rice | Takahashi et al. (2001) |
| Rf- I | Gene | Fertility restorer | Rice | Komeri et al. (2004) |
| Sh4 | QTL | Grain shattering | Rice | Li et al. (2006) |
| Pto | Gene | Bacterial speck blight resistance | Tomato | Martian et al. (993b) |
| Chlornerva | Gene | Iron uptake | Tomato | Ling et al. (1999) |
| Fw 2.2 | QTL | Fruit weight | Tomato | Faray et al. (2000) |
| Brix 9–2–5 | QTL | Sugar content | Tomato | Fridman et al. (2000) |
| Mlo | Gene | Fungal board spectrum resistance | Barley | Buschges et al. (1997) |
| Rpg I | Gene | Fungal stem rest resistance | Barley | Brueggeman et al. (2002) |
| Ppd- HI | QTL | Photoperiod response/flowering time | Barley | Turner et al. (2005) |
| VRNI | Gene | Vernalization /flowering | Triticum monococcum | Yan et al. (2003) |
| VRN2 | Gene | Vernalization /flowering | Triticum monococcum | Yan et al. (2004) |
| Lrl0 | Gene | Fungal leaf rest resistance | Wheat | Feuilet et al. (2003) |
| Lr21 | Gene | Fungal powdery mildew resistance | Wheat | Huang et al. (2003) |
| Pm 3b | Gene | Fungal powder mildew resistance | Wheat | Yahiaoui et al. (2004) |
| NAC | QTL | Protein zinc and iron contents | Wheat | Uauy et al. (2006) |

linked markers needed for chromosome landing in plants, it is clear that additional methods are needed by which new markers can be selectively isolated from specific regions of a genome. The probability of identifying one or more markers within a specified physical distance of a target gene decreases as one gets closer to the target. Ideally one would like to identify a marker at a physical distance from a target that is less than the average insert size of the genomic library from which one expects to isolate the gene. For example, for a YAC library with an average insert size of 200 kb, it would be desirable to identify at least one marker within a 400-kb window containing the target (200 kb on each side). For species with large genome of 1 billion base pairs, the probability would be $4 \times 10^2/10^6 = 4 \times 10^{-4}$. To have a 90% chance of identifying at least one marker within a 400-kb interval, ca. 6000 markers would need to be sampled. It is not likely that many plant species will have molecular linkage maps of this density in the foreseeable future.

Several strategies allowing one to screen a large number of random, unmapped molecular markers in a relatively short time and selecting just those few markers that reside in the vicinity of the target gene have been developed. These high-volume marker technologies that have shown efficacy are RFLP, STS (sequence tagged sites), EST (expressed sequence tag), SSCP (single strand conformation polymorphism), microsatellites, minisatellites, RAPD, SCAR (sequence characterized amplified regions), CAPs (cleaved amplified polymorphic sequences), AFLP, etc. These methods rely on two principles: (i) generate hundreds or even thousands of potentially polymorphic DNA segments and visualize rapidly from single preparations of DNA; and (ii) use of genetic stocks to identify among these thousands of DNA fragments, those few that are derived from a region adjacent to the targeted gene.

In the past few years, thousands of loci scattered throughout the genome have been assayed in just weeks or months by using one or more of these high volume markers technologies. The next problem is to determine which of the amplified loci lie near the targeted gene. Two strategies have proved effective: (i) nearly isogenic lines and (ii) bulk segregant analysis. These are described in Chapter 22.

High-volume marker techniques such as those based on RAPDs, RFLPs, and AFLPs can be used to compare the genetic profiles of pairs of NIL or pairs of bulks to search for DNA markers near the target gene.

In segregating populations of a modest size, a target gene can usually be placed within an interval of 5–10 cM with a high degree of certainty. The markers defining this interval can then be used to screen a large segregating population and identify individuals derived from one or more gametes containing a crossover in the given interval. Only such individuals are useful in orienting other markers closer to the target gene. Once identified, these individuals can be analyzed in relation to all molecular markers within the region to identify those closest to the target. This flanking marker approach was recently used to refine the map position of the *Cf-2* gene, which encodes resistance to *Cladosporium fulvum* in tomato to a 40-kb region between two DNA markers. Through the chromosome landing approach, a tomato gene *Pto*, which confers resistance to the bacterial pathogen *Pseudomonas syringae* pv. *tomato*, has been isolated (Martin *et al.*, 1993). Some of the gene(s) or QTL(s) isolated by map based cloning are described in Table 21.1.

## REFERENCES

Collins, F.S. and Weissman, S.M. 1984. Directional cloning of DNA fragments at a large distance from an initial probe: a circularization method. *Proc. Natl. Acad. Sci. USA* **82**: 6812–6816.

Daelen, R., Van der Lee, T., Diergaarde, P., Groenendijk, J., Topsch, S., Vos, P., Salamini, F. and Schilze-Lefer, T.P. 1997. The barley *Mio* gene: A novel control element of plant pathogen resistance. *Cell* **88**: 695–705.

De Vries, H. 1903. Fertilization and Hybridization. In: Intracellular pangenesis including a paper on Fertilization and Hybridization. Open Court Publishing Co.; Chicago, IL, USA, pp. 217–263 (Translated from Befruchtungund Barstardierung).

Feuillit, C., Travela, S., Stein, N., Albar, L., Nulat, A. and Keller, B. 2003. Map- Based isolation of the leaf rust disease resistance gene *Lr 10* from the hexploid wheat (*Triticum aestivum* L.) genome. *Proc. Natl. Acad. Sci. USA* **100**: 15253–15258.

Fire, A., Xu, S.Q., Montgomery, M.K., Kostas, S.A., Driver, S.E. and Mello, C.C. 1998. Potent and specific genetic interference by double-stranded RNA in *Caenorhabdites elegans*. *Nature* **391**: 806–811.

Frary, A., Nesbitt, T.C., Frary, A., Grandillo, S., Van der Knaap, E., Cong, B., Liu, J., Meller, J., Elber, R., Alpert, K.B. and Tanksley, S.D. 2000. *fw 2.2*: A quantitative trail locus key to the evolution of tomato fruit size. *Science* **289**: 85–88.

Fridman, E., Pleban, T. and Zamir, D. 2000. A recombination hotspot delimits a wild-species quantitative trait locus for tomato sugar content to 484 bp within an invertase gene. *Proc. Nat. Acad. Sci. USA* **97**: 4718–4723.

Gulick, P.J. and Dvorak, J. 1990. Selective enrichment of cDNAs from salt stressed induced genes in the wheat grass, *Lophopyrum elongatum* by the formamide–phenol emulsion reassociation technique. *Gene* **95**: 173–177.

Hunag, L., Brooks, S.A., Li, W., Fellers, J.P., Trick, H.N. and Gill, B.S. 2003. Map-based cloning of leaf rust resistance gene *Lr 21* from the large and polyploid genome of bread wheat. *Genetics* **164**: 655–664.

Komori, T., Ohta, S., Murai, N., Takakura, Y., Kuraya, Y., Suzuki, S., Hiei, Y., Hidemasa Imaseki, H. and Nitta, N. 2004. Map-based cloning of a fertility restorer gent Rf- 1, in rice (*Oryza sativa* L.). *Plant J.* **37**: 315–325.

Li C., Zhou, A. and Sang, T., 2006. Rice domestication by reduced shattering. *Science* **311**: 1936–1939.

Ling, H.Q., Koch, G. and Bumlein Ganal, M.W. 1999. Map-based cloning of chloronerva a gene involved in iron uptake of higher plants encoding nicotainamine synthase. *Proc. Natl. Acad. Sci. USA* **96**: 70–98–7103.

Martin, G.B., Brommonschenkel, S.H., Chunwonges, J., Frary, A., Ganal, M.W., Spiyey, R., Wu, T., Earel, E.D. and Tanksley, S.D. 1999b. Map-Based cloning of a protein kinase gene conferring disease resistance in tomato. *Science* **262**: 1432–1436.

Morgan, T.H. and Catell, E. 1912. Data for the study of sex-linked inheritance in Drosophila. *J. Exp. Zool.* **13**: 79–101.

Poustka, A. and Lehrach, H. 1986. Jumping libraries and linking libraries: the next generation of molecular tools in mammalian genetics. *Trends Genet.* **2**:174–179.

Poustka, A., Pohl, T.M., Barlow, D.P., Frischauf, A.M. and Lehrach, H. 1987. Construction and use of human chromosome jumping libraries from *Not* I digested DNA. *Nature* **325**: 353–355.

Steffenson, B. and Kleinhosf, A. 2002. The barley stem rust-resistance gene *Rpg1* is a novel diseases-resistance gene with homology to receptor kinases. *Proc. Natl. Acad. Sci. USA* **99**: 9329–9333.

Sun, T., Goodman, H.M. and Asubell, F.H. 1992. Cloning the Arabidopsis Ga1 locus by genomic subtraction. *Plant Cell* **4**: 119–128.

Takahashi, Y., Shomura, A., Sasaki, T. and Yano, M. 2001. Hd6, a rice quantitative trait locus involved in photoperiod sensitivity, encodes the á subunit of protein kinase *CK2 Proc. Natl. Acad. Sci. USA* **98**: 7922–7927.

Tanksley, S.D., Ganal, M.W. and Martin, G.W. 1995. Chromosomal landing: a paradigm for map based gene cloning in plants with large genomes. *Trends Genet.* **11:** 63–68.

Turner, A., Beales, J., Faure, S., Dunford, R.P. and Laurie, D.A. 2005. The pseudo- response regulator Ppd- H1 provides adaptation to photoperiod in barley. *Science* **310:** 1031–1034.

Uauy, C., Distelfeld, A., Fahima, T., Blechl, A. and Dubcovsky, R. 2006. A NAC gene regulating senescence improves grain protein, zinc, and iron content in wheat. *Science* **314:** 1298–1301.

Yahiaoui, N., Srichumpa, P., Dudler, R. and Keller, B. 2004. Genome analysis at different ploidy leaves allows cloning of the powdery mildew resistance gene Pm 3b from hexaploid wheat. *Plant J.* **37:** 528–538.

Yan, L., Loukianov, A., Blechl, A., Tranquilli, G., Ramakrishna, W., SanMiguel, P. Bennetzen, J.L., Echenique, V. and Dubcovsky, J. 2004. The wheat *VRN2* gene is a flowering repressor down-regulated by vernalization. *Science* **303:** 1640–1644.

Yan, L., Loukianov, A., Tranquilli, G., Helelguera, M., Fahima, T. and Dubcovsky, J. 2003. Positional cloning of the wheat vernalization gene *VRN1*. *Proc. Natl. Acad. Sci. USA* **100:** 6263–6268.

Yano, M., Yuichi Katayose, Y., Ashikari, M., Yamanouchi, U., Monna, L., Fuse, T., Balba, T., Yamamoto, K., Umehara, Y., Nagamura, Y. and Saskdi, T. 2000. *Hd1*, a major photoperiod sensitivity quantitvive trait locus in rice, is clowely related to the *Arabidopsis* flowering time gene *CONSTANS*. *Plant Cell* **12:** 2473–2484.

## FURTHER READING

Gibson, S and Sommerville, C. 1993. Isolating plant genes: A review. *Trends in Biotech.* **2:** 306–312.

Gupta, P.K. 1999. *Elements of Biotechnology*. Rastogi Publishers, Meerut, India.

Kole, C. and Abbott, A.G. (Eds.) 2008. *Principles and practices of Plant Genomics, Vol. 1: Genome Mapping*, Science Publishers, Enfield, NH, USA.

Miesfeld, R.L.1999. *Applied Molecular Genetics*. Wiley-Liss, New York.

Old, R.W. and Primrose, S.B. 2001. *Principles of Gene Manipulation*. Blackwell Science, UK.

# 22

# Molecular Markers and Marker-assisted Selection

The science of plant genetics traces back to Mendel's classical studies on garden peas. Since then time, researchers have been identifying, cataloging and mapping single gene markers in many species of higher plants. In the early part of the 20th century, scientists discovered that Mendelian factors controlling inheritance, which we now call genes, were organized in linear order on cytogenetically defined structures called chromosomes. Clues to the location of a gene can come from comparing the inheritance of a mutant gene with the inheritance of marker(s) of known chromosomal location. Coinheritance or genetic linkage of a gene of interest (e.g. resistance gene) and marker suggests that they are physically close together on the chromosome. Linkage is a familiar concept in genetics and dates back to early studies on *Drosophila*. It was shown that combination of genes tended to be inherited as groups, linked together because they are close to each other on the same chromosome. In fact the first chromosome map was produced by Sturtevant with segregation data derived from studies on *Drosophila* (Crow and Dove, 1988). The markers on the first genetic map were phenotypic traits scored by visual observation of morphological characteristics of the flies. Markers are "characters" whose pattern of inheritance can be followed at the morphological (e.g., Flower color), biochemical (e.g., protein and/or isozymes), or molecular (DNA) levels. They are so called because they can be used to elicit, albeit indirectly, information concerning the inheritance of "real" traits. A marker must be polymorphic, that is, it must exists in different forms so that chromosome carrying the mutant gene can be distinguished from the chromosome with normal gene by the form of marker it also carries, This polymorphism in marker can be detected at three levels of phenotype (morphological), differences in proteins (biochemical) or differences in the nucleotide sequence of DNA (molecular).

## MORPHOLOGICAL MARKERS

Morphological markers generally correspond to the qualitative traits that can be scored visually. They have been found in nature or as the result of mutagenesis experiments. Morphological markers are usually dominant or recessive. Until recently, however, the genetic markers used to develop maps in plants have been those affecting morphological characters, including genes for dwarfism, albinism, and altered leaf morphology, etc.

Scientists have long theorized about the use of genetic maps and markers to speed up the process of plant and animal breeding. In 1923, Sax proposed identification and selection of minor genes of interest by linkage with major genes, which could be scored more easily. This idea has resurfaced many times and many workers have extended the concept. Unfortunately, there are constraints on the use of morphological markers.

(i) Morphological markers cause such large effects on phenotype that they are undesirable in breeding programs.

(ii) Morphological markers mask the effects of linked minor gene(s), making it nearly impossible to identify desirable linkages for selection.

(iii) Morphological markers are highly influenced by environment. The morphology/ phenotype is the result of genetic constitution and its interaction with environment. Due to varying levels of G x E interaction it is not appropriate to compare the morphological data of varieties that have been collected across different years and/or locations.

(iv) Many morphological traits exhibit dominant intragenic (inter-allelic) interactions. Therefore homozygous dominant and heterozygous individuals cannot be distinguished.

(v) Many morphological traits show intergenic (non-allelic) interactions leading to deviated dihybrid segregation ratios that are not useful for linkage analysis.

(vi) Morphological markers cover only a limited fraction of the genome of an organism.

## BIOCHEMICAL MARKERS

Protein markers, particularly, isozymes (=isoenzymes) have been used very successfully as biochemical markers in certain aspects of plant breeding and genetics as nearly neutral genetic markers (Markert and Moller, 1959). Biochemical markers are proteins produced by gene expression. These proteins can be isolated and identified by electrophoresis and staining. Isozymes, the different molecular forms of the same enzyme that catalyze the same reaction, are proteins. They are revealed on electrophoregrams through a colored reaction associated with enzymatic activity. They are the products of the various alleles of one or several genes. Monomer and dimer isozymes are most often used because analysis of their segregation is easier. Isozymes are generally codominant.

Isozymes are also phenotypic markers and they show tissue variability. Isozymes can be affected by the tissue, growth stage and conditions of plant growth. Some isozymes are better expressed in certain tissues such as roots whereas others in leaf tissue. Protein systems lack adequate polymorphism, genome coverage is low and there is a requirement for the gene to be expressed at the phenotypic level to make the detection possible. To date, only 40–50 reagent systems have been developed that permit the staining of a particular protein in starch. Furthermore, not all of these reagent systems work efficiently with plant species. Therefore, for many species only 15–20 loci can be mapped. As a result, even with the use of isozymes as genetic markers, the full potential of genetic mapping in plant breeding is yet to be realized.

## MOLECULAR MARKERS

A molecular marker is a DNA sequence which is readily detected and whose inheritance can easily be monitored. The use of molecular markers is based on naturally occurring DNA polymorphism, which forms the basis for designing strategies to exploit for applied purposes. A marker must be polymorphic, that is, it must exist in different forms so that chromosome carrying the mutant gene can be distinguished from the chromosome with the normal gene by a marker it also carries. Genetic polymorphism is defined as the simultaneous occurrence of a trait in the same population of two or more discontinuous variants or genotypes.

The first such DNA markers to be utilized were fragments produced by restriction digestion—the restriction fragment length polymorphism (RFLP) based genetic markers. Consequently, several marker systems have been developed. One of the many aspects of this technology is that linkage between molecular markers and traits of interest can be detected in a single cross. With morphological and biochemical markers, a separate cross is required to test linkage with each new marker. The major advantage of the molecular over the other classes of markers are that their number is potentially unlimited, their dispersion across the genome is complete, their expression is unaffected by the environment and their assessment is independent of the stage of plant development. A molecular marker should have some desirable properties.

i. It must be polymorphic as it is the polymorphism that is measured for genetic diversity studies.

ii. Codominant inheritance: The different forms of a marker should be detectable in diploid organisms to allow discrimination of homo- and heterozygotes.

iii. A marker should be evenly and frequently distributed throughout the genome.

iv. It should be easy, fast and cheap to detect.

v. It should be reproducible.

vi. High exchange of data between laboratories.

Unfortunately there is no single molecular marker which meets all these requirements. A wide range of molecular techniques are available which detect polymorphism at the DNA level. These have been grouped into following categories with respect to basic strategy and some major ones have been described.

**Non-PCR based approaches**: Those based on DNA-DNA hybridization between a DNA or RNA probe and total genomic DNA, e.g. Restriction fragment length polymorphism (RFLP).

**PCR based techniques**: Those based on the PCR amplification of genomic DNA fragments: Random amplified polymorphic DNA (RAPD), microsatellites or simple sequence repeat polymorphism (SSRP), amplified fragment length polymorphism (AFLP), arbitrarily primed PCR (AP-PCR), SCAR, SNP etc. However, some assays combine features from both (e.g., retrotransposon based insertional polymorphism—RBIP).

**Targetted PCR and sequencing**: Those based on sequencing: Sequence tagged sites (STS), sequence characterized amplified region (SCARs), sequence tagged microsatellites (STMs), cleaved amplified polymorphic sequences (CAPS), etc.

**History of evolution of molecular markers**: The first DNA based genetic markers were the restriction fragment length polymorphism (RFLP) markers used by Botstein et al. (1980). These were used to construct the earliest genome-wide linkage maps (Helentjaris et al.,1986) and identify the first QTL (Edwards et al., 1987; Paterson et al., 1987). During the 1990s, emphasis was switched to assays based on the polymerase chain reaction (PCR), which brought about a new class of DNA markers, the random amplified polymorphic DNA (RAPD) marker (Williams et al., 1990; Welsh and McClelland et al., 1990). The development of simple sequence repeats (SSRs) (Akkaya et al., 1992; Senior and Heun 1993], amplified fragment length polymorphisms (AFLPs) (Vos et al., 1995) and single nucleotide polymorphism (SNPs) (Jordan and Humphries, 1994) opened the doors for the large-scale deployment of marker technology in genome and progeny screening. There are a large number of marker systems available and each one has advantages and disadvantages. Chronological evolution of molecular markers is given in Table 22.1. The cost sensitivity and suitability of each marker type vary substantially depending on the final application, plant species and instrumentation available at the DNA. The technology related to the detection and analysis of variation at the DNA level is fast-changing. The driver for these developments is, in part, directed at making the technology more efficient, cheaper and robust.

**Table 22.1:** Chronological evolution of molecular markers (adapted from Doveri *et al.*, 2008)

| Acronym | Nomenclature | Reference |
|---------|--------------|-----------|
| RFLP | Restriction Fragment Length Polymorphism | Gordizcker *et al.*, 1974 |
| VNTR | Variable Number Tandem Repeat | Jeffreys *et al.*, 1985 |
| SSCP | Single Stranded Conformational Polymorphism | Orita *et al.*, 1989 |
| STS | Sequence Tagged sites | Olsen *et al.*, 1989 |
| STMS | Sequence Tagged Micosatellite sites | Beckmann and Soller 1990 |
| RAPD | Random Amplified Polymorphic DNA | Williams *et al.*, 1990 |
| AP-PCR | Arbitrarily Primed Polymerase Chain Reaction | Welsh and McClelland 1990 |
| DAF | DNA Amplification Fingerprinting | Caetano-Anolles *et al.*, 1991 |
| SSR | Simple Sequence Repeat | Akkaya *et al.*, 1992 |
| CAPS | Cleaved Amplified Polymorphic Sequence | Akopyanz *et al.*, 1992 |
| MAAP | Multiple Arbitrary Amplicon Prefiling | Caetano–Anolles *et al.*, 1993 |
| SCAR | Sequence Characterized Amplified Region | Paran and Michelmore 1993 |
| SNP | Single Nucleotide Polymorphism | Jordan and Humphries 1994 |
| SAMPL | Selective Amplification of Microsatellite Polymorphic Loci | Morgante and Vogel 1994 |
| ISSR | Inter-Simple Sequence repeat | Zietkiewicz *et al.*, 1994 |
| ALFP | Amplified Fragment Length Polymorphism | Vos *et al.*, 1995 |
| S-SAP | Sequence-Specific Amplified Polymorphism | Waugh *et al.*, 1997 |
| RBIP | Retrotransposon-Based Insertional | Flavell *et al.*, 1998 |
| IRAP | Inter-Retrotransposon Amplified Polymorphism | Kalendar *et al.*, 1999 |
| REMAP | Retrotransposon-Microsatellite Amplified Polymorphism | Kalendar *et al.*, 1999 |

# NON-PCR BASED APPROACHES

## Restriction fragment length polymorphism (RFLP)

RFLP was the first technology which enabled the detection of polymorphism at the DNA sequence level. The genetic information, which makes up the genes of higher plants, is stored in the DNA sequences. Variation in this DNA sequence is the basis for the genetic diversity within a species. In most cases, DNA sequence variation is phenotypically neutral because the majority of mutations altering the order of nucleotides are only maintained during evolution if they are silent.

In this method DNA is digested with restriction enzyme(s) that cuts the DNA at specific sequences, electrophoresed, blotted on a membrane and probed with a labeled clone. Polymorphism in the hybridization pattern is revealed and attributed to sequence difference between individuals. The DNA sequence variation detected by this method was termed restriction fragment length polymorphism

(Botstein *et al.*, 1980). Grodzicker first described this phenomenon for mutant strains of adenovirus. Botstein, White, Skolnick and Davis realized that RFLPs were not necessarily associated with specific genes but were distributed throughout the genome, and they made a detailed proposal for using RFLPs to map all the genes in the human genome. By definition RFLP variation is environmentally independent. Variation in one DNA fragment obtained with a specific enzyme is treated as one RFLP. Plants are able to replicate their DNA with high accuracy and rapidity, but many mechanisms causing changes in the DNA are operative. Simple base pair changes may occur, or large-scale changes as a result of inversions, translocations, deletions or transpositions. This will result in loss or gain of a recognition site and in turn lead to restriction fragments of different lengths. Genomic restriction fragments of different lengths between genotypes can be detected on Southern blots and by a suitable probe. The majority of single nucleotide changes are innocuous, although they may alter a restriction enzyme site so that an enzyme no

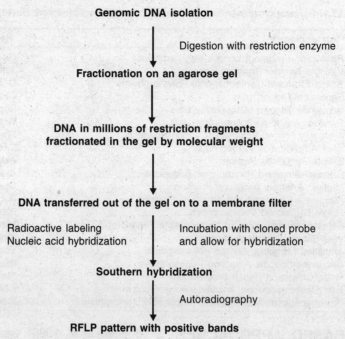

**Genomic DNA isolation**

Digestion with restriction enzyme

**Fractionation on an agarose gel**

**DNA in millions of restriction fragments
fractionated in the gel by molecular weight**

**DNA transferred out of the gel on to a membrane filter**

Radioactive labeling          Incubation with cloned probe
Nucleic acid hybridization          and allow for hybridization

**Southern hybridization**

Autoradiography

**RFLP pattern with positive bands**

**Fig. 22.1:** An outline of conventional RFLP analysis.

longer cuts DNA at that site. For example, the restriction endonuclease *Eco*RI cuts DNA at a specific sequence

↓
... G A A T T C ...   ... G          A A T T C ...
... C T T A A G ...   ... C T T A A  ............ G ...
            ↑

When one or more nucleotides in the endonuclease recognition sequence are altered, *Eco*RI will fail to cut the DNA strands at the altered site and a longer DNA fragment is produced. When a given restriction enzyme site is present in the DNA molecule of one chromosome but absent from the DNA molecule of the other homologous chromosome, a shorter fragment is produced from the chromosome with the site, and a longer fragment from the other chromosome. We are now able to distinguish the two chromosomes in such an individual on the basis of this restriction fragment length polymorphism. The individual is heterozygous for this RFLP and, in the jargon of the geneticist, is said to be informative at that locus. Because

the RFLP is inherited just like a gene, we can follow the individual chromosomes as they pass from generation to generation by tracing the inheritance of the marker fragments.

*Procedure*

RFLP analysis comprises the following basic steps:

 i. DNA isolation.
 ii. Cutting DNA into smaller fragments using restriction enzyme(s).
 iii. Separation of DNA fragments by gel electrophoresis.
 iv. Transferring DNA fragments to a nylon or nitrocellulose membrane filter.
 v. Visualization of specific DNA fragments using labeled probes.
 vi. Analysis of results.

Genomic DNA of an organism is isolated and digested with a particular restriction endonuclease (Fig. 22.1). It will cut the DNA into a large number of fragments. Since DNA has a large number of phosphate groups, which are negatively charged at neutral pH, it will migrate towards the anode in an electric field. In a porous medium of agarose or polyacrylamide, DNA fragments will migrate towards the anode at a rate which is proportional to its molecular weight. Agarose gel electrophoresis has already been explained in Chapter 14.

Relatively small DNAs of chloroplasts will usually produce about 40 discreet restriction fragments when digested with *Eco*RI and can be separated on agarose gels. These are stained with ethidium bromide and pattern can be seen under UV light. But digestion of nuclear DNA yields millions of DNA fragments in a continuous range of sizes the gel appears to run as a smear. However, individual restriction fragments are still well resolved in the gel and RFLPs are still present between DNAs from different organisms.

Since RFLPs of nuclear DNA cannot be directly visualized, the usual procedure is to use small pieces of chromosomal DNA as probes to detect individual restriction fragments. Using the high specificity of DNA-DNA hybridization, such probes can detect individual restriction fragments in the complex mixture of fragments of nuclear DNA present in a restriction digest. To use this technique, a set of chromosomal DNA fragments are prepared as probes. For preparation of probes, DNA is isolated from the species of interest, digested with a restriction enzyme to generate relatively small fragments usually of 0.5–2.0 kb. Individual restriction fragments are ligated into a bacterial plasmid and the plasmid is transformed into a bacterial cell. By growing these transformed bacteria one obtains a large supply of a single plant DNA restriction fragment that is suitable for use as a hybridization probe. Such a set of probes is called a library. Southern blots are prepared from the digested DNA as explained earlier and then probed with one of the cloned probes from the library. For more details read Southern hybridization detection technique in Chapter 14.

The potential of RFLPs as diagnostic markers became evident from studies of the human globin genes, where a direct correlation between the sickle cell mutation carried by a specific β-globin allele and the presence of certain RFLP fragments was evident. The molecular basis of sickle cell anemia is now precisely known. This is not the case for most heritable traits, particularly for those showing quantitative inheritance. For such traits, genetically closely linked markers that do not influence the phenotypic expression of the trait can be identified and they can serve as diagnostic tools. When a genetic linkage map saturated with marker loci is available, the genome of a species can be systematically scanned for markers cosegregating with a specific trait of interest. Because of their abundance, RFLPs are the first class of genetic markers allowing the construction of highly saturated linkage maps. This was first suggested for the human genome by Botstein *et al.* (1980). The potential of this approach in plant breeding was recognized few years later, and its feasibility was experimentally tested in maize and tomato. Since then, an impressive amount of RFLP data has been accumulated in several plant species including important crops. Extensive RFLP linkage maps have been constructed and several genes contributing to agronomically relevant traits have been mapped by RFLP markers.

*Construction of RFLP genetic maps*

In order to discuss the utility of RFLP markers in genetic mapping, it is useful to think of genetic mapping in a slightly different fashion. Consider a genetic cross in a diploid organism in which two homozygous parents are crossed to produce an $F_1$, and the $F_1$ is then selfed to produce $F_2$ population. It is clear that the $F_1$ hybrid from two such parents will have one set of chromosomes from each parent and that the two members of a homologous chromosome pair will be different from one another to the degree to which the parents differ in DNA base sequence.

When the $F_1$ plant undergoes meiosis to produce gametes, its chromosomes will undergo recombination by crossing over. This recombination process is the basis of conventional genetic mapping. The chromosome segments on the same (homologous) chromosome will undergo recombination according to a function which depends on their physical distances. Chromosome segments which are close together (closely linked) will undergo recombination less frequently than those segments which are further apart. Thus, genetic distance or map distance is defined as a function of the recombination which occurs during gamete formation. Previously, the only way to follow chromosome segments and observe their recombination during gamete formation was to observe the phenotype caused by the action of genes. By observing a phenotype such as flower color, plant height, insect resistance etc., the behaviour of chromosome segments has been utilized to construct genetic maps of several plant species. However, the method is cumbersome and extremely time consuming.

Fortunately, RFLP markers provide a way to directly follow chromosome segments during recombination and greatly simplify the construction of genetic map. Instead of looking at the phenotype caused by the presence of a gene on a chromosome segment, one looks directly at RFLP markers on the segments itself. When one considers the inheritance pattern of RFLP markers, it will be seen that they segregate exactly in the same way as do conventional gene markers and follow strict Mendelian rules. This means that segregation pattern of RFLP markers can be analyzed by conventional Mendelian methods and that maps of RFLP markers can be constructed in the same fashion as maps of conventional markers (Fig. 22.2).

To construct RFLP map a fairly well defined series of steps outlined below is followed. For several crop species such as rice, tomato, barley, maize, etc., RFLP maps are already available and one can simply use them without having to construct a new one.

**1. *Select the parent plants*:** As parent plants in a cross for RFLP map construction, one wants to select plants which are genetically divergent enough to exhibit sufficient RFLPs but

not so far distant as to cause sterility of progeny. It would be advantageous also to select a cross in which some desirable agronomic traits are segregating, but this has not always been possible. Since one does not know *a priori* how much RFLP variation will be present in a crop species, it is necessary to make a survey using a random selection of cloned probes. Such a survey should include cultivars or plant introductions thought to represent the spectrum of variability present in the taxon. Related wild species that can be crossed to the cultivar should also be included. One should utilize any available information to select the plants to be surveyed.

After selecting, a range of plants to be surveyed, DNA is isolated and individual plants of each of the accessions are digested with restriction enzymes and screened for polymorphisms by the usual method. Each accession can then be stored for RFLP alleles present, and two accessions showing a usable amount of polymorphism can be selected for crossing.

**2. *Produce a mapping population*:** The selected parent plants are crossed to produce an $F_1$ plant(s). In the case of a selfing plant such as rice, wheat, barley etc., all $F_1$ plants should be alike and heterozygous for all RFLPs exhibited between the parent plants. A mapping population can be derived by selfing the $F_1$ to produce an $F_2$, which is then scored for segregation of the RFLPs or by backcrossing the $F_1$ to one of the parents and observing the segregation in the first backcrosss generation. It is better to use $F_2$ population if this is possible, since more information can be gained than from a backcross population of comparable size. A mapping population of about 50 $F_2$ or backcross plants is sufficient for a fairly detailed map. With plants in which plant breeding studies are being actively carried out, suitable mapping populations might already be available as remnant seed from previous crosses. If this is the case, it may be possible to make a map without having to make any crosses. This can greatly speed up the process. A mapping

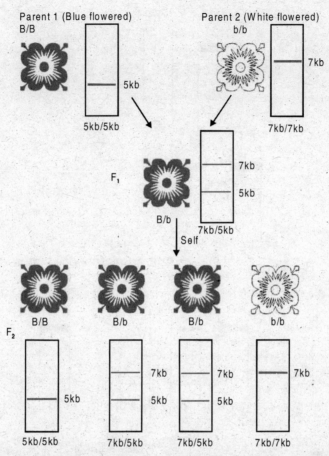

• **Fig. 22.2:** A diagrammatic representation of inheritance of RFLP marker as compared to inheritance of conventional single gene marker controlling flower color.

population can be $F_2$ derived $F_3$ ($F_{2:3}$) population. This ($F_{2:3}$) population is obtained by selling $F_2$ individuals. The $F_3$ families are termed $F_2$ derived $F_3$ ($F_{2:3}$), since each family pedigree traces back to a single $F_2$ individual. This $F_{2:3}$ populations are suitable for mapping of quantitative traits and qualitative traits where replication is needed for accurate phenotypic scoring.

Recombination frequencies can also be estimated from doubled haploids (DH) derived from the pollen of $F_1$ plants. DH population is generated via chromosome doubling of F1 anther or microspore derived haploid plants. DHs are used to make completely homozygous recombinant lines in a single generation.

Advantages of DH lines are: (i) fixed homozygous state is achieved in only a few generations; (ii) DH populations consist of completely homozygous, true breeding individuals and thus represent an immortalized mapping population that can be maintained without any genotypic changes; (iii) scoring of genetic marker data is simpler, since only the two homozygous genotypic classes exist

In maize, recombinant inbred lines (RILs) have also been produced from a number of crosses, through continuous selfing or sib mating in $F_2$ plants. Single seed descent and selfing is repeated through $F_2$ to $F_7$ to $F_{10}$ generations, at which point the RILs are true breeding and fixed. The main advantage is that multiple rounds of

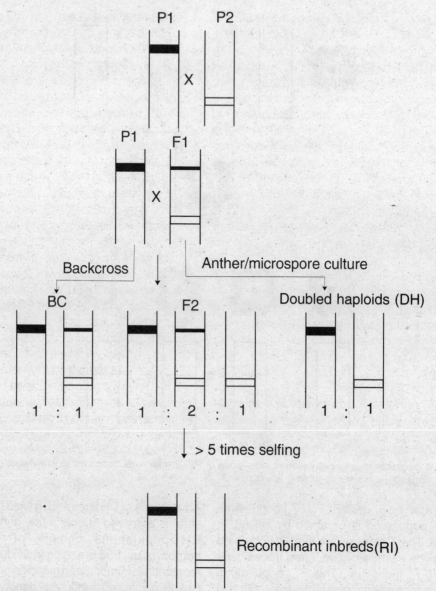

**Fig. 22.3:** Crossing scheme for segregation analysis.

selfing erodes linkage disequilibrium and allows for greater mapping resolution. The crossing scheme has been outlined in Fig. 22.3.

**3. *Scoring of RFLPs in the mapping population***: Once the mapping population is obtained, DNA is isolated from each individual plant in the population. It is important to realize that the chromosome of each plant in the mapping population contains a unique array of parental chromosome segments. To derive a RFLP map it is necessary to determine which parent's chromosome segments would be destroyed by recombination if the $F_2$ or backcross plants were allowed to reproduce sexually. Every effort should be made to keep the individual plants in the mapping population alive so that repeated DNA extractions can be made, and a large amount of DNA can be accumulated. Mapping proceeds by sequentially scoring RFLPs in the individual plants of the mapping population by the following series of procedures:

a. *Screening for polymorphism*: Probes are selected sequentially from the library and tested against the parents in an effort to determine which restriction enzymes will detect a polymorphism between the parents. The number of enzymes required will depend on the amount of variation present in the parent plants. Very polymorphic plants such as inbred lines of maize will usually require only one or two enzymes to detect polymorphism with most probes. In rice, 11 enzymes were used and about 75% of the probes were polymorphic with these.

b. *Scoring*: When a polymorphic probe/enzyme combination is detected, it can then be scored in the mapping population. To accomplish this, a series of agarose gels must be prepared and filters prepared from them for DNA hybridization. These filters are generally called $F_2$ survey or mapping filters. There must be one filter set for each restriction enzyme used in the mapping project, and each filter set must contain digested DNA from each individual plant in the mapping population. Thus one set of mapping filters would contain DNA from each of the $F_2$ plants digested with *Eco*RI, and another set would contain DNA from each of the plants digested with *Pst*I. If a polymorphism between the parent plants has been detected with a given probe/enzyme combination, then that probe will be scored on the corresponding mapping filter set. For example, if probe No. 36 is found to be polymorphic between the parents using *Pst*I, then it could be scored on the *Pst*I mapping filter set. In an $F_2$ mapping population scored for a single RFLP, only three types of plants can be present. There are two homozygous (one similar to each of the parents) and one heterozygous type. Thus, each plant in the

mapping population can be scored as a heterozygote or as one of the two possible types of homozygotes when tested with each library probe.

c. *Linkage analysis*: Data accumulated from scoring the mapping population sequentially with probes from the library is used to construct the linkage map. Linkage analysis is based on the degree to which probes tend to cosegregate. The first probe scored in a mapping study will, of course, provide no information on linkage, but beginning with the second probe one can determine whether linkage is indicated. If the second probe is linked to the first one, they will tend to cosegregate ($F_2$ plants which are heterozygous for the first probe will also tend to be heterozygous for the second probe; plants homozygous for the allele found in one of the parents for the first probe will also tend to be homozygous for the second probe). If no linkage is indicated, distribution of homozygotes and the heterozygotes for the first two probes will tend to be random. Simple statistical tests, such as a Chi-square analysis, will test for randomness of segregation and hence linkage. The first few probes, since they are randomly selected, are unlikely to be linked and hence to show cosegregation. As one sequentially adds probes, however, linkages will be detected. At first there will be many more linkage groups than there are chromosomes, but the two numbers will tend to converge as more markers are added. As each probe is screened, it is tested for possible linkage to all the other markers which have been mapped before. Thus, when one is testing probe No.100, one can test it against the 99 other probes that have been scored before to see if it is linked to any of them. This is essentially like making a 100-point cross. It is evident that a great amount of data is rapidly accumulated in an RFLP mapping project, and a computer is required for efficient data storage and analysis. When a

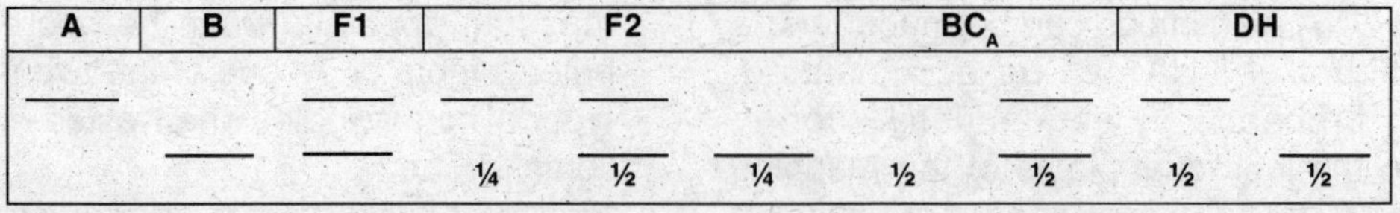

| A | B | F1 | F2 | | | BC$_A$ | | DH | |
|---|---|---|---|---|---|---|---|---|---|
| — | — | — | — | — | — | — | | — | |
| | | | ¼ | ½ | ¼ | ½ | ½ | ½ | ½ |

**Fig 22.4:** Segregation patterns and theoretical ratios with a RFLP codominant marker. A and B are the two parents; BC$_A$: back cross with parent A; DH: doubled haploids.

new probe is scored, if linkage with one or more markers is indicated by the data, the map distance is determined by computer analysis using an algorithm such as maximum likelihood. Recombination data can be converted to map distances using any of the commonly used mapping functions.

## Uses of RFLP

1. It permits direct identification of a genotype or cultivar in any tissue at any developmental stage in an environment independent manner.
2. RFLPs are codominant markers, enabling heterozygotes to be distinguished from homozygotes (Fig. 22.4).
3. It has a discriminating power that can be at the species/population (single locus probes) or individual level (multi locus probes).
4. The method is simple as no sequence-specific information is required.
5. *Indirect selection using qualitative traits*: The greatest potential use of RFLP is its role as indirect selection criteria. It is useful in conjunction with analysis of conventional markers. RFLP maps can be used to supplement regular plant breeding protocols when one wants to select for a conventional gene, but directly selecting for that gene would be expensive, difficult, or time consuming. With indirect selection, one does not directly select for the gene of interest, but rather for one more closely linked RFLP markers. If the RFLP markers are indeed closely linked, they will remain associated with the gene of interest during segregation. This allows one to select for the RFLP marker

with confidence that the conventional gene will also be present, since only relatively rare recombination events would separate the two.

Indirect selection for an RFLP marker has advantages in several plant breeding scenarios. One such situation might involve selection for a recessive gene. Suppose, one was trying to introgress a recessive gene into a cultivar using a backcross breeding program. Backcross protocols involve alternate backcross and selection phases. It is necessary to select progeny bearing the desired gene at several points in the backcross cycle. If a gene is dominant it can be directly selected, but recessive genes will not be expressed in any of the backcross plants, and it would be necessary to carry out progeny testing (probably by selfing the backcross plants to test for the presence of the recessive gene). A recessive allele could be indirectly selected, however, by selecting for a linked RFLP marker. No progeny testing would be necessary, and the process could be greatly speeded up.

6. *Tagging of monogenic traits with RFLP markers*: In inbreeding species, the availability of nearly isogenic lines (NILs) facilitates detection of RFLP markers linked to monogenic disease resistance genes. NILs are obtained by crossing a donor parent carrying the resistance gene with a susceptible recipient parent, followed by repeated backcrosses to the recipient parent under continuous selection for the resistance trait. The donor genome is therefore progressively diluted in the introgressed lines until, in the ideal situation, only a short

**Table 22.2:** Some examples of molecular markers associated with resistance traits in crop plants

| Host species | Pathogen | Resistance gene | Marker |
|---|---|---|---|
| Rice | Pyricularia oryza | Pi-2(t), Pi-4(t) | RFLP |
|  |  | Pi–10(t) | RAPD |
|  | Xanthomonas oryzae | Xa2l | RAPD |
|  |  | Xa3, Xa4, Xa10 | RFLP & RAPD |
|  | Orseolia oryzae | Gm2, Gm4t | RAPD |
| Wheat | Puccinia recondita | Lr9, Lr 24 | RFLP & RAPD |
|  | Erysiphe graminis | Pm1, Pm2, Pm3 | RFLP |
|  | Hessian fly | H21 | RAPD |
| Maize | Leaf blight | rhm | RFLP |
| Barley | Stem rust | Rpg 1 | RFLP |
|  | Barley yellow mosaic | ym4 | RFLP |
|  | Rhyncosporium secalis | — | RFLP |
|  | and barley mild mosaic virus | — |  |
|  | Erysiphe graminis | — | RFLP |
| Brassica napus | Leptosphaeria maculans | — | RFLP |
| Pea | Erysiphe polygoni | er | RFLP |
| Mungbean | Bruchid Callosobruchus | — | RFLP |
| Tomato | Fusarium oxysporum | I2 | RFLP |
|  | Cladosporium fulvum | $Cf_2/Cf_5$ | RFLP |
| Potato | Cyst nematode | H1 | RFLP |

chromosomal segment is retained around the resistance gene. When the parents and the NILs derived from them are compared with molecular probes, only those that are linked to the resistance locus are expected to be polymorphic in the NILs. The linkage and the genetic distance between RFLP markers and the resistance locus are then verified by segregation analysis in normal mapping populations. This technique has been successfully employed for tagging genes of economic value in several crops and some of the examples are given in Table 22.2.

7. *Indirect selection using quantitative trait loci (QTL)*: Several characters of plant species, among which are traits of agronomic importance, are inherited quantitatively. This type of genetic variation is due to multiple factors acting collectively on the expression of a trait. These loci have been designated with the acronym QTL (for quantitative trait locus). Each of the individual genes of such a polygenic system contributes a small positive or negative effect to the trait of interest. Clear dominance is not exhibited and the phenotype has a large environmental component. All these characteristics conspire to make quantitative traits very difficult to analyze. Conventional Mendelian methods of analysis which are suitable for single gene traits cannot be applied to the analysis of quantitative traits, and one is forced to utilize biometrical methods and extensive testing over different years and in different environments in an effort to advance toward the desired state. QTLs are inherited in the same way as the genes of major effect. They segregate, recombine and exhibit linkage theory of association of marker loci with QTLs.

Fundamental advances in this area of plant breeding seemed unlikely before the advent of RFLP mapping techniques. As one can use RFLP markers to simultaneously follow the segregation of all chromosome segments during a cross, the basic idea is to look for correlation between the quantitative trait of interest and specific chromosome segments marked by

RFLPs. If a correlation exists, then the chromosome segment must be involved in the quantitative trait (one or more genes determining the trait must be on that chromosome segment). The difficult part in the procedure is establishing a correlation between the trait and specific chromosome segments. The RFLP markers are easily scored, but the quantitative trait must be scored in the conventional fashion.

RFLP-QTL linkage relationship can be studied in self-pollinated species and among inbred lines in cross-pollinated species. Parents that differ for a number of RFLPs as well as mean values of the quantitative traits are first selected. These parents are crossed and the $F_2$ population is obtained. In the $F_2$ population, a chromosomal segment representing linkage between a RFLP marker and a QTL will be present in the background of random genetic variation due to independent assortment and recombination. By growing samples of $F_2$ population in different environments, one may also study the linkage under different environments. In these cases if mean values of a particular quantitative trait are determined in two groups of plants representing alternative alleles of RFLP, a significant difference in means (of two groups) for quantitative traits will indicate a RFLP-QTL linkage. Individual marker alleles can be assigned breeding values according to the realized effect of the QTL to which they are linked. The realized effect of the QTL is a function of how large an effect the QTL has, and how tightly it is linked to the marker. The selection can then be practiced simultaneously for a number of markers that will have the effect of selecting for QTLs with a positive effect on the quantitative trait. Lander and Botstein (1989) have elaborated a method for mapping QTLs using RFLP maps, which is aimed at identifying promising crosses for QTL mapping, exploiting the full power of complete linkage maps by the approach of interval mapping of QTLs and decreasing the number of progenies to be genotyped. Interval mapping assesses the effects of each genome segment, located between a pair of marker loci, rather than the effect of a QTL associated with a single RFLP.

The interval mapping can be analyzed by maximum likelihood method. When the estimated phenotypic effect $b$ of a single allele substitution at a putative QTL is known, maximum likelihood equations (MLEs) can be calculated which maximize the probability that the observed data will have occurred. These MLEs are compared to MLEs obtained under the assumption that no QTL is linked ($b = 0$), and the evidence that a QTL is existing is indicated by the LOD score (log of the score of odds for the presence of a linkage against the odds of not having this linkage, e.g. if the score is 100:1 in favor of linkage). The LOD score is the ratio of the probability that the data have arisen assuming the presence of a QTL to the probability that no QTL is present. If the LOD score exceeds a predetermined threshold T, a QTL is accepted as being present. Interval mapping can be analyzed by regression method also (Haley and Knott, 1992). Although these two methods may provide similar estimates of QTL parameters, the ML method has several attractive statistical properties such as consistency and efficiency and hence it has great potential for the precise estimation of QTL parameters. Furthermore, the ML method has better interpretability than the regression model in terms of genetic model, suggesting its applicability to practical genetic mapping problems.

Linkage between RFLP and QTL can also be used for the prediction of combining ability. Since RFLPs that are positively correlated for important traits can be identified, inbred lines with RFLPs that should give high general combining ability can be selected.

*Problems*

1. Conventional RFLP analysis requires relatively large amount of highly pure DNA.
2. A constant good supply of probes that can reliably detect variation are needed.
3. It is laborious and expensive to identify suitable marker/restriction enzyme combinations from genomic or cDNA libraries

where no suitable single-locus probes are known to exist.

4. RFLPs are time consuming as they are not amenable to automation.

5. RFLP work is carried out using radioactively labeled probes and therefore requires expertise in autoradiography.

## PCR BASED TECHNIQUES

### Random amplified polymorphic DNA (RAPD) markers

RAPD analysis is a PCR based molecular marker technique. Here, single short oligonucleotide primer is arbitrarily selected to amplify a set of DNA segments distributed randomly throughout the genome. A number of closely related techniques based on this principle were developed almost at the same time and are collectively referred to as Multiple Arbitrary Amplicon Profiling (MAAP). These techniques are:

RAPD    Randomly Amplified Polymorphic DNA
DAF     DNA Amplifying Fingerprinting
AP-PCR  Arbitrarily Primed Polymerase Chain Reaction

All these techniques refer to DNA amplification using single random primers and share the same principle with some differences in the experimental details. Of these, RAPD was the first to become available and is most commonly and frequently used. Williams et al. (1990) showed that the differences as polymorphisms in the pattern of bands amplified from genetically distinct individuals behaved as Mendelian genetic markers. Welsh and McClelland (1990) showed that the pattern of amplified bands so obtained could be used for genomic fingerprinting.

RAPD amplification is performed in conditions resembling those of polymerase chain reaction using genomic DNA from the species of interest and a single short oligonucleotide (usually a 10-base primer). The DNA amplification product is generated from a region which is flanked by a part of 10-bp priming sites in the appropriate orientation. Genomic DNA from two different individuals often produces different amplification patterns [randomly amplified polymorphic DNAs (RAPDs)]. A particular fragment generated for one individual but not for other represents DNA polymorphism and can be used as a genetic marker. Using different combinations of nucleotides, many random oligonucleotide primers have been designed and are commercially available. Such primers can be synthesized in an oligonucleotide synthesizing facility based on sequences chosen at random. No separate information is required from the plant to be studied. The choice of single primer (or RAPD primer) to use is done operationally. Since each random primer will anneal to a different region of the DNA, theoretically many different loci can be analyzed.

The PCR reaction typically requires cycling among three temperatures, the first to denature the template DNA strands, the second to anneal the primer and the third for extension at temperatures optimal for Taq polymerase. This cycle is usually repeated 25 to 45 times (Fig. 22.5).

It is usually sufficient to heat the reaction mixture at 94°C for 30 to 60 s. However the initial denaturation step for the plant DNA should be for at least 3 min. Insufficient denaturation is a common problem leading to the failure of the PCR reaction.

The temperature for the annealing step will depend to some extent on base composition and the length of primers. Generally, an annealing temperature of 35–40°C is used in PCR reactions for RAPD analysis. Annealing may require only a few seconds, but this will depend on a variety of factors such as interference from secondary structure and primary concentration. A time in the range of 30 s to 1 min is usually sufficient.

The extension temperature depends on length and concentration of the target sequence, and in general 72°C temperature is used. As a thumb rule 1 min per kilo base is probably ample. After cycling, an extended extension period of

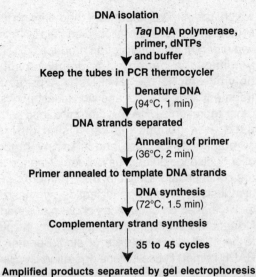

**DNA isolation**

↓ *Taq* DNA polymerase, primer, dNTPs and buffer

**Keep the tubes in PCR thermocycler**

↓ Denature DNA (94°C, 1 min)

**DNA strands separated**

↓ Annealing of primer (36°C, 2 min)

**Primer annealed to template DNA strands**

↓ DNA synthesis (72°C, 1.5 min)

**Complementary strand synthesis**

↓ 35 to 45 cycles

**Amplified products separated by gel electrophoresis**

**Fig. 22.5:** An outline of RAPD analysis.

5–10 min is included to ensure that all annealed templates are fully polymerized.

Most genomes from bacteria to higher plants to human beings contain enough perfect or imperfect binding sites for a short primer of arbitrary sequence. These markers cover the entire genome, i.e. single to repeated sequences. Binding of primer to the complementary sequences allows the amplification of several bands. A DNA sequence difference between individuals in a primer binding site may result in the failure of the primer to bind, and hence in the absence of a particular band among amplification products. The reaction products are conveniently analyzed on agarose gels, stained with ethidium bromide and seen under UV light.

Presence of a RAPD band corresponds to a dominant allele against absence of band that corresponds to a recessive allele. Thus, heterozygous and homozygous dominant individuals cannot be differentiated with RAPD markers (Fig. 22.6). The segregating $F_2$ population may therefore be scored as follows:

Band present    AA (homozygote)
                                   Aa (heterozygote)
Band absent    aa (recessive homozygote)

In $F_2$, these two classes would be expected to segregate in a 3:1 ratio. This causes a loss of information relative to RFLP markers which show codominance. Backcross, recombinant inbred, and the doubled haploid populations do not suffer this loss of mapping information, because the complete information available from a backcross can be obtained from scoring the presence or absence of a polymorphic marker. Keeping in mind this fact, it is easy to select the populations best suited for the construction of genetic maps with RAPD markers

A comparison of the properties of RAPDs in relation to other markers has been shown in Table 22.3. The following characteristics of RAPD markers make them relevant for marker assisted selection.

*Advantages*

1. Need for a small amount of DNA (15–25 ng) makes it possible to work with populations which are inaccessible for RFLP analysis.
2. It involves nonradioactive assays.
3. It needs a simple experimental set up requiring only a thermocycler and an agarose assembly.
4. It does not require species-specific probe libraries; thus work can be conducted on a large variety of species where such probe libraries are not available.

| A | B | F1 | F2 | | BC$_A$ | BC$_B$ | | DH | |
|---|---|---|---|---|---|---|---|---|---|
| — | | — | — | | | — | — | — | |
| | — | — | — | | — | — | | — | |
| | | ¾ | ¼ | 1 | ½ | ½ | ½ | ½ | |

**Fig. 22.6:** Segregation patterns and theoretical ratios with a RAPD dominant marker. A and B are the two parents; BC$_A$: back cross with parent A; BC$_B$: back cross with parent B; DH: doubled haploids.

**Table 22.3:** A comparison of properties of restriction fragment length polymorphisms and randomly amplified polymorphic DNAs

| Characteristic | RAPD | RFLP | AFLP | Sequence tagged microsatellites |
|---|---|---|---|---|
| Principle | DNA amplification | Restriction digestion | DNA amplification | DNA amplification |
| Detection | DNA staining | Southern blotting | DNA staining | DNA staining |
| DNA required- quanlity | Crude | Relatively pure | Relatively pure | Crude |
| DNA required- quantity | Low | High | Medium | Low |
| Primer requirement | Yes (random primer) | None | Yes (selective primer) | Yes (selective primer) |
| Probe requirement | None | Set of specific probes | None | None |
| Use of radioisotopes | No | Yes | Yes-no | Yes-no |
| Part of genome surveyed | Whole genome | Generally low copy coding region | Whole | Whole |
| Dominant/ codominat | Dominant | Codominant | Dominant (Codominant) | Codominant |
| Polymorphism | Medium | Medium | Medium | High |
| Automation | Yes | No | Yes | Yes |
| Reliability | Intermediate | High | High | High |
| Recurring cost | Low | High | Medium | Low |

5. It provides a quick and efficient screening for DNA sequence based polymorphism at many loci.

6. It does not involve blotting or hybridization steps.

*Application*

RAPD markers have been successfully used for the following applications:

1. Construction of genetic maps: Their potential as genetic markers for map construction has been exploited and maps consisting largely or exclusively of RAPD markers have been developed in *Arabidopsis*, pine, *Helianthus*, etc.

2. Mapping of traits: The use of RFLPs for selection of markers linked to monogenic and polygenic traits has been discussed. Likewise RAPD markers may also be used for indirect selection in segregating populations during plant breeding programs. This technique has been employed for tagging genes of economic value, examples of which are given in Table 22.2.

3. Analysis of the genetic structure of populations.

4. Fingerprinting of individuals.

5. Targeting markers to specific regions of the genome.

6. RAPDs have been utilized for the identification of somatic hybrids (See Chapter 9).

7. RAPDs have been recommended as a useful system for evaluation and characterization of genetic resources. Some success has already been achieved in characterization of collections from tree genera *Gliricidia* and *Thenobroma*.

*Limitations*

1. RAPD polymorphisms are inherited as dominant-recessive characters. This causes a loss of information relative to markers which show codominance.

2. RAPD primers are relatively short, a mismatch of even a single nucleotide can often prevent the primer from annealing, hence there is loss of band.

3. The production of nonparental bands in the offspring of known pedigrees warrants its use with caution and extreme care.

4. RAPD is sensitive to changes in PCR conditions, resulting in changes to some of the amplified fragments.

## DNA amplification fingerprinting (DAF)

A single arbitrary primer, as short as 5 bases, is used to amplify DNA using PCR. DAF requires careful optimization of parameters, however, it is extremely amenable to automation and fluorescent tagging of primers for early and easy determination of amplified products. DAF differs from RAPD in following respects:

i. Higher primer concentrations are used in DAF.
ii. Shorter primers of 5–8 nucleotides are used in DAF.
iii. Two temperature cycles in DAF as compared to three in RAPD.
iv. DAF usually produce very complex banding pattern.
v. Products of DAF are analyzed on acrylamide gels and detected by silver staining.

## Arbitrarily primed polymerase chain reaction (AP-PCR)

This is a special case of RAPD, wherein discreet amplification patterns are generated by employing single primers of 10–50 bases in length in PCR of genomic DNA. In the first two cycles, annealing is under non-stringent conditions. AP-PCR differs from RAPD in following respects:

i. In AP-PCR the amplification is in three parts with its own stringency and concentration of constituents.
ii. In AP-PCR higher primer concentrations are used in the first PCR cycles.
iii. Primers of variable length, and often designed for other purposes, are arbitrarily chosen for use (M13 universal sequence primer).
iv. Products of AP-PCR are mostly analyzed on acrylamide gels and detected by autoradiography.

## Amplified fragment length polymorphism (AFLP)

This is a highly sensitive method for detecting polymorphism throughout the genome and is becoming increasingly popular. It is essentially a combination of RFLP and RAPD methods, and it is applicable universally and is highly reproducible. It is based on PCR amplification of genomic restriction fragments generated by specific restriction enzymes and oligonucleotide adapters of few nucleotide bases (Vos *et al.*, 1995). It is a novel DNA fingerprinting technique. DNA fingerprinting involves the display of a set of DNA fragments from a specific DNA sample. Fingerprints are produced without prior sequence knowledge using a limited set of genetic primers. AFLP technique uses stringent reaction conditions for primer annealing and combines the reliability of RFLP technique with the power of PCR technique.

In principle it involves the following steps:

1. DNA is cut with restriction enzymes (generally by two enzymes) and double stranded (ds) oligonucleotide adapters are ligated to the ends of the DNA fragments.
2. Selective amplification of sets of restriction fragments is usually carried out with $^{32}$P labeled primers designed according to the sequence of adaptors plus 1–3 additional nucleotides. Only fragments containing the restriction site sequence plus the additional nucleotides will be amplified.
3. Gel analysis of the amplified fragments. The amplification products are separated on highly resolving sequencing gels and visualized using autoradiography. Fluorescent or silver staining techniques can be used to visualize the products in cases where radiolabeled nucleotides are not used in the PCR.

A schematic representation of the technique has been shown in Fig. 22.7. The restriction fragments for amplification are generated by two restriction enzymes, a rare cutter and frequent cutter. The figure shows that fragments are generated from *Eco*RI and *Mse*I. Oligonucleotides are generally used as adapters and primers. AFLP adapters consist of a core sequence and an enzyme specific sequence.

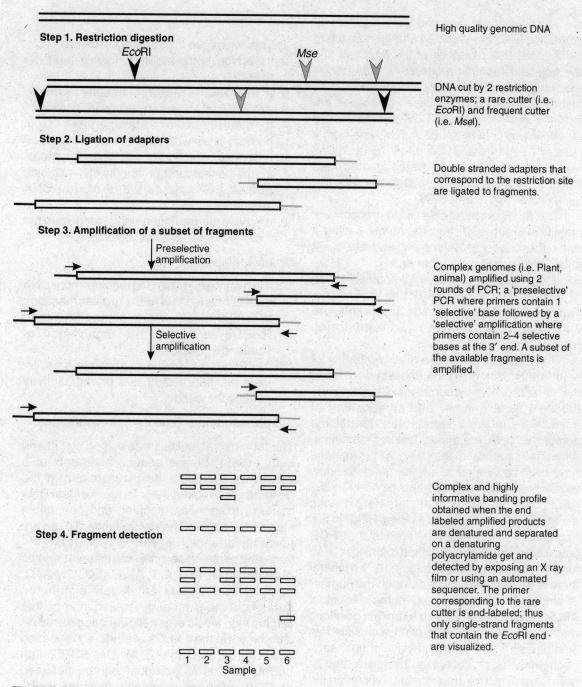

**Step 1. Restriction digestion**

*Eco*RI

*Mse*

High quality genomic DNA

DNA cut by 2 restriction enzymes; a rare cutter (i.e. *Eco*RI) and frequent cutter (i.e. *Mse*I).

**Step 2. Ligation of adapters**

Double stranded adapters that correspond to the restriction site are ligated to fragments.

**Step 3. Amplification of a subset of fragments**

Preselective amplification

Selective amplification

Complex genomes (i.e. Plant, animal) amplified using 2 rounds of PCR; a 'preselective' PCR where primers contain 1 'selective' base followed by a 'selective' amplification where primers contain 2–4 selective bases at the 3′ end. A subset of the available fragments is amplified.

**Step 4. Fragment detection**

Complex and highly informative banding profile obtained when the end labeled amplified products are denatured and separated on a denaturing polyacrylamide gel and detected by exposing an X ray film or using an automated sequencer. The primer corresponding to the rare cutter is end-labeled; thus only single-strand fragments that contain the *Eco*RI end are visualized.

1   2   3   4   5   6
Sample

**Fig. 22.7:** Schematic representation of AFLP technique.

The restriction fragments with their 5′ protruding ends are obtained and then adapters are ligated. The sequence of the adapters and the adjacent restriction site serves as primer binding sites for subsequent amplification of the restriction fragments. Selective nucleotides are included at the 3′ends of PCR primers, which therefore can only prime DNA synthesis from a subset of the restriction sites. Only those restriction fragments in which nucleotides flanking the restriction site match the selective nucleotides will be amplified.

The AFLP procedure results in predominant amplification of restriction fragments having a rare cutter sequence on one end and a frequent cutter sequence on the other end.

The method allows specific coamplification of a high number of restriction fragments. Typically 50–100 restriction fragments are amplified and detected on denaturing polyacrylamide gels.

This method generates a large number of restriction fragment bands, thereby facilitating the detection of polymorphism. Choosing the different base number and composition of nucleotides in the adapters can control the number of DNA fragments that are amplified. Although not many maps using AFLP markers have been developed so far, this method is now widely used for developing polymorphic markers. AFLP is a technique that detects genomic restriction fragments and resembles RFLP in this respect, with the major difference that PCR amplification instead of Southern blotting is used for detection of restriction fragments. In general there is a linear correlation between the number of amplified fragments and genome size. Most AFLP fragments correspond to unique positions on the genome and hence can be exploited as landmarks in genetic and physical maps, each fragment being characterized by its size and its primers required for amplification. A comparison of the properties of AFLPs in relation to other markers has been shown in Table 22.3.

## Advantages

1. This technique is extremely sensitive.
2. It has high reproducibility, rendering it superior to RAPD.
3. It has wide-scale applicability, proving extremely proficient in revealing diversity.
4. It discriminates heterozygotes from homozygotes when a gel scanner is used.
5. It permits detection of restriction fragments in any background or complexity, including pooled DNA samples and cloned DNA segments.
6. It is not a simple fingerprinting technique but it can be used for mapping also.

## Disadvantages

1. It is highly expensive and requires more DNA than is needed in RAPD (1μg per reaction).
2. It is technically more demanding than RAPDs as it requires experience of sequencing gels.
3. AFLPs are expensive to generate as silver staining, fluorescent dye or radioactivity detects the bands.

## Simple sequence repeats (Microsatellites)

The term microsatellites was coined by Litt and Lutty (1989). Simple sequence repeats also known as microsatellites are present in the genomes of all eukaryotes. These are ideal DNA markers for genetic mapping and population studies because of their abundance. These are tandemly arranged repeats of mono-, di-, tri-, tetra- and penta-nucleotides with different lengths of repeat motifs (e.g. A, T, AT, GA, AGG, AAAC, etc.). These repeats are widely distributed throughout the plant and animal genomes that display high levels of genetic variation based on differences in the number of tandemly repeating units at a locus (Table 22.4). This SSR length polymorphisms at individual loci are detected by PCR using locus specific flanking region primers where the sequence is known

**Table 22.4:** List of microsatellites or simple sequence repeats in plants

| Species | Repeats |
|---------|---------|
| *Arabidopsis* | AT, AG, GA, CT, TC, CA |
| Barley | AT, CT, TG, TCT, CTT, ATTT |
| Maize | CT, CA, AC, AG, GA, AG |
| Rice | CA, GT, GA, AT, GGT |
| Sorghum | AG, AC, AAC, AAG |
| Wheat | GA, GT, CT, CA, AT, GT |
| Soybean | AT, CT, TA, ATT, AAT, TAT, TAA |
| Sunflower | CA |
| Tomato | GT, GA, ATT, GATA |
| Grape vine | GA, GT, AG |
| Citrus | ATT, TAT, TAG |
| Yam | CT, AT, TA, CA |

(Discussed in the following section under sequence tagged microsatellites). Thus STMs require precise DNA sequence information for each marker locus from which a pair of identifying flanking markers are designed. This is impractical for many plant and animal slpecies that are not well characterized genetic systems. Nevertheless high levels of polymorphism inherent to SSRs has prompted an alternative approach: the detection of SSR derived polymorphisms without the need of sequencing. The approaches described below all involve PCR amplification fingerprinting with oligonucleotide primers corresponding directly to simple sequence repeat target sites. Some of these SSR-based methods have been collectively termed microsatellite-primed PCR (MP-PCR) and shown in Fig. 22.8.

***Microsatellite directed PCR: unanchored primers***: This approach is also known as SPAR (single primer amplification reaction). Oligonucleotides composed wholly of defined, short tandem repeat sequences and representing a variety of different microsatellite types have been used as generic primers in PCR amplification based application. Most often, a SSR oligonucleotide is used in a single primer amplification reaction (Fig. 22.8). Microsatellite repeat primed reaction is similar to RAPD reaction in that exponential amplification will occur from these single primer reactions only when the particular repeat used as a primer is represented in multiple copies, which are closely spaced (<2–3 kb) and inversely oriented in the template DNA. These reactions are multiplexed, multiple products are simultaneously co-amplified, and therefore multiple loci can be detected from a genome using a single PCR reaction. Apart from the SSR sequence used as a primer, there is no other DNA sequence information required to design and perform these amplifications. The banding pattern after amplification can be resolved on low resolution agarose gels using ethidium bromide staining. The same set of primers can be used across species and even kingdoms, and therefore constitute a general purpose primer set. It has been reported that MP-PCR reactions can be more specific than RAPD reactions since the longer SSR based primers (15-mers to 20-mers) enable higher stringency amplifications. This may result in fewer problems with reproducibility, a complaint frequently leveled against the low stringency RAPD assays with shorter primers.

This approach is simple to perform and require little up-front locus specific or species specific information. The unanchored SSR primers appear to behave during PCR much like a RAPD primer with considerable dependence on specific amplification conditions. SPAR appears to be subject to irreproducibility effects and requires careful optimization of the conditions for specific primer genome combinations to ensure that heterogeneity of banding patterns observed is the result of true genetic polymorphism at fixed loci.

***Microsatellite-directed PCR: anchored primers***: This method is known variously as anchored microsatellite primed PCR (AMP-PCR), inter-microsatellite PCR and inter SSR amplification (ISA or ISSR). In this approach radiolabeled di- or tri-nucleotide repeats (SSR primers) are modified by the addition of either 3′ or 5′-anchor sequences of 2–4 nucleotides composed of nonrepeat bases (Fig. 22.8; Zietkiewicz *et al.*, 1994). The anchored primers

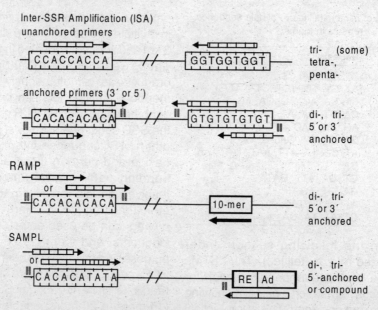

**Fig. 22.8:** Schematic representation of different microsatellites primed PCR methods. RE—restriction enzyme; Ad—synthetic adaptor; 10-mer—arbitrary sequence decamer.

can be $(CA)_8RG$, $(CA)_8RY$, $(CA)_7RTCY$ (where Y = pyrimidine; R = purine). One of the first examples of the use of anchored simple sequence repeat primers was for first strand cDNA synthesis with oligo-d(T) primers carrying one or two non-T bases at the 3′ end. The anchor in the primer serves to fix the annealing of the primer to a single position at each target site on the template, such that every new polymerization event initiates at the same target position. Thus, there is little or no chance for primer slippage on the template. It has been reasoned that since dinucleotide repeats are most abundant in eukaryotic genomes than other types of microsatellites and at least (CA)n: (GT)n repeats appear to be ubiquitously represented in genomes, a small defined set of primers representing these repeats can be developed that would produce complex finger prints.

***Random amplified microsatellite polymorphisms (RAMP)*:** This approach termed random amplified microsatellite polymorphism (Wu *et al.*, 1994) is based on the random distribution of nucleotide sequences immediately flanking a simple sequence repeat. Amplification is performed between a 5′-anchored mono-, di- or trinucleotide repeat primer and a random sequence decamer (RAPD primer) to generate coamplified products capable of carrying codominant length polymorphism that may be present at the SSR target site (Fig. 22.8). The RAPD primer binding site serves as an arbitrary endpoint for the SSR based amplification product and therefore the products obtained are not as restricted by the relative genomic position of SSRs in the genome as they are with ISA. Individual combinations of different SSR based primers with different decamers allows for the generation of nearly limitless numbers of unique amplification products, all of which are bordered at one end by a microsatellite repeat. The SSR primer is [32]P labeled and the amplification products are resolved on polyacrylamide gels. The RAMP method was shown to produce polymorphic coamplification products in *Arabidopsis*. Both dominant and codominat polymorphisms were detectable .

*Selective amplification of microsatellite polymorphic loci (SAMPL)*: It is an SSR based modification of AFLP approach, termed selective amplification of microsatellite polymorphic loci (Vogel and Morgante, 1995). AFLP is a PCR based method for detecting differences in a selected set of restriction fragments and involves the following steps: (i) restriction digestion of total genomic DNA with two endonucleases (one rare cutter and other frequent cutter); (ii) ligation of adapters of known sequences to the two ends of each fragment and (iii) selective amplification of a set of anonymous restriction fragments by using primers complementary to the sequences including adapter sequence + restriction site sequence + selective nucleotides at 3′ end. Amplification may involve two steps, the first involving pre-amplification using primers with fewer (0 or 1) selective oligonucleotides and second involving amplification using primers with 3 selective nucleotides. In SAMPL approach amplification is performed with one labeled SSR primer (anchored through the use of compound repeats) and one unlabeled adapter primer (Fig. 22.8). The SAMPL method uses the same adapter modified restriction fragment templates as are used for AFLP, but the amplification is performed using a $^{32}$P or $^{33}$P-labeled SSR primer combined with an unlabeled adaptor primer. In SAMPL during second amplification, one AFLP primer with three selective nucleotides is used in combination with a SAMPL primer that is at least in part complementary to microsatellite sequences. Generally the SAMPL primer (18–20 nucleotides long) is based on the sequences of two different adjacent SSRs and an associated intervening sequence, known to be found in compound repeats.

Many different combinations of restriction endonucleases, SSR primers and selective adapter primers are possible with SAMPL, allowing for the detection of nearly limitless number of SSR based polymorphisms in the genome. Since SSRs are ubiquitously present in eukaryote genomes, complex fingerprints have been reported in both plant and animal genomes. Both dominant and codominant polymorphisms were detected in plant species.

It is not possible to state definitely which of the methods would be most suitable for every application. Unanchored SSR primer approach requires a heavy note of caution, since tri-, tetra- and penta-nucleotide based oligonucleotides have been shown to exhibit a great deal of nonspecificity, and hence the bands generated may not be SSR based and may not be reproducible. Inter-repeat amplifications with anchored primers are well suited for fingerprinting and germplasm discrimination, but the bands produced are always limited to SSRs within amplifiable distances of one another in the genome. For map construction and trait marking then either of the two methods RAMP and SAMPL involving amplification between a single SSR and an arbitrary fixed site will allow for detection of nearly limitless numbers of SSR based markers. SAMPL may be well suited for detecting large amounts of diverse SSR based codominant polymorphism, although the assay itself is more technically demanding than any other method.

### Retrotransposon-based Markers

Retrotransposons are important vehicles for genome change and are frequently the causal agents of mutation, gene duplication and chromosome rearrangements. The insertion of retrotransposon into new sites are relatively stable because they replicate through reverse transcription of their RNA and integration of the resulting 'cDNA' into a new locus. This mechanism of replication is shared with retroviruses, with difference that retrotransposons do not form infectious particles that leave the cell to infect other cells (Havecke *et al.*, 2004). This replicative mode of transposition can rapidly increase the copy number of elements and can thereby greatly increase plant genome size. Retrotransposons have been found in all eukaryotic genomes analyzed to date. They are distributed as interspersed repetitive sequences almost throughout the length of all host chromosomes and they can constitute a substantial fraction of the nuclear DNA. In maize, for example, 50% of the genome are retrotransposon in origin (San Miguel *et al.,* 1996) and the 10,000–20,000 copies of BARE–1 constitute up to 5% of the barley genome (Manninen and Schulman, 2000). Investigations of genome structure in flowering plants indicated that most variation in the size of nuclear genomes is caused by the differential amplification and/or retention of retrotransposon that contain long terminal repeats (LTRs) (San Miguel *et al.,* 1996; Bennetzen 2002).

The LTR retrotransposons (including retroviruses) are one class of sequences that replicate and transpose via an RNA intermediate (retroelements). Non-LTR retrotransposons, LINEs (long dispersed repetitive element) and SINEs (short interspersed repetitive elements) are also retroelements and together these sequences are found in high copy number in plant species. LTR retrotransposons are located largely in intergenic regions and are the single largest components of most plants genomes. They have been discovered in plants as sources of both spontaneous and induced mutations in maize and tobacco. LTR retrotransposons are further subclassified into *copia*-like and the *gypsy*-like groups that differ from each other in both their degree of sequence similarity and their structure (i.e. the order of functional domains along the element-encoded polycistronic RNA). Please see Chapter 20.

There are several advantages of using retrotransposon sequences as molecular markers. They are ubiquitous, present in moderate to high copy number as highly heterogenous populations, widely dispersed in chromosomes and show insertional polymorphisms both within and between species in plant. As they are stably inherited, retrotransposon positions in the genome act as temporal genetic markers that help in establishing pedigrees, phylogenies and for

genetic diversity studies. Furthermore, the replicative mode that retrotransposons share with retroviruses means that the two LTRs that bound each element is derived form a single LTR, mutations begin to accumulate after transposition and thus divergence between these gives insight to the age of the element.

The exploitation of retrotransposons as marker in plants was first described by Lee *et al.* (1990) where the members of a *copia*-like element were used as RFLP and PCR-based marker in pea. Subsequently, several molecular markers based on retrotransposons have been developed and exploited to study biodiversity in maize, pea and barley and to generate genetic linkage maps in barley (Waugh *et al.* 1997) and pea (Ellis *et al.,* 1998).

***Sequence-Specific Amplified Polymorphism (S-SAP)***: S-SAP was first described by Waugh *et al.* (1997). The method exploits the unique junction created between retrotransposon and non-retrotransposon DNA by anchoring one end of a restriction DNA fragment to and adapter in a similar fashion to AFLP. Unlike AFLP, only one restriction enzyme is required, a frequent cutter (*Mse*I or *Taq*I), and adapters ligated to the ends of the fragments. PCR is performed using two primers, one derived from the sequence near the end of the retroelement, usually within the LTR, and the other from the adapter. Depending on the number of copies in the genome, selective bases can be added to the anchored primer to reduce the number of bands when a retrotransposon family contains too many members, and thus can be optimized to individual retrotransposon families (Leigh *et al.,* 2003). By labeling the LTR derived primer, only LTR related products are visualized. The method is so similar to AFLP that it is possible to use the same restriction-ligation reactions, even though it has been digested using two enzymes (Waugh *et al.,* 1997; Taylor *et al.,* 2004).

S-SAP, like AFLP, is capable of efficiently generating a large amount of markers distributed over the whole genome. The polymorphic information content of S-SAP markers, a measure of their genetic diversity, are higher than those for AFLP. The band intensities are fairly uniform unlike for AFLP, where organellar and repetitive DNAs give rise to intense, often monomorphic bands. The limiting step in the development of the markers system is the availability of terminal sequences from retrotransposons from which to design primers. S-SAP has been applied in barley (Kumar et al.,1997; Waugh et al., 1997), peas (Ellis et al., 1998), wheat (Gribbon et al., 1999), Medicago (Proceddu et al., 2002), tobacco (Melayah et al., 2001) and grapevine (Pelcy and Merdinoglu, 2002).

The high density of *BARE–1* retrotransposons in the cereal genomes has allowed the development of other retrotransposon PCR-based marker that do not require the restriction-ligation of DNA.

***Inter-Retroransposon Amplified Polymorphism (IRAP)***: The IRPA method amplifies DNA fragments between two LTR sequences sufficiently close to one another in the genome to permit PCR amplification of the region (Kalendra et al., 1999). In individuals where one or other of the two LTRs are missing, they will not amplify the corresponding fragment. Insertional polymorphisms are thus generally visualized as polymorphic, dominant bands by electrophoresis on either sequencing or high-resolution agarose gels. The amplified products range from under 100 bp to over several kilobase pairs, with the minimum size depending on the placement of the PCR priming site with respect to the ends of the retrotransposons.

This method was first implemented in *Hordeum* by Kalendar et al. (1999) and since then has been used by Kalendar et al. (2000) and Vicient et al. (2001) for fingerprinting and biodiversity analysis, and for mapping of agronomically important traits (Manninen et al., 2000; Boyko et al., 2002).

***Retrotransposon-Microsatellite Amplified Polymorphism (REMAP)***: REMAP is conceptually very similar to IRAP but instead of amplifying retrotransposon-retrotransposon fragments, it exploits the association between microsatellite sequences and retrotransposons (Kalendar et al., 1999). This method combines LTR primers with SSR primers to generate amplicons when the two sequences are sufficiently close in the genome which allows amplification.

Both IRAP and REMAP require sequence families of sufficient copy numbers to make it probable that they are close enough in the genome to produce amplicons in PCR. To date they have been predominantly used for cereals.

***Retrotransposon-Based Insertional Polymorphism (RBIP)***: RIBP is a method to survey whether a retrotransposon is present at a given site. Two primers flanking the insertion site will amplify when there is no element present. The presence of a retrotransposon displaces the two primers and the product is too large to be amplified. A third primer is included in the reaction that is derived from the element and will only be amplified with one of the external primers when the element is present (Flavell et al., 1998). Since both empty and filled sites will generate different products, the markers are co-dominant. Each PCR reaction surveys a single locus but as detection can be performed using hybridization assays rather than gel-based detection, RBIP can be automated to increased throughput using a Taq Man™ or DNA chip technology in order to increase sample throughput (Favell et al., 2003). The major drawback to this technique is the requirement to isolate and sequence insertion sites.

# TARGETED PCR AND SEQUENCING

An approach opposite to the arbitrarily amplicon profiling procedure is to design primers to target specific regions of the genome. This approach is a targeted PCR and sequencing. The targeted product can be compared to the corresponding product from another individual on an agarose gel, but the resolution achieved will only detect differences in length of fragment resulting from many base pair changes. To resolve all the possible sequence differences, sequencing the entire fragment becomes necessary. A sequence tagged site (STS) is the general term given to a marker which is defined by its primer sequences.

## Sequence tagged sites

STS is a short unique sequence (60 to 1000 bp) that can be amplified by PCR, which identifies a known location on a chromosome (Olson *et al.*, 1989). Specific PCR markers that match the nucleotide sequence of the ends of a DNA fragment can be derived from primers, e.g. an RFLP probe or an expressed sequence tag. To date, all STSs that have been used in mapping projects have been derived from well characterized probes or sequences. STSs are the physical DNA landmarks and PCR is the experimental method used to detect them. STS maps simply represent the relative order and spacing of STSs within a region of DNA. Using this technique, tedious hybridization procedures involved in RFLP analysis can be overcome. STSs have been extensively used for physical mapping of genome. Examples of STSs are:

Sequence tagged microsatellites (STMs).
Sequence characterized amplified regions (SCARs).
Cleaved amplified polymorphic sequence (CAPS).

### Sequence tagged microsatellites (STMs)

The term microsatellites was coined by Litt and Lutty (1989). Simple sequence repeats also known as microsatellites are present in the genomes of all eukaryotes. These are ideal DNA markers for genetic mapping and population studies because of their abundance. These are tandemly arranged repeats of mono-, di-, tri- and tetranucleotides with different lengths of repeat motifs (e.g. A, T, AT, GA, AGG, AAAC, etc.). A motif is A, AT, AGG, etc. and repeat number is denoted by n. Thus a repeat $(AT)_9$ means AT nucleotides are tandemly arranged one after another nine times. In a genome of a particular species when this repeat is identified in a gene, which constitutes a microsatellite, the gene is sequenced with its flanking sequences to design primers for amplification of microsatellites. The regions flanking the microsatellite are generally conserved among genotypes of the same species. PCR primers to the flanking regions are used to amplify the SSR containing DNA fragment. Length polymorphism is created when PCR products from different individuals vary in length as a result of variation in the number of repeat units in the SSR. Genebank sequence data have also been used for designing primers for amplification of microsatellites. Thus SSR polymorphism (SSRP) reflects polymorphism based on the number of repeat units in a defined region of the genome.

Poly G and poly A are the simplest of microsatellites. It now appears that any short nucleotide motif can be found as a microsatellite in the eukaryotic genome. The number and composition of microsatellite repeats differ in plants and animals. The frequency of repeats longer than 20 bp has been estimated to occur every 33 kb in plants, whereas in mammals it is found to occur every 6 kb. In humans AC or TC are very common repeats; in plants AT is more common followed by AG or TC. In general plants have about ten times less SSRs than humans. Microsatellites have been identified in several plant species such as *Arabidopsis thaliana*, *Brassica napus*, rice, maize, wheat, barley, etc. Some of the SSRs in plants have been listed in Table 22.4 (rev Mohan *et al.*, 1997).

### Procedure

A specific microsatellite contained within a stretch of DNA can be amplified by PCR using flanking primer sequences, and then analyzed on Metaphor agarose or polyacrylamide gels. The gels are stained with ethidium bromide and seen

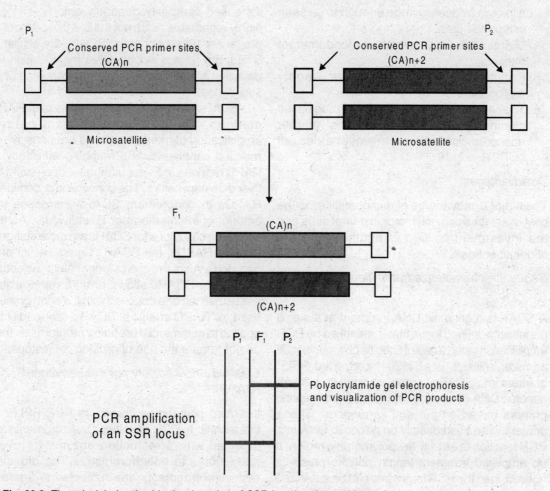

P₁

Conserved PCR primer sites
(CA)n

Microsatellite

P₂

Conserved PCR primer sites
(CA)n+2

Microsatellite

F₁

(CA)n

(CA)n+2

P₁    F₁    P₂

PCR amplification
of an SSR locus

Polyacrylamide gel electrophoresis
and visualization of PCR products

• **Fig. 22.9:** The principle involved in the detection of SSR length polymorphism.

under UV light. The principle underlying the detection of SSR length polymorphism among parents and their $F_1$ hybrid is illustrated in Fig. 22.9. The variation in length of PCR product is a function of the number of SSR units.

This is a relatively new technique and is specifically useful in inbreeding crops like wheat, barley that are characterized by low levels of RFLP variation. A comparison of the properties of STMs in relation to other markers has been shown in Table 22.3.

*Advantages*

1. This technique is very simple and easy to use.
2. Assay is PCR based. It is sufficient to merely separate the amplification products by electrophoresis to observe the results. This reduces the time required considerably to obtain a result comparable to the methods that are based on Southern blotting.
3. The use of radioisotopes can be avoided because the size polymorphism between

alleles is frequently large enough to be seen in agarose gels.

4. Microsatellites segregate as codominant markers.
5. Polymorphism generated is highly reproducible.
6. They are perfectly suited for use in map based cloning because identification of marker anchored clones is conveniently carried out by PCR.

### Disadvantages

The major disadvantage of microsatellites is the cost of establishing polymorphic primer sites and the investment is high in synthesizing the oligonucleotides.

### Sequence characterized amplified regions (SCARS)

A SCAR is a genomic DNA fragment at a single genetically defined locus that is identified by PCR amplification using a pair of specific oligonucleotide primers. Williams *et al.* (1991) converted RFLP markers into SCARs by sequencing two ends of genomic DNA clones and designing oligonucleotide primers based on the end sequences. These primers were used directly on genomic DNA in a PCR reaction to amplify the polymorphic region. If no amplified fragment length polymorphism is noticed, then the PCR fragments can be subjected to restriction digestion to detect RFLPs within the amplified fragment. SCARs are inherited in a codominant fashion in contrast to RAPDs which are inherited in a dominant manner. Paran and Michelmore (1993) converted RAPD markers into SCARs. Amplified RAPD products are cloned and sequenced. The sequence of primers derived from the termini of a band is identified as a RAPD marker. Two 24-base oligonucleotide primers corresponding to the ends of the fragment (the 5′ ten bases are the same as the original 10-mer used in the RAPD reaction and 14 internal bases from the end) have been synthesized. These primers with their increased specificity generally amplify a single highly repeatable band, not the 5–10 bands for the progenitor 10 base primers. SCARs are similar to STSs, but do not involve DNA hybridization for detection and can therefore contain repeated DNA sequences.

SCARs have several advantages over RAPD markers. RAPDs show dominant nature, amplification of multiple loci and are sensitive to reaction conditions. The mapping efficiency of RAPD markers in $F_2$ populations is decreased by their dominant nature. The conversion of dominant RAPDs to codominant SCARs increases the amount of information per $F_2$ individual. As the annealing conditions for SCARs are more stringent than for RAPDs, the SCAR primers detect only one locus. Also, the use of longer oligonucleotide primers for SCARs allows a more reproducible assay than the one obtained with the short primers used for RAPD analysis. SCARs can readily be applied to commercial breeding programs as they do not require the use of radioactive isotopes.

### Cleaved amplified polymorphic sequences (CAPS)

It is also sometimes known as PCR-RFLP. In this technique PCR amplified DNA fragments are digested with a restriction enzyme to reveal restriction site polymorphisms. The digested restriction fragments are subjected to agarose gel electrophoresis followed by staining with ethidium bromide for visualization of bands. Konieczny and Asubel (1993) first adapted the CAPS procedure for genetic mapping by developing a set of CAPS markers for use with *Arabidopsis*. They designed 18 sets of primers which amplified products ranging in size from 0.316 to 1.728 kb from Columbia and Landsberg ecotypes. These amplified products were digested with a panel of restriction endonucleases to identify restriction enzymes that generated ecotype specific patterns.

## Advantages

1. The sizes of the cleaved and uncleaved amplification products can be adjusted arbitrarily by the appropriate placement of the PCR primers.
2. CAPS markers can be readily assayed using standard agarose gel electrophoresis.
3. CAPS markers are codominant.
4. A small quantity of DNA is required to assay CAPS markers.
5. CAPS mapping procedure is technically simple but the results are robust because an amplification product is always obtained.
6. CAPS markers can be assayed relatively quickly.

## Sequence related amplification polymorphism (SRAP)

The SRAP assay involves use of primers designed to amplify the open reading frames (ORFs) of genes (Li and Quiros, 2001). It uses two primers of 17 or 18 nucleotides in length comprising of the following elements: core sequences, which are 13 to 14 bases long, where the first 10 or 11 bases starting at the 5′ end are sequences of no specific constitution ('filler' sequences); followed by the sequence CCGG in the forward primer and AATT in the reverse primer. The core is followed by three selective nucleotides at the 3′ end. The filler sequences of the forward and reverse primers must be different from each other and can be 10 or 11 bases long. For the first five cycles the annealing temperature is set at 35°C; the following 35 cycle are run at 50°C. The amplified DNA fragments are separated by denaturing acrylamide gels. The purpose for using the 'CCGG' sequence in the core of the first set of SRAP primers is to target exons or ORFs. This rational is based on the fact that the exons normally in GC-rich regions. SRAP profiling was shown to target expressed regions of the *Brassica* genome through its use in *Olea europaea* (Reale *et al.,* 2006) gave rise to an excess of retrotransposon sequences.

## Advantages of PCR targeted and sequencing approaches

1. The fragment in which polymorphisms are studied is of known identity, therefore avoiding ambiguities of analyzing RAPD and AFLP bands or random probes.
2. Data based on DNA sequences or restriction site mapping provide both the means of classifying individuals into different classes and assessing relationships among the classes.
3. For population studies, the use of an organellar sequence in complementation with a nuclear sequence can provide useful data with respect to mechanism of differentiation, gene flow and dispersal.
4. The development of genealogical based analytical methods coupled with studies of DNA sequence variation within and among populations is likely to reveal immense information on demographic processes.

## Disadvantages

1. Unless the frequency of variants is high enough to be easily detected by PCR-RFLP or other sensitive gel assays, sequencing is required which in turn necessitates investment of adequate resources and experienced researchers.
2. Although the quality of data is high because the approach is often resource intensive, the coverage of the genome is highly restricted, usually to only one sequence and therefore, to one point of comparison.
3. There is difficulty in identifying sequences that are reliably variable enough at the population level in any system under study.

## Single Nucleotide Polymorphism (SNP)

Single base changes and short insertion/deletion represent the most abundant source of DNA polymorphisms in organisms and they are classified as single nucleotide polymorphisms or SNPs. They represent a defined position at a chromosomal site at which the DNA sequence of two individuals differs by a single base. SNPs were first described by Jordan and Humphries (1994). They have become the marker of choice by virtue of their genome coverage and the parallel testing procedures that enable thousands of loci to be assessed within a single experiment.

In principle, SNPs could be bi-, ti-, or tetra-allelic polymorphism. However, tri-allelic and tetra-allelic SNPS are rare almost to the point of non-existence and so SNPs are sometimes simply referred to as bi-allelic marker (or di-allelic to be etymologically correct). In humans Kwok *et al.* (1996) found an average of one SNP every 1,000 bp. The frequency and nature of SNPs in plants is beginning to receive considerable attention. Tenaillon *et al.* (2001) found an average of one SNP every 104 bp in maize between two randomly sampled sequences. Similar results have been obtained by surveying sequence polymorphisms in eight lines of *Beta vulgaris* (Schneider *et al.*, 2001). These polymorphisms could be used as simple genetic markers that may be identified within or near every gene.

SNP discovery relies on finding differences between two sequences. Direct sequencing of DNA fragments (amplified by PCR) from several individuals is the most direct way to identify SNP polymorphisms. PCR primers are designed to amplify 400–700 bp segments of DNA that are frequently derived from genes of interest or ESTs from the database. PCR is performed on a set of diverse individuals that represent the diversity in the species /population of interest. The PCR products are sequenced directly in both directions and the resulting sequences are aligned and polymorphisms identified.

In addition, the dramatic increase in the number of DNA sequences submitted to the database has made it possible to identify SNPs for several crops by electronic mining (e-mining:) without the need for sequencing (Taillon-Miller *et al.*,1998; Somers *et al.*, 2003). This approach consists in the identification and alignment of sequences from the same locus from different sources (genotypes) allowing the detection of SNPs along these DNA sequences. The prerequisite is to have ample sequences for their alignment to identify polymorphisms and to distinguish real genetic changes from those generated by sequencing errors. The availability of EST database makes it possible to target polymorphisms to functional regions of the genomes and even to specific genes.

Validation of SNPs can be performed by a number of different protocols. The choice of the method for a particular assay depends on many factors, including cost, throughput, equipment needed, difficulty of assay development and potential for multiplexing. Typically, genotyping protocols start with target amplification and follow with allelic discrimination and product detection/identification.

Method employed for SNP genotyping includes: denaturing high-pressure liquid chromatography (DHPLC) (Underhill *et al.*, 1997; Kuklin *et al.*, 1998; O'Donovan *et al.*, 1998), tetra-primers ARMS-PCR (Ye *et al.*, 2001), single-strand conformational polymorphism (SSCP) (Orita *et al.*, 1989a,b), enzymatic cleavage methods (CAPS) (Cotton, 1993; Mashal *et al.*,1995; Youil *et al.*, 1996), MALDI-TOF mass spectrometry-based system, pyrosequencing (Tsuchihashi and Dracopoli, 2002), single-based extension (SBE) or single nucleotide extension (SNuPE) assays, ligase chain reaction (LCR) (Landegren *et al.*, 1988), Taqman™ assays, hybridization assays-allele-specific hybridization (Connor *et al.*, 1983), microarray (Wang *et al.*, 1998) and molecular inversion probes (Hardenbol *et al.*,2003).

Single nucleotide polymorphism has opened the doors for the large scale deployment of marker technology in genome and progeny screening. Today the fingerprinting at 1500 SNP loci of 384 maize lines can be completed in two weeks at a cost of around US$ 35,000, equaling less than $0.10 per data point (Illumina Technology) (Walt, 2005).

## Molecular markers and plant diversity

Variation in the genome sequence and epigenetic activities such as gene silencing and altered chromatin structure within or across a crop species is referred to as "plant genetic diversity". This diversity provides the reason why, for example, some crop varieties are tall and some are short, why some species can survive insect attack while other closely related ones cannot, etc. For millennia, farmers have taken advantage of genetic diversity by selecting the most favorable material for the following season.

Crop breeders need plant genetic diversity in their breeding programs to achieve genetic gain. They can exploit the allelic diversity present in their breeding materials by chromosomes re-assortment and/or recombination, and/or they can import new alleles by hybridizing elite germplasm with exotic materials. The amount of genetic diversity present in a particular collection of individuals can be derived from comparisons between DNA fingerprints (the identification of a living organism). Such estimates of the amount of genetic diversity present can be used to define phylogenetic relationship among germplasm entries, and to evaluate genetic distances separating them. A sufficient characterization of the genetic relationships among germplasm is particularly useful, as it indicates the combination of breeding parents, which is most likely to offer a high degree of allelic complementarity, contributing a large potential for genetic gain in the offspring. Thus, one of the most powerful applications of markers in breeding is the guiding of parental combinations before making crosses. Although most fingerprinting assays are based on anonymous markers whose function is not known. The increasing pace of gene discovery and elaboration of gene function is driving a shift towards markers associated with specific genes. The advantage of such markers is that they can allow selection of parental lines based on specific traits. Fingerprinting is also valuable for establishing varietal identity and purity, and variety protection, which are real issues for the breeder, both in the commercial and the pubic sectors.

# FINGERPRINTING

DNA fingerprinting term was coined for multilocus DNA profiles detected by using multilocus hypervariable probes, which simultaneously score genotypes at several loci without recognizing the allelic relationships of the individual DNA fragments (Jeffreys et al., 1985). DNA typing methods have been widely used in criminal investigation and for determining the pedigree and parentage of an individual. The technique of classical DNA fingerprinting is a derivative of RFLP analysis, but differs from the latter by the type of hybridization probe applied to reveal genetic polymorphisms. For a typical multilocus DNA fingerprint, probes are used which create complex banding patterns by recognizing multiple genome loci simultaneously.

For characterization of genome (finger-printing) a variety of molecular marker techniques are available which are either hybridization based or PCR based. The markers are RFLP, RAPD, AP-PCR, DAF, microsatellites, AFLP, etc. These molecular marker techniques are used as such or these have been modified to reveal polymorphism. It has been reported that techniques which give high information content (estimated as either the polymorphic information content or genetic diversity index or expected heterozygosity) should be used.

RFLP approach has been widely used in fingerprinting of individuals. Southern blot restriction fragment length polymorphism method of DNA typing and population databases are available. Although PCR based databases are also fast coming up for DNA typing.

AFLP approach is widely used for developing polymorphic markers. The high frequency of identifiable AFLPs coupled with high reproducibility makes this technology an attractive tool for detecting polymorphism and fingerprinting. This technique is extremely useful in detection of polymorphisms between closely related genotypes. AFLP analysis depicts unique fingerprints regardless of the origin and complexity of the genome.

A hybridization based approach is oligonucleotide fingerprinting. It reveals multiple genome loci simultaneously. Each of these loci is characterized by more or less regular arrays of tandemly repeated DNA motif that occur in different numbers at different loci. Two categories of such multilocus probes are mainly used. The first category comprises cloned DNA fragments or synthetic oligonucleotides which are complementary to minisatellites (tandem repeats of a basic motif about 10 to 60 bp long). The second category of probes is short, synthetic oligonucleotides which are complementary to microsatellites (tandem repeats of one to five bp). RFLP analysis with microsatellite complementary oligonucleotides is also referred to as oligonucleotide fingerprinting or hybridization based microsatellite fingerprinting. Both minisatellite and microsatellite probes have been applied to RFLP fingerprinting of numerous plant and animal species. High levels of polymorphism between related genotypes are often observed and the technique is being used for genome analysis, paternity testing, genotype identification and population genetics. To exemplify the procedure for oligonucleotide fingerprinting of plants is given below.

1. Isolation of genomic DNA suitable for RFLP analysis.
2. Complete digestion of genomic DNA with an appropriate restriction enzyme.
3. Electrophoretic separation of restriction fragments in agarose gels.
4. Denaturation and immobilization of the separated DNA fragments within the gel or blotting the DNA fragments on a membrane.
5. Hybridization of the dried gel or the membrane to radioactive or non-radioactive labeled microsatellite complementary oligonucleotide probes.

6. Detection of hybridizing fragments (i.e. fingerprints) by autoradiography or by chemiluminescence.
7. Evaluation of banding patterns and documentation by photography.

This technique of in-gel hybridization offers several advantages over Southern hybridization and has been increasingly utilized to reveal hypervariable target regions in a variety of plant materials. The quality, intensity and number of fragments that hybridize with synthetic oligonucleotides vary considerably from probe to probe and from species to species. Since SSRs are abundant and uniformly distributed in the genome, a large number of hybridized fragments, differing in size (few hundred base pairs to as many as 8–10 kb) are generally obtained. Therefore, they may be used as multilocus probes for identification and characterization of varieties, cultivars, breeding lines. Multilocus probes have the advantage of revealing polymorphism at many loci simultaneously, so that a small collection of probes is enough to cover a representative part of the genome.

Microsatellite complementary oligonucleotides are also useful for banding patterns when hybridized to PCR products generated by arbitrary primers. This novel strategy combines arbitrarily or microsatellite primed PCR with microsatellite hybridization. In this technique genomic DNA is amplified with either a single arbitrary 10-mer primer (as in RAPD analysis) or a microsatellite complementary 15-mer or 10-mer primer. PCR products are then electrophoresed, blotted and hybridized to a $^{32}$P or digoxigenin labeled mono-, di-, tri- or tetranucleotide SSR probes [e.g. $(CA)_8$, $(GA)_8$, $(GTG)_5$, $(GCGA)_4$]. Subsequent autoradiography reveals reproducible, probe dependent fingerprints which are completely different from the ethidium bromide staining patterns, and polymorphic at an intraspecific level. This provides for speed of the assay along with high sensitivity, so that high level of polymorphism is detected. This method was termed RAMPO (random amplified microsatellite polymorphism), RAHM (random amplified hybridization micosatellite) or RAMS (randomly amplified microsatellites).

*Evaluation of fingerprints for oligonucleotide fingerprinting*

A fingerprinting pattern obtained from a particular DNA sample is rarely informative on its own. The patterns originating from different samples have to be compared to each other, individual bands are assigned to particular positions and different lanes are screened for matching bands. This multilocus DNA profiles are usually translated into discreet characters i.e. "0" for absence and "1" for presence of band at a particular position. These "0/1 matrices" can then be exploited for different types of studies. Since bands are generally inherited in a Mendelian fashion, an application of RFLP fingerprinting is the estimation of relatedness between parents and offspring. For paternity testing, the fingerprint patterns of mother, father and offspring are compared. All offspring bands must either be derived from mother or father. A large number of non parental bands will exclude paternity. For the evaluation of higher order relationships, a similarity index (*F*) can be calculated from band sharing data of each of the fingerprints according to the formula:

$$F = \frac{2m[ab]}{m[a] + m[b]}$$

where *m*[a] and *m*[b] represent the total number of bands present in lanes [a] and [b] respectively, and *m*[ab] the number of bands which both lanes share. *F* can acquire any value between 0 and 1, where 0 means no bands in common and 1 means patterns are identical. A distance matrix can be prepared from the similarity values, which can be further transformed into a dendrogram.

The application of RAPDs and the related modified marker AP-PCR in variability analysis

and individual specific genotyping has largely been carried out. A modification of RAPD is DAF which has been shown to generate reliable fingerprints, because many more bands are resolved due to short primers. Thus DAF is useful for detecting polymorphisms even between organisms that are closely related. ISSRs or ISA (inter SSR amplification) which has been explained is a suitable marker for fingerprinting.

For fingerprint evaluation obtained by PCR approaches, quantification of pairwise similarities or differences between fingerprints can be done using different algorithms. Most commonly a similarity index is calculated as depictor of band sharing. For example, Jacard's similarity coefficient (*J*) takes into consideration only those matches between bands that are present (Jacard, 1908). Whereas Nei and Li's coefficient (*N*) measures the proportion of bands shared as the result of being inherited from a common ancestor and represents the proportion of bands present and shared in both samples divided by the average of the proportion of bands present in each sample (Nei and Li, 1979). These similarity coefficients are calculated by:

$$J = \frac{sp}{sp + (1-s)} \qquad N = \frac{2sp}{2sp + (1-s)}$$

Where *p* is the proportion of shared bands present in both samples and *s* is the proportion of the total number of bands that are shared either present or absent in both samples. The *N* coefficient has been recommended for analysis of multilocus fingerprinting.

These DNA fingerprints can also be used for genetic mapping. Bands which are polymorphic between two parents of a cross can be analyzed for linkage to each other as well as to other types of markers. A variety of computer programs are available that calculate recombination values and map distances and give a most likely order of markers within linkage groups and thus create genetic maps (e.g. LINKAGE-1, MAPMAKER, GMENDEL, JOINMAP).

# MARKER ASSISTED SELECTION (MAS)

In this technique, linkages are sought between DNA markers and agronomically important traits such as resistance to pathogens, insects and nematodes, tolerance to abiotic stresses, quality parameters and quantitative traits. Instead of selecting for a trait, the breeder can select for a marker that can be detected very easily in the selection scheme. The essential requirements for marker assisted selection in a plant breeding program are:

1. Marker(s) should cosegregate or be closely linked (1cM or less is probably sufficient for MAS) with the desired trait.
2. An efficient means of screening large populations for the molecular marker(s) should be available. At present this means relatively easy analysis based on PCR technology.
3. The screening technique should have high reproducibility across laboratories.
4. It should be economical to use and be user friendly.

Several strategies have been developed that allow one to screen a large number of random, unmapped molecular markers in a relatively short time and to select just those few markers that reside in the vicinity of the target gene. These high volume marker technologies that have shown efficacy are RAPD, AFLP, RFLP, microsatellites, etc. These methods rely on two principles: (i) to generate hundreds or even thousands of potentially polymorphic DNA segments and rapidly visualized from single preparations of DNA; and (ii) use of genetic stocks to identify among these thousands of DNA fragments, those few that are derived from a region adjacent to the targeted gene.

In the past few years, by using one or more of these high volume markers technologies, thousands of loci scattered throughout the genome have been assayed in a matter of weeks or months. The next problem is to determine which of the amplified loci lie near the targeted gene. Two strategies have proved effective.

1. *Nearly isogenic line (NIL) strategy*: Breeders have developed NIL genetic stocks and have been maintaining these inbreds lines that differ at the targeted locus. Nearly isogenic lines are created when a donor line (P1) is crossed to a recipient line (P2). The resulting $F_1$ hybrid is then backcrossed to the P2 recipient to produce the backcross 1 generation ($BC_1$). From $BC_1$, a single individual containing the dominant allele of the target gene from P1 is selected. Selection for the target gene is normally made on the basis of phenotype. This $BC_1$ individual is again backcrossed to P2, and the cycle of backcross selection is repeated for a number of generations. In the $BC_7$ generation, most if not all of the genome will be derived from P2, except for a small chromosomal segment containing the selected dominant allele, which is derived from P1. Lines homozygous for the target gene can be selected from the $BC_7F_2$ generation. The homozygous $BC_7F_2$ line is said to be nearly isogenic with the recipient parent, P2. The banding pattern by high volume marker techniques has also been shown Fig 22.10.

2. *Bulk segregant analysis (BSA)*: This method is more generally applicable, and relies on the use of segregating populations (Michelmore *et al.*, 1991). It requires the generation of populations of bulked segregants (bulks). When P1 and P2 are hybridized, the $F_2$ generation derived from the cross will segregate for alleles from both parents at all loci throughout the genome. If the $F_2$ population is divided into two pools of contrasting individuals on the basis of screening at a single target locus, these two pools (Bulk 1 and Bulk 2) will differ in their allelic content only at loci contained in the chromosomal region close to the target gene. Bulk 1 individuals selected for

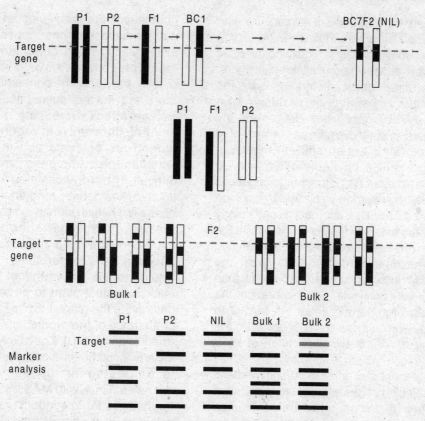

• **Fig. 22.10:** Pattern of markers based on RAPDs/ AFLPs can be used to compare the genetic profiles of NILs and bulks.

recessive phenotype will contain only P2 alleles near the target, while Bulk 2 plants selected for dominant phenotype will contain alleles from both P1 and P2 (dominant homozygote and heterozygote). Bulk 1 and Bulk 2 will contain alleles from both P1 and P2 at loci unlinked to the target.

High volume marker techniques such as those based on RAPDs, RFLPs and AFLPs can be used to compare the genetic profiles of pairs of NIL or pairs of bulks to search for DNA markers near the target gene. DNA fragments indicated by arrows that are observed in the NIL and P1 populations, but not in P2, or that are unique to either Bulk 1 or Bulk 2, should be derived from a locus tightly linked to the target gene.

A large volume of data on plant QTLs has been generated which underline the efforts made to identify phenotype/genotype associations in experimental crosses. In parallel several marker assisted selection schemes have been proposed and tested. The power of DNA markers is their ability to select for genotypes carrying a favorable allelic composition at all marker loci. Favorable QTL alleles can be used to transfer one or more discreet genomic segments from a donor to an elite cultivar by backcrossing, or to conduct marker assisted population improvement by stacking favorable alleles into individuals selected on the basis of marker genotypes.

Marker assisted backcrossing (MABC), in which a chosen allele at a marker locus is transferred from a donor to a recipient line, has been widely used to introgress favorable alleles into elite material which lacks a specific characteristic. A suite of genome-wide markers helps to expedite the progress of the backcrossing process, since it allow the simultaneous selection of the donor allele at the target locus and the rapid recovery of the recurrent parent alleles elsewhere. Although a number of parameters influence the choice of selection strategy, the design of a workable MABC program is relatively straightforward, and genetic gain can be predicted by simulation. MABC is an efficient means of transferring a single favorable allele (e.g., a transgene or a major QTL) into a range of genetic backgrounds, or of improving a particular genotype for a given trait. This latter approach is particularly important when breeding for foodstuffs, where the development of new germplasm is challenging because the new product needs to fit the requirements of local consumers and be better than products already available. For this reason, the introgression of superior alleles via MABC to improve popular cultivars for a specific trait is perhaps the most suitable application of MAS (Rebaut, 2007).

For marker-assisted population improve-ment, individual selected from a segregating population based on their marker genotype are inter-mated at random to produce the following generations, at which point the same process can be repeated a number of times. A second approach aims direct recombination between selected individual as part of a breeding scheme seeking to generate an ideal genotype or ideotype. The ideotype is pre- defined on the basis of QTL mapping. This variety development approach is commonly referred to as marker assisted recurrent selection (MARS), or genotype construction.

Marker assisted selection has become a promising and potent approach for integrating biotechnology with conventional and traditional breeding. The plant breeders interest on molecular markers revolve around certain basic issues which have been illustrated below.

1. *Resistance breeding*: At present breeding for disease and pest resistance is conducted on the segregating populations derived from crosses of susceptible cultivars with resistant donors. These populations are then selected either under natural disease or pest hot spots or under artificially created conditions. Although these procedures have given excellent results, they are time consuming. Besides, there are always susceptible plants that escape attack. Screening of plants with several different pathogens and their pathotypes or pests and their biotypes simultaneously or even sequentially is difficult, if not impossible. Availability of tightly linked genetic markers for resistance genes will help in identifying plants carrying these genes simultaneously without subjecting them to pathogen or insect attack in early generations. The breeder would require a low amount of DNA from each individual plant to be tested without destroying the plant, and see the presence or absence of the product of PCR reaction (marker band) on the gel. Only materials in the advanced generations would be required to be tested in disease and insect nurseries. Thus, with MAS it is now possible for the breeder to conduct many rounds of selection in a year without depending on the natural occurrence of the pest or pathogen as well. Some illustrative examples of molecular markers with resistance genes are given in Table 22.2.

2. *Pyramiding of major/minor genes into cultivars for development of durable resistance/multiple resistance*: Pathogens and insects are known to overcome resistance provided by single genes. Single gene resistances are fragile and often broken down easily. Therefore, breeders intend to accumulate several major and minor resistance genes into one cultivar to achieve durable resistance. Durability of resistance has been increased by developing multilines and by pyramiding of resistance genes. MAS for resistance genes can be useful in these approaches. Based on host pathogen or host insect interaction alone it is often not possible to discriminate

the presence of additional resistance (R) gene(s). With MAS, new R gene segregation can be followed even in the presence of the existing R gene(s) and hence R genes from diverse sources can be incorporated in a single genotype for durable resistance. Pyramiding of bacterial blight resistance genes Xa1, Xa3, Xa4, Xa5 and Xa10 in different combinations using molecular markers has been reported in rice (Yoshimura et al., 1995).

3. *Improvement of qualitative characters*: RFLP markers have been linked to linolenic acid content Fan locus in soybean (Brummer et al., 1995). Not only this, RAPD markers that control somatic embryogenesis in alfalfa have been identified (Yu and Pauls, 1993).

4. *Molecular markers for hybrid vigor*: Hybrids in crops such as maize, sorghum, rice pearl millet, cotton and several vegetable crops have contributed greatly towards increasing the yield potential of these crops. The use of molecular markers has been investigated in maize and rice. Using molecular markers on a set of diallel crosses among eight elite parental lines widely used in Chinese hybrid rice production program, high correlation was found between specific heterozygosity and midparent heterosis. Photoperiod sensitive genic male sterility (PSGMS) genes in rice designated pms-1 and pms-2 were located on chromosomes 7 and 2 respectively (Zhang et al., 1994).

5. *MAS for traits difficult to evaluate*: The MAS is especially useful for the traits that are arduous and/or expensive to evaluate such as male fertility restoring genes for cytoplasmic male sterility in which the presence of the fertility restorer genes (Frs) in the breeding lines cannot easily be detected by conventional breeding techniques as they involve careful and extensive evaluation and analysis of innumerable segregants. The multiple marker alleles at the RG-532 locus closely linked to $Fr_3$ locus offer a possibility of marker assisted screening for more effective sterility. Zhang et al. (1997) have identified RAPD markers linked to a restorer gene for WA gene - cytoplasmic male sterility. In fact, the marker assisted identification of the effective alleles and the use of MAS in breeding can facilitate commercial hybrid rice production. In sugarbeet, selected polymorphic RAPD fragments from PCR fingerprint patterns of mitochondrial DNA from two nearly isogenic cytoplasmic male sterile and male fertile lines were detected, cloned and sequenced by Lorenz et al. (1997).

6. *Molecular markers and abiotic resistance*: Using molecular markers the biochemists and physiologists have identified specific traits beneficial in improving drought responses such as osmotic adjustments, water use efficiency and efficient root system. All these add to the yield improvement in crop plants. In rice and maize, QTL for root traits have been identified and are being used to breed high yielding drought resistant rice and maize genotypes. In many crops, RAPD markers are population specific. MAS has helped to improve yield performance under drought in beans, soybean and peas.

There was general belief that the level of MAS efficiently was too poor for polygenic traits and the common consensus could be summarized by the statement that despite numerous reports of QTL in crops, little has been published on the implementation in breeding programs of MAS based on these QTL. Fortunately, the situation has since changed, and the number of papers reporting successful MAS has increased significantly in recent years. As expected, the use of markers in breeding programs has been adopted primarily for the manipulation of simply inherited traits, for which a limited number of the most significant QTL can impact the phenotypic variance. Some of the important achievements of molecular breeding and some of the examples of genes isolated through map based cloning which is based on molecular markers are shown in Tables 22.5 and 22.6 respectively (adapted from Varshney *et al.*, 2006).

# PROTOCOL

## 22.1: Protocol for RAPD Analysis

1. Isolate the genomic DNA.
2. Add the reagents in a 0.2 ml microcentrifuge tube placed on ice

| Stock | Final | 25 µl |
|---|---|---|
| *Taq* DNA polymerase buffer (10x) | 1x | 2.5 µl |
| dNTPs (10 mM each) | 0.2 mM each | 0.5 µl each |
| Random primer (5 pmoles) | 0.2 µM | 1.5 µl |
| DNA (10–50 ng) | 10–50 ng | 1 µl |
| *Taq* DNA polymerase (5 units/ µl) | 1U | 0.2 µl |

Sterile water to make final volume to 25 µl

3. Prepare cocktail of components. Keep one tube with only cocktail and water (without DNA) always as a control.

4. Mix well and spin briefly.
5. Overlay the reaction with a 25 µl of mineral oil to prevent evaporation of reagents at high temperatures (in those PCR models where hot lid is not provided).
6. Place the tubes in thermocycler block and set the machine for following cycling conditions.

| | |
|---|---|
| Step 1 | 94°C, 2 min |
| Step 2 | (25–30 cycles) |
| | 94°C, 1 min |
| | 40°C, 1 min 30 sec |
| | 72°C, 2 min |
| Step 3 | 72°C, 10 min |
| Step 4 | 4°C end |

After amplification, add 2.5 µl of tracking dye in the sample and store at 4°C.

7. Prepare agarose gel of 1.4 to 1.8%.
8. Load the samples carefully in the wells.
9. Add suitable molecular weight, e.g. 1 kb ladder in one of the lanes.
10. Perform electrophoresis, stain the gel with ethidium bromide and photograph it.

**Table 22.5:** Some important achievements of molecular markers in cereal genetics and breeding (Adapted from Varshney *et al.*, 2006)

| Cereal species | Notable examples of MAS | Reference |
|---|---|---|
| Barley | Release of US variety Tango in 2000 that contains two QTL for adult resistance to stripe rust. | Toojinda *et al.* (1998) |
| | Advancement of a 'Sloop type' variety with CCN (cereal cyst nematodes) resistance for commercial release | CRCMPB Annual Report (2001/2002) |
| | Introgression of *Yd2* gene conferring resistance to barley yellow dwarf virus (BYDV) into a BYDV susceptible background through two cycles of marker-assisted backcrossing | Jefferies *et al.* (2003) |
| Maize | Development of quality protein maize (QPM) through marker-aided transfer of opaque2 gene in backcross programs | Dreher (2000) |
| | Backcross marker-assisted selection for drought tolerance and recurrent selection for grain moisture and precocity | Ribaut and Ragot (2007) |
| Pearl millet | Advance in marker- assisted selection for drought tolerance | Serraj *et al.* (2005) |
| Rice | Release of two Indonesian rice cultivars 'Angke' and 'Conde', in which MAS was used to introduce *xa5* into a background containing *xa4* | Toenniessen *et al.* (2003) |
| | Pyramiding of disease resistance genes in rice against blight | Sanchez *et al.* (2000); Singh *et al.* (2001) |
| | Pyramiding of disease resistance genes in rice against blast | He *et al.* (2004) |
| | Pyramiding of disease resistance genes in rice against blight and blast | Narayanan *et al.* (2002) |
| | Pyramiding of insect and blight resistance | Singh *et al.* (2001) |
| | Quality character | Joseph *et al.* (2004) |
| | The pyramiding of blight resistance with Basmati quality characters | Joseph *et al.* (2004) |
| | Introgression of QTL controlling root traits into Indian upland variety | Steele *et al.* (2006) |

**Table 22.6:** Some examples of genes isolated through map based cloning using molecular markers (Adapted from Varshney *et al.*, 2006)

| Cereal species | Examples of genes isolated through Map based cloning | References |
|---|---|---|
| Barley | Powdery mildew resistance gene - *Mlo* | Buschges *et al.* (1997) |
| | Powdery mildew resistance gene - *Mla* | Wei *et al.* (1999) |
| | Powdery mildew resistance gene - *Rar* | Shirasu *et al.* (1997) |
| | Stem rust resistance - *Rpgl* | Brueggeman *et al.* (2002) |
| | Barley yellow mosaic virus resistance - *rym*3 and /or *rym*4 | Stein *et al.* (2005) |
| Maize | Leaf rust resistance - *Rpl* D | Collins *et al.* (1999) |
| | Flowering time QTL - *Vgt*1 | Salvi *et al.* (2002) |
| Rice | Bacterial blight-resistance gene - *Xa*1 | Yoshimura *et al.* (1998) |
| | Bacterial blight-resistance gene - *Xa*5 | Iyer and McCouch. (2004) |
| | Bacterial blight-resistance gene - *Xa* 21 | Song *et al.* (1995). |
| | Bacterial blight-resistance gene - *Xa*26 | Sun *et al.* (2004) |
| | Rice blast-resistance gene -*PiB* | Wang *et al.* (1999) |
| | Rice blast-resistance gene -*Pi ta* | Bryan *et al.* (2000) |
| | Rice blast-resistance gene -*Pi 5* (t) | Jeon *et al.* (2003) |
| | Rice blast-resistance gene -*Pi 9* | Qu *et al.* (2006) |
| | Plant architecture gene Dwarf1 | Ashikari *et al.* (1999) |
| | A timekeeper of leaf initiation PLASTOCHRONI | Miyoshi *et al.* (2004) |
| | Leaf spotted leaf gene - *Sp*17 | Yamanouchi *et al.* (2002) |
| | Semi-dwarf gene (*sd*1) | Spielmeyer *et al.* (2002) |
| | Seed shattering gene - (*qSHI*) | Konishi *et al.* (2006); Li *et al.* (2006) |
| | QTLs for heading - *Hd*1 | Yano *et al.* (2000) |
| | QTLs for heading - *Hd*3a | Kojima *et al.* (2002) |
| | QTLs for heading - *Hd*4 and *Hd*5 | Lin *et al.* (2003) |
| | QTLs for heading - *Hd*6 | Takahashi *et al.* (2001) |
| | QTLs for heading - *Ehd*1 | Doi *et al.* (2004) |
| | QTL forgrain production, *Gn*1a | Ashikari *et al.* (2005) |
| | QTL for salt tolerance | Ren *et al.* (2005) |
| Sorghum | A major aluminum tolerance gene -*Alt*SB | Magalhaes *et al.* (2006) |

# REFERENCES

Akkaya, M.S., Bhagwat, A.A. and Cregan, P.B., 1992. Length polymorphisms of simple sequence repeat DNA in soybean. *Genetics* **132:** 1131–1139.

Akopyanz, N., Bukanov, N.O., Westblom, T.U. and Berg, D.E. 1992. PCR-based RFLP analysis of DNA sequence diversity in the gastric pathogen *Helicobacter pylori*. *Nucl. Acids Res.* **20:** 6221–6225.

Ashikari, M. *et al.* 1999. A rice gibberellin insensitive dwarf mutant gene Dwarf 1 encodes the alpha subunit of GTP binding protein. *Proc. Natl. Acad. Sci. USA* **96:** 10284–10289.

Ashikari, M. *et al.* 2005. Cytokinin oxidase regulates rice grain production. *Science* **309:** 741–745.

Beckmann, J.S. and Soller, M. 1990. Toward a unified approach to genetics mapping of eukaryotes based on sequence tagged microsatellite sites. *Bio. Technology* **8:** 930–932.

Bennetzen, J.L. 1998. The structure and evolution of angiosperm nuclear genomes. *Curr. Opin. Plant Biol.* **1:** 103–108.

Botstein, D., White, R.L., Skolnik, M. and Davis, R.W. 1980. Construction of a genetic linkage map in man using restriction fragment length polymorphism. *Am. J. Human Genet.* **32:** 314–331.

Boyko, E., Kalendar, R., Korzun, V., Fellers, J., Korol, A., Sculman, A.H. and Gill, B.S. 2002. A high-density cytogenetic map of file *Aegilops tauschii* genome incorporating retrotransposons and defense—related genes: Insights into cereal chromosome structure and function. *Plant Mol. Biol.* **48:** 767–790

Brueggeman, R. *et al.* 2002. The barley stem rust resistance gene Rpgl1 is a novel disease resistance gene with homology to receptor kinases. *Proc. Natl. Acad. Sci. USA* **99:** 9328–9333.

Brummer, E.C., Nickell, A.D., Wilcox, J.R. and Shoemaker, R.C. 1995. Mapping the fan locus controlling linolenic acid content in soybean oil. *J. Hered.* **86:** 245–247.

Bryan, G.T. *et al.* 2000. A single amino acid difference distinguished resistant and susceptible alleles of the rice blast resistance gene Pita. *Plant Cell* **12:** 2033–2046.

Buschges, R. *et al.* 1997. The barley mlo gene: A novel control element of plant pathogen resistance. *Cell* **88:** 95–705.

Caetano-Anolles, G., Bassam, B.J. and Gresshoff, P.M. 1991. DNA amplification fingerprinting using very short arbitrary oligonucleotide primers. *Biol. Technology* **9:** 553–557.

Caetano-Anolles, G., Bassam, B.J. and Gresshoff, P.M. 1993. Enhanced detection of polymorphic DNA by multiple arbitrary amplicon profiling of endonuclease-digested DNA: identification of marker tightly linked to the super nodulation locus in soybean. *Mol. Gen. Genet.* **241:** 57-64.

Collins, N. *et al.* 1999. Molecular characterization of the maize Rp1 D rust resistance haplotype and its mutants. *Plant Cell* **11:** 1365–1376.

Connor, B.J., Reyes, A.A., Morin, C., Itakura, K., Teplitz, R.L. and Wallace, R.B., 1983. Detection of sickle cell BS—globin allele by hybridization with synthetic oligonucleotides. *Proc. Natl. Acad. Sci. USA* **80:** 278–282.

Cotton, R.C. 1993. Current methods of mutation detection. *Mut. Res.* **285:**125–144.

CRCMPB Annual Report 2001/2002. Putting plant breeding into fast-forward. Cooperative Research Centre for Molecular Plant Breeding, University of Adelaide, Australia, p. 51.

Crow, J.F. and Dove, W.F. 1988. A diamond anniversary: the first chromosome maps. *Genetics* **118:** 389–391.

Doi, K. *et al.* 2004. Ehd1, a B type response regulator in rice, confers short day promotion of flowering and controls FT like gene expression independently of Hd1. *Genes Dev.* **18:** 926–936.

Doveri, S., Lee, D., Maheswaran, M. and Powell, W. 2008. Molecular markers: History, Features and Applications. In: *Principles and Practices of Plant Genomics* Vol. 1: *Genome Mapping* (Kole, C. and Abbott, A.G. eds), Science Publishers, Enfield (NH), USA.

Doveri, S., O'Sullivan, D.M. and Lee, D. 2006. Non-concordance between genetic profile of olive oil and fruit: a cautionary note to the use of DNA markers for provenance testing. *J. Agri. food chem.* **10:**1021 (in press).

Dreher, K. *et al.* 2000. Is marker assisted selection cost effective compared to conventional plant breeding methods? The case of quality protein maize. In Proc 4th Ann. Conf. Intern. Consortium on Agricultural Biotechnology Research (ICABR), The Economics of Agricultural Biotechnology, Ravello, Italy.

Dwivedi, S.L., Crouch J.H., Mackill, D., Xu, Y., Blair, M.W., Ragot, M., Upadhyaya, H.D. and Rodomiro, O. 2007a. Molecularization of pubic sector plant breeding: a synthesis of progress and problems. *Advance in Agronomy* **95:** 163–318.

Edwards, M.D., Stuber, C.W. and Wendel, J.F. 1987. Molecular-marker-facilitated investigation of quantitative-trait loci in maize. 1. Numbers, genomic distribution and types of gene action. *Genetics* **116:** 113–125.

Ellis, T.H.N., Poyser, S.J., Knox, M.R., Vershinin, A.V. and Ambrose, M. J. 1998. Polymorphism of insertion sites of Ty1-copia class retrotransposons and its use for linkage and diversity in pea. *Mol. Gen. Genet.* **260:** 9–19.

Flavell, A.J., Knox, M.R., Pearce, S.R. and Ellis, T.H.N. 1998. Retrotransposon-based insertion polymorphisms (RBIP) for high throughput marker analysis. *Plant J.* **16:** 643–650.

Flavell, A.J., Bolshakov, V.N., Booth, A., Jing, R., Russell, J., Ellis, T.H.N. and Isaac, P. 2003. A microarray-based high throughput molecular marker genotyping method: the tagged microarray marker approach. *Nucl. Acids Res.* **31:** e115.

Gilles, P.N., Wu, D.J., Foster, C.B., Dillon, P.J. and Chanock, S.J. 1999. Single nucleotide polymorphic discrimination by an electronic dot blot assay on semiconductor microchips. *Nature Biotechnology* **17:** 365–370.

Gribbon, B.M., Pearce, S.R., Kalendar, R., Schulman, A.H., Paulin, L., Jack, P., Kumar, A. and Flavell, A.J. 1999. Phylogeny and transpositional activity of Ty-copia group retrotransposons in cereal genomes. *Mol. Genet. Genom.* **261:** 883–891.

Grodzicker, T., Williams, J., Sharp, P. and Sambrook, J. 1997. Physical mapping of temperature sensitive mutations. *Cold Spring Harbor Symp. Quat. Biol.* **39:** 439–446.

Haley, C.S. and Knott, S.A. 1992. A simple method for mapping quantitative trait loci in line crosses using flanking markers. *Heredity* **69:** 315–324.

Hardenbol, P., Baner, J., Jain, M., Nilsson, M., Namsaraev, E.A., Karlin-Neumann, G.A., Fakhrai-Red, H., Ronaghi, M.,Willis, T.D., Landegren, U. and David, R.W. 2003. Multiplexed genotyping with sequence-tagged molecular inversion probes. *Nat. Biotechnol.* **21:** 673–678.

Havecker, E.R., Gao, X. and Voytas, D.F. 2004. The diversity of LTR retrotransposons. *Genome Biol.* **5:** 225.

He, Y. *et al.* 2004. Gene pyramiding to improve hybrid rice by molecular marker techniques. In New Directions for a Diverse Planet: Proc. Sci. Cong. Brisbane, Australia (http://www.cropscience.org.au/icsc2004/).

Helentjaris, T., Slocum, M., Wright, S., Schaefer, A. and Nienhuis J. 1986. Construction of genetic linkage maps in maize and tomato using restriction fragment length polymorphisms. *Theor. and Applied Genetics* **72:** 761–769.

Iyer, A.S. and McCouch, S.R. 2004. The rice bacterial blight resistance gene xa5 encodes a novel form of disease resistance. *Mol. Plant Microbe Inter.* **17:** 1348–1354.

Jaccoud, D., Peng, K., Feinstein, D. and Kilian, A. 2001. Diversity arrays: a solid state technology for sequence information independent genotyping. *Nucl. Acids Res.* **29:** e25.

Jeffreys, A.J., Wilson, V. and Thein, S.L. 1985. Hypervariable minisatellite regions in human DNA. *Nature* **314:** 67–73.

Jefferies, S.P. *et al.* 2003. Marker assisted backcross introgression of the yd2 gene conferring resistance to barley yellow dwarf virus in barley. *Plant Breed.* **122:** 52–56.

Jeon, J.S. *et al.* 2003. Genetic and physical mapping of Pi5(t), a locus associated with broad spectrum resistance to rice blast. *Mol. Genet. Genomics* **269:** 280–289.

Jordan, S.A. and Humphries, P. 1994. Single nucleotide polymorphism in Exon 2 of the BCP Gene on 7 931-935. *Hum. Mol. Genet.* **3**: 1915.

Joseph, M. *et al.* 2004. Combining bacterial blight resistance and Basmati quality characteristics by phenotypic and molecular marker assisted selection in rice. *Mol. Breed.* **13**: 377–387.

Kalendar, R., Grob, T., Suoniemi, A. and Schulman, A.H. 1999. IRAP and REMAP: Two new retrotransposon – based DNA fingerprinting techniques. *Theor. Appl. Genet.* **98**: 704–711.

Kalendar, R., Tanskanen, J., Immonen, S., Nevo, E. and Schulman, A.H. 2000. Genome evolution of wild barley (*Hordeum Spontaneum*) by Bare-1 retrotransposon dynamics in response to sharp microclimatic divergence. *Pro. Natl. Acad. Sci. USA* **97**: 6603–6607.

Kojima, S. *et al.* 2002. Hd3a, a rice ortholog of the Arabidopsis FT gene, promotes transition to flowering downstream of Hd1 under short day condition. *Plant Cell Physiol.* **43**: 0961105.

Konishi, S. *et al.* 2006. An SNP caused loss of seed shattering during rice domestication. *Science* **312**: 1392–1396.

Kuklin, A., Munson, K., Gjerde, D., Haefele, R. and Tylor, P. 1998. Detection of single nucleotide polymorphism with WAVE (tm) DNA fragment analysis system. *Genet Test* **1**: 201–206.

Kumar, A., Pearce, S.R., McLean, K. and Heslop- Harrison, J.S. 1997. The Ty1-copia group of retrotransposons in plant: genomic organisation, evolution, and use as molecular markers. *Genetics* **100**: 205–217.

Kwok, P.Y., Deng, Q., Zakeri, H. and Nickerson, D.A. 1996. Increasing the information content of STS-based genome maps: identifying polymorphisms in mapped STSs. *Genomics* **31**: 123–126.

Landegren, U., Kaiser, R., Sanders, J. and Hood, L. 1998. A ligase- mediated gene detection technique. *Science* **241**: 1077–1080.

Lee, D., Ellis, T.H.N., Turner, L., Hellens, R.P. and Cleary, W.G. 1990. A copia-like element in *pisum* demonstrates the uses of dispersed repeated sequence in genetic analysis. *Plant Mol. Biol.* **15**: 707–722.

Leigh, F., Kalendar, R., Lea, V., Lee, D., Donini, P. and Schulman, A.H. 2003. Comparison of the utility of barley retrotransposon families for genetic analysis by molecular marker techniques. *Mol. Genet. Genom.* **269**: 464–474.

Li, C. *et al.* 2006. Rice domestication by reducing shattering. *Science* **311**: 1936–1939.

Li, G. and Quiros, C.F. 2001. Sequence—related amplified polymorphism (SRAP), a new marker system based on a simple PCR reaction: its application to mapping and gene tagging in *Brassica*. *Theor. Appl. Genet.* **103**: 455–461.

Lin, H. *et al.* 2003. Fine mapping and characterization of quantitative trait loci Hd4 and Hd5 controlling heading date in rice. *Breed. Sci.* **53**: 51–59.

Litt, M. and Lutty, J.A. 1989. A hypervariable microsatellite revealed by in vitro amplification of a dinucleotide repeat within the cardiac muscle actin gene. *Am. J. Hum. Genet.* **44**: 397–401.

Lorenz, M., Werhe, A. and Borner, T. 1997. Cloning and sequencing of RAPD fragments amplified from mitochondrial DNA of male sterile and male fertile cytoplasm of sugar beet (*Beta vulgaris* L.). *Theor. Appl. Genet.* **94**: 273–278.

Magalhaes, J.V. *et al.* 2006. High resolution mapping and cloning of AltSB, a major aluminum tolerance gene in sorghum. In Plant and Animal Genome Conference X1V, P61(http://www.intlpag.org/14/abstracts/PAG14_P61.html).

Manninen, I. and Schulman, A.H. 2000. BARE-1 a copia-like retroelement in barley (*Hordeum vulgare* L.) *Proc. Natl. Acad. Sci. USA* **97**: 6603–6607.

Manninen, O., Kalendar, R., Robinson, J. and Schulman, A.H. 2000. Application of BARE1 retrotransposon marker to the mapping of major resistance gene for net blotch in barley. *Mol. Genet. Genome* **264**: 325–334.

Markert, C.L. and Moller, F. 1959. Multiple forms of enzymes. *Tissue, ontogenetic and species specific patterns* **45**: 753–763.

Mashal, R.D., Koontz, J. and Sklar, J. 1995. Detection of mutation by cleavage of DNA heterduplexes with bacteriophage resolvases. *Nat. Genet.* **9**: 177–183.

Melayah, D., Bonnivard, E., Chalhoub, B., Audeon, C. and Grandbastien, M.A. 2001. The mobility of the tobacco Tnt1 retrotransposon correlates with its transcriptional activation by fungel factors. *Plant J.* **28**: 159–168.

Michelmore, R.W., Paran, I. and Kesseli, R.V. 1991. Identification of markers linked to disease resistance genes by bulked segregants analysis: a rapid method to detect markers in specific genomic regions by using segregating populations. *Proc. Natl. Acad. Sci. USA* **88**: 9828–9832.

Miyoshi, K. *et al.* 2004. PLASTOCHRONI, a time keeper of leaf initiation in rice, encodes cytochrome P450. *Proc. Natl. Acad. Sci. U.S.A.* **101**: 875–880.

Mohan, M., Nair, S., Bentur, J.S., Prasada Rao, U. and Bennet, J. 1997. RFLP and RAPD mapping of rice Gm2 gene that confers resistance to biotype 1 of gall midge (*Orseolia oryzae*). *Theor. Appl. Genet.* **87**: 782–788.

Morgante, M., Vogel, J. 1994. Compound microsatellite primers for the detection of genetic polymorphisms. US Plant Appl 08/326456.

Mullis, K. 1990. The unusual origin of the polymerase chain reaction. *Scientific American,* April: 56–65.

Narayanan, N.N. *et al.* 2002. Molecular breeding for the development of blast and bacterial blight resistance in rice cv. IR50. *Crop Sci.* **42**: 2072–2079.

O'Donovan, M.C., Oefner, P.J., Roberts, S.C., Austin, J., Hoogendoorn, B., Guy, C., Speigh, G., Upadhyaya, M., Sommer, S.S. and McGuffin, P. 1998. Blind analysis of denaturing high-performance liquid chromatography as a tool for mutation detection. *Genomics* **52**: 44–49.

Olsen, M., Hood, L., Cantor, C. and Botetein, D. 1989. A common language for physical mapping of the human genome. *Science* **245**: 1434–1435.

Olsen, M., Hood, L., Cantor, C. and Doststein, D. 1989. A common language for physical mapping of the human genome. *Science* **254:** 1434–1435.

Orita, M., Iwahana, H., Kanazawa, H., Hayashi, K., Sekiya, T. 1989. Detection of polymorphism of human DNA by gel electrophoresis as single-strand conformation polymorphisms. *Proc. Natl. Acad Sci. USA* **86:** 2766–2770.

Paran, I. and Michelmore, R.W. 1993. Development of reliable PCR based markers linked to downy mildew resistance gene in lettuce. *Theor. Appl. Genet.* **85:** 985–993.

Pelcy, F. and Merdinoglu, D. 2002. Development of grape molecular markers based on retrotransposons. *In:* Plant, Animal and Microbe Genome X Conf., San Diego, CA, USA.

Peterson A.H., Lander E.S., Hewitt J.D., Peterson S., Lincoln S.E. and Tanksley S.D. 1988. Resolution of quantitative traits into Mendelian factors by using a complete linkage map of restriction fragment length polymorphisms. *Nature* **335:** 721–726.

Porceddu, A., Albertini, E., Barcaccia, G., Marconi, G., Bertoli, F.B. and Veronesi, F. 2002. Development of S-SAP markers based on an LTR-like sequence from *Medicago sativa* L. *Mol. Genet. Genom* **267:** 107–114.

Qu, S.H. *et al.* 2006. The broad-spectrum blast resistance gene Pi9 encodes a nucleotide binding site leucine rich repeat protein and is a member of a multigene family in rice. *Genetics* **172:** 1901–1914.

Ren, Z.H. *et al.* 2005. A rice quantitative trait locus for salt tolerance encodes a sodium transporter. *Nature Genet.* **37:** 1141–1146.

Ribaut, J.M. 2007. The challenges of biotechnologies to improve plant breeding efficiency. Annual Foundation day lecture for Barwale Foundation, IARI, Delhi.

Ribaut, J.M. and Ragot, M. 2007. Marker Assisted selection to improve drought adaptation in maize: The backcross approach, perspectives, limitations, and alternatives. *Journal of Experimental Botany* **58:** 351–360.

Salvi, S. *et al.* 2002. Towards positional cloning of Vgt 1, a QTL controlling the transition from the vegetative to the reproductive phase in maize. *Plant Mol. Bio1.* **48:** 601–613.

Sanchez, A.C. *et al.* 2000. Sequence tagged site marker assisted selection for three bacterial blight resistance genes in rice. Crop Sci. **40:** 792–797.

SanMiguel, P., Tikhonov, A., Jin, Y.K., Motchoulskaia, N., Zakharov, D., Melake-Berhan, A., Springer, P.S., Edwards, K. J., Lee, M., Avramova, Z. and Bennetzen, J.L. 1996. Nested retrotransposons in the intergenic regions of the maize genome. *Science* **274:** 765–768.

Schneider, K., Weisshaar, B., Borchardt, D.C. and Salamini, F. 2001. SNA frequency and allelic haplotype structure of *Beta vulgaris* expressed genes. *Mol. Breed.* **8:** 63–74.

Senior, M.L. and Heun, M.N. 1993. Mapping maize microsatellites and polymerase chain reaction confirmation of the target repeat using a CT primer. *Genome* **36:** 884–889.

Serraj, R. *et al.* 2005. Recent advances in market- assisted selection for drought tolerance pearl millets. *Plant Production Science* **8:** 334–337.

Shirasu, K. *et al.* 1997. A novel class of eukaryotic zinc binding protein is required for disease resistance signaling in barley and development in *C. elegans*. *Cell* **99:** 355–366.

Singh, S. *et al.* 2001. Pyramiding three bacterial blight resistance genes (xa5.xa13 and za21) using marker-assisted selection into indica rice cutivar PR 106. *Theor. Appl. Genet.* **102:** 1011–1015.

Somers, D.J., Kirkpatrick, R., Moniwa, M. and Walsh, A. 2003 Mining single-nucleotide polymorphism from hexaploid wheat ESTs. *Genome* **49:** 431–437.

Song, W.Y., Wang, G.L., Chen, L.L., Kim, H.S. and Pi, L.Y. 1995. A receptor kinase like protein encoded by the rice disease resistance gene, *Xa21*. *Science* **270:** 1804–1806.

Spielmeyer, W. *et al.* 2002. Semi dwarf (sd 1), "green revolution" rice, contains a defective gibberellin 20 oxidase gene. *Proc. Natl. Acad. Sci. U.S.A.* **99:** 9043–9048.

Steele, K.A. *et al.* 2006. Marker assisted selection to introgress rice QTLs controlling root traits into an Indian upland rice variety. *Theor. Appl. Genet.* **112:** 208–221.

Stein, N. *et al.* 2005. The eukaryotic translation initiation factor 4E confers multiallelic recessive bymovirus resistance gene in *Hordeum vulgare* (L.). *Plant J.* **42:** 9123–922.

Sun, X. *et al.* 2004. Xa6, a gene conferring resistance to Xanthomonas oryzae pv.oryzae in rice, encodes an LRR receptor kinase like protein. *Plant J.* **37:** 517–527.

Taillon-Miller, P., Gu, Z., Li, Q., Hillier, L. and Kwok, P.Y.1998. Overlapping genomic sequence: a treasure trove of single-nucleotide polymorphisms. *Genome Res.* **8:** 748–754.

Takahashi, Y. *et al.* 2001. Hd6, a rice quantitative trait locus involved in photoperiod sensitivity, encodes the alpha subunit of protein kinase CK2. *Proc. Natl. Acad. Sci. U.S.A.* **98:** 7922–7927.

Taylor, E.J.A., Konstantinova, P., Leigh, F., Bates, J.A. and Lee, D. 2004. Gypsy-like retrotransposons in *Pyrenophora* species: an abundant and information class of molecular markers. *Genome* **47:** 519–525.

Toenniessen, G.H. *et al.* 2003. Advances in plant biotechnology and its adoption in developing countries. *Curr. Opin. Plant Biol.* **6:** 191–198.

Tenaillon, M.I., Sawkins, M.C., Long, A.D., Gaut, R.L., Doebley, J.F. and Gaut, B.S. 2001. Patterns of DNA sequence polymorphism along chromosome 1of maize (*Zea mays* L.). *Proc. Natl Acad. Sci. USA* **98:** 9161–9166.

Toojinda, T. *et al.* 1998. Introgression of quantitative trait loci (QTLs) determining stripe rust resistance in barley: an example of marker assisted line development. *Theor. Appl. Genet.* **96:** 123–131.

Underhill, P.A., Jin, L., Lin, A.A., Mehdi, S.Q., Jenkins, T., Vollrath, D., Davis, R.W., Cavalli-Sforza, L.L. and

Oefner, P.J. 1997. Detection of numerous Y chromosome biallelic polymorphisms by denaturing high-performance liquid chromatography. *Genome Res.* **7**: 996–1005.

Varshney, R.K., Hoisington, D.A. and Tyagi, A.K. 2006. Advances in cereal genomics and application in crop breeding. *Trends in Biotechnology* **24**: 490–499.

Vicient, C.M., Jaaskelainen, M., Kalendar, R. and Schulman, A.H. 2001. Active retrotransposons are a common feature of grass genomes. *Plant Physiol.* **125**: 1283–1292.

Vos, P., Hogers, R., Bleeker, M., Reijans, M., Van de Lee T., Hornes, M., Fritjers, A., Pot, J., Peleman, J., Kuiper, M. and Zabeau, M. 1995. AFLP: a new technique for DNA fingerprinting. *Nucl. Acids Res.* **23**: 4407–4414.

Walt, D. R. 2005. Miniature analytical methods for medical diagnostics. *Science* **308**: 217–219.

Wang, D.G., fan, J.B., Siao, C.J., Berno, A., Young,P., Sapolsky, R., Ghandour, G., Perkins, N., Winchester,E., Spencer, J., Kruglyak, L., Stein, L., Topaloglou, T., Hubbell, E., Robinson, E., Mittmann, M., Morris, M.S., Shen, N., Killburn, D., Rloux, J., Nusbaum, C., Rozen, S., Hudson, T.J., Lipshutz, R., Chee, M. and Lander, E.S. 1998. Large-scale identification, mapping and genotyping of single-nucleotide polymorphisms in the human genome. *Science* **280**: 1077–1082.

Wang, Z.X. *et al.* 1999. The Pib gene for rice blast resistance belongs to the nucleotide binding and leucine rich repeat class of plant disease resistance genes. *Plant J.* **19**: 55–64.

Waugh, R., McLean, K., Flavell, A.J., Pearce, S.R., Kumar, A., Thomas, B.T. and Powell, W. 1997. Genetic distribution of BARE1 like retrotransposable elements in the barley genome revealed by sequence-specific amplification polymorphisms (S-SAP). *Mol. Gen. Genet.* **253**: 687-694.

Wei, F. *et al.* 1999. The mla (powdery mildew) resistance cluster is associated with three NBS LRR gene families and suppressed recombination within a 240 kb DNA interval on chromosome 5S (HIS) of barley. *Genetics* **153**: 1929–1948.

Welsh, J. and McClelland, M. 1990. Fingerprinting genomes using PCR with arbitrary primers. *Nucl. Acids Res.* **8**: 7213–7218.

Williams, J.G.K., Kubelik, A.R., Livak, K.J., Rafalsh, J.A. and Tingey, S.V. 1990. DNA Polymorphisms amplified by arbitrary primers are useful as genetic markers. *Nucl. Acids Res.* **18**: 6531–6535.

Williams, M.N.V., Pande, N., Nair, S., Mohan, M. and Bennet, J. 1991. Restriction fragment length polymorphism analysis of polymerase chain reaction products amplified from mapped loci of rice (*Oryza sativa* L.) genomic DNA. *Theor. Appl. Genet.* **82**: 489–498.

Yamanouchi, U. *et al.* 2002. A rice spotted leaf gene, Sp17, encodes a heat stress transcription factor protein. *Proc. Natl. Acad. Sci. U.S.A.* **99**: 7530–7535.

Yano, M. *et al.* 2000. Hd1, a major photoperiod sensitivity quantitative trait locus in rice, is closely related to the Arabidopsis flowering time gene CONSTANS. *Plant Cell* **12**: 2473–2484.

Ye, S., Dhillon, S., Ke, X., Collins, A. and Da, I. 2001 An efficient procedure for genotyping single nucleotide polymorphisms. *Nucl. Acids Res.* **29**: 17:e88.

Yoshimura, S. *et al.* 1998. Expression of Xa1, a bacterial blight gene in rice is induced by bacterial inoculation. *Proc. Natl. Acad. Sci. U.S.A.* **95**: 1663–1668.

Yoshimura, S., Yoshimura, A., Iwata, N., McCouch, S., Abenes, M.L., Baraidon, M.R., Mew, T. and Nelson, R.J. 1995. Tagging and combining bacterial blight resistance genes in rice using RAPD and RFLP markers. *Mol. Breed.* **1**: 375–387.

Youil, R., Kemper, B. and Cotton, R.G. 1996. Detection of 81 known mouse beta- globin promoter mutations with T4 endonuclease VII—The EMC method. *Genomics* **32**: 431–435.

Yu, K. and Pauls, K.P. 1993. Identification of a RAPD marker associated with somatic embryogenesis in alfalfa. *Plant Mol. Biol.* **22**: 269–277.

Zhang, G., Angeles, E.R., Abenes, M.L.P., Khush, G.S. and Huang, N. 1994. Molecular mapping of a bacterial blight resistance gene on chromosome 8 in rice. *Rice Genet. Newsl.* **11**: 142–144.

Zhang, G., Bharaj, T.S., Lu, Y., Virmani, S.S. and Huang, H. 1997. Mapping of the RF-3 nuclear fertlity restoring gene for WA cytoplasmic male sterility in rice using RAPD and RFLP markers. *Theor. Appl. Genet.* **94**: 118–127.

Zietkiewicz, E., Rafalaki, A. and Labuda, D. 1994. Genome fingerprinting by simple sequence repeat (SSR)—anchored polymerase chain reaction Amplification. *Genomics* **20**: 176–183.

## FURTHER READING

Caetano-Anolles, G. and Gresshoff, P.M. (eds.) 1997. *DNA markers-Protocols, applications and overview*. Wiley-VCH, New York.

Gebhardt, C. and Salamini, F. 1992. Restriction fragment length polymorphism analysis of plant genomes and its application to plant breeding. *Int. Rev. Cytol.* **135**: 201–235.

Kochert, G. 1990. Introduction to RFLP mapping and plant breeding applications. The Rockfeller Foundation-International Program on rice Biotechnology 1–14.

Kole, C. and Abbott, A.G. (eds.) 2008. *Principles and practices of Plant Genomics*, Vol. 1: *Genome Mapping*, Science Publishers, Enfield, NH, USA.

Mohan, M., Nair, S., Bhagwat, A., Krishna, T.G., Yano, M., Bhatia, C.R. and Sasaki, T. 1997. Genome mapping, molecular markers and marker assisted selection in crop plants. *Mol. Breed.* **3**: 87–103.

# 23

# Gene Transfer in Plants

Gene transfer or DNA uptake refers to the process that moves a specific piece of DNA (usually a foreign gene ligated to a bacterial plasmid) into cells. In plant breeding, techniques involving gene transfer through sexual and vegetative propagation are well established, the aim being to introduce genetic diversity into plant populations, to select superior plants carrying genes for desired traits and to maintain the range of plant varieties. The application of these conventional or classical techniques has produced significant achievements in the yield improvement of major food crops. However, this takes a long time. For example, it usually takes 6–8 years to produce a new variety of wheat or rice by sexual propagation. With the rapid development of genetic engineering techniques based on the knowledge of gene structure and function, plant breeding has been dramatically broadened.

The directed desirable gene transfer from one organism to other and the subsequent stable integration and expression of a foreign gene into the genome is referred as genetic transformation. The transferred gene is known as **transgene** and the organisms that develop after a successful gene transfer are known as **transgenics**. Transgenic plants are the plants that carry the stably integrated foreign genes. These plants may also be called transformed plants.

*In vitro* gene transfer is the laboratory technique of transferring refined desirable genes across taxonomic boundaries into plant and animals from other plants, animals and microbes, or even to introduce artificial, synthetic or chimeric genes into plants. The development of gene transfer techniques in plants has been very rapid. The techniques presently available rely upon natural plant vectors as well as vectorless systems, which include directed physical and chemical methods for delivering foreign DNA into plant cells. Some terms and expressions that will be used in the gene transfer methods have been explained before the actual discussion on gene transfer methods.

## TRANSIENT AND STABLE GENE EXPRESSION

One of the main considerations for any gene transfer technique is the potential of the recipient cell to express the introduced gene. The transferred DNA may be expressed for only a short period of time following the DNA transfer process and this is called **transient expression**. Only a small fraction of the DNA introduced into the cell by direct gene transfer methods becomes stably integrated into the chromosome of the cell. The DNA introduced into majority of the cells is lost with time and cell division. Fortunately, this transient DNA is expressed in the cell and forms the basis of extremely useful transient assays. Transient assays are commonly used for the analysis of gene expression and rapid monitoring of gene transfer. Different variables associated with gene transfer can be optimized using transient expression.

Stable transformation occurs when DNA is integrated into the plant nuclear or plastid genomes, expression occurs in regenerated plants and is inherited in subsequent generations. For stable transformation, the developmental potential of the transformed cells is very important. Ideally, the transformed cells should be capable of regeneration into fertile plants. In general, the smaller the number of cells required for regeneration, the better the culture system is for gene transfer. Therefore, a small cell cluster size gives a larger ratio of transformed to nontransformed cells.

## MARKER GENES

Monitoring and detection of plant transformation systems in order to know whether the DNA has been successfully transferred into recipient cells is done with the help of a set of genes. The marker gene(s) is/are introduced into the plasmid along with the target gene. Marker genes are categorized into two types:

i) Reporter or scoreable or screenable genes
ii) Selectable marker genes

The advantages of using these genes lie in their easy assay, which require no DNA extraction, electrophoresis or autoradiography. Some genes are used both as selectable markers and reporter genes, whereas some are used either as selectable marker or as reporter genes.

### Reporter Genes

A number of screenable or scoreable or reporter genes are available which show immediate expression in the cells/tissue. A reporter gene is a test gene whose expression results in quantifiable phenotype. A reporter system is useful in the analysis of plant gene expression and standardization of parameters for successful gene transfer in a particular technique. An assay for a reporter gene is usually carried out at the level of protein quantification. The principle of using reporter genes in studying molecular processes in a living cell means that in the natural gene, a synthetic modification is introduced (or the protein coding sequence is deleted and replaced by another gene) in order to either simplify the detection of the gene product or to distinguish it from similar or identical genes in the genome. The use of reporter genes requires a method of gene transfer - either transient or stable. An ideal reporter should have the following features:

i. Detection with high sensitivity.
ii. Low endogenous background.
iii. There should be a quantitative assay.
iv. Assay should be nondestructive.
v. Assay should require a minimal amount of effort and expense.

The most commonly used reporter genes have been described below and also summarized in Table 23.1.

**Table 23.1:** Reporter genes used as scoreable markers and their assays

| Reporter gene | Substrate and assay | Identification |
|---|---|---|
| Octopine synthase (*ocs*) | Arginine + pyruvate + NADH | Electrophoresis, Chromatography |
| Nopaline synthase (*nos*) | Arginine + ketoglutaric acid + NADH | Electrophoresis, Chromatography |
| Chloramphenicol acetyl transferase (*cat*) | $^{14}C$ Chloramphenicol + acetyl CoA: TLC separation | Autoradiography |
| β-glucuronidase (*gus or uidA*) | Glucuronides (X-Gluc, 4-MUGluc, PNPG, NAG, REG) | Fluorescence detection, colorimetric, histochemical |
| Bacterial luciferase (*Lux F2*) | Decanal and $FMNH_1$ | Bioluminiscence |
| Firefly luciferase (*luc*) | $ATP + O_2 +$ luciferin | Bioluminiscence |
| Green fluorescent protein (*gfp*) | No substrate required; UV light irradiation | Fluorescence detection |
| Anthocyanin regulatory genes | No substrate required | Visual |

## Opine Synthase

The octopine or nopaline (the two most common opines) synthase genes (*ocs*, *nos*) are present in the T-DNA of Ti (tumor inducing) or Ri (root inducing) plasmids of *Agrobacterium*. The presence of opines in any plant material clearly indicates the transformed status of the plant cells. Furthermore, opines can be detected in some cases before the appearance of any visual symptoms related to the transformed status of the plant material. Opine synthase genes in principle constitute good reporter genes since no natural equivalent to their gene products has been found in plant cells. The value of opines as screenable marker resides in the simplicity of protocols for opine detection, as enzymes are stable, enzymatic assay is inexpensive and easy to perform.

There are basically two ways to assess the presence of opines in a transformed plant tissue. The first one aims at detecting the enzymatic activities responsible for the synthesis of opines. This technique involves a simple protein extraction from the plant sample. The extract is then incubated with the appropriate opines precursors, arginine, pyruvate and NADH for *ocs* and arginine, ketoglutaric acid and NADH for *nos,* and the reaction products are separated by paper chromatography (Fig. 23.1), whereas the second technique aims at directly detecting the presence of opines in the plant tissue. This involves extraction of the plant metabolite from the tissue, the separation of these compounds (opines) by paper electrophoresis and the detection of opines by appropriate staining treatments.

Arginine + Pyuvate + NADH ( *ocs* )

or

Arginine + Ketoglutaric acid + NADH ( *nos* )

↓ *Extract from plant sample*

Reaction products separated by paper chromatography

**Fig. 23.1:** Action of opine synthase reporter gene.

## Chloramphenicol Acetyl Transferase

The gene for chloramphenicol acetyl transferase (*cat*) is one of the most commonly used reporter genes in eukaryotic organisms. Two different *cat* genes have been identified, but the most commonly used reporter was found in the transposable element Tn9. The structural gene codes for chloramphenicol acetyl transferase (CAT) enzyme that carries out acetylation of chloramphenicol. The enzymatic assay for CAT is very sensitive and relatively simple using acetyl CoA and ($^{14}$C) chloramphenicol to convert chloramphenicol into its acetylated derivatives. This is done by incubating the labeled chloramphenicol with a crude extract of plant tissue in a suitable buffer. The presence of acetyl chloramphenicol is detected through autoradiography (Fig. 23.2). The cat gene was the first bacterial gene to be introduced into plant cells (Herrera-Estrella *et al.*, 1983).

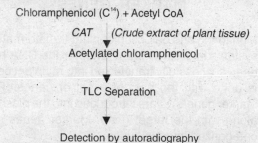

Chloramphenicol (C$^{14}$) + Acetyl CoA

CAT    (*Crude extract of plant tissue*)
↓
Acetylated chloramphenicol
↓
TLC Separation
↓
Detection by autoradiography

**Fig. 23.2:** Action of chloramphenicol acetyl transferase reporter gene.

## Beta Glucuronidase (GUS)

During the last few years the bacterial gene *uid*A, encoding β-glucuronidase (GUS), has become the most frequently used reporter gene for the analysis of plant gene expression. *uid*A encodes a soluble β-glucuronidase enzyme, a homotetramer of approximately 68 kd mass, which requires a pH optimum of 7–8. β-glucuronidase breaks down glucuronide substrate to give a colored reaction, so that its presence can be detected *in situ*. Its wide acceptance has mainly resulted from the availability of a highly sensitive nonradioactive assay using fluorogenic substrate 4-MU Gluc and of a histochemical assay using X-Gluc, that allows a quantitative analysis of cell and tissue specific expression. The major advantage of this reporter is that it does not require DNA extraction, electrophoresis or autoradiography.

β-glucuronidase gene from *E. coli* can be used as a selectable marker gene and a glucuronide derivative of cytokinins, benzyladenine as selective agent. Benzyladenine N-3 glucuronide is inactive as cytokinin but upon hydrolysis by GUS becomes active cytokinin, stimulating growth and regeneration of transformed cells in vitro. The GUS gene serves as both selectable and screenable marker.

## Bacterial Luciferase (Lux F2)

Bacterial luciferases have originated from *Vibrio harveyi* and *V. fischerii*. In these organisms luciferase, is expressed as a heterodimeric flavin monooxygenase that is responsible for catalyzing a light emitting reaction. The *V. harveyi* luciferase (Lux) is encoded by the *lux*A and *lux*B cistrons that are part of a longer operon. The *lux* genes in plant cells can be expressed as separate subunits or as a fused polypeptide. The subunits are correctly assembled in plant cells and the fused protein has been designated as Lux F2. The activity of luciferase is normally measured as the initial maximum light intensity upon mixing the enzyme with its aldehyde and reduced flavin mononucleotide substrates (FMNH$_2$) in the presence of oxygen. Light is measured as photons or quanta per second (qu/s), and if calibrated against a known light standard, the absolute number of enzyme molecules in the reaction can be calculated. Light emission can be monitored visually, photographically, or electronically. A simple low cost luminometer can detect *ca.* 23$^{-5}$ qu/s, corresponding to about 10$^6$ luciferase molecules.

## The Firefly Luciferase (Luc)

Luciferase isolated from the North American firefly, *Photinus pyralis* catalyzes the oxidative decarboxylation of luciferin, a 6-hydroxy-benzothiazole, to oxyluciferin in the presence of ATP, oxygen and magnesium ions. Luminescense in terms of light production can be recorded in a luminometer.

## Green fluorescent protein (GFP)

The green fluorescent protein (*gfp*) gene is an ideal reporter and selectable marker for gene expression analysis and plant transformation. Gene for *gfp* has been obtained from jellyfish *Aequorea victoria*. GFP is a small barrel-shaped protein of 238 amino acids surrounding a fluorescent chromophore. This owes its visible absorbance and fluorescence to a β-hydroxy benzylidine imidazolinone chromophore. It emits green fluorescent light in the blue to ultraviolet range. The formation of green fluorescence can be detected under fluorescent microscope. Visual detection is possible at any time in living cells without any destruction and without the addition of any cofactor or external substrate. In addition gfp gene product does not adversely affect cell growth, regeneration and fertility of transformed plants

## Anthocyanins

Anthocyanins are prominent red and purple pigments of flowering plants. These anthocyanins have been used as visible marker for transformation. At least ten genes that encode regulatory or structural proteins of the biosynthetic pathway control anthocyanin pigmentation within the plant. The regulatory genes C$_1$ and B/R co-ordinately regulate the activity of structural genes. Chimaeric gene constructs have been prepared with C$_1$ and B/R regulatory genes which when introduced into the plant cells produce purple red *color*ation. These red spots can be visually or microscopically identified. These regulatory genes have been used as a visible marker in maize and wheat transformation. This reporter has advantages that a tissue is not sacrificed, transient expression can be seen in the position of target and it does not require addition of substrates.

## Selectable Markers

Selection of transformed cells is a key factor in developing successful methods for genetic transformation. This is done by certain selectable marker genes which are present in the vector along with the gene of interest. Selectable markers are an integral part of plant transformation strategies. The selectable marker genes enable the transformed cells to survive on media containing toxic levels of selection agent while nontransformed cells get killed. A large number of such selectable marker genes

have become available viz. antibiotic, antimetabolite and herbicide resistance genes, hormone biosynthetic genes and genes conferring resistance to toxic levels of amino acids or amino acid analogs. The usefulness of a particular resistance marker depends upon (i) the characteristics of selection agent (ii) the resistance gene, and (iii) the plant material. The selection agent must be toxic to plant cells, though not so toxic that products from dying, no-transformed cells kill adjacent, transformed cells. Different selectable marker genes and their substrates have been summarized in Table 23.2.

## Antibiotic resistance markers

### Neomycin phosphotransferase II (NPT II)

NPT II or aminoglycoside 3'-phosphotransferase II (APH (3')-II) enzyme is the most widely used marker. The enzyme is encoded by

**Table 23.2:** Selectable marker genes used for gene transfer

| Selectable marker gene Reference | Substrate used for selection | Enzyme/Mode of action | |
|---|---|---|---|
| Neomycin phospho-transferase (nptII) | G418, kanamycin, neomycin, Paromycin | Neomycin Phospho-transferase type II (NPT II)/detoxification | Bevan et al. (1983) |
| Hygromycin phospho-transferase (hpt) | Hygromycin B | Hygromycin phospho-transferase/detoxification | Elzen et al. (1985) |
| Bleomycin resistance gene | Bleomycin/phleomycin | Binding protein/detoxification | Hille et al. (1986) Perez et al. (1989) |
| Sulphonamide resistance gene (sulf) | Sulfadiazine/asulam | Dihydropteroate synthase/detoxification | Guerineau et al. (1990) |
| Gentamycin acetyl transferase (AAC-3) | Gentamycin | Gentamycin acetyl transferase/detoxification | |
| Streptomycin phospho-transferase (spt) | Streptomycin | Streptomycin phospho-transferase/detoxification | |
| Dihydrofolate reductase (dhfr) | Methotrexate | Dihydrofolate reductase/insensitive | Herrera-Esterella et al. (1983); Eichholtz (1987) |
| Phosphinothricin acetyltransferase (bar) | L-phosphinothricin (PPT), bialaphos | Phosphinothricin acetyl transferase/detoxification | De Block et al. (1987) |
| 5-enolpyruvyl shikimate 3-phosphate (EPSP) synthase (aroA) | Glyphosate (Roundup) | EPSP synthase/sensitive and insensitive | Shah et al. (1986); Comai et al. (1985) |
| Acetolactate synthase mutant form (csr1–1) | Sulphonylurea, imidazolinones | Acetolactate synthase/insensitive | Haughn et al. (1988) |
| Bromoxynil nitrilase(bxn) | Bromoxynil | Nitrilase/detoxification | Stalker et al. (1988) |
| psbA gene | Atrazine | Q protein insensitive | Cheung et al. (1988) |
| Dehalogenase (deh1) gene | Dalapon | Dehalogenase/detoxification | Buchanan et al. (1992) |
| tfd A gene | 2,4-D | 2,4-dichlorophenoxy acetate monooxygenase/detoxification | Streber and Willmitzer (1989) |

the *npt*II gene, which has been derived from Tn5 transposon. It inactivates a number of aminoglycoside antibiotics such as kanamycin, neomycin, geneticin (G418) and paromycin. Kanamycin is the most commonly used selective agent with concentrations ranging from 50 to 500mg/l.

Resistance to kanamycin in putative transformants and their progeny can be checked by applying kanamycin solution locally or sprayed on soil grown plants to select for kanamycin resistant progeny.

## Hygromycin phosphotransferase

Hygromycin phosphotransferase gene (*hpt*) was originally derived from *E. coli*. Hygromycin B is an aminocyclitol antibiotic that interferes with protein synthesis, is more toxic than kanamycin and kills sensitive cells faster. Hygromycin resistance can also be checked by a nondestructive callus induction test and segregation of *hpt* gene in the progeny of transgenic plants can be scored by seed germination assay.

## Gentamycin acetyltransferase

Three genes encoding aminoglycoside-3-N-acetyltransferases (AAC (3)) have been used as selectable markers in combination with gentamycin antibiotic selection. The genes are AAC (3)-I, AAC (3)-III and AAC (3)-IV.

## Streptomycin and spectinomycin resistance

Two dominant genes, streptomycin phospho-transferase (*spt*) gene from Tn5 and the aminoglycoside-3′-adenyltransferase gene (*aad* A) conferring resistance to both streptomycin and spectinomycin, have been described for plant transformation. Streptomycin and spectinomycin resistance markers differ from other markers in that they allow differentiation by color. Sensitive plant cells bleach but do not die, whereas resistant cells remain green under appropriate conditions.

## Antimetabolite marker

### Methotrexate insensitive dihydrofolate reductase (dhfr)

Methotrexate is an antimetabolite that inhibits the enzyme dihydrofolate reductase (DHFR) and thus interferes with DNA synthesis. A mutant mouse *dhfr* gene encoding an enzyme with very low affinity to methotrexate has been isolated. This *dhfr* gene fused to CaMV promoter yields a methotrexate resistance marker which is used for plant transformation.

## Herbicide resistance markers

### Bar gene

Bialaphos, phosphinothricin (PPT), and glufosinate ammonium are nonselective herbicides that inhibit glutamine synthase (GS) enzyme. GS is necessary for the assimilation of $NH_3$ and in the regulation of nitrogen metabolism in plants. Inhibition of GS results in the accumulation of $NH_3$ which causes death of plant cells. The *bar* gene codes for phosphinothricin acetyl transferase (PAT) enzyme which converts PPT/bialaphos into a nonherbicidal acetylated form by transferring the acetyl group from acetyl CoA on the free amino group of PPT and confers resistance to the cell. This gene has been cloned from two different strains of *Streptomyces hygroscopicus* and *S. viridochromogenes*.

### Enolpyruvylshikimate phosphate (EPSP) synthase (aro A) gene

Glyphosate (N-phosphonomethyl-glycine) is a broad-spectrum herbicide, that inhibits the essential photosynthesis process. It works by competitive inhibition of the enzyme EPSP synthase (5-enolpyruvyl-shikimate 3-phosphate). This enzyme catalyzes the synthesis of 5-enolpyruvylshikimate 3-phosphate from phosphoenolpyruvate and shikimate 3-phosphate, a step which is essential for the synthesis of phenylalanine, tyrosine and tryptophane. In plants, much of the EPSP synthase activity is encoded by nuclear genes, but enzyme is localized in chloroplasts. Mutants of the bacterium *Salmonella typhimurium* resistant to glyphosate have been obtained. These map at *aro* A locus that encodes EPSP synthase.

### Acetolactate synthase (als) gene

The herbicidal action of sulphonylureas and imidazolinones is based on their ability to noncompetitively inhibit acetolactate synthase (ALS) enzyme in the pathway leading to the production of branched chain amino acids *viz.* valine, leucine, isoleucine. Mutant genes have been isolated from *E. coli* and yeast. Mutation to resistance leads to loss of herbicide binding to ALS. This mutant gene has been cloned and used as a selectable marker in transformation experiments.

### Bromoxynil nitrilase (bxn)

Bromoxynil, a nitrilic herbicide is an inhibitor of photosystem II. Bromoxynil is used as a sole nitrogen source by *Klebsiella ozaene.* Bromoxynil nitrilase gene has been isolated which can convert bromoxynil herbicide into 3,5-dibromo-4-hydroxybenzoic acid. This gene has been successfully used as a selectable marker to select for transformed plants.

### New markers

### β-glucuronidase gene

In the traditional selection system using antibiotics of herbicides, the transgenic cells convert the selective agent to detoxified compound that may still exert a negative influence on plant cells. Further, the release of toxic metabolites by dying adjacent cells may also inhibit the growth of transformed cells. In contrast, a new set of markers known as positive selection markers are being developed, which can overcome some of the limitations encountered by the traditional selection system. One such selection system has been established

which uses the β glucuronidase gene from *E. coli* as a selectable gene and a glucuronide derivative of the cytokinins, benzyladenine as selective agent. Benzyladenine N-3 glucuronide is inactive as cytokinin but upon hydrolysis by GUS becomes active cytokinin, stimulating growth and regeneration of transformed cells *in vitro* (Fig. 23.3). The *GUS* gene served as both selectable and screenable markers. The efficiency of the transformation was reported to be about two-fold higher than with kanamycin.

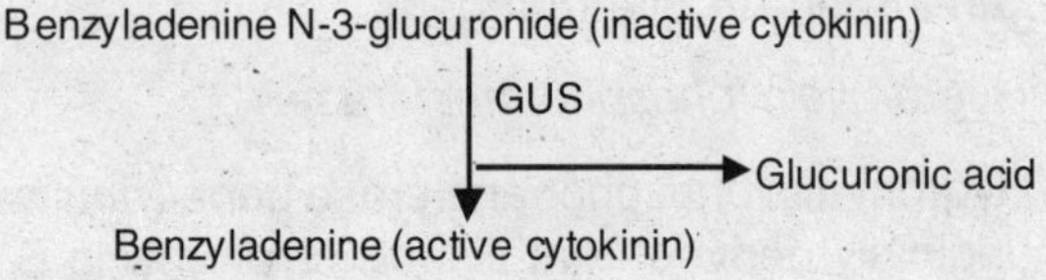

Benzyladenine N-3-glucuronide (inactive cytokinin)

$$\downarrow \text{GUS}$$

Benzyladenine (active cytokinin) $\longrightarrow$ Glucuronic acid

**Fig. 23.3:** Hydrolysis of inactive cytokinin to active cytokinin by GUS gene.

### Gene encoding enzymes in the hormone pathway -ipt

Genes encoding enzymes of the hormone pathways originating from *Agrobacterium* have been successfully used for the selection of the transformed plants, although the presence or activity of the respective gene had to be eliminated or turned down in order to avoid detrimental effects of hormone overdoses on plant development. In 1997, Ebinuma and co-workers developed a multi-auto-transformation (MAT) vector system in which the selectable marker isopentenyl transferase (*ipt*) gene is inserted into the maize transposable element Ac. The *ipt* gene is one of the tumor-inducing genes from *Agrobacterium tumefaciens*, which codes for ispentenyl transferase (*ipt*) and is involved in cytokinin synthesis in plant cells. Introduction of the *ipt* gene under CaMV 35S in plants results in marked increase in endogeneous cytokinin and the production of the extreme shooty phenotype, *ipt* shooty, which exhibits the loss of apical dominance and root formation. As the *ipt* selectable gene is combined with Ac in MAT vector system, phenotypically normal transgenic plants can be obtained subsequently by the removal of *ipt* gene by Ac, achieving a goal of marker-free transformation. However, one of the major limitations of using this system is the low frequency of marker-free transgenic plants, as most of the modified transposable elements (containing *ipt* gene) reinsert elsewhere in the genome shortly after their excision and thus only cells with transposition errors would generate phenotypically normal plants. To overcome this problem, a new MAT vector has been created in which the Ac for removing the *ipt* the gene is exchanged with the site-specific recombination system *Rlrs* isolated from *Zygosaccharomyces rouxii*. The *Rlrs* system comprises a *R* recombinase gene and two *rs* recombination site sequences. The *ipt* combined with the (R) gene was placed within two directly-oriented recognition sites to remove it from transgenic cells after transformation. The improved MAT vector is used to generate marker-free transgenic plants efficiently. Such a system can be applied to woody plants or vegetatively propagated species to produce marker-free transgenic plants as well as providing the basis for the development of an inducible plant transformation system. Expression of *ipt* gene under dexamethasone-inducible promoter led to the recovery of lettuce and tobacco transformants under inducing conditions.

### rol gene of A. rhizogenes

The *rol* gene derived from *A. rhizogenes* is responsible for the proliferation of hairy roots, which spontaneously regenerate into transgenic plants with abnormal phenotype such as wrinkled leaves, shortened internodes and reduced apical dominance. Such plants are easily identified, thus necessitating eviction of selectable markers.

### Phosphomannose isomerase (PMI) gene

This PMI and xylose isomerase *(xyl A)* marker genes are based on the concept of favoring transgenic cells while starving rather than killing the non-transgenic cells. The *man*A gene encoding Phosphomannose isomerase (PMI) has been isolated from *E. coli* and mannose acts as a selective agent. After uptake by plant cells, mannose is phosphorylated by hexokinase to mannose-6-phosphate that results in phosphate and ATP starvation, thus depleting energy from critical functions such as cell division and elongation. Mannose-6-phosphate toxicity in plant cells has been shown to cause apoptosis and programmed cell death. The non-transgenic PMI negative cells are unable to survive on media containing mannose as sole carbon source. Transformed cells with *man*A gene encoding PMI can convert the mannose selective agent to easily metabolizable compound, fructose-6-phosphate, thus improving the energy status of transformed cells. Novartis Agribusines Biotech Research Inc., has licensed the PMI gene selection system, has found the marker to be effective in selection of wheat and maize transgenics. Novartis scientists have reported that PMI protein is safe and readily digested by mammals. However, this system may not work well in those plants, which contain an endogenous level of phosphomannose isomerase.

### Xylose isomerase (xyl A) gene

Gene for xylose isomerase (*xyl*A) isolated from *Streptomyces rubiginosus* has also been employed as a selectable marker and xylose as the selective agent. The enzyme xylose isomerase catalyses the isomerization of D-xylose to D-xylulose. The non-transformed cells cannot utilize the D-xylose as a sole carbon source. But cells which are transformed with *xyl*A growing on xylose have the ability to convert D-xylose to D-xylulose and utilize it as a C source. The xylose isomerase selection system has been tested in potato, tobacco and tomato. Use of *xyl* as a selectable marker in plant transformation is quite safe as xylose isomerase is used in food industry on a commercial scale.

### Betaine aldehyde dehydrogenase gene

The betaine aldehyde dehydrogenase (BADH) gene isolated from spinach has been used as a selectable marker (Daniell *et al.*, 2001). This enzyme is present only in the chloroplasts of a few plant species (members of Chenopodiaceae, Poaceae, etc.). The selection system involves the conversion of toxic betaine aldehyde (BA) by the chloroplast BADH enzyme to a nontoxic glycine betaine, which also serves as an osmoprotectant for enhancing drought and salt stress tolerance. Chloroplast transformation efficiency was 25 fold higher in BA selection than spectinomycin, which is widely used for chloroplast transformation.

### Dihydropicolinate synthase and desensitized aspartate kinase

Regulatory enzymes of aspartate family pathway, which leads to the synthesis of branched chain amino acids, lysine, threonine, methionine and isoleucine from aspartate could be used as selectable markers and lysine plus threonine as selective agent (Perl *et al.*, 1992). Aspartate kinase (AK) the first enzyme of the aspartate family pathway consists of several isoenzymes which are feedback inhibited by lysine and threonine. Another regulatory enzyme of this pathway dihydropicolinate synthase (DHPS) is also sensitive to feedback inhibition by lysine and is the major limiting factor for the synthesis of lysine. Bacterial genes for AK and DHPS enzymes are less sensitive to lysine inhibition than their counterparts in plants and have been used as selectable markers for genetic transformation of tobacco, potato and barley. The plants transformed with bacterial desensitized AK genes were selected for resistance to the presence of lysine plus threonine in the regeneration medium and those transformed with DHPS gene were selected for resistance to the toxic lysine analogue, S-aminoethyl L-cysteine. The transgenic plants showed higher activity of AK and DHPS enzymes, which resulted in overproduction of lysine, threonine and methionine. Thus in plants where accumulation of these essential amino acids is desired, it is possible to obtain both selection and introduction of agronomically desirable traits.

## CHIMERIC GENE VECTORS

The nuclear plant gene consists of different regions, each involved in different functions of transcription and translation of mRNA. Starting with the 5′ end there is a promoter region that is involved in the initiation of transcription, together with enhancer/silencer regions that confer regulation of expression, a transcriptional start or cap site, and the so called CAAT and TATA boxes which help in binding RNA polymerase. One or more untranslated or intron regions are present within the transcribed region. The end of the translation region is determined by a stop codon and followed by a terminator at the 3′ end with polyadenylation signal.

In the majority of cases, whether a direct vectorless delivery system or natural vectors are used for transfer of marker genes, reporter genes and the genes of interest are modified prior to the transfer by different methods. These are placed under the control of different promoter, terminator and enhancer sequences. These are called chimaeric or transgene constructs, since they consist of components derived from different origins (Fig. 23.4). Plants are usually transformed with relatively simple constructs in which the gene of interest is coupled with an appropriate promoter, 5′ leader and a 3′ terminator sequences to ensure efficient transcription, stability and translation of mRNA. The promoter can be of plant, viral or bacterial origin. Some promoters confer constitutive expression, while others may be selected to permit tissue specific expression or

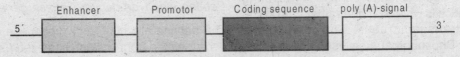

**Fig. 23.4:** Sequence components of a chimeric gene construct.

environmentally inducible expression. The promoters of bacterial origin are *lac, trp,* and *tac,* while phage promoters are T3, T7 and SP6. The cauliflower mosaic virus (CaMV) 35S RNA promoter is often used because it is a plant virus promoter, and plant viruses are dependent on transcription and translation factors of plants. It directs high levels of expression in most tissues. Others such as maize ubiquitin I promoter, *rbc*S ribulose bisphosphate carboxylase small subunit; *Adh*1 alcohol dehydrogenase; *nos* nopaline synthase; and the rice actin promoter/ intron sequence are often preferred for expression in monocots.

The control of gene expression plays an important part in plant development and it is obvious that consideration has to be given to the regulation of genes introduced artificially. As the DNA constructs become more sophisticated, they will contain complex chimeric nucleotide sequences that are a combination of enhancer and silencer sequences, transcription promoters and terminators, protein coding open reading frames possibly with organelle targeting or signal sequences, selectable markers, reporter genes and vector sequences. The DNA constructs can be introduced as naked DNA or as a plasmid, or it can be cloned into a specialized vector for use with *Agrobacterium*. Thus, it is the chimeric gene constructs that are used for expression of transferred genes.

# GENE TRANSFER METHODS

Various gene transfer techniques are grouped in two main categories:
1. Vector mediated gene transfer.
2. Vectorless or direct DNA transfer.

## Vector mediated gene transfer

The term plant gene vector applies to potential vectors both for the transfer of genetic information between plants and the transfer of genetic information from other organisms (bacteria, fungi and animals) to plants. Plant gene vectors being exploited for transfer of genes are plasmids of *Agrobacterium*, viruses and transposable elements.

### *Agrobacterium mediated transformation*

*Agrobacterium* system was historically the first successful plant transformation system, marking the breakthrough in plant genetic engineering in 1983. The breakthrough in gene manipulation in plants came by characterizing and exploiting plasmids carried by the bacterial plant pathogens *Agrobacterium tumefaciens* and *A. rhizogenes*. These provide natural gene transfer, gene expression and selection systems. In recent times, *A. tumefaciens* has been treated as nature's most effective plant genetic engineer.

### *Classification of Agrobacteria*

The bacteria of the genus *Agrobacterium* are well known as gene vectors for plant cells. *Agrobacterium* are gram-negative rods that belong to the bacterial family of Rhizobiaceae. These are often found near soil level at the junction of plant stem and root. The old system of classification according to their phytopathogenic characteristics is as follows:

*A. tumefaciens*: It induces crown gall disease.
*A. rhizogenes*: It induces hairy root disease.
*A. radiobacter*: It is an avirulent strain.

Recently, agrobacteria have been re-classified according to their growth patterns. In this way, three different groups (biotypes) were found. Strains belonging to biotypes 1 and 2 are used in plant genetic engineering experiments.

**Biotype 1:** These strains can be distinguished from other bacteria by the production of ketolactose from lactose. Well known *A. tumefaciens* strains such as Ach5, A6, B6, C58, T37, Bo542 and 23955 belong to biotype 1. They can proliferate at temperatures up to 37°C, but they are usually grown in the laboratory at 28 or 29°C because certain Ti plasmids are somewhat unstable at higher temperatures. Classic *E. coli* media such as LB (Luria Bertini), NB (nutrient broth) and TY (tryptone yeast extract) can be used for growth, but incubation in a defined minimal media such as MM or in a medium such as YMB (yeast maltase broth) also permits growth.

**Biotype 2:** These are identified by their ability to grow on erythritol as a carbon source is a phenotype by which they can be distinguished from other agrobacteria. Laboratory strains of *A. rhizogenes* include the biotype 2 strains NCPPB 1855, ATCC 23834 and A4. The biotype 2 strains proliferate poorly at temperatures higher than 30°C and therefore they are grown at 28 or 29°C. They do not grow well in the classic *E. coli* media, but can instead be cultured on media such as TY + Ca, YMB or in a defined minimal medium (MM).

*Tumor inducing principle and the Ti plasmid*

Smith and Townsend showed in 1907 that a bacterium is the causative agent of crown gall tumors. Experiments of Braun and his coworkers showed that the continued presence of viable bacteria is not required for tumor maintenance (Braun and Stonier, 1958). The bacteria do not penetrate into the plant cells that are converted into tumor cells. Rather, bacteria penetrate into the intercellular spaces and into injured cells and attach themselves only to the wall of healthy plant cells. Attention thus focussed on the identification of a putative 'tumor inducing principle'. Zaenen *et al.* (1974) first noted that virulent strains of *A. tumefaciens* harbor large plasmids and the virulence trait is plasmid borne. Virulence is lost when the bacteria are cured of the plasmid.

*A. tumefaciens* induces tumors called crown galls, while *A. rhizogenes* causes hairy root disease. Large plasmids in these agrobacteria are called tumor-inducing plasmid (Ti plasmid) and root inducing plasmid (Ri plasmid) respectively which confer on their hosts the pathogenic capacity. Both sorts of diseases result from transfer and functional integration of a particular set of the Ti or Ri plasmid into plant chromosomes. It was later revealed that only a small part of the Ti plasmid, the T-DNA or transferred DNA, is transferred to and integrated into the host plant nuclear genome (Chilton *et al.*, 1977).

Crown gall disease caused by *A. tumefaciens* is characterized by unlimited plant cell proliferation (gall formation). Naturally transformed crown gall tumor tissue differs from nontransformed tissue in its ability to grow autonomously on synthetic media in the absence of hormones, i.e. auxins and cytokinins. The plant cells in the tumor acquire two new properties:

1. These show phytohormone independent growth.
2. These contain one or more of unusual amino acid derivatives known as opines. Opine synthesis is a unique characteristic of tumor cells; normal plant tissues do not usually produce these compounds. The ability to produce these opines was specified by the transferred bacterial DNA and not by the host plant genome. They do not appear to have any known useful function in the plant cells and are not directly responsible for the tumorous state of the cells. However, opines are useful to the bacterium since they serve as a source of carbon and nitrogen. In addition, some of the opines also induce conjugation between bacteria. Opines distinguish agrobacteria from other bacteria in the rhizosphere. Agrobacteria are able to utilize opines, thus providing a competitive advantage in the crown gall rhizosphere to the inciting agrobacteria over other soil organisms. This phenomenon has been termed genetic colonization.

It was earlier observed that there were two families of bacteria and the crown galls they established differed by producing two different amino acid derivatives, subsequently called octopine and nopaline. The octopine family has carboxyl-ethyl derivatives of arginine while nopaline family has related compounds that are dicarboxypropyl derivatives of arginine. For the synthesis of octopine and nopaline, the corresponding enzyme octopine synthase (*ocs*)

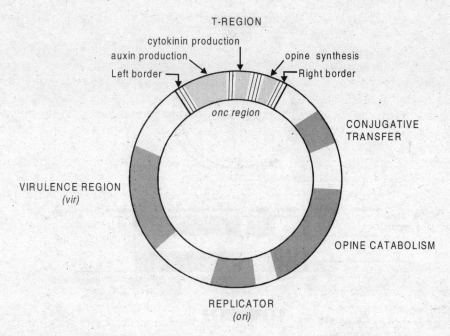

**Fig. 23.5:** Genetic map of an octopine Ti plasmid.

and nopaline synthase (*nos*) are coded by T-DNA. Now, Ti plasmids are defined as octopine, nopaline, succinamopine, agropine or leucinopine depending on the type of opine produced in the tumors.

*Organization of Ti plasmid*

The Ti plasmids of *Agrobacterium* are large circular DNA molecules, up to 200kb in length with molecular weights of about $1.2 \times 10^8$ (3 to 8%) of the *Agrobacterium* chromosome. They exist in the bacterial cells as independent replicating genetic units. The Ti plasmids have major regions for virulence, origin of replication, conjugation, oncogenicity and catabolism of opines. Genetic organization of these regions has been shown in octopine Ti plasmid DNA (Fig. 23.5).

Studies on *A. tumefaciens*—mediated gene transfer system has revealed that the following elements are required for T-DNA transfer to plants.

**T-DNA**: T-DNA is the transferred DNA or transforming DNA of Ti plasmid. The length and

number of T-DNA region vary in Ti and Ri plasmids. The nopaline type Ti plasmid has a single 23-kb T-DNA region (e.g. pTiC58), while octopine type Ti plasmid (pTiAch5) has two closely adjoining regions of 13-kb left-hand piece ($T_L$) and 8-kb right hand piece ($T_R$ DNA). The nopaline T-DNA contains 13 large ORFs whilst there are eight and six large ORFs in the octopine $T_L$ and $T_R$ DNA respectively. The transcripts of the octopine $T_L$ DNA are functionally equivalent to those on the right hand side of the nopaline T-DNA. The T-DNA carries genes for opine synthesis (*nos* or *ocs*) and phytohormone biosynthesis (Fig. 23.6). The T-DNA encodes for both auxin and cytokinin biosynthesis in the plant genome, which alters with the hormone balance in the plant cell and leads to the transformed gall phenotype. Both the hormones synthesized by the T-DNA genes are synonymous with the plant hormones, and although the pathway for cytokinin biosynthesis utilized by A. *tumefaciens* is probably similar to those present in plants, the biosynthesis of auxin from tryptophan is not normally present in plants.

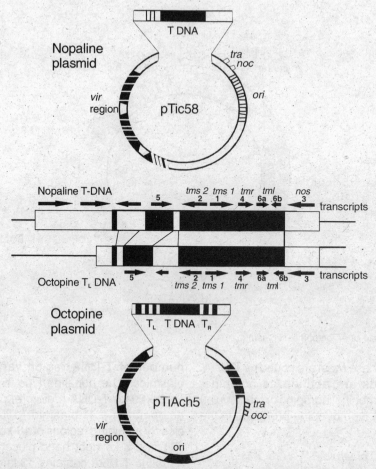

**Fig. 23.6:** Genetic organization of nopaline (pTiC58) and octopine (pTiAch5) plasmid DNAs. Tumor morphology is controlled by expression of genes 1, 2 and 4. *nos* and *ocs* control tumor opine synthesis.

The functionally equivalent genes in the octopine and nopaline T-DNA have been numbered according to the size of their transcript with 6a and 6b encoding similar sized transcripts.

An oncogenic (*onc*) region contains rooty locus (root inhibition, *roi* or tumor morphology root, *tmr*) and results in tumors developing abnormal roots and a shooty locus (shoot inhibition, *shi* or tumor morphology shoot, *tms*) results in tumors with abnormal shoot tissue. Mutations of a further morphogenetic locus, *tml*, produces larger than normal tumors. Two genes comprise the *shi* or *tms* shooty locus and one gene *tmr* represents rooty locus. These *tms*1 and

*tms*2 are responsible for the synthesis of the phytohormone indole 3-acetic acid from tryptophan. Rooty locus *tmr* gene encodes isopentyl transferase that catalyzes the formation of isopentyl adenosine 5′ monophosphate (cytokinin) (Fig. 23.7). The phytohormones in turn alter the developmental program leading to the formation of crown gall.

Gene 4 encodes the enzyme dimethylally1-pyrophosphate: AMP transferase (DMA transferase). The enzyme catalyses the covalent linkage of dimethyl allyl pyrophosphate to the $N_6$ of AMP, yielding isopentany1adenosine 5′-monophosphate, which is the first cytokinin in

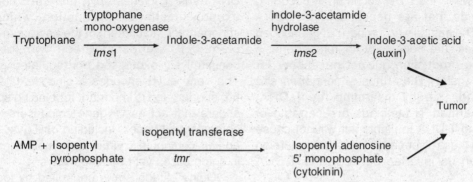

**Fig. 23.7:** Pathway of auxin and cytokinin biosynthesis in *Agrobacterium* induced tumor tissue formation.

the metabolic pathway of this phytohormone. The function of gene 5 in transformed tissue is not clear but its activity is reported to be tissue specific and activated by auxin. Gene 5, especially in combination with mutation at *iaa*H or *iaa*M can affect tumor development or morphology. It modulates auxin response by auto regulated synthesis of a growth hormone antagonist in plants.

Gene 6a is reported to be involved in the secretion of opines from the transformed plant cell. The function of gene 6b is not fully understood, but galls resulting from transformation with mutants in this region of the octopine T-DNA resulted in increased tumor size. The 6b gene has been reported to reduce the response of the plant tissue to cytokinins. This can be achieved by the gene product in several ways, including enhanced auxin production or by increasing plant sensitivity to endogenous auxins. The 6b gene is located close to the right border of the T-DNA. Genetic analysis of Ti plasmid and the T-DNA in particular, has demonstrated that gene 6b has a clear effect on the tumorous phenotype in some hosts but not on others. This may be due to the fact that the gene 6b either inhibits auxin activity or activates the cytokinin activity.

In has been suggested that gene I, gene 2 and gene 4 together give rise to an overproduction of both IAA and cytokinins, which in cooperation with other T-DNA genes like the

*ila* gene (gene 5) and gene 6b lead to the unorganized cell division that produces the crown gall tumour.

Opine synthesis is a unique characteristic of tumour cells. Particular *Agrobacterium* strains induce tumors which produce and secrete specific opines that the strain, or related strain can catabolise with genes present in the Ti plasmid. The induction of tumors, which synthesize unique opines is a central feature in the pathogenic relationship between *Agrobacterium* and the plant. An opine synthesis *os* region is responsible for the synthesis of unusual amino acid or sugar derivatives, which are collectively given the name opines. Opine synthesis is controlled by enzyme octopine synthase and nopaline synthase at loci called *ocs* and *nos* on octopine and nopaline plasmids respectively.

The third component of T-DNA region on all Ti and Ri plasmids is a 25-bp highly conserved directly repeated DNA sequence that flanks and delimits the T-DNA. These sequences act as borders and are referred to as left and right borders. These borders play an important role in the transfer process. In fact, any DNA sequence that is located between these borders is transferred to the plant, and this feature has been exploited in genetic manipulation experiments. The nucleotide sequence of the border repeats is similar in nopaline and octopine Ti plasmids. There are no genes within the T-DNA of naturally occurring Ti plasmids that are necessary for T-DNA transfer. The naturally occurring T-DNA encoded "oncogenes" are necessary for tumorigenesis and opine production but not necessary for T-DNA transfer. Thus, these oncogenes can be deleted, thereby disarming the T-DNA. Such disarmed Ti plasmids are capable of transferring T-DNA to plants but will not cause tumors. In these disarmed plasmids novel foreign genes can be inserted between the T-DNA borders.

**Virulence genes:** The second gene system necessary for T-DNA transfer consists of the virulence (*vir*) genes encoded by the Ti plasmid in a region outside the T-DNA. The 40 kb *vir* region of the octopine Ti plasmid embraces 24 genes involved in virulence. These genes are present in 8 operons called *vir* A–*vir* H, which are coregulated and thus form a regulon. Most of the operons except *vir* A, *vir* F and *vir* G contain several genes, i.e. they are polycistronic. With the exception of the *vir* A and *vir* G genes, the *vir* operons are not transcribed during normal vegetative growth. The genes for virulence are silent till they are induced by certain plant factors. *vir* A and *vir* G make up a two-component regulatory system important for transcriptional activation of the remaining *vir* genes and to some extent their own increased activation.

The remaining *vir* genes are involved in the processing of the T-DNA from the Ti plasmid, and in T-DNA transfer from the bacterium to the plant cell. Certain Vir proteins may also be involved in the targeting of the T-DNA to the plant nucleus, and perhaps in T-DNA integration into plant DNA. Thus, genetic manipulation that stimulates *vir* gene induction (or increase Vir protein activity) may result in increased T-DNA transfer to the plant.

**Chromosomal genes:** The third component necessary for T-DNA transfer consists of a number of genes encoded by the *A. tumefaciens* chromosome. Some of these genes viz. *chv*A, *chv*B, and *psc*A (*exo*C) present on the *A. tumefaciens* chromosome are essential for T-

DNA transfer. Genes *chv*A and *chv*B code for exopolysaccharide production, which are important for attachment of the bacterium to the plant cell. Genes at these two loci are constitutively expressed nonregulated genes. The gene *chv*E encodes a glucose/galactose transporter, and is important for binding specific sugars that act as *vir* gene coinducers. Other chromosomal genes, including *chv*D, *ilv*, *mia*A, and *att*, contribute to virulence in more minor and in some cases yet undefined ways.

Opine catabolism is controlled by loci *noc* and *occ* for nopaline and octopine catabolism respectively. It lies on the right limb of the molecule near to locus *tra* specifying conjugative transfer of the Ti plasmid to avirulent *Agrobacterium* strains. These *noc* and *occ* genes do not form a part of T-DNA.

*T-DNA transfer*

T-DNA transfer begins with the introduction of bacteria into a plant wound. Wounding is a necessary event in the process and may, at least in part, be required for the synthesis by the plant certain compounds that induce the expression of the *vir* genes. Two of the most active substances identified are acetosyringone and β-hydroxy-acetosyringone. The *vir* A product is a periplasmic membrane protein that senses specific phenolic compounds (such as acetosyringone and related molecules) synthesized by wounded plant cells. It is most likely that Vir A protein also interacts with another protein encoded by *chv*E gene present on bacterial chromosome and is important for the binding of sugar coinducers. The presence of these phenolic and sugar inducers results in the autophosphorylation of Vir A, which then transfers the phosphate group to Vir G protein, thus activating Vir G. The activated Vir G protein acts as a transcriptional activator of the other *vir* genes (most likely by binding to specific DNA sequences, the "*vir* box" that precedes the *vir* genes). An early event in the T-DNA transfer is the nicking of Ti plasmid (Fig. 23.8). Two of the proteins coded by *vir* D, Vir $D_1$ and $D_2$, provide

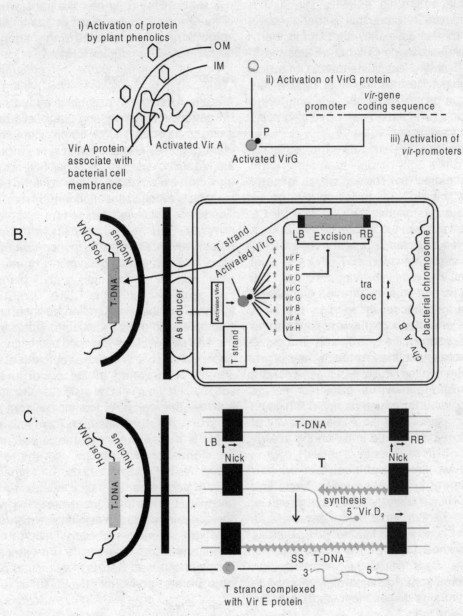

**•Fig. 23.8:** Schematic summary of the events leading to the transformation by *Agrobacterium*; **A**–activation of the vir genes via VirA and VirG; **B**–key events in the induction of vir genes and T-DNA transfer, **C**–key events in the transfer of T-DNA to plant cell. OM–outer membrane; IM–inner membrane; AS–acetosyringone

an endonuclease that initiates the transfer process by nicking T-DNA at a specific site. Nicking takes place between the third and fourth base of the bottom strand of each 25-bp repeat. A nick at the right border 25bp repeat provides a priming end and synthesis of new DNA strand begins from this nick in 5′→ 3′ direction. New strand displaces the old T-DNA single strand. Recent evidence indicates that T-strand retains the Vir $D_2$ protein covalently attached to the 5′ terminus. The presence of Vir $D_2$ makes the 5′ end of the T-strand less vulnerable to an attack by exonuclease. Besides, the Vir $D_2$ protein may also act as a pilot to direct the T-strand to the nucleus of the transformed cell, since it contains a nuclear targeting sequence.

Outside the T-DNA, but immediately adjacent to the right border, is another short sequence called overdrive, which greatly stimulates the transfer process. Overdrive functions like an enhancer. Vir C1 and Vir C2 may act at the overdrive sequence. Vir E2 is a 69 kDa protein that binds single stranded T-DNA and is involved in transporting T-DNA into plant cell nucleus. The remaining Vir proteins are not involved in regulation of expression or T-strand expression; only those encoded by the vir B operon are essential for virulence. Most of the proteins predicted for the vir B operon are located in the membrane and these proteins may form a conjugal pore or pilus structure in the membrane through which the T-DNA is delivered into the plant cell. The vir $B_{11}$ gene has an ATP binding site and the protein has an ATPase activity. It may therefore be involved in delivering energy required for T-DNA transfer. The virH operon consists of two genes and it has a role in the detoxification of certain plant compounds that might otherwise adversely affect the growth of Agrobacterium. Enhanced tumorigenicity was observed for bacteria having the virH genes as compared to those lacking these in certain hosts. vir F plays a role in T-DNA delivery rather than symptom formation. Presence of vi rF gene in octopine Ti strains makes them vastly superior in transferring DNA.

The transfer of T-DNA closely resembles the events involved in bacterial conjugation when the E. coli chromosome is transferred from one cell to another in single stranded form. A difference is that the transfer of T-DNA is usually limited by the boundary of the left repeat, whereas transfer of bacterial DNA is indefinite. It is still not very clear how the transferred DNA is integrated into the plant genome. At some stage the newly generated single strand must be converted into duplex DNA.

### Agrobacterium vectors

It is a fact that the Ti plasmid is a natural vector for genetically engineering plant cells because it can transfer its T-DNA from bacterium to the plant genome. However, wild type Ti plasmids are not suitable as general gene vectors because they cause disorganized growth of the recipient plant cells owing to the oncogenes in the T-DNA. The tumor cells which result from integration of normal T-DNA (wild-type) have proved recalcitrant to attempts to induce regeneration, either into normal plantlets, or into normal tissue which can be grafted onto healthy plants. The oncogenic function of the T-DNA is responsible for inhibiting the regeneration ability of a cell to a complete plant. So it must use vectors in which the T-DNA has been disarmed by making it non-oncogenic. This is most effectively achieved simply by deleting all its oncogenes and substituting them with an insert (desirable gene) between the regions of the left and right border.

Thus, the Agrobacterium Ti plasmid system exploits the elements of the Agrobacterium transformation mechanism. But the following properties of Ti plasmids do not allow their direct use: (i) large size; (ii) tumor induction property and (iii) absence of unique restriction enzyme sites. Agrobacterium plasmids are disarmed by deleting naturally occurring T-DNA encoded oncogenes and replacing them with foreign genes of interest. Besides, restriction enzyme sites are also added into the plasmids.

### Cointegrate vectors

Vectors that recombine via DNA homology into a resident Ti plasmid are often referred to as integrative or cointegrate vectors. In a cointegrate vector, the disarmed Ti vector is covalently linked to donor vector with gene of interest-T-DNA border sequences present in a

*A. tumefaciens* strain to act as one unit. In a cointegrate vector, much of the wild type T-DNA (especially the hormone biosynthetic genes and not the border regions) is replaced with a segment of DNA common to many *E. coli* cloning vectors. Homology between the *E. coli* plasmid-based segment of the modified T-DNA and the donor vector with identical sequences provides a site for recombination to occur which is catalyzed by native *Agrobacterium rec* functions, resulting in the formation of a hybrid or cointegrate Ti plasmid. Because, most cointegrates result from a single recombination event, the ensuing product places the gene of interest along with the entire donor vector with its marker genes between the native T-DNA borders of the disarmed Ti plasmid. The formation of cointegrate vector using the disarmed vector pGV3850 is illustrated in Fig. 23.9. This vector has pBR322 DNA in the T-DNA region, which provides a region of homology with most other cloning vectors. In this example, donor vector pLGVneo1103 carries a plant selectable marker *nos-npt*-II hybrid gene and a bacterial selectable marker, the *kan* gene. Recombination between the homologous pBR322 DNA in the two plasmids produces the co-integrate.

The standard cointegrate vector contains: (i) convenient sites for insertion of the gene of interest, (ii) antibiotic selectable marker gene or genes active in both *E. coli* and *A. tumefaciens*, (iii) a plant functional selectable marker gene, and (iv) *E. coli* functional origin of replication that does not operate in *Agrobacterium*. Research has not demonstrated any specific advantage to retaining the T-DNA and *vir* functions on a single replicon.

*Binary vectors*

The binary vector system consists of two autonomously replicating plasmids within *A. tumefaciens*: a shuttle (more commonly referred to as a binary) vector that contains gene of interest between the T-DNA borders and a helper Ti plasmid that provides the *vir* gene products to facilitate transfer into plant cells. Disarmed helper Ti plasmids have been engineered by removing the oncogenic genes while still providing the necessary *vir* gene products required for transferring the T-DNA to the host plant cell.

The standard components of a binary vector are: (i) multiple cloning site, (ii) a broad host range origin of replication functional in both *E. coli* and *A. tumefaciens* (e.g. RK2), (iii) selectable markers for both bacteria and plants, and (iv) T-DNA border sequences (although only the right border is absolutely essential). A typical binary vector system is shown in Fig. 23.10. The helper Ti plasmid pAL4404 is a derivative of the octopine plasmid pTiAch5 that has the complete T-DNA region deleted but contains *vir* genes. The mini T-DNA donor vector pBin19 contains a truncated T-DNA from the nopaline plasmid pTiT37 comprising the left and right border sequences and a *nos-npt*II hybrid gene as a selectable marker. A *lacZ* polylinker is inserted in the mini T-DNA to provide unique cloning sites and a color test for insertion. The linker carries the galactosidase that produces blue colonies on IPTG/X-GAL indicator plates and white colonies if it is disrupted by an insertion of foreign gene, i.e. pBin 19.

There are a number of advantages associated with using a binary system as compared to cointegrate vector:

1. Binary vectors do not need *in vivo* recombination whereas cointegrate vectors require a recombinational event for stable maintenance within the target *A. tumefaciens* strain.
2. Binary vectors require only that an intact plasmid vector be introduced into the target bacterium, making the process of bacterial transformation both more efficient and quicker (2–3 d versus 4–7d).
3. Binary vector systems with plant ready genes in agrobacteria are easily and efficiently obtained.
4. In binary system, the binary plasmids exist as separate replicons, thus copy number is not strictly tied to that of Ti plasmid. Because

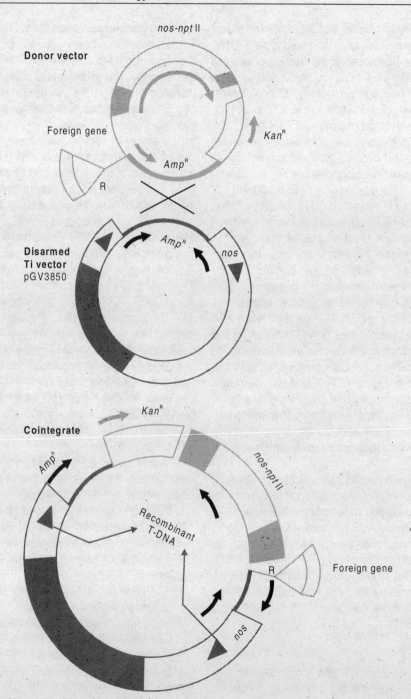

• **Fig. 23.9:** Producing a recombinant T-DNA by cointegrate formation. The donor vector recombines with the disarmed vector by homologous recombination to produce the cointegrate.

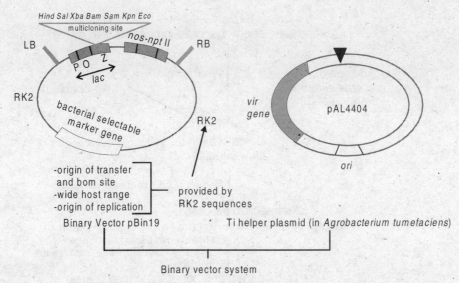

**Fig. 23.10:** A typical binary Ti vector system.

of this, in most cases, confirmation of the transformation event is accomplished via agrobacteria mini preps as compared to Southern hybridization used for cointegrate plasmids.

Most of the recently developed plant transformation vectors are binary, largely due to the ease of both *in vivo* and *in vitro* DNA manipulation and their higher transformation efficiencies, allowing the use of direct *Agrobacterium* transformation techniques.

*Transformation Techniques Using Agrobacterium*

*Agrobacterium*-mediated plant transformation technique is now extensively utilized to generate transgenic plants. The important requirements for *Agrobacterium*-mediated gene transfer in higher plants are:

i. The plant explants must produce acetosyringone or other related compounds for activation of *vir* genes or *Agrobacterium* may be preinduced with synthetic acetosyringone.

ii. The induced agrobacteria should have access to cells that are competent for transformation.

iii. Transformation competent cells/tissues should be able to regenerate into whole plants.

A general scheme for transformation using *A. tumefaciens* as a vector has been shown in Fig. 23.11.

The explant used for inoculation or cocultivation with *Agrobacterium* carrying the desired cointegrate or binary vectors include cells, protoplasts, callus, tissue slices, whole organs or sections etc. There are three general approaches for inoculation of *Agrobacterium*.

*i) Infection of wounded plants*: *In vivo* inoculation of seedlings or well established plants that have been propagated *in vitro* under aseptic conditions is the traditional way to obtain *Agrobacterium* transformed cells. Seedlings are decapitated and the freshly cut surface wound is inoculated with an overnight culture of *Agrobacterium*. The tumor produced is excised and grown as a callus culture. The transforming calli are picked off and regenerated.

*ii) Co-cultivation*: Protoplasts are isolated and during cell wall reformation stage, these are incubated for 24–40 h in a suspension of *Agrobacterium* at about 100 bacteria per

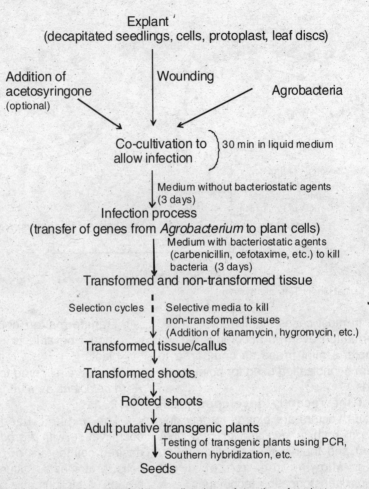

**Fig. 23.11:** *Agrobacterium tumefaciens* mediated transformation of explants.

protoplast. Transformation occurs during the subsequent few days of cocultivation and exposure to selective agent.

iii) **Leaf disc method**: In practice, this procedure can be performed with any tissue explant that provides a good source for initiation of whole plant differentiation. In this regard, newly emerged cotyledons have provided useful material. In the disc, or explant method, a tissue segment is excised and tissue is allowed to incubate in *Agrobacterium* suspension for few hours to three days and then cultured on a media for bacterial growth to take place. Tissue explants are then moved into a media containing a bacteriostatic agent that eliminates the bacteria. Then explants are moved to a media which is designed for the selection of transformed plant cells, and therefore, contains the appropriate antibiotics. The disc method is technically simple and produces transgenic plants very rapidly and is therefore the favored procedure at present.

*Procedure*

A particular *Agrobacterium* strain is first grown in solid media plates (e.g. YEP medium) containing required antibiotics as per bacterial selectable markers. *Agrobacterium* suspension for cocultivation/inoculation is prepared by picking a single colony from media plate and inoculating in 5–10 ml of liquid medium (YEP) containing appropriate antibiotics. If *Agrobacterium* is to be preinduced by acetosyringone (e.g. in monocots), then it should be added in the liquid medium. After 2 days of growth in dark at 28°C, the OD of the culture is measured at 600 nm. The suspension is diluted or concentrated to bring the value at 0.1 OD for cocultivation. Explants are then inoculated/cocultivated with *Agrobacterium* suspension for few hours to few days. Explants are then blot dried on a sterile filter paper and transferred to nutrient culture medium of explant without bacteriostatic agent for 3 days. This allows the *Agrobacterium* to grow, multiply and transfer the T-DNA. The explants are then transferred to media containing bacteriostatic agent like cefotaxine, carbenicillin, etc. to kill the bacteria. These explants are then transferred to a selection medium containing appropriate selection agent (antibiotics, herbicides, etc.) depending upon the plant selectable marker genes employed for transformation. Resistant/tolerant tissues are continuously grown and selected on selection agent followed by regeneration. These transgenic plants are then tested for stable integration and expression of genes by PCR method or by Southern hybridization. These are $T_0$ transgenic plants and seeds obtained from these plants are $T_0$ seeds. Plants obtained from these $T_0$ seeds are $T_1$ plants.

## Agrobacterium mediated virus infection—Agroinfection

The delivery of viral or viroidal sequences to plants using bacterium as a route is referred to as "agroinfection" or "agroinoculation". It has now been shown that plant virus genomes that are cloned into Ti plasmids can be introduced into plants from *Agrobacterium* and they will then initiate a typical infection. This *Agrobacterium*-mediated virus infection has been termed 'agroinfection'. Successful transfer of a single viral genome can be detected as a result of massive amplification during viral replication.

Grimsley *et al.* (1987) constructed a vector that contained the cloned double stranded copy of the maize streak virus (MSV) genome, a gemini virus. This virus has a single stranded DNA genome that replicates in the plant nucleus via DNA intermediates that are autonomous. The naked viral DNA and cloned DNA are not infectious and are transmitted in nature by insects. A head-to-tail dimer of the MSV genome was inserted into a mini-T-DNA vector between border sequences. The plasmid was mobilized into *A. tumefaciens* in a triparental mating and the host agrobacteria were inoculated into leaf wounds of maize seedlings. The maize plants developed all the symptoms of MSV infection. This provides a sensitive assay for T-DNA transfer to monocots.

*Advantages*

*Agrobacterium* mediated transformation has the following advantages:

1. It is a natural means of transfer and it is therefore perceived as a more acceptable technique to those who feel natural is the best.
2. *Agrobacterium* is capable of infecting intact plant cells, tissues and organs. As a result, tissue culture limitations are much less of a problem. Transformed plants can be regenerated more rapidly.
3. *Agrobacterium* is capable of transferring large fragments of DNA very efficiently without substantial rearrangements.
4. Integration of T-DNA is a relatively precise process.
5. The stability of gene transferred is excellent.

*Disadvantages*

Although the introduction of new traits into plants via the *Agrobacterium* system is now a common practice, there are still shortcomings in the system.

1. It has the limitation of host range; some important food crops cannot be infected with *Agrobacterium*. Although lately, much progress has been achieved to overcome this limitation by the development of highly virulent strains of *Agrobacterium*.

2. Sometimes, cells in a tissue that are able to regenerate are difficult to transform. It might be that embryogenic cells are in deep layers to be reached by *Agrobacterium*, or simply are not targets for T-DNA transfer.

### Viruses mediated gene transfer

Vectors based on viruses are desirable because of the high efficiency of gene transfer that can be obtained by infection and the amplification of transferred genes that occurs via viral genome replication. Also, many viral infections are systemic so that gene can be introduced into all the cells in a plant. Viruses provide natural examples of genetic engineering since viral infections of a cell results in the addition of new genetic material which is expressed in the host. Additional genetic material incorporated in the genome of a plant virus might be replicated and expressed in the plant cell along with the other viral genes. Once the biological characteristics of a virus have been suitably selected and manipulated, new genes must be incorporated into the virus in such a way that they are expressed in the plant. Vectors for transferring genes into plants are based on DNA or RNA molecules that naturally express their genetic information in plant cells. The replicating genomes of plant viruses are nonintegrative vectors as compared to those vectors based on the T-DNA of *A. tumefaciens*, which are integrative gene vectors.

The nonintegrative vectors as plant virus vectors do not integrate into the host genome; rather they spread systemically within a plant and accumulate to high copy numbers in their respective target cells. In most cases, viral genomes have been modified to accommodate the insertion of foreign sequences, which are transferred, multiplied and expressed in a plant

as part of the recombinant virus genome. Apart from their pathogenicity, recombinant viruses are designed to mimic their wild type counterparts in all other respects. A viral vector should possess the following characteristics:

1. It should have a broad host range, virulence, ease of mechanical transmission and rate of seed transmission.

2. It should have the potential to carry additional genetic information since strict packaging limitations are there. Thus viruses whose capsid is filamentous or rod-shaped, or viruses that possess a multipartite genome or a helper or satellite component, offer the potential for carrying extra nucleic acid.

3. Virus suitability as a vector depends on the fact that genetic material must be able to be manipulated and be infectious.

At present three groups of viruses are being studied for vector development and transformation.

### Caulimoviruses

Among the plant viruses type, the virus of the caulimovirus group, cauliflower mosaic virus (CaMV), is cited as the most likely potential vector for introducing foreign genes into the plants. This is mainly because caulimoviruses are unique among plant viruses in having a genome made of double stranded DNA, which of course lends itself more readily to the manipulations. They were the first plant viruses to be manipulated by the use of recombinant DNA technology. The caulimoviruses group consists of 6–19 viruses, each of which has a limited host range. The commonest are carnations etched virus (CERV), cauliflower mosaic virus (CaMV), dahlia mosaic virus (DaMV), mirabilis mosaic virus, strawberry vein banding virus. The best known CaMV infects many members of Cruciferae and *Datura stramonium*. The virus is naturally transmitted by aphids but can very easily be transmitted mechanically. It has been found that virion DNA alone or cloned CaMV DNA are infectious when

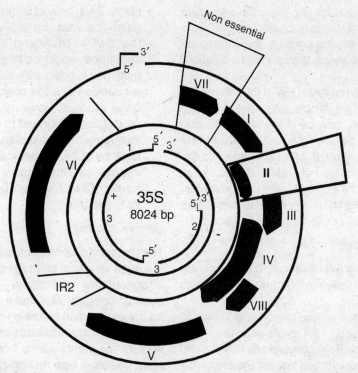

**Fig. 23.12:** A functional map of the cauliflower mosaic virus genome. The coding regions are shown as dark boxes. The different reading frames are indicated by the position of the boxes. The thin line in the centre is the DNA and outside is 35S transcript. Concept of the figure was taken from *Applied Molecular Genetics*, Roger L. Miesfeld, by kind permission of John Wiley & Sons, Inc., 605 Third avenue, New York, USA.

they are simply rubbed on the surface of susceptible leaves. Virus particles can be found at a high copy number (up to $10^6$ virions per cell) and spread rapidly in 3–4 weeks in a systemic manner throughout the entire plant. Virus accumulates in the cytoplasm as inclusion bodies, which consist of a protein matrix with embedded virus particles. The virus particle is spherical, isometric, about 50 nm in diameter, and may be isolated from the inclusion body using urea and nonionic detergents. The DNA exists in linear, open circular and twisted or knotted forms; however, none of the circular forms is covalently closed due to the presence of site-specific single stranded breaks (Fig 23.12).

The genome of CaMV consists of a relaxed circular molecule of 8 kb. DNA sequence analysis revealed that six major and two minor

open reading frames (ORF) are present on one coding strand in a very tightly packed arrangement. The caulimoviruses as well as badnaviruses replicate via reverse transcription of a more than genome length RNA. The mechanism of replication is as follows: The infecting CAMV DNA enters the plant nucleus, where the single stranded overlaps are digested and gaps ligated to give a supercoiled minichromosome. The function of this minichromosome is to act as a template for plant nuclear RNA polymerase II. The transcript thus formed is transported to the cytoplasm where it is either translated or replicated by reverse transcription. The RNA transcript is then copied into the minus strand DNA. Synthesis of the plus strand DNA starts at two primer binding sites near gaps 2 and 3. From gap 2 synthesis proceeds to the 5′ end of the minus strand DNA, whereas synthesis from gap 3 continues to gap 2. This DNA molecule gets packaged into virus particles or re-enters the nucleus and undergoes another round of transcription and/or translation/ replication.

Two genomic regions, ORF II that codes the insect transmission factor and ORF VII with unknown function, can be dispensed with and replaced by the gene of interest. There is one report on the insertion of a foreign gene into the plant describing the use of CaMV. Brisson *et al.* (1984) achieved the expression of foreign protein encoded by the bacterial dihydrofolate reductase gene in turnip plant cells by a CaMV vector.

### Uses as a vector

1. Naked DNA is infective, being able to enter plant cells directly if rubbed on to a leaf with a mild abrasive.
2. As a DNA virus whose genome is known to be packaged in nucleosomes and transcribed by RNA polymerase II, it is more suited for exploitation as an experimental tool than any other plant virus.

### Problems

1. The genome is so tightly packed with the coding regions that there is little room for insertion of foreign DNA. Most deletions of any significant size destroy virus infectivity. The theoretical packaging capacity is of 1000 nucleotides, but attempts to propagate and stably express genes larger than 500 bp have not been successful.
2. CaMV-derived vectors are restricted to members of Cruciferae that can be infected by viral DNA. Recently, some mutant strains of CaMV infecting species of Solanaceae have also been described.
3. CaMV DNA has multiple cleavage sites for most of the commonly used restriction endonucleases. This would limit the usefulness of wild isolates of CaMV.
4. Infectivity with CaMV is another problem because once established it becomes systemic, spreading throughout the whole

plant. This lack of inheritance through the germline may be advantageous in that the CaMV DNA and any inserted gene sequence would be highly amplified in the host plant cells, potentially permitting the expression of large quantities of the foreign gene product. However, it appears that to propagate CaMV and to allow its movement throughout the vasculature of the plant, the DNA must be encapsidated, and this would impose serious constraints on the size of foreign DNA that can be inserted into the viral genome

### Gemini Viruses

The gemini viruses replicate and cause diseases in a wide variety of plant species, including many of agricultural importance. These viruses have some properties that make them ideal vectors for the expression of foreign genes in plants. The potential of gemini viruses as gene cloning vectors for plants stems from work on several plant diseases now recognized as being caused by these agents. Both the curly top virus (CTV) and maize streak virus (MSV) cause important diseases.

The gemini viruses are characterized on the basis of their unique virion morphology and possession of single stranded (ss) DNA. The gemini virus group takes its name from the unusual twin icosahedral capsid structure of its members. These have a small capsid size, 18–20 nm $\times$ 30 nm, and geminate (paired particles) morphology, which makes them different from all other classes of viruses. They have covalently closed circular (ccc) topography of the ssDNA which is in the molecular weight range of 7 $\times$ $10^5$–9 $\times$ $10^5$. All gemini viruses have a single major coat protein subunit in the range 2.7–3.4 $\times$ $10^4$ daltons.

The genome of Gemini viruses consists of either one or two circular, single stranded DNA molecules. The single stranded viral DNA, 2.6–3.0 kb long, is converted into a double stranded replicative form in the nucleus of plant cells. Many copies of the replicative form of a Gemini virus genome accumulate inside the nuclei of infected cells. There is no evidence to date for a reverse transcription step in Gemini virus replication.

Bean golden mosaic virus (BGMV) DNA was found to be 2510 nucleotides long, and if this was the complete genome it would be less than half the length of any other known autonomously replicating plant virus. By comparing the ssDNA of the virus particles with the viral dsDNA found in infected plants, it was found that the nucleotide sequence had twice the complexity expected on the physical size of the viral DNA. This indicates that BGMV DNA is heterogeneous, the virus having a divided genome consisting of two DNA molecules of approximately the same size but different genetic content.

The genomic ssDNA replicates in the nucleus through doublestranded DNA (dsDNA) intermediates, most likely by a rolling circle mechanism. Despite an overall similarity in capsid and genome structure, the Gemini virus group is diverse and can be divided into at least three subgroups.

The genomes of all group members contain two elements that must be part of any functional Gemini virus vector system, although these elements need not be physically present on the same molecule. These elements are an origin of replication and an essential viral replication protein.

Most Gemini virus vectors described to date are relatively primitive and simply permit the investigator to replace the coat protein coding sequence with a reporter or other gene of interest. Gemini virus expression vectors may be used to deliver, amplify and express foreign genes in several different systems of protoplasts and cultured cells, leaf discs, and plants.

The cereal Gemini virus, wheat dwarf virus, is under development as a vector for introducing genes into cereals. It is capable of accommodating and replicating gene inserts up to about 3 kb in length. Three bacterial genes inserted into the wheat dwarf virus genome have been successfully replicated and expressed after transfer into cultured cells of *Triticum monococcum* and *Zea mays*.

## Uses as a cloning vector

1. These viruses contain single stranded DNA that appears to replicate via a double stranded intermediate and thus makes *in vivo* manipulation in bacterial plasmids more convenient.
2. An attractive feature is the ability of bipartite Gemini viruses to contain a deletion or a replacement of virus coat protein sequences by foreign genes without interfering with the replication of the virus genome.

## Problems

1. These are not readily transferred by mechanical means from plant to plant, but transmitted in nature by insects in a persistent manner.
2. The small particle size may present packaging problems for modified DNA molecules and any useful genetic modifications will have to solve the problems of a vector which in its natural state causes severe disease in susceptible plants.

### RNA Viruses

RNA viruses have advantages and disadvantages, which make them the ideal choice for certain applications as vectors. Their assets include ease of use, perhaps the highest levels of gene expression, and the vector's ability to infect and spread in differentiated tissue.

There are two basic types of single stranded RNA viruses. The monopartite viruses have undivided genomes containing all the genetic information and are usually fairly large, e.g. tobacco mosaic virus (TMV). The multipartite viruses, as the name suggests, have their genome divided among small RNAs either in the same particle or separate particles, e.g. brome mosaic virus (BMV) contains four RNAs divided between three separate particles. The RNA components of multipartite genomes are small and appear to be able to self-replicate in plants. With some members of the group, the genes

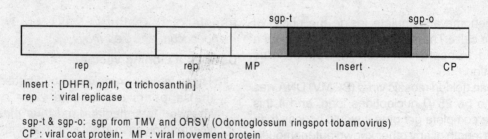

Insert : [DHFR, *npt*II, α trichosanthin]
rep   : viral replicase

sgp-t & sgp-o: sgp from TMV and ORSV (Odontoglossum ringspot tobamovirus)
CP : viral coat protein;  MP : viral movement protein

**Fig. 23.13:** Genomic map of TMV: TB2 plant RNA viral vector.

encoding the coat protein may be dispensable, as their loss does not affect viral DNA multiplication.

The second group, subgenomic RNAs (e.g. RNA IV of BMV), is unlikely to find application as cloning vectors as it is unable to self-replicate in infected plants.

The third group, satellite RNAs has perhaps the greatest potential as it is totally dispensable to the virus. Satellite RNAs vary in size from 270 bases (tobacco ringspot virus satellite) to 1.5 kb (tomato black ring virus satellite). These satellites appear to share little homology with the viral genomic RNAs, templates for their own replication, and utilize the machinery for replication set up by the virus. They are not required for virus replication, but are capable of altering the pathogenicity of the viral infection. These satellite RNAs have a number of other unusual properties, including the ability to code for proteins and stability in the plant in the absence of other viral components.

Plant viruses have only few genes, but these are expressed to high levels by a variety of means. High gene product yields and activities are associated with RNA viral vectors. Plant viral genomes contain a gene for movement through the plant, a gene for the protein making up the protective coat and a gene for replication. Coat proteins are the most abundant viral gene product. One example of a gene replacement construction is the replacement of RNA viral vector coat protein gene of BMV by chloramphenicol acetyl transferase gene (French *et al.*, 1986). Expression of cDNA in

barley protoplasts suggests that RNA virus may be a useful vector for gene manipulation.

A second common type of vector construction is the insertion of a foreign gene into an intact viral genome. An example is the TMV vector TB2 that was the first plant RNA viral vector able to spread systemically in whole plants (Donson *et al.*, 1991). In TB2, the foreign gene is inserted at 3′ end of the movement protein, which is the normal position for the TMV coat protein ORF (Fig 23.13). Thus, foreign gene expression is driven by the native coat protein subgenomic promoter (sgp) and a sgp from related virus ORSV—odontoglossum ringspot tobamovirus drives the expression of ORF of coat protein.

Most RNA viral vectors carrying large or small inserts replicate stably in protoplasts and/ or inoculated leaves. Thus, experiments can be carried out in either protoplasts or differentiated tissue using the present generation of plant viral vectors.

A variety of DNA and RNA viral vectors can be used to express genes in protoplasts but the choice of vector is limited for use in inoculated leaves of whole plants. The other disadvantages are limitation of insert size and the induction of symptoms in the host.

**Vectorless or Direct DNA transfer**

Direct gene transfer has proved to be a simple and effective technique for the introduction of foreign DNA into plant genomes. It has been further sub-divided into three categories:

1. Physical gene transfer methods.

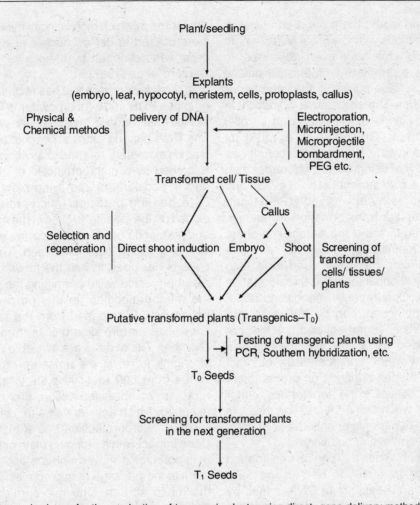

**Fig. 23.14:** A general scheme for the production of transgenic plants using direct gene delivery methods.

2. Chemical gene transfer methods.
3. DNA imbibition by cell, tissue and organs.

**Physical gene transfer methods**

The species and genotype-independent transformation methods wherein no natural vector is involved but which are based on the direct delivery of naked DNA to the plant cells have been grouped under this category. This is also referred to as DNA mediated gene transfer (DMGT). A general scheme for the production of transgenic plants using various direct DNA

delivery methods has been shown in Fig. 23.14. The various methods have been described in detail.

*Electroporation*

Electroporation is the process where electrical impulses of high field strength are used to reversibly permeabilize cell membranes to facilitate uptake of large molecules, including DNA. The electroporation method is based on Neumann *et al.* (1982) for animal cells. It uses a relatively high initial field strength (1–1.5 kV) with a low capacitance and therefore a short decay time. Other methods with low initial field strength and long decay time have been described. In this procedure, a sample of protoplasts is pulsed with high/low voltage pulses in the chamber of an electroporator. The chamber is cylindrical in form with a distance of 1cm between parallel steel electrodes. The pulse is applied by discharge of a capacitor across the cell. It has been reported that using linear DNA rather than circular DNA, a field strength of 1.25 kV/cm and employing polyethylene glycol (PEG) can increase protoplast transformation frequency. PEG is believed to assist the association of the DNA with the membrane.

Electroporation has been used for a long time for transient and integrative transformation of protoplasts. Only recently, conditions under which DNA molecules can be delivered into intact plant cells of sugarbeet and rice that are still surrounded by a cell wall have been standardized. Further, transformability of intact plant cells or tissues depends on pretreatment of the cells or tissues to be transformed, either by mechanical wounding or by treating the cells or tissues with hypertonic or enzyme (e.g. 0.3% macerozyme) containing solutions.

The range of tissues that can be transformed by electroporation seems to be narrower. For tissues that are susceptible to DNA uptake by electroporation, this method is convenient, simple, fast, low cell toxicity and inexpensive to obtain transient and stable transformation in differentiated tissues. The disadvantage of the technique is the difficulty in regenerating plants from protoplasts.

*Particle bombardment/microprojectile/ biolistics*

The technique of particle bombardment also known as biolistics, microprojectile bombardment, particle acceleration, etc. has been shown to be the most versatile and effective way for the creation of many transgenic organisms, including microorganisms, mammalian cells and plant species. The procedure in which high velocity microprojectiles were utilized to deliver nucleic acids into living cells was described by Klein *et al.* (1987) and Sanford *et al.* (1987).

The basic system that has received attention employs PDS 1000 (gun powder driven device) or the PDS-1000/He (helium driven particle gun). The DNA bearing tungsten or gold particles (1–3 μm in diameter), referred to as microprojectiles, is carried by a macroprojectile or macrocarrier and is accelerated into living plant cells. The DNA bearing particles (microprojectiles) are placed on the leading surface of the macrocarrier and released from the macrocarrier upon impact with a stopping plate or screen. The stopping plate is designed to halt the forward motion of macroprojectile while permitting the passage of the microprojectiles. In this procedure when helium gas is released from the tank, a disc known as rupture disc blocks its entry to the chamber. These discs are available with various strengths to resist the pressure of gas, which varies from 500 to 1700 p.s.i. When the disc ruptures, compressed helium gas is suddenly released, which accelerates a thin plastic sheet carrying microprojectiles into a metal screen. Upon impact with a stopping plate or screen, the macroprojectile movement is stopped, but this permits the passage of microprojectiles through the mesh screen. The microprojectiles then travel through a partial vacuum until they reach the target tissue. The partial vacuum is used to reduce the aerodynamic drag upon the microprojectiles and decrease the force of the shock wave created when the macrocarrier impacts the stopping plate. See Fig. 23.15 for a schematic diagram of the PDS–1000/He.

The use of particle bombardment requires careful consideration of a number of parameters. These can be classified into three general categories:

*Physical parameters*

*Nature, chemical and physical properties of the metal particles utilized to carry the foreign DNA*: Particles should be of high mass in order to

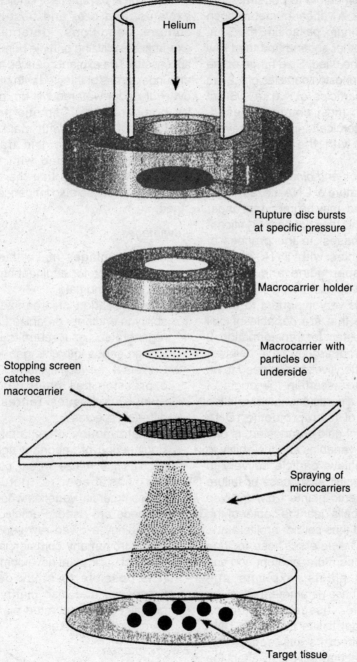

Helium

Rupture disc bursts
at specific pressure

Macrocarrier holder

Macrocarrier with
particles on
underside

Stopping screen
catches
macrocarrier

Spraying of
microcarriers

Target tissue

•Fig. 23.15: Schematic representation of the particle bombardment process.

possess adequate momentum to penetrate into the appropriate tissue. A variety of metals such as tungsten, gold, platinum, palladium, rhodium, iridium and possibly other second and third row transition metals can be used. Size range of the particles is *ca.* 1 μm. Tungsten particles of 1.2 μm diameter and gold particles of 1.0 to 1.6 μm diameter have been most commonly used. Metals should be chemically inert to prevent adverse reactions with the DNA or cell components.

*Nature, preparation and binding of DNA on to the particles*: The nature of DNA, i.e. as single versus double stranded, may be important under some conditions but this may not be a significant variable in specific cases. In the process of coating metal particles with DNA, certain additives such as spermidine and calcium chloride appear to be useful.

*Target tissue*: It is very important to target the appropriate cells that are capable of cell division and are competent for transformation. It is apparent that different tissues have different requirements, thus extensive studies need to be performed in order to ascertain the origin of regenerating tissue in a particular transformation study. Penetration of microprojectile (DNA coated to inert metal of gold or tungsten) is one of the most important variables and the ability to tune a system to achieve particle delivery to specific layers may result in success or failure in recovering transgenic plants from a given tissue. Generally the cells near the center of the target are injured and thus cannot proliferate.

*Environmental parameters*: These include variables such as temperature, photoperiod and humidity of donor plants, explants and bombarded tissues. These parameters effect the physiology of the tissues, influence receptiveness of target tissue to foreign DNA delivery and also affect its susceptibility to damage and injury that may adversely affect the outcome of the transformation process. Some explants may require a healing period after bombardment under special regimes of light, temperature and humidity.

*Biological parameters*: Choice and nature of explants, and pre- and post-bombardment culture conditions determine whether experiments utilizing particle bombardments are successful. The explants derived from plants that are under stress or infected with bacteria or fungi, over- or under-watered will be inferior material for bombardment. Osmotic pre- and post-treatment of explant with mannitol has been shown to be important in transformation. Experiments performed with synchronized cultured cells indicate that the transformation frequencies may be also influenced by cell cycle stage.

*Advantages*

Several advantages make this technique a method of choice for engineering crop species.
1. It is clean and safe.
2. Transformation of organized tissue: The ability to engineer organized and potentially regenerable tissue permits introduction of foreign genes into elite germplasm.
3. Universal delivery system: Transient gene expression has been demonstrated in numerous tissues representing many different species.
4. Transformation of recalcitrant species: Engineering of important agronomic crops such as rice, maize, wheat, cotton, soybean, etc. has been restricted to a few noncommercial varieties when conventional methods are used. Particle bombardment technology allowed recovery of transgenic plants from many commercial cultivars.
5. Study of basic plant development processes: It is possible to study developmental processes and also clarify the origin of germline in regenerated plants by utilizing chromogenic markers.

*Disadvantages*

1. In plants gene transfer leads to non-homologous integration into the chromosome, and is characterized by multiple copies and some degree of rearrangement.

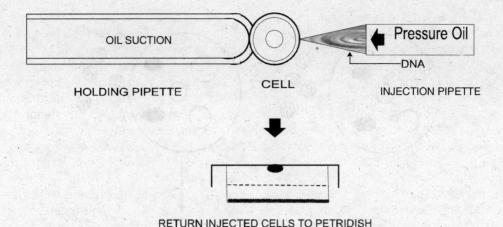

**Fig. 23.16:** Microinjection by holding pipette method.

2. The emergence of chimeral plants.
3. Lack of control over the velocity of bombardment, which often leads to substantial damage to the target cells.

*Macroinjection*

It is the injection of DNA solution (5–10 µl) by micropipettes into the developing floral side shoots (tillers) of plants. Within the floral tillers are archesporial cells that give rise to pollen in the developing sac by two meiotic cell divisions. It was suggested that such cells might also be able to take up large molecules such as DNA. Thus a plasmid encoding kanamycin resistance and under the control of a promoter was injected into the tillers of rye plants. It resulted two plants showing resistance to kanamycin (De la Pena *et al.*, 1987). However, these experiments could not be repeated.

*Microinjection*

Microinjection is the direct mechanical introduction of DNA under microscopical control into a specific target. A target can be a defined cell within a multicellular structure such as embryo, ovule, and meristem or protoplasts, cells or a defined compartment of a single cell. As a direct physical method, microinjection is able to penetrate intact cell walls. It is host-range independent and does not necessarily require a protoplast regeneration system. This method has been proposed for the transfer of cell organelles and for the manipulation of isolated chromosomes. When cells or protoplasts are used as targets in this technique, glass micropipettes of 0.5–10.0 µm diameter tips are used for transfer of micromolecules into the cytoplasm or the nucleus of a recipient cell or protoplast. Recipient cells can be immobilized using methods such as agarose embedding, agar embedding, poly-lysine treated glass surfaces and suction holding pipettes (Fig. 23.16). Once injection has been achieved, the injected cell must be cultured properly to ensure its continued growth and development. With this method the operator has the ability to manoeuver the cell and thus more accurately target the nucleus. Transgenic chimeras have actually been obtained in tobacco and *Brassica napus* by this approach. The main disadvantage of this technique is the production of chimaeric plants with only a part of the plant transformed. The microinjection process is slow, expensive and requires highly skilled and experienced personnel.

*Liposome-mediated transformation*

Liposomes are artificial lipid vesicles surrounded by a synthetic membrane of phospholipids, which have been used in mammalian tissue culture to

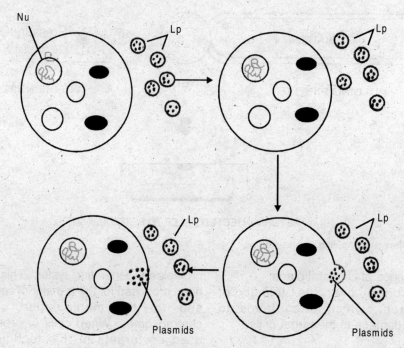

**Fig. 23.17:** Fusion of plasmid filled liposomes with protoplast. Nu–nucleus; Lp–liposomes.

deliver drugs, proteins, etc. into the cells. These can be induced to fuse with protoplasts using PEG and have therefore been used for gene transfer. In this method, DNA enters the protoplasts due to endocytosis of liposomes and it involves the following steps:

i) Adhesion of liposomes to the protoplast surface.

ii) Fusion of liposomes at the site of adhesion.

iii) Release of plasmids inside the cell.

Uptake of liposomes containing DNA molecules has been demonstrated in protoplasts of tobacco and carrot (Fig. 23.17). Liposome-mediated transformation showed higher efficiency when it was used in conjunction with PEG. Liposomes offer many advantages for protoplast transformation including (i) Protection of nucleic acids from nuclease digestion prior to introduction into cellular environment, (ii) Reduced levels of cell toxicity, (iii) Applicability to a wide variety of cells types, (iv) High degree of reproducibility, and (v) Stability and storage of nucleic acids once encapsulated.

*Silicon carbide fiber-mediated transformation*

Silicon carbide fibers (SCF) average 0.6 µm in diameter and 10–80 µm in length. It has been demonstrated that these fibers have the capability to deliver DNA into plant cells. The method involves vortexing a mixture of plasmid DNA encoding a selectable or screenable marker gene, silicon carbide fibers and the explant tissue in an eppendorf tube. Silicon carbide fibers have great intrinsic hardness with sharp cutting edges. DNA delivery in this system is presumably due to cell wall penetration by DNA-coated silicon carbide fibers during vortexing of SCF with explant and DNA adhering to fibers might enter the cells. It is possible that SCF function as numerous needles facilitating DNA delivery into the cells. During the mixing process, DNA penetrates the cell to become stably integrated into the nuclear genome. Advantages of the silicon carbide fiber-mediated transformation method over other procedures includes the ability to transform walled cells thus avoiding protoplast isolation, relative ease of the procedure and very low equipment costs. The disadvantage is that silicon carbide has some carcinogenic properties.

The feasibility of this technique has been demonstrated in fertile transgenic plants of maize and tobacco.

### Ultrasound-mediated DNA transformation

Ultrasound is used for stimulating uptake of foreign DNA by plant protoplasts and leaf segments of tobacco. The procedure involves immersion of explant (leaves/protoplasts) in sonication buffer containing plasmid DNA and sonication with an ultrasonic pulse generator at $0.5 \text{ w/cm}^2$ acoustic intensity for 30 min. Samples are rinsed in a buffer solution and then cultured for growth and differentiation. This technique has the advantages of being simple, inexpensive and multifunctional. One can use standardized conditions and there is no requirement for tissue culture expertise. Little success has been achieved by this technique.

### DNA transfer via pollen

Pollen has been suggested as a vector for gene transfer by various workers. It has been reported that introduction of DNA into gametes followed by fertilization and zygotic embryogenesis will result in gene transfer. This kind of approach would be simpler, faster and cheaper than the *in vitro* methods. Since ovules are difficult to isolate and the injection of DNA into embryo sac *in situ* seemed to be too tedious and unpredictable, it was logical to favor the use of pollen as a vector for DNA. The use of DNA treated pollen as a vector for pollinating fertile plants of maize has been studied. There are various reports which indicate that DNA can be either taken up by germinating pollen and integrated into sperm nuclei or reach the zygote via pollen tube pathway. The main problems are the presence of cell wall and action of nucleases on the DNA. *Agrobacterium* has also been used to transfer the T-DNA via pollen as a vector to overcome the nuclease action on DNA. Both

these approaches have been tried and some success has been achieved.

### Chemical gene transfer methods

This involves plasma membrane destabilizing and/or precipitating agents. Protoplasts are mainly used which are incubated with DNA in buffers containing PEG, poly L-ornithine, polyvinyl alcohol or divalent ions. The chemical transformation techniques work for a broad spectrum of plants.

### PEG mediated gene transfer

The first conclusive demonstration of uptake and integration of isolated Ti plasmid DNA into plant protoplasts was reported in *Petunia* and tobacco in the presence of poly L-ornithine or polyethylene glycol (PEG). The presence of Ti DNA in the plant genome was demonstrated both by the phenotype of hormone auxotrophic growth, the production of the expected opine and by Southern blot analysis of DNA from the transformants.

In a general procedure protoplasts are isolated and a particular concentration of protoplast suspension is taken in a tube followed by addition of plasmid DNA (donor and carrier). To this 40% PEG 4000 (w/v) dissolved in mannitol and calcium nitrate solution is slowly added because of high viscosity, and this mixture is incubated for few minutes (*ca.* 5 min). As per the requirements of the experiment, transient or stable transformation studies are conducted.

Among the most important parameters that affect the efficiency of PEG-mediated gene transfer, are the concentration of magnesium or calcium ions in the incubation mixture, and the presence of carrier DNA. The linearized double stranded plasmid DNA molecules are more efficiently expressed and integrated into the genome than supercoiled forms. Integration of foreign DNA into the nuclear genome occurs predominantly at random sites.

The main application of the technique, apart from analyzing the transformation process itself, is in introducing foreign genes to plant cells. This can be accomplished by constructing a molecule containing a selectable marker and the gene of interest, or more easily by simply mixing DNA of the gene of interest with the selectable marker plasmid in a molar ratio of approximately 3:1 to 10:1, transforming, selecting for the marker and analyzing transformants for the presence of the second gene. The method can also be applied with DNA from different sources. Manipulation of nucleic acids prior to transformation is also possible and there are no host range limitations.

The advantage of the method is that the form of the DNA applied to the protoplasts is controlled entirely by the experimenter and not by an intermediate biological vector. The main disadvantage is that the system requires protoplasts and a functional system for regeneration of these protoplasts to calluses and whole plants. It is therefore not applicable to many plant systems. In addition, the relatively random way in which the DNA is integrated into the genome means that, for the introduction of nonselectable genes, a thorough characterization of the transformants by Southern blot analysis is necessary to confirm the nature of the integration event.

### Calcium phosphate co-precipitation

In this method, DNA is mixed with calcium chloride solution and isotomic phosphate buffer to form DNA-CaPO$_4$ precipitate. The precipitate is allowed to react with actively dividing cells for several hours, washed and then incubated in fresh culture medium. Giving them a physiological shock with DMSO can increase the efficiency of transformation to a certain extent. Relative success depends on high DNA concentration and its apparent protection in the precipitate.

### The polycation DMSO technique

It involves the use of a polycation, polybrene, to increase the adsorption of DNA to the surface followed by a brief treatment with 25–30% DMSO to increase membrane permeability and enhance the uptake. The major advantage of polybrene is that it is less toxic than other polycations and a high transformation efficiency requires very small quantities of plasmid DNA to be used.

### DEAE dextran procedure

Transformation of cells with DNA complexed to the high molecular weight polymer diethyl amino ethyl (DEAE) dextran is used to obtain efficient transient expression. The efficiency increases to 80% when DMSO shock is given. But this technique does not produce stable transformants.

### Advantages

Direct gene transfer for cell transformation circumvents the host range limitation. It is possible to introduce DNA sequences directly into cells and organs by biolistic or electroporation approaches. Cereals, the world's most important crops, have been transformed by direct gene transfer methods.

### Disadvantages

i. Direct gene transfer usually depends on a culture system that allows regeneration of mature plants from protoplasts/cells/organs, which is not usually available for every desired plant species. The problem may be overcome by further improvement of plant tissue culture methods.

ii. These methods show unpredictable pattern of foreign DNA integration. During their passage into the nucleus, the DNA is subjected to nucleolytic-cleavage resulting in truncation, recombination, and rearrangement or silencing.

## DNA imbibition by cells, tissues, embryos and seeds

Incubation of cells, tissues and organs with DNA for transformation has met with little or no success and hence has not resulted in any proven case of integrative transformation, though reporter gene has been expressed from imbibed dry seed and embryos. In all these cases, the plant cell walls not only work as efficient barriers, but are also efficient traps for DNA molecules.

# STATUS AND EXPRESSION OF TRANSFERRED GENES

DNA is introduced into plant cells and these cells are then grown *in vitro* to regenerate plants. Selection and growth of plant cells on selective media provide initial phenotypic evidence for transformation. This includes resistance to antibiotics, herbicides, etc. Several studies have analyzed the inheritance and expression of transferred DNA. The location of the genes and temporal and spatial aspects of their expression may also be important. In general, when transgenic plants are produced by a direct DNA transformation system, more screening has to be done to select lines which contain a limited number of integrated copies. In most cases it will not be possible to separate these copies by crossing, because they will often be tightly linked. The transgenic plants so obtained are self-fertilized. DNA introduced by methods other than *Agrobacterium* mediated transformation integrates into the genome by as yet unknown mechanisms. Generally the DNA integrates at a single locus, but it usually consists of rearranged multimers of the donor DNA. Molecular evidences (e.g. PCR analysis, Southern hybridization) are essential to confirm integration of transferred genes. Seeds are analyzed in the $T_1$ (transgenic first) generation for segregation data. The simple pattern of inheritance (3:1) indicates a single site of insertion and this can be confirmed by Southern blotting. The copy number is usually low, between 1 and 5 copies per cell. The DNA is transmitted through meiosis as a simple Mendelian trait. Occasionally, however, more complex patterns of integration are seen.

In order to apply gene technology successfully to modern agriculture, it is essential to understand transgene expression. Sometimes when a transgene is introduced into an organism it may not show its expression. This is known as gene silencing.

## Gene silencing

Partial or complete inactivation of gene(s) is known as gene silencing. Presence of multiple copies of a transgene in a plant nucleus can also lead to silencing of some or all copies of the gene(s) including the endogenous gene having homology with the transgene. It has also been established that in all such cases of transgene silencing, loss of expression of the transgenes is not due to the loss of these genes but due to their inactivation. Before considering the real causes of gene silencing it is must to know how a foreign DNA sequence is recognized by cell?

**Detection of intrusive DNA:** The ability to recognize 'self' from 'non-self' is a characteristic feature that occurs at cellular level. The observations from genetically engineered plants suggest that an ability to recognize 'self' from 'non-self' exists at nucleic acid level. The possible mechanisms by which intrusive DNA is detected by cell are: (i) Foreign sequence has different base sequence composition from that of endogenous chromosomal environment; (ii) Gene transfer via direct DNA delivery method has no control on copy number. These copies may form independent domains in the cell and work independently with the result they get recognized by genome scanning machinery; (iii) Every cell has its own modification and restriction system that shows xenophobic effect, which does not allow cell to contaminate its chromosomal environment by any foreign sequence.

Genomes are made of isochores that are very long stretches of DNA with high compositional homogeneity. On this basis a concept of gene space has been given. According to this concept if a GC-rich transgene is integrated in to a GC-rich gene space (GC isochore) or AT rich transgene is integrated in to AT-rich gene space (AT isochore), it is normally transcribed. But if GC-rich transgene gets integrated in to AT-rich gene space or vice versa, it is inactivated as there is no compositional homogeneity with neighboring sequence. A powerful initial response by the cell for the presence of foreign DNA may be fragmentation of invasive DNA via the action of cytoplasmic nucleases.

**Causes of gene silencing:** Silencing of a gene can be complete or partial. It may occur at transcriptional or translational level. After the entry of transgene into a receptive genome, there is an increase in methylation pattern. This is a major cause of gene silencing leading to inactivation of gene(s). There are various causes of gene silencing.

**1. *DNA methylation*:** The inactivation of transgene is often accompanied by an increase in DNA methylation and inactivation very frequently correlates with number of copies of integrated transgenes. In plants when foreign DNA is introduced, the modification and restriction system of cell causes methylation of foreign sequence to make it inactivated. DNA methylation is imposed on cytosine residues within symmetrical target sequences. In plants an inverse correlation between gene transcription and cytosine methylation has been observed for certain controlling elements as methylated DNA prevenhts transcription by directly hindering trans-acting factors and basal transcription machinery accessibility. The presence of methylated cytosine residue in the opposite strand provides the information to methylate C-residue in newly synthesized complementary strand with the help of methyl transferase enzyme and S-adenosyl methionine as cofactor. On the basis of DNA methylation gene silencing may be: (i) transcriptional silencing, which is linked to methylation of promoter region; (ii) post-transcriptional silencing that involves inactivation of the coding sequence by methylation.

Transgene Transcriptional gene silencing (TGS) is almost always associated with promoter methylation and both symmetric and asymmetric methylation of cytosine residues is known to occur. The repression of methylated promoters probably results from recruitment of chromatin modifying factor (such as histone deacetylases) and remodeling factors (such as SNF2 helicases) through methylated DNA binding proteins (such as MeCP2) that prevent access of DNA to the transcription machinery, yielding a heterochromatin-like promoter status. The transgene architecture, copy number and genomic position play an important role in determining whether a promoter sequence will be methylated and repressed. Nevertheless, even when the transgene insert is present in multiple (perhaps rearranged) copies and contains repeat sequences, many studies have reported expression. This implies that the induction of TGS is a multicomponent process.

Induction of *de novo* methylation of transgenes involves different modes:

i) DNA methylation via DNA-DNA pairing: When multiple transgene copies integrate as concatamer and DNA coiling occurs, then these copies come in front of each other just like homologous sequences. Due to their mutual suppressing effect there is increase in methylation pattern that leads to inactivation of genes.

ii) Transgene recognition: Transgene get integrated properly but due to older age of transgene(s) or certain environmental stresses the transgene get hypermethylated and inactivated.

iii) Insertion into hypermethylated genomic regions: When a transgene get integrated into a hypermethylated region, then there may be spreading of hypermethylation pattern that leads to inactivation of transgene.

**2. *Homology dependent gene silencing*:** Homologous sequences not only affected the stability of transgene expression but that the activity of endogenous genes could be altered after insertion of transgene into genome. This homology dependent inactivation may involve different modes:

i) *Inactivation of homologous transgenes*: When transgenic plants were retransformed with constructs that are partly homologous with the integrated transgene. In the presence of second construct, the primary transgene becomes inactivated and hypermethylated within the promoter region. *Trans*-inactivation occurs when one transgene locus (that is itself silent) exerts a dominant repressive effect on other loci (which may be linked) that typically include sequence homologies in promoter regions. Sequences as short as 90 bp have been shown to be sufficient to mediate silencing (Vaucheret, 1993). Silencing is also influenced by position of interacting sequences (linked copies are more efficiently silenced). *Trans*- inactivation thus requires interaction of the silencing locus with the target sequence. This has been proposed to be typically caused by ectopic DNA-DNA pairing between the loci, resulting in transfer of the silenced state from one locus to another either by transfer of repressive chromatin states to targets or *de novo* methylation of target sequences. Since accurate pairing of like DNA sequences is intrinsic to vital cellular processes such as meiosis, it should not be surprising that effective systems exist within the genome to mediate recognition of identical sequences.

ii) *Paramutation*: It is interaction of homologous plant alleles that leads to heritable epigenetic effects. Thus combination of two homologous alleles that differed in their state of methylation results in paramutation phenomenon.

iii) *Co-suppression*: Expression of endogenous genes can be inhibited by the introduction of a homologous sense construct that is capable of transcribing mRNA of same strandedness as transcript of the gene. In co-suppression there is suppression of both the transgene and homologous resident gene or inactivation of either of two. This phenomenon was first described in Petunia when a chalcone synthase gene (*chs*) involved in floral pigment biosynthesis was introduced in a deep violet flowering line of *P. hybrida*. 42% of the transgenic plants had white or variegated flowers. It was shown that co-suppression was due to post-transcriptional event, and was independent of promoter, which was necessary for many different transgenes, and a single transgene can cause co-suppression of two or more endogenous genes.

**3. Suppression by antisense genes**: It is post-transcriptional inhibition of gene expression. Introduction of antisense and sense transgenes into plants has been widely used to generate mutant phenotypes. Since 1980 a large number of transgenic plants containing antisense and sense genes have been generated. One of the first successful applications of antisense gene approach was the inhibition of fruit softening by antisense polygalacturonase gene, which lead to first commercial transgenic tomato 'Flavr Savr' variety.

However, it is of major concern that antisense RNA may block RNA production along different pathways. Antisense gene often leads to reduced levels of target mRNA and can potentially form a double stranded structure with complementary mRNAs. Antisense RNA may interfere with the following processes:

i. Antisense transcripts affect the target gene directly in the nucleus thereby preventing synthesis of mRNA i.e. transcription.

ii. Antisense RNA may block processing of mRNAs by masking the sequences recognized by splicing and the polyadenylation apparatus. Antisense RNA may disturb the normal transport of mRNAs out of the nucleus by forming a hybrid with their target mRNAs and disturbs the regular flow of transcripts.

iii. Many antisense RNAs complementary to ribosome binding site have been shown to inhibit translation initiation.

iv. Antisense RNA prevents the accumulation of target mRNA. Sense and antisense RNAs form a double RNA intermediate that is rapidly degraded by ds RNA specific ribonucleases.

Silencing of a *gus* gene driven by a barley aleurone-specific lipid transfer protein (*ltp*) promoter (and a *nos* terminator) was reported in transgenic rice plants derived by protoplast electroporation (Morino *et al.*, 1999). The *ltp-gus* plants typically had from 4 to 12 copies of the insert and at least two loci were implicated in the triggering of silencing. One was a rearranged locus, which yielded a complex RNA transcript containing both sense and antisense *gus* sequences. As this fragment was observed in several silenced transgenic plants, the authors concluded that the silencing was post-transcriptional. This aberrant RNA was hypothesized to interact with the full-length *gus* RNA present at another locus and to cause PTGS.

**4. *Silencing by RNA interference*:** During mid-1990s it was shown that large part of DNA in eukaryotes is actually used for the synthesis of non-coding RNA (ncRNA), which plays an important role in regulating the expression of genes at the post-transcriptional level. It was shown that the ncRNA gives rise to double stranded RNA (dsRNA), which is responsible for gene silencing. The phenomenon was described as RNA interference (RNAi). This discovery of RNAi involving gene silencing was first reported by Andrew Fire and Craig C. Mello (Fire *et al.*, 1998) while working for gene expression studies in *Caenorhabditis elegans*. They were awarded Noble prize for their discovery in 2006.

RNA interference (RNAi) is a phenomenon in which the introduction of double-stranded RNA (dsRNA) endogenously synthesized or exogenously applied into a diverse range of organisms and cell types causes degradation of the complementary mRNA (Fig. 23.1). In the cell, long double-stranded RNAs (dsRNAs typically>200 nucleotides) can be used to silence the expression of target genes in a variety of organisms and cell types (e.g., worms, fruit flies, and plants). Upon introduction, the long dsRNAs enter a cellular pathway that is commonly referred to as the RNA interference (RNAi) pathway. First, the dsRNAs get processed into 20–25 nucleotide (nt) dsRNAs with 2- to 3-nt 3′ overhangs. These dsRNA fragments are called small interfering RNAs (siRNAs) which are generated from long dsRNA by a ribonuclease III like enzyme called Dicer or Dicer like (initiation step). Dicers are multicomponent RNase (nuclease III) associated with siRNA. The siRNA duplex must have 2–3 nucleotide overhanging 3′ ends for efficient cleavage of its target sites. Such 3′ overhangs are characteristics of the product of cleavage reactions catalyzed by RNase III. The siRNAs subsequently associate with ~250 to 500-kD endoribonuclease-containing complex known as RNA-induced silencing complex (RISC). One of the RNA strands is eliminated but the other (antisense strand) remains bound to the RISC complex and serves as a probe to detect the target mRNA molecules. Activated RISC then binds to complementary transcript by base pairing interactions between the siRNA antisense strand and the mRNA where they cleave and destroy the mRNA (effector step). Cleavage of mRNA takes place near the middle of the region by the siRNA strand which renders the mRNA susceptible to other degradation pathways. The sequence specific degradation of mRNA results in RNA silencing. It was also shown that although dsRNA is needed to trigger RNAi, the chemical composition of antisense RNA (relative to that of sense strand) was more important. A feature of RNA silencing is also its ability to spread across cells within a tissue and then to entire organism. Silencing can thus be initiated locally but manifested throughout the organism. RNA dependent RNA polymerases (RdRP) are implicated in the amplification of siRNAs, so that the production and spread of the signal persists for a long time.

Studies in *Caenorhabditis elegans* have shown that double stranded (ds) RNA species

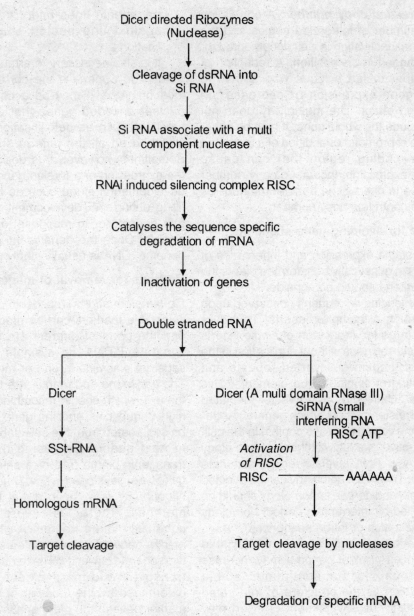

**Fig. 23.18:** A Mechanism of RNAi mediated post transcriptional gene silencing.

can trigger a high level of silencing that specifically targets homologous sequences This dsRNA mediated silencing is much more effective than that of antisense RNA and appears to function through a co-suppression mechanism (Montgomery and Fire, 1998).

**5. *Position effect*:** Whenever a transgene get integrated into an improper region (hypermethylated, heterochromatic, telomeric, compositionally different genomic region) then due to suppressing effect of the adjacent region or environment, it get inactivated.

**6. *Increased copy number*:** A correlation between number of integrated copies and the frequency of inactivation is well documented for copies arranged in *cis* position. A reduction of copy number inside a locus was shown to increase gene expression or decrease the suppressing effect. The multiplicity does not need to include the whole gene. It may be either duplication of promoter or addition of a truncated transgene coding region that can cause decreased expression. Increased copy number may be due to direct gene transfer methods as there is no control on copy number.

## Strategies for avoiding gene silencing

To obtain stable expression and inheritance of transgenes in genetically transformed plants the following criteria should be considered.

i. Gene silencing is frequently observed upon integration of complex inserts. Thus it is better to go for integration of single inserts of transferred gene without duplication in the form of tandem or inverted repeats and consisting of single unique elements. This can be achieved by using vector mediated gene transfer method, which generally introduce one copy (rarely two copies) in the cell.

ii. Irrespective of exact mechanism homology at DNA or RNA levels seem to stimulate silencing events. If homology can be avoided or length and degree of homology should be controlled by interrupting perfect homology with mismatch or intron sequences.

iii Integration of single copy, unmethylated sequences of plant genome may increase the probability for continuous stable expression as the structure of integrated DNA itself, the environment of insert may influence the stability of gene expression.

iv. Gene silencing occasionally becomes evident only after transmission of the gene to next generation. In some cases transgene expression decreased progressively over subsequent generations. Therefore continuous monitoring of expression levels in progeny of even well established transgenic lines might be a precaution against unexpected epigenetic modifications.

It is also necessary to arrange for the gene product to appear in the correct sub-cellular location. Also, it may be acceptable to have nuclear encoded genes that are involved in functions such as detoxification or resistance switched on all the time in some situations. Sometimes, however, it is desirable to ensure the expression of a foreign gene at a particular developmental stage such as flowering, grain filling, during seed development, tuber formation, fruit ripening or in response to environmental signals. Once the gene is integrated into the genome, DNA is usually stably maintained.

## Methods for removal of marker genes

*Co-transformation*: Transgenic plants free from selectable markers can be produced if marker and the gene of interest are placed on two separate T-DNAs in a single plasmid or on separate plasmids in one or more agrostrains. The transgene and marker are thus inserted at the loci, which should recombine at reasonably high frequencies and the gene of interest can be segregated from the selectable marker gene in next generation. Using this technique, transgenic plants free of a selectable marker gene were developed in rice, *Brassica napus* and *Nicotiana*. Likewise in direct transformation methods there is more than 50% cotransformation frequency when selectable marker gene and gene of interest are placed on two separate plasmids. The frequency with which transgene and marker gene segregate is determined by the location of transgene and marker gene in the plant genome and the type of agrostrain used.

*Site-specific recombination*: In this approach the selectable marker gene is flanked with direct repeats of recognition sites for a site-specific recombinase, which allows the enzyme to excise the marker gene from the plant genome by enzyme mediated site specific recombination. Bacteriophage P1 site specific recombination system of *cre/lox* is one which consists of 38-kDa product of *cre* gene (the Cre recombinase) and the asymmetric 34 bp *lox* sites, composed of two 13 bp inverted repeats with 8 bp asymmetric core region that gives directionality to the sites. No additional factors are required for Cre catalyzed recombination between the *lox* sites. The *cre/lox* system has been used to direct the site-specific integration of plasmids at the *lox* sites previously placed in the genome by direct gene transfer or *Agrobacterium* mediated transfer. *cre-lox* system thus offers a method for precise insertion of single copy DNA into genomic targets. Two recombination lox sites of the same orientation flank the integrated DNA. Introduction of Cre results in excision of the internal sequence (insert), if the *lox* repeats are in direct orientation. The *cre* gene can be introduced into *lox* containing plant by either transformation or sexual crossing. Upon transformation, a marker gene cloned between two *lox* sites is eliminated from about 95% of the secondary transformants. Tobacco and *Arabidopsis* transgenic plants free from a selectable marker gene were recovered using *cre/lox* recombination system. The Cre catalyzed excision events in the plant genome are precise and without loss of alteration of nucleotides in the recombination site. The other single chain recombinases, which have also been found to be useful for the removal of marker genes are the FLP/*frt* system of 2 μm plasmid of *Saccharomyces cerevisiae* and *R/rs* system of pSR1 plasmid *of Zygosaccharomyces rouxii*.

*Removal of marker genes by transposases*: Transposable elements have been employed for the dissociation of marker and desirable genes in two ways: I. The marker gene is placed on mobile element, which is lost after transposition. Marker-free transgenic tobacco and aspen plants have been generated by inserting the selectable *ipt* gene into transposable element *Ac*. II. The transgene by itself is mobile, the activation of transposase allows the relocation of the desired transgene to a new chromosomal position. Genetic crosses and/or segregation will

dissociate the two transgenes. Relocation of the transgene at different genomic positions results in differential levels of expression.

## Transplastomics to overcome pollen gene transfer

Transplastomics are the transgenic plants with pure population of transformed chloroplasts (integration of transgenes into plastid genome instead of nuclear genome. Escape of transgenes from transplastomic crops poses a negligible risk of pollen mediated movement of transgenes as transplastomes are not transmitted by pollen.

## Modulation of marker gene expression

Antisense technology could be applied to prevent the expression of marker genes. When a marker gene that is cloned in its reverse orientation is introduced into a cell, it encodes antisense RNA that is complementary to the mRNA of the original marker gene and forms an RNA duplex, which renders the mRNA inactive. Catalytic RNA cleaves either itself or other RNA molecules. Examples of catalytic RNA are ribozymes, which can act specifically on target RNA molecules to render them inactive. Catalytic RNA has created new opportunities to repress the expression of selectable marker genes. The expression of a ribozyme gene directed towards the target mRNA of kanamycin selectable marker in plants resulted in a reduction of the kanamycin gene product. Genetically modified plants containing selectable markers could be further modified with genes that encode antibodies against the marker gene protein product.

## Selectable markers and safety

Current methods of generating transgenic plants employ a "selectable marker gene" which is transferred along with any other genes of interest usually on the same DNA molecule. The presence of a suitable marker is necessary to facilitate the detection of genetically modified plant tissue during development.

The safely assessment for the consumption of food and food products derived from genetically modified crop plants needs to focus attention on the properties of the new gene products. Consideration needs to be given to any possible toxicity, nutritional effects or allergenicity arising from the presence of a selectable marker gene (See Chapter 27 on Biosafety concerns and regulatory measures).

# PROTOCOLS

### 23.1: *Agrobacterium* mediated transformation

**Plant material:** Leaf discs of *Nicotiana tabacum*

**Bacterial strain:** LBA4404 (pBI121). The binary vector pBI121 contains *gus* reporter gene and neomycin phospho-transferase conferring resistance to kanamycin as a selectable marker

1. Prepare *Agrobacterium* culture. Take an appropriate *Agrobacterium* strain containing binary vector pBI121 contained kanamycin, streptomycin and rifampicin as bacterial selectable marker genes.

2. Prepare YEP medium without antibiotics. The constitution of medium is as follows:

   Yeast extract           10 g/l
   Peptone                10 g/l
   NaCl                    5 g/l

   Adjust pH to 7.0 with 1N NaOH
   For solid medium, agar 23 g/l is to be added.

   Dissolve all the constituents of YEP medium in water and adjust pH to 7.0. Aliquot 50 ml each of medium in conical flasks and autoclave at 23 p.s.i. pressure, 121°C for 20 min. Cool the flasks to 50°C and add antibiotics. Prepare YEP medium in petri plates and also maintain liquid broth without agar in flasks.

YEP broth/agar         100 ml
Kanamycin sulphate    0.5 ml of 10
@ 50 µg/ml             mg/ml stock
Streptomycin sulphate   1.0 ml of 10
@ 100 µg/ml           mg/ml stock
Rifampicin @ 50 µg/ml   0.5 ml of 10
                         mg/ml stock

3. Inoculate *Agrobacterium* strain from glycerol stock on a petri dish using a sterile inoculation needle by streaking method. Incubate at 28°C.

4. Next day pick a single colony from the petri dish and inoculate into 23–20 ml of YEP liquid medium contained in 100–230 ml flask. Incubate on a rotary shaker with speed of 120 rpm at 28°C.

5. For culture of leaf discs prepare the following MST medium:
   MS +1.0 mg/l BAP + 0.1 mg/l NAA

6. Collect young healthy green leaves from tobacco plants to be transformed. Cut into small pieces avoiding mid vein portion.

7. Bring it to laminar air-flow unit and transfer them into a beaker containing sodium hypochlorite (1–2% active chlorine) for 10 min.

8. Wash the leaf sample 3–4 times with sterile distilled water. Cut the leaf into small discs of 1cm diameter.

9. Take the overnight grown culture of *Agrobacterium* and measure the OD at 600 nm. For cocultivation, centrifuge the *Agrobacterium* culture at 2500 rpm for 15–20 min. Discarded the supernatant and resuspend the bacterial pellet in MS medium to bring the OD to 0.1.

10. Transfer the *Agrobacterium* suspension in to a sterile petri dish and put the cut leaf discs in the *Agrobacterium* suspension for 30 min.

11. Take the leaf discs out after 30 min and dry them on a sterile Whatman filter paper.

12. Inoculate these leaf discs on MST medium. Incubate for 3 d at 16 h photoperiod.

13. After 3 d transfer the discs individually to MST1 medium containing bacteriostatic agent (MST + 300 mg/l carbenicillin or cefotaxime) and culture again under the same conditions.

14. After 3 days transfer the leaf discs individually to MST2 medium containing selective agent as per the selectable marker used in transformation experiments. For example, (for pBI121–MST1 + 100 mg/l kanamycin). Culture under the same condition.

15. Take out the samples from time to time or select the regenerating transformed shoots and assay for GUS activity.

16. Composition and preparation of X-Gluc stain:

| Stock solution | Final solution | Volume for 10 ml |
| --- | --- | --- |
| 0.1 M K⁺ ferricyanide | 5 mM | 500 µl |
| 0.1 M K⁺ ferrocyanide | 5 mM | 500 µl |
| 0.6% Triton X–100 | 0.3% | 5 µl |
| 0.5 M phosphate buffer (pH 7.0) | 0.1M | 2 µl |
| X-Gluc | Fresh | 30 mg |

Dissolve 30 mg X-Gluc in 500 µl DMSO, before adding the chemical components. Make up to the final volume with distilled water. Store in aliquots of 1ml at –20°C wrapped with aluminium foil.

17. Place the explant material in wells of ELISA plates with X-Gluc stain till it is completely submerged. Incubate the plates at 37°C for 36 h. Wash and keep the stained tissues in 95% ethanol for preservation.

18. Presence of blue spots indicates *gus* gene expression.

## 23.2: Gene transfer by biolistic method: Transient expression

**Plant material:** Immature embryos of any cereal *viz.* wheat barley, rice, maize.

1. Excise immature embryos as explained earlier in Protocol 5.2.

2. Place the embryos with scutellar side up on MS medium supplemented with 2 mg/l 2,4-D. Up to 20 embryos can be placed in the center of a small petri dish containing callus induction medium. Wrap the plates with parafilm and incubate in darkness at 25°C for 1–2 days.

3. Preparation of gold particles: Take 30 mg gold particles of a size of 1 µm in 500 µl of sterile distilled water. Wash it 3–4 times with sterile water, sonicate or vortex well to assure complete mixing of the suspension of particles and distribute 50 µl aliquotes containing 3 mg gold particles into microfuge tubes.

4. Plasmid DNA: Take 5 µg of plasmid DNA. Plasmid DNA isolation procedure has been explained in Protocol 14.1. For transformation experiments DNA should have a *gus* reporter gene and a selectable marker with constitutive promoters.

5. Particle preparation for the PDS–1000/He

   i. Prewash 50 µl aliquot of gold particles are coated with 5 µg plasmid DNA (1 µg/1 µl stored at –20°C) and keep them on ice. Add 5 µg DNA to the gold particle suspension in 1.5 ml microfuge tube.

   ii. Add 10 µl of 0.1 M spermidine (stored at 4°C) with vortexing.

   iii. Add 50 µl of 2.5 M calcium chloride (stored at 4°C) with vortexing.

   iv. Allow the preparation to sit for about 10 min.

   v. Remove supernatant. Resuspend particles in 200 µl ethanol.

   vi. Gently centrifuge and then allow the particles to settle.

   vii. Remove supernatant and repeat once more washing with ethanol.

   viii. Remove supernatant and resuspend particles in 30 µl ethanol with vortexing.

   ix. Remove 5 µl of this suspension with a pipette and spread on the center of each sterile macrocarrier disc, previously positioned in a macrocarrier holder.

x. Air-dry for about 2 min to allow ethanol to evaporate. The macrocarriers are now ready for bombardment. To obtain the best results, use the prepared macrocarriers as soon as possible.

6. Operating procedure with gene gun PDS 1000/He: Follow the procedure as reported by the manufacturer of the gun in the operation manual.

7. Loading the rupture disc: Unscrew the rupture disc cap from the acceleration tube. Place the rupture disc of desired burst pressure (e.g. 1100 p.s.i.) in the recess of the rupture disc retaining cap and screw the retaining cap on to the gas acceleration tube and hand tighten the cap.

8. Load the macrocarrier launch assembly with a sterile stopping screen on the stopping screen support. Place the macrocarrier launch assembly in the second slot from the top in the sample chamber.

9. Positioning of sample: Place the petri dish containing the sample on the petri dish holder. Place the petri dish holder at the desired level (2nd or 3rd shelf from the bottom). Close and latch the sample chamber door.

10. Firing the gun: Follow the procedure as mentioned in the operation manual.

11. Transfer the bombarded material to growth cabinet and kept in dark at 25°C for 1 d.

12. Take 5 embryos from each plate and stain in GUS solution. Follow the procedure as reported in Protocol 23.1.

## REFERENCES

Braun, A.C. and Stonier, R. 1958. Morphology and physiology of plant tumours. *Protoplasmatologia* **10**: 1–93.

Daniell, H., Wiebe, P.O. and Fernandez-San Milan, A. 2001. Antibiotic-free chloroplast genetic engineering-an environmentally friendly approach. *Trends Plant Sci.* **6**: 237–239.

Ebinuma, H., Sugita, K., Matsunaga, E. and Yamakado, M. 1997. Selection of marker-free transgenic plants using the isopentenyl transferase gene as a selectable marker *Proc. Natl. Acad. Sci. USA* **94**: 2117–2121.

Fire, A. Xu, S.Q.., Montgomery, M.K. Kostas, S.A., Driver, S.E. and Mello, C.C. 1986. Potent and specific gene interference by double-stranded RNA in *Caenorhabditis elegans. Nature* **391**: 806–811.

Hille, J., Verheggen, F., Roelvink, P., Franssen, H., Kammen, Avan. and Zabel, P. 1986. Bleomycin resistance gene as dominant selectable marker for plant cell transformation. *Plant Molecular Biology* **7**: 171–176.

Jaiwal, P.K., Sahoo, L, Singh, D.L. and Singh, R.P. 2002. Strategies to deal with the concern about marker genes in transgenic plants: Some environment friendly approaches. *Curr. Sci.* **83**: 128–136.

Jefferson, R.A., Kavanagh, T.A. and Bevan, M.W. 1987. GUS fusions: B-glucuronidase as a sensitive and versatile fusion marker in higher plants. *EMBO Journal* **6**: 3901–3907.

Joersbo, M and Okkels, F.T. 1996. Selectable marker to recover transgenic maize plants (Zea mays L.) via Agrobacterium transformation *Plant Cell Reports* **16**: 219–221.

Koncz, C., Olsson, O., Langridge, W.H.R., Schell, J. and Szalay, A.A. 1987. Expression and assembly bacterial luciferase in plants. Proc. *Natl. Acad. Sci. USA* **135**:

Millar, A.J., Short, S.R., Hiratsuka, K., Chua, N.H., Kay, S.A. 1992. Firefly luciferase as a reporter gene for expression in higher plants. *Plant Molecular Biology Reporter* **10**: 324–337.

Montgomery, M. K. and Fire, A. 1998. Double-stranded RNA as a mediator in sequence-specific genetic silencing and co-supression. *Trends Genet.* **14**: 255–258.

Morino, K., Olsen, O.A. and Shimamoto, K. 1999. Silencing of an aleurone-specific gene in transgenic rice is caused by a rearranged transgene. *Plant J.* **17**: 275–285.

Neumann, E., Schaeffer-Ridder, M., Wang, Y. and Hofschneider, P.H. 1982. Gene transfer into mouse lymphoma cells by electroporation in high electric fields. *EMBO J.* **1**: 841–845.

Perez, P., Tiraby, G., Kallehoff, J. and Perret, J. 1989. Phleomycin resistance as a dominant selectable marker for transformation. *Plant Molecular Biology* **13**: 365–373.

Perl, A., Galili, S., Shaul, O., Ben-Tzvi, I. and Galili, G. 1992. Bacterial dihydrodipicolinate synthase and desensitized aspartate kinase: Two novel selectable markers for plant transformation relying on overproduction of lysine and threonine *Bio/technology* **11**: 715–718.

Stalker, D.M., McBride, K.E. and Malyj, L.D. 1988. Herbicide resistance in transgenic plants expressing bromoxynil nitrilase detoxification gene. *Science* **242**: 419–423.

Streber, W.R. and Willmitzer, L. 1989. Transgenic tobacco plants expressing a bacterial detoxifying enzyme for 2,4-D. *Bio/Technology* **7**: 811–816.

Vaucheret, H. 1993. Identification of a general silencer for 19S and 35S promoters in a transgenic tobacco plant:90 bp of homology in the promoter sequence are sufficient for trans- inactivation. *C.R. Acad. Sci. Paris, Sci. Vie/Life Sci.* **316:** 1471–1483.

Zaenen, I., Van Larebeke, N., Teuchy, H., Van Montagu, M. and Schell, J. 1974. Supercoiled circular DNA in crown gall inducing *Agrobacterium* strains. *J. Mol. Biol.* **86:** 107–127.

Zambryski, P. 1992. Chronicles from the *Agrobacterium* plant cell DNA transfer story. *Ann. Rev. Plant Physiology and Plant Molecular Biology* **43:** 465–479.

## FURTHER READING

Brown, T.A. 1992. *Genetics: A Molecular Approach*. Chapman & Hall, London.

Chawla, H.S. 1998. *Biotechnology in Crop Improvement*. International Book Distributing Co., Lucknow.

Gupta, P.K. 1996. *Elements of Biotechnology*. Rastogi Publishers, Meerut.

Grierson, D. and Covey, S.N. 1988. *Plant Molecular Biology*. Blackie, Glasgow.

Kingsman, S.M. and Kingsman, A.J. 1988. *Genetic Engineering: An Introduction to Gene Analysis and Exploitation in Eukaryotes*. Blackwell Scientific Publication, Oxford.

Mantell, S.H., Mathew, J.A. and McKee, R.A. 1985. *Principles of Plant Biotechnology: An Introduction to Genetic Engineering in Plants*. Blackwell Scientific Publications, Oxford.

Old, R.W. and Primrose, S.B. 2001. *Principles of Gene Manipulation: An Introduction to Genetic Engineering*. Blackwell Scientific Publications, Oxford.

# 24

# Chloroplast and Mitochondrion DNA Transformation

Plants possess three major genomes, carried in nucleus, chloroplast and mitochondrion. Transformation of nuclear genome has already been discussed. The transgenic plants developed in such cases involved insertion of genes into the nuclear genome, not in the organelles. Thus transformations of two organellar genomes of chloroplast and mitochondrion have been discussed separately.

## Chloroplast transformation

Chloroplasts are intracellular organelles present in higher plants and algae, which contain the entire machinery for the process of photosynthesis and their own genetic system. Chloroplasts also participate in the biosynthesis of amino acids, nucleotides, lipids and starch. Chloroplast consists of three main parts; envelope, the thylakoid (membrane structures) and stromas (soluble parts) Fig.24.1. The thylakoid is involved in photochemical reactions to produce chemical energy from light. Stromas include enzymes for dark reactions ($CO_2$ fixation) in photosynthesis, genome replication/ expression and various additional metabolisms. The chloroplast genome is known as plastome. The chloroplast genome (Cp) from higher plants consists of multiple copies of homogeneous circular double-stranded DNA molecules of 120 to 180 kbp in size. In any particular species, all plastid types carry identical, multiple copies of the same genome. In a leaf cell, there are ~100 chloroplasts and each with ~100 copies of the plastid genome will give rise to ~10,000 copies of the plastid genome per cell. There may be significant species specific deviation from their mean value. The gene content and arrangement of chloroplast DNAs in higher plants are relatively uniform from species to species. Chloroplasts are estimated to contain 2000–5000 protein species, and 3500 putative gene products are predicted to possess putative chloroplast signal peptides in Arabidopsis. Chloroplasts contain also sets of non-coding RNA molecules (stable or structural RNAs), typically rRNAs and tRNAs. A total of 30 different tRNA species are known to be encoded in the chloroplast genome from tobacco, rice and so on. All 61 possible codons are used by the chloroplast genes for encoding of polypeptides.

A group of researchers from the United States has produced the complete chloroplast genome sequence of cassava. The genome composed of approximately 164,450 base pairs and includes about 128 genes. Forty nine percent of the genes code for proteins. The *atpF* intron (non-coding region), which was first thought to be conserved in land plants, was

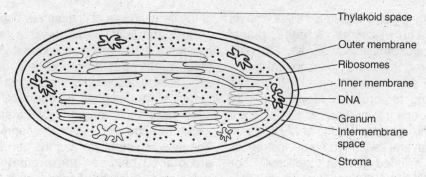

- Thylakoid space
- Outer membrane
- Ribosomes
- Inner membrane
- DNA
- Granum
- Intermembrane space
- Stroma

**Fig. 24.1:** A structure of chloroplast.

found to be absent in the cassava chloroplast genome. The scientists found out that the intron loss is a characteristic of most species belonging to the cassava family (spurges).

In chloroplasts, gene expression is light-dependent, and is developmentally regulated. The mechanism of regulation is mostly post-transcriptional affecting RNA processing and translation initiation. Most chloroplast transcripts are extensively modified to produce functional RNA molecules, the processes of which require precise site recognitions. Products of transgenes integrated in chloroplast genomes would accumulate in leaves and other green parts since they contain chloroplasts, and in fruits that were green before ripening.

The first successful chloroplast trans-formation was reported by Boynton *et al.* (1988) in *Chlamydomonas reinhardii* and the first successful chloroplast transformation in a higher plant was carried out by Svab *et al.* (1990) to introduce spectinomycin resistance into tobacco.

## The Plastid chromosome

The plastid DNA or chromosome is a circular, double-stranded molecule of 120 to 180 kilobase pairs (kbp). Each plastid contains tens to hundreds of copies of the molecule, organized into several nucleoids. These nucleoids are readily observed in chloroplasts stained by DAPI (4, 6-diamidino-2-phenylindole) using fluorescent microscopy. Nucleoids appear interconnected in young and mature chloroplasts. A small number of nucleoids are

present in proplastids but this number increases in young plastids. At this stage of development, plastids are actively dividing and nucleoids are associated with the inner membrane of the plastid envelope. A 130 kDa DNA binding protein which recognizes specific sequences of plastid DNA has been identified in pea. This protein named PEND (for plastid envelope DNA binding) is certainly involved in the binding of nucleoids to the envelope of young plastids and might be involved in DNA replication. Interestingly, the PEND protein is not detected in mature chloroplasts in which nucleoids are more likely attached to thylakoids. In later stages of development the nucleoids are dispersed and associated with grana thylakoids to which they are attached

The plastid chromosome exists as a negatively super coiled molecule. The analysis of DNA conformation by pulse-field electro-phoresis showed that molecules are present as monomers, dimers, trimers and tetramers in a relative amount of 1, 1/3, 1/9 and 1/27, respectively (Deng *et al.*, 1989).

The complete plastid DNA sequences from different plant genera are known and it can be concluded that the chromosome organization is highly conserved. The plastid chromosomes from plants, with some exceptions in the Fabaceae species, contain two inverted repeat (IR) regions separating a large and a small single copy (LSC and SSC, respectively) regions.

The A+T content is not evenly distributed in the plastome. It is higher in non-coding regions and is lower in regions coding for tRNA and for the rRNAs. The plastome of higher plants contains 4 ribosomal RNA genes, 30 tRNA genes, more than 72 genes encoding polypeptides and several conserved reading frames coding for proteins of yet unknown function. In contrast to prokaryotic tRNA genes, no plastid tRNA gene codes for its 3′-CCA end, although in several cases the first C is present. This element is added post-transcriptionally. The set of 30 tRNAs is sufficient to read all amino acid codons. It is important to state that no RNA, even of small size, is imported into chloroplasts. The plastid genes coding for polypeptides can be classified into several categories: genes coding for the prokaryotic RNA polymerase core-enzyme, genes coding for proteins of the translational apparatus and for the photosynthetic apparatus, and genes encoding subunits of the NADH dehydrogenase (*ndh*). Many genes are interrupted by intronic sequences. In contrast to algae, plastid genes of higher plants contain single introns, with the exception of the *rpl2* gene which contains two introns and requires trans-splicing. It is known that vast majority of the plastid proteins are really coded by nuclear genes. In view of this, introduction of novel proteins, or enzymes into plastids can be achieved by two routes: (i) insertion of a chimeric foreign gene construct (containing the gene of interest fused to sequences coding for a plastid transit peptide plus appropriate transcriptional control signals) into the nuclear genome. In this case, the gene product after its synthesis in the cytoplasm is targeted to the chloroplast; (ii) insertion of a foreign gene directly into the chloroplasts for expression by the plastid's own protein synthesis machinery, which is chloroplast transformation. It depends on the integration of transgenes in to the chloroplast exclusively by homologous recombination. Therefore, the foreign gene that is being introduced must be flanked by sequences homologous to the chloroplast genome (Staub and Maliga, 1992). Greater than 400 bp of homologous sequence on each side

of the gene construct is generally used to obtain chloroplast transformants at a reasonable frequency. Chloroplast genes are transcribed by chloroplast specific promoters, chloroplast specific termination signals, and most chloroplast genes are transcribed as operons. Thus, two open reading frames can be inserted into a vector in sequence under the same promoter. The selectable marker and the gene of interest are placed between the promoter and the terminator, which are flanked, by the 5′ and 3′ untranslated regions.

## Selectable markers

The following selectable markers are available for chloroplast transformation:

**Dominant antibiotic resistance markers (Positive selection markers):** Resistance to antibiotics of spectinomycin, streptomycin and kanamycin are dominant primary positive selection markers. The *aadA* gene encodes aminoglycoside adenyl transferase (AAD) which confers resistance to aminoglycoside type antibiotics such as spectinomycin by inactivation of antibiotic (Goldsmith-Clermont, 1991). Spectinomycin is mainly used for chloroplast transformation because it is a prokaryotic translational inhibitor and affects only chloroplasts with little effect on plant cells. It is a non-lethal marker at the cellular level, but is a selective marker at the plastid level. It is possible to distinguish between sensitive and resistant cells on the basis of color. Sensitive cells get bleached whereas resistant cells remain green when grown on a selective medium. AAD also inactivates streptomycin and was used to select transplastomic clones in tobacco. The neomycin phosphotransferase (*npt*II) gene which confers kanamycin resistance has also been used for chloroplast transformation (Carrer *et al.*, 1993)

**Recessive selectable markers (Positive selection markers):** Recessive antibiotic resistance markers are provided by genes encoding antibiotic insensitive alleles of ribosomal RNA genes. Initial transformation vectors carried a plastid 16S rRNA (*rrn*16) gene

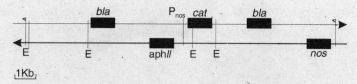

**Fig. 24.2:** Structure of T-DNA construct used for chloroplast transformation, Pnos-cat = chimeric gene for CAT with *nos* promoter, bla–β lactamase gene; *aph*II–aminoglycoside phospho transferase gene; *nos*–nopaline synthase gene.

with point mutation that prevents binding of spectinomycin or streptomycin to the 16S rRNA. The *rrn*16 target site mutations are recessive and were 100 fold less efficient than the currently used dominant *aadA* gene. Likewise *rps*12 mutated ribosomal protein gene was also used which prevents binding of streptomycin. Another category of markers is provided by genes that restore photoautotrophic growth by complementing non-photosynthetic mutants. However, these can be used with a limited set of mutant strains of plants. Other alternative selectable markers are also being studied. The betaine aldehyde dehydrogenase (BADH) encoding gene produces an enzyme that converts toxic betaine aldehyde to non-toxic glycine betaine and serves as an effective selectable marker.

**Secondary positive selection markers:** Secondary markers are genes that confer resistance to the herbicides phosphinothricin (PPT) or glyphosate or to the antibiotic hygromycin (based on expression of the bacterial hygromycin phosphotransferase). Use of secondary selective markers is dose dependent, which are not suitable to select transplastomic clones when only a few ptDNA copies are transformed, but will confer a selective advantage when most genome copies are transformed.

**Negative selection markers:** The ability to identify loss-of-function of a conditionally toxic gene forms the basis of negative selection. Bacterial cytosine deaminase (CD) enzyme encoded by the *codA* gene catalyzes deamination of cytosine to uracil, enabling use of cytosine as the sole nitrogen and pyrimidine

source. CD is present in prokaryotes and in many eukaryotic microorganisms, but is absent in higher plants. 5-fluorocytosine is converted to 5-fluorouracil, which is toxic to cells. This negative selection scheme has been utilized to identify seedlings on 5-fluorocytosine-medium from which *codA* was removed by the CRE-*loxp* site-specific recombinase.

**Transformation methods**

Chloroplast transformation was considered at one stage to be an impossible thing because of double membrane of the organelles acting as a physical barrier. Also there are no viruses or bacteria which could infect chloroplasts and act as vector for gene transfer. However, lately some successes have been achieved in chloroplast transformation. Some of the methods used for gene transfer in to plant nuclei can also bring about transformation of chloroplasts which have been described below.

1. ***Agrobacterium* mediated transformation:** A Ti plasmid vector was constructed using a chimeric gene construct with different T DNA sequences as shown in Fig. 24.2. (i) promoter region of nopaline synthase gene (*Pnos*); (ii) coding sequence of chloramphenicol acetyltransferase (CAT) gene (*cat*); (iii) complete gene for nopaline synthase (*nos*); (iv) gene for amino-glycoside phosphotransferase type II or APH II (*aph*II) and (v) the gene for beta lactamase (*bla*). The gene construct was introduced into tobacco leaf protoplasts by co-cultivation with A. tumefaciens containing the hybrid Ti plasmid. Calli were selected on a medium containing chloramphenicol. From the chloramphenicol resistant (*Cm*r) calli, only 8%

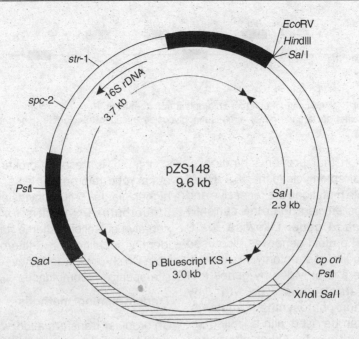

**Fig. 24.3:** Transformation vector pZS148.

had nopaline, suggesting that nos gene was lost or inactivated in majority of cases. Of 20 plantlets regenerated from Cm$^r$ nos$^-$ calli, only eight (40%) were able to root on medium containing chloramphenicol. One of these plants exhibited the presence of CAT activity and APH II activity in chloroplast fraction obtained after cell fractionation. In reciprocal crosses with wild plants, it was found that cat gene was transmitted through the ovule and not through the pollen (maternal inheritance), confirming location of *cat* gene in the Cp genome. The presence of cat gene in chloroplast genome was further confirmed, when Southern blots derived from *Xhol* restriction digests of nuclear, mitochondrial and chloroplast DNAs were hybridized with a probe having *cat* gene (only Cp DNA showed hybridization).

**2. Particle gun method**: Particle gun method also known as biolistic method has been found to be one of the most successful methods for transformation of the chloroplast genome. A variety of vectors have been used for plastid transformation which carry some regions

homologous to a fraction of chloroplast genome. A representative vector pZS148 which contains the different DNA sequences is shown in Fig. 24.3. (i) 16S rRNA gene cloned as a 3.7 kb *Sacl-EcoRV* fragment derived from tobacco genotype (SPC2) which carries resistance to both streptomycin (*str*1) and spectinomycin (*spc*2) due to mutations in 16S rDNA; and also had a *PstI* site, (ii) A 2.9 kb *SalI* fragment carrying *cp-ori* region needed for Cp-DNA replication; (iii) A 3.0 kb segment of pBluescript KS+ phagemid vector for cloning in *E. coli*. This vector has been designed as a shuttle vector to replicate both in *E. coli* and chloroplasts. The vector was introduced in to tobacco leaves by particle gun method and 3 transformants were recovered.

Stable chloroplast transformation depends on the integration of the foreign DNA into the chloroplast genome by homologous recombination, therefore the foreign gene that is being introduced must be flanked by sequences homologous to the chloroplast genome (Staub and Maliga, 1992). Greater than 400 bp of homologous sequence on each side of the gene construct is generally used to obtain chloroplast transformants at a reasonable frequency.

Chloroplast genes are transcribed by chloroplast specific promoters and chloroplast specific termination signals. As mentioned earlier most chloroplast genes are transcribed as operons, hence two open reading frames of gene of interest and selectable marker are to be inserted into a vector in sequence between the promoter and the terminator, which are flanked, by the 5′ and 3′ untranslated regions.

In the procedure a leaf is bombarded by DNA coated tungsten particles, which carries 20 to 50 copies of the donor plasmid vector with the gene of interest. Generally, one or a few plastid genomes in an organelle are transformed and 16 to 17 cell divisions are needed to achieve homoplastomic state. Selection of transformed cells is achieved in the presence of spectionmycin using spectinomycin resistance marker encoded by mutant 16S rRNA gene and a chimeric *aad*A gene encoding aminoglycoside-3-adenyltransferase. Selected cells are regenerated into whole plants, which can be tested for the presence of desired gene in the plastid genome.

The transforming DNA in the vector is incorporated in to the chloroplast genome by two homologous recombination events

i. *Gene replacement*: In the case of gene replacement, a cloned plasmid DNA fragment from an *E. coli* vector is incorporated in to the recipient chloroplast genome, resulting in complete or nearly complete replacement of the homologous region of the resident genome by the donor DNA (Fig. 24.4). For example Transgenic clone PT69D was selected by 16S rDNA encoded spectinomycin resistance (SR) on plasmid pJS75 that carries a 6.2 kb plasmid DNA fragment. Inverted repeat sequences are paired such that large and small single copy regions form loops. Initial interaction of the recipient WT plastid genome (ptDNA) with pJS75 leads to replacement of wild type

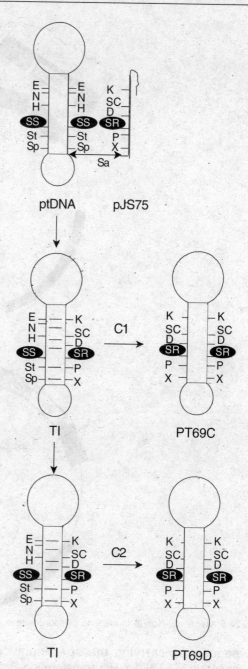

**Fig. 24.4:** Gene replacement in the inverted repeat region of tobacco plastid genome. SS–spectinomycin sensitive allele; SR–spectinomycin resistant allele; TI–Transformation intermediate; E–*Eco*RV, N–*Nci*I; D–*Dra*I, H–*Hind*I, K–*Kpn*I, P–*Pst*I, Sa–*Sal*I, St–*Sca*I, Sp–*Sph*I, St–*Sty*I X–*Xba*I are RFLP restriction enzyme marker sites.

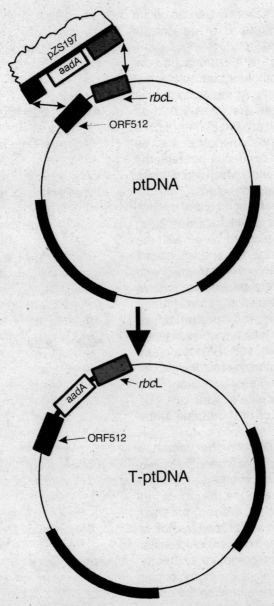

**Fig. 24.5:** Gene insertion in to tobacco plastid genome.

sequences carrying the spectinomycin sensitive (SS) allele with transforming DNA in one of the repeats. The transforming intermediate (TI) replicates and there can be complete copy correction and genome stabilizes for all the markers yielding genome PT69C. There can be differential copy correction in the TI genome which removes the transgenic *Pst*I (P) site but preserves all other markers, yielding the genome PT69D.

ii. **Gene insertion**: Insertion of foreign gene can be obtained by two homologous–recombination events via flanking plastid DNA sequences (Fig. 24.5). For example gene insertion into plastid genome of tobacco has been shown. An *aad*A gene under the control of plastid expression signals was cloned between the *rbc*L and *ORF*512 genes for targeting the insertion of chimeric gene. Plasmid pZS197, a pBluescript KS$^+$ vector with the *rbc*L/*aad*A/*ORF*512 insert was introduced into plastids by particle gun method. Transplastomic lines were selected by spectinomycin resistance encoded in the *aad*A gene. Two homologous recombination events lead to the incorporation of *aad*A gene between the *rbc*L and *ORF*512 of the recipient wild type plastid genome (ptDNA) yielding the T-ptDNA transplastome (Maliga, 1993).

Transformation of *Nicotiana tabacum* with pZS 148 vector DNA (carrying resistance against spectinomycin as well as streptomycin) resulted in development of three transplastomic clones. Screening for *Pst*I site confirmed the replacement of 16S rDNA region in leaves by that of engineered 16S rDNA using SPC2. The transplastomic plants obtained as above, were selfed and crossed to wild type plants. It was shown that in plant raised from seeds obtained due to selfing, there was no segregation, while in crosses the resistance was maternally inherited.

Generally, the transformation of plastids by particle bombardment is 100 fold less efficient despite the high number of Cp DNA copies in a cell (a cell may have up to 100 plastids with 1000 to 10,000 copies of Cp DNA molecules). Compared to this *Chlamydomonas* has 80 Cp DNA copies per cell, but the efficiency is as high as that for nuclear genes (2–100 transplastomic clones per bombardment). Further, the selection for a non-lethal marker (e.g. neomycin, kanamycin, spectinomycin resistance) is critical for obtaining the transplastomic lines, because it allows sufficient time for the resistant plastid genome copies to increase in number to allow phenotypic expression. This explains the low efficiency of transformation of Cp DNA, since eventually all transformed plastids may be derived from a solitary plastid or from a solitary DNA molecule that is transformed.

**3.** ***Other possible methods for introduction of DNA into plastids***: Several other methods have been suggested for DNA delivery into chloroplasts. These methods include the following.

**(a)** ***PEG-mediated transformation***: This approach requires preparation of protoplasts and target species used must be regenerable from protoplasts. Protoplasts take up DNA in the presence of PEG and changes in the plasma membrane allow DNA to penetrate and move into the cytoplasm. The DNA is transported by some unknown means from the cytoplasm into the chloroplast where it may be integrated into the genome. PEG-mediated transformation is less efficient than the particle gun approach.

**(b)** ***Galistan expansion femtosyringe***: It involves the microinjection of DNA into chloroplasts using a small syringe. The heat-induced expansion of a liquid metal galistan, within a syringe forces the plasmid DNA through a capillary tip with a diameter of approximately 0.1 mm in to the chloroplasts. This method is not used now.

**(c)** ***DNA uptake by chloroplasts in vivo***: DNA is injected into the protoplasts either by (i) microinjection; (ii) electroporation or (iii) agitation of chloroplasts in the presence of glass beads and DNA. The DNA available in cytoplasm will be taken up by the chloroplasts. This may be facilitated, if a DNA sequence is identified, which may work as a signal and is recognized by a receptor in the chloroplast, thus facilitating internalization of DNA.

**(d)** ***Microinjection of DNA into chloroplast in vivo***: Microinjection of DNA directly into chloroplasts should also be possible in future although there may be technical problems. For instance, in higher plants, each cell has many plastids and each plastid has several genomes. Therefore, number of plastids per cell may have to be artificially reduced and the optimal number chloroplast to be injected per cell will have to be determined. This will be difficult, because the fate of chloroplasts and their genomes during regeneration of protoplasts is not fully understood. However, microinjection as a method for transformation of *Chlamydomonas* chloroplasts should be a distinct possibility, due to its large size and the presence of a single chloroplast per cell.

(e) *DNA uptake by chloroplasts in vitro*: Introduction of DNA in isolated chloroplasts from higher plants will help in overcoming the following problems faced in DNA uptake *in vivo*: (i) a large number of chloroplasts per cell may not take up the DNA; (ii) the concentration of DNA in the cytoplasm, in the vicinity of chloroplasts, will be low, thus reducing the chances of DNA uptake by chloroplasts. The major problem in this method, however, will be the return of manipulated chloroplasts back to plant cells. There are reports of successful uptake of chloroplasts by plant protoplasts, but stable incorporation of transformed chloroplasts and subsequent regeneration of these transplastomic plants from protoplasts has not been achieved.

## Homoplastomic lines development

A transformed plastid genome is a transplastome. The transformation of plastids means stable integration of transforming DNA into the plastid genome that is transmitted to the seed progeny. Primary chloroplast transformation events involve the transformation of only one or a few genome copies within a single plant cell resulting in cells that contain a mix of transformed and wild type chloroplast genomes. The cells are referred to as **heteroplastomic** and are genetically unstable. Heteroplasmy is of two types: (i) Interplastidic heteroplasmy is when a cell contains chloroplasts with wild type genomes and

chloroplast with transformed genomes and (ii) Intraplastidic heteroplasmy is when wild type and transformed genomes are located within the same chloroplast. Usually heteroplastomic cells are resolved spontaneously to a homoplastomic condition, where the chloroplasts within a cell are either all transgenic or all wild type. Thus, homoplasmy refers to a pure population of plastid genomes. This is achieved by random genome segregation during chloroplast division and random chloroplast segregation during cell division. Homoplasmy can be achieved in chloroplast transformation studies by allowing for a sufficient number of cell divisions under high concentrations of the selective agent usually spectinomycin. Plantlets go through a series of selection and regeneration steps with selective agent (e.g. spectinomycin). Interplastidic heteroplasmy is more likely to disappear rapidly as chloroplast containing only wild type genomes are sensitive to the selection agent and will not survive. Intraplastidic heteroplasmy is more difficult to eliminate as the spectinomycin resistance gene functions as a dominant selectable marker and only few copies are sufficient to confer resistance. Homoplastomic transgenic shoots are typically obtained after 2–4 regeneration cycles under high selection pressure. Assays have been developed to verify homoplasmy of transformed genomes in the shoots and include large-scale assays and PCR-based tests that amplify wild-type genomes (Bock, 2002).

To date, researchers have yet to discover an efficient and inexpensive technique to transform chloroplasts in a wide range of crop species. These methods have only been optimized for the transformation of tobacco. In order for transplastomic technologies to make an impact they must be adapted to other major crop species. Work is currently being done in tomato, Arabidopsis, potato and rice. Research in this field is making rapid progress and plastid transformation systems for agronomically important plant species will likely be revealed in the near future.

## Advantages

1. *Elimination of risk of transgene dispersal:* Chloroplast genomes are inherited maternally through the egg cell in most angiosperm and pollen does not contain plastids of any sort. Therefore, chloroplast genes are only transmitted through the egg to the embryo and there will be no risk of dispersal of transgene through pollen. Maternal inheritance of novel genes is highly desirable in situations where out-crossing between crops and weeds or among crops is a biosafety concern (Daniell *et al.*, 1998). Several agronomically important genes reside in the chloroplast genome and are best expressed in chloroplasts only.

2. *Biosafety:* Biosafety issues are associated with transgenic plants. The risk of transgene escape through pollen referred to as pollen escape will be rare if the gene of interest is inserted into the chloroplast genome. However, with transplastomic plants this problem is not faced due to maternal transmission of transgene. Thus, it excludes the risk of transfer of transgene to other species growing in the adjacent plots.

3. *Elimination of toxic effect of transgene on insects:* There have been some reports that insects (Butterfly larvae) feeding on Bt corn fields showed high mortality because of toxic effect of Bt crystal protein present in the pollen. In transplastomic plants this effect is eliminated because pollen does not transmit the chloroplast genes. In addition the Cry protein is not produced in fruits or pollen grains because chloroplast gene is down regulated in non-green parts of plants thus eliminating the risk of toxic effect of protein on animals or human beings.

4. *High levels of transgene expression:* Very high level of transgene expression has been observed in chloroplasts. The high level of transgene product can be attributed to the fact that there are many chloroplasts within a plant cell whereas there is only one nucleus. DNA inserted into chloroplasts is copied 5,000–10,000 times whereas it is copied only 1–4 times in the nucleus. Higher levels of expression of insect and herbicide resistance gene products were observed in transplastomic plants than in plants containing transformed nuclei. Chloroplasts are ideal expression factories for high-yield protein production with up to 40% of the soluble protein in the cell due to multiplicity of chloroplast genome in a cell and also due to stability of protein in the chloroplasts.

5. *Protein synthesis machinery in the chloroplasts:* Chloroplasts have 70S ribosomes as in the case of prokaryotes. Thus, chloroplasts can express bacterial genes better because transcription and translation in chloroplasts is prokaryotic in nature. Whereas protein synthesis machinery in the nucleus and cytoplasm is not suitable for the microbial genes intended to be transferred in the transgenic crops.

6. *Stable transgene expression:* Transgene expression is more stable in transplastomic plants than in nuclear transformants. Transgenes are integrated into chloroplast genomes by homologous recombination and are not affected by gene silencing. All transplastomic plants in an experiment should be genetically and phenotypically identical. Nuclear transformation in plants occurs by random integration of transgenes into unpredictable locations in the genome by non-homologous recombination and can result in varying levels of expression and in some cases, gene silencing (Kooter *et al.*, 1999).

7. *Multiple gene transfer:* Multiple genes as an operon can be transferred to the chloroplast genome which has the capacity to express multiple genes from a polycistronic mRNA. Thus a complete biosynthetic pathway gene can be transferred as a single transformation event. This is also referred to as transgene stacking. This pyramiding of genes helps to decrease the risk of promoting resistance in pest organisms. Multiple genes associated

with a complete pathway when transferred to the nuclear genome involves generally introduction of individual gene with a promoter and results in variable expression.

8. *Gene silencing*: Gene silencing generally observed in transgenic plants is however minimal or not observed in transplastomic plants.

## Applications

**Transplastomic lines for agronomic traits:** Stable integration of *aadA* selectable marker into tobacco chloroplast genome was reported by Svab and Maliga (1993). In field crops, the most common transgenic traits are resistance to insects and herbicides expressed from nuclear genes. Both types of genes have been successfully engineered for plastid expression. Expression of the Bt insecticidal protein from nuclear genes required construction of synthetic codon-modified genes to improve translation, protect the mRNA from degradation and prevent early translation termination. In contrast the Bt insecticidal protein genes were expressed in plastids from bacterial coding segments. Plants resistant to Bt sensitive insects were obtained by integrating the *cryIAc* gene into the tobacco chloroplast genome (McBride *et al.,* 1995). Plants resistant to *Bt* resistant insects were obtained by hyper-expression of the *cryIIA* gene into the tobacco chloroplast genome (Kota *et al.,* 1999). They have also shown over expression of the Cry2Aa2 protein in tobacco chloroplasts as a model system, and as a possible solution to the evolution of Bt resistance observed in the field.

Cheng *et al.* (2008) successfully transferred into the cabbage chloroplast genome *aadA* and *cry*1Ab genes. The genes were inserted into the pASCC201 vector and driven by the P*rrn* promoter. The cabbage-specific plastid vectors were transferred into the chloroplasts of cabbage via particle gun mediated transformation. Regenerated plantlets were selected by their resistance to spectinomycin and streptomycin. According to antibiotic selection, the regeneration percentage of the two cabbage cultivars was 4–5%. The results of PCR, Southern, northern hybridization and western analyses indicated that the *aad*A and *cry*1Ab genes were not only successfully integrated into the chloroplast genome, but functionally expressed at the mRNA and protein level. Expression of Cry1Ab protein was detected in the range of 4.8–11.1% of total soluble protein in transgenic mature leaves of the two species. Insecticidal effects on *Plutella xylostella* were also demonstrated in *cry*1Ab transformed cabbage.

Plants have also been genetically engineered via the chloroplast genome to confer herbicide resistance and introduced foreign genes were shown to be maternally inherited, overcoming the problem of out-cross with weeds (Daniell *et al.,* 1998). The chloroplast vector pZS-RD-EPSPS contains the 16S rRNA promoter (P*rrn*) driving the *aadA* (aminoglycoside adenyltransferase) and EPSPS gene with the psbA 3′ region (the terminator from a gene coding for photosystem II reaction centre components) from the tobacco chloroplast genome. This construct integrates the EPSPS and *aadA* genes into the spacer region between the *rbc*L, (the gene for the large subunit of Rubisco (ribulose bisphosphate carboxylase) and *orf*512 gene (code for the *acc*1*)* gene) of the tobacco chloroplast genome. Field-level tolerance to glyphosate was obtained by expression in plastids of prokaryotic EPSPS genes; the required protein levels were higher (5% of TSP) than when EPSPS was expressed from nuclear genes. An interestingly split EPSPS gene was developed with gene containment in mind. With one half of the protein encoded in a nuclear gene and the second half in a plastid gene. Pollen transmission from this crop can lead to the transfer of only part of the herbicide resistance gene, which is insufficient to confer glyphosate tolerance.

Ye *et al.* (2003) reported the use of (*CP4* or *bar*) herbicide resistance marker alongwith non lethal selection marker of *aadA* gene that confers resistance to spectinomycin and streptomycin in tobacco. To overcome the apparent herbicide lethality to plastids, a "transformation segregation" scheme was developed that used two independent transformation vectors for a cotransformation approach and two different selective agents in a phased selection scheme. One transformation vector carried an antibiotic resistance (*aadA*) marker used for early nonlethal selection, and the other transformation vector carried the herbicide (*CP4* or *bar*) resistance marker for use in a subsequent lethal selection phase. They reported a plastid cotransformation frequency of 50% to 64%, with a high frequency (20%) of these giving rise to transformation segregants containing exclusively the initially nonselected herbicide resistance marker. Thus, indicating high degree of persistence of unselected transforming DNA, providing useful insights into plastid chromosome dynamics.

**Expression of Recombinant Proteins:** To meet the demands for production capacity of recombinant proteins, there is significant interest in plant based production of vaccines and antibodies. Transgene expression in tobacco plastids reproducibly yields protein levels in the 5% to 20% range. A salient feature of plastid expression is the importance of post-transcriptional regulation and from the same promoter proteins may accumulate in a 10,000 fold range. Transplastomic tobacco plants were produced which carried a gene for human somatotropin (hST).

## Applications in Basic Sciences

i. *Gene Knockouts*: Targeted knockout of plastid genes involves construction of a transformation vector in which a selectable marker in the vector (S1) replaces the target gene (G2) in a larger ptDNA fragment (Fig. 24.6). Selection for antibiotic resistance results in replacement of the target gene with the selective marker (S1) in the ptDNA.

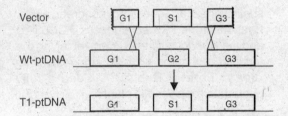

**Fig. 24.6:** Targeted deletion of a plastid gene (G2) by replacement with market gene S1 in vector.

ii. *Over-expression*: The advantage of over expression as a research tool was shown by the unexpected discovery of site-specific RNA editing trans factors. Over-expression of an edited RNA segment and a *clPl*-specific mRNA maturation factor was based on over-expression of the *clP1* 50-UTR in a chimeric transcript. Over-expression was used to probe the plastid *accD* function by replacing the weak *accD* promoter with the strong *rrn* promoter. Another example for over-expression involved relocating the nuclear anthranilate synthase gene to plastids to boost tryptophan production.

iii. *Transcription*: The field of plastid gene transcription also benefits from plastid transformation. The plastid *rpoA* and *rpoB* tobacco knockout plants played a critical role in recognizing that the plastid encoded plastid RNA polymerase (PEP) and the nuclear-encoded phage-type RNA polymerase (NEP) transcribe distinct groups of genes.

iv. *Photosynthesis*: Many photosynthetic genes are encoded in the plastid genome and thus plastid transformation is a necessary tool to probe and improve photosynthesis. Studies in higher plants focused on photosynthetic gene knockouts and Rubisco engineering. Rubisco in higher plants is composed of plastid encoded large *rbc*L and nucleus encoded small subunit *rbc*S genes. The plastid-encoded tobacco *rbc*L gene was replaced with a heterologous sunflower gene. Genes from nongreen algae, Cyanobacteria (*Synechococcus* PCC6301) and the *Rhodospirillum rubrum* replacing the tobacco *rbc*L gene with the *Rhodospirillum rubrum* Form II Rubisco yielded a fully photoautotrophic fertile plant.

## Limitations of Chloroplast Transformation

Chloroplast transformation is much more difficult than nuclear transformation and some of the difficulties encountered during chloroplast transformation are outlined below.

1. Due to unknown reasons chloroplast transformation frequencies are much lower than those for nuclear transformation.
2. The recovery of transplastomic cells is particularly dependent on efficient and prolonged selection procedures. Typically, 2–4 regeneration and selection cycles under high selection pressure are required for the recovery of homoplastomic shoots.
3. Transgene transfer methods into chloroplasts are limited.
4. The transformation systems are far more successful with tobacco than with other plant species. In fact, very few other plant species, e.g. tomato, are at present amenable to chloroplast transformation. Thus, there is an urgent need to develop more generally applicable and efficient chloroplast transformation protocols.
5. Expression of chloroplast genes is mostly light-induced. Therefore, products of transgenes ordinarily would accumulate in green plant parts, e.g., leaves and in such tissues as fruits that were green before ripening. Transgene products accumulate at very low levels in such tissues as potato tubers, etc.

## Mitochondrion transformation

Mitochondria are organelles, found in the cytoplasm of all eukaryotic cells. Their most immediate function is to convert energy found in nutrient molecules (substrates) and store it in the form of adenosine triphosphate (ATP).

Mitochondria vary considerably in shape and size, but all have the same basic architecture. There is a smooth outer membrane, surrounding a very convoluted inner membrane. The convolutions form recognizable structures called cristae. The two membranes have very different properties. Together they create two compartments namely the inter-membrane space (the compartment between the membranes), and the matrix (the very interior of the mitochondria). The matrix contains a highly concentrated mixture of hundreds of enzymes, mitochondrial ribosomes, tRNAs and several copies of the mitochondrial DNA (mt DNA) genome. Mitochondria possess their own genetic material and the machinery to manufacture their own RNAs and proteins.

The smallest mitochondrial genome known among higher plants is that of *Brassica hirta* at 218 kb and the largest known is that of *Cucumis melo* at 2300 kb. Great size variations in mitochondrial DNAs exist among closely related species. For example, species in the Cucurbitaceae possess chloroplast and nuclear genomes of similar sizes, but their mitochondrial genomes range from one of the smallest (watermelon at 230 kb) and to the largest known mitochondrial (melon at 2300 kb) genome among angiosperms. The plant mitochondrial genome is also a circular double-stranded DNA molecule that encodes rRNAs, tRNAs, ribosomal proteins, and a portion of the enzymes used in respiration. Many mitochondrial enzymatic subunits are nuclear encoded, cytoplasmically translated, and imported into the mitochondria. In contrast to the chloroplast genome, the mitochondrial DNA possesses direct repeats spread throughout the genome. Pairing and recombination among these direct repeats produces smaller circular DNA molecules. Continued pairing and recombination among other direct repeats can shift the relative arrangements among coding regions quickly producing polymorphic molecules among relatively closely related plants. Whereas the chloroplast genomes of most higher plants are similar in size and structure, both the nuclear and mitochondrial genomes show great size differences even among relatively closely related species. Among angiosperms, both the chloroplast and mitochondrial genomes predominately show maternal transmission, although occasional biparental transmission is well established and can be under nuclear control. Mitochondria are almost always present in the generative and sperm cells of the male gametophyte.

Mitochondrial transformation has been reported for single-celled *Chlamydomanas* and yeast but there is no routine method to transform the higher-plant mitochondrial genome. The main hurdles to overcome for transformation are:

i. The introduction of foreign DNA into the mitochondrion.
ii. Incorporation of transgene into the mitochondrial DNA.
iii. Absence of selectable markers for mitochondria.
iv. Relatively large number of mitochondria per cell and mitochondrial genomes per mitochondrion.
v. RNA editing may render foreign genes ineffective that are introduced into the mitochondrial genome.

Once selectable markers for mitochondrial transformation become available, the technique could then be applied to higher plants showing maternal transmission of mitochondria. The feasibility of a selection system based on resistance to antimycin and myxothiazol is being tested. These compounds inhibit mitochondrial electron transport through the cytochrome b (cob) protein, which is a mitochondrial gene product. Methods of transgene incorporation into the mitochondrial genome could include particle bombardment of cell cultures, PEG treatments of protoplasts, or microinjection of modified mitochondria to plant cells with a transformation cassette carrying the selectable marker, the gene of interest and flanking mitochondrial regions, followed by treatment with the selective agent.

Mitochondrial transformation in yeast by biolistic method has been reported. A non-reverting strain, which is respiratory deficient because of a deletion in the mitochondrial *oxi3* gene, was bombarded with tungsten microprojectiles coated with DNA bearing sequences that could correct the *oxi3* deletion. Respiratory-competent transformants were obtained in which the introduced *oxi3* DNA is integrated at the homologous site in the mitochondrial genome.

Examples of mitochondrially encoded plant phenotypes include cytoplasmic-genic male sterility (CMS), the non-chromosomal stripe *(ncs)* mutations of maize, the mosaic *(msc)* mutations of cucumber, the *chm*-induced mutations of *Arabidopsis,* and the plastome mutator phenotypes of *Oenothera*. With the exception of CMS, most of these mutant phenotypes are due to deletions or chimeric rearrangements involving mitochondrial coding regions and are maintained in plant population by heteroplasmy.

## Targetting of Transgene Product into Chloroplast and Mitochondria

At present techniques for specific introduction of transgenes into chloroplasts/mitochondria are in the development stage. Therefore, a practical approach is to integrate a transgene into the nuclear genome and target the transgene product into chloroplasts or mitochondria. This targeting is achieved by fusing DNA sequences encoding specific signal peptides or transit peptides with the concerned transgenes. The signal peptides encoded by these sequences direct the associated proteins into the chloroplasts/mitochondria. For example, the transit peptide sequences of the small submit of Rubisco (ribulose bisphosphate carboxylase) guides the fused transgene proteins into chloroplasts. Other examples which are targeted by this signal peptide includes, the two enzyme involved in PHB synthesis, β-glucuronidase, yeast mitochondrial superoxide dismutase, light harvesting chlorophyll binding protein, brome mosaic virus coat protein, etc. Even chloroplast encoded large sub unit of Rubisco could also be targeted using a signal sequence.

Chloroplast targeting has been used to develop glyphosate resistant transgenic tomato and tobacco. Glyphosate inhibits the chloroplast enzyme EPSPS (5-enolpyruvate shikimate-3-phosphate synthase), which is involved in aromatic amino acid biosynthesis. The gene *aroA* of *Salmonella typhimurium* and *E. coli* encodes a glyphosate insensitive EPSPS enzyme. The bacterial gene *aroA* was fused with the 5′-transit peptide sequence from the cDNA of *Petunia* EPSPS and the gene construct was transferred into tobacco and tomato. The transgene product was guided into the chloroplasts of transgenic plants and imparted glyphosate resistance.

Similarly, appropriate signal sequences have been used to guide transgene products into the mitochondria. One such signal sequence is the N-terminal sequence (90 amino acids) of tobacco (β *F1* ATPase), which has been used to target CAT (chloramphenicol acetyl-transferase) and glutamine synthetase into mitochondrial matrix. However, no example of any practical value is available so far.

## REFERENCES

Bock, R. 2002. Transformation technologies for higher plant cell organelles. Available at http://www.biologie.uni-freiburg.de/data/bock/transfor.html

Boynton, J.E., Gillham, N.W., Harris, E.H., Hosler, J.P., Johnson, A.M. and Jones, A.R. 1988. Chloroplast transformation in *Chlamydomonas* with high velocity microprojectiles. *Science* **240**: 1534–1538.

Carrer, H., Hockenberry, TN., Svab, Z. and Maliga, P. 1993. Kanamycin resistance as a selectable marker for plastid transformation in tobacco. *Mol. Gen. Genet.* **241**: 49–56.

Cheng, W.L., Chin, C.L., Jinn, C.Y., Jeremy, J.W.C. and Menq, J.T. 2008. Expression of a *Bacillus thuringiensis* toxin (*cry1Ab*) gene in cabbage (*Brassica oleracea* L. var. *capitata* L.) chloroplasts confers high insecticidal efficacy against *Plutella xylostella*. *Theor. Appl. Genet.*, at http://www.springerlink.com/content/4033416h305833u4/ ?p=58d678bc0ffd4e3b9fcbc4ce98dd16bf&pi=5

Daniell, H., Datta, R., Varma, S., Gray, S. and Lee, S.B. 1998. Containment of herbicide resistance through genetic engineering of the chloroplast genome. *Nature Biotechnol.* **16**: 346–348.

Deng, X.W., Wing, R.A. and Gruissem, W. 1989. The chloroplast genome exists in multimeric forms. *Proc. Natl. Acad. Sci. USA*, **86**: 4156–4160.

Goldschmidt-Clermont, M. 1991.Transgenic expression of aminoglycoside adenyl transferase in the chloroplast: a selectable marker for site-directed transformation of *Chlamydomonas*. *Nucl. Acids. Res.* **19**: 4083–4089.

Kooter, J.M., Matzke, M.A. and Meyer, P. 1999. Listening to the silent genes: transgene silencing, gene regulation and pathogen control. *Trends Plant Sci.* **4**: 340–347.

Kota, M., Daniell, H., Verma, S., Garezynski, F., Gould, F. and Moar, W.J. 1999. Overexpression of the *Bacillus thuringiensis* Cry2A protein in chloroplasts confers resistance to plants against susceptible and *Bt*-resistant insects. *Proc. Natl Acad. Sci. USA* **96**: 1890–1845.

Maliga, P. 1993. Towards plastid transformation in flowering plants. *TIBTECH* **11**: 101–107.

McBride, K.E., Svab, Z., Schaaf, D.J., Hogen, P.S., Stalker, D.M. and Maliga, P.1995 Amplification of a chimeric *Bacillus* gene in chloroplasts leads to extraordinary level of an insecticidal protein in tobacco. *Bio/Technology* **13**: 362–365.

Staub, J.M. and Maliga, P. 1992. Long regions of homologous DNA are incorporated into the tobacco plastid genome by transformation. *Plant Cell* **4**: 39–45.

Svab, Z. and Maliga, P. 1993. High frequency plastid transformation in tobacco by selection for a chimeric *aadA* gene. *Proc. Natl. Acad. Sci. USA* **90**: 913–917.

Svab, Z., Hajdukiewicz, P. and Maliga, P. 1990. Stable transformation of plastids in higher plants. *Proc. Natl. Acad. Sci. USA* **87**: 8526–30.

Ye, G.N., Colburn, S.M., Xu, C.W., Hajdukiewicz, P.T.J. and Staub, J.M. 2003. Persistence of Unselected Transgenic DNA during a Plastid Transformation and Segregation Approach to Herbicide Resistance. *Plant Physiol.* **133**: 402–410.

## FURTHER READING

Gupta, P.K. 2005. Elements of Biotechnology, Rastogi Publishers, Meerut, India, pp. 602.

Maliga, P. 1993. Towards plastid transformation in flowering plants. *TIBTECH*, **11**: 101–107.

Mache, R. and Mache, S.L. 2001. Chloroplast genetic system of higher plants: Chromosome replication, chloroplast division and elements of the transcriptional apparatus. *Curr. Sci.*, **80**: 217–224.

Singh, B.D. 2006. Plant Biotechnology, Kalyani Publishers, India, pp. 755.

# 25

# Transgenics in Crop Improvement

Transfer of genes between plant species has played an important role in crop improvement for many decades. Plant improvement, whether as a result of natural selection or the efforts of plant breeder, has always relied upon evolving, evaluating and selecting the right combination of alleles. Useful traits such as resistance to diseases, insects and pests have been transferred to crop varieties from non-cultivated plants. Since 1970, rapid progress is being made in developing tools for the manipulation of genetic information in plants by recombinant DNA methods. The overall process of genetic transformation involves introduction, integration and expression of foreign gene(s) in the recipient host plant. Plants which carry additional, stably integrated and expressed foreign genes transferred (transgenes) from other genetic sources are referred to as transgenic plants. The development of transgenic plants is the result of integrated application of rDNA technology, gene transfer methods and tissue culture techniques. These techniques have enabled the production of transgenic plants in food, fiber, vegetable, fruit and forest crops. In recent times, plant biotechnology has become a source of agricultural innovation, providing new solution to the age-old problems. Plant genes are being cloned, genetic regulatory signals deciphered and genes transferred from entirely unrelated organisms (notably bacteria and virus) to confer new agriculturally useful traits on crop plants. Genetic transformation facilitates introduction of only specifically desirable genes without co-transfer of any undesirable genes from donor species which normally occurs by conventional breeding methods.

The capacity to introduce and express diverse foreign genes in plants was first described in tobacco by *Agrobacterium* mediated (Horsch *et al.*, 1984; De Block *et al.*, 1984) and vectorless approach (Paszhkowski *et al.*, 1984). The list of plant species that can be transformed by vector mediated (*Agrobacterium*) and vectorless methods has been growing continuously and at present transformation capability has been extended to more than 120 species in at least 35 families. Success includes most major economic crops, vegetables, ornamental, medicinal, fruit, tree and pasture plants. Gene transfer and regeneration of transgenic plants are no longer the factors limiting the development and application of practical transformation systems for many plant species. To illustrate this, monocotyledonous species were earlier considered outside the host range of *A. tumefaciens* which led to the development of direct DNA introduction or vectorless methods for transformation. Thus, most successes in development of transgenic plants in monocots have been due to vectorless systems. However, recently *Agrobacterium* mediated transformation has been reported in monocot species including most important food crops viz. rice (Hiei *et al.*, 1994), maize (Ishida *et al.*, 1996) and wheat (Cheng *et al.*, 1997).

The first generation application of genetic engineering to crop agriculture has been targeted towards the generation of transgenic plants expressing foreign genes that confer resistance to viruses, insects, herbicides or post harvest deterioration and accumulation of useful modified storage products. These have been discussed under various heads.

# RESISTANCE TO BIOTIC STRESSES

Genetic transformation has led to the possibility of transforming crops for enhanced resistance to insects and pathogens and it is rapidly moving towards commercialization. These advances form the basis of a chemical-free and economically viable approach for pest and disease control.

Resistance to biotic stresses has been discussed under the following headings.
1. Insect resistance
2. Viral resistance
3. Fungal and bacterial disease resistance

**Insect Resistance**

Progress in engineering insect resistance in transgenic plants has been achieved through the use of insect control protein genes of *Bacillus thuringiensis.* Insect resistance was first reported in tobacco (Vaeck *et al.*, 1987) and tomato (Fischhoff *et al.*, 1987). Today insect resistant transgenes, whether of plant, bacterial or other origin, can be introduced in to plants to increase the level of insect resistance. Approximately 40 different genes conferring insect resistance have been incorporated into crops. Genes conferring insect resistance to plants have been obtained from microorganisms: *Bt* gene from *Bacillus thuringiensis*; *ipt* gene isopentyl transferase from *Agrobacterium tumefaciens*, cholesterol oxidase

gene from a streptomyces fungus and *Pht* gene from *Photorhabdus luminescens*. Resistance genes from higher plants can be grouped into two categories (i) proteinase and amylase inhibitors and (ii) lectins-snowdrop lectin (GNA), pea lectin, jacalin, rice lectin, etc.). Resistance genes of animal origin are serine proteinase inhibitors from mammals and tobacco hornworm (*Manduca sexta*).

*Resistance genes from microorganisms*

*Bt toxin gene*

*Bacillus thuringiensis* (*Bt*) is an entomocidal bacterium that produces an insect control protein. Bt genes code for the Bt toxin, which differ in their spectrum of insecticidal activity. Most Bt toxins are active against Lepidopteran larvae but some are specific for Dipteran and Coleopteran insects. The insect toxicity of *Bt* resides in a large protein. The toxins accumulate as crystal proteins ($\delta$-endotoxins) inside the bacteria during sporulation. They are converted to active form upon infection by susceptible insect, thereby killing the insect by disruption of ion transport across the brush borders/membranes of susceptible insect.

Several genes encoding Lepidopteran type toxins have been isolated. One such gene from *B. thuringiensis* subsp. *Kurstaki* HD-1 contains an open reading frame of 3468 bp encoding a protein of 1156 amino acids. Chimeric *B. thuringiensis Kurstaki* genes, containing the CaMV 35S promoter and a sequence coding for an active truncated variant as well as the full-length gene, have been constructed and expressed in tomato plants (Fischhoff *et al.*, 1987). The level of insecticidal protein was sufficient to kill larvae of *Manduca sexta*, *Heliothis virescens* and *Heliothis zea*. Analyses of the progeny of transgenic plants showed that the *B. thuringiensis Kurstaki* gene segregated as a single dominant Mendelian marker. A second toxin gene has been cloned from *B. thuringiensis* strain *Berlkiner* 1715. It produces a 1155 amino acid Bt2 protein. Analysis of the level of expression of a Bt2-based chimaeric gene in tobacco plants (Vaeck *et al.*, 1987) showed a correlation between the quality of the toxin and insecticidal activity. Transgenic tobacco plants were protected from feeding damage by larvae of *Manduca sexta*. Bt toxin is not harmful to beneficial insects, mammals and humans. This gene segregates as a single dominant gene.

*Bt* strains contain a great diversity of δ-endotoxins encoding genes. The cloning and sequencing of the first insecticidal protein encoding genes was published in 1981. Today more than 100 crystal protein gene sequences have been published. Each of the crystal proteins has a specific activity spectrum. For example, the Cry1Ab protein is highly active against the European corn borer and is used in current Bt corn hybrids. The Cry1Ac protein is highly toxic to both tobacco budworm and cotton bollworm larvae and is expressed in the Bt cotton varieties, while the Cry3A protein is expressed in Bt potato varieties and provides protection against Colorado potato beetles.

Both the full length and truncated forms of Bt δ-endotoxins have been introduced in plants, conferring demonstrable resistance to tobacco pests (*M. sexta*), tomato pests (*Heliothis virescens* and *Helicoverpa zea*) and potato pests (*Phthorimeae operculella*). The first plants produced were capable of synthesizing the entire protoxin, but expression of the gene was weak and the resulting small quantity of δ-endotoxin gave little or no insect resistance. Further development ultimately led to optimization of the *cry* gene expression in plants. The first generation insecticidal plants have been introduced to the market for commercial purposes.

**Other microorganism derived resistance genes**: Cholesterol oxidase (CO) protein present in the *Streptomyces* culture filtrate showed acute toxicity to boll weevil larvae This gene has been engineered into tobacco. Isopentenyl transferase (*ipt*) gene from *Agrobacterium tumefaciens* codes for a key enzyme in the cytokinin biosynthetic pathway. Expression of *ipt* in tobacco and tomato by a wound inducible promoter has resulted in a decrease in leaf consumption by the tobacco hornworm (*M. sexta*) and reduced survival of the peach potato aphid (*Myzus persicae*).

### Resistance genes from higher plants

With Bt toxins being successfully engineered into crops, efforts are directed towards discovery of non-*Bt* toxin genes having insecticidal activity. A number of non Bt insecticidal proteins interfere with the nutritional needs of the insect. Currently there are two major groups of plant-derived genes that are used to confer insect resistance on crop plants by retarding insect growth and development.

### Proteinase inhibitors

It has been known since 1938 that plants contain peptides acting as protease inhibitors (PIPs). The different proteinases are serine, cysteine, aspartic and metallo proteinases. They catalyze the release of amino acids from dietary protein, thereby providing the nutrients crucial for normal growth and development of insects. The proteinase inhibitors deprive the insect of nutrients by interfering with digestive enzymes of the insect. Two such proteinase inhibitor genes have been described.

### Cowpea trypsin inhibitor gene (CpTl)

CpTl, found in cowpea (*Vigna unguiculata*), is the most active inhibitor identified to date. This inhibitor gene produces antimetabolite substances that provide protection against the major storage pest Bruchid beetel (*Callosobruchus maculatus*). Besides, this gene is also harmful to various Lepidopteran insects (*Heliothis virescens*), Spodopteran insects (*Manduca sexta*), Coleopteran insects (*Callosobruchus, Anthonomus grandis*), Orthopteran insects (*Locusta migratoria*) but is not harmful to mammals. *CpTl* gene has been cloned and constructs containing the CaMV 355 promoter and a full-length cDNA clone 550 bp long were used to transform leaf discs of tobacco. The bioassay for insecticidal activity of transgenic tobacco plants was done with cotton bollworm (*Helicoverpa zea*). Insect survival and plant damage were clearly decreased in transgenic plants compared with control.

### α-Amylase inhibitor

Genes for three α-amylase inhibitors have been expressed in tobacco but the main emphasis has been on transferring the gene of α-amylase inhibitor (αAI-Pv) isolated from adzuki bean (*Phaseolus vulgaris*). It works against *Zabrotes subfasciatus* and *Callosobruchus chinensis*. This α-amylase inhibitor protein blocks the larval feeding in the midgut. The larvae secrete a gut enzyme called α-amylase that digests the starch. By adding a protein that inhibits insect gut α-amylase, the weevil can be starved and it dies.

### Lectins

Lectins constitute another large family of proteins that can be used as insect toxins for genetic engineering of insect resistance. These are plant glycoproteins. Recent interest has mainly concentrated on the lectin from snowdrop (*Galanthus nivalis*), also known as GNA, because it has shown activity against aphids (Down *et al.*, 1996). The gene for this protein has been successfully used in genetic engineering studies and expressed in different species including potato, oilseed rape and tomato. Laboratory tests with modified potato showed that GNA did not increase the mortality but considerably reduced fecundity. An important feature of this protein is that it also acts against piercing and sucking insects. However, one disadvantage is that the protein works well only when it is ingested in large quantities, i.e. when insects are exposed to microgram levels in diet incorporation bioassays. Hence, although a number of lectin encoding genes (wheat germ agglutinin, jacalin and rice lectin) have been expressed in transgenic plants, their insecticidal performance is still very low to make them effective.

### Resistance genes from animals

Resistance genes involved are primarily serine proteinase inhibitors from mammals and the tobacco hornworm (*Manduca sexta*). Based on *in vitro* screening of inhibition of proteolysis by midgut extracts of a range of lepidopteran larvae, bovine pancreatic trypsin inhibitor (BPTI), α-antitrypsin ($\alpha_1$AT) and spleen inhibitor (SI) have been identified as promising insect resistance proteins and have been transferred to a range of plants. However, initial results with potato tuber moth in transgenic potato plants are not encouraging. But *Manduca sexta* derived proteinase inhibitors *viz*. anti-chymotrypsin and anti-elastase expressed in cotton and chitinase in tobacco were found to reduce reproduction of *Bemisia tabaci* and *Heliothis virescens* respectively.

A few insect resistant genes used for generating transgenic plants have been shown in Table 25.1. Insect tolerant transgenic crops covered 18% of the total 114.3 million hectares of the global area under transgenics in 2007. It was mainly due to Bt cotton with 10.8 million hectares (9%) and Bt maize with 9.3 million hectares (7%) (James, 2007).

### Virus Resistance

The development of molecular strategies for the control of virus diseases has been especially successful because of the relatively small genomic size of plant viruses. There are a number of different strategies for using molecular technology to integrate or create new resistance factors in plant virus systems. The approach is to identify those viral genes or gene products, which when present at an improper time or in the wrong amount, will interfere with the normal functions of the infection process and prevent disease development.

### Coat protein mediated cross protection

This is based on the concept of cross protection which is the ability of one virus to prevent or inhibit the effect of a second challenge virus. If

**Table 25.1:** A list of few transgenic plants conferring resistance against insects

| Crop | Gene transferred | Insect(s) controlled |
|---|---|---|
| Tobacco | Bt from *B. thuringiensis* | *Manduca Sexta* |
| | Truncated *cryl* from *B. thuringiensis* | *M. Sexta* |
| | Bt | *M. Sexta* |
| | *CpTl* | *Heliothis armigera* |
| | *CpTl* | *M. Sexta* |
| | Insecticidal protein from *Streptomyces* | Boll weevil *(Spodoptera litura)* |
| | Sweet potato trypsin inhibitor gene | *Spodoptera litura* |
| Tomato | Bt | *Heliothis armigera* |
| Potato | *cry III* from *B. thuringiensis* | Colorado potato beetle *(Leptinotarsa decenlineata)* |
| | Modified *cry III* gene | *L. decenlineata* |
| | Snowdrop lectin (GNA) | Tomato moth *(Lacanobia oleracea)* |
| Cotton | Bt | *Spodoptera* |
| | Bt | Cotton boll worm |
| Pea | *αAl* from bean | Bruchus beetle |
| | *αAl* from bean | Pea weevil |
| Rice | Bt | - |
| | Bt *cryl* gene of *B. thuringiensis* | Striped stem borer |
| | Corn cysteine gene | Coleopteran *(Sitophilus zeamis)* |
| Maize | Bt *cry* II | European corn borer |
| Sugarcane | Bt *cry* I | Sugarcane borer *(Diatracea saccharis)* |

*Bt*–*B. thuringiensis*; *CpTl*–Cowpea trypsin inhibitor; *αAl*–alpha amylase inhibitor.

susceptible strain of a crop is inoculated with a mild strain of a virus, then the susceptible strain develops resistance against more virulent strain. Powell-Abel *et al.* (1986) first demonstrated that transgenic tobacco expressing tobacco mosaic virus (TMV) coat protein showed resistance similar to that occurring in viral mediated cross protection. Since then, a number of coat protein genes from different virus groups have been found to provide resistance when expressed in transgenic plants. Coat protein mediated resistance in many systems is correlated with the inhibition of virus replication at the initial point of infection. This resistance takes the form of reduced numbers of infection sites on inoculated leaves, suggesting that an initial step in the virus life cycle has been disrupted. It has been demonstrated that TMV cross protection may result from the coat protein of the protecting virus preventing uncoating of the challenge virus RNA. Most of the systems in which coat protein mediated resistance has been reported are directed against plus-sense RNA viruses with a single capsid protein. This approach has been used in several crops such as tobacco, tomato,

potato, alfalfa, melons, sugarbeet, squash, rice, maize etc. Kouassi *et al.* (2006) showed resistance in rice against rice yellow mottle virus by expressing its coat protein gene (Table 25.2).

An important negative strand virus is tomato spotted wilt virus (TSWV). In this virus, genomic RNA is tightly associated with nucleocapsid (N) protein. This protein helps in wrapping of viral RNA and also in regulation of transcription-to-replication switch during the infection cycle. Transgenic plants have been developed in tobacco and tomato.

*Non-structural protein mediated resistance*

Viruses encode non-structural proteins that are necessary for replication. Recently, several of these non-structural 'replicase' proteins have been found to provide a high degree of resistance to virus infection when expressed in transgenic plants. Golemboski *et al.* (1990) first demonstrated this phenomenon by expressing the 54 kDa open reading frame (ORF) of TMV in transgenic tobacco. Transgenic tobacco has been developed against pea early browning virus (PEBV) and potato virus X (PVX). Pinto *et al.*

**Table 25.2:** Virus resistant transgenic plants obtained in various species

| Resistance due to coat protein gene | | |
|---|---|---|
| Tobacco | CP-TMV | Corresponding virus |
|  | **C**P-AMV and TRV | ,, |
|  | CP CMV | ,, |
|  | CP-TVMV | TVMV,TEV,PVY |
| Potato | CP-PVX | Corresponding virus |
|  | CP-PVX and PVY | ,, |
|  | CP-papaya ring spot | ,, |
| Tomato | CP-TMV | ,, |
|  | CP-AMV | ,, |
|  | CP-TYLCV | ,, |
| Soybean | CP-BMV | ,, |
| Rice | CP-RSV | ,, |
|  | CP-RYMV | ,, |
| Citrus and Orange | CP-CTV | ,, |
| Maize | CP-MDMV | MDMV and MCMV |
| Melons | CP-CMV | CMV, ZYMV, Water melon mosaic virus |
| *Lactuca sativa* | CP-LMV | LMV |
| *Prunus domestica* | CP-PPV | PPV |
| **Resistance due to nucleocapsid gene** | | |
| Tobacco | NC-TSWV | TSWV |
| Tomato | NC-TSWV | TSWV |

AMV–Alfalfa mosaic virus; BMV–Bean mottle virus; CMV–Cucumber mosaic virus; CP–coat protein; CTV–Citrus tristeza virus; LMV–Lettuce mosaic virus; MDMV–Maize dwarf mosaic virus; NC–Nucleo-capsid; PLRV–Potato leaf roll virus; PPV–Plum pox virus; PVX–Potato virus X; PVY–Potato virus Y; RSV–Rice strip virus; RYMV–Rice yellow mottle virus; TEV–Tobacco etch virus; TMV–Tobacco mosaic virus; TRV–Tobacco rattle virus; TSWV–Tomato spotted wilt virus; TVMV–Tobacco vein mottling virus; TYLCV–Tomato yellow leaf curled virus; ZYMV–Zucchini yellow mosaic virus.

**Table 25.3:** Virus resistant transgenic plants generated in various crops for pathogen derived resistance

| Crop | Transgene | Transgene mode of action | Virus | Reference |
|---|---|---|---|---|
| Potato | Replicase protein | Competition for enzyme | PVY | Audy *et al.* (1994) |
| Pea | Replicase protein | Competition for enzyme | PSbMV | Jones *et al.* (1998) |
| Rice | Replicase protein | Competition for enzyme | RYMV | Pinto *et al.* (1999) |
| Tobacco | Movement protein | Interference with transport | TMV | Cooper *et al.* (1995) |
| Potato | Movement protein | Interference with transport | PLRV, PVX, PVY | Tacke *et al.* (1996) |
| Tobacco | Transport protein | Interference with transport | TEV | Cronin *et al.* (1995) |
| Potato | Viral protease | Polyprotein processing | PVY | Vardi *et al.* (1993) |
| Potato | Antisense RNA | Blocks viral RNA and prevents translation | PLRV | Kawchuk *et al.* (1991) |
| Tomato | Ribozyme | Cleaves viral RNA | CEVd | Atkins *et al.* (1995) |
| Tobacco | Satellite RNA | Competes for capsids | CMV | Harrison *et al.* (1987) |
| Potato | Antiviral protein ribonuclease | Degrades ds-RNA (viroids) | PSTV | Sano *et al.* (1997) |
| Tobacco | Pokeweed antiviral protein | Inhibits rRNA of 60S subunit | TMV, PVX | Wang *et al.* (1998) |
| Tobacco | 2′,5′ oligoadenylate antiviral protein | Degrades ds-RNA | CMV | Ogawa *et al.* (1996) |

CEVd–Citrus exocortis viroid; CMV–Cucumber mosaic virus; PLRV–Potato leaf roll virus; PSbMV–Pea seed borne mosaic virus; PSTV–Potato spindle tuber viroid; PVX–Potato virus X; PVY–Potato virus Y; RYMV–Rice yellow mosaic virus; TEV–Tobacco etch virus; TMV–Tobacco mosaic virus.

(1999) developed rice yellow mosaic virus resistant transgenic plants of rice by expressing replicase gene. Blocking of viral movement through the expression of defective viral movement protein to impede viral infection has been tested in transgenic tobacco (Cooper et al., 1995) and other crops. A similar approach was used to inhibit long distance transport of potyvirus, tobacco etch virus (TEV), by engineering helper component proteinase in tobacco (Cronin et al., 1995). Some of the examples belonging to different categories of resistance besides coat protein mediated resistance are given in Table 25.3.

## Antisense and sense mediated resistance

Another pathogen derived strategy that has been investigated for the control of plant viruses is the transgene expression of antisense and more recently sense segments of viral RNAs. The principle of antisense strategy is to bind up viral RNA with complementary RNA sequences expressed by the plant. Inappropriate RNA-RNA base pairing would potentially prevent accessibility of the viral RNA for replication or gene expression. Thus, antisense and sense constructs could be used to block initial steps important in the establishment of viral infection. Antisense protection has been demonstrated in tobacco expressing complementary RNA to the coat protein

RNA mediated resistance has been demonstrated using positive strand, sense defective constructs. The start codon of the nucleocapsid gene of the negative stranded tospovirus, TSWV was altered so that the mRNA could no longer be translated. Transgenic plants expressing this construct were found to be resistant to TSWV infection at a comparable level to those expressing the translatable mRNA.

## Satellite RNA protection

Satellite RNAs are a class of small (approximately 300 nucleotides), single stranded RNA molecules that are dependent upon a helper virus for replication and virion packaging to cause infection elsewhere. Therefore, satellite RNA depends on virus for its replication and transmission, even though it is unrelated to viral genome. These satellite RNA species have been associated with several different viruses. A number of satellite RNAs have been shown to modulate the replication and symptoms of their helper viruses. Changes in symptom development range from severe necrosis to almost complete symptom attenuation, depending on the associated satellite RNA. Thus, satellite RNAs that attenuate symptoms can be potentially used to reduce the disease severity of the helper virus. Hence its use in transgenics to confer resistance in crops finds an important place. Tien et al. (1987) and Tien and Gusui (1991) demonstrated that the deliberate inoculation of a mild strain of CMV (Cucumber mosaic virus) with a symptom attenuating satellite RNA successfully protected tobacco, pepper, and tomato and cucumber plants from a virulent strain of CMV and reduced yield losses. Tien and Gusui (1991) have reported that 121 transgenic tomato plants expressing an attenuating CMV satellite RNA gave 50% increase in yield over control plants when injected with a severe strain of CMV. This strategy is limited to those virus systems in which attenuating satellite RNAs are found. Kim et al. (1997) generated hot pepper (Capsicum annuum) plants that express CMV satellite RNA. Symptom attenuation in the offspring was confirmed upon inoculation with CMV-Y or CMV-Korea strains.

## Defective interfering RNAs

Defective interfering RNAs (DI RNAs) are similar to satellite RNAs in that they are dependent on the helper virus for replication, use the helper virus coat protein, and can cause an attenuation of symptoms. However, DI RNAs do shares homology to the helper virus as they are directly derived from the viral genome, but have discontinuities in the sequence. Transgenic plants producing DI molecules reduce viral replication and symptom development e.g.

African cassava mosaic virus (ACMV). When a cloned subgenomic form comprising half DNA B of ACMV was introduced into the *Nicotiana benthamiana* genome as a tandem repeat, the transgenic plant was tolerant to ACMV infection and symptoms were reduced on subsequent transfer of virus. The amelioration of symptoms was associated with the release and replication of the subgenomic (defective) DNA B. It was suggested that this mobilized DI is amplified at the expense of full-sized DNA B. This phenomenon was specific to ACMV and other gemini viruses like beet curly top virus.

### Pathogen targeted resistance

*Anti-viral proteins*: A class of polypeptides variously called anti-viral or ribosome-inactivating proteins (RIPs) have been identified in a number of plant species (Stirpe *et al.*, 1992) of which the best known source is pokeweed (*Phytolacca americana*). Three distinct pokeweed antiviral proteins (PAPs) have been identified. The ribosome inhibiting function is due to their ability to modify ribosomal RNA and thereby to interfere with polypeptide translation. Lodge *et al.* (1993) generated transgenic plants of tobacco and potato with PAP isolated from *Phytolacca* which were resistant to PVX, PVY and CMV. Transgenic tobacco expressing pokeweed antiviral protein was found to be resistant to TMV (Wang *et al.*, 1998).

*Mammalian oligoadenylate synthase gene*: In mammals, virus infections are fought via induction of the interferon system. Interferons induce additional proteins that defend the animal directly against viruses. This has been exploited to confer broad based resistance to viral infection in plants. Interferons are secreted by animal cells during cell proliferation, in response to immunological challenges and particularly in response to viral infections, but they do not themselves possess antiviral activity. Interferons induce the synthesis of additional proteins that directly lead to inhibition of virus multiplication. One of these proteins is $2' \rightarrow 5'$ oligoadenylate synthase. This enzyme is activated by double stranded RNA (dsRNA), the replication intermediate of RNA viruses. After activation, the enzyme polymerizes ATP to $2' \rightarrow 5'$ oligoadenylate (2–5A) which in turn activates a latent endoribonuclease (RNase L), an enzyme that degrades ribonucleic acid. A similar enzyme is detectable in plants after TMV infection. Truve *et al.* (1994) showed that expression of $2' \rightarrow 5'$ oligoadenylate synthase from mammalian rat could confer considerable protection against potato virus X in potatoes. They subsequently confirmed their results in field studies and also discovered that their technique confers broad resistance against other RNA viruses also. Ogawa *et al.* (1996) generated transgenic tobacco expressing the mammalian $2' \rightarrow 5'$ oligoadenylate synthase for resistance against cucumber mosaic virus.

*Ribozyme mediated resistance*: Ribozymes are essentially RNA based RNA restriction enzymes capable of catalytically cleaving RNA molecules at specific sites. Although the cleavage is intramolecular, the catalytic domain (a hairpin or hammerhead structure respectively) and flanking antisense arms can be designed to cleave a specific target RNA in trans (before or after a GUC triplet, respectively). The hammerhead will also cleave 3' of GUA or GUU. Ribozymes can be visualized as warheaded antisense RNAs. The ability to direct ribozyme cleavage provides a potentially useful strategy to control plant virus diseases, especially since the majority of agriculturally important plant viruses have RNA genomes. Thus, transgene expression of ribozymes designed to cleave viral RNAs could be used to disrupt viral replication and disease development.

Several different types of ribozymes with different sequences and structures have been identified. Edington and Nelson (1992) have tested the ability of a ribozyme to confer resistance to virus infection *in vivo* in a protoplast system. Results demonstrated that a ribozyme directed against the replicase ORF of TMV was effective at reducing viral accumulation in protoplasts by as much as 90% in the first 24 h post-infection. Steinecke *et al.* (1992) were able to show that a ribozyme can be employed to suppress the expression of a model gene in plant cells. Further experiments in tobacco have shown that this is possible in transgenic plants (Wegener *et al.*, 1994). The ability of ribozyme to cleave target RNA has also been utilized to develop citrus exocortis viroid (CEVd) resistant transgenic tomato (Atkins *et al.*, 1995). Control of viroids has also been achieved by expressing a double stranded RNA specific ribonuclease from yeast in transgenic potatoes for resistance to potato spindle tuber viroid (PSTV) (Sano *et al.*, 1997). Han *et al.* (2000) reported ribozyme mediated resistance to rice dwarf virus in rice plants.

## Disease Resistance

A large number of plant defense response genes encoding anti-microbial proteins have now been cloned. Most of these are transcriptionally activated in response to infection or exposure to microbial elicitor macromolecules. The products of defense response genes may include (i) hydrolytic enzymes, e.g. chitinase, 1–3 β-D glucanase and other pathogenesis related (PR) proteins, (ii) ribosome inactivating proteins (RIPs), (iii) antifungal proteins (AFPs), (iv) biosynthetic enzymes for the production of anti-microbial phytoalexins, (v) wall-bound phenolics, osmotins, thionins, lectins etc. and (vi) hydrogen peroxide.

### Pathogenesis related proteins

These are low molecular weight proteins, which accumulate to significant levels in infected plant tissues. The major classes of PR proteins are tobacco PR-1, PR-2 (β 1–3 glucanase), PR-3 (chitinases), PR-4 (hevein like) PR-5 (thaumatin like and osmotin), etc. The ability of hydrolytic enzymes to break chitin and glucan in the cell walls of fungal pathogen has been exploited to develop crop resistance to pathogens. Various chitinase genes of plants have been isolated and characterized. The first report was in tobacco where a bacterial chitinase gene obtained from soil bacteria (*Serratia marcescens*) was stably integrated and expressed in tobacco leaves (Jones *et al.*, 1988). A basic chitinase gene of bean (*Phaseolus vulgaris*), under the control of strong constitutive promoter of CaMV 35S has been constitutively expressed to high levels in transgenic plants of tobacco and *Brassica napus* (Broglie *et al.*, 1991). This expression resulted in significant protection of the plants from post-emergent damping off caused by the pathogen *Rhizoctonia solani*. In the case of *B. napus*, although the protection was a delay rather than complete inhibition of symptoms, it was concluded that the level of protection was sufficient to be of economic significance in field situations.

There have been no reports of increased resistance from expression of 1–3 β-D-glucanase gene alone in transgenic plants. But glucanase gene when present with chitinase showed fungal resistance in tobacco, tomato, carrot, etc.

### Anti-microbial proteins

Plants and other organisms may contain anti-microbial proteins which are not necessarily associated with induced defense response, but the presence of these proteins exhibit resistance to pathogens. These are ribosome inactivating proteins (RIPs), cysteine rich proteins such as lectins, defensins; thionins, lysozyme, polygalacturonase inhibitors, etc. Transgenic plants showing resistance to pathogens have been generated in various species (Table 25.4).

Barley α-thionin gene when transferred to tobacco showed resistance against *Pseudomonas syringae* pv. *tabaci* and *P. syringae* pv *syringae* (Anzai *et al.*, 1989). Antibacterial proteins of non-plant origin include lytic peptides, lysozymes and iron sequestering glycoproteins. Lytic peptides are small proteins with an amphipathic-α-helical structure whose effect is to form pores in bacterial membranes (e.g. cecropin, attacin, etc.) Cecropins have been expressed in transgenic potato and tobacco (Mourgues *et al.*, 1998) and attacins in apple

**Table 25.4:** Transgenic plants generated in various crops for resistance to fungal and bacterial diseases

| Crop | Gene transferred | Controlled pathogen |
|---|---|---|
| **PR proteins** | | |
| Tobacco | Bacterial chitinase from *Serratia marcescens* | *Alternaria longipes* |
| | Bean chitinase gene | *Rhizoctonia solani* |
| | PR-1-a gene | *Peronospora tabacina Phytophthora parasitica* var. *nicotianae* |
| | Chitinase | *Sclerotinia sclerotiorum* |
| | Chitinase | *Rhizoctonia solani* |
| | Chitinase and 1,3-β glucanase | *Cercospora nicotinae* |
| Tomato | Chitinase and 1,3-β glucanase | *Fusarium oxysporum lycopersici* |
| *Brassica napus* | Chitinase | *Rhizoctonia solani* |
| *Brassica napus* var. *oleifera* | Chitinase | *Cylindrosporium concentricum; Phoma lingam; Sclerotinia sclerotiorum* |
| Rice | Chitinase | *Rhizoctonia solani* |
| Carrot | Chitinase and 1,3-β glucanase | *Alternaria dauci, Alternaria radicina, Cercospora carotae, Erysiphe heraclei* |
| Potato | PR5 | *Phytopthora infestans* |
| Potato | 1,3-β glucanase | *Phytopthora infestans* |
| Kiwi fruit | 1,3-β glucanase | *Botrytis cinerea* |
| **Anti-microbial proteins** | | |
| Tobacco | Barley RIP (ribosome inactivating protein) | *Rhizoctonia solani* |
| Tomato | Prohevein from *Hevea brassiliensis* | *Trichoderma hamatum* |
| Tobacco | Defensin - Rs AFP2 from radish | *Alternaria longipes* |
| Tobacco | Barley α thionin gene | *Pseudomonas syringae* pv. *tabaci; P. syringae* pv. *syringae* |
| Tobacco | Cecropin | *P. syringae* pv. *tabaci* |
| Rice | Cecropin | Bacterial pathogen |
| Potato | Bacteriophage T4 lysozyme | *Erwinia carotovora* subsp. *Atroseptia* |
| Tobacco | Hen egg white lysozyme (HEWL) | *Botrytis cinerea, Verticillium albo-atrum, Rhizoctonia solanum* |
| Tobacco | Lysozyme from human being | *Pseudomonas syringae* pv. *tabaci; Erysiphe cichoracearum* |
| Potato | $H_2O_2$ gene for glucose oxidase | *Verticillium dahlea, Phytophthora; Erwinia carotovora* |
| Tobacco | Cryptogein from *Phytophthora cryptogea* | *Erysiphe cichoracearum, Botrytis cinerea* |
| **Phytoalexins** | | |
| Tobacco | Stilbene synthase | *Botrytis cinerea* |
| *Brassica napus* | Stilbene synthase | — |
| Rice | Stilbene synthase | *Pyricularia oryzae* |

plants (Norelli *et al.*, 1994, 1996). Cecropin gene casette when transferred to tobacco and rice showed resistance against *P. syringae* and bacterial pathogens. Expression of a bacterial gene, *bacterioopsin* (*bo*) encoding a proton pump in transgenic tobacco resulted in complete resistance to *P. syringae* pv. *tabaci* (Mittler *et al.*, 1995).

Bacteriophage T4 lysozyme gene trans-ferred to potato showed resistance to *Erwinia carotovora* subsp *atroseptia* (During *et al.*, 1993). The human lysozyme, which is assembled by stepwise ligation was introduced into the tobacco. The transgenic tobacco plants showed enhanced resistance against *Erysiphe cichoracearum* and *P. syringae*. Thus, the introduction of a human lysozyme gene is an effective approach to protect crops against both fungal and bacterial diseases.

Active oxygen species (AOS) including $H_2O_2$ also has defense mechanism. Transgenic potato expressing a $H_2O_2$ generating fungal gene for glucose oxidase was found to have high levels of $H_2O_2$ and enhanced levels of resistance both to fungal and bacterial pathogens (Wu *et al.*, 1995). Xiangbai *et al.*, (2008) reported development of transgenic oilseed rape (*Brassica napus*) showing resistance to *Sclerotinia sclerotiorum* by introduction of oxalate oxidase (OXO) gene. Oxalic acid (OA) secreted by the pathogen is a key pathogenicity factor. Oxalate oxidase (OXO) can oxidize OA into $CO_2$ and $H_2O_2$. Transgenic oilseed rape (sixth generation lines) constitutively expressed wheat OXO and enhanced resistance to *S. sclerotiorum* (with up to 90.2 and 88.4%) disease reductions compared with the untransformed parent line were obtained. The enhanced resistance of the OXO transgenic oilseed rape to *Sclerotinia* is probably mediated by OA detoxification.

Molina *et al.* (1997) purified lipid transfer protein 2 from barley and the transgenic tobacco plants with this gene showed retarded growth of *P. syringae* as compared to non-transformed controls.

A different approach is offered by the fungal endo α 1,4-D polygalacturonases, which are partly responsible for dissolution of the plant cell wall and are presumably essential for efficient colonization. All dicotyledons that have been investigated so far possess special inhibitors of these endopolygalacturonases (polygalacturonase inhibiting proteins, PGIPs), which show no activity against the endogenous pectinases of the plant. The cloning of a bean PGIP (Toubart *et al.*, 1992) offers the possibility of overexpression in transgenic plants and of potential inhibition of fungal development.

### Engineering toxin insensitivity

The molecular targets of several fungal or bacterial toxins from plant pathogens are now known. Toxin inactivating enzymes have been successfully used to engineer resistance. The bacterial halo blight pathogen of bean, *Pseudomonas phaseolicola*, produces a tripeptide toxin, phaseolotoxin, which causes the chlorotic halos. Phaseolotoxin inhibits the enzyme ornithine transcarbamylase (OC). Bacteria have been selected which contain a phaseolotoxin-insensitive OC and the gene encoding this enzyme has been cloned and transferred to tobacco, where its expression has been shown to prevent the symptoms caused by application of the toxin (De la Fuente *et al.*, 1992). Recently transgenic sugarcane was developed against leaf scald disease. The systemic, xylem invading pathogen, *Xanthomonas albilineans* produces a family of low molecular weight toxins (albicidins) that selectively block prokaryote DNA replication and causes the characteristic chlorotic symptoms by blocking chloroplast development. Transgenic sugarcane expressing an albicidin detoxifying gene (*alb*) from *Pantoea dispersa*, a bacterium that provides biocontrol against leaf scald disease under the control of maize ubiquitin promoter developed resistance to leaf scald disease (Zhang *et al.*, 1999).

### Phytoalexins

Phytoalexins are low molecular weight anti-microbially active secondary metabolites synthesized by the plant in response to an infection, which contribute to the resistance of plants to diseases. During infection, the stored phytoalexins (usually present in special cells or organelles or in a conjugated inactive form) are mobilized, while genes for biosynthetic pathways are induced and the synthesis of more phytoalexin begins. Resveratrol is one of the commonest stilbene (phytoalexin) synthesized in some species. The key enzyme in resveratrol synthesis is resveratrol synthase (often known as stilbene synthase or STS).

Hain *et al.* (1990) demonstrated that introduction of a stilbene synthase (STS) gene from peanut to tobacco resulted in measurable production of peanut stilbene resveratrol and proved the fungitoxic action of resveratrol in plants. STS gene has also been transferred to rice and *Brassica napus.*

*Manipulation of disease resistance genes*

Attempts to isolate disease resistance and avirulence genes have gained momentum in the past few years primarily because of the development of map based cloning and gene tagging strategies. HM1 gene from maize, which confers resistance to *Cochliobolus carbonum,* has been cloned by transposon tagging (Johal and Briggs, 1992). This gene encodes NADPH dependent HC-toxin reductase that inactivates the fungal HC toxin. Resistant genes like *Arabidopsis Rps2* and *RPM*1; *Pto, Cf9, Cf2, Cf4* from tomato; tobacco *N* gene; flax *L6* and rice *Xa*21 have been cloned. A number of avirulence genes have also been cloned *viz.* A*vr*9 and *Avr*4 of *Cladosposium fulvum, NIP*1 of *Rhynchosporium secalis,* etc.

Introduction of resistance (R) gene from a plant variety resistant to a certain pathogen into susceptible varieties is one of the strategies. Martin *et al.* (1993) developed tomato plants with *Pto* resistance gene that confers resistance against *P. syringae* pv *tomato.* Tobacco plants transgenic for *Pto* were resistant to *Pseudomonas syringae* pv *tabaci* expressing *avrPto.* Rice *Xa*21 gene confers resistance to over 30 distinct strains of the bacterium *Xanthomonas oryzae* pv o*ryzae* that causes leaf blight in rice. Song *et al.* (1995) generated rice transgenic plants resistant to *Xanthomonas* expressing *avrXa*21.

The strategy of genetic engineering hypersensitive cell death in response to fungal pathogen attack at the site of infection has been employed successfully. Barnase-barstar system has been used in the introduction of cytoplasmic male sterility. Strittmatter and co workers (1995) developed a strategy for disease resistance in which barnase, a cytotoxic protein with RNase activity, and its inactivator (barstar) are used. The genes encoding these proteins are derived from *Bacillus amyloliquefaciens.* The barnase gene was placed under the control of potato *prp*1–1 promoter, which is activated upon pathogen attack while barstar gene under the CaMV 35S constitutive promoter. Cells are killed only if the barnase activity is higher than barstar activity. Transgenic potato lines expressing both genes showed reduction in symptom development of *Phytophthora infestans.*

Induction of hypersensitive reaction with the production of an elicitor is response to infection by pathogen was reported by Keller *et al.* (1999). Transgenic tobacco plants were generated harboring a fusion between the pathogen inducible tobacco *hsr203J* gene promoter and a *Phytophthora cryptogea* gene encoding the highly active elicitor cryptogein. Under non-induced conditions the transgene was silent but upon infection by virulent fungal pathogen, cryptogein production was stimulated which coincided with the fast induction of several defense genes at and around the infection site. Induced elicitor production resulted in a localized necrosis that restricted further growth of the pathogen. The transgenic plants displayed enhanced resistance to fungal pathogens that were unrelated to *Phytophthora* species such as *Thielaviopis basicola, Erysiphe cichoracearum* and *Botrytis cinerea.*

# RESISTANCE TO ABIOTIC STRESSES

Almost all the abiotic stresses such as drought, low temperature, salinity and alkalinity adversely influence growth and induce senescence, leading to cell death or reduced crop yield. Several abiotic stresses such as drought, salinity and extreme temperatures have a common consequence of causing cellular water deficit or osmotic stress. The response of plants to water deficit is therefore the synthesis and accumulation of low molecular weight compounds called as osmoprotectants. These osmoprotectants within cells lower the osmotic potential and helps maintain turgor. Compatible solutes form a diverse group of compounds, encompassing inorganic ions, organic ions, soluble carbohydrates including polyols (sugar, alcohols), amino acid (proline) and quaternary ammonium compounds such as glycine betaine. Several plant genes that encode key enzymes of osmolytes biosynthetic pathways including sugars, alcohols, glycine betaine and proline have been cloned.

A strong correlation has been reported between proline accumulation with tolerance to drought and salinity stress conditions. γ-Pyrroline-5-carboxylate synthetase (P5CS) is the limiting enzyme in the synthesis of proline. Transgenic tobacco plants were generated with P5CS enzyme and they showed enhanced expression of this enzyme and 10–18 fold more proline. The increase in proline concentration correlated with enhanced growth under salt and drought conditions (Kishore *et al.*, 1995). The ability to synthesize and accumulate glycine betaine is widespread among angiosperms and contributes towards drought and salt tolerance (Fig. 25.1). In plants, glycine betaine is synthesized by the two-step oxidation of choline via the intermediate betaine aldehyde, the reaction being catalyzed by choline monooxygenase (CMO) and betaine aldehyde dehydrogenase (BADH). Transgenic plants of tobacco, *Arabidopsis* and rice with BADH have shown increased salt resistance (Table 25.5).

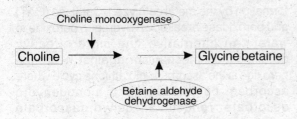

**Fig. 25.1:** Pathway of synthesis of glycinebataine.

An *E. coli* bacterial gene *mtl*D (mannitol-1-phosphate dehydrogenase) involved in the biosynthesis of mannitol was introduced into tobacco and the transgenic plants were tested for salt resistance. After 30 days of exposure to salinity, transgenic plants producing mannitol had a greater shoot height and initiated new and longer roots, while those of the control plants turned brown and did not elongate or branch. Although the improved resistance was not significant enough to be applicable to agriculture, this was the first example of a transgenic plant, altered with a microbial gene, showing a greater resistance to osmotic stress (Tarczynski *et al.*, 1993). Another alteration in osmotic stress resistance has been demonstrated in tobacco plants engineered to express a bacterial gene for fructan biosynthesis (Pilons-Smits *et al.*, 1995). Transgenic plants accumulating fructans exhibited faster growth rates when hydroponically cultured with 10% PEG mediated drought stress. Osmotic stress induces the accumulation of a set of low molecular weight proteins known as dress proteins in plant tissues such as LEAS (late embryogenesis abundant). A barley *lea* gene HVA1 showed stress protection in transgenic rice (Xu *et al.*, 1996). Research has been conducted to identify dehydration responsive transcription factors (DREB) that mediates transcription of several genes in response to cold and water stress. DREB genes which are induced by cold and water stress encode transcription factors that bind to the *cis* acting promoter element (DRE) of stress related genes and turn on their expression (Smirnoff and Bryant, 1999). The binding initiates synthesis of gene products implicated in plant acclimation responses to low temperature and water stress. Overexpression of a fusion of a DRE containing promoter from a dehydration induced gene (*rd29A*) with a DREB gene (*DREB1A*) in *Arabidopsis* resulted in marked increase in the transgenic plants tolerance to freezing, water stress and salinity (Kasuga *et al.*, 1999).

Lee *et al.* (1995) produced high temperature stress tolerant transgenic *Arabidopsis* plants by altering the level of expression of heat shock proteins (HSPs) through change in the expression levels of *Arabidopsis* heat shock protein transcription factor (AtHSF). It was shown

**Table 25.5:** Transgenics produced for various abiotic stresses

| Plant | Gene | Transgenic over-expressing | Claim |
|-------|------|----------------------------|-------|
| Tobacco | Mannitol 1-phosphate dehydrogenase (*mtl*D) from *E . coli* | Mannitol | Tolerance to salinity |
| *Arabidopsis* | *Mtl*D from *E. coli* | Mannitol | Tolerance to salinity |
| Tobacco | *Sac*B from *bacillus subtilis* | Fructan | Resistance to drought |
| Tobacco | TPS1 subunit encoding trehalose synthase from *E. coli* | Trehalose | Tolerance to drought |
| Tobacco | γ - Pyrroline-5-carboxylate synthetase | Proline | Tolerance to osmotic stress |
| Rice | Barley *lea* gene (*HVA*I) | LEA | Tolerance to water deficit and salinity |
| Tobacco | *Bet*A from *E. coli* encoding choline dehydrogenase | Glycine betaine | Tolerance to salinity |
| Rice | *Cod*A from *Arthrobacter globiformis* encoding choline oxidase | Glycine betaine | Tolerance to salinity |
| *Arabidopsis* | *Cod*A from *Arthrobacter globiformis* encoding choline oxidase | Glycine betaine | Tolerance to salinity and cold stresss |
| Tobacco | Δ9-desaturase (Des9) from *Anacystis nidulans* | Desaturase enzyme | Tolerance to cold |
| Tobacco | ω-3 fatty acid desaturase (*Des9*) | Desaturase enzyme | Chilling stress |
| Tobacco | O-3 fatty acid desaturase (*Fad7*) | Desaturase enzyme | Tolerance to cold |
| Tobacco | Cu/Zn superoxide dismutase from rice | Scavenge toxic oxygen species | Oxidative and chilling stress |
| Tobacco | Mn-SOD from *N. plumbaginifolia* | Scavenge toxic oxygen species | Oxidative stress |
| Tobacco | Antifreeze protein (*Afp*) from Winter flounder fish | Inhibits the formation of ice | Freezing stress |
| Tobacco | Hemoglobin (*VHb*) gene from *Vitroscilla stercoraria* | Promote glycolytic flux through NADH oxidation | Hypoxia and anoxia stress |
| Maize | Hemoglobin (Hb) gene from barley | Promote glycolytic flux through NADH oxidation | Hypoxia stress |

that AtHSF of *Arabidopsis* is constitutively expressed but its activity for DNA binding, trimer formation and transcriptional activation of *Hsp* genes is repressed at normal temperature. Malik *et al*. (1999) have reported increase in thermotolerance in transgenic carrot cell lines and plants by constitutive expression of carrot *hsp17.7* gene driven CaMV35S promoter.

Overproduction of active oxygen species in plant cells is a general consequence of drought and salt stresses. Hence, a plant's ability to scavenge these toxic oxygen species is considered to be critical for abiotic stress tolerance. Plants have evolved various protective mechanisms. Among the enzymatic antioxidants, superoxide dismutase (SOD) catalyses the dismutation of superoxide to

hydrogen and molecular oxygen, which maintains a low steady state concentration of superoxide and minimizes hydroxyl radical formation by the metal catalyzed. Catalase (CAT) and peroxidase breakdown hydrogen peroxide to water. An alternative and more effective detoxification mechanism against hydrogen peroxide is the ascorbate-glutathione cycle where ascorbate peroxidase (AP), monodehydro ascorbate reductase, dehydroascorbate reductase and glutathione reductase (GR) work together. Genes or cDNAs encoding AP, CAT, SOD and GR from several crops have been isolated.

Overexpression of mitochondrial MnSOD in chloroplasts conferred tobacco paraquat tolerance (Bowler *et al.*, 1991). Similarly, potato plants overexpressing tomato chloroplast Cu/ZnSOD (Perl *et al.*, 1993), and tobacco plants overexpressing pea chloroplastic Cu/ZnSOD (Sen Gupta *et al.*, 1993) or MnSOD (Schake *et al.*, 1995) showed improved tolerance to oxidative stress. McKersie *et al.* (1996) reported that transgenic alfalfa expressing MnSOD suffered reduced injury from water deficit stress. Enhancement of oxidative stress tolerance in transgenic tobacco plants overproducing Fe-superoxide dismutase in chloroplast has been demonstrated.

Transgenic plants overexpressing cytosolic AP, but not chloroplastic AP are tolerant to paraquat (Pitcher *et al.,* 1994). Similar results are also reported in transgenic tobacco overexpressing a pea cytosolic AP (Webb and Allen, 1995). Transgenic tobacco plants with enhanced GR activity possessed increased resistance to photooxidative stress, whereas transgenic GR deficient plants were less tolerant to paraquat than the control plants (Aono *et al.*, 1995).

Scavenging of hydroxyl radicals: Protection of sensitive metabolic reactions through maintaining the structures of protein complexes or membranes by an increased capacity for hydroxyl radical scavenging, may be an important strategy for engineering tolerance to water stress. Ferritin provides iron for the synthesis of iron proteins, such as ferredoxin and cytochromes, and prevents damage from the free radicals produced by iron/oxygen. Goto *et al.* (1999) developed transgenic rice plants with iron storage ferritin protein which were more tolerant to oxidative damages.

Oxygen deprivation, whether complete (anoxia) or partial (hypoxia) is detrimental to most species of higher plants. This can occur during periods of flooding, when the roots are covered by water. During the following post-anoxic stages several toxic oxidant products are frequently formed. Efforts are directed towards introduction of genes encoding proteins that increase the intracellular oxygen concentration and/or its transport. Constitutive expression of non-symbiotic hemoglobin gene from barley (*Hb*) and bacteria *Vitroscilla* (*VHb*) in maize (Sowa *et al.*, 1998) and tobacco (Holmberg *et al.*, 1997) resulted faster germination, higher growth rates and modified formation of secondary metabolites. Expression of *Hb* in maize helped in maintaining energy status under anaerobic conditions. This result showed that nonsymbiotic hemoglobin acts in plants to maintain the energy status of cells in low oxygen environments by promoting glycolytic flux through NADH oxidation resulted in increased substrate level phosphorylation (Bulow *et al.*, 1999).

Kodama *et al.* (1994) have also reported genetic enhancement of cold tolerance by expression of a gene for chloroplast w-3 fatty acid desaturase in transgenic tobacco. Ishizaki-Nishizawa *et al.* (1996) introduced a desaturase gene from cyanobacteria into tobacco, which led to the introduction of a double bond at the 6-carbon position of linoleic acid to produce linolenic acid. This resulted in increased chilling tolerance in tobacco seedlings. Expression of fish anti-freeze proteins in transgenic plants has been demonstrated, but without a correlation to increased freezing tolerance.

## HERBICIDE RESISTANCE

The use of herbicides to control weeds plays a pivotal role in modern agriculture. A major effort has been devoted in several laboratories to engineer herbicide resistant plants. More progress has been achieved in herbicide resistance as single genes govern the resistance. Three approaches have been followed: (i) Over-production of a herbicide sensitive biochemical target; (ii) Structural alteration of a biochemical target resulting in reduced herbicide affinity, and (iii) Detoxification-degradation of the herbicide before it reaches the biochemical target inside the plant cell.

Resistance to glyphosate and sulfonylurea herbicides has been obtained by using genes coding for the mutant target enzymes 5-enolpyruvylshikimate-3-phosphate synthase

**Table 25.6:** Herbicide resistant genes and their action

| Herbicide | Inhibition | Gene | Product/action |
|---|---|---|---|
| Bialaphos/Basta/ Glufosinate | Glutamine synthase | *bar* | Phosphinothricin $\xrightarrow{\text{PAT}}$ Acetyl phosphinothricin |
| Glyphosate | EPSPS | *aroA* | Mutated EPSPS resistant to glyphosate |
| | | *gox* | Glyphosate $\longrightarrow$ Amino ethyl phosphate (Detoxification) |
| Sulfonylurea and imidazolinone | ALS | *Hra, C3, csr1–1, ahas3r* | Mutated ALS resistant to herbicide |
| Bromoxynil | Photosystem II | *bxn* | Bromoxynil $\longrightarrow$ 3,5, dibromo 4-hydroxy benzoic acid |

EPSPS–5 enol pyruvyl shikimate 3 phosphate synthase; ALS–Aceto lactate synthase; PAT–Phosphinothricin acetyl transferase.

(EPSPS) and acetolactate synthase (ALS) respectively. These two enzymes are involved in amino acid biosynthesis pathway. Resistance to glyphosate has been achieved by using *gox* gene (glyphosate oxidase), which detoxifies the herbicide. This gene has been isolated from Achromobacter bacterial strain. Plants resistant to glufosinate ammonium have been obtained by using genes derived from bacteria that encode phosphinothricin acetyltransferase (PAT), which converts phosphinothricin into its acetylated form. Different herbicide resistant genes and their mode of action have been shown in Table 25.6.

Transgenic plants against various herbicides such as phosphinothricin (bialaphos), glyphosate, sulfonylurea, imidazolinones, bromoxynil, atrazine, 2,4-D, sethoxydim, etc. have been generated in different food crops, vegetables, horticultural and ornamentals species. Herbicide resistant transgenic plants have been reported in various crops, but in cotton, flax, canola, corn and soybean these have already been released for commercial cultivation, and the list is growing rapidly. Herbicide tolerant transgenic crops covered 63% of the total 114.3 million hectares of the global area under transgenics in 2007. It was mainly due to herbicide tolerant soybean (51%), maize (6%), canola (5%) and cotton (1%) (James, 2007).

# TRANSGENICS FOR QUALITY

## Transgenics for improved storage

The first approval for commercial sale of a food product was a transgenic tomato 'Flavr Savr' with delayed ripening, developed by Calgene, USA in 1994. Improved storage or long shelf life of tomato that is suitable for food processing can be developed by two approaches using (i) antisense RNA technology or (ii) using gene for 1-aminocyclopropane-1-carboxylic acid (ACC) deaminase, which degrades ACC to ethylene.

*Antisense RNA technology*: An antisense gene is produced by inverting or reverting the orientation of protein coding region of a gene in relation to its promoter. The RNA produced by this gene has the same sequence as the antisense strand of the normal gene (except for T in DNA and U in RNA) and is thus referred to as antisense RNA. When sense (endogenous gene) and antisense gene are present in the same nucleus, the transcription of the two genes yields antisense and sense RNA transcripts, respectively, which are complementary to each other and thus would pair to form double stranded RNA molecules. It would result: (1) mRNA unavailable for translation; (2) double stranded RNA molecules are attacked and degraded by double stranded RNA specific RNases; (3) It may lead to methylation of

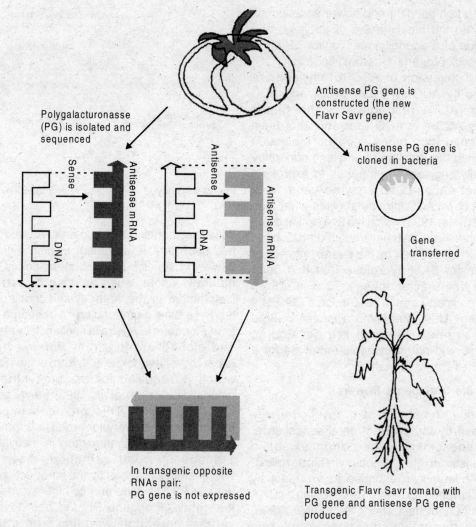

Polygalacturonasse (PG) is isolated and sequenced

Sense

Antisense

DNA

DNA

Antisense mRNA

Antisense

Antisense mRNA

Antisense PG gene is constructed (the new Flavr Savr gene)

Antisense PG gene is cloned in bacteria

Gene transferred

In transgenic opposite RNAs pair:
PG gene is not expressed

Transgenic Flavr Savr tomato with PG gene and antisense PG gene produced

• **Fig. 25.2:** Production of Flavr Savr tomato using antisense RNA gene.

promoter and coding regions of normal genes resulting in silencing of endogenous gene.

Calgene used antisense RNA technology in tomato. An enzyme polygalacturonase (PG) degrades pectin, a component of fruit cell wall which leads to softening of fruit and deterioration in the fruit quality. On the basis of PG gene sequence, an antisense PG gene was constructed and tomato plants were transformed. These transgenic tomato plants produced both sense and antisense mRNA for the PG gene resulting in RNA-RNA pairing. This would result

in non-production of PG gene product, and thus preventing the attack of PG gene upon pectin in the cell wall of ripening fruit and thereby preventing softening of fruit (Fig. 25.2). Calgene gave the brand name MacGregor to its transgenic tomato and as per the company's remark, it can stay on the market shelf for approximately two weeks longer without softening.

***Use of gene for 1-aminocyclopropane-1-carboxylic acid (ACC) deaminase***: The phytohormone ethylene plays a major role in regulating fruit ripening and flower senescence process. In the biosynthesis of ethylene, the conversion of S-adenosyl methionine (SAM) to 1-aminocyclopropane-1-carboxylic acid (ACC) and the conversion of ACC to ethylene are catalyzed by ACC synthase (ACS) and ACC oxidase (ACO) respectively. The genes encoding ACS and ACO (*acs* and *aco*) have been cloned from many species. Monsanto approached the problem of ethylene control by genetically engineering tomatoes to express a gene that encodes an enzyme capable of breaking down ACC into metabolites other than ethylene (Klee, 1993) which led to a similar delay in fruit ripening. Ethylene production has been reduced in tomato fruit by the overexpression of the gene for SAM hydrolase that had been isolated from bacteriophage T3 (Fig. 25.3). Several companies such as Zeneca, DNA Plant Technology USA, Monsanto, Pioneer Hi-Bred and Limagrain's Vilmorin are developing transgenic and non-transgenic tomatoes for a longer shelf-life fresh market.

### Longer life transgenic flowers

The post harvest life of many flowers is determined by the onset of petal senescence. Petal senescence in carnation exhibit a characteristic "in rolling" behavior. This "in rolling" is also observed in response to exogenously supplied ethylene (Cornish, 1993). Using a cDNA clone for ACC oxidase (*aco*), Savin *et al.* (1996) produced transgenic carnations expressing antisense ACC oxidase and producing flowers with little detectable ethylene and a marked delay in petal senescence. This extended vase life of flowers by 200%. The genetically engineered cut flower carnation (*Dianthus carophyllus*) for longer life containing an antisense *aco* gene has been granted approval in Australia and also in European Union (Tanaka *et al.*, 1998).

### Transgenic for flower color and shape

As consumers of flowers are attracted to novel products, flower color has been the target for biotechnology. Anthocyanins are the major

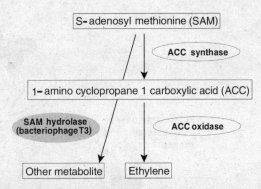

**Fig. 25.3:** Delayed ripening by overproduction of SAM hydrolase.

flower pigments in higher plants. A major target has been the development of blue pigmented flowers in species in which these colors do not naturally occur, such as rose or carnation. Delphinidin is the anthocyanin that normally leads to blue pigmentation. A research group from Florigene in Australia successfully identified and cloned a gene from *Petunia hybrida* encoding a flavonoid 3′ 5′-hydroxylase (F3′5′H), which is required for the biosynthesis of delphinidin (Holton *et al.*, 1993; Stevenson and Cornish, 1993). This provides important information necessary for producing transgenic plants with blue pigmentation in their flowers. Transgenic violet carnations have been successfully produced by the introduction of the petunia F3′5′H gene and its high expression in the petals. Florigene Pty Ltd has developed various carnation varieties with modified flower color namely Moonlite, Moonshade, Moonshadow and Moonvista varieties which also showed herbicide tolerance have been released for commercial cultivation in Australia, Japan and European Union.

Modification of intensity of flower color has also been achieved. White and pink flower varieties have been obtained by introducing sense and antisense chalcone synthase (*chs*) transgenes in to petunia, chrysanthemum, rose, carnation, gerbera, torenia (Suzuki *et al.*, 1997). Flower color has been changed by the introduction of pathway specific transcriptional activators that control the structural genes in anthocyanin biosynthesis. Expression of maize transcriptional factor 'R' (Lc allele) under the control of CaMV 35S promoter elevated the amount of anthocyanin in transgenic Arabidopsis and tobacco (Lloyd et al., 1992). Constitutive expression of Lc gene in petunia resulted in change of leaf color to purple due to accumulation of anthocyanin (Bradley et al., 1998). This approach may allow novel color patterns to be introduced into ornamental species by the controlled expression of anthocyanins and flavonols.

The size and shape of ornamental plants can be modified by alteration of the ratio of cytokinin: auxin in transgenic plants by expressing oncogenes from Ti and Ri plasmids from Agrobacterium tumefaciens and A. rhizogenes respectively. Constitutive expression of the rolC gene encoding a cytokinin β glucosidase from the Ri plasmid into potato leads to dwarf and bushy phenotype transforming potato into an ornamental plant (Schmulling et al., 1988). Similarly introduction of this gene into roses have made them bushy (Tanaka et al., 1998).

## Transgenics for male sterility

Male sterility in plants is inherited both in a nuclear and cytoplasmic manner. Cytoplasmic male sterility (cms) is due to defects in the mitochondrial genome. Generally cms is associated with defective functioning of the tapetum of the anthers that supply nutrients to the developing pollen grains. The female components of these plants maintain fertility. Transgenics with male sterility and fertility restoration are produced in Brassica napus (Mariani et al., 1990). In tobacco also, male sterility was introduced using mitochondrial mutated gene (Hernould et al., 1992). This was achieved by introducing a ribonuclease gene. A hybrid gene with a ribonuclease coding sequence and a tapetum specific promoter was constructed. Mariani et al. (1990) formed a gene construct having an anther specific promoter (from TA29 gene of tobacco) and bacterial coding sequence for a ribonuclease (barnase gene from Bacillus amyloliquefaciens). Barnase gene encoded a product ribonuclease enzyme, which was cytotoxic and killed only the tapetal cells, thus preventing pollen development and ultimately leading to male sterility (Fig. 25.4). Using this gene construct (TA-29-RNase), male sterile transgenics have been produced in tobacco, lettuce, cauliflower, cotton, tomato, corn, etc. But there was a limitation to this approach as the above mentioned transgenic plants were always male sterile. Crossing with a restorer line could not restore male fertility in them. It was subsequently reported that male fertility in these transgenic plants could be restored by crossing them with a second set of transgenic plants which expressed a chimaeric ribonuclease inhibitor gene under the control of tapetum specific promoter. Mariani et al. (1992) used another gene construct barstar from

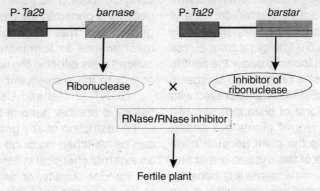

Fig. 25.4: A fertility restorer system in Brassica.

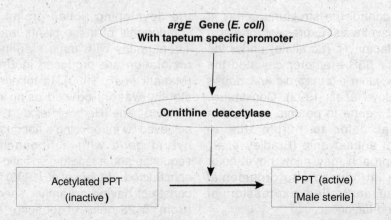

**Fig. 25.5:** A fertility restorer system by the use of herbicide.

*Bacillus amyloliquefaciens* with *TA-29* promoter for development of transgenic male fertile plants in *B. napus*. Barstar gene product is an inhibitor of ribonuclease which forms a complex with ribonuclease and neutralizes the cytotoxic properties. When male sterile plants (barnase gene) were crossed to male fertile plants (barstar gene), the F₁ plants were fertile due to fertility restoration by suppression of cytotoxic ribonuclease activity in the anther by the formation of cell specific RNase/RNase inhibitor complex. This system was used for hybrid seed production and commercial canola varieties have been released which were developed by Aventis Crop Science and Bayer Crop Science.

Kriete *et al.* (1996) developed another strategy for inducing male sterility. In this system a bacterial gene *argE* encoding N-acetyl ornithine deacetylase is used. This enzyme deacetylates the inactive herbicide (acetyl bialaphos) and makes it an active herbicide by deacetylation (Fig. 25.5). The gene from *E. coli* has been introduced in tobacco under the control of a tapetum cell specific promoter. Transgenic plants expressing *argE* gene when sprayed with acetylated (inactive) form of bialaphos showed the normal development of plants but anther tissue collapsed. Thus the plant became male sterile. The advantage of this system is that one need not maintain the male sterile line because plants can be propagated sexually without any

problem since *argE* does not affect the development of pollen. Moreover since male sterility is induced by an inactive herbicide in normal plants, no genetic restorer of male sterility is required. This strategy can be applied to other plants where the hybrid production is difficult.

## Transgenics for terminator seed

The technology that terminates the viability/fertility of seeds after a given time is known as terminator technology and the gene involved is popularly known as terminator gene. The terminator gene technology has generated a lot of attention due to a patent (No. 5,723,765) on "Control of Plant Gene Expression" which was granted on March 3, 1998 by the United States Patent and Trademark Office to United States Department of Agriculture and the Delta and Pine Land Co, USA. This patent was not for any new gene as is generally the case, but was for known genes and has been given for control of mechanisms to terminate the expression of desired traits either in the first generation of plant or only to the subsequent generations.

GURT or Genetic Use Restriction Technology is another general term used that refers to the restriction of any genetic trait in plant that can be switched on or off by the application of an external chemical inducer. This could include the trait for sterility, or any other trait such as color, ripening, cold tolerance, etc. T-GURT refers to the restriction of a specific trait in a plant also referred to as 'traitor technology'. V-GURT refers to restriction of the variety by engineering plants whose seeds will not germinate if replanted, which is terminator technology.

The terminator technology is based on the use of a suitable lethal gene which makes the second-generation seeds infertile. The seed company sells the first generation seeds which are fully developed, normal and fertile to produce healthy plants bearing seeds or fruits that can be used as food but will not germinate if planted. This will force the farmers to buy fresh seeds from the seed company to grow next season crop, since they cannot use the harvested seeds for next season crop.

The terminator technology employs three stretches of DNA (genes) which carry the necessary genetic information into the plants (Fig. 25.6).

1. *Terminator (lethal gene)*: It could be any gene that produces a protein which is toxic to plants and does not allow the seeds to germinate. A lethal gene codes for ribosome inhibiting protein (RIP). Lethal gene encoding RIP interferes in the synthesis of all proteins in the plant cell, without being toxic to other organisms. Thus expression of RIP gene in the cells of embryo would prevent germination of the seeds. It is attached with a particular type of promoter one that is active only in the late stages of seed development. LEA (late embryogenesis abundance) promoter is ideal when the aim is to express the trait in the second generation of seed onwards. Such a promoter would be active only after the first generation plant has completed vegetative growth. To prevent the expression of lethal RIP gene in the first generation seed, a blocking sequence is placed in between the LEA promoter and the lethal gene. The blocking sequence itself is flanked by specific excision sequences (LOX sites). When the blocking sequence is excised out by site specific excision at flanking LOX sites, the lethal gene comes in direct contact with the promoter and thus show expression in all the subsequent generations during late embryogenesis stages.

2. *Recombinase gene*: The second gene construct consists of a gene coding for an enzyme called recombinase. This enzyme is able to recognize the excision (LOX) sequences and removes these sequences along with the blocking sequence from the first gene construct by a process of recombination. A preferred recombinase excision sequence system is a bacteriophage CRE/LOX system where the CRE protein (recombinase) performs site-specific recombination of DNA at LOX sites. This recombinase gene is placed behind a repressible promoter specific for a repressor encoded by the third gene. This promoter can be repressed and thus recombinase enzyme will not be produced if a particular protein is present. This site-specific recombination takes place during germination of first generation seed on sowing and thus removes the excision and blocking sequences from the first gene construct.

3. *Repressible gene*: A third gene produces a protein called repressor protein which represses the promoter of the recombinase gene in the second gene construct. The repressor protein itself becomes inactive when it binds to a specific chemical, i.e. tetracycline. The inactive repressor (i.e. repressor-tetracycline complex is not able to repress the promoter attached to recombinase gene, thus allowing the synthesis of recombinase enzyme.

*Mechanism in pure line seed production*

It is reported that plant cells will be genetically modified with these strips of DNA and the plants regenerated through tissue culture methods. During the first generation, i.e. when companies are producing the seed, the plants with these

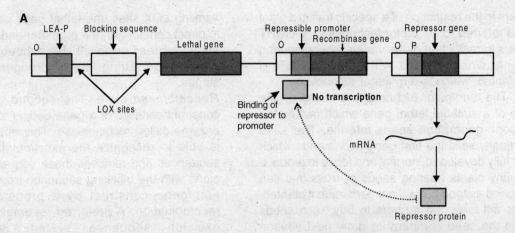

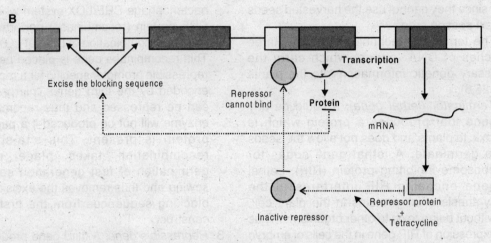

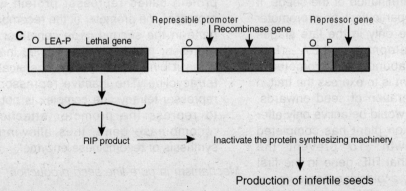

Production of infertile seeds

• **Fig. 25.6:** A schematic diagram showing different gene sequences involved in terminator seed technology. **A**–Action of repressor gene on repressible promoter; **B**–Action of recombinase gene to excise blocking sequence; **C**–Synthesis of lethal gene product.

stretches of DNA will be normal. The blocking sequence is firmly present between the reporter and lethal gene. Seeds are therefore formed without any trouble.

When the first generation seeds mature, these seeds will be exposed to a certain chemical (tetracycline) and sold in the market to the farmers. The repressor protein is being produced by the third gene which in the presence of tetracycline becomes inactive and cannot bind on the repressible promoter site. Thus, recombinase gene will become active on the second strip of DNA. The recombinase promptly removes the excision and blocking sequences from the first gene construct. At this stage LEA promoter is in direct contact with the lethal gene. But the lethal gene is not expressed, because the promoter has been chosen to be active only at a particular stage of seed development i.e. late embryonic stage. As a result the seed germinates properly to produce healthy second generation plant in the farmer's field.

When the second generation plant starts producing seeds, in the late embryogenesis stage, the LEA promoter becomes active and produces a large amount of ribosome inactivating proteins, which in turn inactivate the protein synthesizing machinery of cells, i.e. ribosomes. This results in the production of infertile second generation seeds. These seeds can be used as food, but will not germinate if planted to grow as subsequent generation plants.

Zeneca, U.K has developed a system called improved plant germplasm invention, dubbed as verminator technology by RAFI, Canada. This technology uses a gene from the fat tissue of rat, the product of which interferes with the growth of plant. If the crop is sprayed with a specific chemical, this gene is inactivated and the plant growth is normal. When a farmer buys a 'verminator seed' from the company, the crop raised from this seed is normal, but the seed produced from this crop gives rise to plants that can grow normally only when the plants are sprayed with a specific chemical. Thus, the farmers are forced to buy specific chemical if they have to raise the crop from the saved seed or buy fresh seed.

The companies argue that genetic trait control will offer farmers a menu of traits that can be turned on or off depending on their needs. Unfortunately, GURT (genetic use restrictive technology) is a confusing terminology, and the gene giants are using this to their advantage in intergovernmental negotiations. Using the GURT, the seed industry argues that T-GURT could have potential benefits for farmers and agricultural productivity, but dodges the clear-cut case against terminator technology and the calls to ban it. Industry is hiding behind GURT, thus making it more difficult for government negotiators to take decisive action against terminator technology. In a self-serving but well-reasoned memo, UPOV, the international body that coordinates plant breeders' rights, 'has considerable disadvantages for society'. Stung by negative publicity related to the escape of DNA from genetically modified (GM) plants, industry continues to 'greenwash' terminator technology by promoting it as terminator technology by promoting it as a biosafety tool for containing unwanted gene flow from GM plants.

*Terminator technology for use in hybrid seed production*

An alternative technique to using terminator gene is producing a pair of genetically engineered plants and then hybridizing them. One transgenic parent will contain the gene construct containing LEA promoter, excision sequences and the lethal gene. The second transgenic parent will have the gene construct containing the recombinase gene linked to a germination specific promoter. A cross between these two transgenic parent plants will produce hybrid seeds carrying genes from both the parents. When the hybrid seed is planted, although the recombinase gene will become active during germination and will remove the excision and blocking sequences from the first gene construct, yet it will produce healthy second generation plants. The lethal gene will express only when the plant tries to produce the second generation seeds on the farmer's field, which will result in the production of sterile seeds. This technique is particularly suitable for use with plants that are being planted as a yearly crop from seeds.

# TRANSGENIC PLANTS AS BIOREACTORS

Through genetic engineering techniques, compounds of commercial interest that were previously produced only from exotic sources of plants, animal and microbial origin are now produced in domesticated crops. Transgenic plants can act as living bioreactors for the inexpensive production of chemicals and pharmaceuticals, and this is also known as **molecular farming**. It has been possible to produce carbohydrates, fatty acids, polypeptides, vaccines, industrial enzymes and biodegradable plastics using the molecular farming approach (Table 25.7).

## Carbohydrates

Starch is one of the components of plant cells. Tailor-made starches with reduced level of amylose or increased amount of starch have been development in potato. Starch precursors have also been rerouted into the biosynthetic pathways of other storage carbohydrates. Non-fructan storing tobacco and potato plants have been transformed with fructosyl transferase gene from *Bacillus subtlis* to accumulate fructan. Production of novel compounds has also been attempted through genetic engineering. Tobacco plants have been transformed with the mannitol-1-phosphate dehydrogenase gene (*mtl*D) from *E. coli*. These transgenics synthesized more mannitol and were also tolerant to high salinity (Tarczynski *et al.*, 1993). Similarly myo-inositol o-methyl transferase gene from *Mesembryanthemum crystallimum* (ice plant) has been incorporated into tobacco to produce pinnitol a myo-inositol derived cyclic sugar alcohol. Trehalose is an additive that improves processed and dried food by making them tastier. The plant biotechnological companies MOGEN (Leiden, The Netherlands) and Calgene (Davis, USA) have reported the synthesis of small quantities of trehalose in transgenic tobacco plants.

Sucrose, a disaccharide of linked glucose and fructose is the most preferred sweetner. The key enzyme in sucrose synthesis is sucrose phosphate synthase (SPS). Overexpression of SPS holds promise for achieving the goal of increased sucrose synthesis in plants. Maize SPS has been overexpressed in tomato (Laporte *et al.*, 1997). In the transgenic plants sucrose level increased with a concomitant reduction in starch.

## Lipids

Monounsaturated fatty acid level in plants is desirable so that the nutritional value of oils is enhanced. When a rat desaturase gene was introduced into tobacco, level of palmitoleic acid (16:1) and oleic acid (18:1) levels increased (Grayburn *et al.*, 1992). For the manufacture of detergents and for specialized nutritional applications, thioesterase gene from California bay tree was incorporated into *Arabidopsis* for the accumulation of medium chain 12-carbon fatty acid (lauric acid) as storage lipids (Voelker *et al.*, 1992). The first non-food products of plant bioengineering in commercial production is a genetically engineered rape seed plant variety (Laurical) that was modified to produce lauric acid, a 12-carbon fatty acid used to make soaps and detergents (Table 25.7). The researchers needed to introduce only one gene (lauroyl-ACP thioesterase), which came from the California bay tree (ACP–Acyl carrier protein). That gene shut off fatty acid synthesis after 12 carbons rather than allowing the acids to grow to 18-carbon length normal for the plant, while having little effect on the productivity. *Brassica napus* with modified oil content was developed by Calgene Inc. and released in USA and Canada. Some of the transgenic lines showed up to 60% lauric acid in their oils. Besides soybean with

**Table 25.7:** The use of plants as bioreactors for the production of lipids, carbohydrates and proteins

| Compound | Origin of gene(s) | Application | Plant species used |
|---|---|---|---|
| **Lipids** | | | |
| Medium chain fatty acids | California bay tree (Umbellularia californica) –Thioesterase | Food, detergent, industrial | Oilseed rape |
| Mono-unsaturated fatty acids | Rat-desaturase | Food | Tobacco |
| Poly-hydroxybutyric acid | *Alcaligenes eutrophus* | Biodegradable plastics | *Arabidopsis*, oilseed rape, soybean |
| Saturated fatty acids | *Brassica rapa* | Food, confectioneries | Oilseed rape |
| γ-Linolenic acid | Desaturase *(Synechocystis)*; *Mortierella alpina* | For reducing coronary diseases | Tobacco; canola |
| Ricinoleic acid | Castor oleate hydroxylase | Nylon paints, varnishes, lubricants | *Arabidopsis* |
| **Carbohydrates** | | | |
| Amylose free starch | *Solanum tuberosum*(GBSS) | Food, industrial | Potato |
| Cyclodextrins | *Klebsiella pneumoniae-* Cyclodextrin glucosyl transferase | Food, pharmaceutical | Potato |
| Fructans | *Bacillus subtlis-* Fructosyl transferase | Industrial, food | Tobacco, potato |
| Increased amount of starch | E. coli (*glgC16*) | Food, industrial | Potato |
| Trehalose | E. coli | Food stabilizer | Tobacco |
| Sucrose | Maize sucrose phosphate synthase (SPS) | Sweetner | Tomato |
| Ononitol | *Mesembryanthemum crystallinum* (ice plant: gene *imt I*) | Industrial, food | Tobacco |
| **Proteins** | | | |
| Alpha-trichsantin | Chinese medicinal plant | Inhibition of HIV replication | *Nicotiana benthamiana* |
| Angiotensin converting enzyme inhibitor | Milk | Anti-hypersensitive effect | Tobacco, tomato |
| Antibodies | Mouse | Various | Mainly tobacco |
| Antigens | Bacteria, viruses | Orally administered vaccines | Tobacco, tomato, potato, lettuce |
| Antigens | Pathogens | Subunit vaccine | Tobacco |
| Enkephalin | Human | Opiate activity | Oilseed rape, *Arabidopsis* |
| Hirudin | Synthetic | Thrombin inhibitor | Oilseed rape |
| **Enzymes** | | | |
| Alpha amylase | *Bacillus licheniformis* | Liquefaction of starch; industrial application | Tobacco |
| Phytase | *Aspergillus niger* | Increased phosphate utilization from feed | Alfalfa, Tobacco |
| Manganese dependent lignin peroxidase | *Phanerochaete chrysosporium* | Bleaching and pulping of paper | Alfalfa, Tobacco |
| Beta-(1,4) xylanase | *Clostridium thermocellum*; *Cryptococcus albidus* | Animal feed, paper and pulp, baking | Tobacco, Canola |

Table adapted from Goddjin and Pen (1996).

modified oil content developed by DuPont Canada Agricultural Products has also been released for commercial cultivation.

γ-Linolenic acid (GLA) is the first intermediate in the bioconversion of linoleic acid to archidonic acid. GLA is important in alleviating hypercholesteromia and many coronary heart diseases. It is not produced in oil seed crops. This conversion is catalyzed by $\Delta^6$ desaturase enzyme and the gene encoding this enzyme has been cloned from cyanobacterium, *Synechocystis*. Constitutive expression of this gene in transgenic tobacco resulted in the production of GLA (Reddy and Thomas, 1996). This desaturase gene has been cloned from filamentous fungi (*Mortierella alpina*) and transferred to canola and tobacco respectively (Knutzon *et al.*, 1998; Sayanova *et al.*, 1997).

Hydroxy fatty acid, ricinoleic acid has industrial applications for the manufacture of nylon paints, varnishes, resins, lubricants and cosmetics. Transgenic *Arabidopsis* expressing castor oleate hydroxylase gene accumulated ricinoleic acid up to 17% of the total lipid fraction. Similarly, a oleate 12-hydroxylase: desaturase gene has also been transferred from *Lesquerella fendleri* to *Arabidopsis* for production of hydroxylated fatty acid (Broun *et al.*, 1998).

## Protein quality

Considerable progress has also been made in the improvement of nutritional quality of legume storage protein by expression of gene for methionine-rich protein. Brazil nut (*Bertholletia excelsa*) seed protein, 2S albumin of Brazil nut under the promoter of phaseolin gene was introduced into tobacco. The accumulation of 2S albumin resulted in a 30% increase in the methionine content in the transgenic tobacco seed (Altenbach *et al.*, 1989). Similarly, expression and accumulation of Brazil nut 2S albumin protein in transgenic *Arabidopsis*, rape seed, soybean, French bean seeds and potato have also been reported. Unfortunately, Brazil nut 2S albumin was found to be allergenic (Nordlee *et al.*, 1996) which raised the safety concern of development of transgenic plants expressing this protein. Sunflower seed albumin rich in methionine is not an allergen has been used for development of transgenic plants. The sunflower albumin gene under the control of pea vicilin gene promoter was transferred to lupin and the methionine content of seed protein of transgenic plants increased by two fold. Rat feeding experiments demonstrated the superior nutritive value of transgenic lupin (Molvig *et al.*, 1997). Another seed storage protein gene (*Ama1*) isolated from Amaranthus is a good candidate for introduction into crop plants as the protein has well balanced amino acid composition (Raina and Datta, 1992). This gene has been introduced into potato and the level of accumulation of *Ama1* was found to be 0.3% of total soluble protein (Chakraborty *et al.*, 1998). Soybean glycinine gene has been introduced into rice to increase its protein content and to make it more digestible. In addition some important amino acids like lysine lacking in quantity in normal rice were replished (Momma *et al.*, 1999). Gene encoding a human milk β-casein was expressed in transgenic potato under the control of an auxin inducible promoter (Chong *et al.*, 1997). These findings open the way for reconstitution of human milk proteins in plant foods.

7S legume β-phaseolin seed storage protein gene driven by rice *gt1* (gltutelin 1) gene promoter was transferred in rice. Transgenic rice plants expressed the gene in their endosperm and showed up to 4% of their total proteins to be β-phaseolin. Likewise 11S legumin protein gene driven by *gt1* promoter has also been transferred and expressed in rice endosperm. Du Pont scientists have patented a gene encoding a protein CP3–5 with 35% lysine and 22% methionine. This CP3–5 gene was coupled with seed specific promoters and transferred to maize. Maize with improved lysine content has been released for commercial cultivation in Canada, Mexico, USA and Philippines.

## Enzymes

Phytase enzyme produced in plant tissues of alfalfa and tobacco serves a good example for increased phosphate utilization from feed for animals (Table 25.7). Enzyme phytase hydrolyzes phytate to inorganic phosphate and inositol, thereby making the phosphorus readily available to animals like chicken. Phytase is encoded by a gene from *Aspergillus niger*. It enhances phosphorus utilization by chickens from their feed to the extent that phosphate supplement in the feed is not required. Transgenic tobacco seeds expressing the gene coding for phytase were fed to broiler chickens for a period of 4 weeks. This produced a gain in body weight that was comparable to those obtained with feed supplemented with phosphate or with *A. niger* phytase.

## Vitamin and mineral

Vitamin A content in the diet plays a major role in preventing disorders such as color blindness. Rice, which is the staple food for billions of people does not contain any provitamin A in the endosperm. Immature rice endosperm is capable of synthesizing the early intermediate geranyl geranyl diphosphate (GGPP), which can be used to produce the uncolored carotene phytoene by expressing the enzyme phytoene synthase in rice endosperm. The synthesis of beta carotene requires three additional plant enzymes: (i). phytoene desaturase and (ii). ξ-carotene desaturase each catalyzing the introduction of two double bonds and (iii). lycopene β-cyclase, encoded by the *lcy* gene. Ye *et al*. (2000) used *Agrobacterium* mediated transformation method in rice to introduce the entire β carotene biosynthetic pathway into rice endosperm. Plant phytoene synthase (*phy*) gene originated from daffodil (*Narcissus pseudonarcissus*), bacterial phytoene desaturase (*crtl*) originating from *Erwinia uredovora* placed under the control of endosperm specific glutelin (*Gt*1) and the constitutive CaMV35S promoter respectively. The phytoene synthase cDNA contained a 5'sequence coding for a functional transit peptide and the *crtl* gene contained the transit peptide (*tp*) sequence of the pea Rubisco small subunit. These two direct the formation of lycopene in the endospeum plastids, the site of geranyl geranyl diphosphate formation. A third gene lycopene β-cyclase from *Narcissus pseudonarcissus* controlled by rice glutelin promoter which carried a functional transit peptide allowing plastid import and selectable marker *aphIV* were introduced. The transgenes providing these enzyme activities were transferred into rice using *Agrobacterium* mediated transformation. The resulting transgenic rice plants produced sufficient β-carotene which can be converted to vitamin A and the rice has been named as golden rice.

Shewmaker *et al*. (1999) have introduced phytoene synthase gene into *Brassica napus* that resulted in increased content of carotenoides. This work has been conducted by Monsanto to enhance the carotenoid levels of oilseed crops with a focus on the accumulation of beta-carotene in the seed of canola.

The carotenoid content of tomato fruit was enhanced by producing transgenic lines containing a bacterial carotenoid gene (*crtl*) encoding phytoene desaturase, which converted phytoene to lycopene (Romer *et al*., 2000). Expression of this gene did not increase total carotenoid levels but it increased B carotene level up to 45% of the total carotenoid content.

A yeast gene, encoding S-adenosyl methionine decarboxylase (ySAMdc; *SPE 2*) fused to fruit specific and ripening-inducible $E_8$ promoter, was expressed in tomato, which led to an increase in the level of polyamines spermidine and spermine (Mehta *et al*., 2002) This enhanced level of polyamines led to an increase in lycopene content of tomatoes, prolonged vine life and better fruit juice quality.

The major micronutrient deficiencies worldwide concern iron, with 24 percent of the world's population (up to 60 percent in developing countries) or 1.4 billion women suffering from iron deficiency anemia and vitamin A deficiency, affecting approximately 400 million children, or seven percent of the world population. Iron deficiency is the consequence of (a) a very low amount of iron in the endosperm,

(b) a high concentration of phytate (the major cause for inhibition of iron resorption in the intestine) and, (c) lack of high sulphur containing proteins enhancing iron resorption.

Consequently Potrykus and his coworkers at Switzerland aimed to (a) an increase of iron content with a ferritin transgene from *Phaseolus vulgaris*, (b) reduce phytate in the cooked rice with a transgene for heat stable phytase from *Aspergillus fumigatus*, and (c) increase the resorption-enhancing effect from a transgenic sulphur rich metallothionin like protein from *Oryza sativa*. All transgenes are under endosperm specific regulation. Goto *et al.* (1999) and Vasconncelos *et al.* (2003) analyzed a series of transgenic rice plants for all the genes mentioned and have achieved so far a two-fold increase in iron content and a high activity of the *A. fumigatus* phytase reducing phytate completely after one hour of cooking. Expression of metallothionin like protein led to an increase in cystein of approximately 25% above control. Kim *et al.* (2005) manipulated calcium content of rice grain with $H^+/Ca^{2+}$ transporter gene in order to increase its nutritional value.

### Biodegradable plastic

Commercial plastic producing plants are also several years away, although researchers have been making progress. Polyhydroxy butyrate (PHB), an aliphatic polyester with thermoplastic properties, was initially synthesized in plants by introducing the relevant genes acetoacetyl-CoA reductase and PHA synthase from the bacterium *Alcaligenes euthophus* into *Arabidopsis thaliana*. The original gene transfer worked, although the resulting plants were sickly and made very little PHB. The problem was solved by adding a sequence to the gene that causes the enzymes they make to be targeted to storage vesicles called plastids. This both provides the enzymes with high levels of a key starting compound for PHB synthesis and protects the rest of the plant cell from the possible harmful effects of PHB accumulation. As a result, PHB synthesis showed a 100-fold increase with no significant

ill effects on plant growth or seed yield (Poirier *et al.*, 1995).

### Proteins, peptides and vaccines

New developments in plant biotechnology have suggested that it may be possible to use genetically engineered plants and plant viruses to produce vaccines against human diseases, ranging from tooth decay to life threatening infections such as bacterial diarrhea, cholera, and AIDS. Conventional vaccines consist of attenuated or inactivated pathogens. But, in many cases, the genes encoding a critical antigen has been isolated and expressed in other bacteria/animals, and the recombinant protein so produced is used as a vaccine, which is known as recombinant vaccine. Thus, recombinant protein is encoded by a natural or synthetic gene present in a recombinant DNA construct and is expressed in a host other than the species to which the gene belongs. For such recombinant vaccines, generally the entire protein molecule is not necessary because antigenic property is localized to a small part of molecule for effective immunization. Therefore, either a constituent polypeptide or a small part of a polypeptide may be used as a vaccine, such vaccines are known as subunit vaccines. These are produced using recombinant DNA technology for cloning and expressing the appropriate DNA sequences encoding antigenic polypeptides or polypeptide fragments in suitable hosts.

One of the earliest examples of the production of pharmaceutical polypeptides in plants exploited the natural high level expression of seed storage protein. The neuropeptide [leu] enkephalin was produced in *B. napus* as part of the seed storage protein 2S albumin (Krebbers and Kerckhove, 1990). Viral vectors have been used to produce chimaeric coat proteins in transgenic plants. Turpen *et al.* (1995) have described a method for engineering the capsid protein of TMV as either internal or C-terminal fusions with peptides carrying epitopes derived from malarial sporozites. Both internal and

C-terminal fusion constructs yielded high titers of genetically stable recombinant virus which produced appropriate monoclonal anti-malarial antibodies in infected tobacco plants. The zona pellucida ZP3 protein has been fused with TMV capsid protein. ZP3 protein of mammalian oocyte has been a target for immune contraception and an epitope of 13 amino acids from murine ZP3 has now been expressed in plants as a fusion with TMV capsid protein. Mice immunized with the recombinant TMV particles developed antibodies against ZP3 (Beachy et al., 1996). Theoretically, vaccination with the modified viral coat protein might work as a form of birth control, because antibodies to zona pellucida could prevent fertilization of egg. The same is true for epitopes derived from humal immunodeficiency virus which, are expressed as alfalfa mosaic virus coat protein fusion products (Yusibov et al., 1997). Cowpea mosaic virus (CMV) has also been engineered as an expression system for the production of foreign peptides such as epitope derived from human rhinovirus 14 and human immunodeficiency virus (HIV-1). The modified plant viruses elicit the production of mouse antibodies that neutralize virus in test tube experiments (McLain et al., 1995). It indicates the possibility of producing a preventive HIV vaccine. Fusion coat proteins may represent a cost-effective way for generating vaccines (Table 25.7).

## Production of edible vaccine antigens

Transgenic plants show the promise for use as low-cost vaccine production system. Researchers are taking several tacks to making plant based vaccines, but the edible vaccines now under development by a team led by Charles Arntzen of Texas A & M University in USA are likely to be the cheapest and easiest to administer. The idea behind the edible vaccines is to have people take their dose by eating, as part of their diet, the plant that produces the vaccine. Arntzen's team started work aimed at developing oral vaccines to prevent enteric diseases, including cholera and diarrhea caused by bacteria such as E. coli, Shigella and Salmonella. Bacterial diarrheas are a leading cause of infant deaths in the developing world. The first report of the concept of using a plant expression system for production of an edible vaccine appeared in a patent application under International Patent Cooperation Treaty by Curtiss and Cardineau (1990), which described a means to express a surface protein (spa A) from Streptomyces mutans in tobacco to 0.02% of total leaf proteins. Since then various antigenic determinants against viral and bacterial pathogens have been expressed in transgenic plants.

*Escherichia coli* **Labile Toxin:** *Escherichia coli* labile toxin (LT), which is responsible for causing diarrhoea, is composed of a 27 kDa A subunit and five 11.6 kDa B subunits that pentamerise. The B pentamer (LT-B) has been used as a vaccine component, as antibodies against this would block toxin activity. LT-B was expressed in transgenic potato (Haq et al., 1995). However, the maximum expression level achieved represented 0.01% of total soluble protein. Mice fed with such transgenic tuber developed an oral immune response. Mason et al. (1998) created a synthetic gene for LT-B that contained plant-preferred codons. As a result, the expression level increased to 0.19% of the total soluble protein. In a clinical trial, transgenic tubers expressing LT-B, when fed to human volunteers, showed the development of serum and mucosal immune response to LT-B. In another study, the recombinant LT-B, produced in potato tubers, was found immunogenic and, on oral administration, elicited a systemic and local IgA response. A problem with potatoes is that they must be cooked before they are eaten. The heat used in cooking will cause the vaccine protein to denature, reducing or eliminating its ability to elicit immunity. A synthetic gene encoding, a variant of LT-B, was also expressed in transgenic corn and oral administration of transgenic corn elicited serum and mucosal immune responses in mice (Steatfield et al., 2001).

***Vibrio cholerae* Toxin:** Cholera toxin (CT) is very similar to *E. Coli* LT. The genes encoding CTA and CTB were amplified by PCR and then cloned into plant expression vectors. Hein *et al.* (1996) generated tobacco plants expressing CT-A or CT-B subunits of the toxin. Expression on these genes was controlled by the CaMV 35S promoter and 35S enhancer. In potato tubers, CTB was expressed at the level of 0.3% of total soluble protein (Arakawa *et al.*, 1998). The transgenic potato tubers when fed to mice, serum and intestinal CTB specific antibodies were induced. CTB was also expressed in tobacco chloroplasts at a level ranging between 3.5 and 4.1% of the total soluble protein (Daniell *et al.*, 2001).

**Hepatitis B Virus Surface Antigen:** Hepatitis B virus infection causes acute or chronic hepatitis, and hepatocellular carcinoma. Mason *et al.* (1992) developed the first vaccine consisting of hepatitis B surface antigen (HbsAg). This antigen is produced in large amounts in liver cells of infected individuals. Tobacco plants were genetically transformed with the gene encoding HbsAg, which was driven by CaMV 35S promoter with double enhancer linked to tobacco etch virus (TEV) 5′ non-translated leader sequence. HbsAg expression levels ranged up to 66 ng/mg of soluble protein. The immunogenicity of tabacco derived HBsAg was tested intraperitoneally in mice. HbsAg-specific antibodies of all IgG subclasses as well as IgM antibodies were produced. The plant derived HbsAg could also prime T cells *in vivo* that could be recalled *in vitro* to proliferate upon stimulation with the immunogen. Lupin and lettuce were created for expressing HbsAg at levels of 150 ng/g fresh weight in lupin callus and 5.5 ng/g fresh weight in lettuce leaf (Kapusta *et al.*, 1999). Mice fed with transgenic callus elicited HbsAg specific IgG response. Human volunteers, fed with transgenic lettuce, developed specific serum IgG rsponse to HbsAg. Transgenic potatoes were also developed expressing the hepatitis B surface antigen (Richter *et al.*, 2000). Mice fed with transgenic tuber developed a primary immune response, which was boosted by intraperitoneal delivery of a single subimmunogenic dose of commercial HbsAg.

The potato problem, however, may be nearing a solution through a switch to a carrier that is eaten raw: bananas. Arntzen and his colleagues have introduced a foreign gene into banana plants and shown its expression. They did not use a gene for a potential vaccine protein in these experiments, but plan to put the *E. coli* enterotoxin gene in bananas. If that succeeds, Arntzen wants to add genes for additional vaccine proteins. Another antigen Norwolk virus capsid protein (NVCP) has also been expressed. The first step was to show that proteins made in plants could elicit immune response in animals. They achieved the goal by introducing the gene encoding a surface protein from the hepatitis B virus (HBV) into tobacco plants. Not only did the plants make the viral protein, but also when injected into mice, it triggered production of antibodies that recognize the hepatitis B protein.

Similarly, a rabies virus coat glycoprotein gene has been expressed in tomato plants. Carrillo *et al.* (1998) expressed structural protein, VP1, of foot and mouth disease virus in *Arabidopsis*. The mouse that was immunized intraperitoneally with a leaf extract elicited immune response to synthetic peptides carrying various epitopes of VP1, or to complete VP1. Furthermore, all the mice immunized with leaf extract were protected against challenge with virulent foot and mouth disease virus.

### Antibody Production in Plants (Plantibody)

Many groups have been working on the expression of antibodies in plants with a view to exploiting plants as bioreactors for the large scale production of antibodies, for the following reasons.

1. Plants can assemble heavy and light chains into complete antibodies.

2. Plants permit appropriate post-translational-modification for the production of antibodies.

3. Several groups have expressed complete antibody by targeting the antibody via endoplasmic reticulum to apoplast, the extracellular aqueous region in which hydrolytic degradation is minimal and antibodies secreted into it can accumulate in a relatively stable environment. The extraction of antibody is also simpler and can be achieved by conditions milder than those required for proteins located elsewhere.

4. Genetically stable seed stocks of antibody-producing plants can be produced and stored indefinitely at low cost. The seed stock can be converted into a harvest of large quantity of antibody within one growing season.

Production of a fully functional antibody in plants is not a very straightforward task because of the multi-subunit structure of antibody molecule. Moreover, expression of a complete antibody may not be required for many applications. For example, production of antigen binding domain as present in single chain Fv or Fab molecules may be enough to block binding of a pathogen or a virulence factor secreted by the pathogen and thus may limit the spread of the infection. However, a monovalent antibody fragment may have reduced affinity for binding to antigens as compared to a bivalent $F(ab)_2$ fragment or a complete antibody. Further, divalent nature may be required for aggregation of cells or bacteria in some cases. Production of these fragments without complete constant region might be sufficient for the conditions where attenuation of the function of the antigen is required. Different antibody fragments have been expressed in plants for blocking actions of phytochrome A (Owen et al., 1992) abscisic acid (Longstaff et al., 1998), as well as for sequestration of organic pollutants (Artasenko et al., 1995). Holmberg et al. (1997) introduced the gene for Vitreoscilla hemoglobin (VHb) in tobacco. Vitreoscilla synthesizes elevated quantities of a novel homodimeric hemoglobin

(VHb) in oxygen limiting environment. Expression of VHb in E. coli increases growth rate and final cell density in oxygen limited cultures. When VHb was expressed in transgenic tobacco it exhibited enhanced growth, on an average 80–100% more dry weight. Transgenic plants contained on an average 30–40% more chlorophyll and 34% more nicotine. Thus, the introduction of this gene could lead to enhanced plant growth and yield. Hood et al. (1998) purified avidin from high expressing lines of transgenic maize and it made commercially available. Avidin is primarily localized in the embryo and represents 72% of aqueous soluble extracted protein from dry seeds.

Ma et al. (1995) produced functional, high molecular weight secretory immunoglobulin in transgenic tobacco, which acts against adhesion protein of Streptococcus mutans known to cause tooth decay. They had cloned the heavy and light chains of murine antibody, murine joining chain and a rabbit secretory component into separate transgenic tobacco plants. Hybrid plants obtained by multiple cross-pollination events co-expressed all components and produced a functionally active secretory antibody. The Secretory antibodies were found to be more stable because of their dimeric nature and thus protected teeth from infection for a longer period as compared to murine IgG. The First human trial of a monoclonal secretory antibody produced in transgenic plants was conducted by planet Biotechnology, Inc (Mountain View, CA). Their product CaroRx™ was recombinant sIgA/G, purified from mature tobacco plants by ammonium sulphate precipitation and protein G immunoaffinity chromatography. The clinical efficacy of both plant sIgA/G and murine IgG monoclonal antibodies was tested by their application directly to teeth for three weeks with two applications per week. In both cases there was no recolonization by Streptococcus mutans in saliva or teeth till day 118 of the experiment (Ma et al.1998, Giddings, 2001).

Potato tubers have also been used as a biofactory for high level production of a

**Table 25.8:** Expression of certain antibodies and antibody-derived fragments of therapeutic and diagnostic value in plants

| Target | Antibody and antibody derived fragments | Plant | Reference |
|--------|------------------------------------------|-------|-----------|
| *Streptococcus mutans* | Guy's 13 in 3 forms:<br>Plant G13: Guy's13 IgGl<br>Plant GI/A: Guy's 13 IgG/IgA hybrid heavy chain consisting of variable $\gamma 1$–$\alpha 2$–$\alpha 3$ domains<br>Plant G2/A:Guy's 13 IgG/IgA hybrid heavy chain consisting of variable $\gamma$ 1–$\gamma$ 2–$\alpha 2$–$\alpha 3$ domains | Tobacco | Ma *et al.* (1994) |
| Human creatine kinase | MAK 33 IgG and Fab fragment | *Arabidopsis thaliana* | De Wilde *et al* (1996) |
| CD-40, Non Hodgkin's lymphoma | Single chain immunotoxin composed of bryodin 1 fused to scFv region of anti-CD40 antibody G28-5 | Tobacco | Francisco *et al.* (1997) |
| Herpes Simplex Virus (HSV) | Humanized anti-herpes simplex virus 2 (HSV-2) monoclonal antibody (IgG) | Soybean | Zeitlen *et al.* (1998) |
| Non Hodgkin's lymphoma | ScFv of IgG from, mouse B cell lymphoma | Tobacco | McCormick *et al.* (1999) |
| Zearalenone produced by members of *Fusarium* | ScFv | *Arabidopsis thaliana* | Yuan *et al.* (2000) |
| Carcinoembryonic antigen (CEA) | ScFv Ab (scFvT84.66) | Wheat and rice | Stoger *et al.* (2000) |
| Carcinoembryonic antigen (CEA) | T84.66/GS8 diabody | Tobacco | Vaquero *et al.* (2002) |
| Hepatitis B surface antigen | ScFv | Tobacco | Ramirez *et al.* (2002) |
| Human anti-rhesus D | IgGl | *Arabidopsis thaliana* | Bouquin *et al.* (2002) |
| Human chorionic gonadotropin (hCG) | ScFv, diabodies and chimaeric antibodies (mouse variable domain and human immunoglobulin constant domain) | Tobacco | Kathuria *et al.* (2002) |

recombinant single chain Fv (SCFv) antibody. Recombinant antibodies accumulated up to 2% of the total soluble tuber protein (Artasenko *et al.*, 1998).

Several other antibodies and antibody-derived fragments of therapeutic and diagnostic value have been expressed in plant. A few examples are cited in Table 25.8.

For Production of complete antibodies in plants to a level of 0.055 to 5% of the total cellular protein, following three strategies have commonly been applied.

1. Separate transgenic plants are obtained with individual heavy and light chain expressing vectors. Plants expressing high level of functional antibody are produced in the progeny after cross-pollination between plants containing heavy/light chain.

2. As plant expression vectors are generally large and contain only one promoter and one polylinker region, one vector each is used for every immunoglobulin gene. Therefore, two separate vectors, one encoding heavy chain and the other encoding light chain, are used to obtain transgenic plant. The level of expression achieved is generally low.

3. A single vector with genes encoding both the heavy and light chains is used to obtain a transgenic plant.

## Production of Pharmaceutically Important Proteins in Plants

Glucocerabrosidase (GC) catalyses the degradation of complex glycosylceramide lipids in humans. Inherited deficiency in GC causes Gaucher's disease and leads to bone marrow expansion, bony deterioration and visceromegaly. Ceredase is one of the most expensive drugs of the world. For a 50 kg person, the drug alone costs in the range of $ 70,000 to $ 300,000 per year. Successful production of human GC (hGC) in plants can clearly denote a dramatic example of the potential of plant-based system for the cost-effective bioproduction of human pharmaceutical proteins. The gene of hGC was engineered into tobacco using *Agrobacterium* mediated transformation. Such plant produced hGC was found to be enzymatically active, which indicated that the gene product was correctly folded in the plant cell. This result represents one of the first active human enzymes produced in transgenic plants. Further, in the crude extract of transgenic tobacco leaves, hGC represents 10% of the total soluble protein.

Therefore, transgenic plants show promise for use as low cost vaccine production systems. Attempts are being made to express many proteins of immunotherapeutic use at high levels in plants and to use them as bioreactors of the modern era (rev. Sharma *et al.*, 1999).

# REFERENCES

Altenbach, S.B., Pearson, K.W., Meeker, G., Staraci, L.C. and Sun, S.S.M. 1989. Enhancement of methionine content of seed proteins by the expression of a chimeric gene encoding a methionine rich protein in transgenic plants. *Plant Mol. Biol.*, **13:** 513–522.

Anzai, H., Yoneyama, K. and Yamaguchi, I. 1989. Transgenic tobacco resistant to a bacterial disease by the detoxification of a pathogenic toxin. *Mol. Gen. Genet.*, **219:** 492–494.

Aono, M., Saji, H., Sakamoto, A., Tanaka, K., Kondo, N. and Tanaka, K. 1995. Paraquat tolerance of transgenic *Nicotiana tabacum* with enhanced activities of glutathione reductase and superoxide dismutase. *Plant Cell Physiol.*, **36:** 1687–1691.

Arakawa, T., Chong, D.K.X., and Langridge, W.H.R. 1998. Efficacy of a food plant-based oral cholera toxin B subunit vaccine. *Nat Biotechnol.* **16:** 292–297.

Arakawa, T., Chong, D.K.X., Merritt, J.L., and Langridge W.H.R. 1997. Expression of cholera toxin B subunit oligomers in transgenic potato plants. *Transgenic Res.* **6:** 403–413.

Artasenko, O., Peisker, M., zur. Nieden, U., Fiedler, U., Weiler, E.W. *et al.* 1995. Expression of a single-chain Fv antibody against abscisic acid creates a wilty phenotype in transgenic tobacco. *Plant J.* **8:** 745–750.

Atkins, D., Young, M., Uzzell, S., Kelly, L., Fillatti, J. and Gerlach, W.L. 1995. The expression of antisense and ribozyme genes targeting citrus exocortis viroid in transgenic plants. *J. Gen. Virol.* **16:** 1781–1790.

Audy, P., Palukaitis, P., Slack, S.A. and Zaitlin, M. 1994. Replicase mediated resistance to potato virus Y in transgenic tobacco plants. *Mol. Plant Microbe-Interact.* **7:** 15–22.

Beachy, R.N., Fitchen, J.H. and Hein, M.B. 1996. *Ann. N. Y. Acad. Sci.* **792:** 43–49.

Bouquin, T., Thomsen, M., Nielsen, L.K., Green, T.H., Mundy, J. *et al.* 2002. Human anti-rhesus D IgG antibody produced in transgenic plants. *Transgenic Res.* **11:** 115–122.

Bowler, C., Slooten, L., Vandenbranden, S., De Rycke, R., Botterman, J., Sybesma, C., Van Montagu, M. and Inze, D. 1991. Manganese superoxide dismutase can reduce cellular damage mediated by oxygen radicals in transgenic plants. *EMBO J.* **10:** 1723–1732.

Bradley, J.M., Davies, K.M., Deroles, S.C., Bloor, S.J. and Lewis, D.H. 1998. The maize Lc regulatory gene upregulates the flavonoid s biosynthetic pathway of petunia. *Plant J.* **13:** 381–392.

Broglie, K., Chet, I., Holliday, M., Crossman, R., Biddle, P., Knowlton, S., Manain, I.J. and Broglie, R. 1991. Transgenic plants with enhanced resistance to fungal pathogen *Rhizoctonia solani. Science* **254:** 1194–1197.

Broun, P., Boddupalli, S. and Somerville, C. 1998. A bifunctional oleate 12-hydroxylase desaturase from *Lesquerella fendleri. Plant J.* **13:** 201–210.

Bulow, L., Holmberg, N., Lilius, G. and Bailey, J.E. 1999. The metabolic effects of native and transgenic hemoglobin on plants. *Trends Biotechnol.* **17:** 21–24.

Carrillo, C., Wigdorovitz, A., Oliveros, J.C., Zamorano, P.I., Sadir, A.M., Gomez, N., Salinas, J., Escribano, J.M. and Borca, M.V. 1998. *J. Virol.* **72:** 1688–1690.

Chakraborty, N., Chakraborty, S., Kesarwan, M., Azam, M. and Datta, A. 1998. Proceedings of IUBS symposium, Taiwan, Taipei, 1997, pp. 125–131.

Cheng, M., Fry, J.E., Pang, S., Zhou, H., Hironaka, C.M., Duncan, D.R., Conner, T.W. and Wan, Y. 1997. Genetic transformation of wheat mediated by *Agrobacterium tumefaciens. Plant Physiol.* **115:** 971–980.

Chong, D.K., Roberts, W., Arakawa, T., Illes, K., Bagi, G., Slatter, C.W. and Langridge, W.H. 1997. Expression of human milk protein beta casein in transgenic potato plants. *Transgenic. Res.* **6:** 289–296.

Cooper, B., Lapidot, M., Heike, J.A., Dodds, J.A. and Beachy, R.N. 1995. A defective movement protein of TMV in transgenic plants confers resistance to multiple viruses whereas functional analogs increases susceptibility. *Virology* **206:** 307–313.

Cronin, S., Verchot, J., Haldeman-Cachill, R., S. and Carrington, J.G. 1995. Long distance movement factor: A transport function of the potyvirus helper component proteinase. *Plant Cell* **7**: 549–559.

Curtiss, R.I., and Cardineau, C.A. 1990 Oral immunization by transgenic plants world patent application WO 90/02484.

Daniell, H., Lee, S.B., Panchal, T., and Weibe, P.O. 2001. Expression of the native cholera tozin B subunit gene and assembly as functional oligomers in transgenic tobacco chloroplasts. *J Mol. Biol.* **311**: 1001–1009.

De Block, M., Herrera-Estrella, L., van Montagu, M., Schell, J. and Zambryski, P. 1984. Expression of foreign genes in regenerated plants and their progeny. *EMBO. J.* **3**: 1681–1689.

De Haan, P. *et al.* 1992. Characterization of RNA mediated resistance to tomato spotted wilt virus in transgenic tobacco plants. *Biotechnology* **10**: 1133–1137.

De la Feunte, H.M., Mosqueda-Cano, G., Alvarez-Morales, A. and Herrera-Estrella, L. 1992. Expression of a bacterial phaseolotoxin resistant ornithyl transcarbamylase in transgenic tobacco plants confers resistance to *Pseudomonas syningae* pv *phaseolicola*. *Bio/Technology* **10**: 905.

De Wilde, C., De Neve, M., De Rycke, R., Bruynss, A.M., De Jaeger, G. *et al* . 1996. Intact antigen-binding MAK33 antibody and Fab fragment accumulate in intercellular spaces of *Arabidopsis thaliana, Plant Sci.* **114**: 233–241.

Down, R.E., Gatehouse, A.M.R., Hamilton, W.D.O. and Gatehouse, J.A. 1996. Snowdrop lectin inhibits development and decreases fecundity of the glass house potato aphid (*Aulacorthum solani*) when administered in vitro and via transgenic plants both in laboratory and glass house trials. *J. Insect Physiol.* **42**: 1035–1045.

During, K., Porsch, P., Flauddung, M. and Lorz, H. 1993. Transgenic potato plants resistant to the phytopathogenic bacterium *Erwinia carotovora*. *Plant J.* **3**: 587–598.

Edington, B.V. and Nelson, R.S. 1992. Utilization of ribozymes in plants: plant viral resistance. In: *Gene Regulation- Biology of antisense RNA and DNA.* (eds. Erickson,. R.P. and Izant, J.G.) Raven Press, New York

Fischhoff, D.A., Bowdish, K.S., Perlak, F.J., Marrone, P.G., McCormick, S.M., Niedermayer, E.J., Rochester, E.J., Rogers, S.G. and Fray, R.T. 1987. Insect tolerant transgenic tomato plants. *Bio/Technology* **5**: 807–813.

Francisco, J.A., Gawlak, S.L., Miller, M., Bathe, J., Russell, D. *et al.* 1997. Expression an characterization of bryodin I and a bryodin I-based single-chain immunotoxin from tobacco cell culture. *Bioconjug chem.* **8**: 708–713.

Giddings, G. 2001. Transgenic plants as protein factories. *Curr. Opin. Biotechnol.* **12**: 450–454.

Golemboski, D.B., Lomonossoff, G.P. and Zaitlin, M. 1990. Plants transformed with a tobacco mosaic virus non-

structural gene sequence are resistant to the virus. *Proc. Natl. Acad. Sci. USA* **87**: 6311.

Goto, F., Yoshihara, T., Shigemoto, N., Toki, S. and Takaiwa, F. 1999. *Nature Biotechnol.* **17**: 282–286.

Grayburn, W.S., Collins, G.B. and Hildebrand, D.F. 1992. *Bio/Technology* **10**: 675–678.

Hain, R., Biessler, B., Kindl, H., Schroeder, G. and Stocker, R. 1990. Expression of a stilbene synthase gene in *Nicotiana tabacum* results in synthesis of phytoalexin resveratrol. *Plant Mol. Biol.* **15**: 325.

Han, S. Wu, Z, Yang, H., Wang, R, Yie, Y, Xie, L. and Tien, P. 2000. Ribozyme mediated resistance to rice dwarf virus and transgene silencing in the progeny of transgenic rice plants. *Transgenic Research* **9**: 195–203.

Haq T.A., Mason, H.S., Clements, J.D. and Arntzen, C.J. 1995. Oral immunization with a recombinant bacterial antigen produced in transgenic plants. *Science* **268**: 714–716.

Harrison, B.D., Mayo, M.A. and Baulcombe, D.C. 1987. Virus resistance in transgenic plants that express cucumber mosaic virus satellite RNA. *Nature* **328**: 799–801.

Hein, M.B., Yeo, T.C., Wang, F. and Sturtevant, A. 1996. Expression of cholera toxin subunits in plants. *Ann N Y Acad Sci.* **792**: 50–55.

Hernould *et al.* 1992. *Bulletin di Information* **3–4**: 17–18.

Hiatt, A., Cafferkey, R. and Bowdish, K. 1998. Production of antibodies in transgenic plants. *Nat (Lond)* **342**: 76–78.

Hiei, Y., Ohta, S., Komari, T. and Kumashiro, T. 1994. Efficient transformation of rice (*O. sativa* L.) mediated by *Agrobacterium* and sequence analysis of the boundaries of the T-DNA. *The Plant J.* **6**: 271–282.

Holmberg, N., Lilius, G., Bailey, J.E. and Bulow, L. 1997. Transgenic tobacco expressing *Vitroscilla* hemoglobin exihibits enhanced growth and altered metabolite production. *Nat. Biotechnol.* **15**: 244–247.

Holton, T.A., Brugliera, F., Lester, D.R., Tanaka, Y., Hyland, C.D., Menting, J.G.T., Lu, C.Y., Farcy, E, Stevenson, T.W. and Cornish, E.C. 1993. Cloning and expression of cytochrome p450 genes controlling flower color. *Nature* **366**: 276–279.

Hood, E.E., Howard, J.A. and Tannotti, E.L. 1998. Transgenic corn: a new source of valuable industrial products. In: *Proc. Corn utilization and Tech., Conf.,* Missouri, USA, pp. 101–104.

Horsch, R.B., Fraley, R.T., Rogers, S.G., Sanders, P.R., Lloyd, A. and Hoffman, N. 1984 Inheritance of functional foreign gene in plants. *Science* **223**: 496–498.

Ishida, Y., Saito, H., Ohta, S., Hiei, Y., Komari, T. and Kumashiro, T. 1996. High efficeiency transformation of maize (*Zea mays* L.) mediated by *Agrobacterium tumefaciens*. *Nature Biotech.* **14**:745–750

Ishizaki-Nishizawa, O., Fuji, T., Azuma, M., Sekiguchi, K., Murata, N., Ohtani, T. and Toguri, T. 1996. Low temperature resistance of higher plants is significantly enhanced by a non-specific cyanobacterial desaturase. *Nature Biotech.* **14:** 1003–1006.

James, C. 2007. Global Status of commercialized Biotech/GM Crops:2007. ISAAA Brief 37.

Johal, G.S. and Briggs, S.P. 1992. Reductase activity encoded by the HM1 disease resistance gene in maize. *Science* **258:** 985.

Jones, A.L., Johansen, I.E., Bean, S.J., Bach, I. and Maule, A.J. 1998. Specificity of resistance to pea seed borne mosaic potyvirus in transgenic peas expressing the viral replicase (*Nib*) gene. *J. Gen. Virol.* **79:** 3129–3137.

Jones, J.D.G., Dean, C., Gidoni, D., Bond-Nutter, D., Lee, R., Bedrock, J. and Dunsmuir, P. 1988. Expression of bacterial chitinase protein in tobacco leaves using pro photosynthetic promoters. *Mol. Gen. Genet.* **212:** 536–542.

Kapusta, J., Modelska, A., Figlerowicz, M., Pniewski, T., Ltellier, M. *et al.* 1999. A plant derived edible vaccine against hepatitis B virus. *FASEB J.* **13:** 1796–1799.

Kasuga, M., Liu, Q., Miura, S., Yamaguchi-Schinozaki, K. and Shinozaki, K. 1999. Improving plant drought, salt and freezing tolerance by gene transfer of a single stress inducible transcription factor. *Nat. Biotechnol.* **17:** 287–291.

Kathuria, S.R., Nath, R., Pal, R., Singh, O., Fischer, R. *et al.* 2002. Functional recombinant antibodies against human chorionic gonadotropin expressed in plants. *Curr Sci.* **82:** 1452–1457.

Kawchuk, L.M., Martin, R.R. and Pherson, J.Mc. 1991. Sense and antisense RNA mediated resistance to potato leaf roll virus in Russet Burbank potato plants. *Mol. Plant-Microbe Interact.* **4:** 247–253.

Keller, H., Pamboukdjian, N., Ponchet, M., Poupet, A., Delon, R., Verrier, J.L., Roby, D. and Ricci, P. 1999. Pathogen induced eleicitin production in transgenic tobacco generates a hypersensitive response and nonspecific disease resistance. *Plant Cell* **11:** 223–235.

Kim, W.S. Kim, J. Krishnan, H.B. and Nahm, B.H. 2005. Expression of *Escherichia coli* branching enzyme in caryopses of transgenic rice results in amylopectin with an increased degree of branching. *Planta* **220:** 689–695.

Kishore, P.B.K., Hong, Z., Miao, G.H., Hu, C.A.A. and Verma, D.P.S. 1995. Overexpression of D1-pyrroline-5 carboxylate synthetase increases proline production and confers osmotolerance in transgenic plants. *Plant Physiol.* **108:** 1387–1394.

Klee, H.J. 1993. Ripening physiology of fruit from transgenic tomato (*Lycopersicon esculentum*) plants with reduced ethylene synthesis. *Plant Physiol.* **102:** 911–916.

Knutzon, D.S., Thurmond, J.M., Huang, Y.S., Chaudhary, S., Bobik, E.G. Jr., Chan, G.M., Kirchner, S.J. and Mukherji, P. 1998. Identification of D5 desaturase from *Mortierlla alpina* by heterologous expression in bakers yeast and canola. *J. Biol. Chem.* **273:** 29360–29366.

Kodama, H., Hamada, T., Horiguchi, G., Nishimura, M. and Iba, K. 1994. Genetic enhancement of cold tolerance by expression of a gene for chloroplast 3 fatty acid desaturase in transgenic tobacco. *Plant Physiol.* **105:** 601–605.

Kouassi, N.K., Chen, L., Sire C., Bangratz-Reyser, M., Beach, R.N., Fanquet, C.M. and Brugidou, C. 2006. Expression of rice yellow mottle virus coat protein enhances virus infection in transgenic plants. *Archives of Virology* **151:** 2111–2122.

Krebbers, E. and van de Kerckhove, J. 1990. Production of peptides in plant seeds. *Trends Biotechnol.* **8:** 1–3.

Kriete, G., Neuhaus, K., Perlick, A.M., Puhler, A. and Broer, I. 1996. Male sterility in transgenic tobacco plants induced by tapetum specific deacetylation of externally applied non-toxic compound N-acetyl L-phosphinothricin. *Plant J.* **9:** 809–818.

Laporte, M.M., Galagon, J.A., Shapiro, J.A., Boersig, M.R., Shewmaker, C.R. and Sharkey, T.D. 1997. Sucrose phosphate synthase activity and yield analysis of tomato plants transformed with maize sucrose phosphate synthase. *Planta* **203:** 253–259.

Lee, J.H., Hubel, A. and Schoffl, F. 1995. Depression of the activity of genetically engineered heat shock factor causes constitutive synthesis of heat shock proteins and increased thermotolerance in transgenic *Arabidopsis Plant J.* **8:** 603–612.

Lindbo, J.A., and Doughery, W.G. 1992 Untranslatable transcripts of the tobacco etch virus coat protein gene sequence can interfere with tobacco etch virus replication in transgenic plants and protoplasts. *Virology* **189:** 725–733.

Lloyd, A.M., Walbot, V. and Davies, R.W. 1992. *Arabidopsis* and *Nicotiana* anthocyanin production activated by maize regulators R and C1. *Science* **258:** 1773–1775.

Lodge, J.K., Kaniewski, W.K. and Tumer, N.E. 1993. Broad spectrum virus resistance in transgenic plants expressing pokeweed antiviral proteins. *Proc. Natl. Acad. Sci. USA* **90:** 7089–7093.

Longstaff, M., Newell, C.A. Boonstra, B., Strachan, G., Learmonth, D. *et al.* 1998. Expression and charactrisation of single chain Fv antibody fragments produced in transgenic plants against the organic herbicides atrazine and paraquat, *Biochin Biophys Acta* **1381:** 147–160.

Ma, J.K.C., Hiatt, A., Hein, M.B., Vine, N., Wang, G. *et al.* 1995. Generation and assembly of secretory antibodies in plants. *Science* **268:** 716–719.

Ma, J.K.C., Hikmat, B.Y., Wycoff, K., Vine, N.D., Chargelegue, D. *et al.* 1998. Characterization of a recombinant plant monoclonal secretory antibody and preventive immunotherapy in human. *Nat Med.* **4:** 601–609.

Ma, J.K.C., Lehner, T., Stabila, P., Fux, C.I. and Hiatt, A. 1994. Assembly of monoclonal antibodies with IgGi and IgA heavy chain domains in transgenic tobacco plants. *Eur J Immunol.* 131–138.

Malik, M.K., Slovin, J.P., Hwang, C.H. and Zimmerman, J.L. 1999. Modified ex-pression of a carrot small heat shock protein gene, hsp17.7, results in increased or decreased thermotolerance *Plant J.* **20:** 89–99.

Mariani, C., de Beuckeleer, M., Truettner, J., Leemans, J. and Goldberg, R.B. 1990. Induction of male sterility in plants by a chimaeric ribonuclease gene. *Nature* **347:** 737–741.

Mariani, C., de Beuckeleer, M., Truettner, J., Leemans, J. and Goldberg, R.B. 1992. A chimaeric ribonuclease–inhibitor gene restores fertility to male sterile plants. *Nature* **357:** 384–387.

Martin, G.B., Brommonschenkel, S.H., Chunwongsee, J., Frary, A., Ganal, M.W., Spivey, R., Wu, T., Earle, E.D. and Tanksley, S.D. 1993. Map based cloning of a protein kinase gene conferring disease resistance in tomato. *Science* **262:** 1432–1436.

Mason, H.S., Haq, T.A., Clements J.D. and Arntzen, C.J. (1998) Edible vaccine protects mice against *Escherichia coli* heat-labile enterotoxin LT-B gene. *Vaccine* **16:** 1336–1343.

Mason, H.S., Lam, D M-K. and Arntzen, C.J. 1992 Expression of hepatitis B surface antigen in transgenic plants. *Proc Natl Acad Sci USA* **89:** 11745–11749.

Mc Cormick, A.A., Kumagai, M.H., Hanley, K., Turpen, T.H., Hakim, I. *et al.* 1999. Rapid production of specific vaccines for lymphoma by expression of the tumor-derived single-chain Fv epitopes in tobacco plants. *Proc Natl Acad Sci USA* **96:** 703–708.

McKersie, B. D., Bowley, S. R., Harjanto, E. and Leprince, O. 1996. Water deficit tolerance and field performance of transgenic alfalfa overexpressing superoxide dismutase. *Plant Physiol.* **111:** 1177–1181.

McLain, L., Portia, C., Lomonossoff, G.P., Durrani, Z. and Dimmock, N.J.1995, AIDS Res. Hum. Retroviruses **11:** 327–334.

Mehta, R.A., Cassol, T., Li, N., Ali, N., Handa, A.K. *et al.* 2002. Engineered polyamine accumulation in tomato enhances phytonutrient content, juice quality, and vine life. *Nat. Biotechnol.* **20:** 613–618.

Mittler, R., Shulaev, V. and Lam, E. 1995. Coordinated activation of programmed cell death and defense mechanism in in transgenic tobacco plants expressing a bacterial proton pump. *Plant Cell* **7:** 29–42.

Molina, A. and Garcia, O.F. 1997. Enhanced tolerance to bacterial pathogens caused by transgenic expression of barley lipid transfer protein LTP2. *Plant J.* **12:** 669–675.

Molvig, L., Tabe, L.M., Eggum, B.O., Moore, A.E., Craig, S., Spencer, D. and Higgins, T.J.V. 1997. Enhanced methionine levels and increased nutritive value of seeds of transgenic lupin (*Lupinus angustifolius*) expressing a sunflower seed albumin gene. *Proc. Natl. Acad. Sci. USA* **94:** 8393–8398.

Momma, K., Hashimoto, W., Ozawa, S., Kawai, S., Katsube, T., Takaiwa, F., Kito, M., Utsumi, S. and Murata, K. 1999. Quality and safety evaluation of genetically engineered rice with soybean glycinin: analyses of the grain composition and digestibility of glycinin in transgenic rice. *Biosci. Biotechnol. Biochem.* **63:** 314–318.

Mourgues, F., Brisset, M.N. and Chevreau, E. 1998. Strategies to improve plant resistance to bacterial diseases through genetic engineering. *Trends. Biotechol.* **16:** 203–210.

Nordlee, J.A., Taylor, S.L., Townsend, J.A., Thomas, L.A. and Bush, R.K. 1996. Identification of a Brazil nut allergen in transgenic soybeans. *New Eng. J. Med.* **334:** 688–692.

Norelli, J.L., Aldwinckle, H.S., Destephano-Beltran, L. and Jatnes, L.M. 1994. Transgenic "mailing 26" apple expressing the attacin E gene has increased resistance to *Erwinia amylovora. Euphytica* **77:** 123–128.

Norelli, J.L., Jensen, L.A., Momol, M.T., Mills, J.Z., Ko, K-S., Aldwinckle, H.S. and Cummis, J.M. 1996. Increasing the resistance of apple rootstocks to fire blight by genetic engineering: A progress report. *Acta Hortic.* **411:** 409.

Ogawa, T., Hori, T. and Ishida, I. 1996. Virus induced cell death in plants expressing the mammalian 2'5' oligoadenylate system. *Nat. Biotech.* **14:** 1566–1569.

Owen, M., Gandecha, A., Cockboun, B. and Whiltelam, G. 1992 Synthesis of a functional anti-phytochrome single-chain Fv protein in transgenic tobacco. Biotechnology. (NY), *Nat. Biotech.* **10:** 790–794.

Paszhowski, J., Shillito, R.D., Saul, M., Mandak, V., Hohn, T. *et al.* 1984. Direct gene transfer to plants. *EMBO J.* **3:** 2712–2722.

Perl, A.R., Perl, T., Galili, S., Aviv, D., Shalgi, E., Malkin, S. and Galun, E. 1993. Enhanced oxidative stress defense in transgenic potato expressing tomato Cu, Zn superoxide dismutases. *Theor. Appl. Genet.* **85:** 568–576.

Pilons-Smits, E.A.H., Ebskamp, M.J.M., Paul, M.J., Jenken, M.J.W., Weisbeek, P.J. and Smeekens, S.C.M. 1995. Improved performance of transgenic fructan accumulating tobacco under drought stress. *Plant Physiol.* **107:** 125–130.

Pinto, Y.M., Kok, R.A. and Baulcombe, D.C. 1999. Resistance to rice yellow mosaic virus (RYMV) in cultivated African rice varieties containing RYMV transgenes. *Nat. Biotechnol.,* **17:** 702–707.

Pitcher, L.H., Repeti, P. and Zilinskas, B.A. 1994. Overproduction of ascorbate peroxidase protects transgenic tobacco plant against oxidative stress (Abstract No. 623). *Plant Physiol.* **105:** S–169.

Poirier, Y., Nawrath, C. and Somerville, C. 1995. Production of Polyhydroxyalkanoates, a Family of Biodegradable Plastics and Elastomers, in Bacteria and Plants. *Bio/Technology* **13:** 142–150.

Powell-Abel, P.A., Nelson, R.S., De, B., Hoffman, N., Rogers, S.G., Fraley, R.T. and Beachy, R.N. 1986. Delay of disease development in transgenic plants that express tobacco mosaic virus coat protein gene. *Science* **232:** 738–743.

Raina, A. and Datta, A. 1992. Molecular cloning of a gene encoding a seed specific protein with nutritionally balanced amino acid composition from *Amaranthus*. *Proc. Natl. Acad. Sci. USA* **89:** 1774–1778.

Ramirez, N., Ayala, M., Lorenzo, D., Palenzuela, D., Herrera, L., *et al*. 2002. Expression of a single-chain Fv antibody fragment specific for the hepatitis B surface antigen in transgenic tobacco plants. *Transgenic Res.* **11:** 61–64.

Reddy, A.S. and Thomas, T.L. 1996. Expression of a cyanobacterial D6 desaturase gene results in gamma linolenic acid production in transgenic plants. *Nat. Biotechnol.* **14:** 639–642.

Richter, L.J., Thanavala, Y., Arntzen, C.J. and Mason, H.S. 2000. Production of hepatitis B surface antigen in transgenic plants for oral immunization. *Nat Biotechnol.* **18:** 1167–1171.

Romer, S., Fraser, P.D., Kiano, J.W., Shipton, C.A., Misawa, N. *et al*. 2000. Elevation of the provitamin A content of transgenic tomato plants. *Nat Biotechnol.* **18:** 666–669.

Sano, T., Nagayama, A., Ogawa, T., Ishida, I and Okada, Y. 1997. Transgenic potato expressing a double stranded RNA-specific ribonuclease is resistant to potato spindle tuber viroid. *Nat. Biotechnol.* **15:** 1290–1294.

Savin, K.W., Baudinette, S.C., Graham, M.W., Michael, M.Z., Nugent, G.D., Lu, C.Y., Chandler, S.F. and Cornish, E.C. 1996. Antisense ACC oxidase delays carnation petal senescence. *HortSci.* **30:** 970–972.

Sayanova, O., Smith, M.A., Lapinskas, P., Stobart, A.K., Dobson, G., Christie, W.W., Shewry, P.R. and Napier, J. 1997. Expression of a borage desaturase cDNA containing an N-terminal cytochrome b5 domain results in the accumulation of high levels of D6 desaturated fatty acids in transgenic tobacco. *Proc. Natl. Acad. Sci. USA* **94:** 4211–4216.

Schake, S. A., Wong-Vega, L. and Allen, R. D. 1995. Analysis of tobacco plants that overexpress pea chloroplastic manganese superoxide dismutase (abstract No. 260). *Plant Physiol.* **108:** S–62.

Schmulling, T., Schell, J. and Spena, A. 1988. Single gene from *Agrobacterium rhizogenes* influence plant development. *EMBO J.* **7:** 2621–2629.

Sen Gupta, A., Heinen, J.L., Holady, A.S., Burke, J.J. and Allen, R.D. 1993. Increased resistance to oxidative stress in transgenic plants that overexpress chloroplastic Cu/Zn superoxide dismutase. *Proc. Natl. Acad. Sci. USA* **90:**1629–1633.

Shewmaker, C., Sheehy, J., Daley, M., Colburn, S. and Ke, D.Y. 1999. Seed-specific expresssion of phytoene synthase: increase in carotenoids and other metabolic effects. *Plant J.* **20:** 401–412.

Shintani, D. and Dellapenna, D. 1998. Elevating the vitamin E content of plants through metabolic engineering. *Science* **282:** 2098–2100.

Smirnoff, N. and Bryant, J.A. 1999. DREB takes the stress out of growing up. *Nat. Biotechnol.* **17:** 229–230.

Song, W.Y, Wang, G.L., Chen, L.L., Kim, H.S. and Pi, L.Y. 1995. A receptor kinase like protein encoded by the rice disease resistance gene, *Xa21*. *Science* **270:** 1804–1806.

Sowa, A.W., Duff, S.M.G., Guy, P.A. and Hill R.D. 1998. Altering hemoglobin levels changes energy status in maize cells under hypoxia. *Proc. Natl. Acad. Sci. USA* **95:** 10317–10321.

Steinecke, P., Herget, T. and Schreier, P.H. 1992. Expression of a chimeric ribozyme gene results in endonucleolytic cleavage of target mRNA and a concomitant reduction of gene expression *in vivo*. The *EMBO J.* **11:** 1525–1530.

Stevenson, T.W. and Cornish, E.C. 1993. Cloning and expression of cytochrome p450 genes controlling flower color. *Nature* **366:** 276–279.

Stirpe, F., Barbieri, L., Battelli, ,M.G., Soria, M. and Lappi, D.A. 1992. Ribosome inactivating proteins from plants: present status and future prospects. *Bio/Technology* **10:** 405–412.

Stoger, E., Vaquero, C., Torres, E., Sack, M. and Nicholson, L. 2000. Cereal crops as viable production and storage systems for pharmaceutical scFv antibodies. *Plant Mol Biol.* **42:** 583–590.

Streatifield, S.J., Jilka, J.M., Hood, E.E., Turner, D.D., Bailey, M.R. *et al*. 2001. Plant based vaccines unique advantage. *Vaccine* **19:** 2742–2748.

Strittmatter, G., Janssens, J., Opsomer, C. and Botterman, J. 1995. Inhibition of fungal disease development in plants by engineering controlled cell death. *Bio/Technol.* **13:** 1085–1089.

Suzuki, K., Zue, H., Tanaka, Y., Fukui, Y., Mizutani, M. and Kusumi, T. 1997. Molecular breeding of flower color of *Torenia fournieri*. *Plant Cell Physiol.* **38:** 538–540.

Tacke, E., Salamani, F. and Rhode, W. 1996. Genetic engineering of potato for broad spectrum protection against virus infection. *Nat. Biotechnol.* **14:** 1597–1601.

Tanaka, Y., Tsuda, S. and Takaaki, K. 1998. Application of recombinant DNA to floriculture. In: *Applied Plant Biotechnology* (eds. V.L.Chopra, V.S. Malik and S.R. Bhatt), Oxford & IBH, New Delhi, pp. 181–235.

Tanaka, Y., Yonekura, K., Fukuchi-Mizutani, M., Fukui, Y., Fujiwara, H., Ashikari, T. and Kusumi, T. 1996. Molecular and biochemical characterization of three anthocyanin synthetic enzymes from *Gentiana triflora*. *Plant Cell Physiol.* **37:** 711–716.

Tarczynscki, M.C., Jensen, R.G. and Bohnert, M.J. 1993. Stress protection of transgenic tobacco by production of the osmolyte mannitol. *Science* **259:** 508–510.

Tien, P, Zhang, X., Qiu, B. and Wu, G. 1987. Satellite RNA for the control of plant diseases caused by cucumber mosaic virus. *Ann. Appl. Biol.* **111:** 143.

Tien, P. and Gusui, N. 1991. Satellite RNA for the biocontrol of plant diseases. *Adv. Virus Res.* **39:** 321.

Toubart, P., Desiderio, A., Salvi, G., Cervone, F., Daroda, L., De Lorenzo, G., Bergmann, C., Darvill, A.G. and

Albersheim, P. 1992. Cloning and characterization of the gene encoding the endo polygalacturonase inhibiting protein (PGIP) of *Phaseolus vulgaris* L. *Plant J.* **2:** 367–373.

Truve, E., Kelve, M., Aaspollu, A., Kuusksalu, A., Seppannen, P. and Sarma, M. 1994. Principles and background for the construction of transgenic plants displaying multiple virus resistance. *Arch. Virol. (Suppl)* **9:** 41–50.

Turpen, T.H., Reinl, S.J., Charroenvit, Y., Hoffman, S.L., Fallarme, V. and Grill, L.K. 1995. Malaria Epitopes Expressed on the surface of Recombinant Tobacco Mosaic Virus *Bio/Technology* **13:** 53–57.

Vaeck, N., Reynaerts, A., Hofte, H., Jansens, S., Beukeleer, M.D., Dean, C., Zabeau, M., Montagu, M.C. and Leemans, J. 1987. Transgenic plants protected from insect attack. *Nature* **328:** 33–37.

Vaquero, C., Sack, M., Schuster, F., Finnern, R., Drossard, J. *et al*. 2002. A carcinoembryonic antigen-specific diabody produced in tobacco. *FASEB J.* **16:** 408–410.

Vardi, E., Sela, I., Edelbaum, O., Livneh, O., Kuznetsova, L. and Stram, Y. 1993. Plants transformed with a cistron of a potato virus Y protease (Nia) are resistant to virus infection. *Proc. Natl. Acad. Sci. USA* **90:** 7513–7517.

Vasconncelos, M. Datta K. Oliva, N. Khalekuzzaman, M. Torrizo, L. Krishnan, S. Oliveira, M. Goto, F. and Datta, S.K. 2003. Enhanced iron and zinc accumulation in transgenic rice with ferritin gene. *Plant Sci.* **164:** 371–378.

Voelker, T.A., Worrell, A.C., Anderson, L., Bleibaum, J., Fanc, C., Hawkins, D.J., Radke, S.E. and Davis, H.M. 1992. Fatty acid biosynthesis redirected to medium chains in transgenic oil seed plants. *Science* **257:** 872–874.

Wang, P., Zoubenko, O. and Tumer, N.E. 1998. Reduced toxicity and broad spectrum resistance to viral and fungal infection in transgenic plants expressing pokeweed antiviral protein II. *Plant Mol. Biol.* **38:** 957–964.

Webb, R.P. and Allen, R.D. 1995. Overexpression of a pea cytosolic ascorbate peroxidase in *Nicotiana tabacum* confers protection against the effects of paraquat (abstract No. 272). *Plant Physiol.* **108:** S–64.

Wegener, D., Steinecke, P., Herget, T., Petereit, I., Philipp, C. and Schreier, P.H. 1994. Expression of a reporter gene is reduced by ribozyme in transgenic plants. *Mol. Gen. Genet.* **245:** 465–470.

Wu, G, Shott, B.J., Lawrence, E.B., Levine, E.B., Fitzsimmons, K.C. and Shah, D.M. 1995. Disease resistance conferred by expression of a gene encoding H2O2 generating glucose oxidase in transgenic potato plants. *Plant Cell* **7:** 1357–1368.

Xiangbai, D., Ruiqin J., Xuelan G., Simon J. F., Hong C., Caihua D., Yueying L., Qiong H. and Shengyi L. 2008. Expressing a gene encoding wheat oxalate oxidase enhances resistance to *Sclerotinia sclerotiorum* in oilseed rape (*Brassica napus*). *Planta*.

Xu, D., Duan, X., Wang, B., Hong, B., Ho, T.D. and Wu, R. 1996. Expression of a late embryogenesis abundant protein gene HVA1 from barley confers tolerance to water deficit and salt stress in transgenic rice. *Plant Physiol.* **110:** 249–257.

Ye, X., Al-Babili, S., Kloeti, A., Zhang, J., Lucca, P., Beyer, P. and Potrykus, I. 2000. Engineering the Provitamin A (Beta carotene) biosynthetic pathway into (carotenoid free) rice endosperm. *Science* **287:** 303–305.

Yuan, Q., Hu, W., Pestka, J.J., He, S.Y. and Hart, L.P. 2000. Expression of a functional antizearalenone single- chain Fv antibody in transgenic Arabidopsis plants. *Appl. Environ. Microbiol.* **66:** 3499–3505.

Yusibov, V., Modelska, A., Steplewski, K., Agadjanyan, M., Weiner, D., Hooper, D.C. Koprowski, H. and Yusibov, V. 1997. Immunization against rabies with plant derived antigen. *Proc. Natl. Acad. Sci. USA* **95:** 2481–2485.

Zeitlen, L., Olmsted, S.S., Moench, T.R., Co, M.S., Martinell, B.J. *et al*. 1998. A humanized monoclonal antibody produced in transgenic plants for immuno protection of the vagina against genital herpes. *Nat Biotechnol.* **16:** 1361–1364.

Zhang, L., Ku, J. and Birch, R.G. 1999. Engineered detoxification confers resistance against a pathogenic bacterium. *Nat. Biotechnol.* **17:** 1021–1024.

## FURTHER READING

Pattanayak, D., Kumar, P.A. 2000. Plant biotechnology: Current advances and future perspectives. *Proc. Ind. Acad. Sci. B.*, **6:** 266–310.

Sharma, A.K., Jani, Dewal and Raghunath, C. and Tyagi, A.K. 2004 Transgenic plants as bioreactors, Ind J. Biotechnology, **3(2):** 274–290.

Singh, B.D. Plant Biotechnology, Kalyani Publishers, Ludhiana, India, pp. 755.

# 26

# Impact of Recombinant DNA Technology

## Agriculture

Biotechnology exploded onto the scene of crop based agriculture in 1984 with the first report of inheritance of a foreign gene engineered into tobacco plants. Prior to this time, many scientists speculated on the use of recombinant DNA techniques for improvement of organisms. However, the list of plant species that can be transformed has been expanding year after year. Success includes most major economic crops, vegetables, ornamental, medicinal, fruit, tree and pasture plants. The initial publication of *Agrobacterium* mediated transformation, and those that followed shortly thereafter sparked a new industry in plant biotechnology. Many companies were quick to jump on the 'biotech wagon' prompted by claims for the potential to modify crop plants. In 1990s there was an upsurge of private sector investment in agricultural biotechnology. Some of the first products were insect and herbicide resistant crop plants and 'Flavr Savr' tomato with delayed ripening. The development of transgenic crops to these agrochemicals was the work of big multinational companies who were producing these chemicals. The companies were looking for their profits and they invested heavily on seed industry to develop transgenic crops which were resistant to their trade mark chemicals (herbicides and pesticides). In other words MNCs were able to sell their improved genetically modified seeds and also their herbicide and pesticide chemicals. No doubt area under these transgenic crops has been increasing steadily in developed countries and also the production level has gone higher and as per reports the benefits have been accrued by the farmers.

## Status of transgenic crops

Transgenic plant products have now been released as commercial varieties in a range of crops (Table 26.1). By all accounts, 1996 was a critical year in the history of plant biotechnology. For the first time farmers planted large areas under transgenic/biotech plants. In 1995 the variety of tomato named as Flavr Savr was given regulatory approval for commercial cultivation which had delayed ripening gene. In the following year, cotton varieties expressing the Bt crystal protein toxic to a number of lepidopteran pests including the tobacco budworm and cotton bollworm were introduced. At the same time Bt corn hybrids with improved resistance to European corn borer became available to the farmers. All these commercial introductions have taken place in USA. Transgenics for herbicide tolerance was the dominant trait in 1997. Now the transgenics have been released for various traits from virus resistance to quality and for industrial use.

**Table 26.1** Commercial releases of transgenic plant varieties (Source: James, C., 2007)

| Trait | Crop | Product name | Organization |
|---|---|---|---|
| **Insect resistance** | | | |
| Bt (*cry1Ac, cry2Ab*) resistance to lepidopteran insects | Cotton | Bollguard | Monsanto Co. |
| Bt (*cry1Ac*) resistance to lepidopteran insects | Cotton | - | Calgene |
| Bt (*cry3A*) resistance to Colorado potato beetle | Potato | Newleaf | Monsanto Co. |
| Bt (*cry1Ab*) resistance to European corn borer | Maize | Yield Guard | Monsanto Co. |
| Bt (*cry3Bb1*) resistance to corn root worm | Maize | | Monsanto Co. |
| Bt (*cry1Ab*) resistance to European corn borer | Maize | Maximizer | Ciba Seeds |
| Bt (*cry3A*) & Coat Protein (CP) gene; resistance to Olorado potato beetle and potato virus (PVY) | Potato | New leaf Y | Monsanto Co. |
| Bt (*cry1Ab*) | Maize | Knock Out | Novartis, Mycogen, Pioneer Hi-Bred Int. Inc |
| Bt | Rice | | Agricultural Biotech Research Institute, Iran |
| Bt (*cry1Ac*) | Tomato | | Monsanto Co. |
| Bt | Poplar | | Research Institute of Forestry |
| **Virus resistance** | | | |
| Virus resistance (coat protein gene) | Squash | Freedom II Destiny, liberator | Asgrow Seeds, Inc. Upjohn, USA |
| Virus resistance (satellite RNA) | Tomato, Capsicum | — | Beijing University |
| Virus resistance (satellite RNA) | Tobacco, | | |
| Virus resistance (coat protein) | Cantaloupe | — | Asgrow Seeds, Inc. |
| Virus resistance | Plum (*Prunus domestica*) | | USDA-Agricultural Research Service |
| Virus resistance (coat protein) | Papaya | Rainbow, Sun up | Cornell University, |
| Virus resistance (coat protein) | Papaya | | South China Agricultural University |
| **Herbicide resistance** | | | |
| Glufosinate (*bar - PAT* gene) | Maize | Liberty Link | AgrEvo Canada Inc. |
| Glufosinate (*bar - PAT* gene) | | Liberty Link | AgrEvo USA Co. |
| Sethoxydim | | — | BASF |
| Glufosinate (*bar - PAT* gene) | | — | DeKalb Genetics Corp. |
| Glyphosate (EPSPS gene) | | Roundup Ready | Monsanto Co. |
| Glufosinate (*bar - PAT* gene) | | — | Plant Genetic Systems |
| Glufosinate (*bar - PAT* gene) | Oilseed rape Canola | Innovator | AgrEvo Canada Inc. |
| Glufosinate (*bar - PAT* gene) | | ,, | AgrEvo USA Co. |
| Glyphosate (EPSPS gene) | | Roundup Ready | Monsanto Co. |
| Glufosinate (*bar - PAT* gene) | | — | Plant Genetic Systems |
| Bromoxynil (*bxn* gene) | | — | Rhone-Poulenc |
| Bromoxynil (*bxn* gene) | Cotton | BXN | Calgene Inc. |
| Sulfonylurea (*als* gene) | — | — | Du Pont |
| Glyphosate (EPSPS gene) | — | Roundup Ready | Monsanto Co. |
| Glyphosate (EPSPS gene) | Soybean | Roundup Ready | Monsanto Co. |
| Glufosinate (*bar - PAT* gene) | — | Liberty Link | Aventis Crop Science |
| Glufosinate (*bar - PAT* gene) | Rice | Liberty Link | Aventis Crop Science; Bayer Crop Science |
| Sulfonylurea (*als* gene) | Flax | Triffid | Univ. of Saskatchewan |
| Glufosinate | Chicory | — | Bejo-Zaden |
| Glyphosate (EPSPS gene) | Sugarbeet | | Monsanto Co. |
| Herbicide | Alfalfa | | Monsanto Co. & Forage Genetics International |

*(contd. Table 26.1)*

*(contd. Table 26.1)*

| Trait | Crop | Product name | Organization |
|---|---|---|---|
| **Quality characters** | | | |
| Vine ripened, shelf life (PG gene) | Tomato | Flavr Savr | Calgene |
| Vine ripened, shelf life (amino cyclopropane -1-carboxylic acid (ACC) synthase gene) | Tomato | Endless Summer | DNA Plant Technology |
| Delayed softening, Paste consistency (PG gene) | Tomato | — | Zeneca |
| Delayed ripening/ longer shelf life and improved taste (S-adenosyl methionine hydrolase-(SAMase) gene) | Tomato | | Agrotope Inc. |
| Delayed ripening | Melon | | Agrotope Inc. |
| Increased vase life/Delayed ripening | Carnation | | Florigene Pty Ltd |
| Modified flower color | Carnation | — | Florigene Pty Ltd |
| Modified oil characteristics | Canola | Laurical | Calgene |
| Modified fatty acid content—high oleic acid (fatty acid desturase (*fad2*) gene) | Soybean | | Du Pont Canada Agricultural Products |
| Increased lysine content | Maize | | Monsanto Co. |
| Increased nicotine content | Tobacco | | Vector Tobacco Inc. |
| Cedar pollen peptide | Rice | | National Inst. Agro-biological Sciences, Japan |
| High cell number | Poplar (*Populus alba*) | | National Inst. Agro-biological Sciences, Japan |
| Insect and virus resistance | Potato | | Monsanto Company |
| **Stacked traits** | | | |
| Herbicide tolerance and modified oil content | Soybean | Du Pont | |
| Herbicide and insect resistance | Maize | | Syngenta Seeds, Monsanto Company, Pioneer Hi-Bred Int. Inc. |
| Herbicide and insect resistance | Cotton | | Syngenta Seeds, AVIPE, Monsanto Company |
| Herbicide tolerance and male sterility/fertility restorer | *Brassica napus* | | Aventis Crop Science; Bayer Crop Science |
| Herbicide tolerance and male sterility/fertility restorer | Maize | | Bayer Crop Science; Pioneer Hi-Bred Int. Inc |
| Herbicide tolerance and fertility restorer | Chicory | SeedLink | Bejo Zaden BV |
| Modified flower color and herbicide tolerance | Carnation | Moonlite, Moonshade, Moonshadow, Moonvista | Florigene Pty Ltd |

There has been dramatic and continuing increase in area planted to transgenic crops. From 1.7 million hectares in 1996, the area increased nearly sixty seven fold to 114.3 million hectares in 2007. The accumulated hectarage during the 12 years period from 1996–2007 has reached almost 700 million hectares, making it the fastest adopted crop technology (Table 26.2). In 2007, the number of countries planting biotech crops totaled 23 which comprised of 12 developing and 11 industrial/developed countries. They are in order of hectarage, USA, Argentina, Brazil, Canada, India, China, Paraguay, South Africa, Uruguay, Philippines, Australia, Spain, Mexico, Colombia, Chile, France, Honduras, Czech Republic, Portugal, Germany, Slovakia, Romania and Poland (Table 26.3). Notably, the first eight of these countries grew more than 1 million hectares each. Results of 2007 year have shown that more than 50 million farmers have adopted this technology so far. In 2007, 43% of the global biotech crop area which is equivalent to 49.4 million hectares, was grown in developing

**Table 26.2:** Global Area of Biotech Crops in the First 12 Years (Source: James, C., 2007)

| Year | Hectares (million) | Acres (million) |
|------|--------------------|-----------------|
| 1996 | 1.7 | 4.3 |
| 1997 | 11.0 | 27.5 |
| 1998 | 27.8 | 69.5 |
| 1999 | 39.9 | 98.6 |
| 2000 | 44.2 | 109.2 |
| 2001 | 52.6 | 130.0 |
| 2002 | 58.7 | 145.0 |
| 2003 | 67.7 | 167.2 |
| 2004 | 81.0 | 200.0 |
| 2005 | 90.0 | 222.0 |
| 2006 | 102.0 | 252.0 |
| 2007 | 114.3 | 282.0 |
| Total | 690.9 | 1,707.3 |

**Table 26.3:** Global Area of Biotech Crops in 2007: by Country (Source: James, C., 2007)

| Country | 2007 (Million Hectares) | % |
|---------|-------------------------|---|
| USA | 57.7 | 50 |
| Argentina | 19.1 | 17 |
| Brazil | 15.0 | 13 |
| Canada | 7.0 | 6 |
| India | 6.2 | 5 |
| China | 3.8 | 3 |
| Paraguay | 2.6 | 2 |
| South Africa | 1.8 | 2 |
| Uruguay | 0.5 | <1 |
| Philippines | 0.3 | <1 |
| Australia | 0.1 | <1 |
| Spain | 0.1 | <1 |
| Mexico | 0.1 | <1 |
| Colombia | <0.1 | <1 |
| Chile | <0.1 | <1 |
| France | <0.1 | <1 |
| Honduras | <0.1 | <1 |
| Czech Republic | <0.1 | <1 |
| Portugal | <0.1 | <1 |
| Germany | <0.1 | <1 |
| Slovakia | <0.1 | <1 |
| Romania | <0.1 | <1 |
| Poland | <0.1 | <1 |
| Total | 114.3 | 100 |

countries where growth between 2006 and 2007 was substantially higher (8.5 million hectares or 21% growth) than industrial countries (3.8 million hectares or 6% growth). Five principal developing countries committed to biotech crops, span all the three continents of the South; they are India and China in Asia, Argentina and Brazil in Latin America and South Africa on the African continent. The USA is one of the six "founder biotech crop countries" having commercialized biotech maize, soybean, cotton and potato in 1996, the first year of global commercialization of biotech crops. The USA continued to be the lead biotech country in

2007 with 57.7 million hectares followed by Argentina (19.1 mhac), Brazil (15.0 mhac), Canada (7.0 mhac), India (6.2 mhac) and China (3.8 mhac).

The four principal transgenic crops in the world are soybean, maize, cotton and rapeseed/canola. Biotech soybean continued to be the principal biotech crop in 2007, occupying 58.6 million hectares (57% of global biotech area), followed by maize (35.2 million hectares at 25%), cotton (15.0 million hectares at 13%) and canola (5.5 million hectares at 5% of global biotech crop area) (Table 26.4). Herbicide tolerance has consistently been the dominant trait since 1996 when transgenic crops were released for commercial cultivation. In 2007, herbicide tolerance, deployed in soybean, maize, canola, cotton and alfalfa occupied 63% or 72.2 million hectares of the global biotech 114.3 million hectares while insect resistant varieties occupied 20.3 million hectares with 18% of global biotech planted area (Table 26.5). The stacked double and triple traits mainly herbicide tolerance with insect resistance or virus resistance trait occupied a larger area (21.8 million hectares, or 19% of global biotech crop area). In the USA two third of biotech maize hectarage featured double or triple constructs of Bt and herbicide tolerance traits whereas over 93% of biotech cotton featured stacked traits of insect and herbicide resistance. The triple gene products in biotech maize featured two Bt genes (to control European corn borer complex and rootworm) and one herbicide trait, which was first commercialized in the USA in 2005. Besides the main biotech crops of soybean, maize, cotton and canola some other crops have been commercialized viz. herbicide tolerant alfalfa grown on less than 0.1 million hectares in USA, Bt rice in Iran grown on 4000 hectares. Biotech virus resistant squash (2000 hectares) and PRSV resistant papaya in Hawaii (2000 hectares) and in China 3500 hectares area under biotech sweet pepper, tomato and poplar.

**Table 26.4:** Biotech crop area as percentage of global area of principal crops in the year 2007 (Source: James, C., 2007)

| Crop | Global area* | Biotech crop area | Biotech crop area as % of global area |
|---|---|---|---|
| Soybean | 91 | 58.6 | 64 |
| Cotton | 35 | 15.0 | 43 |
| Maize | 148 | 35.2 | 24 |
| Canola | 27 | 5.5 | 20 |
| Total | 301 | 114.3 | 38 |

*FAO 2005 hectarage

**Table 26.5:** Dominant Biotech Crops in 2007 (Source: James, C., 2007)

| Crop | Million Hectares |
|---|---|
| Herbicide tolerant Soybean | 58.6 |
| Stacked traits Maize | 18.8 |
| Bt Cotton | 10.8 |
| Bt Maize | 9.3 |
| Herbicide tolerant Maize | 7.0 |
| Herbicide tolerant Canola | 5.5 |
| Stacked traits Cotton | 3.2 |
| Herbicide tolerant Cotton | 1.1 |
| Herbicide tolerant Alfalfa | <0.1 |
| Others | <0.1 |
| Total | 114.3 |

Farmers decided to plant more genetically modified rapeseed, corn, soybean and cotton because of more convenient and flexible crop management, higher productivity and safer environment through decreased use of conventional pesticides. The number of farmers benefiting from biotech crops continued to grow. In 2007 about 12 million farmers were growing biotech crops with 11 million (more than 90%) were small farmers from developing countries. The 11 million small farmers included 7.1 million farmers in the cotton growing provinces of China, 3.8 million farmers in India, 100,000 small farmers growing Bt maize in Philippines and several thousand in South Africa.

## Benefits of transgenic crops

An initial assessment of the net benefits from Bt insect resistant transgenic crops of cotton, corn and potato are estimated to be $ 33, $ 27.25 and $ 19 per acre respectively. The herbicide tolerant gene(s) introduced into different crops have no effect on yield *per se*, but significantly better weed control was achieved which in turn increased yield dependability. Farmers in USA using herbicide tolerant soybean required only one application of herbicide. This decreased requirement of herbicides on herbicide tolerant varieties, translated to herbicide savings of 10 to 40%. It also resulted in better control of weeds and soil moisture conservation, improved yield dependability, no carry-over of herbicide residues and much more flexibility in agronomic management of the crop. Herbicide tolerant canola in Canada lowered herbicide requirement from 570 g to 160 g active ingredient per acre, increased yield up to 20%, with an average increased yield of 9%.

A study conducted by NCFAP (National Center for Food and Agriculture Policy) in 2001 has shown that eight biotech cultivars adopted by US growers increased crop yields by 4 billion pounds, saved growers $ 1.2 billion by lowering production costs and reduced pesticide use by 46 million pounds. These eight cultivars include insect resistant corn and cotton, herbicide tolerant canola, corn, cotton and soybean, and virus resistant papaya and squash. The adopted cultivars provided a net value of $1.5 billion, which was determined by adding any increased value of the crop plus or minus any changes in grower costs.

Among the eight adopted crops, the greatest yield increase came from insect resistant corn (3.5 billion pounds) and insect resistant cotton (185 million pounds). The greatest cost savings was realized in herbicide tolerant soybeans ($ 1 billion), herbicide tolerant cotton ($ 133 million) and herbicide tolerant corn ($ 58 million). The greatest pesticide reduction was seen in herbicide tolerant soybean (28.7 million pounds) and herbicide tolerant cotton (6.2 million pounds).

NCFAP studied 40 cultivars that have been developed or are being developed to manage pests in 27 crops. The 40 pest management case studies were classified according to six pest control criteria: insect resistance, herbicide tolerance, nematode resistance, bacteria resistance, virus resistance and fungus resistance. It demonstrated that biotechnology is having and can continue to have significant impact on improved yields, reduced grower costs and pesticide reduction. If growers adopt all of the cultivars examined by NCFAP, the total net economic impact would be $ 2.5 billion per year, an annual increase in production of 14 billion pounds and a pesticide reduction of 163 million pounds per year.

As expected, in 2007, the global market value of biotech crops, estimated by Cropnosis, was US $ 6.9 billion representing 16% of the US $ 42.2 billion global crop protection market in 2007, and 20% of the ~US $ 34 billion 2007 global commercial seed market (Table 26.6). The US $ 6.9 billion biotech crop market comprised of US $ 3.2 billion for biotech maize (equivalent to 47% of global biotech crop market), US $ 2.6 billion for biotech soybean, US $ 0.9 billion for biotech cotton (13%), and US $ 0.2 billion for biotech canola (3%). Of the US $ 6.9 billion biotech crop market, US $ 5.2 billion (76%) was

**Table 26.6:** The Global Value of the Biotech Crop Market, 1996 to 2007 (Source: James, C., 2007)

| Year | Value (Million of US $ ) |
|------|--------------------------|
| 1996 | 115 |
| 1997 | 842 |
| 1998 | 1,973 |
| 1999 | 2,703 |
| 2000 | 2,734 |
| 2001 | 3,235 |
| 2002 | 3,656 |
| 2003 | 4,152 |
| 2004 | 4,663 |
| 2005 | 5,248 |
| 2006 | 6,151 |
| 2007 | 6,872 |
| Total | 42,344 |

in the industrial countries and US $ 1.6 billion (24%) was in the developing countries. The market value of the global biotech crop market is based on the sale price of biotech seed plus any technology fees that apply. The accumulated global value for the eleven-year period, since biotech crops were first commercialized in 1996, is estimated at US $ 42.4 billion. The global value of the biotech crop market is projected at approximately US $ 7.5 billion for 2008 (James, 2007). In the recent global study on the benefits of biotech crops, Brookes and Barfoot (2008) estimated that USA and Argentina have enhanced farm income from biotech crops by US $ 15.9 billion and US $ 6.6 billion in the first eleven years of commercialization (1996–2006) and the benefits for 2006 alone are estimated at US $ 2.9 billion and US $ 1.3 billion respectively. A study conducted in Argentina showed that approximately 75–77% of the benefits were gained by farmers while 25% of the benefits were gained by technology developers and the Argentine government. Other countries which have gained from biotech crops are China (US $ 5.8 billion), Brazil (US $ 1.9 billion), India (US $ 1.5 billion) and Canada (US $ 1.2 billion). The survey on economic benefits also revealed accumulative reduction in pesticide usage of 289,300 metric tonnes of active ingredient which is equivalent to 15.5% reduction as measured by Environmental Impact Quotient. In addition to the direct savings from insect resistant and herbicide tolerant traits associated with yield improvements, reduced pesticides, fuel and labor there were indirect benefits of no/low till systems and lower fuel consumption. These benefits have contributed to a permanent reduction in carbon dioxide emissions and resulted in high carbon sequestration in soil, with carbon dioxide savings of approximately 9 billion kg in 2005 alone.

## Biotech crops in India

Bt cotton is the only crop till 2007 which has been released for commercial cultivation. Bt cotton confers resistance to important insect pests of cotton was first adopted in India as hybrids in 2002 with approximately 50,000 hectares. Insects pests *Helicoverpa armigera* (American bollworm), *Spodoptera litura* (army worm), *Bemisia tabaci* (whitefly) pests currently reduce cotton yield by 50%. It has been reported that 70% of all pesticide use is in cotton. In India, Department of Biotechnology oversees and monitor the projects in biotechnology while large-scale tests and release were to be overseen by the Ministry of Environment and Forests. The modified DBT guidelines recommend that greenhouse and limited field trials in an open environment be done for atleast one year with minimum of four replications and ten locations in the agro-ecological zone for which material is intended.

---

### Bt cotton release in India—Sequence of events

| | |
|---|---|
| 1990 | Bt cotton importation and biosafety regulations; Monsanto refused permission to backcross bollgard into local varieties to get Bt cotton. |
| 1996 | Mahyco given permission to import and backcross bollgard into local varieties to get Bt cotton. |
| April–May 98 | Monsanto acquires 26% stake in Mahyco. |
| July 1998 | Mahyco given permission to plant Bt cotton trials in AP, Punjab, Karnataka, Haryana and Maharashtra. |
| Nov 1998 | Bt cotton trials become public knowledge. |
| 2nd Dec 1998 | It was reported that no terminator genes are present in Bt cotton. |
| 29th Dec 1998 | US files patent plea for terminator gene in India. |
| 28th Nov 1998 | Protests against Mahyco-Monsanto field trials. Burning of Bt cotton field trial crop in Karnataka. |
| June 2001 | GEAC denies permission to Mahyco and asked for conduct of large trials. |
| Sept. 2001 | Around 500 farmers in Gujarat were found to have planted the seed of Bt cotton on 11,000 acres. Around 480 acres were planted in AP also. Though Bt cotton was not recommended yet for commercial cultivation by GEAC. |
| 26th March 2002 | GEAC headed by Dr A.M. Gokhale approved 3 varieties of cotton for commercial cultivation—Mech 12, Mech 162, Mech 18 with conditions. |

i. Seeds should contain the label of GM on them.

ii. To submit the data of seed selling every year to GEAC including the area.

iii. To avoid pollination, Bt field should be surrounded by non-Bt crop (15 lines) around it.

iv. Every farmer has to maintain the ratio of 80:20:: Bt cotton: non-Bt cotton.

---

On March 26, 2002 M/S Mahyco (Maharashtra Hybrid Company) was the first to receive approval for three Bt cotton hybrids i.e. Mech 12, Mech 162 and Mech 184. Genetic Engineering Approval Committee (GEAC) under the MoEF approved the release of Bt transgenic cotton hybrids for commercial cultivation for a period of three years after extensive field trials. The Bt transgene in the converted Indian inbred lines behaves as a single dominant Mendelian factor and is stably integrated in the plant genome. The Bt trait has been successfully transferred into more than 60 cotton lines by traditional backcrossing method of plant breeding. DBT assessed the biosafety of the *cry*IAc Bt gene in these hybrids and then allowed the field testing through systematic trials under ICAR-AICCIP for two years during 2000 and 2001. M/s Mahyco had undertaken field trials throughout India under different agro-climatic conditions. GEAC imposed the following approval conditions for commercialization of Bt cotton:

i. Valid for three Years: April 02 to March 05

ii. Three hybrids namely Mech 12, Mech162 and Mech184

iii. Provide same non Bt seed to meet refuge requirements

iv. Conduct studies to monitor resistance development

v. Provide information to government on distribution of the seed through its dealers and agents

vi. Labeling requirements such as GEAC number, etc.

vii. Develop Bt based IPM program

viii. Conduct educational and awareness programs

ix. Meet other requirements as stipulated

The performance of the Bt cotton was being regularly monitored both at the state and the central levels by monitoring and evaluation committees. Approximately 72,000 packets of seeds containing Bt cotton hybrids and its non Bt cotton hybrids counterparts for covering one acre each was sold by Mahyco in the *kharif* season of 2002. In the first year Bt cotton was grown in six states covering an area of ~50,000 hectares. The area was doubled to ~100,000 hectares in 2003, increased to four fold in 2004 to reach half a million hectares. In 2005 the area planted to Bt cotton reached 1.3 million hectares and in 2006 it was 3.8 million hectares. However, in 2007 there was a record increase in the adoption of Bt area from 3.8 million hectares to 6.2 million hectares, which represents 66% of the total cotton area (9.4 million hectares) in India (Table 26.7). Although the major states growing Bt cotton are Maharashtra, AP, Gujarat and MP but it has spread to all the states in India. Not only the area has increased but by May 2008, GEAC had approved 131 hybrids for commercial cultivation in all the different agro-climatic zones with four different gene events in different cotton hybrids. GEAC has approved the commercial release of an indigenous cotton variety called *Bikaneri Narma* (BN) Bt expressing Bt *cry* 1Ac protein in the North, Central and South Cotton Growing Zones in India. This Bt cotton variety is the first public sector genetically modified (GM) crop in India. These gene events are: Bollgard-I with *cry*1Ac gene and Bollgard-II with *cry*1Ac and *cry*2Ab stacked genes developed by Mahyco-Monsanto, Event 1 developed by JK seeds featuring *cry*1Ac sourced from IIT Kharagpur and GFM event was developed by Nath Seeds featuring *cry*1Ac and *cry*1Ab sourced from China. India's apex biotech regulatory body, the Genetic Engineering Approval Committee, while reviewing its earlier decision directing the CICR to conduct large scale trials (LST) of Bt BN variety in North Zone,

decided to approve the commercial cultivation of Bt BN variety because the farmers can save the seeds for planting the following season if it was allowed for LST.

**Table 26.7:** Adoption of Bt cotton in India
(Source: James, C., 2007)

| Year | Thousand hectares |
|------|-------------------|
| 2002 | 50 |
| 2003 | 100 |
| 2004 | 500 |
| 2005 | 1,300 |
| 2006 | 3,800 |
| 2007 | 6,200 |

As a result India has become world's third largest producer of cotton behind China and USA. In the year 2001 India planted (21.3 million) nearly twice as many acres in cotton as China (11.8 million), but produced less than half the crop 11.8 million bales versus 24.4 million bales. India ranks first in the world in acres planted but third in production after China and US. By comparison, US cotton farmers who planted 15.8 million acres and produced 20 million bales. In recent years, the Indian crops have been so poor that the country has imported more raw cotton to supply its textile mills. India was ranked among the world's biggest importing countries in 2001 with just 308 kg/ha cotton yield which was less than half the global average. After the adoption of Bt technology, the yield levels have gone to 560 kg/ha in 2007–08 and is now an exporter rather than an importer of cotton. In India the cotton production for the year 2007–2008 was 31 million bales (1 bale = 170 kg) as compared to 15.8 million bales in 2001–02.

It is estimated that approximately 3.8 million small farmers planted on an average 1.65 hectares of Bt cotton in 2007. It has been reported that the use of insecticides on Bt cotton reduces the cost of producing cotton by 30%. It has increased yield by up to 50%, reduced insecticide sprays from 20 to seven, with environmental and health implications, and increased income by up to US $ 225 or more per hectare. At the national level, increased farmer income from Bt cotton in 2006 was estimated at US $ 840 million to US $ 1.7 billion, production has almost doubled. A study by Gandhi and Namboodiri (2006) reported yield gain of 31% due to Bt cotton, 39% reduction in the number of pesticide sprays and 88% increase in profit.

Now the initial obstacles in the access of a new technology have all but gone. All progressive Indian farmers' await the go ahead for GM papaya, tomato, corn, brinjal, cauliflower, mustard and other fruits and vegetables currently under multi-location field trials as approved by GEAC.

**Biotech Crops in Pakistan:** Pakistan's National Biosafety Committee (NBC) in May 2008 granted approval to Monsanto to import 13 biotech cotton seed varieties from India. About 200 gm each of the Bt cotton variety will be tested in different cotton growing areas during the current cotton season. The limited field trials will follow the concept of "Refugia and Buffer Zone" for growing genetically modified (GM) crops under Pakistan's biosafety rules and National Biosafety Guidelines. NBC will control and monitor the whole process from the importation of the seeds to field testing.

## Insect resistant plants through transgenic approach

Insecticides based on the crystalline toxin from *Bacillus thuringiensis* (*B.t*) vividly illustrate the influence of molecular biology and genetic engineering on traditional crop protection. The isolation of insect toxin genes from *B.t* has opened up an area relevant to plant protection. Plant Biotechnology has become a source of agricultural innovation, providing new solutions to age-old problems of crop protection from insect pests. The simple reason for this interest is that crop damage inflicted by phytophagous insects is staggering despite the use of sophisticated crop protection measures, chiefly chemical pesticides. This brings to the conclusion why transgenics for insect resistance. The cost associated with management practices

and chemical control of insects approaches $ 10 billion annually, yet global losses due to insects still account for 20% to 30% of the total production. Every year individual farmers face the possible devastation of their crops by insect infestations. Insecticides to control insects are mainly chemically synthesized and they have a negative or adverse effect on host as well as on environment apart from the cost incurred in their use. Chemical pesticide used for protection from insects has very hazardous effect because:

i. Highly inefficient.
ii. Most of the applied chemicals is wasted either in runoff or in subsequent washing from the plant surfaces, and it has been estimated that as much as 98% of the sprayed pesticides end up in the soil.
iii. It is difficult to deliver pesticides to the most vulnerable part of the plant e.g. roots or inside of stem or fruit.
iv. Pesticides and their residues are often highly toxic to beneficial organisms.
v. A plant species in its natural environment is in balance with the various organisms including insects that depend upon it for food. These insects in turn provide food for predatory insects, birds, fish and mammals.
vi. The natural order is often overlooked when considering crop plants and use of pesticides that may be toxic to non-target organisms.
vii. Soil condition may decline.
viii. Human toxicity is a major concern.
ix. Very strong selection pressure on insect populations is imposed by insecticides causing resistance to such compounds to be rapidly acquired.
x. Overuse of pesticides and herbicides can decrease the vigour of the crop and actually make it more susceptible to insect attack.

In view of this, genetic engineering of insect resistance into crops represents an attractive opportunity to reduce insect damage.

*Advantages*

1. More effective targeting of insects protected within plants.

2. Greater resilience to weather conditions.
3. Season long protection.
4. Insects are always treated at the most sensitive stage.
5. It affords protection of plant tissues which are difficult to treat using insecticide. For example control of larvae of corn root worm (*Diabrotica* spp.) which attack underground tissues of maize and the cabbage seed weevil (*Centrorhynchus assimilis*) which is the major pest of oil seed rape attacking the developing ovules with immature pods.
6. Only crop eating insects are exposed.
7. Fast biodegradability.
8. No need of skilled hand.
9. Reduced exposure of farmers, laborers, and non-target organisms to the pesticides.
10. Net financial benefit to farmers.
11. Increased production of a particular crop.
12. Increased activities of natural enemies of pests, because of reduction in insecticide sprays.
13. Reduced amount of pesticide residues in the food and food products.
14. A safer environment to live because of reduction in pesticide use.

*Limitations/Apprehensions*

1. The secondary pests will no longer be controlled in the absence of sprays for the major pests.
2. Cost of production of transgenic plant is very high.
3. Cost of transgenic seed sold to the farmers is very high with the result that there is increased burden on the farmer.
4. Development of resistance in insect population may limit the usefulness of transgenic crops for pest management.
5. Evolution of new insect biotypes: Due to deployment of transgenic crops there is danger of evolution of new biotypes of insects.
6. Insect sensitivity: There are many species of insects that are not susceptible to currently available Bt proteins. There is need to broaden the pool of genes, which can be effective against insects that are not sensitive to currently available genes.
7. Gene escapes into the environment.
8. Secondary pest problem: Most crops are not attacked by a single pest species, but a complex of insect pests attack on it. In the absence of competition from major pests, secondary pests may assume a major pest status. Effective and timely control measures should be adopted for the control of secondary pests on transgenic crops.
9. Effect on non-target organisms: One of the major concerns of transgenic crops is their effects on non-target organisms, about which little is known at the moment. Bt proteins are rapidly degraded by the stomach juices of vertebrates. Most Bt toxins are specific to insects as they are activated in the alkaline milieu of insect gut. However, Bt proteins can have harmful effects on the beneficial insects. Although, such effects are much less sever than those of the broad spectrum insecticides. [For more details see Chapter 27]

*Strategies for resistance management*

Development of transgenic plants should be based on the overall philosophy of IPM, and consideration of not only gene construct, but alternate mortality factors, reduction of selection pressure, and monitoring of populations for resistance development to design more effective management strategies. The success of *Bt* crops will depend on whether target pests develop resistance to them. Insects have demonstrated a high capacity to develop resistance to a wide array of chemical insecticides. More recently, field populations have been shown to be equally adept at developing resistance to microbial sprays based on *Bt* δ-endotoxins. To increase the effectiveness and usefulness of transgenic plants, it is important to develop resistance management strategies to minimize the rate of development of resistance in insect populations to the target genes.

1. The first step in resistance management is to establish the target pest's baseline susceptibility to the insecticidal protein. The baseline should be determined by geographical location, insect species and selective agent used. The selective agent, the transgenic plants should be precisely characterized as to the nature amount of δ-endotoxin produced. Baselines should be established on field populations, not in domesticated laboratory cultures. Once a baseline has been established, regular monitoring can be used to detect changes in susceptibility that may indicate early stages of resistance in local insect populations.

2. Use of resistance management strategies from the beginning: A large number of conceptual strategies for resistance management have been given. Most of these strategies are based on mixtures of toxins to be deployed for insect control, tissue specific production and induced toxin production. In mixing genes within a plant (gene pyramiding or gene stacking), genes of two or more insecticidal proteins or different genes need to be introduced into the same plant. In tissue specific production, the plants are engineered so that the toxin is produced only in the tissues where the insect feeds. In induced toxin production, the plant is engineered in such a way that it produces the toxin when the insect starts feeding.

3. Gene pyramiding: Many of the candidate genes used in genetic transformation of crops are either too specific or only mildly effective against the target insect pests. Thus, to convert transgenics into an effective weapon in pest control e.g. by delaying the evolution of insect populations resistant to the target genes, it is important to deploy genes with different modes of action in the same plant. Several genes such as trypsin inhibitors, secondary plant metabolites, vegetative insecticidal proteins, plant lectins and enzymes that are selectively toxic to insects can be deployed along with the Bt genes to increase the durability of resistance. Advances have been made in biotechnology for introducing and expressing multiple transgenes in crops.

4. Gene deployment: There is a need to develop appropriate strategies for gene deployment in different crops or regions depending on pest spectrum, their sensitivity to the insecticidal genes, and interaction with the environment. The deployment of different genes and their level of expression should be based on insect sensitivity and level of resistance development.

5. Refugia strategy: Currently the focus is on designing strategies to retard resistance development in the insects. A consensus has developed on the high dose/refuge strategy, which aims to eliminate resistance genes and to dilute resistance development in the insect population. The most effective way to eliminate resistance genes of insects is by targeting the heterozygous resistant insects, since in an insect population the large majority of resistance genes are in a heterozygous condition. Resistance to *Bt* has been proved to be recessive in all cases. Therefore, heterozygous resistant insects are only slightly less susceptible than the fully susceptible insects and both can be effectively controlled by *Bt* plants. For the successful removal of heterozygous resistant insects from a population, it is essential that the plants express a high dose of *Bt* protein during the entire season and for all the developmental stages that feed on the plant. If the levels of *Bt* protein decrease towards the end of growing season, heterozygous resistant insects could survive and consequently increase the frequency of resistance genes in the following generation. Thus, by growing transgenic plants with high *Bt* expression levels only the homozygous resistant insects are left in the field. Then by providing an area

in the vicinity of the *Bt* crop where the susceptible insects can survive, the mating between homozygous resistant individuals and susceptible insects can be promoted, leading to heterozygous offspring that are susceptible to the *Bt* crop. Thus a refugia strategy should be adopted to provide a source of susceptible insects for mating with the selected population to prevent the fixation of resistance. Refuges can naturally occur when a certain percentage of an agricultural acreage is planted with non-transgenic plants. Refuges may also be established by design through the use of seed mixtures or susceptible border rows, or may occur outside the crop for highly mobile insect species with alternate host plants. It has been recommended that refuge plants can be in 4–20% in maize and 20–40% in cotton.

6. Transgenic plants producing additional insecticidal proteins with different modes of action, different targets, or both should be developed.

7. Regulation of gene expression: In most cases resistant genes have been inserted with constitutive promoters such CaMV35S, maize ubiquitin or rice actin 1 which direct expression in most plant tissues. Limiting the time and place of gene expression by tissue specific promoters such as phenylalanine ammonia lyase (PHA-L) for seed specific expression, *RsS*1 for phloem specific expression or inducible promoters such as potato *pin*2 wound-induced promoters might contribute to the management of resistance development, and unfavorable interactions with the beneficial insects.

8. Development of synthetics: One targeted deployment strategy is the development of synthetics. By incorporating various constitutively expressed Cry toxins into lines adapted for specific environments, synthetics can be formed quickly, which are effective against the pest complex and are compatible with the natural enemies. Once released to the farmers, the synthetics can be maintained as narrow based populations at the farm level by removing the plants showing insect damage. Lines with resistance through the conventional method of breeding can also be included as a component in developing the synthetics to increase the durability of resistant germplasm.

9. Destruction of carryover population: Destruction of pupae or the carryover population (that has been exposed to Bt crops in the previous generations) from one season to another is an important component of resistance management. Ploughing the fields immediately after the crop harvest will expose the pupae of insects to biotic and abiotic factors. Destruction of stems or burning of stubbles of cereal crops will help in reducing the carryover of stem borer larvae. Therefore, appropriate agricultural practices need to be followed that reduce the carryover of pests from one season to another, including appropriate crop rotations, and observing a 'close season'.

10. Control of alternate hosts: Removal of alternate hosts is required in case alternate hosts play an important role in pest population build-up. Efforts should be made to remove the alternate hosts of the pests from the vicinity of the crop. This practice will help in reducing the pests density, and low to moderate levels of pest abundance can be effectively controlled by the transformed crops.

11. Use of planting windows: Following a planting window, when the crop can escape pest damage or avoid peak periods of insect abundance, can also be useful in maximizing the benefits from transgenic crops or prolong the life of transgenic crops.

12. Use of economic thresholds and IPM: Crop growth and pest incidence should be monitored carefully so that appropriate control measures can be initiated in time.

Care should be taken to use control options such as natural enemies, nuclear polyhedrosis virus (NPV),. neem or entomopathogenic nematodes and fungi, which do not disturb the natural control agents. Use of pesticide formulations such as soil application of granular systemic insecticides and spraying soft insecticides such as endosulfan may be considered to suppress populations in the beginning of the season. Broad spectrum and most toxic insecticides may be used only during the peak activity periods of the target pest. Efforts should be made to rotate pesticides with different modes of action, and avoid repetition of insecticides belonging to the same group or the insecticides that fail to give effective control of the pests.

## Herbicide tolerant plants through transgenic approach

Herbicide tolerance has been the dominant trait since 1996 when GM crops were released commercially. However, there have been arguments against and favor of this trait in transgenic crops which have been summarized below.

*Arguments in favor of herbicide tolerance in crop plants*

1. Use of highly specific and powerful herbicides in combination with transgenic crops tolerant to specific herbicides will lead to more effective control of weeds and reduced herbicide use.
2. Would promote use of environmentally more benign chemicals that are less likely to leach into ground water.
3. Farmers would get a greater option for weed control and more flexibility in the choice of crops for rotation or for double crop planting.
4. Environmental gains of GM crops: Three independent teams of CAST (Council of Agricultural Science and Technology) reviewed the available scientific literature to compare the environmental impacts of biotechnology derived and traditional crops. The study involved various criteria and was based on herbicide tolerant soybeans, the most widely planted crop and further studies showed similar benefits for corn also. These are:

i. *Soil quality*: No till soybean acreage in US has increased significantly since the introduction of herbicide tolerant soybeans. Use of no till farming results in less soil erosion, dust and pesticide run off as well as increased soil moisture retention.

ii. *Water quality*: Use of biotechnology derived soybeans enables farmers to use a more benign herbicide that rapidly dissipates in soil and water.

iii. *Air quality*: Greenhouse gas emissions from some farm operations decreases by an estimated 88% as a result of biotech soybeans planted in a no-tillage system, which may help slow global warming.

iv. *Biodiversity*: The no-till practices commonly associated with biotech soybeans provide a more favorable habitat for birds and other wildlife. No-tillage systems provide food and shelter for wild life such as pheasants and ducks.

v. *Land use efficiency*: Biotechnology derived soybeans may lead to increased yields through improved weed control and the adoption of narrow row spacing.

*Arguments against herbicide tolerant transgenics in crop plants*

1. Use of herbicide tolerant transgenic crops can lead to transfer of herbicide tolerance genes to sexually compatible wild relatives or weeds, which can be a major potential threat to environment.
2. Transgenic crops can create 'super weeds'.
3. It would actually increase the dependence on a few herbicides rather than reducing herbicide usage.

4. It may increase the problem of weed control if weeds develop resistance to such herbicides through gene flow from transgenic crops.
5. Herbicide tolerance is being sought not only for environmentally comparatively acceptable herbicides but also for older, more toxic and persistent products.
6. Non-chemical means of weed control, such as crop rotation, dense plantings, cover cropping, ridge tillage, and others, however labour intensive for farmers, are preferable than the use of any herbicide at all.
7. Gene flow is the primary risk in releasing transgenic plants.

However, genetically improved crops show no sign of turning into super-weeds according to a 10-year study by a team of scientists in Britain. Researchers planted genetically improved varieties of oilseed rape, potato, maize and sugar beet alongside conventional crops in 12 different habitats in the UK to see whether the plants could invade natural habitats. But the research team found that native wild plants displaced both genetically improved plants and ordinary crops and that conventional crop actually outlived the biotech ones. Scientists said most of the crops died out after 4 years, and after ten years the only survivor was one type of non-biotech potato. When the study began, the kind of biotech plants available were varieties engineered to make them better able to withstand spraying with herbicides or insect attack, not crops engineered to withstand drought or natural pests. It was not expected that herbicide tolerance would give plant an ecological advantage. The results do not mean that other genetic modifications could not increase weediness or invasiveness of crop plants, but they do indicate that arable crops are unlikely to survive for long outside cultivation.

## Plant based pharmaceuticals

Attempts are being made to replace the traditional fermentation procedure for the biopharmaceuticals to plant based production.

The benefits of using plants are the ability to increase production at low cost by planting more acres rather than building fermentation capacity. Plant transgenic technology is being used to produce a plant that will generate a seed that expresses a desired therapeutic protein. This seed can be propagated under right growing conditions to yield plants and seed stock for producing the desired protein. This protein of interest can be extracted from the seed to make a biopharmaceutical. Dow Plant Pharmaceuticals is using corn to grow pharmaceuticals by designing and selecting the plant which will contain the active pharmaceutical within the endosperm and seed compartment.

## Production of biofuels

There are two principal biofuels currently in use, ethanol produced from sugarcane, maize and other starchy grains, and biodiesel principally from rapeseed/canola, soybean and palm oil. To put biofuels into context, global biofuel production in 2004 was only 33 billion litres, equivalent to only 3% of the 1,200 billion litres of gasoline used. Brazil has been the world leader in the production and consumption of sugarcane-based ethanol for the last 25 years, followed by the US where ethanol production from maize increased rapidly from 4 billion litres in 1996 to 44 billion litres in 2005. US has accorded high priority to biofuels and is using 18% of its maize production to generate 2% of the non-diesel transport fuel in 2005, allowed the US to overtake Brazil in ethanol production in 2005. Ethanol is generated through fermentation of sugar and starchy grain crops and biodiesel is produced through esterification of vegetable oil from oilseed crops such as rapeseed or soybean and the addition of methanol. For the short term, goals for biofuel production in the next 10 to 15 years will be met with the "first generation" technologies of ethanol and biodiesel from the food crops sugarcane, maize, rapeseed and soybean and palmoil. However, the long term goals will need a "second generation" of technologies that will feature the production of biofuels from energy crops, rich in lingo-cellulose biomass such as the tall grasses of switch grass (*Panicum virgatum*) and *Miscanthus*, fast growing trees including willow, hybrid poplar and eucalyptus, crop waste products including straw, maize stover, bagasse, sawdust, wood thinnings and organic residues such as municipal solid waste. These "second generation" technologies are required for two reasons. Firstly, to expand by a quantum amount, the volume of biomass feedstocks, and secondly to increase significantly the efficiency and cost-effectiveness of converting biomass to liquid biofuel.

Biotechnology can be used to cost effectively optimize the productivity of energy generation crops with high biomass/hectare. This can be achieved by developing crops tolerant to abiotic stresses especially drought and salinity, biotic stresses and also to raise the ceiling of potential yield per hectare through modifying plant metabolism. There is also an opportunity to utilize biotechnology to develop more effective enzymes for the downstream processing of biofuels.

Several factors have contributed to the recent increases in interest and investments in biofuels. These include the recent rise of the price of oil upto US $ 130/ barrel, growing concern about increasingly high consumption of oil and moreover that the supply of fossil fuels is finite and concentrated in geographical areas which could seriously disrupt supplies in a world of political turmoil.

One of the original issues with biofuels was that the energy required to produce them was more than the energy they generated i.e. a negative energy balance. This was a valid critique of early experiences with maize, however the energy balance is now well over 1.0 as a result of higher "biofuel yield" from improved maize (which could be significantly further enhanced through biotechnology) and also in ethanol refining.

It is emphasized that biofuel targets must not be achieved at the expense of food security.

Thus, a strategy must be in place to address the issues involved in the food versus fuel debate, particularly when it relates to the developing countries where there is a food deficit. Developing countries must look for food security first and then biofuel security.

## Transgenic crops in future

Current cultivars primarily address the problem of biotic stresses caused by pests and diseases and abiotic stress of weeds. This sustainable increase in productivity on the same area of cropland allows biodiversity to be conserved. Initial progress has been made with maize food crop, oilseed crops of soybean and canola, fibre crop of cotton, and biotech papaya and squash. The next wave of cultivars will greatly extend biotechnology into the control and prevention of crop to diseases and abiotic stresses with drought tolerance trait available in next 5 years which will be followed by salt tolerance. Golden rice enriched with pro-vitamin A is expected to be approved by 2012. Biotech rice has already been released temporarily in Iran in 2005, however, large scale cultivation of biotech rice against biotic stresses will be released soon for commercial cultivation in China and other countries. Multilocational field trials of biotech egg plant, tomato, cauliflower, rice are underway in India and expected to be given regulatory approval for commercial cultivation. Other GMOs in near future includes soybean and canola oils containing more unsaturated fatty acids, higher yielding peas that remain sweet longer, bananas and pineapples with delayed ripening qualities, bananas resistant to fungi, high protein rice, tomatoes with higher antioxidant content, fruits and vegetables with higher levels of vitamins.

There is increasing evidence that transgenics can produce healthier food. Another opportunity is to use plants as chemical factories. Thus, the next major phase of the plant revolution will emphasize the engineering of desirable traits in plants. Intensive efforts have also been made to channelize plant metabolism into producing industrial feedstocks, pharmaceuticals and nutraceutical products. The selection of superior strains and the conventional breeding of soybeans have led to lines with reduced levels of antinutritional oligosaccharides, stachyose, raffinose, and galactose. Modification of the genome of soybeans to desirable fats is leading to healthier foods and to useful industrial chemicals. Another effort that succeeded was to use soybean plants to produce vernolic acid and ricinoleic acid, derivatives of oleic acid that are used as hardeners in paints and plastics.

Nobel laureate Norman Borlaug estimates that to meet projected food demands by 2025, average cereal yield must increase by 80% over the 1990 average. Biotechnology, one of the many tools of agricultural research and development, could contribute to food security by helping to promote sustainable agriculture centered on small holder farmers in developing countries. This biotechnology revolution is very relevant to the problems of food security, poverty reduction, and environmental conservation in the developing world. Biotechnology will help us do things that we couldn't do before and do it in a more precise and safe way. Biotechnology will allow us to cross genetic barriers that we were never able to cross with conventional genetics and plant breeding. In the past conventional breeders were forced to bring along many other genes with the genes of interest like insect or disease resistance that we wanted to incorporate in a new variety. These extra genes often had a negative effect, and it took years of breeding to remove them. But with use of biotechnological tools we have the more precise ways to introduce only the desired gene.

## Biotechnology in Process Industry

Recombinant DNA technology helps in the development of improved microorganisms with increased enzyme production. It improves the productivity to cost ratio so that the enzyme productivity in microorganisms is increased and the desired enzyme can be produced in commercial quantities. Today a growing number of enzymes are produced using GMOs with

applications in varied sector such as pulp and paper, food processing, textiles, etc. In 1988, Novo Nordisk was the first company to commercially produce a fat degrading enzyme, phytase using genetically modified fungi. The scientists inserted the gene producing phytase into the *Aspergillus oryzae* fungus, known for its ability to produce high yields of enzyme. As a result, pure phytase could be produced with hundred fold increase in yield as compared to wild organism.

Using rDNA technology it is possible to replace petroleum derived polymers with biological polymers derived from grain or agricultural biomass. DuPont has created high performance polymer Sorona from the bioprocessing of corn sugar and this fibre has been used to create clothing.

Plants and microbes have been genetically modified to produce polyhydroxybutyrate for producing biodegradable plastics. Production of natural protein polymers such as spider silk and adhesives from barnacles through microbial fermentation using rdNA technology is now possible.

## Bioremediation

Application of biotechnology for better environment has also been made by using naturally occurring microorganisms to identify and filter manufacturing waste before it is introduced into the environment. Bioremediation programmes involve the use of microorganisms for cleaning the contaminated air, tracks of land, lakes and waterways. rDNA technology helps in improving the efficacy of these processes. rDNA technology is also being used in development of bioindicators where bacteria have been genetically modified as bioluminescors that gives light in response to several chemical pollutants and hazardous chemicals. For example mercury resistance gene (*mer*) or toluene degradation (*tol*) gene is linked to genes that code for bioluminescence within living bacterial cells, the biosensor cells can signal extremely low levels of inorganic mercury or toluene present in contaminated water and soils by emitting visible light and which can be measured with fibre optic fluorometers.

## Healthcare applications

Since the approval of recombinant insulin as the first recombinant biopharmaceutical, many biotechnology companies as well as traditional pharmaceutical companies have devoted their resources to discover and develop numerous recombinant protein therapeutics or biopharmaceuticals.

Although, the initial efforts were towards development of recombinant forms of natural proteins and biologics derived from natural sources, several monoclonal antibody (mAb) based therapeutics have also been developed now. USA and EU are the leaders in biotechnology research and commercialization and almost all the products are first approval in these countries. Five of the pioneering biotechnology companies such as Amgen, Genentech, Chiron, Genzyme and Biogen have had a spectacular and consistent revenue growth for the past two decades and have multiple biopharmaceuticals in the market, as well as in the pipeline. Several pharmaceutical companies such as Johnson & Johnson, Eli Lilly, Roche, Bayer and Schering Plough are also at the forefront of the recombinant therapeutics revolution primarily thorough their acquisition off products for further development or by becoming marketing partners of various small to medium size biotechnology companies.

It has been reported that about 60 recombinant products have been approved till 2003 and are being used as therapeutics in various forms across the world. The annual global market for biopharmaceutical was estimated at more that US $ 30 billion in 2002, as compared to US $ 12 billion in 1999 and US $ 5 billion in 1989. Notable target indication for approved therapeutics includes diabetes, haemophilia, hepatitis, myocardial infarction and various cancers.

Table 26.8 gives the list of recombinant therapeutics approved in USA and Europe till middle of 2003. These have been produced in various transformed hosts such as *E. coli;* yeasts such as *Saccharomyces cerevesiae, Hansenulla polymorpha, Pichia pastoris,* etc. and mammalian cell lines such as Chinese hamster ovary cell, Baby hamster kidney cell lines etc.

However, it may be noted that all recombinant products are considered new if there is a change in the use of cell lines, major changes in the process techniques and methods of transformation of hosts, based on which it has been reported that nearly 144 products are approved for therapeutic use.

Globally, approximately 500 candidate biopharmaceuticals are undergoing clinical evaluation which are mainly for the treatment of cancer, infectious diseases, autoimmune disorders, neurological disorders and AIDS/HIV related conditions.

Regarding the status of approved rDNA therapeutics in India, 18 therapeutically distinct products have been approved for marketing through imports/indigenous manufacture (Table 26.9).

Most of the companies are only marketing the products by collaborating with foreign companies. The companies producing rDNA therapeutics are: Shantha Biotechnics, Bharat Biotech, Panacea Biotech, Wockhardt, Dr. Reddy's Laboratories, Serum Institute of India, etc. In addition, there is significant research and development activity in both private and public institutions. Some of the products, which are under development, include recombinant anthrax vaccine, recombinant HIV vaccine, recombinant cholera vaccine, streptokinase etc.

**Future:** Biotechnology, for many, raises important questions relating to ethics, intellectual property rights, and biosafety. There have been widespread protests against the spread of agro-biotechnology. Some of the concerns come from scientists who fear that "novel" products will destroy agricultural diversity, thus changing

**Table 26.8:** Approved recombinant therapeutics in USA and EU (Source: Anonymous, 2004)

| Products | Therapeutic indication | Produced in host | Brand name | Company | Years of approval |
|---|---|---|---|---|---|
| Blood factors Factors VIII | Hemophilia Type A | Animal cell | Recombinate | Baxter Health-care/Genetics Institute | 1992 (US) |
| | | CHO | Bioclate | Aventis Behring | 1993 (US) |
| | | | Kogenate | | |
| | | BHK cells | | Bayer | 1993 (US) 2000 (EU) |
| | | | Helixate Nex Gen | | |
| | | | | Bayer | 2000 (EU) |
| | | | Refacto | | |
| | | CHO cells | | Genetics Institute/ Wyeth Europa | 1999 (EU), 2000 (US) |
| Factor VIIa | Some forms of homophilia | BHK cells | NovoSeven | Novo-Nordisk | 1996 (EU),1999(US) |
| Factor IX | Homophilia Type B | CHO cells | Benefix | Genetics Institute | 1997 (US, EU) |
| Anticoagulants Recombinant hirudin | Prevention of venous thrombosis | S. cerevisiae | Revasc | Aventis | 1997 (EU) |
| | Anticoagulation therapy for heparin associated thrombocytopenia | S. cerevisiae | Refludan | Hoechst Marion Roussel Behringwerke AG | 1997 (EU), 1998 (US) |
| **Blood clot dissolvers** | | | | | |
| Tissue plasminogen activator | Acute myocardial | CHO cells | Activase | Genentech | 1987 (US) |
| | | E coli | Retavase | Boerhringer Mannheim/ Centocor | 1996 (US) |
| | | | Rapilysin | Roche | 1996 |
| | | | Ecokinase | Galenus Mannheim | 1996 (EU) |
| | Myocardial infarction | CHO cells | TNKase | Genentech | 2000 (US) |
| | | | Tenecteplase | Boehringer Ingelheim | 2001 (US) |
| Streptokinase | Acute myocardial infarction | | Haverkinase | Cuban company | 1990 |
| **Growth factors and hormones** | | | | | |
| Insulin | Diabetes mellitus | E.coli | Humulin Novolin | Eli Lilly Novo | 1982 (US) |
| | | | Insuman | Nordisk | 1991 (US) |
| | | | | Hoechst AG | 1997 (EU) |
| Insulin Analogues Insulin lispro | | | | Eli Lilly | 1996 (US, EU) |
| Bio Lysprol | | | Liprolog | Eli Lilly | 1997 (EU) |
| Insulin glargine | | | Lantus | Aventis | 2000 (EU, US) |
| | | | Optisulin | Aventis | 2000 (US) |
| Insulin aspart | | S. cerevisiae | NovoRapid | Novo Nordisk | 1999 (EU) |
| Insulin | | | Actrapid/Velosulin/ Monotard/Insulatard/ Protaphane/Mixtard/ Actraphane/Ultratard | Novo Nordisk | 2002 (EU) |

*(contd. Table 26.8)*

*(contd. Table 26.8)*

| Products | Therapeutic indication | Produced in host | Brand name | Company | Years of approval |
|---|---|---|---|---|---|
| Human growth homone | hGH deficiency in children | *E.coli* | Protropin | Genetech | 1985 (US) |
| | | | Humatrope | Eli Lilly | 1987 (US) |
| | | | Nutropin | Genetech | 1994 (US) |
| | | | BioTropin | Savient Pharmaceuticals | 1995 (US) |
| | | | Genotropin | Pharmacia & Upjohn | 1995 (US) |
| | | | Norditropin | Novo Nordisk | 1995 (US) |
| | | | Saizen | Serono Laboratories | 1996 (US) |
| | Treatment of AIDS associated catabolism/wasting | | Serostim | Serono Laboratories | 1996 (US) |
| | Growth failure/ Turner's syndrome | | Nutropin AQ | Schwartz pharma | 1994 (US) 2001 (EU) |
| | Treatment of acromegaly | | Somavert | Pfizer | 2003 (US) 2001 (EU) |
| Glucagon | Hypoglycemia | *S. cerevisiae* | Glucagen | Novo Nordisk | 1998 (US) |
| Calcitonin | Paget disease | *E. coli* | Forcaltonin | Unigene | 1999 (EU) |
| Luteinizing hormone | Some forms of infertility | CHO cells | Luveris | Ares-Serono | 2000 (EU) |
| Thyrotrophin- (TSH) | Detection/ treatment of thyroid cancer | CHO cells | Thyrogen | Genzyme | 1998 (US) 2000 (EU) |
| Choriogona- dotrophin | Used in selected assisted reproductive technique | CHO cells | Ovitrelle | Serono | 2000(US) 2001 (EU) |
| Parathyroid hormone (shortened) | Treatment of osteoporosis in menopausal women | *E. coli* | Forteo | Eli Lilly | 2002 (US) |
| Parathyroid hormone | Treatment of osteoporosis | *E. coli* | Forsteo | Eli Lilly | 2003 (EU) |
| Follicle stimulating hormone | Anovulation and super ovulation Infertility | CHO cells | Gonal F Puregon Follistim | Ares-Serono N.V. Organon N.V. Organon | 1995 (US) 1997 (US) 1996 (EU) |
| Erythropoietin | Treatment of anemia | Mammalian cell line CHO cells | Epogen Procrit Neorecormon | Amgen Ortho Biotech Boehringer Mannheim | 1989 1990 1997 |
| Granulocyte monoclonal stimulating colony factor (GM-CSF) | Autologous bone marrow transplantation Chemotherapy induced neutropenia | *E. coli* | Leukine Neupogen | Immunex Amgen | 1991 1991 |

*(contd. Table 26.8)*

*(contd. Table 26.8)*

| Products | Therapeutic indication | Produced in host | Brand name | Company | Years of approval |
|---|---|---|---|---|---|
| Platelet derived growth factor (PDGF) | Lower extremity diabetic neuropathic ulcers | S. cervisiae | Regranex | Ortho-McNeil Pharmaceuticals Janssen-Cilag | 1997 |
| **Cytokines** Interferons IFN-α2a | Hairy cell leukemia | E. coli | Roferon A | Hoffmann La Roche | 1986 |
| IFN-α, synthetic type I IFN | Chronic hepatitis C | E. coli | Infergen | Amgen Yamanouch Europe | 1997 |
| IFN-α2b | Hairy cell leukemia, genital warts | E. coli | Intron A | Schering Plough | 1986 |
| | Chronic hepatitis C | | Rebetron | " | 1999 |
| | Hepatitis B, C, and various cancers | | Alfatronol | " | 2000 |
| | Hepatitis B and C | | Virtron | " | 2000 |
| IFN-β₁a | Multiple sclerosis Relapsing/remitting multiple sclerosis | E. coli | Betaferon Betaseron | Schering AG Berlex Laboratories and Chiron | 1995 1993 |
| IFN-β₁a | Relapsing multiple sclerosis | CHO cells | Avonex Rebif | Biogen Ares Serono | 1997 1998 |
| IFN-γ₁b | Chronic granulomatous disease | E. coli | Actimmune | Genentech | 1990 |
| **INTERLEUKING** IL-2 | Renal cell carcinoma | E. coli | Proleukin | Chiron | 1992 |
| IL-11 | Prevention of chemotherapy induced thrombocytopenia | E. coli | Neumega | Genetics Institute | 1997 |
| **Vaccines** Hepatitis B surface antigen | Hepatitis B prevention | S. cerevisiae | Recombivax | Merck | 1986 |
| Combination vaccine, containing HbsAg | Vaccination of infants against H. influenzae type B and hepatitis B | S. cerevisiae as one component | Comvax | Merck | 1996 |
| | Vaccination against hepatitis B, diphtheria, tetanus, and pertusis | S. cerevisiae as one component | Tritanrix-HB | Smith Kline Beecham | 1996 |
| | | S. cerevisiae as one component | | | |
| | Immunization against hepatitis A and B. | S. cerevisiae as one component | Twinrix, adult and pediatric forms | Smith Kline Beecham | 1996 |
| | Immunization against diphtheria, tetanus, and hepatitis B | | Primavax | Pasteur Merieux MSD | 1998 |

*(contd. Table 26.8)*

*(contd. Table 26.8)*

| Products | Therapeutic indication | Produced in host | Brand name | Company | Years of approval |
|---|---|---|---|---|---|
| | Immunization against *H. influenzae* type B and hepatitis B | | Procomavax | Pasteur Merieux MSD | 1998 |
| OspA, a lipoprotein found on the surface of *B. burgdorferi* | Lyme disease vaccine | *E. coli* | Lymerix | SmithKline Beecham | 1998 |
| **Additional Products** | | | | | |
| Dornase- α,DNase | Cystic fibrosis | CHO cells | Pulmozyme | Genentech | 1993 |
| β-glucocere-brosidase produced in *E. coli* Differs from native human enzyme by one amino acid, R495?H, and has modified oligosaccharide com | Treatment of Gaucher's disease | | Cerezyme | Genzyme | 1994 |
| TNF-α, | Adjunct to surgery for subsequent tumor removal, to prevent or delay amputation | *E.coli* | Beromum | Boehringer Ingelheim | 1999 |
| IL-2 diphtheria toxin fusion protein that targets cells displaying a surface IL-2 receptor | Cutaneous T-cell lymphoma | | Ontak | Seragen/Ligand Pharmaceuticals | 1999 |
| TNFR-IgG fragment fusion protein | Rheumatoid arthritis | CHO cells | Enbrel | Immunex (US) Wyeth Europa (EU) | 1998 |
| Fomivirsen, an antisense oligonucleotide | Treatment of CMV retinitis in AIDS patients | | Vitravene | ISIS pharmaceuticals | 1998 |

agricultural patterns into unrecognizable and uncontrollable forms. The dominance of a highly concentrated private sector has raised fears of a new phase of comparative disadvantage and increased dependency in the developing world. A country has to set up proper regulatory procedures so that new technology should reach the ultimate users. Biosafety concerns as elucidated in the next chapter should be properly addressed. It should be well remembered that every technology has potential to do both harm and good. Then, it is left to the ultimate user to weigh benefits against the potential harm and take a decision based on sound economics. Ultimately to use or not to use agricultural biotechnology products should remain the farmer's decision.

In Bt cotton case, no damage to the environment has materialized so far in any country where the crop has been grown.

**Table 26.9:** Recombinant therapeutics approved for marketing in India (Source: Anonymous, 2004)

| Molecules | Therapeutic application |
| --- | --- |
| Human insulin | Diabetes |
| Erythropoietin | Treatment of anaemia |
| Hepatitis B vaccine (recombinant surface antigen based) | Immunization against Hepatitis B |
| Human growth hormone | Deficiency of growth hormone in children |
| Interleukin 2 | Renal cell carcinoma |
| Interleukin 11 | Thrombocytopenia |
| Granulocyte Colony Stimulating Factor | Chemotherapy induced neutropenia |
| Granulocyte Macrophage Colony Stimulating Factor | Chemotherapy induced neutropenia |
| Interferon $\alpha$2a | Chronic myeloid leukemia |
| Interferon $\alpha$2b | Chronic myeloid leukemia, Hepatitis B and Hepatitis C |
| Interferon $\gamma$ | Chronic granulomatous disease and Severe malignant osteopetrosis |
| Streptokinase | Acute myocardial infarction |
| Tissue Plasminogen Activator | Acute myocardial infarction |
| Blood factor VIII | Haemophilia type A |
| Follicle stimulating hormone | Reproductive disorders |
| Teriparatide (Forteo) | Osteoporosis |
| $\alpha$-Drerecogin | Severe sepsis |

Furthermore, refuge management practices will assist the bollworm retain its resistance to the organic pesticides as well as prevent gene flow. Finally, if considered otherwise, further research and development can be undertaken to tackle these aspects. Denying the farmers access to a technology, which has the potential to revolutionize agriculture, on the grounds of unfounded and overhyped fears, must no longer be an option available to agriculture.

Biotechnology is now no more considered as fashionable but a fact of application of science for the benefit of humans. The path from inception of the agricultural biotechnology industry to field production of commercial products has been criss-crossed by obstacles ranging from scientific and technological challenges to legal and regulatory hurdles, to economic factors and social concerns. Some expectations have been met with the commercial development of herbicide and insect resistance in crops clearly pointing to a strong future for biotechnology in agriculture. The future prospects in this area are also exciting with strong indications that novel genes can be introduced to generate plants producing materials ranging from pharmaceuticals to biodegradable plastics, vaccines, accumulation of minerals, vitamins, fatty acids, carbohydrates, etc.

## REFERENCES

Brookes, G. and P. Barfoot. 2008. GM Crops: Global Socio-economic and Environmental Impacts 1996–2006, P.G. Economics.

Cao, J., Tang, J.D., Strizhov, N., Shelton, A.M. and Earle, E.D. 1999. *Mol. Breed.* **5**: 131–141.

Gandhi, V. and Namboodiri, N.V. 2006. The adoption and economics of Bt cotton in India: Preliminary results from a study. IIMA Working Paper No. 2006–09–04, pp. 1–27.

James, C. 2007. Global Status of commercialized biotech/ GM Crops:2007. *ISAAA* Brief No. **37** ISAAA, Ithaca, N.Y.

## FURTHER READING

Anonymous, 2004. National consultation on biosafety aspects related genetically modified organisms. By Biotech Consortium India Ltd. and DBT, India.

Sharma, H.C. and Ortiz, R. 2000. Transgenics, pest management, and the environment. Curr. Sci., **79**: 421–437.

James, C. 2007. Global status of commercialized biotech/GM crops: 2007. ISAAA Brief No. 37, ISAAA, Ithaca, N.Y.

# 27

# Biosafety Concerns and Regulatory Framework

After the advent of rDNA technology in 1970s, discussions began within the scientific community about the risks associated with recombinant DNA/genetic engineering experiments. The main concerns were: (i) Molecular biologists may not be well-versed with the laboratory practices needed for such type of work; (ii) 'Hybrid organisms' could be created with biological activities of an unpredictable nature and has the potential for infection of workers; (iii) 'Hybrid organisms' may escape from the laboratory with unpredictable consequences since their survival and behavior in the environments would be largely unknown. National Academy of Sciences, USA in 1974 examined the various issues and made certain recommendations and also established the Recombinant Advisory Committee. In 1975, at Asilomer, California, an International meeting was held which formulated the first set of recommendations on safety of recombinant DNA experiments. These formed the basis of subsequent biosafety guidelines and regulations in USA and other countries. The first National Institute of Health (NIH) guidelines were prepared in 1975. In 1977, the NIH prepared an Environmental Impact Statement (EIS), which estimated the possible undesirable impacts of experiments and the likelihood of their occurrence based on probabilities of creating and dispersing potentially hazardous agents.

Initially the regulations on recombinant DNA and the genetic modification of organisms focused on the technology itself. Therefore, biosafety concerns were primarily focused towards the safety procedures for recombinant DNA work within the laboratory. The emphasis was to ensure that researchers take proper steps to contain organisms that potentially posed a risk to themselves or human health generally. **Biosafety** refers to policies and procedures adopted to ensure environmental safety during the course of development and commercialization of genetically modified organisms (GMOs). As per CBD Biosafety is a term used to describe efforts to reduce and eliminate the potential risks resulting from biotechnology and its products.

Most cloning experiments on *E. coli* K-12, certain strains of *Bacillus subtilis* and *Saccharomyces cerevisiae* were considered exempt from other requirements of NIH guidelines by 1981. Eventually, complete exemption was granted for most of the recombinant DNA research. A major revision of the guidelines was effected in 1982; containment levels were lowered, and experiments, that were previously prohibited, including deliberate release of recombinant DNA-containing organisms into the environment, were changed to category requiring review and approval by NIH. In USA, the NIH guidelines are followed by all federal agencies that fund research on recombinant DNA.

# RISK ANALYSIS

Risk is defined as "the probability of harm". The risk associated with any action depends on the three elements of the following equation.

Risk = hazard X probability X consequences

A hazard is anything that might conceivably go wrong. A hazard does not in itself constitute a risk. The probability associated with a hazard also depends, in part on the management strategy used to control it. Risk can be under estimated if some hazards are not identified and properly characterized, if the probability of the hazard occurring is greater than expected or if its consequences are more severe than expected.

Risk analysis consists of three steps i.e. risk assessment, risk management and risk communication. Risk assessment evaluates and compares the scientific evidence regarding the risks associated with alternative activities. Risk management develops strategies to prevent and control risks within acceptable limits and relies on risk assessment. In addition to the scientific assessment, it also takes into consideration various factors such as social values and economics. Risk communication involves an ongoing dialogue between regulators and the public about risk and options to manage risk so that appropriate decision can be made.

***Risk assessment***: The concept of risk assessment is quite complex and relies on both science and judgment. It involves determination of potential and anticipated effects of rDNA research to concerned workers and of the products of research on human health and environment if they are released deliberately or accidentally in the environment or as a result of their consumption. Risk assessment should be carried out in a scientifically manner and be in accordance with recognized assessment techniques. A lack of scientific knowledge should not be interpreted as an absence of risk or an acceptable risk, but further information may be sought. During laboratory research work risk can be assessed as (i) initial risk assessment and (ii) comprehensive risk assessment. Initial risk assessment is made by the investigator on the basis of the risk group (RG) to which organism, on which he proposes to conduct experiments. Organisms are classified in to four risk groups (1–4) on the basis of their potential effects on a healthy human adult. After the initial risk assessment, a comprehensive risk assessment should be made where an appropriate level of containment for an experiment should be decided. This assessment should consider organism factors such as virulence, pathogenicity, infection dose, environmental stability, spread, etc. and type of manipulations planned.

Risk assessment procedures may vary from case to case but for GMOs it generally covers the following:

1. Characterization of the host, donor and gene transfer process including the molecular and phenotypic characterization. It requires knowledge on introduced genes, its expression, the characteristics, concentration and localization of expressed products and consequences of expression. Genetic stability of introduced gene is also to be ascertained.

2. The modified organism is analysed for the risks of pathogenicity, allergenicity, teratogenicity, etc. Appropriate assessment procedures and criteria will vary for the genetic modifications in different organisms and products thereof. The pathological and taxological assessments mainly deal with the protein expression studies of the inserted genes. However appropriate *in vitro* and *in vivo* studies are needed to assess the changes in pathogenicity, toxicity and other effects of result of a combination of traits that enhance its possibility of survival by adhesion, invasion and toxigenicity. If the host organism and the genetic material being transferred to the host organism does not contain traits conferring pathogenicity, it is

most likely that the modified organism will also not be pathogenic. The same is true for assessment of toxigenicity and other parameters.

3. Substantial equivalence: Since rDNA technology creates vastly different organisms, it has been felt that safety assessment methodologies may not be sufficient. Thus a concept of substantial equivalence has been promoted, which means that if GMO is substantially equivalent to an existing organism, safety assessment of the product should be made in the context of existing organism. The rationale of substantial equivalence helps in adoption of specific testing procedures. A case by case approach is used in assessment which is very useful particularly in the development of transgenic plants and their products. Three tiers of equivalence have been proposed as per the guidelines of OECD and WHO for the risk assessment of GMOs including food products.

(a) When substantial equivalence has been established for an organism or food product, it is considered to be safe as its conventional counterpart and no further safety evaluation is needed.

(b) When substantial equivalence has been established apart from certain defined differences, further safety assessment should focus on these differences: a sequential approach should focus on the new gene product(s) and their structure, function, specificity and history of use. If a potential safety concern is indicated for the new gene product(s), further in vivo and / or *in vitro* studies may be appropriate.

(c) When substantial equivalence cannot be established, this does not necessarily mean that the food produce is unsafe. Not all such products will require extensive safety testing. The design of any testing program should be established on a case-by-case basis… with the implication that further studies, including animal feed trials, may be required, especially when the new organism/product is to be used as food and is intended to replace a significant part of the diet.

4. Effects related to gene transfer and marker genes: Transfer of genetic information may take place by one or more mechanisms such as transformation of released genetic material, transduction (viral transfer of DNA into bacteria) or conjugation by cell to cell contact and exchange of plasmid. Risk assessment also looks into the possibilities of the same, although, the probability of occurring any of the above is extremely low. Further, it should also be assessed whether such gene transfer will confer some selective advantage to new host such as virulence, adherence, substrate utilization or production of bacterial antibiotics. There have been some concerns about the use of antibiotic resistance genes as markers but the same has been largely discounted by conducting extensive risk assessment.

5. Ecological effects: When organism is released into the environment risk assessment is totally different from that for contained use. Almost all countries have developed a series of questions regarding the organism, vectors and procedures and the site into which it is to be released. An example of such questions in case of a transgenic crop is as follows:

i. What parts of the engineered plant will contain the new protein?

ii. What non-target species will be exposed to it?

iii. What is its toxicity to those species?

iv. What is their expected level of exposure?

v. What are the likely biological effects of exposure?

In many cases, particularly with the release of genetically modified plant, microbes for bioremediation, or animals, the major difference is that the modified organism has been designed to survive in the environment, and in particular circumstances has been designed to be fitter for the environment than the wild-type. Examples might be heat or cold tolerance, drought or saline-tolerance in plants, and tolerance to heavy metals or organic chemicals for both plants and microbes. This implies many different approaches to risk assessment than that for contained use. The ecological risk assessment model developed by the US Environmental Protection Agency (EPA) prescribes a systematic approach comprising distinct steps that occur in phases.

**Risk management**: It is the use or application of procedures and means to reduce the negative consequences of a risk to an acceptable level. It is employed during the development and evaluation of an organism in a systematic fashion in the laboratory, through stages of field testing to commercialization. If GMO is to be produced then identification of DNA sequences encoding the desired trait, choice of marker gene, regulatory sequences for the expression of transgene, transformation method, etc. should be considered. For example, concerns have been expressed about the use of antibiotic resistant gene a marker gene, although several studies unanimously concluded that the risk was immeasurably low. Therefore, ongoing efforts are undergoing to identify other types of genes useful as markers and to develop methods for removing marker genes before GM products are commercialized. In addition, a number of specific promoter gene sequences have been identified, which can turn on gene expression in specific tissue/situations. For example, a leaf specific promoter, directing toxin production in the leaves but not roots, stems or flowers could control a gene encoding a toxin active against a leaf-attacking pest. The method of transformation should be such so as to avoid including any extraneous DNA sequence e.g. direct gene transfer by a gene gun in plants, avoids the potential for inserting unnecessary vector DNA because no vector is used. On the other hand, cells transformed by *Agrobacterium* mediated DNA transfer methods usually contain extra pieces of DNA coming from the vector in addition to the desired gene or genes. Such

sequences have generally been proven as safe but it is better to avoid, if possible. It may be noted that right from the initiation of research, risk assessment and management consideration be kept in mind and integrated appropriately into the research plan for production of GMOs.

Containment facilities at different levels of physical and biological and for GM plants, green house containment and risk management of field trials are very important and hence discussed separately.

**Risk monitoring**: In view of the speculations about potential harm from GMOs introduced into the environment, there has been considerable focus on the monitoring to follow the fate of these organisms and the transgenes they carry and to be vigilant about the unanticipated consequences. Monitoring programs have been classified into three categories i.e. experimentation, tracking and surveillance corresponding to progressive scale up in field test and commercialization. In view of different monitoring objectives for each successive stage, there is a need to consider larger geographic sampling areas and longer term observation regimes. The development of the monitoring plan will consider the objectives, available knowledge about the organisms, environment, conditions of release and potential risk as determined in a risk assessment and the regulatory requirements. The designing of the plan include specific sampling regimes and testing procedures. It is important to keep the monitoring plan dynamic so as to incorporate modifications in response to changing conditions or unanticipated problems that might develop during the implementation of the plan.

**Risk Communication**: Whereas risk assessment and management procedures are intended to identify and minimize potential negative effects on human health and the environment, risk communication is an integral part of biosafety procedures to ensure public acceptance of GMOs. It is important to interact with public at large about the specific risks and actions taken to alleviate them before announcing GMO field tests and commercialization. Insufficient or inaccurate information needs to misperceptions of risk resulting in adverse public opinion. For example, risk controversies like the current debate over GM food can be divided into technical and non-technical components. The technical components are generally regarding the scientific hazards evaluated in a risk assessment and the management options arising from the assessment. The non-technical components include the cultural and ethical issues generally raised by non-experts, allegations about secretive regulatory decisions etc.

It is a valuable exercise to have effective risk communication and some of the communication strategies are as follows:

- Accept and involve the public as a legitimate partner and treat adversaries with respect
- Coordinate, collaborate and provide information through credible sources
- Be honest, frank and open, don't keep secrets and acknowledge mistakes made
- Listen to and acknowledge people's concerns
- Be proactive and speak clearly with a balanced and realistic information strategy
- Meet the needs of the media and identify and train communicators

Public opinion about biotechnology is based on misperceptions of risk fueled by insufficient or inaccurate information. More fully informed opinions can arise only when people have a better and more realistic understanding of how biotechnology will affect their immediate lives and the environment in which they live. Risk communication is thus an important first step towards public dialogue concerning the development and use of GMOs.

## CONTAINMENT

Biological safety in laboratories is achieved by adopting good laboratory practices and containment strategies. The term **Containment** is used in describing the safe methods for managing infectious agents in the laboratory environment where they are being handled or maintained. It is a combination of laboratory procedures, equipment and installations, and host-vector systems designed to minimize accidental release of organisms, their dissemination and survival in the environment and accidental infection of laboratory workers and of persons outside the laboratory. The basic biosafety requirement in the development of a GMO is to limit the spread of GMO and its genetic material. The purpose of containment is to reduce exposure of laboratory workers, other persons, and outside environment of potentially hazardous agents. Containment can be of two types: physical and biological. Containment facilities for different Risk Groups as per the recommendations of World Heath organization (WHO) and as applicable are described.

**Biological containment (BC):** It is concerned with making changes in GMOs so that hazards are reduced when they are released deliberately or accidentally into the environment. The guidelines require that *E. coli* strains and vectors to be used should be highly debilitated so that they are not able to survive outside the laboratory, transfer the foreign gene into another organism and not able to infect human beings. This can be achieved by the use of: (i) auxotrophic mutants of *E. coli*; (ii) strains lacking recombination ability (*recA-*); (iii) plasmid vectors that are non self-transmissible and non-mobilizable; and (iv) no transposons or antibiotic resistance genes in *E. coli* strains.

In consideration of biological containment, the vector (plasmid, organelle, or virus) for the recombinant DNA and the host (bacterial, plant or animal cell) in which the vector is propagated in the laboratory will be considered together. Any combination of vector and host which is to provide biological containment must be chosen or constructed to limit the infectivity of vector to specific hosts and control the host-vector survival in the environment.

**Physical Containment (PC):** When there are real physical barriers to prevent the escape of biological where the organism is designed so that it is not able to survive in any environment other than that of laboratory. The objective is to confine recombinant organisms thereby preventing the exposure of the researcher and the environment to the harmful agents. The protection of personnel and the immediate laboratory environment from exposure to infectious agents is provided by good microbiological techniques and the use of appropriate safety equipment, (primary containment). The protection of the environment external to the laboratory from exposure to infectious materials is provided by a combination of facility design and operational practices (secondary containment). There are three elements of containment viz. laboratory practice and technique, safety equipment and facility design. Physical containment was grouped in to four categories which were designated as P1, P2, P3 and P4 but they are now referred to as Biosafety levels

## Biosafety levels

These are combination of laboratory practices and techniques, safety equipment and laboratory facilities appropriate for the operations performed and the hazard posed by the infectious agents. The proposed safety levels for work with recombinant DNA technique take into consideration the source of the donor DNA and its disease-producing potential. The four biosafety levels (BL) corresponds to (P1<P2<P3<P4) facilities approximate to 4 risk groups assigned for etiologic agents.

*Biosafety Level 1 (BL1)*: It is appropriate for undergraduate and secondary educational training and teaching laboratories and for other facilities in which work is done with defined and characterized strains of viable microorganisms not known to cause disease in healthy adult human.

*Biosafety Level 2*: It is applicable in clinical, diagnostic, teaching and other facilities in which work is done with the broad spectrum of indigenous moderate-risk agents. Laboratory

workers are required to have specific training in handling pathogenic agents and to be supervised by competent scientists. Accommodation and facilities including safety cabinets are prescribed, especially for handling large volume or high concentrations of agents when aerosols are likely to be created. Access to the laboratory is controlled.

*Biosafety level 3*: It is applicable to clinical, diagnostic, teaching, research or production facilities in which work in done with indigenous or exotic agents where the potential for infection by aerosols is real and the disease may have serious or lethal consequences. Personnel are required to have specific training in work with these agents and to be supervised by scientists experienced in this kind of work. Specially designed laboratories and precautions including the use of safety cabinets are prescribed and the access is strictly controlled.

*Biosafety level 4*: It is applicable to work with dangerous and exotic agents which pose a high individual risk of life-threatening disease. Strict training and supervision is required and the work is done in specially designed laboratories under stringent safety conditions, including the use of safety cabinets and positive pressure personnel suits. Access is strictly limited.

## Good laboratory practices

The most important rules are listed below, not necessarily in order of importance:

1. Mouth pipetting should be prohibited.
2. Eating, drinking, smoking, storing food, and applying cosmetics should not be permitted in the laboratory work area.
3. The laboratory should be kept neat, clean and free of materials not pertinent to the work.
4. Work surfaces should be decontaminated at least once a day and after any spill of potentially dangerous material.
5. Members of the staff should wash their hands after handling infectious materials and animals and leaving the laboratory.

6. All technical procedures should be performed in a way that minimizes the creation of aerosols.

7. All contaminated liquid or solid materials should be decontaminated before disposal or reuse; contaminating materials that are to be autoclaved or incinerated at a site away from the laboratory should be placed in the durable leak proof containers, which are closed before being remove from the laboratory.

8. Laboratory coats, gowns, or uniforms should be worn in the laboratory; laboratory clothing should not be worn in non laboratory areas; contaminated clothing should be disinfected by appropriate means.

9. Safety glasses, face shields, or other protective devices should be disinfected by appropriate means.

10. Only persons who have been advised of the potential health hazards and meet any specific entry requirements (immunization) should be allowed to enter the laboratory working areas; laboratory doors would be closed when work is in progress; access to animal houses should be restricted to authorized persons, children are not permitted in laboratory working areas.

11. There should be an insect and rodent control program.

12. Animals not involved in the work being performed should not be permitted in the laboratory.

13. The use of hypodermic needles and syringes should be restricted to parenteral injection and aspiration fluids from laboratory animals and diaphragm vaccine bottles. Hypodermic needles and syringes should not be used as a substitute for automatic pipetting devices in the manipulation of infectious fluids. Cannulas should be used instead of sharp needles wherever possible.

14. Gloves should be worn for all procedures that may involve accidental direct contact with blood, infectious materials, or infected animals. Gloves should be remove aseptically and autoclaved with other laboratory wastes before disposal. When disposable gloves are not available, reusable gloves should be used. For removal they should be cleaned and disinfected before re-use.

15. All spills, accidents and overt or potential exposure to infectious materials should be reported immediately to the laboratory supervisor. A written record should be prepared and maintained. Appropriate medical evaluation, surveillance, and treatment should be provided.

16. Baseline serum samples may be collected from and stored for all laboratory and other at risk personnel. Additionally serum specimens may be collected periodically depending on the agents handled or the function of the facility.

17. The laboratory supervisor should ensure that training in laboratory safety is provided. A safety or operations manual that identifies known and potential hazards and that specifies practices and procedures to minimize or eliminate such risks should be adopted. Personnel should be advised of special hazards and required to read and follow standard practices and procedures.

*Greenhouse Containment*

Greenhouse facilities should be made as per the specifications required for the category of transgenic plant being handled. Conventional greenhouses designed to keep insects and animals out and plant and plant parts are in can be made suitable for GMOs by structural upgrades. For higher level of containment facilities have to meet specifications such as controlled and filtered airflow, systems to control and disinfect water leaving the facility, autoclave for on-site sterilization of plant material and equipment, disinfecting the facility after experiments, strict limits on whom is allowed to enter, and staff and worker training.

## Risk Management in Field Trials

The procedures vary depending on the nature and magnitude of identified risks, which in turn depend on basic characterization of the organism, the nature of genetic modification and most important, the site where the GMO is to be released. The local ecosystem of the site should be carefully examined before planning the release of a GMO. Some of the risk management strategies are explained below:

For preventing and minimizing the unintentional spread of GMO or a genetic material, measures should be taken to confine them within a site/zone having designated borders/limits. This can be used by both physical and biological means.

*Physical strategies for confinement*

i. Physical means to confine GMOs particularly in case of plants and animals consist of geographical or spatial isolation by the use of structures such as fences, screens, mesh etc. The access to the site should be controlled.

ii. In case of plants, appropriate isolation distance should be worked out to control the fertilization of sexually compatible species growing in the vicinity by transgenic pollen. It is essential to collect information about the presence and distribution of cross-fertile wild or weedy relatives of cultivated species near the proposed site.

iii. In case when spatial isolation is not possible due to paucity of land, the following procedures are also used to reduce or prevent the spread of GMO or transgene through pollen or seeds.

a. Border rows of the non-GM variety may be planted around the test plot to "trap" pollen from the GMO.

b. Flowering structures may be covered to screen out pollinating insects and /or prevent pollen spread by insect vectors, wind, or mechanical transfer.

c. Female flowers may be covered after pollination to prevent loss or dissemination of GM seed.

d. In case where research objectives do not require seed production for analysis or subsequent planting, flower heads may be removed before pollen and seed production.

e. Plant material of experimental interested may be harvested before sexual maturity.

f. Test plots may be located surrounded by roads or buildings.

*Biological strategies for confinement*

A common method of biological confinement is reproductive isolation, which can be achieved by adopting some of the following strategies.

i. GM plants may be grown in an area where sexually compatible wild or weedy species are not found.

ii. All plants of sexually compatible wild or weedy species found within the known effective pollinating distance of the GM crop may be removed.

iii. Flowers may be cover with bags to screen out insect pollinators or prevent wind pollination.

iv. Production of viable pollen may be prevented by using genetic male sterility, applying a gametocyte, or removing all reproductive structures at an early stage of development.

v. Tubers, rhizomes, storage roots, and all tissues capable of developing into mature plants under natural conditions may be recovered.

vi. Differences in flowering time may be exploited so that GM pollen is not shed at the time when sexually compatible plants nearby are receptive.

*Other procedures*

i. Incorporating genes into chloroplast DNA instead of chromosomal DNA.

ii. Genetically engineer transgenic plant to produce sterile seeds. This technology was developed as a "technology protection" system to secure intellectual property rights for the improved seed (the so-called Terminator gene). It is highly effective for risk-management purposes, but has raised ethical questions regarding seed saving and the role of multinational corporations in controlling seed and therefore food supplies in developing countries. This technology is not preferred and also not recommended for release of any transgenic plant containing this technology or terminator gene sequences.

iii. Environmental conditions such as temperature, water supply and humidity can also be manipulated to limit reproduction, survival or dissemination of GMOs outside the experimental area.

iv. Chemicals such as herbicides, fungicides, insecticides and disinfectants can also be used to limit survival and reproduction of GMOs outside the trial area.

v. At the end of the experiment, the whole experimental area can be sterilized or treated with appropriate chemicals.

It may be noted that these measures will be applicable on case-by-case basis on organism with novel traits. Generally the risks are acceptably low at the time of field-testing, due to earlier extensive research being conducted during research and development.

### General precautions

Careful records of all experiments need to be kept as they provide documentation of the genetic modification and verification data, observed phenotype and any other unexpected observations. This information is necessary not only for documenting performance in the field but also to ensure compliance with risk management and redress in the case of accidental release.

The monitoring procedures are applied in such a way that appropriate measures can be taken in case of unexpected effects during or after the release.

It is also important to have appropriate termination procedure to ensure that the GMOs are effectively removed from the experimental area. The required measures are determined by the type of organism, their natural means of spread and the environment in which testing was carried out. Some form of disinfecting is necessary for microorganisms whereas harvesting seeds and ploughing in or burning residual plant material are usually effective in case of plants. Detailed guidelines have been developed for the same, which should be followed.

It may be noted that the biosafety risks to human health and environment can be reduced to acceptable levels by careful management. In addition to the science based issues, the risk management options also take into consideration the policies of the regulatory authorities and the measures that are possible scientifically and economically. The costs of risk management are borne by the applicant in essentially every country and therefore it is important to ensure that the necessary risk management requirements are worked out in such a way that both private as well as public funded research organizations can undertake the same.

### National Biosafety Regulatory Framework in India

Govt. of India has evolved a comprehensive regulatory mechanism for development and evaluation of GMOs and rDNA research work. The Ministry of Environment and Forests (MoEF) is the nodal agency for release of GMOs in the country. The Ministry has enacted Environment and Protection Act (EPA), 1986, rules 1989, to provide for protection and improvement of environment and the related matters. The rules and regulations cover the areas of research as well as large scale applications of GMOs and products made there from. The rules also cover the application of hazardous microorganisms which may not be genetically modified. Department of Biotechnology (DBT)

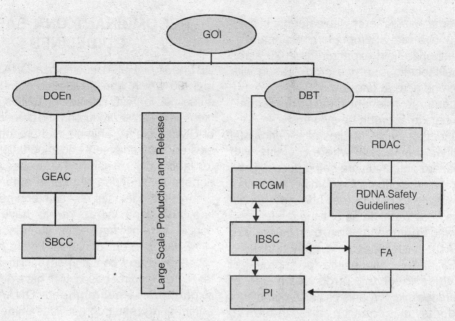

**Fig. 27.1:** GOI–Government of India; DBT–Department of Biotechnology; RDAC–Recombinant DNA Advisory Committee; IBSC–Institutional Biosafety Committee; RCGM–Review Committee on Genetic Manipulation; DOEn–Department of Environment; GEAC–Genetic Engineering Approval Committee; SBCC–State Biotechnology Coordination Committee; PI–Principal Investigator (R & D/Industry/Others); FA–Funding Agency (Govt./Private and Public Institutions)

had formulated recombinant DNA Guidelines in 1990 which were revised in 1994. There are six competent authorities for the regulatory mechanism as described below and shown in Fig. 27.1. The salient features are:

1. **Recombinant DNA Advisory Committee (RDAC):** A committee constituted by DBT referred as RDAC, take note of developments in biotechnology at national and international levels and recommends safety regulations for research and applications.

2. **Institutional Biosafety Committee (IBSC):** It is the nodal point for interaction within the institution for implementation of guidelines. For this, institution carrying out research activities on genetic manipulation should constitute IBSC with one DBT nominee. The main activities are: (i) to note and to approve r-DNA work; (ii) to ensure adherence of r-DNA safety guidelines of government; (iii) to prepare emergency plan according to guidelines; (iv) to recommend to RCGM about category III risk or above experiments and to seek RCGM's approval; (v) to act as nodal point for interaction with statutory bodies; (vi) to ensure experimentation at designated locations, taking into account approved protocols.

3. **Review Committee on Genetic Manipulation (RCGM):** DBT has next higher level of body known as RCGM which has the following functions:

i. To bring out manuals of guidelines specifying procedures for regulatory process on GMOs in research, use and applications including industry with a view to ensure environmental safety.

ii. To review all the work going on r-DNA projects involving high risk category and controlled field experiments.

iii. To lay down procedures for restriction or prohibition, production, sale, import and use of GMOs both for research and applications.

iv. To permit experiments with category III risks and above with appropriate containment.

v. To authorize field experiments in 20 acres in multi-locations in one crop season with up to one acre at one site.

vi. To generate relevant data on transgenic materials in appropriate systems.

vii. To undertake visits of sites of experimental facilities periodically where projects with biohazard potentials are being pursued and also at a time prior to the commencement of the activity to ensure that adequate safety measures are taken as per the guidelines.

4. **Genetic Engineering Approval committee (GEAC):** It functions as a body under the MOEF and is responsible for approval of activities involving large scale use of hazardous microorganisms and recombinant products in research and industrial production from the environment angle. It also has the following functions:

i. To permit the use of GMOs and products thereof for commercial applications.

ii. To adopt procedures for restriction or prohibition, production, sale, import and use of GMOs both for research and applications under EPA, 1986.

iii. To authorize large scale production and release of GMOs and products thereof into the environment.

iv. To authorize agencies or persons to have powers to take punitive actions under the EPA.

5. **State Biotechnology Coordination committee (SBCC):** In each state there is a State Biotechnology Coordination committee (SBCC) headed by the Chief Secretary where research and applications of GMOs are contemplated.

6. **District Level Committee (DLC):** DLC is the district level committee headed by district collector as an authoritative unit to monitor safety regulations. Both SBCC and DLC work along with RCGM in the inspection and monitoring of the experiments at the field sites.

# RECOMBINANT DNA SAFETY GUIDELINES

DBT had formulated recombinant DNA Guidelines in 1990 which were revised in 1994. It include guidelines for R&D activities on GMOs, transgenic crops, large-scale production and deliberate release of GMOs, plants, animals and products into the environment, shipment and importation of GMOs for laboratory research. The issues relating to genetic engineering of human embryo, use of embryos and fetuses research and human germ line, and gene therapy areas have not been considered while framing the guidelines.

Four different biosafety levels have been recognized and containment facilities for each level are recommended which have already been explained. Recombinant DNA/ genetic engineering research activities guidelines have been grouped under three categories:

*Category* I: It includes: (i) Routine cloning of defined genes, defined non-coding stretches of DNA and open reading frames in defined genes in *E. coli* or other bacterial/fungal hosts which are generally considered as safe to human, animals and plants. (ii) Transfer of defined cloned genes into *Agrobacterium*. (iii) Use of defined reporter genes to study transient expression in plant cells and to study genetic transformation conditions. (iv) Molecular analysis of transgenic plants grown *in vitro*. It also includes those experiments which involved self cloning using strains and also interspecies cloning belonging to organisms in the same group. The experiments need intimation to IBSC in the prescribed proforma.

*Category* II: It include experiments carried out in the Lab and green house/net house using defined DNA fragments non-pathogenic to human and animals for genetic transformation of plants, both model species and crop species and the plants are grown in green house/net house for molecular and phenotypic evaluation. Permission to perform experiments will be provided by IBSC. It would be intimated to RCGM before execution of experiments.

*Category III*: It includes experiments having high risk where the escape of transgenic traits into the open environment could cause significant alterations in the biosphere, the ecosystem, plants and animals by dispersing new genetic traits the effects of which cannot be judged precisely. Further this also includes experiments conducted in green house and open field conditions not belonging to the above category II types. Such experiments could be conducted only after the clearance from RCGM and notified by DBT. It also includes experiments on toxin gene cloning, cloning of genes for vaccine production, etc.

The controlled release of GMOs should be done under appropriate containment facilities to ensure safety and to prevent unwanted release in the environment.

Pre-release tests of GMOs in agriculture should include elucidation of requirements for vegetative growth and persistence and stability in small plots and experimental fields.

# REVISED GUIDELINES FOR RESEARCH IN TRANSGENIC PLANTS, 1998

DBT brought out separate guidelines for research in transgenic plants in 1998 which include the guidelines for toxicity and allergenicity of transgenic seeds, plants and plant parts. The genetic engineering experiments on plants have been grouped under three categories. Category I includes routine cloning of defined genes, defined non-coding stretches of DNA and open reading frames in defined genes in *E. coli* or other bacterial/fungal hosts which are generally considered as safe to human, animals and plants. The category II experiments include experiments carried out in lab and greenhouse/net house using defined DNA fragments non-pathogenic to human and animals for genetic transformation of plants, both model species and crop species. Category III includes experiments having high risk where the

escape of transgenic traits into the open environment could cause significant alterations in the biosphere, the ecosystem, plants and animals by dispersing new genetic traits the effects of which cannot be judged precisely. Further this also includes experiments conducted in green house and open field conditions having risks mentioned above.

Risk management is employed during the development and evaluation of an organism in a systematic fashion in the laboratory, through stages of field testing to commercialization. Right from the initiation of research, risk assessment and management considerations should be kept in mind. Planned field experiments with transgenic plants are permitted only after a step-wise (laboratory to growth chamber and greenhouse) evaluation, either in India or elsewhere, to generate data on the following: (i) Characteristics of the donor organisms providing the target nucleic acids; (ii) Characteristics of the vectors used; (iii) Characteristics of the transgenic inserts which includes specific functions coded by the inserted nucleic stretches including the marker gene inserts, expression of nucleic acid products and their activities, toxicity and allergenicity of nucleic acid products to human and animals sequence, (iv) regulatory mechanism utilized in the expression of cassette, (v) cell lines used for shuttling and amplification of the cassette (vi) Characteristics of the transgenic plants which include the methods for detection of transgenic plants, methods of detection and characterization of the escaped transgenic traits in the environment, toxicity and pathogenicity of transgenic plants and their fruits to other plants in the ecosystem and environment, possibility of and the extent of transgenic escape and pollen transfer to wild near relatives and pathogenicity, toxicity and allergenicity of the transgenic plants and their fruits to human and animals. In addition, the following is also insisted upon.

(a) Laboratory data to show that protein products of transgenes are safe to the environment and human beings.

(b) Isolation distance as applicable to foundation seed of the crop be provided to transgenic crops when grown in the field.

(c) A few rows of the same crop as the transgenic one should be planted beyond the isolation distance to act as pollen trap.

(d) Non-transgenic plants should be grown within the isolation distance at 1 or 5 m intervals to determine the distance of pollen escape.

(e) All the vegetative plants and left-over seeds must be destroyed by burning after the conduct of experiments.

(f) After the experiment, the land may be left fallow and all plants that emerge must be destroyed.

(g) The experimental field may be visited by the company authorized personnel only, and all records of visits are to be maintained.

(h) Full account of transgenic seeds produced is to be kept and no part of this seed lot can be transacted or further propagated without authorization as per the guidelines.

To monitor over a period of time, the impact of transgenic plants on the environment, a special Monitoring cum Evaluation Committee (MEC) has been set up by the RCGM. The committee undertakes field visits at the experimental sites and suggests remedial measures to adjust the trial design, if required, based on the on-spot situation. This committee also collects and reviews the information on the comparative agronomic advantages of the transgenic plants and advises the RCGM on the risks and benefits from the use of transgenic plants and advises the RCGM on the risks and benefits from the use of transgenic plants put into evaluation. Trials will be done for at least one year with minimum of four replications and ten locations in the agroecological zone. The biological advantage of transgenic will be judged and communicated by RCGM to GEAC for consideration of release in to the environment.

The guidelines include complete design of a contained green house suitable for conducting research with transgenic plants. Besides, it provides the basis for generating food safety information on transgenic plants and plant parts.

A separate section (No. 6) in the Seed Policy, 2002 on transgenic plant varieties has been put in place. Under this all genetically engineered crops/varieties will be tested for environment and biosafety before their commercial release. Seeds of transgenic plant varieties for research purposes will be imported only through National Bureau of Plant Genetic Resources as per EPA Act, 1986. Clearance for import of transgenic material for research purposes would be provided by RCGM. Transgenic varieties will be tested for 2 seasons for their agronomic values under All India coordinated project trials of ICAR, in coordination with the tests for environment and biosafety clearance as per EPA Act before any variety is commercially released. Transgenic plant varieties cannot be protected by patent regime in India but can be registered under Protection of Plant Varieties and Farmers' Rights (PPV&FR) Act, 2001 in the same manner as non-transgenic plant variety.

Besides, Drug and Cosmetic rules (8th amendment), 1988 and Drug Policy, 2002 for all recombinant products considered to be new products, and guidelines for generating preclinical and clinical data for rDNA therapeutics, 1999 are under operation.

## Cross border movement of transgenic germplasm for research purposes

As per the revised guidelines for research in transgenic plants, 1998, clearance for import of transgenic material, for research purposes would be provided by the RCGM to the concerned importer applicant. The RCGM will issue an import certificate after looking into the documents related to the safety of the material and the national need. The RCGM will take into consideration the facilities available with the importer for in-soil tests on the transgenic material.

The importer of a transgenic material may import the material accompanied by an appropriate phyto-sanitary certificate issued by the authority of the country of export. However, such import may be routed through the Director, NBPGR on the basis of the import permit issued by the RCGM of DBT. The import certificate would be cancelled if NBPGR is not provided the phyto-sanitary certificate. NBPGR will provide information on the time that is required for phyto-sanitary evaluation. These evaluations will be done in a time- bound manner in presence of the agents of the institutes or the commercial organizations that are importing the material, if they so desire. Parts of the seed material will be kept at NBPGR in double lock system in the presence of the importer. This lot of seed will act as a source material in case of any legal dispute.

### Biosafety concerns

Conventional breeding involves crossing related species to develop plants with desired characteristics which are selected among the progeny for reproduction and the selection is repeated over many generations. Genetic engineering bypasses reproduction altogether. It transfers gene horizontally from one individual to another as opposed to vertical gene transfer from parent to offspring. It makes use of infectious agents as vectors or carriers of genes so that genes can be transferred between distant species that would never interbreed in nature. Due to horizontal gene transfer in the development of GMOs there have been conflicting reports on the benefits of GMOs, risks, apprehensions, environmental concerns, and social concerns for the release of GMOs. Some researchers argue that transgenic organisms pose potentially no risks than the conventionally modified organisms. Some argue that technologically modified genes cannot be reliably assessed with the present incomplete knowledge about living systems. Research is being conducted on field tests of transgenic plants and data is generated and the potential risks from the use of GMOs and products thereof have been described. Risk, as explained

earlier, means different things to different people in different situations. It can be thought of as a combination of probability and consequence i.e. the likelihood of an event multiplied by the impact of the event. Formal risk assessment usually considers: (i) what can go wrong? (ii) how likely is it to happen? and (iii) what are the consequences if it does not happen? It is important to recognize that just because an event may happen, does not mean it will happen. Once the chance of a risk occurring has been calculated, political, social, cultural and economic considerations will determine whether people believe it to be acceptable. Discussions on GMOs have shown that risk perceptions differ dramatically, even between experts, depending on individuals, their motives and values. Broadly these risks have been categorized under risks to human health, environmental concerns and social and ethical grounds.

## RISK TO HUMAN HEALTH

### Risk of toxicity

The risk of toxicity may be directly related to the nature of the product whose synthesis is controlled by the transgene or the changes in the metabolism and the composition of the organisms resulting from gene transfer. In some cases organism may contain inactive pathways, with the addition of new genetic material which, could reactivate these inactive pathways or otherwise increase the level of toxic substances within the plants. This could happen if the on/off signals associated with the introduced gene were located on the genome in places where they could activate the previously inactive genes. Further, the modified metabolism due to introduction of tolerance to chemical substances such as herbicides may also lead to appearance of novel metabolites in the cell. For example herbicide resistant varieties have been released to permit the use of glyphosate and bromoxynil herbicides for weed control. It has been reported that bromoxynil causes birth defects in animals and is also toxic to fish. Thus it may cause

---

**Box 1: GM potato feeding experiment on rats**

Arpad Pusztai, a scientist in Scotland came to international attention when he announced to the media that eating genetically modified potatoes depressed rat immune systems and caused changes in their intestinal tract. The potatoes were modified to contain and express a gene for snowdrop lectin, called Galanthus nivalis agglutinin (GNA), which acts as an insecticidal protein as well as it is an antimetabolite i.e. it slows down growth. Dr Pusztai and colleagues compared rats fed with GM modified potatoes and non-modified potatoes. There were difficulties with the use of raw potatoes diet as rats did not like to eat raw potatoes and a standard 110 day trial had to be abandoned after 67 days because the rats were starving. The GM potatoes affected the rat immune system and the structure of their internal organs. They asserted that the outcome was the result of the way the lectin gene had inserted into the potato genome, rather than the expression of lectin by the potatoes. After considerable deliberations and review of all available data, it was concluded that evidence did not support Dr Pustzai's conclusions. An important point was that genetic modification introduced in potato was a gene to make lectin, a toxin. The public, relying on the media accounts, were not told the rats were being forced to eat a toxic lectin, and that this toxin, not the method used to introduce it, might have been responsible for their illness.

---

serious hazards to farm workers. Glyphosate is also reported to be toxic to soil organisms and to fish. It also accumulates in the fruits and tubers since plants cannot degrade it. Many of the substances involved in plant defence are also known to produce toxic effect e.g. lectin encoding genes. Wheat germ agglutinin (WGA) lectin gene is of plant origin and the product is heat stable and resistant to proteolytic digestion. But in rats it causes loss of weight and hypergrowth of small intestine. Thus it points to a very serious observation that although gene is of plant origin, from a food crop but it does not guarantee the food safety of the gene product [See Box 1]. Toxic effects of some of the genes are shown in Table 27.1. In view of the above, every GMO needs to be carefully evaluated for toxicity to human and animals. Most of such toxicity risks can be assessed using scientific methods both qualitatively and quantitatively. The controversy of Starlink corn affecting the human health is mentioned in Box 2.

### Risk of allergies

Production of GMOs sometimes includes the introduction of newer proteins from the organisms which have not been consumed as foods and may cause allergy. All known food allergens are proteinaceous in nature, mostly heat stable and resistant to proteolytic digestion but most proteins do not elicit any allergenic effects on humans. It may be noted that allergies can be developed by any individual to even common foods such as egg, milk, fish, soybean, wheat, etc. Some of the proteins involved in plant defence reactions are known food allergens e.g. peanut lectin, soybean Kunitz trypsin inhibitor, wheat germ agglutinin, barley 15 kDa alpha amylase/trypsin inhibitor (Table 27.1).

The consequences of allergenic reactions depend on the quantity of allergen ingested and duration of consumption as reaction becomes more severe with increasing dose. Also, there is no evidence that GM products pose more risk than conventional products. Allergenicity assessments generally focus on the characterization of any protein produced as a result of transfer of a transgene. The source of the protein, its history of safe use, the function of the gene/protein, its digestibility, stability to heat and other processes, are all used in comparison of the protein with known allergens.

**Table 27.1:** Toxic and allergenic effects of some compounds (Reviewed in Franck Oberaspach and Keller, 1997)

| Compound | Role in Plant defense | Toxic effects | Allergenicity |
|---|---|---|---|
| Lectins | Involved in fungus and insect resistance | Weight loss and hyper-growth of small intestine in rats | Allergenic |
| Proteinase inhibitors | Insect resistance | Pancreas enlargement, hyperplasia and adenoma | Allergenic |
| Thionins | Antifungal | Toxic on intravenous injection nontoxic orally (insufficient data) | — |
| Alkaloids[1] | Insect resistance | Toxicity problems1 | — |
| Phytoalexins[2] | Fungus and insect resistance | Toxicity problems2 | — |
| Virus coat protein | Virus resistance (transgenic) | Most likely nontoxic (toxicity not reported) | — |
| Cry protein | Insect resistance | Most likely nontoxic (toxicity not reported) | — |

1. A potato variety derived from a cross with *Solanum chalcoense* had up to 650 mg/kg of glycoalkaloids as compared to 20–150 mg/kg of glycoalkaloids in normal varieties. Variety was withdrawn from the market due to acute toxicity to humans.
2. A conventionally developed insect resistant variety of celery had 8 fold higher psoralen (phytoalexin) content. Psoralen is known to be toxic, mutagenic and carcinogenic.

---

**Box 2: The Starlink corn incident**

The US Environment Protection Agency in 1998 approved for use as animal feed a corn modified by insertion of the *cry9C* gene from Bt encoding for an insecticidal crystal protein endotoxin by Aventis Crop Science. The corn was marketed as StarLink™. Because of concern that the protein Cry9C could be allergenic, it was not approved for human food. In September 2000, a coalition of environmental and food safety groups announced that *cry9C* DNA fragments had been found in a popular brand of taco shells sold in the United States. In addition, the Cry9C protein was discovered in some non-StarLink™ seed corn and used as human food. As a result, there was a voluntary recall of corn-derived food products in the United States by manufacturing companies, some of whom took steps, such as mandatory testing requirements, to ensure no further contamination of corn used in human food with Starlink corn.

Late in 2000, a further review of the potential allergenicity of Cry9C, and of mechanisms for assessing suspected allergenic reactions to StarLink™ corn concluded that the Cry9C protein had a medium likelihood of providing to be a potential allergen and that seven out of 34 reactions to a meal containing corn products were probably allergic.

The presence of Cry9C protein in seed corn was due to physical contamination, or it can be due to cross pollination from Starlink corn. Thus this incident raised many issues:

i. The difficulties of restricting a GM food for animals or industrial purposes from that of human food when unmodified crop is used for humans.
ii. The difficulty of preventing accidental physical contamination
iii. The difficulty of ensuring adherence to separation requirements to prevent cross pollination of GM modified and unmodified crop.
iv. The need for appropriate labeling and for post market monitoring to identify allergenic reactions rapidly and correctly.

However, Aventis Crop Science voluntarily took many steps that this Starlink corn is not contaminated with corn used by humans and also withdrew this corn from market even for animal use.

Any potential safety issue is identified and a decision is taken whether to proceed further with the transgenic under review or not.

The possibility of transferring allergens with genetic engineering came to light when a methionine-rich 2S storage albumin protein producing gene from the Brazil nut (*Bertholletia exceisa*) was incorporated into soybean to enhance its sulphur containing amino acids in the seed storage protein. The process was experimented by Pioneer Hi-Bred in the USA. The transgenic soybean was meant for use as an animal feed. Pioneer Hi-Bred considered the possibility that nut protein might finds its way into the human nutrition and therefore the tests were commissioned. Unfortunately the Brazil nut 2S gene was found to be allergenic which raised the biosafety concerns. The tests conducted by the scientists on allergens confirmed that consumption of the transgenic soybean could trigger an allergic response in those sensitive subjects who were allergic to Brazil nut. The company, therefore, decided not to release the transgenic soybean for sale. The allergenic results showed that the undesirable properties of the Brazil nut protein were transferred when the gene was moved into a different species. In general, if a gene produces an allergenic protein in one species, it will likely do so in a new species.

However, sometimes a non-immunogenic protein could become immunogenic when it is expressed in another species. Transgenic pea line has been developed with alpha amylase inhibitor gene which gives protection from pea weevil (*Bruchus pisorum*), but animal feeding experiments revealed that it caused an immune response in mice including inflammation of lungs (Prescott *et al.*, 2005). This was not expected from alpha amylase inhibitor gene but it was found that subtle changes occurred when it was expressed in pea. Thus, there is a strong argument in favour of case by case biosafety assessment of transgenic plants regarding allergenicity assessment.

Specific guidelines have been developed by WHO, FAO, US Food and Drug Authority. Based on these guidelines DBT, regulatory agency in India has formulated a set of guidelines.

**Antibiotic resistance**

Current methods of generating transgenic plants employ a "selectable marker gene" which is transferred along with any other gene of interest usually on the same DNA molecule. The presence of a suitable marker is necessary to facilitate the detection of genetically modified plant tissue during development. The most widely used selectable marker is a bacterial gene for neomycin phosphotransferase (*nptII*), besides *aadA* gene for streptomycin and spectinomycin resistance and the *hpt* gene for resistance to hygromycin. The use of these antibiotic resistance markers has raised the concerns that eating foods carrying antibiotic resistance marker would reduce the effectiveness of antibiotic to fight disease when these antibiotics are taken with meals. The antibiotic resistance gene produce enzyme that can degrade antibiotics. Therefore, theoretically if a transgenic tomato with an antibiotic resistance gene is eaten at the same time as an antibiotic, it could destroy the antibiotic in the stomach. This issue was first raised during the approval process of Calgene's Flavr Savr tomato and Ciba-Geigy's Bt corn 176. A possible concern on the transfer of resistance to gut microorganisms and the potential for transfer of resistance to potentially hazardous microorganisms has been raised.

From various studies it has been concluded that these selectable markers basically have no effect and following points have emerged which are summarized below:

1. The transfer of *nptII* gene from plant material to gut microflora is extremely unlikely because there is no evidence that such transfer can occur through horizontal gene transfer (HGT). HGT is defined as the transfer of genetic material from one species

to another species that is not usually sexually compatible or in other words the spread of genetic information from a plant to microorganism via non-sexual processes is referred as horizontal gene transfer. For plant DNA to be transferred by HGT a whole set of conditions must occur: (i) Available DNA from the plant must be free from the cells, of sufficient length and persist long enough for uptake. DNA in dying plant cells is generally rapidly degraded but it can survive in some soils, aquatic environments or the digestive tract of mice long enough to be available for uptake. (ii) A bacterial recipient must be in a suitable state for DNA uptake (competent) and a mechanism for uptake needs to be in place. How often this occurs in bacteria in natural surroundings is unknown but competence can be induced in the laboratory? (iii) The recipient cell needs to incorporate, maintain and use the incoming DNA. Integration will depend on sequence homology and a gene will only be useful if it can be read by the recipient's cell machinery.

Some studies have shown that gene transfer to plant associated fungi has taken place but there is no evidence for stable integration and subsequent inheritance. This conclusion is supported by studies, which demonstrated that horizontal gene transfer from plants to microbes did not occur under a variety of test conditions (Nap *et al.*, 1992; Redenbaugh *et al.*, 1994; Schloeter *et al.*, 1995; Prins and Zadoks, 1994). Moreover, there are numerous barriers in the gut which make this event extremely unlikely to occur and the event, if it were ever to occur, would be unlikely to be maintained in the absence of constant selection pressure for resistance due to the extremely limited use of these particular antibiotics.

Although, horizontal transfer of DNA does occur under natural circumstances and laboratory conditions, its probability is extremely rare in the acid environment of the human stomach or even outside environment. Recently, it has been demonstrated that horizontal transfer of DNA containing *npt*II gene (a deletion mutation) can occur at a rate of $1 \times 10^{-8}$ i.e. one transformant per 10 million cells, under strong kanamycin selection pressure. The probability of such transfer would further decrease in natural circumstances.

2. If the event of a transfer occurs by an unknown mechanism from the genome of genetically modified plants or products derived from them to gut micro flora and this event being maintained, this would not add significantly to the inherently large microbial population of kanamycin and neomycin resistant microbes in the gut of either humans (Nap *et al.*, 1992) or animals (McAllan *et al.*, 1973; Nap *et al.*, 1992).

3. The expression of *npt*II gene in genetically modified plants is controlled by a plant specific promoter which is not expected to function in bacteria. In the unlikely event of transfer of the *npt*II gene and stable propagation of the intact gene fragment in bacteria, the gene is unlikely to be expressed and even less likely that a DNA rearrangement event occurs that will place the functional nptII encoding open reading frame in front of a bacterial promoter.

4. *npt*II gene if expressed in intestinal bacteria, antibiotic therapy would not be compromised, as the co-factors necessary for the enzyme to inactivate kanamycin and neomycin are not present at the required concentration range in the gut. Moreover the NPTII protein would be rapidly degraded in the gut.

5. In veterinary there is limited usage of kanamycin and neomycin.

6. The *npt*II gene occurs ubiquitously in nature and resistance to this class of antibiotics is already widespread. Therefore, in the highly unlikely situation that a transfer event does occur in gut microflora in humans or animals, this would not significantly impact the overall

frequency of kanamycin or neomycin resistant bacteria in the gut or rumen.

Therefore, the overall risk is considered to be effectively zero and the therapeutic use of antibiotics in humans or animals will not be impacted by the commercialization of transgenic crop containing antibiotic resistant selectable marker genes. Although, the ideal situation would be to develop strategies to remove a selectable marker from a transgenic plant prior to commercialization. There are a number of methods for generating marker-free transgenic plants. The strategies being developed generally rely on high transformation efficiencies and transformation with suitable excision vectors, which will allow the removal of the selectable marker gene.

### Eating Foreign DNA

There have been apprehensions about danger from eating the foreign DNA in GM foods i.e. the pieces of DNA that did not originally occur in that food plant. DNA being present in all living things such as plants, animals, microorganisms and is eaten by human beings with every meal. Most of it is broken down into more basic molecules during the digestion process whereas a small amount that is not broken down is either absorbed into the blood stream or excreted. In an experiment on feeding mice with a harmless detectable DNA sequence, its progress was tracked through the gastrointestinal tract and the body. About 5% of DNA was detectable in small intestines, large intestines and faeces up to eight hours after the meal, 0.05% in the blood stream up to eight hours; very small fragments in liver and spleen up to 18 hours and no foreign DNA after 42 hours. It has been reported that even if foreign DNA finds it way into tissues of an organism, it is destroyed by body's normal defense system. The DNA of the modified crop will usually be processed and broken down by the digestive system in the same way as that of conventionally bred, or otherwise modified crops. So far there is no evidence that DNA from GMOs including transgenic crops is more dangerous

to human health than DNA from conventional crops, animals or associated microorganisms that are normally eaten.

According to a FAO/WHO document on safety aspects of GMOs of plant origin, the amount of DNA which is ingested varies widely, but it is estimated to be in the range of 0.1 to 1.0 gram per day. Novel DNA from a GM crop would represent less than 1/250,000 of the total amount consumed. This means that the possibility of the transfer of genes that have been introduced through genetic modification is extremely low. In fact, it would require that all of the following events would probably occur: (i) the relevant gene(s) in the plant DNA would have to be released probably as linear fragments; (ii) the gene(s) would have to survive harvesting, preparation and cooking, and also nucleases action in the plant; (iii) the gene(s) would have to compete for uptake with other dietary DNA; (iv) the recipient bacteria or mammalian cells would have to be able to take up the DNA and the gene(s) would have to survive enzyme digestion; (v) gene(s) would have to be inserted into the person's DNA by very rare recombinant events and the gene would have to stably integrate and express in new host.

The probability of above events, actually taking place is extremely rare or virtually negligible.

### Use of promoters of virus origin

Concerns have been expressed regarding the harm to human health by the use promoters of virus origin e.g. 35S promoter of cauliflower mosaic virus (CaMV). This promoter causes cauliflower mosaic disease in several vegetables, such as cauliflower, broccoli, cabbage and canola. It was suspected that CaMV promoter might be harmful if it invades human cells and turns on certain genes. However, a multi step chain of events would have to occur for the CaMV promoter to escape the normal digestive breakdown process, penetrate a cell of the body and insert itself into a human chromosome. Such a probability is extremely low (virtually impossible), as it has been explained earlier that the normal body defense eliminate any stray fragments of foreign DNA that enter into the bloodstream from the digestive tract.

### Changes in nutritional level

Concerns about the accidental changes in the nutritional components of transgenic crops while incorporating other traits have been raised. For example isoflavone levels in Roundup-Ready herbicide resistant transgenic soybean showed minor differences in comparison to conventional varieties. A controversy about increase in phytoestrogen levels in herbicide tolerant soybean has been reported in Box 3.

## RISK TO ENVIRONMENT

### Gene flow or dispersal from transgenics

Genetic traits with selective advantages in agricultural or natural systems are liable to spread beyond the cultivated variety in which they were originally introduced. Accidental cross breeding between GMO plants and traditional varieties through pollen transfer can contaminate the traditional local varieties with GMO genes resulting in the loss of traditional varieties for the farmers. The wind, rain and insect pollinators can contribute to the spread of pollen resulting in the contamination of local varieties through cross pollination/breeding. In Europe, if contamination by the transgenic pollen exceeds the limit of 1%, then the farmer will not be able to get the label of GM free. Creation of new weeds or increasing the problems posed by existing weeds, due to transfer of modified genes by hybridization or cross-pollination with uncultivated plants or wild relatives, has been suggested as the primary risk in releasing transgenic plants.

The likelihood of the transgenes conferring herbicide tolerance, pest resistance or disease resistance being transferred by cross-pollination to sexually compatible species, and the possibility of producing persistent weeds or invasive plant populations, have motivated researchers to study pollen movement in various crops and also sexual compatibility among crops and related species (Table 27.2). Inter-varietal gene flow which occurs frequently and inter-specific gene flow which is a much rarer event are by no means limited to transgenic traits. However, transgenic plants are considered as 'special cases' due to two reasons: (a) many transgenic characters correspond to acquisition of one or several foreign genes, for which no equivalent character is present or would have appeared by spontaneous or induced mutation; and (b) unlike most characters obtained by empirical selection, transgenic traits are dominant and monogenic, and may thus be readily transmittable by out-crossing.

---

**Box 3**

The increase in phytoestrogen levels in herbicide tolerant soybeans can cause breast cancer A letter from SAG, Basler Appell gegen Gentechnologie, Basle Appeal against Genetic engineering was sent to Swiss Confederate Councillor on Feb. 4, 1997 that we fear Roundup Ready soybean produces large quantities of pseudoestrogens (substances that occur in natural environment and influence or mimic the functions of hormones) when it is sprayed with Roundup herbicide. Today it is assumed that estrogen hormones play an important role in the emergence of breast cancer.

SAG report was based on 1988 report which claimed that conventional French beans produces an estrogenically effective isoflavonoid (coumestrol) after the application of glyphosate (a.i. of Roundup). Hoever, investigations revealed no indication that GM soybeans exhibit any raised concentrations of phytoestrogens following treatment with Roundup herbicide.

**Table 27.2:** The extent of pollen dissemination from transgenic varieties to other varieties of the same crops (Reviewed by Messeguer, 2003)

| Transgenic Crop | Gene/trait | Rate of pollen dispersal (as per cent seed of non-transgenic variety expressing the transgene) |
|---|---|---|
| Brassica napus | bar/ Glufosinate tolerance | (i)1.5% at 1 m; 0.00033% at 47 m<br>(ii) 0.15% at 200 m; 0.00038% at 400 m |
| Zea mays | bar/ Glufosinate tolerance | 1% at 18 m |
| Oryza sativa | pat/ glufosinate tolerance | 0.1% at 1m; 0.1% at 5 m |
| Gossypium hirsutum | tfdA/ 2,4-D resistance;<br>Bt/ insect resistance | 11.2% at 1 m;<br>0.03% at 50 m |

Further, some of the important factors that may influence gene flow are:

i. Proximity of the transgenic with compatible wild relatives
ii. Sexual compatibility between crop plant and species or weed
iii. Mating system and mode of pollination
iv. Synchronization of flowering of crop and wild relative or weed
v. Relative fitness of weed-crop hybrid
vi. Mode of seed dispersal
vii. Nature of transgenic character itself

Studies on gene flow from transgenics irrespective of distance have been conducted. Although out-crossing frequency appears to be strongly influenced by the distance of the recipient plant from the test plot but some studies showed that distance is not the barrier. Hybrids between a transgenic *B. napus* expressing an oil quality modification gene and the weedy *B. campestris* were superior to *B. campestris* in terms of seedling establishment. Transgene introgression can occur from *Beta vulgaris* in to their wild relatives in variable frequencies (Table 27.3). Arias and Rieseberg (1994) reported that physical distance alone will be unlikely to prevent gene flow between cultivated and wild population of sunflower. Skogsmyr (1994) on transgenic potatoes (*Solanum tuberosum* cv. Desiree), containing marker systems like NPTII and GUS, indicated that gene dispersal could occur over long distances and to a higher extent than had been previously shown. Ellstrand (1992) using cultivated radish (*Raphanus sativus* L.), which is out-crossing and insect-pollinated, showed that some hybrids containing marker gene could be found even 1km away. A report on corn contaminated by GMO genes in Mexico is reported in Box 4.

However, studies also showed that transgenics pose little or no risk through gene flow. Field trials of transgenic rapeseed *(Brassica napus)* with GUS, kanamycin and asulam resistance markers by Paul *et al.* (1995) and

**Table 27.3:** Examples of gene transfer from transgenic crops to their wild relatives (Reviewed by Messeguer, 2003, and Malik and Saroha, 1999)

| Transgenic crops | Gene/trait | Wild relative | Remarks |
|---|---|---|---|
| Brassica napus | Oil quality modification gene | Brassica campestris | Hybrids have advantage over B. campestris during seedling establishment |
| | | Wild radish (Raphanus raphanistrum | Hybrids produced with frequency of ~ 5x10⁻⁴ |
| Oryza sativa | Herbicide resistance | Red rice (weedy form) | Red rice seeds showing herbicide resistance ranged from (i) 0.8 to 19.7% and (ii) 0.002 to 0.6% in different studies. |
| Beta vulgaris | | Wild beet | Gene flow occurs; transgenic progeny found> 200 m behind a strip of hemp plants planted for containment |
| Solanum tuberosum | — | Solanum dulcamara and Solanum nigrum | No evidence of transgene transfer to wild relatives. |

---

### Box 4: A report on corn contaminated by GMO genes in Mexico

David Quist and Ignacio Chapela in November 2001 reported that transgenic DNA constructs–35S promoter sequence together with sequences from alcohol dehydrogenase had been found in a number of creole maize varieties in two remote mountain locations in Oaxacan state of Mexico. The report was widely circulated in media, drawing attention to risks of food security and threats to the genetic diversity. CIMMYT conducted research on 28 accessions from Oaxacan landraces for a study on gene flow. Their initial study could not detect the 35S promoter either in historical accessions from its extensive seed bank or in samples collected recently from the field. In a second phase, the CIMMYT researchers took seeds of 15 maize accessions from gene bank and eight of the accessions were from the state of Oaxaca. DNA was amplified using a primer corresponding to CaMV35S promoter, a fragment of DNA found in most commercial transgenic maize and not known to exist naturally in the maize genome. To further ensure that the reactions are working correctly, all DNA samples were amplified using a primer corresponding to a fragment of DNA known to exist naturally in the maize genome (SSR marker phi076). All positive controls amplified correctly while none of the 15 accessions of gene bank amplified the CaMV35S promoter sequence, indicating that that there is no CaMV35S promoter sequence in the accessions. This was followed by another study on seeds of 42 samples of maize varieties collected from the fields of Oaxaca state. All positive controls (SSR marker phi022) amplified correctly while samples from Oaxaca state did not show the presence of CaMV35S sequence.

As is known maize is an out crossing species and the farmer's varieties are not the same as they were two years ago. Work at CIMMYT has shown that creole maize varieties planted by small farmers in Mexico are constantly changing, both as a result of the biology of plant and traditional practices of farmers. Oaxaca, which is the center of maize diversity, a land race is not stable, uniform or distinct like a plant variety. Farmer's deliberately use external sources of seed to maintain vigour. Gene flow is constant and the real question is whether it makes any difference if one of the genes that have flowed is a transgene. Substitution of one variety for another in an agricultural field does not threaten biodiversity but it is the conversion of native and wild land to agriculture in the first place

Nature, the leading scientific journal disowned the above mentioned paper on April 5, 2002 it published in the 2001 about the environmental safety of GM crops. It followed protests by more than 100 leading biologists who spotted mistakes in the research by two American scientists and attacked Nature for giving respectability to inaccurate results. After studying criticisms of the paper and obtaining new information from its authors, Philip Campbell, the Editor of Nature, agreed that it should not have passed its process of peer review. Author of one of the refutation papers Dr Matthew Metz said the original study relied on flawed techniques "The discovery of transgenes fragmenting and promiscuously scattering throughout the genomes would be unprecedented and is not supported by Quist and Chapela's data.

---

studies by Luby and McNicol (1995) on 80 raspberry populations, using a recessive marker gene *spineless(s)*, and an experiment on a tomato line containing the *anthocyaninless* (*an*) genetic marker by Groenewegen *et al.* (1994), have indicated the following: (i) limited gene dispersal may occur following large-scale gene deployment; (ii) gene flow events are probably infrequent and appear to be strongly influenced by genotype of the immediately adjacent plants; and (iii) spread is localized for genes having probable selective neutral value. McPartlan and Dale (1994) provided evidence that the extent of gene dispersal from transgenic to non transgenic potatoes fall markedly with increasing distance, and was negligible at 10 m. There was also no evidence of transgene movement from potato to wild species, *Solanum dulcamara* and *S. nigrum,* under field conditions.

Thus, a safe policy would be not to grow transgenic crops where wild relatives occur in proximity of cultivated fields.

### Strategies to Prevent Gene Flow of Transgene Escape

Some strategies suggested by researchers to avoid the possible risk of gene dispersal from transgenic plants:

*Isolation zone*: An isolation zone devoid of vegetation discourages gene flow from transgenic plant to wild or weedy species or to other local cultivars. The isolation distance depends on the mode of reproduction (self or cross-pollination), and also the natural agents promoting cross-pollination (insects/wind, etc.). There are established isolation distances for different crops, mainly used for seed production under controlled conditions which are being followed for experimental studies on gene flow, but their utility in farmers' fields is doubtful.

*Trap crop*: Use of trap crops, which are non-transgenic varieties of the same crop, planted adjacent to the transgenic plot, can cleanse emigrating pollinators of transgenic pollen, thus preventing gene flow. An experiment conducted by M/s Calgene Inc. in California and Georgia with transgenic canola/rapeseed *(B. napus),* using kanamycin resistance marker, showed that effectiveness of trap crop depends on the width of the isolation zone. Gene escape was reduced when the two varieties were separated by 8 m, but escape increased across a 4 m isolation zone. In India, the recommended isolation distance is the minimum isolation required for "foundation seed" production of non-transgenic varieties of the concerned crop. At the end of isolation distance, few rows of non-transgenic variety should be planted as pollen trap.

*Male sterility*: Male sterility may be engineered to crops that will not produce pollen or pollen that is inactive, is another suggested mechanism to prevent gene flow from transgenic to sexually compatible species through pollen. Linking of the engineered gene to gene that is lethal in pollen is a mechanism that can provide effective male sterility.

*Chloroplast transformation*: A highly potential strategy that may be widely applied in future to effectively circumvent the problem of gene flow is 'plastid transformation' or 'chloroplast transformation'. In this strategy the gene construct is introduced into chloroplasts and a selection strategy is adopted that allows cells to retain only transformed chloroplasts. This is in contrast to 'nuclear transformation', which has so far led to the development of various transgenic varieties that have been commercialized. Since chloroplasts are inherited through cytoplasm, and pollen from transgenic plants usually does not carry any significant cytoplasm, thus there are negligible chances for 'transgene escape' or gene flow through this strategy (See Chapter 24 on Chloroplast and Mitochondrion transformation).

*Other strategies*: The other proposed strategies are: (i) Removal of flowers from transgenic plants; (ii) Removal of sexually compatible species. But these manual strategies may not find application at a commercial level; (iii) Mechanisms like terminator technology and RBF (recoverable block of function) cause embryo lethality and prevent pollen mediated transgene dispersal. Both these techniques are not desirable. As per Indian Protection of Plant Varieties and Farmers Rights Act, 2001 no variety can be registered for commercial exploitation if that contain terminator gene technology or GURT (genetic use restriction technology). These technologies have already been explained (See Chapter 25). RBF system consists of a "blocking" sequence, e.g., *barnase* gene, linked to the transgene of interest and a recovering sequence, *e.g., barstar* gene in this case. All the three genes are present in one transformable construct. In tobacco, the *barnase* gene was driven by a germination-specific promoter, while the *barstar* gene was under the control of a heat shock promoter. Under natural conditions, the *barstar* gene is not expressed because the heat shock promoter cannot be induced; as a result, expression of *barnase* in germinating seeds kills the young seedlings.

It is now believed that whenever the crop and weed are different forms of the same species as in sunflower, squash, radish, etc., crop to wild species/weed gene flow can occur when these species happen to grow nearby. Thus, the possibility of gene flow must be carefully examined on a case by case basis.

### Resistance/tolerance of target organisms

The potential benefits of planting insect-resistant transgenic crops include decreased insecticide use and reduced crop damage. However, the innate ability of insect populations to rapidly adapt to environmental pressures poses the development of strains which are resistant to these genes is a serious threat to the long-term efficacy of insect resistant transgenics. Adaptation by insects and other pests to pest protection mechanisms can have environmental and health impacts. For example, adaptation by insect populations to an environmentally benign pest control technique could result in the use of chemical pesticides with higher toxicity. Laboratory studies have indicated evolution of insect strains resistant to GM crops. However, no conclusive evidence has so far been reported.

### Generation of new live viruses/ super viruses

Concerns have been raised that due to GM virus resistant plants it could encourage viruses to grow stronger or give rise to new or stronger variants that can infect plants. There are two mechanisms by which this can happen: (i) recombination and (ii) transcapsidation.

*Recombination*: In transgenics, viruses recombination can occur either by copy choice or cleavage and ligation. It can occur between GMO produced viral genes and closely related gene of any incoming virus infecting that GMO. Such recombination may produce viruses that can infect a wider range of hosts or that may be more virulent than the parent viruses. This may cause increase in pathogenicity, although there are very few known cases. RNA recombination could also take place between the transcribed transgene and an infecting virus (Greene and Allison, 1994). Evidence about recombination between different viruses has not been collected, but phylogenetic studies of plant RNA virus evolution indicate recombination is common, and some of the new viruses could survive. However, Falk and Bruening (1994) argued that recombination could take place in the field between viruses with different host ranges during mixed infections, but new viruses did not appear to be the result. There is a need for selection pressure to promote recombination. The frequency of recombination between transgene RNA and viral genomic RNA is unlikely to be higher than that occurring naturally between viral genomic RNAs; and the viability of any new virus is unlikely to be higher than that of existing viruses throughout the infection cycle.

*Transcapsidation (heteroencapsidation)*: Transcapsidation is partial or full coating of the nucleic acid of one virus with a coat protein of a differing virus. Thus in a GMO it involves the encapsulation of the genetic material of infecting virus by the GMO produced viral proteins. Such hybrid virus could transfer viral genetic material to a new host plant that it could not otherwise infect. Except in the rare circumstances this would be a one time effect only, because the viral genetic material carries no genes for the foreign proteins within which it was encapsulated and would not be able to produce a second generation of hybrid viruses. Interactions that might result in transcapsidation rarely occur in natural joint infections, and thus are unlikely in transgenic situations. As the coat protein region, which determines vector specificity, is not important for protection against viruses in transgenic plants, vectors which do not pose a risk of such interactions can be designed (Hull *et al.*, 1994). Transcapsidation with transgenic plants would occur only slightly above the existing background of natural events, and its significance in viral spread has not been fully understood.

So far in the field trials with potato carrying the gene for potato leaf roll virus (PLRV) coat protein and potato viruses S, X and Y failed to acquire PLRV coat protein. Thus it points to the fact that the fears of development of new viruses may not be realistic.

# ECOLOGICAL CONCERNS

## Increased weediness

The perceived agricultural concerns of GM crops are that these may become weeds or they may invade natural habitats. Although there is no consensus on how to define a weed or the attributes that indicate a plant is likely to become a weed. Weeds do, however, tend to have a preference for disturbed habitats and high physical variability, allowing continuous adaptation to changing environments. Alternatively, a plant without these attributes may become a serious weed if it finds itself in an environment for which it is suited that lacks enemies such as herbivores or diseases. Thus weediness or invasiveness means the tendency of the plant to spread beyond the field where it was first planted, causing undesirable ecological changes. For example, "Kudzu" was introduced into the Southern United State to control soil erosion, but now it has became a major invasive weed in the region. There are apprehensions about GM crops becoming weeds. For example, a salt tolerant GM crop if escapes into marine areas could become a potent weed there.

Many weedy attributes such as dormancy, physical variability, continuous flowering and seeding have been bred out of crop plants over thousands of generations. These traits are now introduced into crop plants using genetic engineering. Some forage grasses, legumes and oilseed rape have a shorter history of domestication and are more likely to become weeds.

One way to assess potential weediness of GM crop is to calculate its finite rate of increase. This calculation is based on the processes affecting population growth, including fecundity, seed survival, seed germination, and seedling survival to maturity. The calculation can take into account any effects of the newly introduced DNA on any step in the process. This analysis has been applied to GM oilseed rape with resistance to the herbicide glufosinate. It has been found that oilseed rape was no more invasive in disturbed or undisturbed habitats than non GM-rape. Calculations of the potential weediness of GM crops, based on population growth studies for oilseed rape and other GM crops showed that these were no more invasive or persistent than their conventional counterparts in a range of environments.

## Creation of super-weeds

GM crops may hybridize with related weedy relatives and the transgenes (for example herbicide resistance) may be transferred resulting in the creation of superweed. The potential for a crop to hybridize with a weed is highly dependent on sexual compatibility and relatedness. Even if a crop plant and a weed were sufficiently compatible, the survival of any resulting plants would depend on overcoming a number of barriers. Traditional plant breeders use many techniques to encourage plants to hybridize but the process cannot be guaranteed to occur naturally.

One situation where hybridization is more likely to occur is where a crop species is growing alongside its wild relative. In these situations there has always been the potential for crop genes to transfer to the relative. It is important to consider, therefore, whether GM crops are more prone to transferring genes than their non-GM counterparts. For most transgenic traits, GM crops are not likely to transfer either their transgenes, or any other gene, to other species than crop cultivars have done in the past.

Transgenes can spread to wild relatives by cross-pollination, thus creating superweeds. This has occurred in oilseed rape and sugarbeet, creating potential superweeds. Spread of genes by cross-pollination is to be expected, whether the plants are transgenic not. However, a recent report suggests that transgenes may be up to 30 times more likely to escape than the plant's own genes. A large proportion of transgenic varieties under commercial cultivation have herbicide resistance. There is a fear about the development of super-weeds i.e. a weed that acquires the herbicide tolerant gene due to genetic contamination or through horizontal gene transfer.

Environmental organizations believe that weeds will become more resistant through the use of new herbicide resistant transgenic cultivars and their commercialization. Farmers will apply high doses of herbicides over several seasons, rather than limiting their use or using them in rotation with other herbicides, thus increasing chances of development of herbicide resistant weeds. The transgenic plants themselves are already turning up as volunteer plants after the harvest, and have to be controlled by additional sprays of other herbicides. The use of glyphosate with genetically engineered resistant plants will encourage the evolution of glyphosate resistance in weeds and other species, even without cross-pollination. A ryegrass highly resistant to glyphosate has already been found in Australia. Resistance evolves extremely rapidly because all cells have the capability of mutating their genes at high rates to resistance if they are exposed continuously to sub-lethal levels of toxic substances including herbicides, pesticides and antibiotics. This is inherent to the "fluidity" of genes and genomes that has been documented within the past 20 years (Ho, 1998). It will render resistant plants useless after several generations, as the herbicide is widely applied. At the same time, resistant weeds and pathogens may become increasing abundant. Additional herbicides will then have to be used to control the resistant weeds.

Thus there is a need for case by case approach as herbicide resistance may be safe in one location but riskier in another.

### Loss of biodiversity/reduction of cultivars

There have been concerns about reduction in the genetic diversity in cropping systems (i.e. *in situ*) by the development and global spread of improved crop varieties to the green revolution. This genetic erosion has occurred as the farmers have replaced the use of traditional varieties with monocultures. This is expected to further intensify as more and more transgenic crops are introduced which bring in considerable economic benefits to the farmers. The relative rate of susceptibility to any unforeseen infections or destructive situation increases when single varieties are used in cropping system in place of multiple varieties. However, it is argued that there is always a continuous and localized experimentation going on for the development of more effective crops which helps in maintaining genetic diversity. Thus conservation of land races should be done both *in situ* and *ex situ*. Already germplasm banks have been established in various countries.

Herbicide resistant transgenic crops are incompatible with sustainable agriculture. Many studies within the past 10 to 15 years have shown that sustainable organic agriculture can improve yields and regenerate agricultural land degraded by the intensive agriculture of the green revolution. Sustainable organic agriculture depends on maintaining natural soil fertility as well as on mixed cropping and crop rotation. This has been reversing due to the destructive effects of intensive agriculture that have led to falling productivity since 1980s. Glyphosate resistant plants requires application of glyphosate. It is highly toxic to fish and earthworm. It also harms mycorrhizal fungi symbiotically associated with the roots of plants, which are now found to be crucial for maintaining both species diversity and productivity of ecosystems. The depletion of mycorrhizal fungi in intensive agriculture could therefore decrease both plant biodiversity and ecosystem productivity, while increasing ecosystem instability. "The present reduction in biodiversity on Earth and its potential threat to ecosystem stability and sustainability.can only be reversed or stopped if whole ecosystems, including ecosystem components other than plants are protected and conserved."

Concerns have been raised that GM plants expressing antimicrobial proteins or Bt toxins could affect soil microbial communities. Bt toxins in soil are estimated to have a half life of 10 to 30 days. The rate of degradation is highly dependent on soil types. Clay particles can bind and inactivate Bt irreversibly. Bt is not taken up and accumulated by other parts and research has shown that microbes near the roots of GM plants are unchanged. The only exceptions reported so far are GM peroxidase producing alfalfa and GM tobacco modified for decreased lignin. It is too early to conclude whether GM crops can have a negative impact on agricultural and natural ecosystems by means of secondary ecological effects. Few examples of secondary effects have been found to date are negative enough to result in problems at an ecosystem level.

## Non-target effects

Non-target or unintended effects is another perceived risk as transgenics growing in a particular ecosystem can cause direct or non-target effects on certain microbes or insects growing in a particular ecosystem. GM crops containing insect-resistant Bt (*Bacillus thuringiensis*) toxin have been comprehensively studied. Possible environmental effects include direct effect on non-target insects due to exposure to GM plant material and also any indirect effect on non-target insects via so-called multi-trophic food chains.

For example when Bt cotton was released for commercial cultivation, extensive analysis was carried out on the effect of Bt crystal proteins on various non-target insect populations (honey bees, green lacewing, ladybird beetle, parasitic wasp, etc.), mammals (goat, sheep, buffalo, cow, rabbit, etc), birds and other organisms within the environment. Bt gene product (Cry proteins) is rapidly degraded by the stomach juices of vertebrates, but they could have harmful effects on non-target insect species. Any non-target insects that are vulnerable to Bt toxin will be affected if they eat any part of the GM crop. There was considerable media coverage when Monarch butterfly caterpillars that were fed only on Bt maize pollen in a laboratory experiment died (See Box 5). Another species that may be affected directly by Bt crops is the honey bee (*Apis mellifera*). At high doses Bt is toxic to bees but pollen from GM plants is unlikely to reach the doses required. Research with most of the widely grown commercial crops has found no effect on colony performance.

Studies in general did not show the effect of Bt-endotoxin on non-target organisms. Bt protein is relatively unstable so it will not remain or build-up in the food chain. Commonly, predators and parasites reared on insects feeding on GM plants do not grow to the same weight as those reared on insects feeding on non-GM plants. This is probably because there are lesser insects available on Bt-defended plants or because Bt-exposed insects are nutritionally poor for the predator/parasite.

One study, with lacewing, found the larvae died when they were fed prey raised on Bt maize. The researchers suggested Bt, which is normally not toxic to lacewings, had become toxic during processing by the lacewings' prey. It seems much more likely the prey used (*Spodoptera littoralis*) was not a good food for lacewing. Not all interactions will result in negative impacts. Bt maize, for example has reduced levels of insect damaged tissue, therefore it is less infestated by fungi that produce the mycotoxins that are toxic to humans and domestic animals.

Yet another transgenic plant has been shown to harm beneficial insects up the food-chain. Ladybirds fed on aphids that have eaten transgenic potato with snow-drop lectin lived half as long, laid 38% fewer eggs that were 4 times

## Box 5: Bt pollen effect on monarch butterfly

Losey *et al.* (1999) reported that monarch butterfly (Danaus plexippus) larvae reared on leaves of milkweed that were dusted with Bt maize pollen grew slowly and mortality rate was high as compared to larvae fed on normal maize pollen. This study raised concerns on conservation of monarch butterflies in USA and was widely and incorrectly interpreted to mean that GM crops were threatening non-pest insects. But this report was contradicted when follow up studies showed the effect of GM pollen on non-target insects, including the monarch butterfly to be negligible under real life conditions. In the first place, corn pollen does not fly long distances because it is pretty heavy. Also, most monarchs are moving at different times of the season when there's no corn pollen. It is argued that some of them might get killed by Bt corn pollen, but how many get killed when they are sprayed with insecticides. Further the scientific community discounted the original report because (1) it was conducted under artificial lab conditions, (2) the larvae were allowed to eat only corn pollen (which they don't often encounter in the open environment) and (3) there was no comparison group of larvae fed on ordinary corn pollen sprayed with regular Bt insecticide. The question is how much Bt corn pollen does it take to cause toxic effects in monarch caterpillars? It has been reported that monarch larvae eating leaves with pollen coating densities below 1000 grains/cm$^2$ had no effect on caterpillars weight or survival rate. Another question is what are the chances for caterpillars to encounter that dose under natural conditions? The answer is on an average less than 30% of the pollen that corn produces ends up on milkweed leaves. Data pooled from various locations showed that average density of Bt corn pollen density was about 170 grains/cm$^2$ and it rarely went above 600 grains/cm$^2$. Thus, given the low toxicity of Bt corn pollen and the low rates of exposure, the effect of Bt corn pollen from common commercial Bt hybrids on monarch butterflies is negligible.

more likely to be unfertilized and 3 times less likely to hatch. This transgenic potato has now been revealed to be highly toxic to rats, and is most probably harmful to small mammals in the wild (Birch *et al.*, 1997)

## Persistence of the transgene or transgene product

The gene transferred into an organism or the resultant product can actually remain in environment leading to environmental problems. For example, in case of Bt crops it was suspected that insecticidal proteins can persist in the environment but experiments have proved that these are degraded in the soil. GM herbicide resistant oil seed rape, maize, sugarbeet and GM potato expressing either Bt toxin or pea lectin as mentioned earlier also did not survive well outside the agricultural field and did not take a weedy character in UK where field trials were conducted over a 10 year period. There are also concerns in case of microorganisms about their capacity to adapt to new environment conditions and persist in the environment as spores. It has been suggested that transgenic volunteers may persist in the field and become weeds due to their increased fitness. However there are no reports from many field trials and large scale cultivation of transgenic crop varieties that a transgenic has become a weed.

## GM crops affect the purity of other crops

Conventional non-GM crops will inevitably receive transgenes from GM crops, resulting in situations that are either undesired or unlawful. This has already happened in the case of GM Starlink maize containing the *cry9C* gene which was found in non-GM maize grains in the US (See Box 2). The organic farming industry is also particularly concerned about genetic mixing through pollen dispersal and mixing of seed. Liability may become a major issue.

### Increased use of chemicals

Most of the transgenic varieties released for cultivation are against a particular herbicide and the farmers have to use a particular herbicide e.g. Roundup (Trade name of Monsanto herbicide) for a transgenic variety Roundup Ready Soybean. It is reported that farmers growing such varieties used a specific Roundup herbicide 2–5 times more in per unit area of land as compared to other weed management programmes. This increased use of herbicide is due to the fact that there is increased tolerance to Roundup of key weed species.

### Effect on Rhizosphere and microflora

Transgenic plants can influence the composition of microflora in their root zones or rhizospheres. Field studies with *B. napus* transgenic expressing barnase and barstar genes and non transgenic lines showed comparable rhizospheres qualitatively and quantitatively. Similarly, there was no difference in the bacterial cell consortia extracted from rhizosphere of non-transgenic and transgenic expressing *cry*1A(b) maize plants (Baumgrarte *et al.*, 2004). But Andreote *et al.* (2004) observed significant differences in the composition of bacterial communities associated with the rhizosphere of non-transgenic and transgenic (expressing genes *cab* and *npt*II) plants of tobacco and eucalyptus.

### Unpredictable gene expression or transgene instability

There is considerable evidence that transgenes show instability because they become deleted or modified following their integration in to the plant genome. The level of transgene expression may also decline with the passage of generations as gene silencing may occur (See Chapter 23). If and when these phenomena occur the crop becomes vulnerable to such stresses for which the transgenes provided protection. Unintended

genomic changes can also occur as a secondary consequence of genetic modification. Such changes can lead to production of new proteins that may be toxic or allergenic or may disrupt or alter metabolic pathway that play a role in making the GMO successful.

## SOCIO-ECONOMIC AND ETHICAL CONSIDERATIONS

One of the most important considerations in commercialization of transgenic crops is its socio-economic implications. Quite often, scientists might come up with techniques which *per se* may look novel and innovative, but from a global perspective may have wide-ranging socio-political and socio-economic ramifications. The development of Genetic Use Restriction Technologies is a prominent example. Though it looks appropriate for private sector but for the farmers of developing countries who rely of farm-saved seed is totally a bad thing and will be disastrous for such countries. India has totally banned the use of GURT in plant varieties for registration under Protection of Plant Varieties and Farmers' Rights Act, 2001. India is a land of small farm holdings. There are about 90 million operational farm holdings in the country, and about 60 percent of them own less than one hectare of land. Thirty five percent own land between one hectare to 4 hectare and only 5% own more than 4 hectares of land. Farming provides a livelihood to nearly 60% of the Indian population. Unless developing countries have policies in place to ensure that small farmers have access to better agricultural input delivery system, extension services, markets, and infra-structure. There will be increased inequality of income and wealth for small farmers if transgenics are introduced as big farmers are likely to capture most of the benefits through early adoption of the technology, expanded production, and reduced unit costs. However, in India farmers have adopted the biotech cotton technology and got benefits. Companies were selling the biotech seed at a much higher rate and the case was filed and hence Indian legal system intervened and the companies reduced the price. If benefits are to flow then regulatory system and legal system is also to be put in proper framework.

Another concern of the developing countries is the open liberalized and globalized markets where competition on plain-leveled field would often determine who is going to prevail or perish. Obviously, the one with efficient, effective and relevant technological intervention would be able to compete in terms of cost and quality of the product. Therefore, it would be the cutting edge of science and technology which decides as to who is going to attain and sustain advantages on short-term and long-term basis in the international markets. Some developing countries may be in a position to share significantly the benefits accruing from biotechnological developments on a long run. For instance, countries such as India, China, Mexico, Brazil, Thailand and Philippines share numerous important attributes including large domestic markets, a strong agricultural production and trading capacity, reasonably strong agricultural research network, a broad human capital base, and good technological and scientific capability in crop biotechnology. But, many other countries in the Asia and Africa are in a potential disadvantage position in effective exploitation of frontier technologies, and consequently, international trade, unless there are significant policy changes.

It is well known that MNCs have a dominant role in biotechnological research and the key technologies are in the 'hands of a few' peoples. Further, there is consolidation of seed industry where large MNCs have either purchased or are in the process of purchasing smaller seed and biotechnology companies. Consequently, the fear is that transgenic crops will prove to be expensive for resource-poor farmers. Considering the fact that crop biotechnology has a heavy involvement of private sector, particularly a few MNCs, these acquisitions and

mergers have raised apprehensions and valid concerns that market considerations and mergers will greatly influence the areas as well as commodities chosen for research in biotechnology, besides strengthening monopolistic tendencies.

Public acceptance is one of the major hurdles for the adoption of the first wave of products of agricultural biotechnology. In many countries, people are naturally cautious about the transgenic cultivars and their products. In general, people around the world appear to accept biotechnology in medical applications more easily than biotechnology in the field of agriculture or food processing. In some countries, the public and the scientific community hold different views as in European Union, commercialization of genetically modified crops have faced stiff resistance whereas public acceptance is much better in the USA. There are several factors that can play a key role in public acceptance of genetically modified crops. Scientific demonstrations of biosafety of transgenic crops and reviews by government agencies are extremely important in gaining public acceptance. What role credible experts will play in communicating the issues to the public in a realistic and effective manner can make a huge difference. Public acceptance is also greatly determined by the kind of information provided by the media to the general public and various organizations concerned about farmers. Misinformed public debates on key issues related to crop biotechnology can result in erosion of public confidence and can create mistrust in the technology and its developers, irrespective whether the developers are from the public or private sector (See Box 6). Clear and understandable consumer information is a very important part of the public acceptance process. Besides media, research organizations and scientific institution concerned with crop improvement must also take up the responsibility in bringing awareness in public about the applications of genetic engineering in agriculture, their potential benefits as well as constraints.

---

**Box 6: Multinational companies disregard the rights of farmers
(e.g., Percy Schmeiser *vs* Monsanto)**

Percy Schmeiser became the focus of international attention after he was taken to court by Monsanto, a biotech seed company, for using their patented Roundup Ready Canola seeds illegally. Schmeiser claimed pollen had drifted onto his property and he had merely replanted its seed next season. The Canadian court studied the report of samples from the 1,030 acres of canola planted and grown by Schmeiser Enterprises, Ltd. It consisted of 95% to 98% pure Roundup Ready plants as determined by independent testing, which could not have come from natural causes such as wind drift, spillage, etc. There were no close neighbors who grew Roundup Ready Canola when Schmeiser's lawyers claimed wind blew it on the enterprises, Ltd. Farms. The closest neighbour was five miles away. The judge explicitly rejected the pollen-drift theory because the facts did not support it. The court ruled that Schmeiser knew or should have known he was illegally using the protected plant variety and ordered him to pay damages of $ 40,000. Percy Schmeiser and his litigation with Monsanto had become familiar to the Commission long before his appearance as a witness for the Bio Dynamic Farming and Gardening Association. Anti GMO campaigners mentioned his case as an example of the perceived evils of genetic modification business.

## INTERNATIONAL PROTOCOLS AND CONVENTIONS ON BIOSAFETY

**Cartagena Protocol on Biosafety:** There have been various conventions, protocols and treaties for safe use of GMOs and their products. A Convention on Biological Diversity (CBD) was adopted in June 1992 which came into force in 1993. The objectives of CBD are "the conservation of biological diversity, the sustainable use of its components and access and benefit sharing arising out of utilization of biological resources. Under this convention contracting parties agreed to consider and develop appropriate procedures to address the safe transfer, handling and use of any living modified organisms (LMOs). CBD recognized that biotechnology inventions may have adverse effects on conservation and sustainable use of biological diversity (Article 19.3 of CBD) and a biosafety protocol is the result of that process. A biosafety protocol named as Cartagena protocol on biosafety to the Convention on Biological Diversity was finalized in February 1999 at Cartagena, Colombia but adopted a year later on 29 January, 2000 in Montreal, Canada.

Protocol entered into force in September 11, 2003. 147 countries have ratified the protocol as of May 2008. India signed the Cartagena Protocol on Biosafety on 23rd January 2003, which aims at ensuring an adequate level of protection in transfer, handling and use of genetically improved organisms, particularly during their trans-boundary movement. The protocol features two separate set of procedures one for LMOs that are to be intentionally introduced into the environment and one for those that are to be used directly as food or feed or for processing. It incorporates the use of precautionary principle, the application of the advance informed agreement (AIA) procedure for import of the organisms, risk assessment, risk management cooperation in preventive and emergency measures, capacity building and exchange of scientific, technical, environmental and legal information regarding the organisms through a biosafety clearing house mechanism. Various elements of the protocol which needed attention are as follows:

**i. Advance informed agreement procedure:** Under the biosafety protocol, the most rigorous procedures are reserved for

GMOs that are to be introduced intentionally into the environment. These include seeds, live fish and other organisms that are destined to grow and that have the potential to pass their modified genes on to succeeding generations. It includes four components: notification by the Party of export or the exporter, acknowledgment of receipt of notification by the Party of import, decision procedure and review of decisions. The purpose of this procedure is to ensure that importing countries have both the opportunity and the capacity to assess risks that may be associated with the LMO before agreeing to its import.

Specifically, the exporter starts by providing the government of importing country a detailed, written description of the LMO in advance of the first shipment. A Competent National Authority in the importing country is to acknowledge receipt of notification. The decision could be (i) approving the import, (ii) prohibiting the import, (iii) requesting additional relevant information, or (iv)extending the 270 days by a defined period of time.

Except in a case in which consent is unconditional, in other cases the Party of import must indicate the reasons on which its decision are based. The absence of response is not to be interpreted as implying consent.

A party of import may, at any time, in light of new scientific information, review its decisions. However, the Protocol's AIA procedure does not apply to certain categories of LMOs, i.e.
- LMOs in transit;
- LMOs destined for contained use;
- LMOs intended for direct use as food or feed or for processing

It should be noted that, while the Protocol's AIA procedure does not apply to certain categories of LMOs, Parties have the right to regulate the importation on the basis of domestic legislation. In this way the AIA Procedure ensures that recipient countries have the opportunity to assess any risks that may be associated with LMO before agreeing to its import.

**ii. Procedures for LMOs intended for direct use as food or feed for processing (LMOS-FFP):** LMOs intended for direct use as food or feed, or processing (LMOs-FFP) and not as seeds for growing new crops represent a large category of agricultural commodities.

Instead of requiring the use of the AIA procedure, the Protocol establishes a simpler system for the transboundary movement of LMOs-FFPs. Under this procedure, governments that approve these commodities for domestic use must communicate this decision to the world community via the Biosafety Clearing-House within 15 days of its decision. They must also provide detailed information about their decision.

Decisions by an importing country on whether or not to import these LMO-FFPs are taken under its domestic regulatory framework. In the absence of domestic regulatory framework a country may declare through the biosafety Clearing House that its decisions on the first import of LMOs-FFP will be taken in accordance with risk assessment as set out in the protocol and timeframe for decisions making. In case of insufficient relevant scientific information and knowledge, the importing country may use precaution in making their decisions on the import of LMOs-FFP.

**iii. Risk assessment:** The protocol empowers governments to make decisions in accordance with scientifically sound risk assessments. These assessments aim to identify and evaluate the potential adverse effects that a LMO may have on the conservation and sustainable use of biodiversity in the receiving environments. They are to be undertaken in a scientific manner using recognized risk assessment techniques. A country considering permitting the import of a GMO is responsible for ensuring that a risk assessment is carried out, but it also has the right to require the exporter to do the work or bear the cost.

**iv. Risk management and emergency procedures:** The Protocol requires each country to manage and control any risks that may be identified by a risk assessment. Key elements of risk management include monitoring systems, research programs, technical training and improved domestic coordination amongst government agencies and services. The protocol also requires each government to notify and consult other affected governments when it becomes aware that LMOs under its jurisdiction may cross international borders due to illegal trade or release into the environment.

This will enable them to pursue emergency measures or other appropriate action. Governments must enable them to pursue emergency measures or other appropriate action. Governments must establish official contact points for emergencies as a way of improving international coordination.

**v. Export documentation:** For GMOs intended for direct introduction into the environment, the accompanying documentation must clearly state that the shipment contains GMOs. It must specify the identity and relevant traits and characteristics of the GMO; any requirements for its safe handling, storage, transport and use; a contact point for further information and the names and addresses of the importer and exporter. In cases where a government agrees to import a genetically modified commodity intended for direct use as food or feed or for processing, the shipment must clearly indicate that it "may contain" living modified organisms and that these organisms are not intended for introduction into the environment.

**vi. The Biosafety Clearing House (BCH):** The Biosafety Clearing House contains information on national laws, regulation, and guidelines for implementing the Protocol. The Biosafety Clearing-House also includes information required under the AIA procedure, summaries of risk assessments and environmental reviews, bilateral and multilateral agreements, reports on efforts to implement the Protocol, plus other scientific, legal, environmental and technical information. The Biosafety Clearing-House and has been developed largely as an Internet-based system and can be found at http://bch.biodiv.org.

**vii. Unintentional transboundary movement of LMOs:** When a country knows of an unintentional transboundary movement of LMOs that is likely to have significant adverse effects on biodiversity and human health, it must notify affected or potentially affected States, the Biosafety Clearly House and relevant international organizations regarding information on the unintentional release. Countries must initiate immediate consultation with the affected or potentially affected States to enable them to determine response an emergency measures.

**viii. Capacity-building and finance:** Parties are encouraged to assist with scientific and technical training and to promote the transfer of technology, know-how, and financial resources.

**ix. Public awareness and participation:** It is clearly important that individual citizens understand and are involved in national decisions on GMOs. The Protocol therefore calls for cooperation on promoting public awareness of the safe transfer, handling and use of GMOs. It specifically highlights the need for education, which will increasingly have to address GMOs as biotechnology becomes more and more a part of our lives. The Protocol also calls for the public to be actively consulted on GMOs and biosafety. Individuals, communities and nongovernmental organizations should remain fully engaged in this complex issue. This will enable people to contribute to the final decisions taken by governments, thus promoting transparency and informed decision-making.

**x. Issue of non-Parties:** The Protocol addresses the obligations of Parties in relation to the trans-boundary movements of LMOs to and from non-Parties to the Protocol. The trans-boundary movements between Parties and non-Parties must be carried out in a manner that is consistent with the objective of the Protocol. Parties are required to encourage non-Parties to adhere to the Protocol and to contribute information to the Biosafety Clearing-House. (Article 24).

**xi. Institutional arrangement at the national level:** Parties are required to designate national institutions to perform functions relating to the Protocol. Each party needs to designate one national focal point to be responsible on its behalf for liaison with the Secretariat. Each Party also needs to designate one or more competent national authorities, which are responsible for performing the administrative functions required by the Protocol and which shall be authorized to act on its behalf with respect to those functions. A party may designate a single entity to fulfill the functions of both focal point and competent national authority.

## WTO and Other International Agreements

Cartagena protocol is the only international agreement which deals exclusively with GMOs, but there are number of other agreements, which address various aspects of biosafety.

i. International Plant Protection Convention (IPPC) protects plant health by assessing and managing the risks of plant pests. The IPPC is in the process of setting standards to address the plant pest risks associated with GMOs and invade species. Any GMO that could be considered as a plant pest falls within the scope of this Treaty. The IPPC allows governments to take action to prevent the introduction and spread of such pests. It established procedures for analyzing pest risks, including impacts on natural vegetation.

ii. Codex Alimentarius Commission addresses food safety and consumer health. On March 8, 2002 at Yokohama, Japan a task force of the Codex Alimentarius Commission has reached an agreement on a final draft of "Principles for the risk analysis of foods derived from Biotechnology". The Commission is also considering the issue of labeling biotech foods to allow the consumer to make an informed choice. The Codex Alimentarius, or the food code, has become the seminal global reference point for consumers, food producers and processors, national food control agencies and the international food trade.

iii. World Organization for Animal Health referred as Office of the International des Epizootics, (OIE), which develops standards and guidelines designed to prevent the introduction of infectious agents and diseases into the importing country during international trade in animals, animal genetic material and animal products.

iv. World Trade Organization (WTO) stipulates a number of WTO agreements, such as the Agreement on the Application of Sanitary and Phytosanitary Measures (SPS) and the Technical Barriers to Trade (TBT) Agreement contain provisions that are relevant to biosafety.

Whereas both the Cartagena Protocol on Biosafety and WTO advocate the use of science based risk assessments as a means to justify trade related measures, there are areas of potential conflicts mainly with respect to application of precautionary approach in the conditions that scientific evidence is insufficient and the socio-economic considerations in the decision of importing LMOs.

For example with regard to a certain biotechnology product, the details needed to be considered for risk assessment under the protocol and the SPS are very different. The SPS does not specify exactly what a risk assessment is, but the protocol elaborates this in detail. Further, the SPS does not mention risk management, but merely risk assessment. The scope of socio-economic considerations under the Protocol is wide, while SPS puts strict limits on the economic considerations. The mandatory labeling of LMOs-FFPs under the Protocol may also be in conflict with the WTO rules.

## Regulatory frameworks in different countries

**USA:** The USDA's Animal and Plant Health Inspection Services (APHIS) is the lead agency for the regulation of genetically engineered plants including the experimental evaluation of these products in confined field trials. The Environmental Protection Agency (EPA) is responsible for assuring the human and environmental safety of pesticidal substances engineered into plants, and the Food and Drug Administration (FDA) is responsible for assuring that foods and drugs derived from genetic engineered are as safe as their traditional counterparts. Products are generally regulated according to their intended use, with some products being regulated under more than one agency e.g. pesticidal plants. All the three regulatory agencies have the legal power to demand immediate removal from the marketplace of any product post commercialization if any new and valid data indicates safety concerns on consumers or the environment. Regarding research and development activities, compliance with National Institute of Health guidelines is mandatory for working with GMOs for all scientists receiving federal funding or working for federal agencies.

**European Union:** The deliberate release of GMOs into the environment is under the Directive 2001/18/EC. The Directive puts in place a step by step approval process on a case by case assessment of the risks to human health and the environment before any GMO or product consisting of, or containing GMOs can be released to the environment or placed on the market. The approval process requires submission of the notification to the competent authority in a member state of European Union (EU) where the GMO will be field tested/marketed. The member state gives a summary/assessment report for the EU commission and competent authorities of all member states. Besides, public is also provided an opportunity to provide comments, which are discussed in an attempt to reach agreement. At the end of review process, the competent authority provides written consent for marketing the GMO for a period of no more than 10 years. The period of validity,

the conditions for marketing the product, the labeling and monitoring requirements are all specified in the consent. The Directive requires that the labeling should include the words 'this product contains genetically modified organisms'.

**Canada:** Canadian Food Inspection Agency (CFIA), Health Canada and Environment Canada co-ordinates the regulations of the biotechnology products. The CFIA is responsible for regulating the import, environmental release, variety registration and use in livestock feeds of plants with novel traits. Health Canada is solely responsible for assessing the human health safety of foods. Environment Canada is responsible for administering the new substances notification regulations and for performing environmental risk assessment of toxic substances, including organisms and microorganisms that may have been derived from biotechnology.

Canada is the only country where regulatory system is operative on the novelty of traits expressed by plants or the novel attributes of foods or food ingredients, irrespective of the means by which the novel traits were introduced. The Canadian regulatory system refers to plants with novel traits (PNT) and novel foods in place of GM plants or GM foods. Under this regime, all agricultural commodities and food products, whether they are produced using conventional technologies or biotechnologies, are governed under the same Acts. Depending on the type of product, the relevant piece of legislation is applied i.e. Seeds Act, Feeds Act, Fertilizers Act, Food and Drugs Act, Health of Animals Act, or the Canadian Environmental Protection Act (CEPA). For example, a herbicide tolerant Canola produced by genetic engineering or mutagenesis (established plant breeding tool) are subject to same environmental or food safety risk assessment although the latter approach has been in use for more than 80 years.

**Australia:** In Australia, research, manufacture, production, commercial release and import of GMOs are regulated under the Gene Technology Act, 2000 by the Gene Technology Regulator (GTR).

**Argentina**: Regulations concerning the environmental release of GMOs were developed by Commission Nacional Asesora de Biotecnologia Agropecuaria (The National Advisory Committee on Agricultural Biosafety or CONABIA) and are enforced by Secretary of Agriculture, Livestock, Fisheries and Food (SAGPyA). The regulatory requirements for GMOs are based on guidelines in the form of non-legislative resolution that are integrated in the overall regulatory system and there is no specific law that makes the resolutions legally binding.

**Asian region**: China's biosafety guidelines were produced by the State Science and Technology Commission in December 1993, under which the administrative responsibility for biosafety of various products has been assigned to the relevant administrative departments. In 2002 China has established rules on GMOs to strengthen the safety and management of GMO products. Besides other detailed procedures, these rules require all GM products to be labeled. Japan uses voluntary guidelines administered through four governmental agencies viz. Ministry of Science & Technology for lab work, Ministry of International Trade and Industry for industrial applications, Ministry of Agriculture, Forestry and Fisheries for safety of animal feeds, feed additives and environmental release of GMOs and Department of Health and Welfare for food and food additives produced by rDNA technology. Philippines and Malaysia have completed their biosafety guidelines. Thailand has already approved field testing of GMOs after finalization of regulations in 1993.

## REFERENCES

Andreote, F.D. *et al.* 2004. Bacterial community in rhizosphere and rhizoplane of transgenic tobacco and eucalyptus (Abstr). Int. Cong. Rhizosphere 2004—Perspectives and Challenges—A tribute to Lorenz Hiltner. 12–17. Sept., 2004, Munich, Germany.

Arias, D.M. and Rieseberg, L.H. 1994. Gene flow between cultivated and wild sunflowers. *Theor. Appl. Genet.* **89**: 655–660.

Baumgarte, S., Trescher, K. and Tebbe, C.C. 2004. Field studies on the environmental fate of the Bt-toxin of transgenic corn (Mon810) and its effect on bacterial communities in soil (Abstr). Int. Cong. Rhizosphere 2004—Perspectives and challenges—A tribute to Lorenz Hiltner. 12–17 Sept., 2004, Munich, Germany.

Birch, A.N.E., Geoghegan, I.I., Majerus, M.E.N., Hackett, C. and Allen, J. 1997. Interaction between plant resistance genes, pest aphid-population and beneficial aphid predators. *Soft Fruit and Pernial Crops*. October 68–79.

Ellstrand, N.C. 1992. Threat of gene escape via hybridization. *Outlook on agriculture* **21**: 228.

Falk, B.W. and Bruening, G. 1994. Will transgenic crops generate new viruses and new diseases? *Science* **263**: 1395–1396.

Franck-Oberaspach, S.L. and Keller, B. 1997. Consequence of classical and biotechnological resistance breeding for food toxicology and allergenicity. *Plant Breed.* **116**: 1–17.

Greene, A.E. and Allison, R.F. 1994. Recombination between viral RNA and transgenic plant transcripts. *Science* **263**: 1423–1425.

Greoenewegen, C., King, G. and George, B.F. 1994. Natural cross pollination in California commercial tomato fields. *HortScience* **29**: 1088.

Ho, M.W. 1998, 1999. *Genetic Engineering Dream or Nightmare? The Brave New World of Bad Science and Big Business,* Gateways Books and Third World Network, Bath and Penang.

Hull R., Gibbs, M., Bruening G., Falk, B.W., Green, A.E. and Allison, R.F. 1994. Risks in using transgenic plants. *Science* **264**: 1649–1651.

Losey, J.E., Rayor, L.S. and Carter, M.E. 1990. 1999. Transgenic pollen harms monarch larvae. *Nature* **399**: 214.

Luby, J.J. and McNicol, R.J. 1995.Gene flow from cultivated to wild raspberries in Scotland: developing a basis for risk assessment for resting and deployment of transgenic cultivars. *Theor. Appl. Genet.* **90**: 1133–1137.

Malik, V.S. and Saroha, M.K. 1999. Marker gene controversy in transgenic plants. *J. Plant Biochem. Biotechnol.* **8**: 1–13.

McAllan, A.B. and Smith, R.H. 1973. Degradation of nucleic acids in the rumen. *British J. Nutrition* **29**: 467–474.

McPartlan, H.C. and Dale, P.J. 1994. An assessment of gene transfer by pollen from field-grown transgenic potatoes to non-transgenic potatoes and related species. *Transgenic Research* **3**: 216–225.

Messeguer, J. 2003. Gene flow assessment in transgenic plants. *Plant Cell Tissue Organ Cult.* **73**: 201–212.

Nap, J.P., Bijvoet, J. and Strikema, W.J. 1992. Biosafety of kanamycin resistant transgenic plants: an overview. *Transgenic Crop* **1**: 239–249.

Paul, E.M., Thompson, C. and Dunwell, J.M. 1995. Gene dispersal from genetically modified oil seed rape in the field. *Euphytica* **81**: 283–289.

Prescott, V.E., Campbell, P.M., Moore, A., Mattes, J., Rothenberg, M.E., Foster, P.S., Higgins, T.J.V. and Hogan, S.P. 2005. Transgenic expression of bean alpha-amylase inhibitor in peas results in altered structure and immunogenicity. *J. Agric. Food Chem.* **53**: 9023–9030.

Prins, T.W. and Zadoks, J.C. 1994. Horizontal gene transfer in plants, a biohazard? Outcome of a literature review. *Euphytica* **76**: 133–138.

Redenbaugh, K., Hiatt, W., Martineau, B., Lindemann, J. and Emlay D. 1994. Aminoglycoside 3′-phosphotransferase II: review of its safety and use in the production of genetically engineered plants. *Food Biotechnology* **8**: 137–165.

Schloeter, K., Futterer, J. and Potrykus, 1. 1995. "Horizontal" gene transfer from a transgenic potato line to a bacterial pathogen *(Erwinia chryanthem)* occurs-if at all- at an extremely low frequency. *Bio/technology* **13**: 1094–1098.

Skogsmyr, I. 1994. Gene dispersal from transgenic potatoes to cospecies: a field trial. *Theoretical and Applied Genetics* **88**: 770–774.

## FURTHER READING

Anonymous 2004. National Consultations on Biosafety aspects related to Genetically Modified Organisms, BCIL, Delhi and DBT, India, pp. 338.

Singh, B.D. 2006. Plant Biotechnology, Kalyani Publishers, India, pp. 755.

Rai, M. and Prasanna, B.M. 2000. Transgenics in Agriculture. Indian Council of Agricultural Research, Delhi, pp. 142.

# 28

# Genomics

Genomics is a rapidly emerging area of research and in the coming years it will revolutionize our understanding of biology. Genomics is the study of how genes and genetic information are organized within the genome, and how this organization determines their function. Genomics has been described either as functional or as structural genomics. *Functional genomics* involves a study of the functions of all specific gene sequences and their expression in time and space in an organism while *structural genomics* involves a study of the structure of all proteins encoded in a fully sequenced genome. Plant genomics is a part of this larger field and embraces the study of whole genome of plants, their physical and molecular organization, evolution and functions of the myriad of constituent genes. What the genome contains and how the genome functions is addressed in this chapter. Genomes of eukaryotes comprise thousands of genes and millions of base pairs, which store the massive data and information not only on genes themselves, but relating also to their expression. The advent of genomics is a consequence of the discovery of elegant and simple procedures of sequencing of nucleic acids. This ability to sequence the genome has aroused tremendous importance because deciphering the sequence of bases would unlock the whole blueprint of the development of an organism. Thus if we

understand the genomes of crops like rice, wheat, maize, *Arabidopsis*, potato, etc. we can ensure a better future with the capability of more perfect and précised genetic manipulation for enhanced yield and survival under adverse conditions. This chapter is devoted to understand the basic steps involved in sequencing of genome, location of genes in DNA sequences and understanding how the genome functions. In the past attention was directed at the expression pathways for individual genes, with groups of genes being considered only when the expression of one gene is linked to that of another. Now the issues have become more general and relate to the expression of the genome as a whole rather than an individual gene function.

## MAPPING OF PROKARYOTIC GENOMES

A number of microbial genomes have already been sequenced, followed closely by simple eukaryotic genomes like yeast and the nematode *Caenorhabditis elegans,* Arabidopsis and rice (Table 28.1). The first organism to have its genome sequenced was *Haemophilus influenzae* (Fleischmann *et al.*, 1995). It has a genome size of 1.8 Mbp encoding 1743 recognized genes. *Bacillus subtilis* with a genome of 4.2 Mbp has 4100 genes (Saier,

**Table 28.1:** A list of organisms whose genome has been sequenced

| Organism | Genome size (Mb) | No. of genes |
|---|---|---|
| *Haemophilus influenzae* | 1.83 | 1743 |
| *Helicobacter pylori* | 1.66 | 1,590 |
| *Methanococcus janaschii* | 1.66 | 1,680 |
| *E. coli* | 4.6 | 4,288 |
| *Bacillus subtilis* | 4.2 | 4100 |
| Yeast (*Saccharomyces cerevisae*) | 13 | 6000 |
| *Drosophila* | 139 | 13,601 |
| *Caenorhabditis elegans* | 97 | 19,099 |
| *Arabidopsis thaliana* | 140 | 27,000 |
| Human | 3200 | 33,000–150,000 (40,000) |

1998). The prototypical bacterium *E. coli* has a genome size of 4.6 Mbp with 4288 recognized genes (Blattner *et al.*, 1997). The genome of one eukaryote (the brewers yeast *Saccharomyces cerevisiae*) has been sequenced.

***E. coli* genome:** In 1997, Fred Blattner and his group at the University of Wisconsin reported the complete nucleotide sequence of nonpathogenic laboratory strain of *E. coli* K-12 called MG1655. Their 15 year effort led to the assembly of one continuous DNA strand containing all the 4,639,321 bp of the circular *E. coli* genome. A similar feat was accomplished shortly thereafter by a consortium of Japanese research laboratories using the W3110 strain of *E. coli* K-12. *E. coli* genome is very compact with a homogeneous distribution of genes throughout the chromosome. Numerous clusters of genes encoding functionally related proteins were identified as physically linked regulatory operons. This compact arrangement of genes is also observed in the genomes of three other bacteria which have been recently sequenced: (1) the *Bacillus subtilis* genome was found to contain 4100 protein coding sequences distributed across 4,214,810 base pairs, (2) *Mycobacterium tuberculosis*, a pathogenic microorganism with 4000 genes distributed across a 4,411,529 base pair genome, and (3) the gastric pathogen, *Helicobacter pylori*, which is responsible for peptic ulcer disease, has a minimum of 1590 proteins in a genome of 1,667,867 base pairs.

*E. coli* DNA sequence identified 4288 actual and proposed gene coding sequences, of which 38% have no attributed function based on homology searches with all available databases. It was found that approximately 88% of the genome encodes proteins or RNAs, ~11% appears to be utilized for gene regulatory functions, and ~1% consists of repetitive DNA sequences. The average distance between *E. coli* genes is only 120 bp. The coding sequences are continuous and lack the noncoding introns that are commonly found in most eukaryotic genes. *E. coli* K-12 is the best characterized free-living organism and research over the past 50 years has provided an abundance of genetic, biochemical and physiological data. Using this information Blattner's group was able to assign function to ~3200 genes as summarized in Table 28.2.

## MAPPING OF EUKARYOTIC GENOMES

Sequencing of entire genome in humans, *Arabidopsis thaliana*, fruit fly, rice and so many other organisms have been completed. The major revolution in the study of genomes of eukaryotes was brought about due to the availability of recombinant DNA and PCR technologies that became available during 1970s and 1980s respectively. These technologies helped in the preparation of molecular maps for a variety of animal and plant genomes. Ideally the objective of genomics research in any species should be to sequence its entire genome and to decipher functions of all the different

**Table 28.2:** Gene functional groups identified in the DNA sequence of *E. coli* K-12 (Source: Miesfeld, R.L., 1999)

| Functional class | Number of genes | Percent of total |
|---|---|---|
| Transport proteins | 427 | 9.9 |
| Energy metabolism | 373 | 8.7 |
| Replication, transcription, translation | 352 | 8.2 |
| Putative enzymes | 251 | 5.9 |
| Cell structure and membrane proteins | 237 | 5.5 |
| Nucleotide, amino acid, fatty acid metabolism | 237 | 5.5 |
| Regulatory functions | 178 | 4.2 |
| Miscellaneous gene functions | 601 | 14.0 |
| Hypothetical, unclassified, unknown | 1632 | 38.1 |
| Total | 4288 | 100 |

coding and non-coding sequences. In crop sciences, genomics has been promoted using *Arabidopsis* and rice as model plants. *Arabidopsis* is a weed plant, found all over the world, easy to grow, and to use for genetic experiments as it has one of the smallest plant genome with 145 million base pairs. *Arabidopsis* is the model plant for dicot plant families while rice has been selected to be the model plant for monocot families. Rice genome is also very simple, only three times larger than that of *Arabidopsis*. Also the genomic organization in cereals is highly conserved irrespective of the plant species, genes are lined up in the same order on the chromosomes of different cereal species. This phenomenon is called *synteny*.

Base sequencing is fundamental to any genome-sequencing project. Sanger's dideoxy chain termination method is now preferred over that of Maxam and Gilbert technique. High throughput techniques using fluorescent dideoxynucleotide terminators, laser excitation of bands of newly synthesized DNA and sensing of their colors by photomultipliers attached to computers are employed. As far as prokaryotes are concerned it is rather easy and simple to sequence the genomes (see chapter 17 for DNA sequencing techniques). But the actual execution of any eukaryote multicellular organism is rather complex because of the great mass of DNA (see Table 13.2). In humans 3.2 billion base pairs of DNA is present. Many plants have comparable or larger genomes. *Arabidopsis*, known to possess the smallest

genome among flowering plants, has 145 Mbp of DNA and around 20,000 to 50,000 genes. The rice has about 420 Mbp while wheat has ~16 billion base pairs of DNA. If sequencing at a time, in a single lane and in a single gel in the electrophoresis apparatus is carried out, one can handle DNA fragments that are only 1–2 kb pairs long while actual sequences may be read for only 200–500 bp. In eukaryotic genomes the problem becomes acute by the fact that if DNA of such eukaryotic organism is digested to such a small size, millions of fragments would be formed, which would make the task not only unmanageable but would leave no exact location in the genome. Thus for any genome sequencing program the following steps are undertaken. In this chapter the relevant points have been mentioned because the techniques and procedures have already been described in the earlier chapters in details.

1. *Construction of linkage maps with molecular markers*: Generally classical genetic linkage maps is a prerequisite for such program. These maps are based on recombination frequencies, the distance between the genes is measured in centimorgans (cM). From a genetic linkage map, one has to proceed for the construction of 'physical map' with landmarks along the whole length of the DNA molecules, and with distance measured in base pairs. In this regard, concept of molecular marker maps was developed by Botstein and coworkers which relied on restriction fragment length polymorphism

(RFLP). The other molecular marker maps with RAPD, microsatellites, AFLP, etc. have also been prepared (see chapter 22). RFLP markers are also positioned initially by linkage analysis and thus only approximately, they represent real molecular markers that can be actually hybridized to cloned DNA fragments and thus allow the work on construction of an actual physical map to begin. The RFLP markers are then hybridized to the vectors containing the DNA fragments that are to be sequenced, enabling the positioning of various cloned DNA fragments along the chromosomes in the cytogenetic map.

2. *Construction of Gene libraries*: An essential step in genome analysis after digestion of DNA in to smaller fragments, whether by restriction endonucleases or by shearing is cloning in to a vector. Cloning enables multiplication of the DNA fragments for further handling or distribution to different laboratories. Significant advances have taken place for the development of vectors which could accommodate a big fragment of DNA. Cosmids allowed 40–55 kb long fragment as compared to 10 kb long fragment to be inserted in a plasmid vector. However, even cosmid vectors were not quite appropriate for the task of sequencing the large genomes of eukaryotes. The development of yeast artificial chromosomes (YACs) allowed up to 1000 kb fragments to be cloned and gave a real impetus to genome studies. Cloning of such large fragments reduces the number of potential gaps in the assembly of contigs. It has been reported that 7500 YAC clones with an average insert size of ~ 400 kb DNA can contain the entire human genome but actually 60–70,000 clones have to be screened by a thousand or more probes. This is because of redundancy in preparing libraries and is essential to ensure that overlapping clones are available for the entire genome. To prepare libraries, after the colonies are individually picked from petri dishes they are first transferred to 96 well microtiter plates for replication and distribution. Sometimes colonies are gridded on special nylon filters by a robot with 96 prong spotting device which enables repetitive stamping on large nylon filters for preparation of replicates. The nylon membranes can then be processed for screening of libraries.

3. *Screening of libraries and constructing contigs*: The nylon membrane with colonies can be made to rest on an agar culture medium from which cells can absorb nutrients and further multiply. But it can also be dried, which greatly facilitates the screening process since the filters can be bathed by a DNA probe such as by a RFLP marker (the cells are lysed *in situ* and yeast DNA denatured before hand). Any spot that hybridizes can be detected by autoradiography and the corresponding yeast clone located in the microtitre plate. However, RFLP markers are few and far between and often there is only one marker per YAC clone. So when the PCR techniques became available, to order the clones precisely the concept of marking them by sequence tagged sites (STS) was developed. STS represents a landmark of a unique short ~300 bp DNA sequence that can be had from any region of the cloned DNA. The critical feature being that it should be possible for probes to be generated by PCR by defining a pair of sites for annealing primers. To generate STS, only single-pass partial sequencing of the clones is required and it is through such probes that an investigator can determine the overlap relationship between clones. It will indicate whether a clone under test, is identical, shorter or longer than a previously tested clone or represents a new clone altogether. STS made it possible to mark YAC clones. Analogous to STSs are the ESTs (expressed sequence tags) which greatly helped in the

sequencing of genomes. Like STSs, ESTs represent single-pass sequences, but unlike STSs, which are derived from genomic clones, ESTs represent short sequences of cDNAs from the 3′ or 5′ ends. ESTs are obtained far more easily and they are very useful because the cDNAs from which they are derived represent genes that are actually expressed and ESTs, therefore greatly facilitate the task of annotation of the final sequence.

When ordered overlapping clones are available, contigs are constructed which represent stretches of contiguous sequence ready clones. Generally some gaps remain and one has to resort to gap closure by chromosome walking for which special probes have to be generated from clones at the ends of contigs by either plasmid rescue or by inverse polymerase chain reaction. These probes are then employed to identify new YAC clones to bridge the gaps.

4. *Sequencing*: After a minimal set of overlapping YAC clones have been identified, the DNA in each clone is further fragmented either randomly or by restriction digestion and then subcloned in cosmids as per standard practice. Overlapping cosmid clones are identified by fingerprinting which is done by restriction digesting the clones, filtering out common bands of vectors with the aid of computers and then comparing the digestion patterns to find contiguous clones. The cosmid clones are then ready for further breakage, which is done randomly either by shearing or by sonication into 1–2 kb fragments and subcloned into pUC plasmids or M13 phages. Finally the DNA from each subclone is extracted, reaction set-up for *in vitro* DNA synthesis as necessary in the Sanger method employing the M13 phage and sequenced.

5. *New vectors BACs and PACs and the shot-gun approach*: Since the storage of YAC colonies require special procedures and YAC clones were often found to be unstable or chimeric (incorporating more than one DNA fragment), efforts were made to develop alternate vectors like BACs or PACs (Chapter 16). Satisfactory BAC vectors have been developed in which ~100 kb long DNA fragments are being directly cloned. Overlapping BACs can be found by fingerprinting—one can then proceed directly to subcloning and sequencing. Since YACs can be multiplied only in yeast cells, BACs are maintained in bacteria, simplifying all presequencing steps. Thus one can do away with the initial ordering of BACs by RFLP or other genetic markers. In fact with the aid of computers and special software, major sequencing effort can largely be restricted to BAC ends alone which provide adequate information for determining overlaps. In this strategy, while the ends of every BAC are sequenced, the number of BACs that need to be sequenced fully is much smaller and is determined later. To begin construction of a contig, one starts with a fully sequenced 'seed' BAC. The database search for BAC ends may, reveal overlaps with about 30 other BACs. When two most important extreme end BACs are located, then sequence them fully, continuously expanding search for new BACs for full sequencing at either end, until one tackles an entire chromosome. This random shot-gun method does require more extensive sequencing and powerful computers, but it has been claimed that it more than compensates for time and money that go in a conventional sequencing program, where considerable effort has to go for construction of linkage maps of molecular markers.

## Gene location in DNA sequences

In genomics study of what genome contains and how the genome functions is undertaken. A genome sequence is not an end to itself. Gene location in DNA sequence is the prime aspect followed by its function. Once a DNA sequence has been obtained then various methods can be employed to locate the genes, which can be simple sequence inspection by computer and the methods that involve experimental analysis.

*Sequence inspection*

Genes are not random series of nucleotides but a specific sequence with specific features, which determine whether the sequence is a gene or not. Though it is not a powerful tool yet but it gives an idea of gene location.

i. Genes are open reading frames: genes that code for proteins comprise open reading frames (ORFs) which consist of a series of codons that specify the amino acid sequence of protein. It begins with an initiation codon (usually ATG) and ends with a termination codon (TAA, TAG, TGA). Searching a DNA sequence with ORFs that begins with an initiation codon and a termination triplet is one way for location of genes. The success to ORF scanning is the frequency with which termination triplets appear in the DNA sequence. If the DNA has random sequence and a GC content of 50% then each of the termination triplets (TAA, TAG, TGA) will appear on an average once every $4^3 = 64$ bp. If the GC content is > 50% then the termination triplets being AT rich will occur less frequently, perhaps every 100–200 bp. On the basis of start of ORF with an initiation codon and number of termination triplets, the random DNA should not show ORF longer than 50 triplets in length. Most genes on the other hand are longer than 50 codons (the average lengths are 317 codons for *E. coli* and 483 codons for *S. cerevisiae*). ORF scanning in its simplest form therefore takes a figure of 100 codons as the shortest length of a putative gene and records positive hits for all ORFs longer than this. With simple bacterial genome ORF scanning is an effective procedure for location of genes, but this is less effective in higher eukaryotes. Because there is more space between real genes, and presence of introns, so do not appear as continuous ORFs in the DNA

sequence. In other words genes of eukaryotes do not appear in their DNA sequences as long ORFs and simple scanning cannot locate them. Three modifications to the simple procedure have been adopted. These are:

a. **Codon bias**: It means that all the codons are not used equally frequently in the genes of a particular organism. The codon bias of the organism being studied is therefore written into ORF scanning software.

b. **Exon-intron boundaries**: The exon-intron boundaries can be searched as these have distinctive sequence features. The consensus sequence of the upstream exon-intron boundary is usually described as:
5′—AG↓GTAAGT—3′
(arrow indicate the precise boundary point) The down stream intron-exon boundary is defined as:
5′—Py Py Py Py Py PyNCAG↓—3′
(Py means pyrimidine nucleotide T or C; N is any nucleotide)

c. **Upstream control sequences**: The upstream control sequences can be used, although these are variable, for location of genes.

Besides these three simple extensions of simple ORF scanning, additional features or strategies are also possible with individual organisms. For example vertebrate genomes contain CpG islands upstream of many genes, these being sequences of approximately 1 kb, in which the GC content is greater than the average for the genome as a whole.

In eukaryotes homology searches are conducted to test whether a series of triplets is a real exon or a chance sequence. In this analysis the DNA databases are searched to determine if the test sequence is identical or similar to any genes that have already been sequenced. The main use of similarity searching is to assign functions to new discovered genes. However, at this point it enables tentative exon sequences location by ORF scanning. If the tentative exon sequence gives one or more

**Table 28.3:** Commonly used software tools for sequence assembly, prediction of coding regions, annotation, detecting repeats and splice sites (Source: Singh , 2000)

| Software | Function |
|---|---|
| TRANSCAN | Detects for RNA coding regions |
| FGENEH | A dynamic programming algorithm that uses linear discriminant functions. Used to identify genes |
| GENSCAN | Prediction program for coding regions and exon/intron splice site |
| Genquest | Program for sequence assembly, analysis and comparison |
| MUMmer | A whole genome alignment tool |
| REPEATMASKER | A repetitive element filter that screen a sequence against a library of repetitive sequences and searches for low complexity regions |
| Annotator | An interactive genome annotation tool |
| PSI-BLAST and variants | The most commonly used program for homology searches |
| Genefinder | Popular tool to detect coding regions and splice sites |
| Grail XGRAIL/Grailant | Predicts exon, putative genes, detect promoter regions, poly-As, CpG islands, similarities in ESTs and repetitive segments |
| Net Plant Gene | Predicts exon/intron splice site |
| AAT | Analysis and annotation tool |

positive matches after a homology search then probably it is an exon. Some of the softwares used for prediction of coding regions are mentioned in Table 28.3.

*Experimental techniques*

These procedures locate genes by examining the RNA molecules that are transcribed from a DNA fragment. Techniques which map the positions of transcribed or expressed sequences in a DNA fragment can therefore be used to locate exons and entire genes. It should be kept in mind that the transcript is usually longer than the coding part of the gene because it begins several nucleotides upstream of the initiation codon and several nucleotides down stream of the termination codon. This analysis indicates that a gene is present in a particular region and it can locate the exon-intron boundaries.

**Hybridization tests**: This is a procedure for studying expressed sequences. RNA molecules can be separated by agarose gel electrophoresis and transferred to a nitrocellulose or nylon membrane by the process of northern blotting (see chapter 14). If a northern blot is probed with a labeled DNA fragment, then RNAs expressed from the fragment will be detected. Northern hybridization is a means of determining the number of genes present in a DNA fragment and

the size of each coding region. There are two problems. (i) Some individual genes give rise to two or more transcripts of different lengths, because some of their exons are optional and may or may not be retained in the mature RNA. (ii) With many species it is not possible to make a preparation from an entire organism so that RNA extract is obtained from a single organ or tissue. So if any gene not expressed in that tissue or organ will not be present in the RNA population and so will not be detected when the RNA is probed with DNA fragment being studied.

A second type of analysis uses DNAs rather than RNAs to overcome the problems with poorly expressed and tissue specific genes. If a DNA fragment from one species is used to probe a Southern blot of DNAs from related species, and one or more hybridization signals are obtained then it is likely that the probe contains one or more genes, is called **zoo-blotting**.

**cDNA sequencing**: Northern hybridization and zoo-blotting enables the presence or absence of genes in a DNA fragment to be determined, but gives no positional information relating to the location of those genes in the DNA sequence. The easiest way is to sequence the relevant cDNAs. The success of cDNA sequencing depends on the frequency of appropriate cDNAs in the cDNA clone library that has been prepared. For this purpose various methods of cDNA capture or cDNA selection have been devised, based around repeated hybridization between the DNA fragment being studied and the pool of cDNAs. Because the cDNA pool contains so many different sequences it is generally not possible to discard all relevant clones by these repeated hybridizations, but it is possible to enrich the pool for those clones that specifically hybridize to the DNA fragment. This reduces the size of the library and then it must be screened under stringent conditions to identify the desired clones.

Completeness of individual cDNA molecules also determines the success or failure of this approach. Traditionally cDNAs are made by copying RNA molecules into single stranded DNA with reverse transcriptase and then converting single stranded DNA into double stranded DNA with a DNA polymerase (see chapter 17 for cDNA cloning). There is always a chance that one or other of the strand synthesis reactions will not proceed to completion, resulting in a truncated cDNA. The presence of intramolecular base pairs in the RNA can also lead to incomplete copying. Truncated cDNAs may lack some of the information needed to locate the start and end points of a gene and its exon-intron boundaries. For location of the precise start and end points of gene transcripts RACE (rapid amplification of cDNA ends) is performed (see chapter 18). In this type of PCR, RNA rather than DNA is the starting material. The first step in this type of PCR is to convert the RNA into cDNA with reverse transcriptase, after which the cDNA is amplified with *Taq* polymerase then followed by RACE for 5' or 3' end depending upon the desired information.

Other method for transcript mapping is heteroduplex analysis. DNA region being studied is cloned as a restriction fragment in an M13 vector from which the DNA obtained is single stranded. When mixed with an appropriate RNA preparation the transcribed sequence in the cloned DNA hybridizes with the equivalent mRNA, forming a double stranded heteroduplex

(Fig. 28.1). The start of this mRNA lies within the cloned restriction fragment so some of the cloned fragment participates in the heteroduplex, but the rest does not. The single stranded regions can be digested by treatment with a single strand specific nuclease S1. The size of the heteroduplex is determined by degrading the RNA component with alkali and electrophoresing the single stranded DNA in an agarose gel. This size measurement is then used to position the start of the transcript relative to the restriction site at the end of cloned fragment.

## Yeast (*S. cerevisiae*) genome

The yeast genome consists of 16 linear chromosomes, each containing a centromeric region required for chromosome segregation and special repeat sequences at the two ends called telomeres, which provide important functions during DNA replication and in chromosome maintenance. Unlike the *E. coli* chromosome, which contains a single origin of replication, eukaryotic chromosomes contain multiple initiation sites for bidirectional DNA replication. The nucleotide sequence of the entire *S. cerevisiae* genome has been determined and found to contain 12,068 kb of DNA. Sequence analysis has identified 5885 potential protein coding genes and another 455 RNA coding genes (tRNA, snRNA, and rRNA genes). Almost 70% of the genome is devoted to protein coding sequences, with the average distance between genes being ~2 kb. Interestingly, unlike most other eukaryotic genes, only 4% of the ~6000 yeast genes have introns, and even then, most of these genes, contain only a single intron near the 5' end of the coding sequence.

## Human genome

On 26th June 2000 successful completion of the sequencing of Human Genome project was announced jointly by two groups, The Human Genome Consortium and Celera Genomics. This was the first draft of Human genome sequence, which is over 3 billion nucleotide long, consisting of 24 chapters, viz., 22 autosomes, X and Y sex

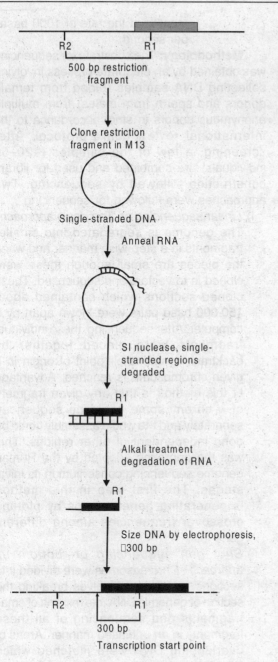

**Fig. 28.1:** Transcript mapping by S1 nuclease. Concept of the figure taken from *Genomes,* T.A. Brown, by kind permission of BIOS Scientific Publishers, Oxford, U.K.

chromosomes. Although the euchromatic regions have been sequenced, the Y chromosome and many gaps in other chromosomes are unlikely to be completed soon due to large regions of repeats.

## Historical facts

1984–86: The idea for a large-scale effort to learn the sequence of the entire human genome was put forward. The practicality of sequencing was discussed at two meetings, one at Santa Cruz, USA and another at Santa Fe, USA.

1988: Two government agencies of USA—Department of Energy (DOE) and the National Institute of Health (NIH) drafted the first proposal to coordinate research and technical activities related to human genome.

1990: The project began in earnest when the projected goals for the first five years of a fifteen year plan were published. Dr. James Watson pledges to decode the genome by 2005 at a cost of $ 3 billion. The project evolved over the 1990s decade and in its final stages had over 16 international centres: USA (8), Germany (3), Japan (2) and France, China, UK with one centre each.

Two major technical advances that made the project feasible were the (i) development of yeast artificial chromosomes (YAC) vectors and (ii) fluorescence based detection of dideoxy terminated fragments that enabled near total automation of sequencing.

1990: Craig Venter, an NIH scientist developed a short cut method to find fragments of human genes based on which whole genes can be identified.

1995: Sequencing of the genome of a bacterium (*Haemophilus influenzae*) by Venter's shot-gun method established.

1995–1996: Publication of maps based on sequence tagged sites (Hudson *et al.,* 1995), human genetic map (Dib *et al.,* 1996), transcript map (Schuler *et al.,* 1996).

1997: Publication of STS based radiation hybrid map.

1998: Physical map of the genome presented (Dekoulas *et al.,* 1998).

May 1998: Craig Venter (NIH scientist) joins a new company that plans to complete the human genome in three years well ahead of US government's target. The company was later named as Celera Genomics, Inc.

December: The first full genome of eukaryote
1998 roundworm animal *Caenorhabditis elegans* sequenced.

1999: Completion of the sequencing of chromosome 22 (Little, 1999; Dunham *et al.,* 1999).

March Two groups led by Craig Venter
2000: and Gerald M. Rubin sequenced the genome of *Drosophila* using Venter's shot gun strategy.

May 2000: Completion of the sequencing of chromosome 21 (Hattori *et al.,* 2000).

26th June Publicly funded Human Genome
2000: Project and Celera Genomics Inc., USA jointly announced the completion of 'rough draft' of human genome. 86% of the whole genome has been evaluated and classified to be highly accurate. The final draft was compiled from a total of 22 billion bases covering the total human DNA sequence almost seven fold. The genome was sequenced in the past 6

months at the rate of 1000 bases per second.

**Methodology:** The substrate for sequencing was obtained by an elaborate process involving collecting DNA samples (blood from female donors and sperm from males) from multiple anonymous donors in strict accordance to the international review board protocol. After screening, a few of the samples (~20–50 individuals) are combined and used for library construction followed by sequencing. Two approaches were followed for sequencing

i) *Linear sequencing or 'Top down approach'*: The genome is segregated into smaller fragments in a step-wise manner and when the pieces are small enough these were cloned in to vectors and sequenced. These cloned sections which contained about 150,000 base pairs were blown apart by a computer. After sequencing, these individual fragments were pieced together by backtracking until their point of origin in a given chromosome is reached. Advantage of this method is that any given fragment of a chromosome could be sequenced separately and the whole assembly could be done independent of other regions. This was the approach adopted by the Human genome sequencing consortium in its initial stages. The first step in this method is generating genetic maps by plotting crossover frequencies among different gene loci.

ii) *Shot gun sequencing or 'bottom up approach'*: Chromosomes were divided into sections and then it involves breaking the section of genomic DNA into millions of small fragments and sequencing of all these fragments in an unbiased manner. Areas of overlapping DNA were matched which formed larger and larger segments. Because of the degeneracy of this process the total fragments generated typically covers the genome 10 times its original size. The overlapping of fragments was exploited in the assembly phase by computational exercise. Although prone to artefacts, especially in repeat regions, this process has been successfully applied for the sequencing of all the microbial genome sequences published and by Celera Genomics Inc. for sequencing of human genome. This has been possible mainly due to the increase in accuracy of the algorithms performing the assembly.

In either approach the actual sequencing is performed in a manner of dideoxy chain termination method proposed by Sanger in 1977. Presently most of the sequencing is done by specially designed high-speed sequencers, which require little human intervention. Each of the four dideoxy terminators is tagged with different fluorescence markers which are read by automated instruments (explained in chapter 17). DNA sequencing machines developed by Perkin Elmer now Applied Biosystems can sequence up to 0.5 million bases per day.

As the sequencing of a given fragment which was inserted into a plasmid, cosmid, YAC, BAC or PAC is completed it is subjected to a set of computational procedures, which align the fragments, do error checking and detect potential protein coding regions. Once a region is predicted to be a coding segment, the features of the gene are analysed (e.g. presence of specific signatures, CpG islands, etc.) and similarity searches run to establish their identity. Some of the programs that are commonly used are listed in Table 28.3. As can be easily made out, it is an exercise that is highly computation intensive. Celera Genomics Inc. announced that during the assembly involved more than 500 million trillion) calculations, making it the biggest exercise in the history of computation biology (Singh, 2000).

Now it is known that 99.8% of the 3.2 billion base pairs between any two humans would be the same and 0.2% different. For every 500 nucleotides there is one nucleotide which varies between two individuals. This means any two of us differ only in 6 million locations out of 3.2 billion locations. Now, if these locations have very little effect then both of us will be identical. More strikingly some might not consider it so, humans are more than 98% similar to chimpanzees. So one has to understand the variation in even few locations in the DNA sequence can lead to a severe disease and variation in humans. The study of variation will also tell us what we are, but also where we came from. It will also tell us what types of diseases we are susceptible to, because it will be possible to find out those variations that are associated with disease and those that are not. This exercise will involve functional analysis of the genome. The aim will be to find out how various genes are expressed in various tissues. This will be done in various tissues using DNA chips, where thousands of genes will be put on a single glass slide or silicon chip against which the messenger RNA isolated from various tissues will be hybridized. From this study it will be possible to pinpoint the expression of which genes has gone up in a tumor cell and expression of which genes has gone down.

*Arabidopsis genome*

*Arabidopsis thaliana*, a small herbaceous species belongs to Crucifer (Brassicaceae) family. This plant species has been adopted as a model species for dicotyledonous crops. The species carries five pairs of chromosomes (2n=10) and the total physical size of the chromosomes together equals 125 Mbp, which in fact is the smallest known genome of a higher plant.

The organization of *A. thaliana* genome is also particularly simple and favorable to molecular genetic and genome experimentation (Fig. 28.2). The content of high repeat sequences (10%) and medium repeat sequences (10%) is singularly low, as these repeatable elements can represent up to 95% of the DNA in the genomes of other plant species. Furthermore, these repeated sequences are not dispersed in the genome, but grouped in specialized regions of the genome: the first group essentially

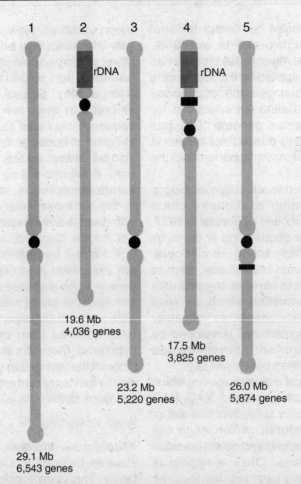

**Fig. 28.2:** The five chromosomes of *Arabidopsis thaliana*. The sequenced regions are represented in light grey. The size of these regions and their gene content is indicated beneath each chromosome. The regions of chromosome 2 and 4 coding for rRNA are represented in dark grey. The heterochromatic regions and centromeres are represented in black.

corresponds to repeated peri-centromeric sequences which surround the centromeres, the second to the two regions of the nucleolar organiser which codes for the ribosomal RNAs, and are localised at the extremity of chromosomes 2 and 4. Around 80% of the genome is therefore constituted of rarely repeated or unique sequences, essentially corresponding to the genic faction. Thus the density of information is exceptionally high, as the sequencing of the genome has confirmed.

The global size of the genetic map of *A. thaliana* is 500–600 cM, implying that on average

a unit of genetic mapping corresponds to a physical size in the range of 200 kb, which is a size compatible with current techniques of cloning and sequencing. Note that is average equivalence can find some local disparity (from 20 to 2,000 kb/ cM within certain regions).

Since 1990, efforts have been put into increasing our knowledge of the genome of this model species. Three American groups, namely Meyerowitz, Somerville and Goodman at three different universities took the first steps leading to the organization of a multinational project on sequencing of the *Arabidopsis* genome and which later has come to be known as AGI (*Arabidopsis* genome initiative). The first RFLP map was prepared by Meyerowitz group and then a second one by Goodman group, which were later integrated (Hauge *et al.*, 1993). Early RFLP maps were prepared from $F_3$ individuals while later maps were prepared by using recombinant inbreds. By 1996 the distance between RFLP markers had been reduced to >1.0 cM which represents a physical spacing of about ~200 kb suitable for chromosome walking. Subsequently other molecular markers like RAPDs, CAPs, AFLPs, SSPs and microsatellite markers have been added. For sequencing of all the chromosomes an international collaborative group comprising 25 laboratories spread over the world under AGI was formed with various chromosomes assigned to different groups. Initially the conventional 'clone-by-clone' was the dominant approach but later on Venter's shot-gun approach for sequencing was followed. Extensive use of cDNA libraries and ESTs was made. More than 700,000 EST tags are available in the dbEST database (www.ncbi.nlm.nih.gov) and this approach has now been used in numerous other plant species.

To overcome a certain number of inherent limitations with the EST approach, the complete sequence of the nuclear genome of *A. thaliana* was put together by an international consortium of laboratories. The final summary of this colossal work was published in December 2000 in *Nature* (AGI, 2000), and the results are available in their entirety in the diverse databases accessible online; a good starting place to consult the data is the international database *The Arabidopsis Information Reserve* (TAIR: www.arabidopsis.org). The complete chloroplast (154 kb, 79 gene) and mitochondrial (367 kb, 58 gene) genomes are also available.

The size of the genome sequence is close to 115 million nucleotides covering all of the five chromosomes with the exception of zones of ribosomal DNA and the centromeres, which represent around 10 Mb in total. This sequence, of high quality, has permitted the discovery of around 26,000 genes potentially coding for proteins. Close to 8% of these predicted proteins seem to be targeted to the chloroplasts compared to 1.5%, which are targeted to the mitochondria. The genome is particularly dense with a gene occurring every 4.5 kb on an average. The length of a gene is on average 1.9 kb and 79% of the genes are interrupted by one or more introns. Of the 26,000 genes identified in the original publication, less than 10% have a biological function defined experimentally. Close to 30% have no sequence similarity with a gene of known function and for the remaining 60%, a function can be predicted based on sequence comparison. A quiet high degree of redundancy has been identified. Thus, 26,000 predicted genes can be regrouped into 11,000 multigenic families and only 35% of these are unique. Repeated and tandem genes are frequent and concerns 17% of the genes. The global analysis has brought to light a great number of duplications, which are more or less recent, 24 duplicated segments of more than 100 kb, representing close to 60% of the genome.

Out of more than 25,000 genes identified in *A. thaliana,* reliable functional information is only available for few thousand amongst them. Functional genomics proposes high throughput approaches permitting fast acquisition of knowledge on gene function. It was an ambitious objective that the international community of *Arabidopsis* biologists fixed for themselves that in the 10 coming years, they would is to associate possible biological function to all the genes identified by sequencing. This requires the gathering of all possible information for each gene and protein on structure, regulation, biochemical activity and sub-cellular localisation, interaction with other molecules, natural allelic variability, etc.

### Rice genome

Rice is one of the most important plant species used by mankind, as it forms the staple diet of more than 3 billion people. Cultivated rice (*Oryza sativa*) behaves as a diploid species and its genome as analysed by cytogenetical methods belongs to the AA group. Rice has one of the smallest genome of cereals. The estimated size of 400–440 Mbp is much smaller than that of maize (2,500 Mbp) or of wheat (16,000 Mbp). The rice genome is distributed in to 12 chromosome pairs (2n = 24).

Rice genome sequencing began with the accumulation of the necessary resources. Several dense saturated genetic maps had been produced in Japan, US, France, Korea, accumulating thousands of markers. A major initial effort was made by the Japanese Rice Genome Program, who constructed numerous cDNA libraries, delivered more than 10,000 ESTs in the early 1990s, prepared YAC (Yeast Artificial Chromosome) and PAC (P1 phage derived Artificial Chromosome) large insert libraries. Another key effort was made in the USA, supported by the Rockefeller and Novartis Foundations, to construct ordered BAC libraries. These BAC libraries partially ordered by fingerprinting and contiging the clones and by BAC end sequencing, were available at the end of 1997 when the IRGSP (International Rice Genome Sequencing Project) was set up with the goal of obtaining the full sequencing of the rice genome by 2007. This consortium decided to use the clone by clone strategy following a minimum tiling path for each chromosome and distributed specific chromosomes to the participating countries. Japan received chromosomes 1, 2, 6, 7, 8, and 9, the US received chromosomes 3, 10 and 11 China received chromosome 4, Taiwan received chromosome 5 and France received chromosome 12. Several additional countries (Thailand, Korea, Brazil and India) sequenced limited portions of one chromosome (Delseny 2007; International Rice Genome Sequencing Project, 2005).

Sequencing indeed started efficiently by the end of 1999 and has been accelerated, due to three outsider competitors, who independently from the consortium, decided to sequence the whole genome by different approaches.

Two agrochemical companies, Monsanto and Syngenta have produced drafts of the sequence of the same *japonica* variety as IRGSP, Nipponbare, and the Beijing Genome Institute produced a draft shot gun sequencing of the two *indica* varieties which are (93–11 and Pei-Ai645) the paternal and maternal parents of the Chinese super hybrid LYP 9. The complete finished sequence was announced by the IRGSP in November 2004 (International Rice Genome Sequencing Project, 2005) and at the same time the Beijing Group reported a considerably elaborated draft of the *indica* sequence (Yu *et al.*, 2005). All these sequences are available from Genbank (http://www.ncbi.nlm.nih.gov/Genbank).

The sequencing phase of the rice genome is essentially finished and the most difficult task, the physical and functional annotation, remains to be carried out. The first reported numbers ranged between 32,000 and 62,000 genes for the shotgun *indica* genome. However the gene number has been reduced and currently stands at 37,544 (International Rice Genome Sequencing Project, 2005). The most updated rice genome annotation can be found on the following websites:

-http:// riceorgaas.dna.affrc.go.jp
-http:// rad.dna.affrc.go.jp
-http:// rice.tigr.org

Structural annotation was improved with the determination of more than one million ESTs, 28,000 full length cDNAs and with the comparison of the sequence with *A. thaliana* genome. More than 85% of the *A. thaliana* genes have homologues in rice. However the gene order between the two species has been almost completely reshuffled. Approximately 71% of the predicted rice genes have homologues in *A. thaliana* and 2,859 genes seem to be unique to rice and other cereals (International Rice Genome Sequencing Project, 2005). About 95% of known and expressed genes of rice have an equivalent in other cereals, thus justifying the use of rice as a model. Gene distribution along the chromosomes is not even. Gene density is usually higher in the distal regions of chromosome arms (18–19 genes/ 100 kb) than in the proximal ones (10–12 genes/ 100 kb).

Amongst the identified genes, there are no major changes in the functional categories in comparison to *A. thaliana*. For instance the same transcription factor families are represented in both genomes in roughly equivalent proportions. Considering flowering time and morphology, some *A. thaliana* genes have no orthologue in rice, which is not so surprising given the biological differences between the two species. An important difference concerns disease resistance (R) genes. The TIR (Toll-Interleukin Receptor- like), - NBS (Nucleotide Binding Site) and - LRR (Leucine Rich Repeat), class which account for 2/3 of the *A. thaliana* R genes is completely absent from the rice genome, and probably from other cereal genomes. The biological significance of this difference is not yet clear (Zhou *et al.*, 2004).

When the physical distances along the chromosomes are compared to genetical distances; a clear cut suppression of recombination is observed in centromeric regions. The ratio between physical and genetic distances thus varies between 210 kbp/ cM in euchromatin, up to 636 kbp/ cM in heterochromatin.

Sequencing of the rice genome revealed several additional important features. It shows the presence of a large number of mitochondrial and chloroplastic DNA insertions along the chromosome. Chromosome 12 thus has 1% of its total DNA represented by such insertions. It shows duplicated nature of the genome. Most of the genes are members of multigene families. Some result from tandem duplications and sometimes make large clusters, such as a group of 10 protein kinase genes on chromosome 1. This organization pattern is superimposed on to large segmental duplications representing several megabases. The rice genome can be organized into 18 pairs of duplicated segments covering nearly 70% of the genome. These duplications, with the exception of one between the short arms of chromosomes 11 and 12, are probably more ancient than the ones undergone by the *A. thaliana* genome, since there are on average only 17% of the gene which are present as pairs within the two segments of a duplication (International Rice Genome Sequencing Project, 2005).

Distribution of heterochromatin is also quite uneven among chromosomes. The chromosome with the largest heterochromatin is chromosome 10. An intriguing issue is the organization of the centromere. The centromere in rice contains many copies of a 155bp satellite DNA and copies of a centromere specific retrotransposon. The number of repeats of this satellite varies from 1,980 on chromosome 1 to 80 on chromosome 8.

The genome of the two varieties of rice has been sequenced and hence it is possible to compare them directly. However, this comparison is difficult, as about 15 % of the two genomes cannot be simply aligned due to the presence of many retrotransposons (Han and Xue, 2003). Nevertheless, with the improvement of sequencing quality, a total of 80,127 polymorphic sites between the two species could be recognized. SNP frequency varies from 0.53 to 0.78% which is 20 times more than that which is observed between the *A. thaliana* Columbia and Landsberg ecotypes (International Rice Genome Sequencing project, 2005). These retrotransposons are extremely helpful to distinguish the two rice types and to date their divergence.

The results of functional genomics of model plants will increase the understanding of basic plant biology as well as the exploitation of genomic information for crop improvement. A series of important traits in valuable crops like cereals, sugarcane, soybeans, cotton, potato, horticultural and ornamental crops can therefore be seen from a general perspective using gene function analysis of model plants. Recently, the isolation of genes determining plant height from *Arabidopsis* led to the identification of orthologous (Genes coding for the same function that are derived from the same ancestral gene. Two orthologous genes thus belong to two distinct species resulting from a common ancestor and these genes may indicate a position preserved in the genome of two species) dwarf genes in rice and other cereals. These were found to be the dwarf genes that were introduced in to modern varieties by conventional breeding in the green revolution. This is the example of conservation of important traits in the plant kingdom and illustrates the potential of gene function discovery in model plants for application in a wide variety of crop plants.

Sequencing an entire genome is not possible in most of the cases. A rapid way to establish an inventory of expressed genes is by determining partial sequences of cDNA called expressed sequence tags (ESTs). Thus genomics technology is being used to develop expressed sequence tags (ESTs) which provide a cost effective and rapid approach for describing all the genes of an organism. The sequences are sufficiently accurate to unambiguously identify the corresponding genes in most cases. Thousands of sequences can thus be determined with a limited investment. EST information is present in public databases for a variety of species. Powerful programs for rapid sequence comparisons with databases are accessible via the internet. The institute of Genome Research has ordered overlapping ESTs into tentative contigs. EST databases have proven to be a tremendous resource for finding genes and for interspecies sequence comparison, and have provided markers for genetic and physical mapping and clones for expression analyses. The relative abundance of ESTs in libraries prepared from different organs and plants in different physiological conditions also provides preliminary information on expression patterns for the more abundant transcripts.

## Functional genomics

Functional genomics involves a study of functions of all specific gene sequences and their expression in time and space in an organism. Once a new gene has been located and sequenced, then the next question is what function it controls. It was thought that from conventional studies on *E. coli* and yeast that most of the genes and their functions have been assigned. But the genomic sequences revealed that in fact there are large gaps in our knowledge. Of the 4288 protein coding genes in the *E. coli* genome sequence, only 1853 (43% of the total) had been previously identified (Blattner *et al.*, 1997). For *S. cerevisiae* the figure was only 30%. Thus there is a great challenge even if every gene in a genome is sequenced. We need to understand how the genome as a whole operates within the cell. Attempts are being made to find out the pattern of gene expressions, which genes are switched on and which are switched off in different tissues during different developmental stages.

The following approaches are used to determine the function of unknown genes.

### Computer analysis

Computer analysis plays an important role in locating genes in the DNA sequences. For this the most powerful tool available is *homology searching*. It locates genes by comparing the DNA sequence under study with all the other DNA sequences in the databases. The basis of homology searching is that the related genes have similar sequences and so a new gene is discovered by virtue of its similarity to an equivalent, already sequenced gene from a different organism. Homologous genes are ones that share a common evolutionary ancestor revealed by sequence similarities between the genes. A pair of homologous genes does not usually have identical sequences because the two genes undergo different random changes by mutation, but they have similar sequences because these random changes have operated on the same starting sequence, the common ancestral gene. Homology searching makes use of these sequence similarities. The basis of the analysis is that if a newly sequenced gene turns out to be similar to a second, previously sequenced gene, then an evolutionary relationship can be inferred and the function of new gene is likely to be same or similar to the function of the second gene. In terms of homology and similarity, if the nucleotide sequences are identical by, for example 80%, then percentage value is given to similarity and referred as the genes show 80% similarity rather than 80% homology.

It has been seen that homology analysis can provide information on the function of an entire gene or of segments within it. A homology search can be conducted either with nucleotide or amino acid sequence. It is best to conduct homology search using amino acid sequence. Because, there are twenty different amino acids in proteins but only four nucleotides in DNA, so genes that are unrelated usually appear to be more different from one another when their amino acid sequences are compared. Thus it is less likely to give spurious results if the amino acid sequence is used. Several software programs like BLAST exist for carrying out this type of analysis. A positive match may give a clear indication of the function of new gene or the implications of the match might be more subtle. In particular, genes that have no obvious evolutionary relatedness might have short segments that are similar to one another. The explanation of this is often that, although the genes are unrelated, their proteins have similar functions and the shared sequence encodes a domain within each protein that is central to the shared function. Although the genes themselves have no common ancestor, the domains do, but with their common ancestor occurring at a very ancient time, the homologous domains having subsequently evolved not only by single nucleotide changes, but also by more complex rearrangements that have created new genes within which the domains are found. An example has been quite often cited is of *tudor* domain, 120 amino acid motif, first identified in *Drosophila melanogaster*. A homology search using the *tudor* domain as the test revealed that several

known proteins contain this domain. These proteins are, RNA transport proteins during *Drosophila* oogenesis, a human protein with a role in RNA metabolism, and others whose activity appears to involve RNA in one way or the other. Thus it suggests that *tudor* domain sequence plays some part in RNA binding, RNA metabolism or some other functions involving RNA. The information from the computer analysis thus points the way to the types of experiment should be done to obtain more clear cut data on the function of the *tudor* domain.

Homology analysis has been conducted on *S. cerevisiae* genome for assigning functions to new genes. Of the 6000 genes in yeast, 30% had already been identified by conventional genetic analysis, before the sequencing project started. 70% of the remaining genes were studied by homology analysis and were put in following categories.

By the homology analysis 30% of the genes in the genome could be assigned functions. About half of these were clear homologs of genes whose functions had been previously established, and about half had less striking similarities, restricted to domains only.

About 10% of all the yeast genes had homologs in databases, but the functions of these homologs were unknown. The homology analysis is thus unable to help in assigning functions to these genes. These yeast genes and their homologs are called *orphan families*.

The remaining yeast genes, about 30% of the total had no homologs in the databases. A small portion (about 7%) were questionable ORFs which might not be real genes, because of unusual codon bias. The remainder 23% look like genes and are unique are called *single orphans*.

*Experimental analysis*

In conventional genetic analysis for study of gene function, starting point is a phenotype and the objective is to identify the underlying gene or genes. Genetic basis of phenotype is usually studied by searching for mutant organisms in which the phenotype has become altered. The mutants might be obtained experimentally or present in a natural population. The gene or genes that have been altered in the mutant organism are then studied by genetic crosses, which can locate the position of gene(s) in the genome and also determine if the gene is the same as the one that has already been characterized. The gene can then be studied by molecular biology techniques and sequencing is done. While in functional genomics the problem addressed is in opposite direction. It starts from the new gene or gene sequence and hopefully leading to the identification of its function i.e. associated phenotype.

The general principle of conventional analysis is that the gene responsible for a phenotype can be identified by determining, which genes are inactivated in organisms that display a mutant version of the phenotype. If the starting point is the gene rather than the phenotype, then the equivalent strategy would be to mutate the gene and identify the phenotypic change that results. Disrupting genes either by the introduction of mutations in to genes or by silencing the gene families using RNA interference approach has been followed. Mutation induction approach has been carried out in a genome wide fashion in a number of plants, but RNAi studies tend to be by each gene family individually. This is the basis of most of the techniques used to assign functions to unknown genes. Mutation libraries have been developed either by chemical mutagenesis or insertional mutagenesis.

**a. Tilling:** Tilling (targeting induced local lesions in genomes) has established itself as powerful tool for functional genomics and reverse genetics. This is a high throughput methodology, the conjunction of a high-throughput molecular reverse genetics technique and mutation induction, allowing for the mining of new alleles in a known locus and can be performed at a very early stage. The end product of Tilling process is a plant (and its offspring)) that has been identified with a change in a specific gene of interest. The plant line is then useful for determining the overall effect/role of that gene on the characteristics of the plant. The tilled population can be produced and then screened as mutant alleles in specific gene.

**b. Insertional Mutagenesis:** Gene knock-outs are where the activity of a gene has been eliminated due to the insertion of a DNA fragment. The two major methods for generating these are either by inserting a T-DNA insertion or a transposon sequence. T-DNA insertion is the most generally applicable method since it can be used for any plant that can be transformed and regenerated. Since each transformant is an independent event with the T-DNA being relatively randomly inserted into the genome a large number of independent transformation event are needed to inactivate every gene. The need for the generation of large numbers of independent transformants therefore limits this technology to plant species or particular lines that are capable of being transformed in a high throughput manner.

The following methods are employed to inactivate specific genes or overexpress genes for determining the function.

*Gene inactivation*: To inactivate a specific gene is to disrupt it with an unrelated segment of DNA which can be achieved by homologous recombination between the chromosomal copy of the gene and a second piece of DNA that shares some sequence identity with the target gene (Fig. 28.3). For present purposes it is enough to know that if two DNA molecules have similar sequences then recombination can result in segments of the molecules being exchanged.

The gene inactivation procedure by homologous recombination has been explained with an example from *S. cerevisiae*. The gene inactivation procedure employs deletion cassette, which carries a gene for antibiotic resistance preceded by the promoter sequences needed for expression in yeast and flanked by

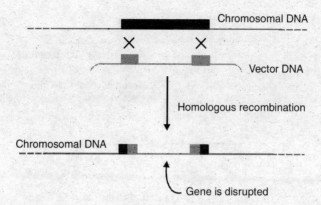

Chromosomal DNA

Vector DNA

Homologous recombination

Chromosomal DNA

Gene is disrupted

**Fig. 28.3:** Gene inactivation by homologous recombination. Concept of the figure taken from *Genomes,* T.A. Brown, by kind permission of BIOS Scientific Publishers, Oxford, U.K.

two restriction sites (Fig. 28.4). Before using the deletion cassette new segments of DNA are attached as tails to either ends. These segments have sequences identical to parts of the yeast gene (start and end segments) that is going to be inactivated. This modified deletion cassette is put in a vector. This modified deletion cassette is then introduced in to a yeast cell. Recombination occurs between the gene segments in the vector (DNA tails) and the chromosomal copy of the target gene in the yeast, replacing the yeast target gene with the antibiotic resistant gene. Cells in which the disruption has occurred are identified because they now express antibiotic resistance gene and so will grow on an agar medium containing geneticin antibiotic. The colonies lack the target gene and their phenotypes can be examined to gain insight into the function

Gene inactivation procedure has been used in mice, which is analogous to yeast. On mice the experimentation is done for equivalent genes to human beings so that functions identified from this organism can be utilized on human genes. In such experiments not one mutated cell but whole mutant mouse is needed so that a complete assessment of the effect of gene inactivation on the phenotype can be seen. To achieve this embryonic stem (ES) cells are used which are totipotent and can give rise to all types of differentiated cells leading to complete

organism formation. The ES cell is engineered and then injected into a mouse embryo, which continues to develop and gives rise to a chimera. These chimeric mice are allowed to mate with one another. Some of the offsprings result from fusion of two engineered gametes and will therefore be non-chimeric as each of their cells will carry the inactivated gene. These are called knockout mice and it is hoped that examination of their phenotypes would provide the desired information on the function of gene being studied.

*Gene overexpression*: In earlier techniques gene inactivation was used to determine the function of a gene but a complementary approach has been devised in which the test gene is much more active than normal gene and to determine what changes, if any, this has on the phenotype. Thus there is a need to distinguish between a phenotype change that is due to specific function of an over-expressed gene and a less specific phenotypic change that reflects the abnormality of the situation. To overexpress a gene, a special type of cloning vector, multi-copy is used which directs the synthesis of as much protein as possible. The vector must also contain a highly active promoter so that each copy of the test gene is converted into large quantities of mRNA leading to high amount of protein. Some genes functions have been identified in mouse using this approach.

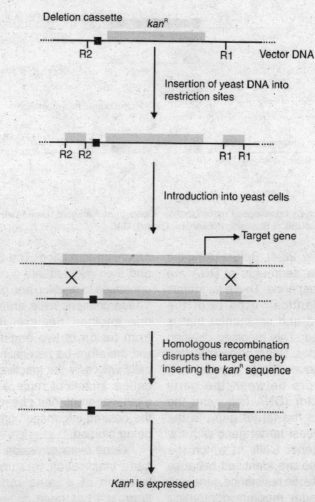

**Fig. 28.4:** Gene inactivation by the use of a yeast deletion cassette. The start and end segments of the target gene are inserted into the restriction sites and the vector introduced into yeast cells. Recombination between the gene segments in the vector and the chromosomal copy of the target gene results in the disruption of target gene. Cells in which the disruption has occurred are identified because they now express the antibiotic resistance gene and will grow on an agar medium containing geneticin, Kan$^R$–Gene for geneticin resistance;—■—: yeast promoter sequences; R$_1$, R$_2$–Restriction sites. Concept of the figure taken from *Genomes*, T.A. Brown, by kind permission of BIOS Scientific Publishers, Oxford, U.K.

Inactivation and overexpression can determine the overall expression of a gene but they cannot provide detailed information on the activity of a protein coded by a gene. For example, it might be that part of a gene codes for an amino acid sequence that directs its protein product to a particular compartment in the cell, or is responsible for its action in response to a particular signal. To test these hypotheses, it would be necessary to delete, mutate or alter a relevant part of the gene and then to determine its effect. This can be achieved by various procedures of site directed mutagenesis (explained in chapter 19). After mutagenesis the gene sequence must be introduced into the host cell so that homologous recombination can replace the existing copy of the gene with the modified version.

One can also abolish expression by antisense DNA or by activating transposons. Considerable amount of work has been done on the insertion and mobilization of the well known transposons of maize, namely the *Ac/Ds*, *Spm/dspm* and *Mu* systems in model plants such as tobacco and *Arabidopsis*. Special strategies to identify transposon tagged lines in large populations have been made. (see chapter 20).

Clues to the function of a gene can be obtained by determining where and when the gene is expressed. If gene expression is restricted to a particular organ or tissue of a multicellular organism, or to a single set of cells within an organ or tissue, then reporter genes and immunochemistry can be used to locate where and. when genes are expressed. The pattern of gene expression within an organism can be studied with a reporter gene (*uid*A, *lux*, GFP, etc.) which have been explained in chapter 21. For the reporter gene to give a reliable indication of where and when a test gene is expressed the reporter is subjected to same regulatory signals as the test gene. This is achieved by replacing the ORF of the test gene with the ORF of the reporter gene, the regulatory signals being contained in the region of DNA upstream of an ORF (Fig. 28.5). The reporter gene will now display the same expression pattern as the test gene which can be determined by examining the organism for the reporter signal.

Immunocytochemistry is used to locate the position of protein within the cell. It makes use of an antibody that is specific for the protein of interest and so binds to this protein and no other. The antibody is labeled so that its position in the cell can be visualized. For low-resolution studies, fluorescent labeling and light microscopy are used.

Gene function can often be determined if protein coded by the gene interacts with other known proteins. If an interaction to a well-characterized protein is identified, then it might be possible to infer the function of unknown protein. There are several methods for studying

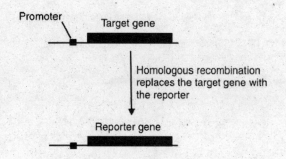

**Fig. 28.5:** A reporter gene. Concept of the figure taken from *Genomes*, T.A. Brown, by kind permission of BIOS Scientific Publishers, Oxford, U.K.

protein-protein interactions but two most useful approaches have been discussed.

**Phage display:** In this approach a special type of cloning vector based on bacteriophage or M13 filamentous bacteriophage is used (Clackson and Wells, 1994). The vector is so designed that a new gene that is cloned into it is expressed in such a way that its protein product is synthesized as a fusion with one of its coat proteins. The phage protein therefore carries the foreign protein into the phage coat, where it is displayed and hence its name 'phage display'. In such an approach cloning vector used for phage display is a bacteriophage genome with a unique restriction site located within a gene for a coat protein (Fig. 28.6). To create a display phage, the DNA sequence coding for the test protein is ligated into the restriction site so that a fused reading frame is produced in which the series of codons continues unbroken from the coat protein gene into the test gene. This is transformed into *E. coli* where recombinant molecule directs synthesis of a hybrid protein, made up of the test protein fused to coat protein. Phage particles produced by these transformed bacteria therefore display the test protein in their coats. To make this approach a powerful strategy, a phage display library is produced by ligating many different DNA fragments into the cloning vector. The library contains a collection of clones displaying a range of proteins. After transformation, the bacteria are plated on to agar medium so that plaques are produced. Each

**(A) Production of a display phage**

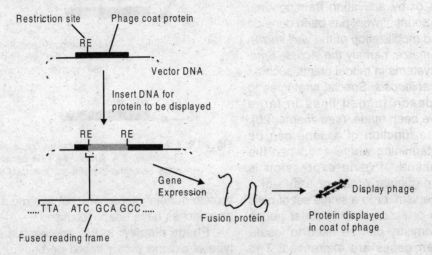

Restriction site     Phage coat protein

RE

Vector DNA

Insert DNA for
protein to be displayed

RE      RE

Gene
Expression

Fusion protein

Display phage

Protein displayed
in coat of phage

......TTA ATC GCA GCC.....

Fused reading frame

**(B) Use of phage display library**

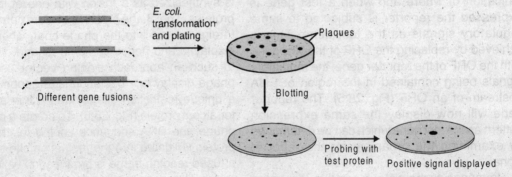

Different gene fusions

E. coli.
transformation
and plating

Plaques

Blotting

Probing with
test protein

Positive signal displayed

**Fig. 28.6:** Phage display approach for studying protein-protein interaction: A.–The cloning vector is a bacteriophage genome with unique restriction sites located within a gene for coat protein. To create display phage, the DNA sequence coding for the test protein is ligated into the restriction site so that a fused reading frame is produced in which the series of codons continues from the coat protein gene into the test gene. After *E.coli* transformation, this recombinant molecule directs synthesis of a hybrid protein made up of the test protein fused to the coat protein. Phage particles therefore display the test protein in their coats. The technique was originally carried out with the gene III coat protein of the filamentous phage f1, but has now been extended to other phages including λ. B.–A phage display library is produced by ligating many different fragments of DNA into cloning vector. After transformation, the bacteria are plated on to an agar medium so that plaques are produced. Each plaque contains a phage displaying a different protein. It is blotted on a nitrocellulose membrane and is probed with a test protein. If the test protein interacts with any of the displayed proteins then it will bind to the membrane. Concept of the figure taken from *Genomes,* T.A. Brown, by kind permission of BIOS Scientific Publishers, Oxford, U.K.

(A) The two-hybrid system

(B) Screening for protein interactions using the two-hybrid system

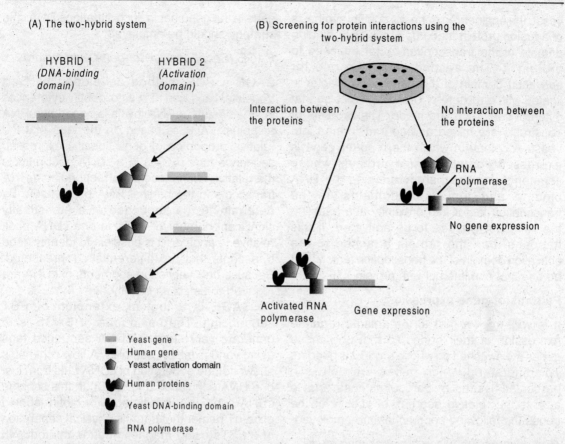

HYBRID 1
(DNA-binding domain)

HYBRID 2
(Activation domain)

Interaction between the proteins

No interaction between the proteins

RNA polymerase

No gene expression

Activated RNA polymerase

Gene expression

- Yeast gene
- Human gene
- Yeast activation domain
- Human proteins
- Yeast DNA-binding domain
- RNA polymerase

**• Fig. 28.7:** Yeast two hybrid system: A.–A gene fpr a human protein is ligated to the gene for the DNA binding domain of a yeast transcription factor. After transformation of yeast, this construct specifies a fusion protein, part human protein and part yeast transcription factor. On the right, various human DNA fragments are ligated to the gene for the activation domain of the transcription factor, which specifies a variety of fusion proteins. B.–The two sets of constructs are mixed and cotransformed into yeast. A colony in which the reporter gene is expressed contains fusion proteins whose human segments interact, thereby bringing the DNA binding and activation domains of the transcription factor into proximity and enabling the RNA polymerase to be activated. Concept of the figure taken from *Genomes,* T.A. Brown, by kind permission of BIOS Scientific Publishers, Oxford, U.K.

plaque contains phage displaying a different protein. The lawn of plaques is blotted on to a nitrocellulose membrane and the test protein applied. If the test protein interacts with any of the displayed proteins then it will bind to the membrane. Binding is directed either because the test protein is labeled or by treating the membrane with a labeled antibody specific for the test protein.

**Yeast two hybrid system:** The two hybrid system makes use of *S. cerevisiae* strain that lacks a transcription factor for a reporter

gene. Transcription factors are responsible for controlling the expression of genes. There are two segments of a transcription factor coded by separate genes, DNA binding domain for binding to a DNA sequence upstream of gene and activation domain that activates the RNA polymerase enzyme to copy the gene into RNA. One of these genes, usually the one for DNA binding domain is ligated to the gene for the protein whose interaction one wants to study (e.g. protein X). This protein can be from any organism (Fig. 28.7). After introduction into yeast, the engineered gene will specify synthesis of a fusion protein made up of the DNA binding domain of the transcription factor attached to protein X. The second gene, the one for activation domain of the transcription factor is ligated with a mixture of DNA fragments so that different constructs are made. The two sets of constructs are mixed and co-transformed into yeast. A colony in which the reporter gene is expressed contains fusion proteins whose segments interact thereby bringing the DNA binding and activation domains of the transcription factor into proximity and enabling the RNA polymerase to be activated. If this happens then the protein attached to the activation domain of the transcription factor must be one that can interact with protein X.

## Patterns of gene expression

It is well known that all organisms regulate expression of their genes so that only those genes are switched on whose product is needed. To understand which genes are active in particular tissues at particular times and to assess the relative degrees of activity of the genes the following approaches are employed.

### Gene expression assay by measuring levels of RNA transcripts

The first stage of gene expression involves copying the gene into an RNA transcript. Identification of a gene transcript is therefore the most direct way of determining whether the gene is switched on. This can be done by immobilizing the DNA copy of the gene on a solid support and hybridizing it with a sample of RNA extracted from the cells being studied. In fact the RNA is usually converted into cDNA prior to being used as the hybridization probe, as cDNA synthesis is the most convenient way of obtaining labeled sequences representing RNA molecules. If transcripts of the gene are present in the RNA extract then a hybridization signal will be seen. In order to assay the entire mRNA content of a cell referred to as **transcriptome,** this basic procedure has to be repeated for every coding

gene in the genome. In such situations DNA chip strategy should be employed.

### SAGE (Serial analysis of gene expression)

SAGE a PCR method developed by Bert Vogelstein and Ken Kinzler to identify differences in steady-state mRNA levels between two RNA samples. SAGE is based on the idea that the relative proportion of gene specific expressed sequence tags (ESTs) in a cDNA pool reflects the relative abundance of corresponding mRNA transcripts in the original RNA preparation. By determining the sequence of a statistically significant number of ESTs in one cDNA pool, relative to another, it is possible to identify gene transcripts that are differentially represented. Because only very short segments of cDNA are required to establish identity.

SAGE is a logical extension of EST sequencing. The basic idea of SAGE is to generate very short 3′ cDNA sequence tags using various mRNA samples. A flow scheme is shown for the generation of cDNA tags from just one RNA population (Fig. 28.8). In this scheme cDNA synthesis is initiated with biotin labeled oligo-dT primer that permits physical separation of 3′ESTs using magnetic beads coated with streptavidin. The two enzymes used for tag delineation are *Nla*III, which recognizes the sequence CTAG, and *Bsm*FI, a type II restriction endonuclease that cleaves 10 bp downstream of the sequence GGGAC. Restriction enzyme digested products are adaptor mediated ligated and PCR amplified to link large number of short ESTs (10–15 bp) into a single array for automated DNA sequencing.

These short tags have proven sufficiently long to unambiguously identify corresponding genes in databases (Velculescu *et al.*, 1997). Expression patterns for different genes are reflected by the relative abundance of individual tags. SAGE patterns have been studied in humans and yeast but the technique has not been applied to plants. A prerequisite for the identification of tags is the availability of large sequence databases for the species under study.

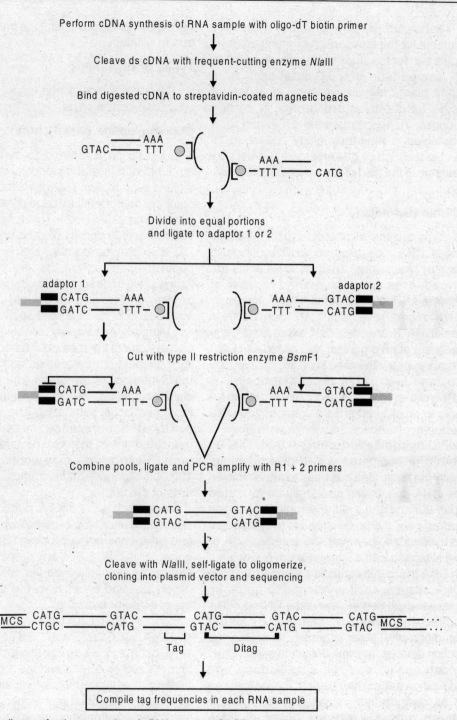

Perform cDNA synthesis of RNA sample with oligo-dT biotin primer

Cleave ds cDNA with frequent-cutting enzyme NlaIII

Bind digested cDNA to streptavidin-coated magnetic beads

Divide into equal portions
and ligate to adaptor 1 or 2

adaptor 1                                                         adaptor 2

Cut with type II restriction enzyme BsmF1

Combine pools, ligate and PCR amplify with R1 + 2 primers

Cleave with NlaIII, self-ligate to oligomerize,
cloning into plasmid vector and sequencing

Tag          Ditag

Compile tag frequencies in each RNA sample

**Fig. 28.8:** A flow diagram for the generation of cDNA tags in a SAGE. It is a DNA sequencing dependent gene analysis method to identify differentially expressed gene transcripts. Concept of the figure taken from Miesfeld: *Applied Molecular Genetics.* Copyright©1999 John Wiley and Sons. This material is used by the permission of John Wiley and Sons.

The technique is powerful but not very convenient for the comparison of many different samples and for the study of the rarer transcripts.

A comparison of EST databases from different plants reveals similarity in genes for specific functions and diversity in coding sequences. To supplement the rDNA and PCR technologies, a new technology described as DNA chip technology has emerged and will play a great role in the field of genomics (Gupta *et al.*, 1999).

### DNA chip technology

DNA chips are the microscopic array of single stranded DNA molecules immobilized on solid surface by biochemical analysis. DNA chips are also called as *gene chips* or *biochips* or *microarrays.* DNA chips contains a high density of DNA microarrays, which are derived from either cDNA (from identified expressed gene sequences) of an organism also called as cDNA microarrays or synthesized short oligonucleotide microarrays prepared by photolithography.

**Principle:** Hybridization of an unknown sample to an ordered array of immobilized DNA molecules of known sequences produce a specific hybridization pattern that can be analyzed or compared with a given sample. Actually the basic idea has been to attach known probe DNA sequences covalently on to a solid support surface and it is the sample RNA which is poured over the probe area for hybridization. Before doing so, however the sample RNA or cDNA is tagged with a fluorescent dye. Because each DNA microchip contains a standardized set of DNA sequences, it is referred to as probe whereas the labeled experimental DNA or RNA is called the target or sample. By using autoradiography, laser scanning, fluorescence detection devices, enzyme detection system, etc. one can easily read the chip surface and hybridization pattern can be both qualitatively as well as quantitatively analyzed. In ordinary northern analysis various types of RNA molecules remain immobilized in the gel and the DNA molecules constituting the probe are initially free in solution.

### Types of DNA chips

i. Oligonucleotide based chips
ii. cDNA based chips

### Oligonucleotide based chips

It is based on a method to synthesize large amounts of different oligonucleotides *in situ* on a glass support using light directed, solid phase, combinatorial chemistry developed by Affymetrix (Santa Clara, CA, USA). This type of DNA chips contains a high density of short oligonucleotides (10–25 mers) microarray which are prepared by photolithographic technique, a process in which light is used to direct the simultaneous synthesis of many different chemical compounds, synthesis occurs on a solid support.

Integration of two key technologies forms the cornerstone of the method. The first technology, light directed combinatorial chemistry, enables the synthesis of hundreds and thousands of discreet compounds at high resolution in precise locations on a substrate. The second laser confocal fluorescence scanning permits measurement of molecular interactions on the array. Such arrays may contain 100,000 to 400,000 oligonucleotides immobilized within an area of 1.6 cm$^2$.

Actually it is not always possible to produce cDNA from each type of mRNA which is a difficult and time consuming process. So, rather than going for cDNA, ESTs are prepared from EST library. As mentioned earlier EST is the short oligonucleotide either from the start or end of the cDNA. For example

5′–TACGGGCAATACGATGGGCC–3′
3′–ATGCCCGTTA–5′ (EST)

So, the EST oligonucleotide will be immobilized on a solid surface that will facilitate binding of sample DNA, which will be detected by different detection techniques.

A key advantage of this approach over non-synthetic methods is that photoprotected version of the four DNA building blocks allow chips to be manufactured directly from sequence databases thereby removing the uncertain and burdensome aspects of sample handling and tracking.

**Production of oligonucleotide microarrays/ chips:** Oligonucleotide microarrays are produced by *in situ* synthesis techniques using phosphoramidites. It utilizes photolithographic deprotection of 5'OH groups at specific sites on the chips, so that individual specific bases may be added at these deprotected sites during chemical synthesis.

**Procedure**: A glass wafer can be taken as a support. This surface can be either non-derivatized glass microscopic slide or aminoalkylsilane derivatized one. The basic purpose of derivatization is to clean the surface of solid support as hybridization is highly sensitive to any foreign bodies present on the surface. The procedure has been depicted in Fig. 28.9. The solid surface is derivatized with bis (2-hydroxyethyl) aminopropylthiethoxy silane. The 5'-O($\alpha$-methyl-6-nitropiperonyloxy carbonyl)-N-2' deoxynucleoside phosphoramidites are then attached to derivatized substrate through a synthetic linker e.g. 4,4'-dimethoxytrityl (DMT)-hexaethyloxy-O-cyanoethyl phosphoramidite [X]. The surface is then selectively activated for DNA synthesis by shining light through a photomask [MI]. Photolithographic masks having apertures are used to cover those regions where a particular nucleotide is not desired to be added. Light with a wavelength of more than 280 nm through microapertures (~800 × 12800 µm) of photolithograpghic masks are used. The glass wafer is then flooded with photoprotected DNA base [A-X] resulting in defined coupling on the chip surface.

*Coupling reaction*: 3'-O-phosphoramidite activated deoxynucleoside (its 5'-hydroxyl is protected with a photolabile group) is presented on the surface that was illuminated, and coupling occurs at sites that were exposed to light. The substrate is then rinsed. A second photomask is used to deprotect defined regions of the wafer for next coupling reaction. Repeated deprotection and coupling cycles enable the preparation of high density oligonucleotide microarrays.

Highly efficient strategies can be used to synthesize any arbitrary probe at any discreet, specified location on the array in a minimum number of chemical steps. For example, the complete set of 4N polydeoxynucleotides of length N, or any subset of this set, can be synthesized in only 4 × N chemical cycles. Thus giving a reference sequence, a DNA chip can be designed that consists of a highly dense array of complementary probes with no restriction on design parameters. The amount of nucleic acid information encoded on the chip in the form of different probes is limited only by the physical size of the array and the achievable lithographic resolution (Fodor, 1997)

*Advantages*

1. The photoprotected versions of the four DNA building blocks allow chips to be manufactured directly from sequence databases.
2. The use of synthetic agents minimizes chip to chip variation by ensuring a high degree of precision in each coupling phase.
3. Photolithography has enabled the synthesis of entirely independent sequences at individual grid positions and allowed miniaturization of the arrays.

*Disadvantage*

It needs photolithographic masks, which are expensive, expensive and time-consuming to design and built.

**Ink jets:** It is a 'drop on demand' delivery approach to manufacture microarrays. A biochemical sample of phosphoramidite solution is loaded into a miniature nozzle equipped with a piezoelectric fitting and an electric current is used to expel a precise amount of liquid from the jet on to the precise location on a surface (Fig. 28.9). This drop on demand technologies allows high-density gridding of virtually any

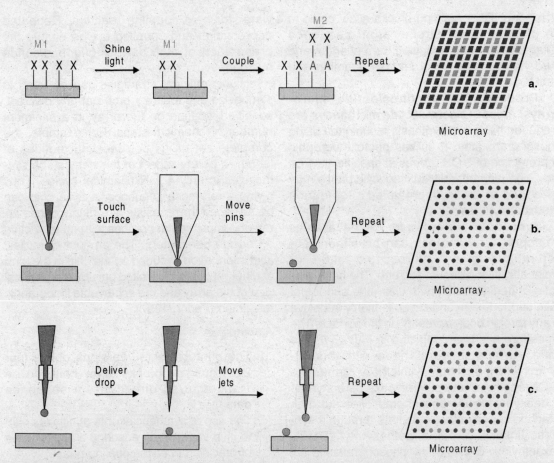

**• Fig. 28.9:** Three approaches to microarray manufacturing are depicted: **a.**–Photolithography- a glasswafer modified with photolabile protecting group (X) is selectively activated for DNA synthesis by shining light through a photomask (M1). The wafer is then flooded with a photoprotected DNA base (A-X), resulting in spatially defined coupling on the chip surface. A second photomask (M2) is used to deprotect defined regions of the wafer. Repeated deprotection and coupling cycles enable the preparation of high density oligonucleotide microarrays. **b.**–Mechanical microspotting–a biochemical sample is loaded into a spotting pin by capillary action, and a small volume is transferred to a solid surface by physical contact between the pin and the solid substrate. After the first spotting cycle, the pin is washed and a second sample is loaded and deposited to an adjacent address. Robotic control systems and multiplexed pinheads allow automated microarray fabrication. **c.**–Ink jetting- a biochemical sample is loaded into a miniature hole equipped with a piezoelectric fitting and an electric current is used to expel a precise amount of liquid from the jet onto the substrate. After the first jetting step, the jet is washed and a second sample is loaded and deposited to an adjacent address. A repeated series of cycles with multiple jets enables rapid microarray production.

biomolecule of interest including cDNAs, genomic DNAs, antibodies and small molecules. This approach has been used to prepare microarrays of single cDNAs at a density of 10,000 spots/cm². Because ink jetting does not require direct surface contact, piezoelectric delivery is theoretically amenable to very high throughput. Piezoelectric based delivery of phosphoramidite reagents has recently been used for the manufacture of high-density oligonucleotide microarrays (Blanchard, 1998).

## cDNA based chips

These are often derived from cDNAs, so also called as cDNA microarrays. These are produced by taking cDNAs or ORFs, which are PCR amplified and immobilized on a solid support of microscopic slide or nylon filter. This procedure originally developed by Davis, Brown and their colleagues at Stanford has been commercialized by Synteni/Incyte Companies in USA. Their attachment on a solid surface is done by delivery/deposition of DNA probe aliquots at specific sites on the surface of chip. The delivery can be performed manually or robotically. The following two delivery technologies are extensively used.

**Mechanical microspotting**: Micropsotting is a miniaturized version of earlier DNA spotting techniques which encompasses a family of related deposition technologies that enable automated microarray production by printing small quantities of pre-made substances onto solid surfaces (Fig. 28.9). Printing is accomplished by direct surface contact between the printing substrate and a delivery mechanism that contains an array of tweezers, pins or capillaries that serve to transfer biochemical samples to the surface. The microarrays currently manufactured by Synteni contain as many as 10,000 groups of cDNAs in an area of ~3.6 cm$^2$. The microspotting is unlikely to produce the densities available by photolithography, but improvements are being made in microspotting technologies for automated production of chips containing 100,000 groups of cDNAs in an area of ~6.5 cm$^2$.

The following steps are involved in microspotting for production of cDNA microarrays:

1. Cloned cDNA fragments are amplified by PCR, which are referred to as probe DNA samples.
2. 1 × 3 inch microslide with chemically modified surface is prepared for microspotting.
3. Probe DNA samples prepared as above are loaded each into a spotting pin by capillary action and a small volume is transferred to chemically modified solid surface at a specified position. This is achieved by a physical contact between the pin and the solid surface.
4. After the first spotting cycle, the pin is washed and the second sample is similarly loaded and deposited at an adjacent position.

These steps are speeded up by robotic control systems and multiplexed pinheads.

**Advantages:** Ease of use, low cost, versatility and affordability are the important advantages.

**Disadvantages:** One disadvantage of microspotting as that each sample must be synthesized, purified and stored prior to microarray fabrication.

**Ink jets:** Ink jets provide another approach to the manufacture of microarrays as explained earlier.

### Hybridization and detection methods

Hybridization of the target DNA to a microarray yields sequence information. The target DNA is labeled and incubated with microarray. If the target DNA has regions complementary to the probes on the array, then the target DNA will hybridize with these probes. For the detection of hybridization patterns on DNA chips, the technique of reverse dot blot is utilized. In dot blots where the target DNA is dot blotted on the membrane and the probes are labeled opposed to this for DNA chips, the probes are anchored in the form of microarrays and the target DNA is labeled. Once hybridization is completed, the hybridization is detected by various methods, which rely on the use of an enzyme system while others use radiolabeling, and/or fluorescence.

Under a fixed set of hybridization conditions, e.g. target concentration, temperature, buffer and salt concentration, etc., the fraction of probes bound to targets will vary with the base composition of the probe and the extent of the target—probe match. In general, for a given length, probes with high GC content will hybridize more strongly than those with high AT content. Similarly probes matching the target will hybridize more strongly than probes with mismatches, insertions and deletions.

Radioactive and non-radioactive detection methods have been discussed earlier in a chapter on gene cloning. Non-radioactive methods involve biotin or digoxigenin labeling while radioactive labeling requires direct detection through autoradiography, gas phase ionization and phosphorimagers (Eggers et al., 1994). However there are drawbacks with the detection methods involving radioactivity. In order to avoid these problems, fluorochromes may be used which will allow direct detection due to fluorescence. In this one target DNA may be labeled with more than one fluorochromes for hybridization of microarray on the DNA chips which would also allow multiplexing.

A target sequence is fluorescently tagged and then injected into the chamber, where the target hybridizes to its complementary sequences on the array. Laser excitation enters through the back of the array, focused at the interface of the array surface and the target solution. Fluorescence emission is collected by a lens and passes through a series of optical filters to a sensitive detector. By simple scanning the laser beam or translating the array, or a combination of both, a quantitative two-dimensional fluorescent image of hybridization intensity is rapidly obtained.

The hybridization patterns can be screened using automatic scanner (e.g. Scan Array 3000). These detection systems are based either on lens based systems (epifluorescent and confocal microscopes) or on charge coupled device (CCD) based systems. Lens based systems are not well suited to detect small quantities. Therefore, CCD detection systems have been developed to detect small quantities of array bound molecules. In this method labeled target DNA is hybridized to an immobilized probe on a silicon wafer which is then placed on the CCD surface and a signal is generated. A fluorescence microscope fitted with a CCD camera and computer is used. This will allow sequencing by hybridization with an oligonucleotide matrix.

Two versions of commercial readers are available. A first generation system from Molecular Dynamics as well as a recently released high performance system from Hewlett-Packard. Chip production is now in a scalable format with production capacity of ~5000 to 10,000 chips per month (Fodor, 1997).

**Double stranded DNA chips:** Presently DNA chip technology has been extended to fabrication of double stranded DNA chips on a solid surface. In one of the approaches single stranded oligonucleotides (up to 40 mers) are immobilized on to a surface via the 3′ end. To these a 16-mer primer is hybridized which can be extended with Klenow fragment, thereby converting the immobilized DNA into a double stranded form (Fig 28.10a). By incorporating fluorescent label into the DNA as a marker, it was shown that double stranded DNA is accessible to protein. An alternative approach is to couple 12-mer oligonucleotides to gold support and then these oligos are hybridized to double stranded DNA with a complementary single stranded end, followed by treatment with DNA ligase which results in surface coupled double stranded DNA (Fig. 28.10b).

## Functions

1. It allows systematic study of DNA-protein interaction.
2. How eukaryotes regulate gene expression?
3. Pinpointing nucleotide amino acid interaction.

## The characteristic features of DNA chips/ microarrays

i. *Parallelism*: Microarray analysis allows parallel acquisition and analysis of massive

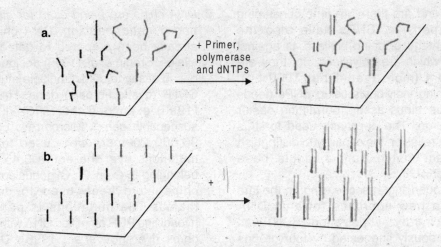

**Fig.28.10:** Approaches for making double stranded DNA arrays. **a.**–Single stranded oligonucleotides are synthesized on a glass surface in the presence of primer, polymerase and dNTPs. Since all the strands are not of uniform length and thus not accessible to the primer and polymerase will not be converted into double stranded DNA of desired length and sequence. **b.**–12-mer oligonucleotides are coupled to a small solid support. Hybridization to double stranded DNA with a complementary single stranded end, followed by treatment with DNA ligase, results in surface coupled double stranded DNA.

data. It allows meaningful comparison between genes or gene products represented in microarrays.

ii. *Miniaturization*: The analysis involves miniaturization of DNA probes and reaction volumes, thus reducing time and reagent consumption.

iii. *Speed*: Microarray analysis is highly sensitive and allows rapid data acquisition.

iv. *Multiplexing*: Multiple samples can be analyzed in a single assay. The labeling and detection methods that involve multicolor fluorescence allow comparisons of multiple samples in a single DNA chip. Multiplexing also increases the accuracy of comparative analysis by eliminating complicating factors such as chip to chip variation, discrepancies in reaction conditions and other shortcomings inherent in comparing separate experiments.

v. *Automation*: Manufacturing technologies permit the mass production of DNA chips and automation lead to proliferation of microarray assays by ensuring their quality, availability and affordability.

vi. *Combinatorial synthesis*: Using the combinatorial synthesis strategy, a set of all $4^k$ oligonucleotides of the length $k$ nucleotides ($k$-mers) can be generated in $4k$ synthesis cycles. For example, the set of all 4-mers (256) can be synthesized in 4 rounds, each round having 4 cycles, thus making a total of 16 cycles.

### Applications of DNA chips

1. Functional genomics: Microarrays for gene expression analysis provided an integrated platform for functional genomics.

2. Single nucleotide polymorphisms and point mutations: Recently single nucleotide polymorphisms (SNPs) as biallelic genetic markers have been extensively used as the markers of choice. Markers like RFLP, simple sequence length polymorphisms (SSLPs) are being utilized but these markers had some drawbacks. They need gel based

assays and are therefore time consuming and expensive. SNPs have also the disadvantage of being biallelic as against SSLPs which are polyallelic, but they are abundant (more than 1 per 1000 bp). Genotyping individuals using SNPs needs only plus/minus assay, permitting easier automation. The approach used for this purpose relies on the capacity to distinguish a perfect match from a single base mismatch.

3. DNA sequencing/ resequencing: In the late 1980s a new approach towards DNA sequencing by hybridization (SBH) was simultaneously suggested by four groups (Southern, 1988; Bains and Smith, 1988; Lysov *et al.*, 1988; Drmanac *et al.*, 1988). The method involves manufacturing the sequencing DNA chips that contain a complete set of immobilized oligonucleotides of a particular size (e.g. 8-mers) and hybridization of the target DNA of unknown sequence (whose sequence is to be determined) on to these DNA chips. The hybridization patterns are then recorded using one of the several suitable devices as discussed earlier. Identification and analysis of the overlapping oligomers that form perfect duplexes with the DNA of interest permits reconstruction of the target DNA sequence.

4. *Characterization of mutant populations exposed to various selection pressures*: This is done to collect information about the fitness value of a variety of alleles for each of the large number of genes in a species. This is particularly useful in those organisms where complete genome sequence is known. In such cases deletions, insertions or substitutions can be introduced and then the fitness value is analyzed. Such an approach is called reverse genetics where study is started from the DNA sequence and concluded with the phenotype.

5. *DNA chips are being used for diagnostics and genetic mapping*. DNA chips have already been prepared to detect mutant alleles. Cronin *et al.* (1996) designed a chip based assay to detect multiple mutations in CFTR (cystic fibrosis) gene. Hacia *et al.* (1996) examined the *BRCA*-1 (cancer susceptible gene). Microarrays containing 96,000 20–mers were used to identify mutations over the entire 3.45 kb DNA belonging to exon 11. Oligonucleotide DNA chips have been used to detect ß-thalassemia mutations in patients by hybridizing PCR amplified DNA with the DNA chips (Yershov *et al.*, 1996). DNA chip technology has also been successfully applied to genotyping of hepatitis C virus in blood samples (Livache *et al.*, 1998).

6. *Proteomics (study of protein–protein interactions*: Protein linkage maps can also be created using genomic sequence information.

7. *Agricultural biotechnology*: DNA chips with ESTs can be used to collect data on expression in an agricultural crop under different conditions. Transgenic plants can also be rapidly analyzed using microarray on DNA chips, and expression patterns under different environmental conditions can be predicted at the gene level. DNA microarray will also be extensively used for the study of DNA polymorphism (e.g. single nucleotide polymorphisms—SNPs) to develop molecular markers tagged to specific economic traits for molecular marker assisted selection in breeding program.

### Proteomics

Proteomics involves a study of the structure of all proteins encoded in a fully sequenced genome which is also referred to as proteomics. Thus it includes investigation of biological processes by the systematic analysis of a large number of expressed proteins for specific properties such as their identity, quantity, activity and molecular interactions. The term proteome was coined to refer to the total protein complement expressed by the genome.

The genome-wide analysis of mRNA expression by microarray approach is providing important clues about expression patterns and thus the function of gene products (discussed in the preceding section). However, for a substantial number of proteins there may by only a loose correlation between mRNA and protein levels. In addition the functions of proteins depend considerably on post-translational modifications and interactions with other proteins. The analysis of proteins is the most direct approach to define the gene function. Thus whether the new sequence obtained corresponds to any functionality in terms of transcription and translation is to identify the proteins which it codes. Therefore, information on mRNA levels is not sufficient to obtain a complete picture of the way gene expression is regulated within the cell. Protein expression data are more informative, but are much more difficult to obtain in a parallel fashion.

To understand the function of genes, efficient approaches for identification of proteins, for determining profiles of protein expression in different tissues and under different conditions, for identification of post-translational modifications of proteins in response to different stimuli and for characterization of protein interactions are critical. Therefore protein analysis or proteomics is turning out to be a major subject of research. There are three major steps in proteome analysis:

Two-dimensional protein gel analysis
Mass spectrometry (MS) for analysis of proteins
Database searching

### Two dimensional protein gel electrophoresis

Two-dimensional PAGE (polyacrylamide gel electrophoresis) is the method of choice to study the abundance and posttranslational modifications of several hundred proteins in parallel. O'Farell made a path breaking advancement in electrophorectic protein separation methods by combining isoelectric focusing (IEF) and sodium dodecyl sulfate (SDS) gel electrophoresis, resulting in the powerful two dimensional gel electrophoresis (2D) technique (see chapter 14). Several innovations in the basic 2D technique have made this technique suitable for a range of different applications. This method has been optimized for the separation of both soluble and membrane protein fractions as well as glycoproteins. 2D separated proteins can be subjected to analysis of amino acid composition, immunological characterization and peptide mapping. The amino acid sequence of 2D separated proteins can be directly determined.

### Mass spectrometry (MS) for analysis of proteins

Typically the MS technique consists of a source to generate ions from the sample and an analyzer to separate and detect these ions according to their mass. The important development in proteomics research is the combination of 2D with mass spectrometry for the analysis of separated proteins. MS has essentially replaced the classical technique of Edman degradation because it is very sensitive and can also deal with protein mixtures and offers much higher throughput. Simple MS such as MALDI-TOF (matrix assisted laser desorption ionization-time of flight) measures only the mass. However, tandem mass spectrometry also allows the amino acid sequence determination. In this approach proteins are separated by 2D and stained spots are excised and subjected to in-gel digestion with trypsin. The resulting peptides are separated by on-line high performance liquid chromatography (HPLC). The eluting peptides are subjected to tandem mass spectrometry in which two stages of mass analysis are linked in series. The peptides are ionized by electrospray ionization and then delivered to first MS in which peptides are identified based on their mass to charge ($m/z$) ratio. The selected peptide is fragmented by collision with an inert gas (e.g. argon) and finally the second MS analysis separates the resulting peptide fragments. The MS spectrum of the peptide gives valuable information on amino acid sequence. The amino acid sequence information helps in homology searching and cloning or database identification of the corresponding gene. Recently developed technique allow automation of in-gel tryptic digestion of all the proteins in the 2D gel, followed by their transfer to a membrane that can be scanned by MS to obtain diagnostic peptide masses for peptide mass finger printing. However, until the complete genome sequence is not available for peptide mass fingerprinting, the amino acid sequence approach of tandem mass spectrometry is the method of choice. The synthesis of 2D + MS is the backbone of the present day proteomics research.

*Database searching*

Database searching: Further support to proteomics research has been lent by the emerging computer technologies. Identification of proteins by mass requires access to protein sequence database (see chapter 29). The most commonly used databases are:

SWISS—2DPAGE for protein identification
NCBI/BLAST—sequence database
SWISS-PROT—sequence database
SWISS-MODEL—for three dimensional structure
PROSITE—for domain structure
GenBank and EMBL—DNA data banks

The science of bioinformatics is a cardinal part of the present day proteomics science as development of sophisticated software for an efficient analysis and storage of data with partially automated comparison of multiple 2D gels is needed in scaling proteomics to meet challenges put by genomics research.

A large number of two-dimensional gel databases exist for many organisms. A database for plasma membrane proteins in *Arabidopsis* is under development, which not only provides information on abundance and posttranslational modification, but also on the intracellular localization of a subset of proteins. This approach can be extended to other organelles, and even multisubunit protein complexes, provided that efficient separation procedures are available.

Other techniques based on MS of complex mixtures of protein fragments, may be developed in the future for the study of total protein complements. The protein interactions can be analyzed directly through proteomics science by performing co-precipitation studies with a bait protein followed by mass spectrometric read-out of the proteins. The post-translational modification of proteins is a key regulatory event in many cellular processes, including signalling, targeting and metabolism. Current developments in proteomics enable the global analysis of post-translationally modified proteins. MS has been employed to characterize function-critical post-translational modifications, including phosphorylation and glycosylation.

Several newer methods of protein analysis such as affinity purification, antibody usage, yeast two hybrid system, phage display (discussed earlier), etc. have been combined with 2D for proteomics research. DNA chips can also be used for the study of proteomics. DNA chip arrays can be used to identify the genes involved in protein–protein interactions. Protein–protein interactions can be studied using the yeast two hybrid system. In this system, two fusion proteins are used for the activation of transcription of a reporter gene in yeast. The first fusion protein contains a DNA binding domain fused to a protein of interest and the second fusion protein carries an acidic transcription activation domain fused to a second protein of interest. Specific interaction between two chimeric proteins leads to transcriptional activation of reporter gene which can be easily scored with color based assays. The identity of two proteins of interest is confirmed by sequence analysis of each clone thus identified. Phage representation libraries can also be used for DNA chip based detection systems. This involves the use of fusion proteins encoded by chimeric sequences of phage viral coat proteins and genes of interest (discussed earlier).

The proteome approach has been used in yeast to study gene function through the generation of knockout or overexpression mutants and for the analysis of changes in protein profiles on two-dimensional gels. In maize comparison of proteomes of lines nearly isogenic for gene *opaque*2 allowed the identification of new targets for the encoded transcription factors. In *Arabidopsis* two-dimensional gel profiles were used to characterize developmental mutants and allowed the hypothesis of the overproduction of cytokinins in one of the mutants, which was confirmed subsequently. Thus proteomics research is proving important for characterization of individuals or lines, estimation of genetic variability within and between populations, establishment of genetic distances to be used in phylogenetic studies and characterization of mutants with localization of genes encoding several proteins. Selective applications of proteomics in animal and microbial systems include discovery of target molecules, designing/discovery of novel biomolecules and proteins (pharmaceuticals, industrial and environmental applications) finding high value peptides/proteins, antibodies, vaccines, enzymes, therapeutic peptides, drug discovery and biomolecular engineering.

Proteomics is becoming a necessity in plant biology for deciphering the function and the role of genes in the plant genome sequencing projects. Some examples where 2D analysis has proved to be of great use in plants are shown in Table 28.4.

## Metabolomics

After DNA sequencing to establish an inventory of the genes of an organism, gene expression analysis (transcriptome–the full complement of activated genes, mRNAs or transcripts in a tissue at a particular time in response to a specific stimulus), and protein analysis (proteomics–the analysis of the functions and interactions of proteins in organisms), the remaining functional genomics challenge is analysis of the metabolome. The term "metabolome" was created by analogy to transcriptome and proteome. The metabolome encompasses all the metabolites that represent the catabolic and anabolic activities being performed by proteins at any given time (Fiehn, 2002), plus compounds that can be produced in tissue through non enzymatic reactions. Like the transcriptome and proteome, the metabolome shows specificity with respect to species, organ developmental stage, and environmental conditions. Because metabolites are the end products of cellular functions, their levels can be viewed as the response of biological systems to environmental or genetic manipulation. Metabolomics denotes comprehensive, nonbiased, high-throughput analyses of the complex metabolite mixtures typical of plant extracts. The ultimate aim of these approaches is to allow a study of the entire range of chemicals of low molecular weight synthesized and metabolized by plants.

Over 50,000 biochemical compounds have already been identified in plants. Nevertheless,

**Table 28.4:** Examples of the use of 2D gel electrophoresis technique in plant systems

| Application |
| --- |
| Identification of somatic embryogenesis related proteins |
| Construction of wheat seed protein map |
| Scoring of polymorphism in *Saccharum* species |
| Identification of marker proteins for distinguishing *indica-japonica* rice |
| Preparation of data file of rice seed proteins |
| Tagging of plasma membrane proteins in *Arabidopsis* |
| Identification of lumenal and peripheral proteins in pea thylakoids |

100,000 and perhaps even over 200,000 compounds are still to be discovered (Fiehn, 2002). This enormous number includes compounds belonging to a wide variety of structurally different metabolic classes, which are all, in one way or another, inter-related. This multitude of chemical components, with many different carbon skeletons, functional groups and physicochemical properties, makes plants an invaluable source of food, pharmaceuticals, health-promoting compounds, flavour and fragrance compounds, biocides and fine chemical, but also a source of toxins and xenobiotics.

The challenges of metabolomics are, first, how to measure the maximum number of these chemicals simultaneously, and second, how to make sense of the vast datasets that are generated by such measurement. Achieving the broadest overview of metabolic composition is very complex and entails establishing a multifaceted strategy for adequate tissue sampling homogenisation, optimal sample extraction, metabolite separation, detection and identification, automated data gathering, handling, analysis and quantification (Hal *et al.*, 2002).

*Terminology*

For metabolite analyses, different analytical approaches are used and Fiehn (2002) classified them under the following four general approaches:

i. *Target analysis*: It is centred on a small number of metabolites. Very often the substrate and the product of an enzymatic reaction are analysed with the aim of characterising a given enzymatic reaction *in vivo* or *in vitro.* So the primary effect of a genetic modification or of a treatment can be studied without the immediate need to analyse a larger number of metabolites.

ii. *Metabolite profiling*: It enlarges the analysis to include all metabolites of a pathway or a given chemical class of metabolites. Here too the analysis is restricted to a selected number of pre-determined metabolites, which again simplifies sample preparation and analysis. Such studies concentrate on a few, well-defined metabolites, which are often part of well-characterized metabolic pathway.

iii. *Metabolomic approaches*: It is the unbiased identification and quantification of all metabolites present in a specific biological sample from a plant grown under defined conditions. The component of the metabolome can be viewed as end products of gene expression that define the biochemical phenotype of a cell or tissue. The determination of large numbers of metabolites thus provides a broad view of the biochemical status of an organism that can be used to monitor and assess gene function. Differentially, expressed metabolites must be chemically identified and correlations can be sought between sets of differentially displayed metabolites. The individual as well as the correlated metabolites are used to identify metabolic pathways or networks that have been affected. These pathways are then used to determine the broader biological significance of the response or the assign gene function.

iv. *Metabolic fingerprinting*: It aims to classify samples by global and fast analysis of a large number of metabolites. Metabolic fingerprinting approaches focus on collecting and analyzing data from crude metabolite mixtures to rapidly classify samples. It is a high-throughput technique which enables a snapshot of the metabolic composition at a given time, where it is not initially necessary or feasible to determine levels of metabolites individually. Metabolic fingerprinting is the rapid classification of samples according to their original biological provenance. Fourier transform infrared spectroscopy (FT-IR) has been used successfully in metabolic fingerprinting technology. FT-IR is a physicochemical method that measures predominantly the vibrations of bonds within functional groups and generates a spectrum that can be regarded as a biochemical or metabolic "fingerprint" of the sample (Johnson *et al.,* 2003). NMR methods also provide metabolite fingerprints with good chemical specificity for compounds containing elements with non zero magnetic moment, such as $^{1}H$, $^{13}C$, $^{15}N$, and $^{32}P$ which are commonly found in most biological metabolites. Analysis of NMR spectra of solvent extracts of plants has the potential to provide a relatively unbiased fingerprint, containing the majority of the metabolites present in the solution.

### Analytical Techniques

A metabolite analysis consists of several steps. Samples must be taken in defined conditions while avoiding effects of harvesting procedures on metabolite contents. Sampling has to be fast, involving efficient and rapid quenching of metabolism in order to block immediately any biochemical reaction and obtain a snapshot of the biochemical status. This involves the arrest of all enzymatic reactions and is very efficient in the case of tissue cultures where the liquid culture can be directly infused with extraction solvent. For thicker higher plant tissue, rapid quenching is possible by freeze clamping. Ideally samples should be frozen at liquid nitrogen temperatures or lyophilized.

Progress in bioanalytical techniques has made it possible to analyse metabolites directly in crude biological extracts and from frozen and freeze dried samples. Samples should then be homogenized prior to extraction. Currently, no comprehensive comparisons of extraction techniques that show high reproducibility, robustness and recovery for all clotseha•g comn published. No single technique currently exists that is appropriate for measurement of all metabolites. The choice of technique depends on the scientific question. Targeted metabolite analysis is focused on specific chemical classes of metabolites and numerous techniques have been used. True metabolomics requires more advanced techniques, and these are still in the process of development and evaluation, especially with regard to comprehensiveness, selectivity and sensitivity (Krapp *et al.,* 2007).

*Nume*rous techniques exist for metabolite detection. Recent developments have favored mass spectrometry (MS) as a detection mode that is sensitive, relatively non-selective, and able to identify compounds with reference to databases of mass spectra obtained for authentic standards. These properties make MS the method of choice for metabolite detection in complex matrices and this technique is particularly powerful when allied to prior chromatographic resolution of such mixtures. Gas chromatography (GC) coupled to mass spectroscopy (GC/MS) and more recently, liquid chromatography (LC) coupled to mass spectroscopy (LC/MS) are able to detect several hundred chemicals, including sugars, sugar alcohols, organic acids, amino acids, fatty acids and a wide range of diverse secondary metabolites. A further step is the quantification and, last but not least, the organization of the results in a database (data processing). Most of the statistical tools are the same as used in genomics. Automated data gathering handling and analysis require considerable computational developments.

### Metabolomics is an important tool for functional genomics: linking genotype and phenotype

Over the last few years, two main strategies have been followed in order to discover the role of a gene: forward genetics *and reverse genetics.* *Forward genetics* consists in searching for interesting phenotypes by exploiting all kinds of natural or induced variation of the genome. The underlying genetic modification has then to be elucidated. Reverse genetics has as its starting point a specific gene whose in vivo activity *is deli*berately modified, e.g. by antisense, mutagenesis, co-suppression, over-expression, site directed mutagenesis (SDM), RNAi approaches, etc. Transgenic or mutant plants provide an excellent means to explore changes in metabolic network connectivity through metabolomic analysis of the effect of the specific perturbation of a gene interest.

Genetic perturbation achieved by mutations or by integration of foreign DNA sequences leading to either over-or under-expression of genes were used to modify expression product and the resultant changes have been assessed to infer gene function. Transgfunction. Transgenic potato plants over-expressing invertase specifically in the tubers were analysed by GC/ MS to establish an unbiased metabolic profiling (Roessner et al., 2000). Studies of sense or antisense plants with progressively decreasing expression of a target enzyme have shown that diagnostic changes in metabolites can often be detected in lines where the alteration of enzyme activity is too small to produce any visual phenotype (Scheible et al., 2000). These technologies have been developed over the past few years for the detailed profiling of various classes of metabolites specific to, or characteristic of, plants. These approaches involves the profiling of sugars, organic and amino acids, flavonoids and their conjugates, caroteniods and related compounds such as tocopherols and plastoquinones, monoterpenes, phenylpropnaoids and lignin formation. Application of the same profiling method to A. thaliana plants treated with different types of herbicides demonstrated different profiles diagnostic of the site of inhibition (Fraser et al., 2000). Further, to overcome problems such as gene redundancy, quantitative analysis is needed rather than qualitative analysis. Moreover, many agronomic traits are under the control of multiple loci and quantitative genetic analysis is the only way to identify genes involved in these important traits. Recombinant inbred lines (RILs) are valuable tool to detect quantitative trait loci (QTL) from allelic variation in two parental lines. The first detection of QTL for biochemical traits, such as enzyme activities and metabolite contents was carried out using maize leaves (Causse et al., 1995). The use of QTL in physiological and biochemical approaches has been reviewed by Prioul et al.

(1997). The complementary use of near isogenic inbred lines (NIL) allowed identification and cloning of a QTL for sugar content in tomato which appeared to be an invertase gene (Fridman et al., 2000). The exploration of the natural variation in accessions or cultivars is another way to validate a candidate gene as a QTL analysis in metabolic studies is more recent in A. thaliana, but offers an enormous potential (Loudet et al., 2003).

## Phenomics

The phenome is the storage, analysis and integration of phenotypic information. This information is generally collated by breeders and agronomists, who have traditionally applied phenotypic information for crop improvement. The international Crop Information System (ICIS) is arguably the most advanced database of the type and enables breeders to load and store information pertaining to genealogy and phenotype (McLaeren et al., 2005). ICIS was initially developed by the International Rice research Institute (IRRI) bioinformatics and biometrics unit and is widely used for rice and wheat is being extended to include information on legumes. Data are predominantly provided by the International Center for Agricultural Research in Dryland Agriculture (ICARDA), International Crop Research Institute for the Semi- Arid Tropics (ICRISAT), as well as the United State Department of Agriculture (USDA). Database may be queried across multiple datasets to identify varieties which have agronomic qualities such as disease resistance, yield, and response to stress.

TASSEL (Trait Analysis by aSSociation, Evolution and Linkage) software applies trait information and evaluates linkage disequilibrium, nucleotide diversity, and trait associations. It works with many different types of diversity data and connects to the Panzea and Gramene data repositories.

The GERMINATE database is a modular system and includes a Data Integration Module to store information on the nature of the plant samples. A passport Module stores Multi Crop Passport Descriptors (MCPDs), standards developed jointly by the Food and Agriculture Organization (FAO) and the International Plant Genetic Resources Institute (IPGRI). Passport data include geographical information about accession collection sites, using a data format, which is consistent with Geographic Information System (GIS) programs. The Datasets Module stores genotypic (marker), phenotypic and manages pedigree and list information, and has been adapted from the Genealogy Management System (ICIS- GMS) module of the ICIS database (Edwards and Batley, 2008). With the volume of genetic and genomic information being produced the technologies described above, there is an increasing ability to utilize this information for the improvement of crops. The linking of the heritable material, the DNA sequence of genes and genomes that is passed from one generation to another, with the traits that are associated with the possession of this DNA, is one of the greatest challenges facing biological scientists today. Bioinformatics plays a key role linking the genome information to the observed phenotype or phenome, bringing together and integrating the information to logical interrogation and knowledge discovery by researchers.

## REFERENCES

Arabidopsis Genome Initiative. 2000. Analysis of the genome sequence of the flowering plant *Arabidopsis thaliana*. *Nature* **408**: 796–815.

Bains, W. and Smith, G.C. 1998. A novel method for nucleic acid sequence determination. *J. Theor. Biol.* **135**: 303–307.

Blanchard, A. 1998. In: *Genetic Engineering, Principles and Methods* (ed. Setlow, J.) **20**: Plenum Press, New York, pp. 111–124.

Blattner, F.R., Plunkett, G., Bloch, C.A., Perna, N.T., Burland, V., Riley, M., Collado-Vides, J., Glasner, J.D., Rode, C.K. and Mayrew, G.F. 1997. The complete genome sequence of *Escherichia coli* K-12. *Science* **227**: 1453–1474.

Causse, M., Rocher, J.P., Henry, A.M., Charcosset, A., Prioul, J.L. and de Vienne, D. 1955. Genetic dissection of the relationship between carbon metabolism and early growth in maize, with emphasis on key-enzyme loci. *Mol. Breeding* **1**: 259–272.

Clackson, T. and Wells, J.A. 1994. *In vitro* selection from protein and peptide libraries. *Trends Biotechnol.* **12**: 173–184.

Cronin, M.T. *et al.* 1996. Cystic fibrosis mutation detection by hybridization to light generated DNA probe arrays. *Hum. Mutat.* **7**: 244.

David, D. and Jacqueline, B. 2008. Bioinformatics: fundamentals and Applications in Plant Genetics, Mapping and Breeding. *Principles and Practices of Plant Genomics* **Vol. 1**: 269–293.

Dekoulas, P., Schuler, G.D., Gyapay, G., Beasley, E.M., Soderlund, C. *et al.* 1998. A physical map of 30,000 human genes. *Science* **282**: 744–746.

Delseny, M. 2007. Rice: A model plant for Cereal Genomics. In: *Functionel Plant Genomics.* (Morot-Gaudry, J.F., Lea, P. and Briat, J.F., eds), Science Publishers, Enfield (NH), pp. 397–408.

Dib, C., Faure, S., Fizames, C., Samson, D., Drouot, N., Vignal, A., Millasseau, P., Marc, S., Hazan, J., Seboun, E., Lathrop, M., Gyapay, G., Morissette, J. and Weissenback, J. 1996. A comprehensive genetic map of the human genome based on 5264 microsatellites. *Nature* **380**: 152–154.

Drmanac, S., Kita, D., Labat, I., Hauser, B., Schmidt, C., Burczak, J.D. and Drmanac, R. 1998. Accurate sequencing by hybridization for DNA diagnostics and individual genomics. *Nat. Biotechnol.* **16**: 54–58.

Dunham, I., Shimizu, N., Roe, B.A. and Chissoe, S. 1999. The DNA sequence of human chromosome 22. *Nature* **402**: 489–495.

Eggers, M., Horgan, M., Reich, R.K., Lamture, J., Ehrlich, D., Hollis, M., Kosicki, B., Powdrill, T., Beatie, K., Smith, S., Varma, R., Gangadharan, R., Malik, A., Burke, B. and Wallace, D. 1994. Biotechniques. *Curr. Sci.* **17**: 516–524.

Fiehn, O. 2002. Metabolomics - the link between genotypes and phenotypes. *Plant Mol. Biol.* **48**: 155–171.

Fleischmann, R.D., Adams, M.D., White, O., Clayton, R.A., Kirkness, E.F., Kerlavage, A.R., Bult, C.J., Tomb, J.F., Dougherty, B.A. and Merrick, J.M. 1997. Whole genome random sequencing and assembly of *Haemophilus influenzae* Rd. *Science* **269**: 496–512.

Fodor, S.P.A. 1997. Massively parallel genomics. *Science* **270**: 393–394.

Fraser, P.D., Pinto, M.E., Holloway, D.E. and Bramley, P.M. 2000. Application of high-performance liquid chromatography with photodiode array detection to the metabolic profiling of plant isoprenoids. *Plant J.* **24**: 551–558.

Fridman, E., Pleban, T. and Zamir, D. 2000. A recombination hotspot delimits a wild-species quantitative trait locus for tomato sugar content to 484 bp within an invertase gene. *Proc. Nat. Acad. Sci.* USA **97**: 4718–4723.

Gupta, P.K., Roy, J.K., and Prasad, M. 1999. DNA chips, microarrays and genomics. *Curr. Sci.* **77**: 875–884.

Hacia, J.G., Brody, L.C., Chee, M.S., Fodor, S.P.A. and Collins, F.S. 1996. *Nature Genet.* **14**: 441–447.

Hall, R., Beale, M., Fiehn, O., Hardy, N., Sumner L. and Bino, R. 2002. Plant metabolomics: the missing link in functional genomics strategies. *Plant Cell* **14**: 1437–1440.

Han, B., Xue, Y. 2003. Genome wide intraspecific variations in rice. *Current Opinion Plant Biol.* **6:** 134–138.

Hattori, M. *et al.* 2000. The DNA sequencing of human chromosome 21. *Nature* **405:** 311–319.

Hauge, P.M. *et al.* 1993.An integrated genetic/RFLP map of the *Arabidopsis thaliana* genome. *Plant J.* **3:** 745–754.

Hudson, D.J. *et al.* 1995. A gene map of the human genome. *Science* **270:** 540–546.

International Rice Genome Sequencing Project. 2005. The map- based sequence of the rice genome. *Nature* **436:** 793–800.

Jonson, H. E., Boardhurst, D., Goodacre, R. and Smith, A.R. 2003. Metabolic fingerprinting of salt-stressed tomatoes. *Phytiochemistry* **62:** 919–928.

Krapp, A., Morot- Gaudry, J.F., Boutet, S., Bergot, G., Lelarge, C., Prioul, J.L. and Nctor, G. 2007. Metabolomics, In: *Functional Genomics* (Morot-Gaudry, J.F., Lea, P. and Briat, J.F., eds), Science Publishers, Endfield, NH, USA.

Little, P. 1999. The book of gene. *Nature* **402:** 467–468.

Livache, T., Brigitte, F., Roget, A., Morchand, J., Bidan, G., Teoule, R. and Mathus, G. 1998. Polypyrrole DNA chip on a silicon device: example of Hepatitis C virus genotyping. *Anal Biochem.* **255:** 188–194.

Loudet, O., Chaillou, S., Krapp, A. and Daniel-Vedele, F. 2003. Quantitative trait loci analysis of water and anion contents in interaction with nitrogen availability in *Arabidopsis thaliana. Genetics* **163:** 211–22.

Lysov, Y.P., Florentiev, V.L., Khorlin, A.A., Karpko, K.V., Shik, V.V. and Mirzabekov, A.D. 1998. Determining DNA nucleotide sequence by means of oligonucleotide hybridization: A new method. *Dokl. Acad. Sci. USSR* **303:** 1508–1511.

McLaren, C.G., Bruskiewich R.M., Portugal, A.M. and Cosico, A.B., 2005. The international rice information system. A platform for meta- analysis of rice crop data. *Plant Physiol* **139(2):** 637–642.

Miesfeld, R.L.1999. *Applied Molecular Genetics.* Wiley-Liss, New York.

Pereira, A. 1999. Plant genomics is revolutionizing agricultural research. *Biotechnology and Development Monitor* **40:** 2–7.

Prioul, J. L., Quarrie, S. Causse, M., Devienne, D. 1997. Dissecting complex physiological functions through the use of molecular quantitative genetics. *J. Exp. Bot.* **48:** 1151–1163.

Roessner, U., Wangner, C., Kopka, J., Trethewey, G.N. and Willmitzer, L. 2000. Simultaneous analysis of metabolites in potato tuber by gas chromatography-mass spectrometry. *Plant J.* **23:** 131–142.

Saier, M.H. 1998. Genome sequencing and informatics: New tools for biochemical discoveries. *Plant Physiol.* **117:** 1129–1133.

Salse, J., Piegu, B., Cooke, R. and Delseny, M., 2004. New *in silico* insight into the synteny between rice (*Oryza sativa* L.) and maize (*Zea mays* L.) highlights reshuffling and identifies new duplications in the rice genome. *Plant J.* **38:** 396–409.

Scheible, W.R., Krapp, A. and Stitt, M. 2000. Reciprocal diurnal changes of phosphoenolpyruvate carboxylase expression and cytosolic pyruvate kinase, citrate synthase and NADP-isocitrate dehydrogenase expression regulate organic acid metabolism during nitrate assimilation in tobacco leaves. *Plant Cell Env.* **23:** 1155–1167. .

Schuler, G.D. Boguski, M.S., Stewart, E.A., Stein, L.D. *et al.* 1996. A gene map of the human genome. *Science* **274:** 540–546.

Singh, S.K. 2000. The Human genome: A first look at the blueprint of human inheritance. *Current Science* **79:** 144–148.

Southern, E.M. 1998. International patent application PCT GB 89/00460.

Velculescu, V.E., Zhang, L., Vogelstein, B. and Kinzler, K.W. 1995. Serial analysis of gene expression. *Science* **270:** 484–487.

Weckwerth, W. 2003. Metabolomics in system biology. *Ann. Rev. Plant. Biol.* **54:** 669–689.

Yershov, G., Barsky, V., Belgovsky, A., Kirillov, E., Kreindlin, E., Ivanov, I., Parinov, S., Guschin, D., Drobishev, A., Dubiley, S. and Mirzabekov, A. 1996. DNA analysis and diagnostics on oligonucleotide microchips. *Proc. Natl. Acad. Sci. USA* **93:** 4913–4918.

Yu, J., Wang, J., Lin, W. *et al* 2005. The genomes of *Oryza sativa:* a history of duplications. *Plos Biol.* **3(2):** e 38.

Zhou, T., Wang, Y., Chen, J.Q., Araki, H., Jing, Z., Jiag, KI, Shen, J. and Tian, D. 2004. Genome-wide identification of NBS genes in *japonica* rice reveals significant expansion of divergent non TIR NBS- LLR genes. *Mol. Gen. Genomics* **271:** 402–415.

## FURTHER READING

Brown, T.A. 2001.*Genomes*, Bios Scientific Publishers Ltd, Oxford, UK.

Maheshwari, S.K., Maheshwari, Nirmala and Sopory, S.K. 2001. Genomics, DNA chips and a revolution in plant biology. *Curr. Sci.,* **80:** 252–261.

Gupta, P.K., Roy, J.K., and Prasad, M. 1999. DNA chips, microarrays and genomics. *Curr. Sci.,* **77:** 875–884.

Dubey, H., Grover, A. 2001. Current initiatives in proteomics research: The plant perspective. *Curr. Sci.,* **80:** 262–269.

Kole, C. and Abbott, A.G. (eds) 2008. *Principles and practices of Plant Genomics, Vol. 1: Genome Mapping,* Science Publishers, Enfield, NH, USA.

Krapp, A., Morot-Gaudry, J.F., Boutet, S., Bergot,G., Lelarge, C., Prioul, J.L. and Noctor, G. 2007. Metabolomics, In: *Functional Genomics* (Morot-Gaudry, J.F., Lea, P. and Briat, J.F. eds), Science Publishers, Enfield, NH, USA.

# 29

# Bioinformatics

In the last few decades, advances in molecular biology and the equipment available for research in this field have increased rapid sequencing of large portions of the genomes of several species. In fact, to date, several bacterial genomes (more than 568), four animal genomes such as *Drosophila,* human being and atleast three plant genomes (*Arabidopsis thaliana* from dicotyledons, Rice from monocotyledons and Poplar from trees) have been sequenced. Several other genomes from microbial, animal and plant kingdom are actively being sequenced. The Human Genome Project, designed to sequence all 24 of the human chromosomes has been sequenced. This has resulted in the growth of the International Nucleotide Sequence Database Collection (National Centre for Biotechnology Information NCBI, European Bioinformatics Institute EBI, and DNA Databank of Japan DDBJ). The volume of data in the sequence databases (including nucleotides, proteins, etc.) has long exceeded the threshold limit of manual analysis. Popular sequence databases, such as GenBank and EMBL (European Molecular Biology Lab), have been growing at exponential rates. This deluge of information has necessitated the careful storage, organization and indexing of sequence information.

Bioinformatics is a new discipline emerging from combination of Biology, Mathematics and information technology.

Bioinformatics is a science associated to database like activities involving persistent sets of data that are maintained in a consistent state over essentially indefinite periods of time. Thus, bioinformatics is the marriage of biology and information technology. The discipline encompasses many computational tools and methods used to manage, analyze and manipulate large sets of biological data especially gene and protein sequences. The definition of bioinformatics and the term itself is not universally agreed upon. Some definitions available in the literature are as follows:

1. Bioinformatics—The science of developing computer databases and algorithms for the purpose of speeding up and enhancing biological research.
2. Bioinformatics—a combination of Computer science, Information Technology and Genetics to determine and analyze genetic information.
3. Bioinformatics—is an integration of mathematical, statistical and computer methods to analyze biological, biochemical and biophysical data.

4. Bioinformatics—can be defined as the storage, manipulation and analysis of biological information via computer science.

5. Bioinformatics—is the science concerned with the development and application of computer hardware and software to the acquisition, storage, analysis and visualization of biological information.

The massive amount of sequence data generated by different sequencing projects requires electronic databases to store, organize, and index it. In addition, the databases need specialized tools for scientists to see and analyze the available information, as well as to submit new or revised sequence data. Though biological databases form a major component of bioinformatics, but along with it there are other important areas of activity, in which it is actively involved, such as DNA and amino acid sequence comparison, elucidation of structure of various genes and proteins (at 1, 2 and 3-D levels), finding their biological functions (so called annotation of genomes), finding evolutionary relationships existing among various organisms based on molecular variation and design of suitable drugs and vaccines to diseases using various *in silico* algorithms and computer techniques.

Variety of available bioinformatics tasks:

Creation of Biological databases
Sequence similarity search
Discovery of coded genes and their structure
DNA and protein function and annotation
Elucidation of protein structure
Drug design
Biochemical (metabolic pathway) analysis
Tools for analysis of High-through put molecular biology experimentation data (Micro arrays, Mass spectrometry, etc)
Comparative genome analysis
Building gene and protein networks
Scientific literature search

The simplest tasks used in bioinformatics concern the creation and maintenance of databases of biological information. Nucleic acid sequences (and the protein sequences derived from them) comprise the majority of such databases. While the storage and organization of millions of nucleotides is far from trivial, designing a database and developing an interface whereby researchers can both retrieve existing information, enabling them to make meaningful discoveries in their day to day research and submit new entries is only the beginning. New bioinformatics techniques based on mathematical and statistical algorithms are being continually developed to help interpret genomic information. For example, they are used to locate a gene within a nucleotide sequence, to predict protein structure or function, and to find relationships or similarities between sequences. Eric Lander, whose initial training was in mathematics, has been a driving force for the development of many of the information technology tools. For example coding genes can be discovered from sequenced raw data using tools such as GRAIL or Genscan or structure of protein sequences coded by these genes can be elucidated and their functions may be validated linking them to known protein domains and folds.

Biotechnology is instrumental in developing databases on genomics and proteomics pertaining to such developments in agriculture as well as other forms of life in general. Scientists are now working to segregate specific databases of different life forms from the available information in different databases. GenBank and SwissProt are such large public domain databases that are repositories of DNA and protein sequences respectively. GenBank is supported by National Centre for Biotechnology Information, United States Department of Health (www.ncbi.nil.org). SwissProt–a repository of protein sequences is supported and maintained by Swiss Institute of Biotechnology. Both these are on-line public databases along with many others that enable free access and down load of available gene and protein sequences. These databases are very essential for current and future biotechnology research.

The rapid advances in computer technology are responsible for the revolution in bioinformatics. This includes the collection of biological data that had once been dispersed in file drawers; now wend their way into vast databases. By virtue of their comprehensive nature and instant cross accessibility, make the information into a commodity more valuable than the sum of its parts. This has brought sweeping changes in the character of bioinformatics. Bioinformatics differs from computational biology and bioinformation structure. The most pressing tasks in bioinformatics involve the analysis of sequence information. **Computational biology** refers to algorithmic tools to facilitate biological analysis, whereas bioinformation structure comprises the entire collection of information management system, analysis tools and communication networks supporting biology. **Computational biology** involves the following:

1. Finding the genes in the DNA sequences of various organisms.
2. Developing methods to predict the structure and/or function of newly discovered proteins and structural RNA sequences.
3. Clustering protein sequences into families of related sequences and the development of protein models.
4. Aligning similar proteins and generating phylogenetic trees to examine evolutionary relationships.

### Development of Bioinformatics

The development of bioinformatics started with the networking of computers and accumulation of data on genes and proteins. The first protein sequence was reported for bovine insulin in 1956 consisting of 51 amino acid residues. The first bioinformatics database was constructed a few years after the first protein sequence became available. Nearly a decade later the first nucleic acid sequence for yeast alanine tRNA with 77 bases was reported. Dayhoff from 1965–1978 gathered all the available sequence data to create the first bioinformatics database as 'Atlas of Protein sequence and Structure'. The

protein Data Bank was set up in 1972 with a collection of ten X-ray crystallographic protein structures, and the SWISSPROT protein sequence database began in 1987. The European Molecular Biology Laboratory (EMBL) established their data library in 1980 to collect, organize and distribute nucleotide sequence data and other related information which is now performed by European Bioinformatics Institute (EBI). In early 1980s National Centre for Biotechnology Information (NCBI) was established in USA, which serves as primary information databank and provider of information. DNA Databank of Japan (DDBJ) established sometime later. The discipline developed further during mid-1980s when use of computers became very popular for data storage, and access by users. It was this period when various tools to store, manage, and access the information through computers were developed. The researchers started relying on computers for experimental data recording, analyzing, storing, browsing technical literature and modeling of proteins and genes. During the 1980s, for example, a software package "PC/GENE" was introduced by Intelli-Genetics, a US company (Moore *et al.*, 1988). This package could translate a given sequence of gene to its protein end product with prediction of its secondary structure. This paved the way for the development of a number of new software packages for biotechnologists. The examples of early packages are the Staden Package for DNA sequences and PROPHET for data management and analysis in life sciences.

After the formation of databases, tools became available to search sequence databases at first in a very simple way, looking for keyword matches and short sequence words, and then more sophisticated pattern matching and alignment based methods. BLAST (Basic Local Alignment Search Tool, http://130.14.29.110/BLAST/) is the most popular tool available for conducting sequence similarity search of biological sequences. Similar algorithm has been the mainstay for rapid search of sequence database since its introduction which is complemented by the more rigorous and slower FASTA and Smith-Waterman algorithms.

**Algorithm** is a set of rules (steps) for calculating or problem solving carried out by a computer program.

Suites of analysis algorithms, written by leading academic researchers became more widely available for basic sequence analysis. These algorithms were typically single function black boxes that took input and produced output in the form of formatted files. Different sequence analysis tools are designed to work either in Windows or UNIX environments based on user friendly Graphical User Interface (GUI) or command prompt style respectively in displaying the search output of all commands which were used to operate the algorithms. With some suites having hundreds of possible commands, each taking different command options and input formats.

Internet gave the colossal power to the computers and made a major impact on the information technology and bioinformatics. During the 1990s, with the introduction of WWW (world wide web) the developmental work in bioinformatics got a boost with vast connectivity and accessibility to databases in a user friendly way. The success of internet came about with the advent of various networking tools and the servers in the early and mid 1990s.

Bairoch (1991) established a moderate database pertaining to protein sequence and structural correlations on the "Net". This database was known as PROSITE, which was further strengthened with a database on sequence analysis and comparison of protein sequences known as SEQANALREE. The internet, which is an information superhighway has practically compressed the world into a cyber colony through local area networks (LAN), wide area networks (WAN), metropolitan area networks (MAN) and other intranets. The development of internet and the emergence of WWW as a common vehicle for communication

and access to databases are an exciting aspect in bioinformatics.

Tools for bioinformatics and computational biology include servers and computation boxes, cross-platform software, web-based interfaces tools and databases both proprietary and public. At present many bioinformatics software and data access tools are available. Strategies for searching the Net can be categorized broadly as useful for cataloguing the WWW directories and search engines. The search engines are non-specific and produce voluminous results while directories are like 'yellow pages' which are browsed by the search engines. A popular scientific information search engine to locate published journal articles is PubMed, available with NCBI central database (http://www.ncbi.nlm.nih.gov/entrez/query.fcgi/DB=pubmed). PubMed is supported by US National Library of Medicine, and is a repository of millions of published scientific publications for free down loads either as abstract or full length article form coupled with several search options. The early 1990s also experienced the development of proteomics related databases on two dimensional polyacrylamide gel electrophoresis, known as SWISS-2DPAGE. Then a fully organized and detailed protein sequence database now famous as SWISS-PROT was introduced. This database helps in providing a high level of annotation, a minimal level of redundancy, and a high degree of integration with other databases. Recent developments in this database include cross references to additional databases. A computer annotated supplement to SWISS-PROT has improved it significantly. This was possible with a variety of new documentation files and improvements to TrEMBL (Translated EMBL). The TrEMBL consists of entries in a SWISS-PROT like format derived from the translation of all coding sequences in the EMBL nucleotide sequence database. Some important databases which are commonly used by biotechnologists are those from Incyte, Pangea systems,

PE-informatics, GCG, NCBI, PDB, SRS, UDB, SWISS-PROT, EMBLnet, ICBnet, Medline, Flybase, Mendelian inheritance in man (MIM), US Patent and Trademark Office (USPTO) and SeqWeb. USDA maintains a crop genome database server at Cornell University at http://ars-genome.cornell.edu/. A number of universal resource locators for specific databases can be easily found on the web. There are various programs and tools available for making searches, algorithms analysis, modeling and the computer graphics of the databases on genomics and proteomics.

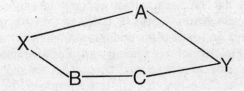

**Fig. 29.1:** A Computer network.

### Role of Internet in Bioinformatics

The **Internet** is an international network of computers derived from an earlier system, **ARPAnet,** developed by the US military in 1969. Biological Information is generally stored in computers all over the world and if one has to access that information the computers must be joined together in a network. A typical computer network is a group of computers that can communicate over a telephone system to allow the exchange of data. For transfer, the data is broken into small packets, which are sent independently and reassembled at the point of destination. A typical network connection is shown in Fig 29.1 where information is sent from computer X to computer Y via two different routes of A or B and C. Different routes are being used if one route fails then the network is still available through another route.

Information transfer over the Internet is governed by a set of protocols called **TCP/IP.** TCP is the **Transmission Control Protocol,** which determines how data is broken into packages and reassembled. IP is the **Internet Protocol,** which determines how the packets of information are addressed and routed over the network. To access the Internet, a computer must have the correct hardware (generally a network card and / or a modem), the appropriate software and permission for network access.

The **World Wide Web (WWW)** is a way of exchanging information over the Internet using a program called a *browser*. A number of browsers are available for working on the WWW, the most widely used of which are *Internet Explorer* and *Netscape Navigator*. Most of the computers are now Internet ready with the appropriate hardware and one or both of these browser programs installed as standard. The WWW was developed in 1992 which allows the display of information pages containing *multimedia objects* (e.g. text, images, audio and video) in a special format called *hypertext*. In a hypertext document, text is displayed normally and can be read and manipulated like any other text document, but some words and objects are highlighted in a different color and these are known as *hypertext links* (or simply **hyperlinks**). Clicking on a hyperlink directs the browser to access another hypertext document, which might be on the same computer or might be on any other computer linked to the Internet. The new document may have its own hyperlinks and thus the process can be repeated allowing the user to move rapidly from computer to computer around the world downloading information as he or she goes (this is commonly known as *surfing the web or surfing the net*).

The WWW works on the basis that each hypertext document has a unique address known as a *uniform resource locator* (*URL*). Every computer on the Internet has an IP address, which is in the form of four integers conventionally separated by dots. Associated with this is a text version of the address, for example http://www.bios.co.uk, which is the publisher's address. The equivalent IP address for the publisher is 195.172.6.15. If a local user tries to contact http://www.bios.co.uk, how does the browser find the correct site? The loca' computer first contacts Internet computers called **Domain Name Servers (DNSs)** that try to understand parts of the address starting with the most significant (right hand) part. For example, most text addresses have a country abbreviation, in this case 'uk' for United Kingdom, but American addresses do not, since the Internet was an American invention. If the computer one is trying to access is providing a service on the WWW, it is known as a web server. This means there are numerous files available for browsing, and each can be identified by a unique URL. Such files are specified by extra characters separated from the main Internet address by a solidus (/). For example, the URL http://www.bios.co.uk/bioiformatics refers to a subdirectory on the publisher's web server that corresponds to the web site accompanying the book. Once the DNS has found the Internet name for the server, it is for the server itself to work out what to do about any extensions to the URL, such as '/bioinformatics'.

## Online

Online is an approach to keep one up-to-date in knowledge at any place and at any time. The strongest reason for adopting this approach is becoming a lifelong necessity and challenge for professionals who want to maintain their competitive advantage in the workforce. With ever increasing pace of knowledge generation, the skills and the information a student acquires through formal education soon becomes obsolete. With the internet and associated information and communication technologies one can get up-to-date information. The online course consists of five components:

1. Web resources: It is an informational material on the course site itself or on external web pages linked to the course.
2. Tests and quizzes: It allows the students to assess their own understanding of the materials or have their understanding evaluated by others.
3. Learning activities offline: It is an individual study by a student from textbooks, labs, CD-ROMs, etc.
4. Home work: It is like assignments, collaborative projects, etc. which a student has to take.
5. Discussion forum: It begins with all the above components by discussion, email conferencing, etc. to interact with one another.

## Browsing working and downloading

Browsing the Internet is simply a case of clicking on the desired hyperlinks and allowing the associated pages to download. Some pages download faster than others, which may be due to content (pages with many images and other large files will take longer to download than pages that contain text alone) or due to the speed of connection (there are bottlenecks in many parts of the Internet, and it is advisable to find a local web server to minimize the number of routers the information has to pass through).

To access a particular web site, it may first be necessary to type in the URL the address bar of the browser. Once a page has been accessed, however, it should not be necessary to type in the URL again.

A number of public search engines are available allowing the user to search for sites of interest using particular keywords, but it may be better to start with some dedicated bioinformatics sites. Some useful bioinformatics web sites have been listed separately in this chapter itself.

## Sequences and Nomenclature

The nomenclature system adopted in bioinformatics is based on recommendations of International Union of Pure and Applied Chemists (IUPAC). The nucleotide and amino acids sequences have been reduced to single letter codes, which has facilitated the understanding and utilization of data generated by diverse research groups. The four bases in

**Table 29.1:** Single letter symbols used to denote base in DNA sequences

| Symbol | Meaning | Logic for symbol | Symbol for the complementary sequence |
|---|---|---|---|
| A | Adenine | Adenine | T |
| C | Cytosine | Cytosine | G |
| G | Guanine | Guanine | C |
| T | Thymine | Thymine | A |
| R | G or A | Purine | Y |
| Y | T or C | Pyrimidine | R |
| M | A or C | Amino (bases having) | K |
| K | G or T | Keto (bases having) | M |
| S | G or C | Strong (nature of base pairing) | S* |
| W | A or T | Weak (nature of base pairing) | W* |
| H | A or C or T | Not G** | D |
| B | C or G or T | Not A** | V |
| V | A or C or G | Not U** | B |
| D | A or G or T | Not C** | H |
| N | A or C or G or T | Nucleotide | N |

\* The same symbol is used for the sequence of complementary strand since G pairs with C (S denotes both) and likewise A pairs with T (W denotes both)

\*\* Not G, the letter H occurs immediately after letter G in The English alphabet; Not A comes after A; Not U–the letter V follows; Not C–letter D follows C

the DNA sequences are denoted by single letters. For ambiguities in the sequence, whether G or A purine, C or T thymine or others, symbols have been developed (Table 29.1). The base sequences of two complementary strands of a DNA molecule are represented using the same symbols. Even those positions that exhibit ambiguity can be represented by this system of symbols. But some symbols are complementary to themselves, e.g. S and W. S represents G or C and C pairs with G and therefore both strands of a DNA duplex will have S at the given position. In databases, sequence of only one strand is listed which runs from 5′ to 3′ direction.

The amino acids have been traditionally represented by three letter symbols. But in bioinformatics they are represented by single letter as in DNA sequences. For ambiguities as in the case of glutamine or glutamic acid, the position is given the symbol Glx and single letter code as Z. The symbol X is used to denote that the position may have any amino acid (Table 29.2). The protein synthesis begins at the N-terminus and ends at C-terminus. Thus amino

**Table 29.2:** Symbols used to represent amino acids in protein sequences

| Single letter code | Amino acid | Three letter code |
|---|---|---|
| A | Alanine | Ala |
| B | Asparagine or Aspartic acid | Asx |
| C | Cystine | Cys |
| D | Aspartic acid | Asp |
| E | Glutamic acid | Glu |
| F | Phenylalanine | Phe |
| G | Glycine | Gly |
| H | Histidine | His |
| I | Isoleucine | Ile |
| K | Lysine | Lys |
| L | Leucine | Leu |
| M | Methionine | Met |
| N | Asparagine | Asn |
| P | Proline | Pro |
| Q | Glutamine | Gln |
| R | Arginine | Arg |
| S | Serine | Ser |
| T | Threonine | Thr |
| V | Valine | Val |
| W | Tryptophan | Trp |
| X | Any amino acid | Xaa |
| Y | Tyrosine | Tyr |
| Z | Glutamine or Glutamic acid | Glx |

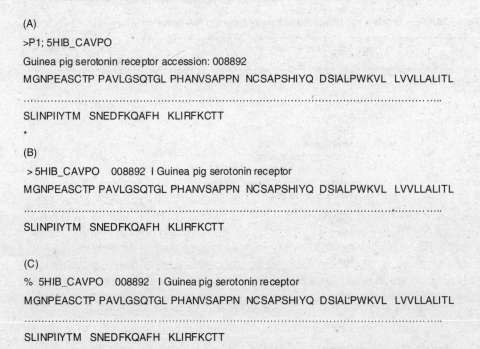

(A)

>P1; 5HIB_CAVPO

Guinea pig serotonin receptor accession: 008892

MGNPEASCTP PAVLGSQTGL PHANVSAPPN NCSAPSHIYQ DSIALPWKVL LVVLLALITL

...........................................................................................................................................

SLINPIIYTM SNEDFKQAFH KLIRFKCTT

*

(B)

 > 5HIB_CAVPO   008892  I Guinea pig serotonin receptor

MGNPEASCTP PAVLGSQTGL PHANVSAPPN NCSAPSHIYQ DSIALPWKVL LVVLLALITL

...........................................................................................................................................

SLINPIIYTM SNEDFKQAFH KLIRFKCTT

(C)

% 5HIB_CAVPO   008892  I Guinea pig serotonin receptor

MGNPEASCTP PAVLGSQTGL PHANVSAPPN NCSAPSHIYQ DSIALPWKVL LVVLLALITL

...........................................................................................................................................

SLINPIIYTM SNEDFKQAFH KLIRFKCTT

**Fig. 29.2:** Protein sequence of a guinea pig serotonin represented in different sequence formats: **A.** NBRF/PIR format; **B**. FASTA format; **C**. GDE format.

acid sequences in databases are listed from the N-terminus at the extreme left of the sequence to the C-terminus at the extreme right of the polypeptide.

## File Formats

If biological data is to be used by computer programs, it must be presented in a standard format that can be read by computer. It is very common to put data in text files. Popular sequence repositories such as GenBank produce output in flat text format. Text files hold (almost) only the text and little auxiliary information about formatting. As different on-line tools are designed using various computer languages and work on suitable platforms they accept user input data in certain formats. Some of the data formats include—Flat files, FASTA, ASN, etc.

Many bioinformatics databases and software applications are designed to work with sequence data, and this requires a standard format for putting in nucleic acid and protein sequence information. Three of the most common sequence formats are NBRF/PIR (National Biomedical Research Foundation/ Protein Information Resource), FASTA and GDE. Each of these formats has facilities not only for representing the sequence itself, but also for inserting a unique code to identify the sequence and for making comments which may include for example the name of the sequence, the species from which it was derived, and an accession number for GenBank or another appropriate database. Fig. 29.2 shows the same protein sequence of a guinea pig serotonin receptor represented in the three sequence formats. In NBRF/PIR format first line begins with '>P1;' which specifies a protein sequence, for nucleic acid sequence >N1; is used. The semicolon is followed by a code. In this case 5HIB_CAVPO is a sequence identifier. Serotonin is also known as 5-hydroxy tryptamine, thus 5H1B identifies the protein as serotonin receptor

1B, while CAVPO identifies its source as guinea pig (*Cavia porcellus*). Then it follows a comment line followed by the sequence itself and is terminated by an asterisk (*). It is conventional to give files in this format the extension '.pir' or '.seq'.

Fig. 29.2. B shows the FASTA format. The first line begins with '>' but there is no designation of protein or nucleic acid sequence. The code is entered next and this is followed (on the same line) by comments, although it is conventional to delimit the comments with 'I' symbol. FASTA files also allow lower-case letters for the amino acids. Files in this format commonly have the extension '.fasta'. Fig. 29.2 C shows the GDE format. This is essentially the same as the FASTA format, but the '>' symbol in the first line is replaced by '%'. Files in this format have the extension 'gde'.

All the three file formats ignore spaces and carriage returns. This allows sequences to be typed out in a manner that is convenient for the user. In Fig. 29.2 for examples a space has been inserted every 10 amino acid residues and a carriage return after every 60, making it much easier to manually count the residues and identify amino acids at specific positions in the sequence.

## Files of Structural data

The raw materials for bioinformatic studies on macromolecular structures are PDB files. These are text files using a format devised by the Protein Data Bank. Such files contain orthogonal atomic co-ordinates together with annotations, comments and experimental details.

## Submission of Sequences

It is essential that the data are submitted in supported format and that the submission is carried out by means of software provided by the database curators. Examples are: 'WebIn' provided by EMBL (www.ebi.ac.uk/embl/Submission) and 'BankIt' provided by the NCBI (http:// www.ncbi/nlm.nih.gov/BankIt/), each of which can be used to submit sequences to the databases over the WWW. A powerful stand-alone software tool, Sequin, is provided by the NCBI and can be used on UNIX, PC/Windows and Macintosh systems for sequence submission for those with no WWW access http://www.ncbi.nlm.nih.gov/Sequin/index.html).

## Databases

A biological database is a large, organized body of persistent data, usually associated with computerized software designed to update, query, and retrieve components of the data stored within the system. A simple database might be containing many records, each of which includes the same set of information. For example, a record associated with a nucleotide sequence database typically contains information such as contact name the input sequence with a description of the type of molecule, the scientific name of the source organism from which it was isolated and often literature citations associated with the sequence.

Most biological databases consist of long strings of nucleotides (guanine, adenine, thiamine, cytosine and uracil) and /or amino acids (threonine, serine, glycine, etc.). Each sequence of nucleotides or amino acids represents a particular gene or protein, respectively. Sequences are represented in shorthand, using single letter designations. This decreases the space necessary to store information and increase processing speed for analysis.

### Primary sequence databases

The primary sequence databases are repositories for raw sequence data, and can be accessed freely over the World Wide Web (WWW). There are three such databases, comprising the International Nucleotide Sequence Database Collaboration. These are GenBank, maintained by the National Center of Biotechnology Information (NCBI), the Nucleotide Sequence Database maintained by the European Molecular Biology Laboratory (EMBL), and the DNA Databank of Japan

(DDBJ). New sequence can be deposited in any of the databases since they exchange data on a daily basis. The databases contain not only sequences but also extensive annotations.

*Subsidiary (Secondary) sequence databases*

The main sequence databases have a number of subsidiaries for the storage of particular types of sequence data. For example, the other divisions of GenBank include dbEST for expressed sequence tags (ESTs), dbGSS for single-pass genomic survey sequences (GSSs), dbSTS for sequence tagged sites (STSs; unique genomic sequences that can be used as physical markers) and the HTG (high-throughput genomic) division, which is used to store unfinished genomic sequence data. Secondary databases house information relating to various gene and protein sequences and also provide easy to analyze tools. For example ProtParam tool enables easy analysis of protein sequence to find its various physico-chemical properties such as amino acid bias, molecular weight and half life time, etc.

*Composite databases*

These house all annotated information about map location, structural and cellular functions of various genes and proteins representing the organism. TAIR- (The Arabidopsis information resource), EnsEMBL–a complete information resource about human genome are some of the examples of composite databases.

Most of the biological databases contain nucleotide and protein sequence information, but there are also databases, which include taxonomic information such as the structural and biochemical characteristics of organisms. The power and ease of using sequence information has however, made it the method of choice in modern analysis.

Computer technology has provided an obvious solution to this problem. Not only can computers be used to store and organize sequence information into databases, but they can also be used to analyse sequence data

rapidly. The evolution of computing power and storage capacity has so far been able to outpace the increase in sequence information being created. Theoretical scientists have devised new and sophisticated algorithms, which allow sequences to be readily compared using probability theories. These comparisons become the basis for determining the gene function, developing phylogenetic relationships and simulating protein models.

The major tasks considered in the design of a database are data storage and enabling easy retrieval of information. For researchers to benefit from the data stored in a database, two requirements must be met:

i. Easy access to the information
ii. A method for extracting only that information needed to answer a specific biological question.

Successful public data services suffer from continually escalating demands from their user communities. It is probably important to realize that these databases will never completely satisfy a very large percentage of the user community. The range of interests in biology varies and virtually there is no end to the depth and breadth of desirable information. In all cases, however, the principal requirement of the biomedical and biotechnology research communities on the public data service includes:

1. Timeliness: The basic data should be available on an internet accessible server within days (or hours) of publication or submission.
2. Annotation: Deep, consistent annotation comprising supporting and ancillary information should be attached to each basic data object in the database.
3. Supporting data: In some instances database users will need to examine the primary experimental data, either in the database itself, or by following cross-references back to network accessible laboratory databases.
4. Data quality: Database users and advisors have repeatedly asserted that data quality

is of the highest priority. However, because the data services, in most cases lack access to supporting data, the quality of the data must remain the primary responsibility of the submitter.

5. Integration: Each data object in the database should be cross-referenced to representations of the same or related biological entities in other databases.

A search of the recent literature has discovered more than 200 databases of molecular, structural, genetic or phenotypic data. The major well-known biological databases are listed in Table 29.3. Some specialized databases, which are perhaps of more interest to biotechnologists and molecular biologists, are listed below.

## Molecular Biology Databases

Most biologists and biotechnologists are familiar with the more well known nucleotide sequence databases such as GenBank or EMBL (European Molecular Biology Lab) as well as the protein related databases of SWISS-PROT, and PDB. However, there are numerous specialized biological databases that have been created out of particular need, either to answer a particular biological question or to better serve a particular segment of the biological community. Some of the molecular biology databases have been described so that one can use these resources in the design and analysis of their experiments. Comprehensive listings of molecular biology databases are available through the following web catalogues:

1. Infobiogen (Table 29.3) provides an online catalogue of molecular biology databases. The catalogue has listing of more than 500 databases, further categorized based on content as—DNA related (87), RNA related (29), protein related (94), genomic (58), mapping (29), protein structure (18), literature (43) miscellaneous (153) as of October 30, 2001.

2. The Molecular Biology Database Collection (Table 29.3) is an online catalogue of key databases of value to the biological community. This collection contains today a list of more than 1000 high-quality online databases. The databases included in this collection are considered particularly relevant for biotechnology research as they provide new value to the underlying data by virtue of curation, new data connections or other innovative approaches.

3. The European Bioinformatics Institute has been maintaining Biocatalogue since 1993 (Table 29.3). The Biocatalogue is a software directory of general interest in molecular biology and genetics. It is maintained by Upasala University. It is also categorized into key areas similar to the Infobiogen catalogue.

## Bioinformatics databases and analysis services

Some of the databases and analysis services are mentioned below.

**GenBank**—Main nucleotide sequence database of NCBI, USA. This database contains the nucleotide sequences of genomic DNA.

**EMBL**—A comprehensive nucleotide (DNA and RNA) sequence database held at EMBL. This is compiled from various sources and there is collaboration between GenBank of NCBI and DDBJ at GenomeNet of Japan.

**dbEST**—This contains EST entries from GenBank, EMBL and DDBJ which is held by NCBI, USA.

**Mito**—This contains mitochondrial genome sequences held by NCBI, USA.

**NCBI's Molecular Modeling Database (MMDB):** An integral part of Entrez information retrieval system, is a compilation of all the PDB three dimensional structures of biomolecules. The difference between the two databases is that the MMDB records, reorganize and validate the information stored in the database in a way that enables cross-referencing between the chemistry and the three dimensional structure of macromolecules. By integrating chemical sequence, and structure information, MMDB is designed to

**Table 29.3:** Major databases on different aspects [Source: Adak and Srivastava (2002)]

| Database Name | Link | Contents |
|---|---|---|
| **Molecular Biology Database Catalogs** | | |
| Infobiogen Catalogue of Molecular Biology Databases | http://www.infoobigen.ft/service/dbcat | |
| The Molecular Biology Database Collection | http://www.oxfordjournals.org/cgi/content/full/29/1/1/DCl | |
| The European Bioinformatics Institute Biocatalogue | http://www.ebi.ac.uk/biocat/ | |
| **Major Nucleotide Sequence Databases** | | |
| NCBI's GenBank | http://www.ncbi.nlm.nih.gov/Genbank | All known nucleotide and protein sequences: International Nucleotide Sequence Data Collaboration |
| EMBL. Nucleotide sequence database | http://www.ebi.ac.uk/embl/ | All known nucleotide and protein sequences: International Nucleotide Sequence Data Collaboration |
| DNA Data Bank of Japan (DDBJ) | http://www.ddbj.nig.ac.ip | All known nucleotide and protein sequences: International Nucleotide Sequence Data Collaboration |
| **Major Protein Databases** | | |
| Protein Information Resource (PIR) | http://pir.georgetown.edu | Comprehensive, annotated, non-redundant protein sequence database |
| Swiss-Prot/ TrEMBL | http://www.expasy.ch/sprot | Curated protein sequences |
| InterPro | http://www.ebi.ac.uk/interpro | Integrated resource for protein families, domains, and sites |
| MetaFam | http://metafam.ahc.umn.edu/ | Integrated protein family information |
| Membrane Protein Database | http://biophys.bio.tuat.ac.jp/ohshima/database/ | Membrane sequences, trans membrane regions and structures |
| TRANSFAC | http://transfac.gbf.de/TRANSFAC/ | Transcription factors and Binding sites |
| **Major Protein Structure Databases** | | |
| Protein Data Bank (PDB) | http://www.rcsb.org/pdb/ | Structure data determined by X-ray crystallography and NMR |
| NCBIs MMDB | http://www.ncbi.nlm.nih.gov/Structure | All experimentally-determined 3D protein structures linked to NCBI's Entrez |
| SCOP | http://scop.mrc-lmp.cam.ac.uk/scop/ | Familial and Structural Protein Relationship |
| **Major Mutation Databases** | | |
| NCBI's dbSNP | http://www.ncbi.nlm.nih.gov./SNP/ | Database of single nucleotide polymorphisms |
| ALFRED | http://alfred.med.yale.edu/alfred/index.asp | Allele frequencies and DNA Polymorphisms |
| OMIM | http://www.ncbi.nlm.nih.gov.OMIM/ | Catalog of human genetic and genomic disorders |
| **Major Plant and Microbial Genome Databases** | | |
| Genoplante | www.genoplante.org | Genomics for plant improvement |
| UK CropNet | http://ukcrop.net/ | Comprehensive gateway to crop genomes |
| NCBI's Microbial Genome Gateway | http://www.ncbi.nlm.nih.gov/PMGifs/Genomes/micr.html | |
| DOE's Microbial Genomics Gateway | http://microbialgenome.org/ | |
| TIGR's Comprehensive Microbial Resources | http://www.tigr.org/tigr-scripts/CMR2/CMRHomepage.spl | Completed Microbial Genomes |

| Database Name | Link | Contents |
|---|---|---|
| **Major Organism Specific Genome Databases** | | |
| Genome Online Databases (GOLD) | http://wit.integratedgenomics.com/ | Information regarding complete and ongoing genome projects |
| Fly base | http://www.fruitfly.org | Drosophila sequences and genomic information |
| Mouse Genome Database (MGD) | http://www.informatics.jax.org/ | Mouse genetics and genomics |
| *Arabidopsis thaliana* genome database | http://www.tigr.org.tdb/e2K1/athl | *Arabidopsis thaliana* genome database |
| *Saccharomyces* Genome Databases (SGD) | http://genome-www.stanford.edu/Saccharomyces/ | *S. cerevisiae* genome information |
| Rice Genome Project | http://rgp.dna.affrc.go.jp/ | Reporting current data in the rice genome project |
| ZmDB | http://zmdb.iastate.edu | Reporting current data in the maize genome project |
| **Major Gene Expression Databases** | | |
| Gene Expression Omnibus | http://www/nchi./nlm.nih.gov./GEO | NCBI's Repository for gene expression (under development) |
| Stanford Micro array Database (SMD) | http://gnome-www4.Stanford.edu | Gateway to microarray data Stanford labs |
| **Integrated bioinformation resources** | | |
| Integrated Genome Database | http://genome.dkfz-heidelberg.de/igd/> | |
| Entrez Browser | http://www.ncbi.nim.nih.gov> | |
| ExPASy | http://expasy.hcuge.ch> | |
| SRS (Sequence Retrieval System) | http://www.ebi.ac.uk/srs/srsc> | |
| **Online directories of WWW-accessible bioinformation and analysis service** | | |
| Bio-wURLD (searchable collection of biology URLs) | http://www.ebi.ac.uk/htbi/bwurld.pl> | |
| BRASS | http://mbisg2.sbc.man.ac.uk/gradschool/bioinf/brass95.html> | |
| Harvard Biological Laboratories | http://golgi.harvard.edu/> | |
| Biology Servers (Lawrence Berkeley Laboratory) | http://genome.lbl.gov/BioSrver.html> | |
| The WWW Virtual Library: Biomolecules | http://golgi.harvard.edu/sequences.html> | |
| Web Resources for Protein scientists | http://www.prosci.uci.edu/ProSciDos/> | |
| Motif Bioinformatics Server (Stanford University) | http://stanford.edu:80/> | |

serve a resource for structure based homology modeling and protein structure prediction. MMDB record have value added information compared to the original PDF entries, including explicit chemical graph information, uniformly derived secondary structure definitions, structure domain information, literature citation matching, and molecule based assignment of taxonomy to each biologically derived protein or nucleic acid chain.

**The conserved Domain Database**—The conserved Domain Database (CDD) is a collection of sequence alignments and profiles representing protein domains conserved in molecular evaluation. It includes domains from Smart and Pfam two popular Web- based tools for studying sequence domains, as well as domains contributed by NCBI researchers. CD Search another NCBI search service, can be used to identify conserved domains in a protein query sequence. CD search uses RPS-BLAST to compare a query sequence against specific matrices that have been prepared from conserved domain alignments present in CDD. Alignments are also mapped to known three-dimensional structures, and can be displayed using Cn3D.

**PDB: The Protein Data Bank**—The PDB was the first "e;bioinformatics'e; database ever built and is designed to store complex three dimensional data. The PDB was originally developed and housed at the Brookhaven National Laboratories but is now managed and maintained by the Research Collaboratory for Structural Bioinformatics (RCSB). The PDB is a collection of all publicly available three-dimensional structures of proteins nucleic acids, carbohydrates, and a variety of other complexes experimentally determined by X-ray crystallography and NMR.

**SWISS-PROT and TrEMBL**—SWISS-PROT and the related database TrEMBL (translated EMBL) are repositories for annotated protein sequences. Characteristic features of SWISS-PROT entries include the DR (reference), KW (key words) and FT (features) fields. It is the presence of these careful and extensive annotations that makes SWISS-PROT so popular with biochemists.

SWISS-PROT provides the most up-to date and extensively annotated information on protein sequence and its quality reflects its active management by human curators. TrEMBL is another database in the same format. The entries in TrEMBL are derived from translation of all coding sequences in the EMBL Nucleotide Sequence Database that are not already in SWISS-PROT. As further data ensures the reliability of annotations, TrEMBL entries are moved to SWISS-PROT. Detailed queries of the text annotation in the databases can be carried and using SRS and Entrez tools.

**BLAST:** The introduction of BLAST, or The Basic Local Alignment Search Tool, in 1990 made it easier to rapidly scan huge databases for homologies, or sequence similarity, and to statistically evaluate the resulting matches. BLAST works by comparing a user's unknown sequence against the database of all known sequences to determine likely matches. Sequence similarities found by BLAST have been critical in several gene discoveries. Hundreds of major sequencing centers and research institution around the country use this software to transmit a query sequence from their local computer to BLAST server at the NCBI via

the Internet. In a matter of seconds the BLAST server compares the user's sequence with up to million known sequences and determines the closest matches.

BLAST enables not only locating the orthologous or homologous sequences derived from a common ancestor but also provides a statistical scores (such as E values and similarity scores) to validate the search results. The sequence similarity search tool BLAST comes in different flavors—BLASTn (for DNA sequences), BLASTp (Protein sequences). Similar to BLAST there are also numerous similarity search tools exist such as FASTA, WUBLAST and PSI-BLAST, which basically differ in efficiency of search. Almost all these tools are based on computer technique called Pattern search employed by algorithms like Smith-Waterman (http://www.med.nyu.edu/rcr/rcr/course/sim-sw.html) and Needleman & Wunsch (http://www.gen.tcd.ie/molevol/nwswat.html) employed in design of various types of BLAST tools.

Not all significant homologies are readily and easily detected. Some of the most interesting are subtle similarities that do not always rise to statistical significance during a standard BLAST search. Therefore, NCBI has extended the statistical methodology used in the original BLAST to address the problem of detecting weak, yet significant, sequence similarities. PSI-BLAST, or Position Specific Iterated BLAST, searches sequence databases with a profile constructed using BLAST alignments, from which it then constructs what is called a position specific score matrix. For protein analysis, the new Pattern Hit Initiated BLAST, or PHI-BLAST serves to complement the profile based searching that was previously introduced with PSI-BLAST. PHI-BLAST further incorporate hypotheses as to the biological function of a query sequence and restrict the analysis to a set of protein sequences that are already known to contain a specific pattern or motif. Specialized BLASTs are also available for human, microbial, and other genomes, as well as for vector contamination, immunoglobulins, and tentative human consensus sequences.

**RPS-BLAST** is a "reverse" version of PSI-BLAST, which has been described above. Both RPS-BLAST and PSI-BLAST use similar methods to derive conserved features of a protein family. However, RPS-BLAST compares a query sequence against a database of profiles prepared from ready-made alignments whereas PSI-BLAST builds alignment starting from a single protein sequence. The programs also differ in purpose: RPS-BLAST is used to identity-conserved domains in a query sequence, whereas PSI-BLAST is used to identify other members of the protein family to which a query sequence belongs.

**VAST**, or Vector Alignment Search Tool, is a computer algorithm developed at NCBI for use in identifying similar three dimensional protein structures. VAST is capable of detecting structural similarities between proteins stored in MMDB even when no sequence similarity is detected.

**VAST Search** is NCBI's structure-structure similarity search service that compares three-dimensional coordinates of newly determined protein structures to those in the MMDB or PDB databases. VAST Search creates a list of structure neighbors or related structures that a user can then browse interactively. VAST search will retrieve almost all structures with an identical three-dimensional fold although it may occasionally miss a few structures or report chance similarities.

## Specialized Resources

**Plant Databases**: The completion of the sequencing of the entire genome of the model plant *Arabidopsis thaliana (Arabidopsis* Genome Initiative, 2000) is hailed as the beginning of a new era by the plant biotechnologists. Various efforts in sequencing of the genomes of major crop plants are underway and will be completed shortly, scientists are now faced with challenge of identifying new plant genes, Understanding the function of newly discovered plant genes and

"reaping the plant gene harvest" by using this new information in improving crop yield. While plant genomics (i.e. unraveling the functions of plant genes based on whole genome sequencing) is still in its infancy, plant biotechnologists have been more active in creating proteomics and metabolic profile databases. Proteomics uses two-dimensional (2D) gel electrophoresis to separate the proteins in a cell or tissue by size and pH characteristics, followed by mass spectrometry to help identify each component of the resulting gel pattern. Using this technology, researchers are building protein expression pattern databases for *Arabidopsis*, rice, maize, and pine trees (see http://www.expasy.ch/ch2d/2d-index.html for links to these databases). The databases provide pictures of 2D gels with links to previous publications on proteins of interest, the plant tissue from which it was derived and sequence data for proteins. However, as an increasing number of plant genomes become available, plant genomics will play an even more significant role in plant biotechnology. Two gateways to the emerging area of plant genomics are reviewed (Table 29.3).

**UK Crop Net**: Comparative genomics (ascribing function through alignment of similar nucleotide or protein sequences) has become a key area of research in plant biotechnology. This is because genomes of closely related plant species have been found to have remarkably similar genes and gene functions. As the vast amount of plant genomic data becoming available, the use of bioinformatics to improve plant varieties is also becoming vital. To make sense of all the genomic data, UK CropNet was established in 1996 with specific aims of developing software and databases that will facilitate the querying of genomic information from different crop species. Particular emphasis has been placed on developing software tools for comparative mapping. UK CropNet has used the AceDB (Durbin & Thierry-Mieg, 1992) database system to create separate databases for each of the UK CropNet Projects with individual databases for *Arabidopsis,* Barley, *Brassica*, Forage grasses, and Millet.

***Genoplante*:** Genoplante is an instance of collaboration between publicly funded institutions and private organization in furthering scientific research in plant genomics and bioinformatics. Genoplante is a major partnership programme in plant genomics, which links public research in France (INRA, CIRAD, IRD, CNRS) and the main private companies involved in crop improvement and protection (Biogemma, Aventis Crop Science, Bioplante). The objective of Genoplante is to create a network of laboratories across Europe, which will pool their combined resources to discover new plant genes, study the genomics of model plant species like *Arabidopsis* and rice, and conduct genome-based research on major crops under cultivation in Europe.

### Bioinformatics Databases and Analysis Services

It is probably best to try the Major sites first. Some of the services highlighted in later section are at the major sites but are mentioned again to help newcomers to the field find their way to the resources. Some are at other sites, or may be at both. Once you reach other sites you will often find they have very good sets of links that help you find more useful services

# BIOINFORMATICS TOOLS FOR ACCESS TO DATABASES

Existing autonomous biological databases contain related data that are more valuable when interconnected with other related databases. However, the expertise to build and populate biological databases does not reside in any single institution. Therefore, biological databases are built by different teams, in different locations, for different purposes and using different data models and supporting database management systems. As a consequence, connecting the related data they contain is not straightforward. Experience with existing biological databases and their interfaces (e.g. the WWW) indicates

that it is possible to form usual queries across these databases, but that doing so usually require expertise in the semantic structure of each source database. Biological databases often provide bioinformatics application as part of the user interface. There are three primary types of bioinformatics tools that are commonly coupled with the databases:

1. Text based databases searching
2. Similarity based database searching
3. Visualization tools

### 1. Text based Database Searching

Most publicly available biological databases provide a text-based search interface that allows retrieval of entries that "match" user specified word(s) or phrases. Advanced search features typically include Boolean searches (combining terms) with or/and/not), wild cards, etc. The quality of results of a text based search depends on the quality of the contents of the database. Hence, in either designing a biological database or conducting a text based search the following things should be kept in mind.

I. Spelling errors can exclude relevant entries in free-form text. Inconsistencies may also result in relevant errors being excluded (e.g. IL2 and IL2 are both used to refer to Interleukin-2).

II. Use of proper key words

III. Use of controlled vocabulary. It is important to understand the organization and hierarchy of such a controlled vocabulary if used in searching (e.g. PubMed uses MeSH–medical subject headings as controlled vocabulary).

### 2. Similarity based Database Searching

It is based on *BLAST* and *FASTA*. With large-scale genome sequencing projects, the flood of DNA sequence data coming into public databases is staggering. Researchers are increasingly relying on inferring the function of putative genes through similarity to well-characterized proteins. It is important to realize that designing an *in silico* sequence similarity search needs to be as carefully designed as wet lab experiment in order to get biologically meaningful results.

Sequence similarity searches use alignments to determine a "match". Alignment of two sequences is matching of the two sequences. Except that they allow the most common mutations, insertions, deletions, and single-character substitutions. The basic considerations in using a sequence-similarity search are:

**a. Global vs. Local Alignment:** Global alignment forces complete alignment of the input sequences whereas local alignment will align the most similar segments. The choice of global vs. local depends on the assumptions made by the user as to whether the sequences are related over their entire length or presumed to share only isolated regions of homology. As similarities will span segment rather than entire sequences, hence local alignment is the most popular database similarity search.

**b. Alignment Algorithms:** There are a variety of relatively efficient alignment algorithms, each of which aim to determine the most optimal alignment. Needleman & Wunsch (1970) first described the Needleman-Wunsch algorithm for optimal global alignment, followed by a slight variant, the Smith-Waterman algorithm (Smith & Waterman, 1981) for optimal local alignment of two sequences. These two methods were developed prior to wholesale genome sequencing. Today the special purpose parallel machines and the massive computation time required by these algorithms have rendered them almost obsolete. Most users prefer BLAST (Altschul *et al.* 1990) or FASTA (Pearson & Lipman, 1988), which rely on heuristic strategies to speed up alignment searches. Promising regions are first determined through rapid match searches, and only then Smith-Waterman algorithm is invoked. This approach permits FASTA or BLAST to run 10 to 100 times faster than conventional Smith-Waterman, at the cost of missing a few alignments. Some of the adjustable parameters described below provide the user the flexibility to trade off between speed and accuracy. BLAST in general, tends to be faster and is more sensitive (detects more alignments) but FASTA returns fewer false hits.

**c. Search Parameters:** The effectiveness of alignment algorithms depends on its parameters and a careful choice is necessary, without which important alignments may be overlooked or too many spurious alignments may be returned. There are three sets of parameters that can be specified by the user to control the results which are as described below:

i. *Alignment parameters:* It includes the choice of substitution matrix and the costs associated with gaps. The substitution matrix is the cost associated with substituting one residue with another in aligning two protein sequences. The most popular substitution matrices are the PAM (Schwart and Dayhoff, 1978) and BLOSUM (Henikoff and Henikoff, 1992) family of matrices. The gap cost parameters involve a cost associated with opening a gap and a lesser cost associated with extension of a gap.

ii. *Algorithmic parameters:* The different algorithmic parameters mostly control the heuristics on which BLAST and FASTA rely and hence allow the user to control the speed and accuracy of their alignment. One should refer to the online manuals available at the BLAST and FASTA sites for a detailed description of these algorithmic parameters.

iii. *Out put parameters:* It includes a threshold for the E-score and the desired number of matches. The E-score is a measure of the statistical significance of an alignment where it combines the raw score from the alignments, the lengths of the query sequences, and the size of the database. The E-score gives the expected number of sequences in the database that would align

**Table 29.4:** BLAST and FASTA variants for different searches [Source: Adak and Srivastava (2002)]

| Programme | Input Sequence | Comparison Database | Common Use |
|---|---|---|---|
| BLASTn/FASTA | Nucleotide | Nucleotide | Align a new DNA sequence to a nucleotide sequence. |
| BLASTp/FASTA | Protein | Protein | Seeks to align an amino acid query sequence to a protein sequence database. |
| BLASTx/FASTx | Nucleotide (translated) | Protein | Analyze new data sequence (translated) to find potential coding regions. |
| TBLASTn/tFASTx | Protein | Nucleotide (translated) | Useful for EST analysis. |
| TBLASTx/tFASTx | Nucleotide (translated) | Nucleotide (translated) | Useful for EST analysis. |

with the given raw score by chance. Typically in a database of the size of Genbank or Swiss-Prot, one expects random matches of 5 to 10 sequences, and thus alignments with E-Scores less than 5 (or 10) are ignored. However, in smaller databases, such as the PDB, it is important to consider smaller E-scores.

Various forms of BLAST and FASTA are used in alignment of different types of biological sequences, some of which are listed in Table 29.4.

Pair wise sequence alignment using BLAST and FASTA has been extended in two ways. (i) for multiple alignment of DNA or protein sequences and (ii) structural alignment for determination of protein structural neighbors, where the extension of BLAST to 3-dimensional coordinates is called VAST (Gibrat *et al.*, 1996). In multiple alignments the most common method called Clustal W (Gibson *et al.*, 1994) creates a multiple alignment of DNA or protein sequences, starting with BLAST or FASTA pair-wise alignment scores.

### 3. Visualization tools

***Protein Structure Visualization:*** RasMol and Kinemage: Most protein structure databases are today equipped with visualization tools. The most frequently used being the freely available RasMol (Sayle and Milner-White, 1995). The name "RasMol" is derived from Raster (the array of pixels on a computer screen) and Molecules. The initial of the author of RasMol and R.A.S. is probably coincidental.

***RasMol:*** This is a olecular graphics software that allows visualization of proteins, nucleic acids and small molecules, for which a 3-dimensional structure is available. In order to display a molecule, RasMol requires an atomic coordinate file that specifies the position of every atom in the molecule, through its 3-dimensional cartesian coordinates. RasMol accepts this coordinate file in a variety of formats, including the PDB format. The visualization provides the user a choice of colour schemes and molecular representations (wire frame, cylinder,stick bonds, alpha-carbon trace, space filling (CPK) spheres, macro-molecular ribbons, hydrogen bonding and dot surface). Additional features such as text labeling for selected atoms, different color schemes for different parts of the molecule, zoom, rotation, etc. have made this the most popular of all visualization tools.

Chime and Protein Explorer are derivatives of RasMol that allow visualization inside web browsers, while RasMol runs independently outside a web browser.

***Kinemage:*** One of the drawbacks of RasMol was that it failed to allow the user to move two molecules or parts of a molecule complex, relative to each other. For example, RasMol cannot show the binding of a substrate to, or its release from an enzyme. This drawback was corrected when kinemage (kinetic images) was developed (Richardson and Richardson, 1989). Kinemages are set up to illustrate a particular idea about a three dimensional object, rather than neutrally displaying that object. They incorporate the authors selection, emphasis and viewpoints, etc.

## Integrated Molecular Biology Database system

A database is repository that provides a centralized and homogenous view of its contents. The repository is created and modified through a database management system (DBMS). Every data item in the database is structured according to a *schema*, which is defined as a set of pre-specified rules through the data definition language. The contents of the database can be typically accessed through a graphical user interface (GUI) that allows browsing through the contents of the repository, which is similar as one may browse through the books in a library. Most databases also allow querying of its contents through a specialized query language. The data definition language and the query language form the data model and define the semantics of the manipulations and operations allowed on the database. For example, the schema of Genome Sequence Data Base (GSDB) and the Mouse Genome Database (MGD) are defined using the data definition language of the Sybase relational DBMS, the structure of the *Arabidopsis thaliana* database (AtDB) as well as numerous other genomic databases have been defined using AceDB (Durbin and Thierry-Mieg, 1992) and the structure of Genome Database (GDB) and the Protein Data Bank (PDB) are defined using an object protocol model (OPM) (Chen and Markowitz, 1995) that allows storage of images on top of a relational database management system. For molecular biology databanks maintained as files, the data definition languages used for defining their structure are not based on a data model *per se* and range from generic notations such as the ASN.1 data exchange format used for Genbank to ad-hoc data definition languages such as that employed for EMBL.

Thus, studies of molecular biology data involve comprehensive exploring of multiple databases. Rather than requiring the user to combine information retrieved from multiple databases, it is better to provide an integration of the databases. But the main hurdle in the integration of multiple biological databases is their inherent heterogeneity. These inherent heterogeneities are caused by:

1. *Heterogeneity of content*: Different databases are used to store a variety of information. For example, protein sequence information is available through Swiss-Prot while protein structure information is available through the Protein Data Bank (PDB.)

2. *Heterogeneity of database management system*: Different data types require different databases management systems. For example, table structured data can be easily stored through relational database management systems (such as those of Oracle), storage of text data as in GenBank is not amenable to relational DBMSs, while image require object database management systems such as that of OPM (Chen and Markowitz, 1995)

3. *Heterogeneity of Data model*: The data model (the schema and the query language) of heterogeneous systems will also vary. For example, relational systems mostly use SQL (structured query language), while special query languages such as OQL (object query language) needed to be used for object databases like the PDB.

In integrating molecular biology databases, the following issues need to be addressed:

i. *Integration of Data*

A basic problem underlying the integration of heterogeneous databases is the autonomy of the sources, which has led to lack of cooperation and non-standardization of format with some notable exceptions. For example, GenBank, EMBL and the DNA Data Bank of Japan (DDBI) cooperate in creating a centralized repository for the human genome sequence data and daily exchange of data is made for the purpose of synchronization. A cooperation/co-licensing agreement is the first step in the creation of integrated systems and the two things are important: (i) as data is exchanged between heterogeneous systems, schema converters are required (which convert data from one schema to either the schema of another database or a global schema); (ii) Data conflicts and errors need to be resolved in a systematic manner.

### ii. *Integration of User Interfaces*

*Browsing Interface*: It presents a unified view of the data in one of the two possible ways (Markowitz & Ritter, 1995): (a) A global schema is created by unifying the schema of the component databases, or (b) Local views of the data which use the "local" schema of the component databases.

*Query Interface*: For integrating intelligence into database query tools are very important. Ideally, such tools would be capable of using knowledge of the schemata of remote databases to construct properly formed queries to each of the relevant remote databases, and capable of integrating the retrieved data into a coherent report for the user. The software modules that perform these functions have been termed 'mediators'. Principal advantage of an integration system based on mediators is that it allows the individual databases to operate autonomously, but to function collectively as a federation. A federated database is a collection of essentially autonomous databases in which each constituent database may be implemented using a different schema and using a different database engine. The federated database provides defined access methods to the distributed databases and allows the user to view them collectively as a single database. While the development of mediator based query systems is a subject of continuing computer science research, recent implementations (e.g. the CPL/Kleisli system) have shown that it is possible to provide access both to relational databases (the Genome Database) and to unconventional sources of biological data [e.g.

the *Caenorhabditis elegans* genetic database (ACeDB), ASN.1 and BLAST] without building a monolithic database or writing a very application specific code for each query.

Each of the component databases may support different types of queries (e.g. free text search, keyword search, accession number search, etc). Integrated querying of multiple databases is still the holy grail of heterogeneous database management systems today. The main hurdles to integrated querying are:

a. Rewriting the queries and query rules in the integrated system to operations on the source databases;

b. Ability to implement join across heterogeneous items from the source databases

c. The ability to identify redundancy

d. The ability to exploit parallelism of different servers (computers) in use for the source databases

### iii. *Integration of Visualization and Analytical Tools*

With integration comes the advantage that bioinformatics applications (analytical and visualization tools) can be developed on a common data model. Some of the tools for bioinformatics and computational biology have been mentioned below as reported by Horton (1998).

A net-work computation server GeneMatcher system by Paracel has been designed for wide variety of supported algorithms that allows researchers to rapidly compare similarities and differences between DNA and protein sequence data. Gene Explorer 1.4 from Molecular Simulations (MSI), a computational environment for molecular biology and bioinformatics is compatible with several major genomic database systems, including those from Incyte, Pangea Systems, PF informatics, GCG, NCBI, PDB SRS, companies involved in bioinformatics.

Gene Explorer 1.4 automatically tells scientists when new database hits for previous experiments are found in database updates. In addition, the program helps researchers identify protein homologs with more sensitivity, using its protein threading technology.

SeqWeb from Genetics Computer Group (GCG) a part of Oxford Molecular Group have developed a web based interface that provides easy access to a core set of programs in the Wisconsin package, which are now accessible through Netscape Navigator or Microsoft Internet explorer. With SeqWeb users can use sequence files from their desktop, access group project data on the server and search a site's standard and proprietary databases through a web browser. SeqWeb links to local databases, as well as those on the internet, such as Medline, PDB, Flybase, Mendelian Inheritance in Man (MIM) and US Patent and Trademark Office patent database. Certain dynamic databases: Synergy from NetGenics, DiscoveryBase from Molecular Application Group; BioMerge3.0 from PE Informatics have been developed. These have been designed for providing centralized access to both proprietary and public genomic data

StackDB, a program from Pangea systems, provides an integrated dataset of the publicly available human ESTs and alignments. The sequence tag alignement and consensus knowledgebase or STACK is another public database of gene sequences expressed in the human genome. It features ESTs organized according to tissues and provides a comprehensive representation of each gene with alignments of its expressed fragments.

## Database integration strategies

There are currently three types of integration strategies being used for molecular biology databases: data warehousing, link-driven federation, and semantic integration. These strategies along with discussion of the integrated molecular biology databases systems that have resulted from these efforts are reviewed as under.

**Data Warehousing:** Data Warehouses represent the materialization of a global schema, i.e. the integrated database is defined by the global schema and loaded with data from the component databases. The steps involved in creating a data warehouse are:
- Downloading of data from the component databases
- Data cleaning (removal of erroneous entries and resolving data conflicts)
- Reformatting the data into the global schema

A data warehouse is often confused with a consolidation of multiple databases. In consolidating multiple databases, the component databases are put together into a larger database and the individual component databases are discarded, whereas in data warehousing, the component databases are not disturbed. Consolidation is far more complex and expensive, requiring consensus on common names, data structures, and policies. Furthermore, existing applications on component databases must be converted in order to function on the consolidated system. The relative advantages and disadvantages of data warehousing systems are listed in Table 29.5. Current data warehousing systems used in molecular biology databases are:

*GUS-Genomics Unified Schema*: Genomics Unified Schema (GUS) was given by Davidson *et al.* (2001) It is a warehousing based data integration system from University of Pennsylvania. GUS uses a relational data model and stores nucleotide, amino acid sequences and annotations in Tables. The data sources already included are GenBank/ EMBL/DDBJ, dbEST and SWISS-PROT. GUS builds and maintains a map between DNA sequence based entries at some sites and gene-based entries at others through its local storage of the necessary data. Its tables hold the conceptual entities that DNA sequences and annotations indirectly represent, which are genes, the RNA derived from these genes and the proteins from those RNAs. While transforming the data into

**Table 29.5:** Data Warehousing [Source: Adak and Srivastava (2002)]

| Advantages | Disadvantages |
|---|---|
| Downloaded source data can be manipulated into suitable formats | High maintenance cost as data needs to be constantly synchronized with component databases |
| Global schema allows browsing of data through a unified view | Large initial costs associated with setup and schema development |
| Execution of queries is usually very fast because all the data is locally available | Storage requirements add to cost |
| System is reliable because there is no outside dependency | System does not scale easily—not easy to add new database |

gene-centric organization, it cleans data to identify erroneous annotations and misidentified sequences.

GUS facilitates data maintenance by tracking how data is generated/accessed from sources and subsequently modified. This helps in learning about knowledge of genes that data items represent. The revisions of original data can be by the source itself, annotations are slowly experimentally verified and predicted values become more accurate with better algorithms. For example, with computationally derived annotations, GUS stores the algorithm used, its implementation version, input parameters and the run time information.

To keep its databases synchronized with external data sources after initial download, GUS retrieves updates and new entries from them based on the source's change schedule or periodically. The changed fields of the modified entries are detected in the database based on a difference operation, and updated accordingly. Both the new and updated entries are subjected to an annotation update process in which the protein and DNA sequence and their annotations are transformed into gene and protein based entries. The user always sees a production version of the database while updates are made to the next development version. When the development cycle is over, the database version is put into production, and a new development version is created.

*Link Driven Federation*

The link driven federation approach has been used successfully by mainly online molecular biology databases to add value to their data. A link connects an object in one database to objects in possibly another database. Links can also be to objects in the same database. The links provided will allow the user to start from a data item of interest in a particular database and then jump to other related data sources through the links. The user has to still interact with individual sources; only the interaction is easier through convenient links and invoking/querying the individual source databases directly is not required. The most widely used integrated molecular biology systems are Entrez (Fig. 29.3) sequence retrieval system, SRS (Etzold & Argos, 1993), and LinkDB (Fujibuchi *et al*, 1998), Integrated Genomic database (IGD), the ExPASyWWW server are examples of this approach.

*Advantages*

1. Point and click links make it convenient for the user to see related sources
2. Links to a variety of information for each entry in a database.

*Disadvantages*

1. Manual link creation difficult for large databases
2. Does not scale easily; for each entry in a database difficult to generate all links to a new database
3. Changes in source database schema may result in links becoming obsolete

**Entrez:** The National Center for Biotechnology Information (NCBI), which is part of the National Institutes of Health, USA is the

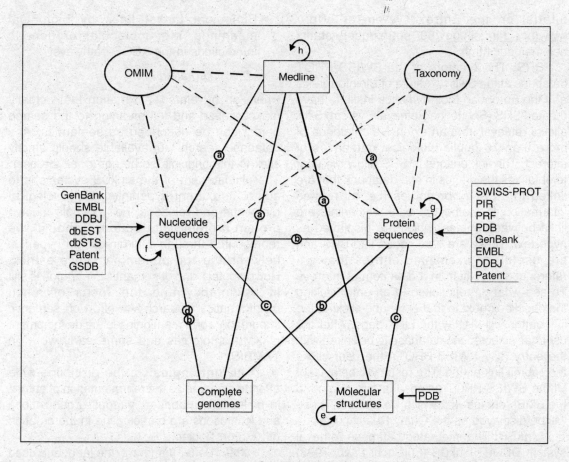

**Fig. 29.3:** The Entrez browser is an integrated set of databases, indexes and access tools produced by NCBI, USA. The original sources of data in the Entrez databases are shown in italics. Connections between the databases are based on: (a) literature citations in sequence or structure; (b) coding region features; (c) primary structure identity with sequence; and (d) alignments. Connections within databases (neighbors) are based on: (e) three-dimensional structural similarity; (f) nucleotide sequence similarity; (g) amino acid sequence similarity; and (h) term frequency statistics. OMIM (the on-line version of Mendelian Inheritance in Man (MIM) is not formally integrated within Entrez, but OMIM entries are linked and offer entry points into the Entrez system.

foremost repository of publicly available genomic and proteomic data. Their integrated information retrieval system, known as Entrez, is perhaps the most utilized of all biological database systems. Entrez uses a link based approach to cross-reference entries from different databases. The nucleotide sequence database GenBank, the medical literature database PubMed, NCBI's protein sequence database, NCBI's protein structure database and NCBI's database of whole genomes (Fig. 29.3). Hard links are applied between the different databases, whenever there is a logical connection between entries. Link out from PubMed citations also provide links to user defined external web pages (for example, the full text journal articles, biological data, sequence centers, etc.) These external resources provide a URL, resource name, brief description of their web site, which PubMed uses to create the links to their sites. For complete review of the features and complexities of Entrez, one may refer to a tutorial on the Entrez system at http://www.ncbi.nlm.nih.gov:80/ entrez/query/static/help/helpdoc.html.

**SRS:** The creators of the SWISS-PROT database at the Swiss Institute of Bioinformatics and the European Bioinformatics Institute have created SRS (Sequence Retrieval System) SRS allows retrieval from an extensive catalogue of more than 75 public biological databases of interest. The link buttons in SRS allow the user to obtain all the entries in one databank that are linked to an entry or entries (i.e. all related information) in another databank. Hyperlinks are links between entries, which are displayed as hypertext. These are hard coded into SRS and are useful for examining entries that are referenced directly from a data item of interest. To see what data is linked to an entry, ticking the checkbox next to that entry and clicking the link button will display the LINK page. After the user has selected the database to be linked with the entry (say SWISS-PROT), the user clicks the submit link button. The result will be a list of all the SWISS-PROT entries that are related to the EMBL entries with which we started. These will be displayed on the Query Result page.

**LinkDB:** The integrated database retrieval system DBGET/LinkDB (Fujibuchi *et al.*, 1998) is the backbone of the Japanese GenomeNet service. DBGET is used to search and extract entries from a wide range of molecular biology databases, while LinkDB is used to search and compute links between entries in different databases. Once an entry is retrieved through DBGET, all links from this entry can be obtained by clicking on the entry name, which causes the search against LinkDB. In addition to the original links provided by the source database, which are embedded in the entry, LinkDB also aims at providing computer generated links, which includes:

• Factual Links: links between database entries which include, e.g. Medline ID and GenBank accession
• Similarity Links: links produced by similarity search e.g. the results of BLAST and FASTA

• Biological Links: links by biological meanings, e.g. molecular or genetic interactions in the KEGG pathways.

*Semantic Integration*

Provision of related information that actually makes sense and full meaning to the search object can be described as semantics. Any database search tool available should strictly stick to this principle so as not to fail the user. The ultimate aim of the LinkDB systems is to add to its current hyperlinks using biological meanings and relations. For example, known protein-protein interactions should be represented through a bi-directional hyperlink between the two protein sequence entries. However, this kind of semantic integration is still in its infancy and the focus of much bioinformatics research. An effort on semantic integration worth mentioning is the development of XML ontologies and some early work on TAMBIS.

**XML ontologies:** In the discipline-wide effort to standardize the representation of entries in biological databases, various organizations and institutions are participating in the creation of XML ontologies.

• **XML:** Better than html (the language used in creation of web pages), XML (eXtensible Markup Language) has emerged as the de facto format for exchanging data in the last few years with abundant development tools and backing from major vendors. XML is ideal for format integration because it can handle semi structured data (unlike the rigid data structures of relational database systems). The wide use of XML is easily seen in the variety of standards being proposed which are shown in Table 29.6.

• **Ontology:** An ontology is a set of concepts (objects, events and relations) that are specified in order to create an agreed upon vocabulary for exchanging information. Specification of ontology involves (1) determination of which concepts are to be included in the ontology (2) assigning English language meaning to the terms; and (3) defining all possible relations

**Table 29.6:** XML—Standards in Biology [Source: Adak and Srivastava (2002)]

| XML Standard | Description |
|---|---|
| AnatML | A language for storing geometric information and documentation obtained as part of the musculoskeletal modeling project |
| Array XML (AXML) | For exchanging and storing data from micro array experiments |
| Bioinformatic Sequence Markup Language (BSML) | A public domain standard for encoding and display of DNA, RNA and protein sequence information. |
| Biopolymer Markup Language (BIOML) | BIOML standard allows the full specification of all experimental information known about molecular entities composed of biopolymers (e.g. proteins and genes) |
| Genome Annotation Markup Elements (GAME) | The goals of GAME at least in the perspective of the bioxml community, are to provide an XML ontology and tools for annotating biosequence "annotation features" |
| CellML | For storage and exchange of computer based biological models |
| Clinical Trial Data Model | FDA safety domain and metadata models with an XML ontology for clinical data |
| Gene Expression Markup Language (GEML) | An open-standard XML format for microarray and gene expression data |
| Gene Ontology Markup Language | The Gene Ontology Consortium is attempting to produce a dynamic controlled vocabulary that can be applied to all eukaryotes, even as we gain more knowledge on gene and protein roles in cells |
| Gene X Gene Expression Markup language (Gene XML) | Part of the Gene X project, a massively distributed gene expression database |
| Molecular Dynamics Markup Language (MoDL) | MoDL (pronounced "Model") is an XML application that allows chemical simulation data visualization over the Web |
| Systems Biology Markup Language (SBML) | Represent models of biological systems common in research on a number of topics including cell signaling pathways, metabolic pathways, biochemical reactions, etc. |
| Taxonomic Markup Language (Taxonomic ML) | Taxonomic ML seeks to standardize (1) the description of the structure (topology) of a biological phylogeny; (2) the presentation of statistical data about the phylogeny and (3) the option of superimposing a Linnaean taxonomy |
| XML Description Language for Taxonomy (XDELTA) | XML file format derived from DELTA (DEscription Language for TAxonomy) standard |

between the concepts in the ontology. Almost all the ontology tools are designed using XML language that exploits features such as detection and text parsing. GOA–Gene ontology annotation (http://www.ebi.ac.uk/GOA/) is the most popular keyword based text search ontology tool that retrieves all information relating to search query.

**TAMBIS**: Transparent Access to Multiple Bioinformatics Information Sources (TAMBIS) is an integration system for molecular biology where a domain dependent ontology is used for information retrieval. The ontology is central to the system and plays a role in query formulation and execution (Goble *et al.*, 2001).

## Detection of genes

Bioinformatics tools have been developed for detection of genes, functions and its annotation. The task is simple in prokaryotes but in

prokaryotes it is difficult due to complexity of genes.

1. In the first instance genomic clone of interest is identified by subjecting to **zoo blot hybridization** (hybridization with the whole genomic DNA of a variety of species). A clone found positive is likely to represent a gene. This clone of interest may be then hybridized with cDNA libraries or used for northern blot assays. A positive assay identifies a gene since only genes are transcribed. Other approaches for gene identification can also be followed.

2. The most important task in gene identification in a DNA sequence is the identification of the correct reading frame. Correct reading frame is assumed to be the longest frame uninterrupted by a stop codon. Such a reading frame is called open reading frame. Location of the 3' end of an ORF is

relatively easy than that of 5′ end because of the presence of Kozak sequence (CCGCCAUGCG) flanking the AUG codon. Detection of genes may follow one of the two strategies:

i. Sequence of known genes, cDNAs, ESTs and proteins present in databases are compared with the genome sequence which is done by homology search tools e.g. BLAST.

ii. Specialized softwares are used to detect genes. For prokaryotes the programs like GENMARK (a modified GenScan algorithm) and Glimmer can identify the genes. Identification of ORFs usually exceeding 300 nucleotides is sufficient to findmost genes in prokaryotic genomes. However, such a simple criteria will miss smaller genes and overlapping genes. GENMARK program resolves this by using algorithms that consider differences in base composition between genes and noncoding DNA.

For eukaryotes various softwares used for gene predictions are GENIE, GENESCAN, GRAIL, Genefinder, HMM Gene, etc. However, an algorithm trained with DNA sequence of one organism e.g. for humans will not perform satisfactorily for *C. elegans* or any plant genome. The gene prediction programs search for gene specific features such as promoters, splice sites, polyadenylation signals or for gene content of ORFs. Gene search programs currently being used employ different search criteria. The gene prediction programs used in eukaryotes use the output of several algorithms to generate a whole gene model. In this model, a gene is defined as a series of exons that are coordinately transcribed. The various features of genes recognized during gene detection include transcriptional and translational controls, e.g. cap sites, TATA box, polyadenylation sites, 5′ and 3′ splice sites, differences in base composition between coding and noncoding DNA, etc.

3. Identification of function of a new gene–The simplest way is to translate the gene sequence in to its amino acid sequence of protein it is expected to encode. This protein sequence is then compared with the sequences in the protein database. A program tBLASTx as mentioned earlier will perform these functions. If the encoded protein is homologous to a protein the database, it will confirm the identification of gene.

4. Identification of functional domains–Several bioinformatics tools for identification of motif and domains in proteins are PRINTS, PROSITE, SMART, BLOCKS, etc.

5. Detection of noncoding RNAs–Different types of RNAs *viz* ribosomal, transfer, small are noncoding and are easiest to find by sequence similarity programs. The program tRNAScanSE for example searches for tRNA sequences.

6. Genome annotation–After the gene identification and prediction of its functions, the data is then used for genome annotation. Standard genome annotation languages have become widely accepted. GAME is a program for describing experimental evidence to support annotation. Likewise DAS (distributed annotation system) is useful for indexing and visualization. Besides software tools like BioPerl 2001, BioJava 2001 are used for storing, manipulation and visualization of genome annotations.

7. Molecular phylogenetics–The DNA and protein sequence data can be used to investigate evolution of genes and their protein products which is called as molecular phylogeny. These can be represented either by dendrogram or a phylogenetic tree.

## Applications

1. Data acquisition: High throughput biology requires automatic capture and conversion of biological data to symbolic digital representation without human intervention.

Increasing the efficiency of current DNA sequencing systems will require base calling (the process of interpreting the signals from a fluorescent DNA sequencing instrument as a stream of nucleotides passing the detector) algorithms that provide increased accuracy, increased read-lengths and confidence estimations. These improvements will permit improvements in the automation of the assembly and finishing of DNA sequences.

2. Assembly of sequence contigs: Large scale sequencing projects require sequence assembly algorithms. Bioinformatics will help in generating contigs because softwares have been developed.

3. Predicting functional domains in genomic sequences: Advances in sequencing technology are expected to make the acquisition of large amount of anonymous sequences, which can be converted to its functions. It is likely that experimental determination of coding regions will be necessary to use computational methods to find all the exons in a genomic sequence. Computational approaches to recognize functional sites based on certain assumptions have been generated.

4. Sequence alignment: Once the sequence of nucleotides is obtained, it is compared to all the known sequences to see if any instructive similarities can be detected. Finding a similar sequence of known function can often be helpful in elucidating the function of a new cloned gene. For example, a quarter of positionally cloned human genes were found to have matches among *Saccharomyces* sequences of known function.

5. Searching structure databases: Newly determined protein structures are increasingly showing structural similarity to previously determined ones, even when no sequence similarity is detectable. New algorithms allow protein structures to be compared with databases of all known structures.

6. Determination of macromolecular structure: X-ray crystallography and nuclear magnetic resonance (NMR) are the major experimental methods used for deducing the macromolecular structure. Both methods produce extremely large amounts of data and are entirely dependent upon the availability of powerful computers and sophisticated processing algorithms.

7. Molecular evolution: Bioinformatics helps in constructing phylogenetic trees. The comparative genomics has provided a path of evolution to researchers through accumulated data on genomics and proteomics.

8. Bioinfomatics in medicine: The amount of data available through sequencing is shifting research in molecular biology and genetics from a purely experimental approach to one in which experiments can be planned in front of a computer. The analysis of data will allow the identification of genetic markers of diseases (like cancer), and the discovery of currently unknown physiological mechanisms which are at the basis of currently incurable diseases, offering clues for therapies including genetic therapy. Biocomputing techniques offer opportunities to biotech companies which want to give a stronger rational basis to the process of drug discovery. Experiments can be designed much more intelligently and the understanding of molecular interactions can be enhanced dramatically. Trial-and-error experiments can be replaced by predictions that allow for the design of custom-made drugs. *In silico* tools such as MODELER, Molecular Environment (MOE), CATH and RasMol are designed based on detection of epitope (or active/binding site) detection using antigen-receptor interactions. These tools not only take in to consideration various topographica, geometrical aspects of various

molecules (from both antigen/drug and protein receptor) but also chemical bond energy parameters required for defining the efficacy of the designed drugs. *In silico* design of drugs relieves pharmaceutical industry from routinely followed laborious and time consuming pharmacological path of drug discovery. For example if development of a drug requires nearly 10 years of time with an approximate investment of $ 1 billion, use of *in silico* techniques would require investment of $ 400 million and time can be reduced to 6 years. Drugs based on rational design are 'smarter' in the sense that they are optimized for the target molecule in question and therefore tend to elicit far fewer side effects, which is a major problem for many therapies (e.g. cancer and AIDS therapies). Smarter design also calls for lower doses of drugs to achieve the desired effect. Furthermore, access to public databanks/resources improve the exchange of information on drug design research, promotes cooperation and reduce waste due to unnecessary competition, duplication and re-inventing the wheel.

9. Bioinformatics in agriculture: Agriculture can exploit the same computational tools, databases and strategies for investigating disease resistance markers, pest resistance genes and so on. As a result, crop harvests can increase and special plants can be engineered to alleviate food shortages in developing countries.

## Challenges

1. Data management: The physical mapping of genomes and large scale sequencing projects require storage of information. In laboratory conditions data of a few 100 megabytes to nearly 100 gigabytes can be stored. Information technology talks in terms of terabytes (trillion bytes), petabytes (1000 terabytes) and exabytes (1000 petabytes). It is stated that all the words ever spoken by human beings amount to about five exabytes.

Now a new, zettabyte term has been coined for such huge databases involving 1000 exabytes and 1000 zettabytes as a yottabyte or $10^{24}$ bytes. To realize the promise of molecular and structural biology will require high performance computers. High throughput biological research requires new and improved tools for managing information both at the laboratory level and at the national/international database levels.

2. Research is also needed for new and improved analytical methods. These methods are essential for turning molecular and structural data into biological knowledge.

3. The searching of biological databases via the WWW is becoming increasingly difficult. There are differences in database structure and nomenclature that hinder research efforts and standardizations have met with much resistance.

4. Audit and control of databases is becoming important because databases are becoming larger and larger. The complex computing environment and resource crunches would make it vital for information technology auditors to have practical guidance on conducting audits and also ensuring security in today's quickly changing computing environment.

5. The development of various strategies to surf WWW, and reaching a consensus on coining uniform definitions and adopting uniform platforms and technologies is another challenge in the practice of bioinformatics.

## REFERENCES

Adak, S. and Srivastava, B. 2002. Bioinformatics Advancing biotechnology through Information Technology. Part 1. Molecular Biology Databases. *Indian J. Biotechnology* 1: 101–116.

Altschul, S.F., Gish, W., Millr, W., Myers, E.W. and Lipman, D.J. 1990. Basic local alignment search tool. *J Mol Biol.* 215: 403–410.

Bairoch, A. 1991. A dictionary of sites and patterns in proteins. *Nucl. Acids Res.* 19: 2241–2245.

Batini, C., Lenzerini, M. and Navathe, S. 1986. A comparative analysis of methodologies for database schema integration. ACM *Comput Sur.* **18:** 323–364.

Chen, I. A. and Markowitz, V.M. 1995. An overview of the object protocol model, (OPM) and OPM data management tools. *Inf Sys.* **20:** 393–418.

Davidson, S., Crabtree, J., Brunk, B., Schug, J., Tennen, V., Overton, C. and Stocckert, C. 2001. K2 /Kleisli and GUS. Experiments in integrated access to genomic data source *IBM Syst. J.* **40:** 512–531.

Dayhoff, M.O., Schwart, R.M. and Orcutt, B.C. 1978. A model of evolutionary change in proteins. In: *Atlas of protein sequence and structure* 5 (ed. M.O. Dayhoff), pp. 345–352. National Biomedical Research Foundation, Washington, DC.

Durbin, R. and Thierry-Mieg, J. 1992. Syntactic definition for the AceDB database manager. Available at http://probe.nalusda.gov.8000/acedocs.

Durbin, R., Krogh, A., Mitchison, G. and Eddy, S. 1999. Biological Sequence Analysis probabilistic models of proteins and nucleic acids Cambridge University Press, Cambridge, UK.

Etzold, T. and Argos, P. 1993. SRS: An indexing and retrieval tool for at file data libraries. *Comput Appl Biosci.* **9:** 49–57.

Fujibuchi, W., Goto S. and Migimatsu, H. 1998. DBGET/LinkDB: an integrated database retrieval sustem. In Pacific Symposium on Biocomputing 1998. pp. 683–694.

Gibrat, J.F., Madej, T., and Bryant, S.H. 1996. Surprising similarities in structure comparison. *Carr Opinion Struct Biol.* **6:** 377–385.

Gibson, T.J. Thompson, J.D. and Higgins, D.G. 1994. CLUSTAL W: improving the sensitivity of progressive multiple sequence alignment through sequence weighting position specific gap penalties and weight matrix choice. *Nucl. Acids Res.* **22:** 4673–4680.

Goble, C., Stenens, R. Ng, G., Bechhofer, S., Paton, N., Baker, P., Peim, M., and Brass, A. 2001. Transparent access to multiple bioinformatics information resources. *IBM Syst. J.* **40:** 532–551.

Henikoff, S., and Henikoff, J.G. 1992. Amino acid substitution matrices from protein blocks. *Proc. Natl. Acad. Sci. USA* **89:** 10915–10919.

Horton, B. 1998. Sequence cyberbiology. *Nature* **393:** 603–606.

Karp, P. 1995. A strategy for database interoperation. *J Computational Biol.* **2:** 573–583.

Markowitz, V.M. and Ritter, O. 1995. Characterizing heterogeneous molecular biology database systems. *J Computational Biol.* **2:** 547–556.

Moore, J., Engelberg, A. and Bairoch, A. 1988. *Biotechnique* **6:** 566–572.

Needleman, S.B. and Wunsch, C.D. 1970. A general method applicable to the search for similarities in the amino acid sequences of two proteins. *J Mol Biol.* **48:** 443–453.

Pearson, W.R. and Lipman, D.J. 1988. Improved tools for biological sequence analysis, *Proc Natl Acad Sci USA* **85:** 2444–2448.

Richardson, J.S. and Richardson, D.C. 1989. Principles and patterns of protein conformation. In: Prediction of protein structure and the principles of protein conformation. (ed. Fasman, G.D.) Plenum Press, New York, pp. 1–98.

Sayle, R.A. and Milner-White, E.J. 1995. RASMOL: biomolecular graphics for all. *Trends in Biochemical Sciences* **20(9):** 374–376.

Smith T.F. and Waterman M.S. 1981. Identification of Common Molecular Subsequences. *Journal of Molecular Biology,* **147:** 195–197.

## FURTHER READING

Adak, Sudeshna and Srivastva, Biplav. 2002. Bioinformatics Advancing Biotechnology through Information Technology Part 1: Molecular Biology Databases. *Ind. J. Biotechnology,* **1:** 101–116.

Benton, D. 1996. Bioinformatics–principles and potential of a new multidisciplinary tool. *Trends in Biotech.* **14:** 261–279.

Singh, B.D. 2006. Plant Biotechnology, Kalyani Publishers, Ludhiana, India, pp. 755.

Tripathi, K.K. 2000. Bioinformatics: the foundation of present and future biotechnology. *Curr. Sci.* **79:**570–575.

Westhead, D.R., Parish, J.H. and Twyman, R.M. 2003. Instant Notes Bioinformatics. BIOS Scientific Publishers Ltd. Oxford, U.K. pp. 257.

# 30

# Intellectual Property Rights

With the advent of biotechnology, one of the issues is the legal characterization of new inventions. Thus, an important part of the biotechnology industry concerns patenting or intellectual property rights. Loosely defined, intellectual property (IP) is a product of the mind. Intellectual property is intangible in contrast to real property (land) or physical property, which one can see, feel and use. With any type of property there are property rights. When IPs are expressed in a tangible form, they can also be protected. Intellectual property rights (IPRs) have been created to protect the right of individuals to enjoy their creations and discoveries. In fact, IPRs can be traced back to the fourteenth century, when European monarchs granted proprietary rights to writers for their literary works.

IPRs have been created to ensure protection against unfair trade practice. Owners of IP are granted protection by a state and/or country under varying conditions and periods of time. This protection includes the right to: (i) defend their rights to the property they have created; (ii) prevent others from taking advantage of their ingenuity; (iii) encourage their continuing innovativeness and creativity; and (iv) assure the world a flow of useful, informative and intellectual works.

The convention establishing the World Intellectual Property Organization (WIPO) in 1967, one of the specialized agencies of the United Nations System, provided that "Intellectual Property" shall include rights relating to

i. Literary, artistic and scientific works;
ii. Performance of performing artists, phonograms and broadcasts;
iii. Inventions in all fields of human endeavor;
iv. Scientific discoveries;
v. Industrial designs;
vi. Trademarks, service marks, and commercial names and designations; and
vii. Protection against unfair competition and all other rights resulting from intellectual activity in the industrial, scientific, literary, or artistic fields.

## PROTECTION OF INTELLECTUAL PROPERTY

Intellectual property is protected and governed by appropriate national legislations. The national legislation specifically describes the inventions that are the subject matter of protection and those which are excluded from protection. For example, methods of treatment of the humans or animals by surgery or therapy, inventions whose use would be contrary to law or morality, or inventions which are injurious to public health are excluded from patentability in the Indian legislation.

With the growing recognition of IPR, the importance of worldwide forums on IPs is realized. Companies, universities and industries want to protect their IPR internationally. In order to reach this goal, countries have signed numerous agreements and treaties and developed organizations.

## World organizations

GATT   General Agreement on Tariffs and Trade
TRIPs  Trade Related Aspects of Intellectual Property Rights
WIPO   World Intellectual Property Organization
UPOV   International Union for Protection of New Plant Varieties
WTO    World Trade Organization

GATT was framed in 1948 and was meant to be a temporary arrangement to settle amicably among countries disputes regarding who gets what share of the world trade. GATT serves as a code of rules for international trade and a forum to discuss and find solutions to trade problems of the member countries. Agriculture as a tradable commodity was included for the first time in the 8th round of GATT negotiations in 1986 at Uruguay. However, the Uruguay round discussions were successfully concluded on December 15, 1993 because of apprehensions and serious concerns of developing countries due to inclusion of agriculture as a tradable commodity. The negotiations were signed in the form of an accord on April 15, 1994 at Marrakesh, Morocco, by about 124 countries and lead to the formation of World Trade Organization (WTO) on 1 January 1995. The Uruguay round covered 15 distinct areas and the Agreement on Agriculture (AoA) forms a part of the final act of the Uruguay negotiations. The long term objective of the agreement is to establish a fair and market oriented agricultural trading system and that a reform process should be initiated through the negotiation of commitments on support and protection and through the establishment of strengthened and more operationally effective GATT rules and disciplines. In this 8th round of negotiations another decision was taken to form WTO as a rule-based body for all trade and trade-related issues. Thus WTO became operative from 1st January 1995. WTO is a multilateral trading system. It has now 153 members. It envisages a comprehensive scheme for protection of IPRs and the establishment of a legitimate reward system for the creative inputs of the inventors of intellectual property under the broad category of one of the WTO agreements i.e. Trade Related Aspects of Intellectual Property Rights (TRIPs). Various articles under TRIPs cover different forms of IP, *viz.* patents, trademarks, copyrights, designs, etc. All these forms of IP have a bearing on biotechnology and agriculture sector.

In 1961 at Paris, a "Union International Pour La Protection Des Obtentions Vegetables" UPOV (International Union for the Protection of New Varieties of Plants) was signed for coordinating the inter country implementation of PBR and it entered into force in 1968 with its headquarters based at Geneva.

## Forms of protection

Usually IPRs are protected by the following legal theories:
- Patents
- Copyrights
- Trademarks
- Trade secrets
- Geographical indications
- Designs
- Layout design of integrated circuits

Of these, patents are the most important forms of protection for research and development organizations. One of the most important examples of IP is the processes and products that result from the development of genetic engineering techniques. Another example of IP is the development of crop varieties, which are protected through Plant Breeder Rights or PBRs. Thus, a better understanding of patents as the form of IP by research scientists and university/institute administrators will increase the pace of research for technological developments in biotechnology. Before describing the patents, the other forms of IP have been explained briefly.

## Copyright

Copyright protects only the form of expression of ideas, not the ideas themselves. The creativity protected by copyright law is creativity in the choice and arrangement of words, musical notes, colors, shapes and so on. Copyright was created to provide protection to composers, writers, authors and artists to protect their original works against those who copy; those who take and use the form in which the original work was expressed by the author. The copyright vest in original work in whatever form it may be. The owner of a registered copyright enjoys the ability of blocking the unauthorized copying or public performance of a work protected by copyright and provides the power to exploit the same for the period prescribed under the Act.

Computer software/program is another mode of expression. A computer program is separately defined in the Act as "set of works expressed in words, codes, schemes or any other form including a machine readable medium capable of causing computer to perform a particular task or achieve a particular result." A computer program is produced by one or more human authors but, in its final mode or form of expression it can be understood directly only by a machine (the computer) not by human readers. In India, The Copyright Act 1957 as amended in 1994 is in force. The Copyright protection of computer software is under the Information Technology Act, 2000.

In biotechnology, the copyright may cover DNA sequence data that may be published. However, an alternative sequence coding for same protein may be prepared using wobble in the genetic code, so that the copyright is not infringed.

The rights bestowed by the law on the owner of copyright in a protected work are described as exclusive rights to authorize others after permission to use the protected works. Copyright protection is generally valid in the country that grants the protection. Many countries respect the copyright of other countries, but if one wants protection in a certain country, it is best to apply for a copyright in that country. The agriculture industry uses copyright protection regularly. Directions on the use of a product and descriptions of products are few examples of copyright use.

It is not necessary under the Indian Copyright Act to register with the Copyright Office to get copyright protection. A work is protected by a copyright at the moment it is created in a tangible form (written copy, recorded music, filmed movie, digital data saved on a computer disk). Registration of the work is however highly recommended because such registration is helpful in an infringement suit. As per the Copyright Act, the register of copyrights (where the details of the work are entered on registration) is a prima facie evidence of the particulars entered therein. The documents purporting to be copies of any entries therein, or extracts from the register which are certified with the seal by the Registrar of Copyrights, are admissible as evidence in all courts. India is a member of both Berne and Universal Conventions and Indian law extends protection to all copyrighted works originating from any of the convention countries.

Copyrights for different aspects have different time periods of protection viz. sixty years for literary, dramatic, musical or artistic work, fifty years for performance rights and 25 years for broadcasting rights

In Geneva on December 20, 1996, The WIPO Diplomatic conference adopted two treaties namely WIPO Performance and Phonogram Treaty (WPPT) and WIPO Copyright Treaty (WCT). Both the treaties address the challenges posed by digital technologies, in particular the dissemination of protected material through internet and therefore these treaties are also referred to as Internet Treaties.

## Trademark

It is a mark capable of being represented graphically and which is capable of identifying and distinguishing the goods or services of one person from those of others and may include any word, name, design or device, shape, number, slogan, smell, sound or anything that helps the consumer to identify the products and/or companies. In short, a trade mark is a brand name. In modern times, trade marks have developed in to identifiers of products from individual companies and are important business assets. Today trade mark has become almost synonymous with 'brand'.

In research, laboratory equipments bear trademarks that are well known to workers in their field. Certain vectors useful in recombinant DNA technology are also known by their trademarks.

Commercial names and designations constitute another category of elements of intellectual property. Trade names are generally names, terms or designations that serve to identify and distinguish an enterprise and its business activities from those of other enterprises, whereas trademarks are used to differentiate between a company's product and all other products. For example slogan "Just Do It" identifies the Nike company.

Trademark law, unlike patent or copyright law, confers a perpetual right. So long as the trademark continues to identify a single source, anyone who uses a very similar mark may be liable for trademark infringement. The perpetual right of trademarks depends on the use. The basic idea of 'use it or lose it' is essential to preserving trademark rights. A company cannot register a trademark and then not use it. The product for which the trademark was registered must be used commercially.

India has a Trade Marks Act, 1999. Overall there are 42 classes distinguishing the goods and services. There are 34 classes of goods and 8 classes of services. Essential features of a trademark are distinctiveness whether inherent or acquired. Only the owner of a trade mark may file an application for its registration. Generally, the person who uses or controls the use of the mark, and controls the nature and quality of the goods to which it is affixed, or the services for which it is used, is the owner of the mark such as companies, individuals, partnerships, trade unions or lawful associations. The duration of a registered trade mark is for a period of ten years. It can be renewed for another ten years from the date of expiration of the original registration by making an application in a prescribed manner within a prescribed period and subject to payment of the prescribed fees.

Trademark rights are so important that multinational companies spend huge amount of money to maintain their respective trademarks around the world. Every country has different trademark law. However, there are agreements to ensure that a company's trademark in one country is protected in another country.

Besides trademarks there are other marks also: (i) Certification marks—a mark capable of distinguishing the goods or services in connection with which it is used in the course of trade that are certified by the proprietor of the mark in respect of origin, material, mode of manufacture of goods or performance of services, quality, accuracy or other characteristics from goods or services not so certified and registrable like AGMARK, FPO, ISO, etc.; (ii) Service marks—means service of any description which is made available to potential users in connection with business of any industrial or commercial matters such as banking, communication, education, financing, insurance, chit funds, real estate, transport, storage, material treatment, processing, supply of electrical or other energy, boarding, lodging, entertainment, amusement, construction, repair, conveying of news or information and advertising, etc. like LIC, SBI, PNB; (iii) Collective marks—a trade mark or service mark used, or intended to be used, in commerce, by the members of a cooperative, an association, or other collective group or organization, including a mark that indicates membership in a union, an association, or other organization e.g. INTUC, Automobile Association (AA), etc.

Rights of registration of trade mark are valid in a country where it was registered. If the products are sold in other countries, one should consider applying for foreign registration. However, there is a provision of international registration also. There is a Madrid Agreement which came onto force in 1989 and was meant for repression of false or deceptive indications of source on goods. A mark to be registered in member states should be first registered at the national level in the country of origin of the applicant. The first registration is called 'basic registration'. The country having given the basic registration can only transmit the request for international filling to the International Bureau of the World Intellectual Property Organization (WIPO) along with the list of the countries in which protection is being sought. There is no provision for direct filing a request under the agreement. A Madrid protocol came into force on December 1, 1995 to remove some of the discrepancies in the Madrid Agreement especially concerning the International Registration of Marks. Besides, there is Nice Agreement concerning the International Classification of Goods and Services for the purpose of the Registration of the Marks (1957) and Vienna Agreement establishing an International Classification of the Figurative Elements of Marks (1973).

### Patent

A patent is a government granted exclusive right to an inventor to prevent others from practicing i.e. making, using or selling the invention. The word patent is derived from the Latin word 'Patere' which means 'to open'. A patent is a personal property, which can be licensed or sold like any other property. As the right is conferred by the State, it can be revoked by the State under very special circumstances even if the patent has

been sold or licensed or manufactured or marketed in the meantime. A patent is territorial in nature i.e. within the boundaries of a particular country, which has given the patent.

The purpose of a patent is to encourage and develop new innovations. The Patent Law recognizes the exclusive right of a patentee to gain commercial advantage out of his invention. There are three criteria to issue a patent for the innovation.

i. *Novelty*: Novelty is assessed in global context. An invention will be considered novel if it does not form a part of the global state of the art. The inventor must establish that the invention is new or novel. The novelty requirement refers to the prior existence of an invention. If an invention is identical to an already patented invention, the novelty requirement is not met, so a patent cannot be issued. An invention will not be novel if it has been disclosed in the public through any type pf publications in the magazines, technical journals, books, newspapers, etc., or in radio, T.V. anywhere in the world before the filing of a patent application. Novelty is determined through extensive literature and patent searches. It should be realized that patent search is the most crucial parameter for ascertaining the novelty.

ii. *Inventiveness (Non-obviousness)*: It is an invention and not merely discovery. It is non-obvious to one skilled in the field. The non-obvious requirement refers to the level of difficulty required to invent the technology. If an invention is so obvious that anyone having an ordinary skill would have thought of it, then it does not meet this requirement.

iii. *Usefulness (Industrial application)*: It has a utility or is useful for the society. The useful requirement refers to the practical use of invention. If an invention provides a product that is required or needed in some manner, then it meets this requirement

In the patent adequate disclosure should be made so that others can also work on it. It should have the features: (i) be a written description; (ii) enables other persons to follow; (iii) adequate and (iv) deposit mechanism.

There are three types of patents: (i) plant patent, (ii) design patent, and (iii) utility or regular patent.

Plant patents are granted for newly developed asexually propagated plants. It provides protection for 20 years. A design patent protects ornamental characteristics. The life span of a design patent is only 14 years. Examples of design patents are the companies dealing with toys, souvenirs and industrial manufacturers. Utility patent is the most common patent used by universities and companies to protect their results of research and development. In USA and most of the countries it has a life span of 20 years from the date of filing an application as it is one of the requirements of TRIPs regulations.

Most biotechnology inventions are filed as utility patents and not as plant patents. As a utility patent, it is possible to protect plant genes and other controlling elements rather than just the plant, and to control the use of genetic material of a number of plants and for multiple uses such as disease resistance, herbicide resistance, or pharmaceutical or oil production.

The purpose of a patent is to promote the progress of science and useful arts. The patent law promotes this progress by giving the inventor the right of exclusion. In exchange for this right to exclude others, the inventor must disclose all details describing the invention, so that when the patent period expires, the public may have the opportunity to develop and profit from the use of invention.

## Patent application

There are different types of patent application for grant of patent:
1. Ordinary application
2. Convention application
3. PCT international application
4. PCT national phase application
5. Application for patent of addition
6. Divisional application.

Normally an ordinary application is filed when no priority is claimed by giving reference to other application under process or any application made in a convention country. A convention application is that application if a country is a member of convention countries which are group of countries or a union of countries or an Inter-governmental organization which are notified by the Ministry of Commerce and Industry, Government of India.

Filing of an application for a patent should be completed at the earliest possible date and should not be delayed until the invention is fully developed for commercial working. Further, a published or disclosed invention cannot be patented. Hence inventors should not disclose their inventions in symposia or as research papers before filing a patent application. The preparation of a patent application is quite complex and generally an attorney is required to draft the application. Basically there are two types of patent documents: Application with provisional specification or complete specification. A **Provisional Specification** application should contain the description of the invention with drawing if required. But there are no claim(s). This document is submitted when inventor(s) is/are still working on the invention. This is being done to claim the priority of an invention. However, the complete specification including claims fairly based on the matter disclosed in the provisional specification should be filed within 12 months from the date of filing of provisional specification. The date of patent will be the date of filing (Priority date) of provisional specification. An application with **Complete Specification** should fully describe the invention with drawing(s) if required disclosing the best method known to the applicant and it ends with claim(s) defining the scope of protection sought. Writing of claim(s) is the most important component of the patent application. Claims are those portions of the patent that describe what can be accomplished with the invention and what is protected in the patent. This is neither to be too general nor to be too narrowly defined.

Patent with complete specification is a document having the following components:

i. Title of invention
ii. Field of invention
iii. Background of invention with regard to drawbacks associated with prior art
iv. Object of invention
v. A summary of invention
vi. A brief description of the accompanying drawing if any
vii. Detailed description of the invention with reference to examples and drawing
viii Claim(s)

Before filing an application, patent search should be made to determine the novelty of an invention. Certain patent search resources are given Table 30.1. A patent is enforced in the country that issues it. For each country a separate application is to be filed in that country where protection is sought if a country is not a member of patent cooperation treaty (PCT).

### International Patenting and Patent Cooperation Treaty

PCT is a multilateral treaty entered into force in 1978. Through PCT, an inventor of member country (contracting state) of PCT can simultaneously seek patent protection for his/her invention in all/any of the member countries, without having to file a separate application in the countries of interest, by designating them in the international PCT application. India became a member of this treaty on Dec 7, 1998. PCT does not provide an institutional authority for granting international patents. There is, in fact, nothing like an international patent or world patent, which may cover all countries or at least all those who are members of the PCT.

The PCT facilitates obtaining patents in several countries relatively quickly. It standardizes the patent search and examination process so that the initial evaluation done in the country of filing is valid in all member countries. The principal objective of the PCT is to simplify the patenting system and to render more effective and more economical services by the offices which have responsibility for administering it in the interests of the users. To achieve its objective, the PCT:

i. Establishes an international system which enables the filing of applications to a single Patent Office in one language having similar effect in each of the countries which are party to the PCT and which the applicant names in his application;
ii. Subjects each international application to an international search which results in a report citing the relevant prior art which may have to be taken into account in deciding whether the invention is patentable; that report is made available first to the applicant and is later published;
iii. Provides for centralized international publication of international applications with the related international search reports, as well as their communication to the designated Patent Offices; and
iv. Provides the option of an international preliminary examination of the international applications to decide whether the claimed

**Table 30.1:** Patent search resources

| Resource | Internet address |
|----------|------------------|
| USPTO—Patent and trademark office | www.uspto.gov/web/menu/search.html |
| EPO—European Patent Office | http://www.european-patent-office.org/online/index.htm |
| Delphion Intellectual Property Network | http://www.delphion.com/ |
| The French Office (INPI) patent database | http://www.inpi.fr/inpi/inbrevet.htm |
| Free Patent Search Site | http://www.inpi.fr/inpi/inbrevet.htm |
| Biotechnology patents | http://www.nal.usda.gov/bic/Biotech_patents/ |
| Indian patent office database | http://www.ipindia.nic.in |

invention meets certain international criteria for patentability.

Further, PCT helps to facilitate and accelerate access by industries and other interested sectors to technical information related to inventions and to assist developing countries on gaining access to technology.

Under PCT, a patent application is first filed either in the home country or another country on the choice of the applicant. The National phase follows the International phase. The applicant has to enter the National phase otherwise International application loses its effect in the designated states. Any application can enter National phase within 31 months from the international filing date. In India Form I A has to be filled for National phase entry. It is not mandatory to submit the documents while entering the National phase since it is obligatory on the part of World Intellectual Property Organization (WIPO) to send the required document to designated office. The applicant has to deposit the national fee, usually the same as the fee required for the filing of a national or regional application.

In the normal case, if applicant is not taking the PCT route or a country is not a member of PCT then an applicant has to do multiple filing in different countries which makes the task difficult, besides expenditure on filing, search, translation, etc. In contrast, PCT establishes an international system which requires filing of only one additional application which is called "International Application" (IA) at a designated patent office available in all PCT member countries and patent can be written in one of the languages approved in the PCT country where the IA is filed.

India is a member state of WIPO which acts as an International Organization responsible for the promotion and protection of intellectual property throughout the world. There is another international treaty known as The Paris Convention which is meant to promote trade among the member countries and has been devised to facilitate protection of industrial property simultaneously in the member countries without any loss in the priority date. All the member countries would provide national treatment to all the applications from other member countries for protection of industrial property rights. The Convention was first singed in 1883. Since then the Convention has been revised several times, in 1900 at Brussels, in 1911 at Washington, in 1925 at The Hague, in 1934 at London, in 1958 at Lisbon and in 1967 at Stockholm. The last amendment took place in 1979. India became a member of the Paris Convention on December 7, 1998. (Readers may note the use of the phrase 'Industrial Property' and not Intellectual Property).

***Use of technical information in patent documents:*** Patent literature has an important use as a source of useful scientific information:
A) It is useful in searching for new technical solutions in the course of research and development and the creation and assimilation of new technology.
B) A scientist or technologist when beginning a new project should first study the patented literature and other information sources to find out what should not be invented, what technical solutions have already been found, when, by whom and where. The timely use of patent information makes it possible to avoid unnecessary expenditure and save time and resources. It also provides information on the trends of the development of knowledge in that field of research.

*Revocation of patent*

A patent is a monopoly granted by the statute in consideration of disclosure of invention. It is a limited monopoly for something that will pass to the public domain after expiry of the term of the patent. A patent granted does not mean that all the claims made in the specifications are valid. The validity of a patent can be challenged even after it is granted. Such kind of objections raised after the grant of patent is called revocation. A revocation can also be filed even after the term of a patent has been expired. Need for revocation of patent arises due to the fact that when patent application is filed, it is difficult to examine whether the innovation is new or it is dressed up. Also public deserves an opportunity to challenge the patent. Generally there are 4 months for opposing the patent and if it was not opposed during that period or any organization misses an opportunity for opposition, then it goes for revocation of patent. There are three modes of revocation of patents in India.

1. Central Government
2. Controller
3. Courts

There are various grounds of revocation of a patent:

i. in the public interest;
ii. nonworking of patent;
iii. patentee not entitled to the patent;
iv. patent wrongfully obtained;
v. failure to disclose the details;
vi. noncompliance with secrecy order;
vii. subject of claim is not an invention;
viii. not a patentable invention;
ix. subject matter lacks novelty;
x. lacks inventiveness;
xi. subject matter not useful;
xii. not sufficiently described;
xiii. claims not clearly defined

### Patenting of biological material

Biotechnology is the synergistic union of the biological sciences and the technologically based industrial arts. It is the utilization of biological processes for the exploitation and manipulation of living organisms or biological systems, in the development or manufacture of a product or in the technological solution to a "real-world" problem. Patent laws in most of the countries were tuned for non-biologic material. The general prerequisites for patentability, namely novelty, inventiveness and utility apply to biotechnology inventions as well. In biotechnology the basic aspect is biological material or biological process or biological

product with industrial application. But the issue of whether living organisms, such as microorganisms, plants or animals, or naturally occurring substances, such as DNA and proteins, cloning and bioinformatics may constitute the subject of an invention is very controversial and hence considered separately.

**Microorganisms**: Louis Pasteur, the famous French scientist, received U.S. Patent No. 141,072 on July 22, 1873, claiming "yeast, free from organic germs of disease, as an article of manufacture." But the US courts later held that the "discovery of some of the handiwork of nature" was unpatentable. With the phenomenal growth of genetic engineering in the late 1970s, the patentability of living microorganisms came in to the scene. In 1980, Dr Chakrabarty got a patent (No. 4259444) on the discovery of oil eating *Pseudomonas* bacteria (*P. aeruginosa* and *P. putida*), thus the microorganism life forms could be patented in USA. USPTO rejected the claims on *Pseudomonas* bacteria as product of nature, but the case went to Supreme Court and decision was in favor of Chakrabarty in a landmark case, *Diamond (USPTO commissioner) vs. Chakrabarty (inventor)*. Chakrabarty's *Pseudomonas* bacteria were manipulated to contain more plasmids controlling the breakdown of hydrocarbons was "a new bacterium with markedly different characteristics from any found in nature". The Supreme Court stated that new microorganisms not found in nature were "manufacture" or "composition of matter" within the meaning of US Patent Act $\gamma 101$ and thus patentable. The "product of nature" objection as mentioned by USPTO for rejection of patent therefore failed and the modified organisms were held patentable. Following this decision, EPO and the Japanese Patent Office (JPO) also started granting patent protection for microorganisms in 1981. A provision of EPC, Article 53(b) is relevant here which states that patents shall not be granted for "plant or animal varieties or essentially biological processes for the production of plants or animals, however the provision does not apply to "microbiological processes or the products thereof" (Chawla, 2005).

The microorganisms and microbiological inventions can be patented in India under Patents Act, 1970 provided the strain is new. This provision of patent protection to microorganisms has been implemented from 20th May 2003 in India. Inventor has to deposit the new strain in any recognized international depositary. Budapest Treaty is an international convention governing the recognition of microbial deposits in officially approved culture collections which was signed in Budapest in 1973 and later on amended in 1980. There are 34 International depositaries for deposition of microbial cultures. India signed the Budapest treaty on 17 December 2001. In India, Microbial Type Culture Collection and Gene Bank (MTCC) at Institute of Microbial Technology (IMTECH), Chandigarh is a recognized international depositary for certain kinds of microorganisms. Before filing the patent application involving microorganism, inventor(s) have to deposit the strain to IDA, which will issue the number and this has to be quoted in the patent application.

**Plant patents:** Plant patents are obtainable in US, South Africa and Japan. The US Plant Patent Act of 1930 (PPA) granted property rights for privately developed plant varieties of asexually reproducing plants. These rights were extended to new and distinct asexually varieties for a period of seventeen years. Advances in breeding technology provided the momentum for the 1970 Plant Variety Protection Act (PVPA). The PVPA provided protection for sexually reproducing plants, including seed germination. In 1980 Diamond vs. Chakrabarty case set in motion the trend towards the legal acceptance of the commodification of germplasm. In 1985, The US Patent and Trademark Office gave a ruling that plants previously under breeder's rights are patentable subject matter under industrial patent law and utility patents could be granted. Among higher plants, "tryptophan-overproducing maize" obtained through tissue culture known as the "Hibberd Patent," was the

first plant patent issued to Molecular Genetics Research and Development, USA. Following this, US utility patents could be granted for other types of plant e.g. genetically modified plants. Plant patents have been granted by EPO from 1989. In 1993, Agracetus Company, USA, claimed broad patent protection over all genetically engineered cotton varieties regardless of how they are produced. The broad-based claim of patent protection over an entire species was unprecedented and was challenged. In 1994, Agracetus Company, USA obtained a similar patent on all transgenic soybean plants in Europe. But in 1995, EPO severely restricted the scope of patent on herbicide resistant plants and allowed claims only on the herbicide resistant gene and the process used in the generation of plants. In Japan, plant patents are allowed, but there are some disputes over territorial rights. Life forms of plants and animals except microorganisms are not patentable in India. In pursuance to the TRIPs agreement, India has enacted "Protection of Plant Varieties and Farmers' Rights" (PPV&FR) Act, 2001, a *sui generis* system of plant variety protection and Plant Breeders Rights have been described in detail separately.

**Animal patents:** The question of whether multicellular animals could be patented was examined by the USPTO in 1980s. In 1987, Ex Parte Allen case the key issue was the patentability of polyploid pacific coast oysters that had an extra set of chromosomes. However, USPTO rejected the patent application on the ground of obviousness On April 12, 1988 USPTO issued the first patent on transgenic non-human animal "Harvard Mouse" (U.S. Patent No. 4,736,866) developed by Philip Leder (Harvard University) and Timothy Stewart. The "Harvard Mouse", was created through a genetic engineering technique of microinjection. To the fertilized egg a gene known to cause breast cancer was injected and then this egg was surgically implanted in to the mother so that she may bring it to the term. The resulting transgenic mice were extremely prone to breast cancer.

After initial reluctance by the EPO, European patent was issued in 1992. By 2002 more than 300 patent applications for transgenic animals have been filed but so far few have been granted by EPO. In Japan animals became patentable subject matter after 1988 when the "Harvard Mouse" patent was issued by the USPTO and JPO issued patents on animals, the majority of them being the products of genetic engineering.

Indian Patents Act, 1970 amendment 2005 has excluded from patentability under section 3(j) "plants and animals as a whole or any part thereof other than microorganisms but including seeds, varieties and species and essentially biological processes for production or propagation of plants and animals" and section 3(i)"any process for medical, surgical, curative, prophylactic (diagnostic, therapeutic), or other treatment of human beings, or any process for a similar treatment of animals to render them free of disease or to increase their economic value or that of their products". This is in pursuance to the TRIPs agreement Article 27.3(a) and (b). Further TRIPs Article 27.2 mentions that States may exclude from patentability inventions, the prevention within their territory of the commercial exploitation of which is necessary, to protect *ordre public* or morality including to protect human, animal or plant life or health or to avoid serious prejudice to the environment. provided that such exclusion is not made merely because the exploitation is prohibited by law. Thus, human beings or their treatment procedures are neither patentable in India nor anywhere else.

*Cloning*: Cloning is the process of transferring nucleus of an adult multicellular organism's cell to an unfertilized egg of the same species while transgenic cloning is when a particular gene is added to the nucleus of an adult organism cell before its transfer to an unfertilized egg of the same species. Dolly the first mammal sheep was created in 1997 by cloning. Creation of animals by cloning is patentable in some countries. However, patenting of human cloning issue varies in different countries. Japan banned human cloning

in 2001 but had permitted researchers to use human embryos that were not produced by cloning. Recently in July 2004 Japan Government Science Council has permitted limited cloning of human embryos for scientific research. Britain and South Korea also allow cloning of human embryos for therapeutic purposes. However, United States prohibits any kind of human embryo cloning but allows patenting of animal cloning.

In the controversial issue of cloning no attempt has been made to implement strict legislation in US, but in Europe in July 1998, a European directive (98/44/EC) was adopted on the legal protection of biotechnology inventions. Another major difference is that US patents on the human embryonic stem cells have been granted while in Europe the ethics of stem cells patentability is still a controversial subject of debate.

*Biological compounds*: Biological compounds, such as DNA, RNA and proteins, are not themselves living, but naturally occur in nature. The ability to isolate genes and produce the proteins they encode has enormous commercial impact. The availability and scope of patent protection on genes and genome-related technologies is considered vital for the survival and success of the biotechnology industry. Under U.S. Patent law, DNA sequences are considered chemical compounds by the USPTO and are patentable as compositions of matter. In its "Utility Examination Guidelines," the USPTO explained that isolated and purified DNA molecule that has the same sequence as a naturally occurring gene is different from the naturally occurring compound as it is processed through purifying steps that separate the gene from other molecules naturally associated with it and hence eligible for patent protection. If a patent application discloses only nucleic acid molecular structure for a newly discovered gene, and no utility for the claimed isolated gene, the claimed invention is not patentable. Since one of the requirements of a patent is utility. In 1992, The US Government's National Institute of

Health (NIH) applied for patents on thousands of human gene sequences. The patent application was denied in 1993 because NIH had no idea how the human gene sequences could be used or what role they will play in the human body. The main issue was whether these sequences are useful as required in the patent statute. This requirement was interpreted to mean that in the process or product to be patented there should be specific benefit in currently available form. Subsequently the word useful was interpreted to mean practical utility. However, EPO differs in this respect of utility or usefulness criteria, which stipulates that for patentability inventor has to show its industrial application for grant of a patent. As per EPC Implementing Regulations of EU directive (98/44/EC) in 1999 the new provisions are summarized as follows: The definition of biotechnological invention, according to Rule 23b, is invention that concerns "a product consisting of or containing biological material or a process by means of which biological material is produced, processed or used." This includes DNA-related inventions, such as an isolated DNA fragment, gene, DNA sequence analysis protocols and its software products. The definition of biological material is "any material containing genetic information and capable of reproducing itself or being reproduced in a biological system." For example plasmid, which is simply a piece of DNA containing a group of genes which cannot reproduce by itself, but it can be reproduced in a biological system, such as bacteria. The biological materials, such as DNA, protein, plasmids are patentable if the materials are isolated from its natural environment or produced by means of a technical process. Rule 23e further pronounces that the simple discovery of one of the elements of the human body, including the sequence or partial sequence of a protein or a gene, cannot constitute patentable inventions if industrial application, i.e. utility, of the claimed gene or protein sequences or a partial sequence is not disclosed in the patent application.

Thus USA and Japan grant patents on all plants of a particular species in to which a specific new gene is inserted by biotechnological means. In this way a gene can be patented along with legal claims over the isolated gene and DNA sequences, the genetic engineering tools that use the sequences and over the plants derived from these tools. The USA and Japan have also granted patents on transgenic plants (Chawla, 2005).

Indian Patents Act, 1970 allows inventions on isolation for a substance like DNA, gene sequences are patentable if function has been ascribed to that gene sequence. The JPO also points out that since "the aim of the patent law is to develop industries, only inventions that are useful or having industrial applicability are patentable".

*Patenting Procedure in India*: An application is required to be filed according to the territorial limits where the applicant or the first mentioned applicant in case of joint applicants for a patent, normally resides or has domicile or has a place of business or the place from where the invention actually originated. In India Head Patent Office is located at Kolkata whereas branch offices for different zones are in Delhi, Mumbai and Chennai. Application may be made by the inventor, either alone or jointly with another inventor, or assignee of the inventor or legal representative of the inventor. A patent application is to be submitted in the prescribed form for one invention only. Every application shall accompany provisional or a complete specification. A provisional specification is a document which contains the description regarding the nature of an invention but it does not contain every detail regarding the invention and also no claims. This type of application is filed to claim the priority date of an invention. The complete specification after provisional must be filed within 12 months from the date of filing the provisional specification. The document that contains the detailed description of invention alongwith the drawings and claims is called patent application with complete specifications. Claims refer to the scope of protection sought.

There are different types of patent applications but most common is an ordinary application which is made in the Patent Office without claiming any priority of application made in a convention country or without any reference to other application under process in the office. Besides, there are applications as patent of addition, convention application, divisional application, patent cooperation treaty (PCT) national phase and international phase application. All the patent offices in India have been designated as receiving offices (RO) for acceptance of PCT applications. This PCT route helps in saving the initial investments towards filing fees, translation, etc. World Intellectual Property Organization (WIPO) office situated at Geneva coordinates all the activities of PCT. Every application shall be made in the prescribed forms.

It is a common experience that through ignorance of patent law, inventors act indiscreetly and jeopardize the chance of obtaining patents for their inventions. The most common of these indiscretions are: a) to publish their inventions in newspapers or scientific and technical journals, before applying for patent; b) to wait until their inventions are fully developed for commercial working, before applying for patents. Delay in making application for a patent involves risks, namely, (i) other inventors might forestall the first inventor in applying for the patent, and (ii) there might be either an inadvertent publication of the invention by the inventor himself, or the publication thereof by others independently of him.

### Geographical Indications

The geographical indications (GIs) for the purpose of TRIPs agreement have been described as indications which could be used to recognize that a good has originated in a particular territory, region or locality, where the given quality, reputation or other characteristic of the good are essentially attributable to its geographical origin. This is another way of protecting a country's biodiversity assets which are uniquely endemic to certain geographical locations in the country and has been accepted in terms of quality and traits to that geographical region is through appropriate national legislations. Protection of GI under the TRIPs agreement is defined in two Articles (Articles 22 and 23). All products are covered by Article 22 which defines a standard level of protection. Article 23 provides a higher or enhanced level of protection for GIs for wines and spirits. Products such as Scotch whisky, Champagne, California wines fall under this category. Protection of GIs may be at the national or regional level. The word 'Tuscany' for olive oil produced in a specific area of Italy, or 'Roquefort' for cheese produced in France is protected in the European Union and in US as US certification mark. A number of countries want to extend this level of protection to a wide range of other products.

GI can be used, when protected in legal terms, to prevent others from misleading the public or creating unfair competition in trade for the goods that have their origin from a particular territory, region or locality. In the TRIPs there is an important provision that if some one uses a GI as a trademark and it misleads the public as to the true place of origin, then registration of trademark is refused or invalidated *ex officio*.

India enacted The Geographical Indication of Goods (Registration and Protection) Act, 1999. Prior to that there was no specific law governing GIs of any agricultural, natural or manufactured goods including food stuff. India being party to the TRIPs agreement, was required to extend protection of GI for goods imported from other countries, which provide for such protection. A law to the effect in the country was necessary because other member countries would be obliged to give protection to the goods only if there are laws to protect the goods in the country of origin. GI as per Indian Act means an indication in relation to goods, which identifies such goods (agricultural goods, natural goods or manufactured goods or any goods of handicrafts or of industry, including food stuff) as originating, or manufactured in the territory of country, or a region or locality in that territory as the case may be, where a given quality, reputation or other characteristics of such goods is essentially attributable to its geographical origin. GI is not registered on the individuals name as in the case of patent or trademark but it is given to any association of persons, producers, organization or authority established by or under the law. The applicants must represent the interest of the producers and their names should be entered in the register of GI as registered proprietor for the GI applied for. A producer includes any person who trades or deals in such production, exploitation, making or manufacturing as the case may be of the goods. The registration of a GI is valid for 10 years which can be further renewed for 10 years by repayment of fees.

GI cannot be a trademark, the only exception is that trademark held prior to the Act or in good faith. According to GI an indication includes any name (excluding the name of origin country, region or locality), geographical or figurative representation or any combination of them. A registered GI can provide the benefits: (i) it confers legal protection to facilitate an action for infringement in India; (ii) it prevents unauthorized use of a registered GI by others; (iii) it boosts export of GIs by providing legal protection; (iv) it promotes economic prosperity of producers; (v) it enables seeking legal protection in other WTO member countries. Since GIs are inherently collectively owned, they act as an excellent tool for regional or community-based economic development. Any region that has a specialty associated with it, where a quality link exists or can be established between the product and the region, should consider the advantages of using a GI to distinguish its product from lower-quality and non-regional competitors.

Different types of goods including handicrafts (Channapatna toys, Mysore rosewood inlay), textiles (Pochampally Ikat,

Chanderi saree, Kota Doria, Mysore silk, Kullu shawls), agricultural (Darjeeling tea, Coorg orange, Mysore betel leaf, Nanjanagud banana, Palakkadan Matta rice) have already been registered as GIs in India.

## Trade Secret

A trade secret is any information that gives a company a competitive edge over competitors and which the company maintains as secret and away from public knowledge. Trade secrets often include private proprietary information. Examples of trade secrets are the Coca Cola company brand syrup formula, Polaroid company instant film chemical formula, etc. The nature or the identity of a product is maintained secret for as long as the company can keep this information from becoming public knowledge.

Trade secret rights are mainly kept and enforced through agreements between employers and employees. These are non-patented. Usually at the time employment begins, an employer makes an employee sign an agreement that grants the employer trade secret protection. The agreements protect the company by preventing its competitors from enticing key personnel since these individuals cannot divulge the trade secret material without incurring severe penalties. Criminal prosecution of an employee who steals trade secrets from their employer is a recognized remedy. Some examples of trade secret misappropriation are given in Boxes 1 to 3.

Trade secrets have an unlimited term, which may be perpetual. The term is as long as it takes the public or a competitor to determine how to make the product and to ascertain the nature and identity of the trade secret. Disclosure of a trade secret and its unauthorized use can be punished by the court and the owner may be allowed compensation. However, if a trade secret becomes public knowledge by independent discovery or other means, then it is not punishable by court. There is no separate law or Act as these are dealt under the civil law.

---

### Box 1: Trade secret misappropriation of Brandon Process

This is an example of trade secret litigation between two pharmaceutical companies and at stake were billions of dollars and thousands of jobs. On one side was Wyeth—one of the world's foremost manufacturers and makers of Premarin, a hormone replacement therapy drug which has been sold in US since 1942. Wyeth makes Premarin from the estrogens found in the pregnant mare urine (PMU) using a closely guarded secret chemical process known as Brandon Process named after the manufacturing facility in Brandon, Manitoba. Brandon Process extracts estrogens from PMU and transforms them in to dry powder—Preserved condensed urine desiccated (PCUD). Brandon Process is neither generally known nor readily ascertainable. For more than sixty years, no company has ever been able to replicate the Brandon Process. Brandon Process is not patented. In 1940's and 1950's Wyeth obtained patents on several methods discovered in connection with its estrogen extraction research and the last patent expired in 1975. On the other side was Natural Biologics—David Saveraid (Agricultural salesman) is the founder and President, a start up company in Albert Lea, Minnesota, where it extract natural conjugated estrogens from PMU in the form of a dry bulk product. It intended to market a generic form of Premarin. An eight year court battle fought over trade secret misappropriation. Finally the Court concluded that measures taken by Wyeth's information protection efforts fell in the upper end of the range of reasonableness for corporate security programs. Natural Biologics was "a company built upon misappropriation". Judge found that the Company's CEO (David Saveraid (Agricultural salesman) had conspired with a former Wyeth scientist (Dr. Irvine, a chemist) to steal the company's secret Brandon Process. Judge ordered natural Biologics shut down, ordered the destruction of all research materials and drug supplies at the company and banned its leaders from ever seeking to re-enter the estrogen replacement market.

---

### Box 2: Imprisonment to Ex-Dupont Chemist for Trade secret misappropriation

Gary Min also known as Yonggang Min, a former chemist of DuPont Co., was sentenced to 18 months imprisonment and a fine of $ 30,000, after he pleaded guilty. Min was accused of stealing $ 400 million worth of confidential DuPont technology to take overseas. Judge Sue Robinson termed the sentence as 'fair', and justified it by stating that there was no evidence of any master plot and that Mr Min's remorse and contrition are beyond question. Moreover, the actual harm to DuPont was found to be negligible. This incident of trade secrets theft surfaced only when Mr. Min started to look for possible job opportunities in July 2005.

---

Trade secrets are much more common in industry where scholarly publication is not required. Trade secrets in the area of biotechnology may include (i) cell lines, (ii) gene transfer parameters, (iii) hybridization conditions, (iv) corporate merchandising plans, and (v) customer lists.

### Designs

The expression design means only the ornamental feature of shape, configuration or pattern applied to any article by any industrial process or means, whether manual, mechanical or chemical, separate or combined, which in the finished article appeal to end are judged solely by the eye. Design means the features of shape, pattern, etc. applied to an article or an object is protected and not the article or object itself. The features are conceived in the author's intellect. He gives those ideas conceived by him a material (visual) form as a pictorial illustration or as a specimen, prototype or model.

---

**Box 3: Jury Awards compensation and damages against Sears for trade secret misappropriation**

The U.S. District Court for the Northern District of Illinois handed down the largest over award of compensation and punitive damages in the case of trade secrets. The dispute arose when Mr Bob Kopras and its company Roto Zip Tool Corp, complained against Sears, Roebuck and Co. for misappropriation of trade secrets. The dispute dates back almost a decade when Roto Zip was one of the major suppliers of Sears. Mr. Kopras, in the year 1999 had disclosed, confidentially, to Sears the draft of a new model of hand held combination saw. No contract was signed between the parties as they failed to agree on the price. With time Mr. Kopras developed on his draft, unknown to the facts that Sears have employed a Chinese company to make a copy of the design against the non disclosure agreement with Mr. Kopras. Kopras launched his finished product, and after a week, in August 2001, Sears introduced the Craftsman Combination Tool. Kopras filed a suit in 2004 complaining of misappropriation of trade secrets. The jury noted that Sears have been selling the device, and that there is a dramatical dip in the sales of Roto Zip's rotary saw. The jury found Sears guilty of willful and malicious trade secret misappropriation and breach of contract.

---

**Box 4: Monsanto Company v. Loren David**

Plaintiff, Monsanto Company has developed Roundup Ready Technology, which involves inserting a chimeric gene into a seed that allows the plant to advantageously continue to break down sugars in the presence of glyphosate herbicide. Crops grown from such seeds are resistant to Roundup and other glyphosate-based herbicides. When, Roundup Ready seeds are planted and used in conjunction with a glyphosate- based herbicide, Roundup Ready plants will survive, while weeds and other plants lacking the Roundup Ready gene will be killed. Plaintiff has claimed this technology in the '605 patent. Plaintiff licenses seed companies to incorporate the Roundup Ready genes into their plants and to sell soybean seeds containing the Roundup Ready gene. All purchasers of such seeds are required to enter into a Technology agreement that grants them the right to use the seeds. The Technology agreement stipulates that buyers may use the seeds for the planting of only single commercial crop, but that no seeds from that crop, may be saved for future harvests. Defendant, Mr. Loren David, a commercial farmer who owns soybean fields in North and South Dakota executed a Monsanto Technology Agreement and planted the contested soybeans at issue in this case. Plaintiff alleged that the seeds that Defendant planted were Roundup Ready soybeans improperly saved from the previous year's harvest and thereby violated the Technology agreement. Plaintiff filed suit for patent infringement, breach of contract, unjust enrichment, and conversion. The District Court held that Defendant had wilfully infringed the '605 patent and breached the Technology agreement by planting saved seed from a prior year's crop and thereby awarded damages of $ 226,214.40 in favor of Plaintiff attorney fees, prejudgment interest, and costs, bringing the total damages award to $ 786,989.43.

---

In India, The Designs Act, 2000 is in force in India, in which the features are protected as design by registration. The act confers exclusive right to apply a design to any article in any class in which the design is registered.

To qualify for registration, the design mustL

i. be a new or original design and

ii. not have been previously published in India.

## Layout designs of integrated circuits

Integrated circuit topographies are the 3D configurations of electronic circuits embodied in integrated circuits, products or layout designs. Today they are at the heart of modern information, communication, entertainment, manufacturing, medical and space technologies. In India layout designs of the integrated circuits are covered under the Semiconductor Integrated Circuits Layout Design Act, 2000.

## Plant Breeders Rights

In order to understand the implication of IPRs on plant varieties and on agricultural production, one has to understand the plant breeders rights (PBRs) and how does it differ from patents. PBR refers to the legal protection provided to a breeder, originator, or owner of a variety. The concept is based on the realization that if commercial plant breeding is to be encouraged for the benefit of agriculture and society, measures have to be taken to allow breeders to profit from their products (Mishra, 1999).

### History

As per Article 27.1 all inventions, whether products or processes, in all fields of technology provided they are new, involve an inventive step and are capable of industrial application shall be eligible for patents. However, Art 27.3 (b) mentions that parties may exclude from patentability plants and animals other than microorganisms, and essentially biological processes for the production of plants or animals other than non-biological and microbiological processes. However, for plants, they were required to provide the protection by an effective *sui generis* system or patents or a combination of patents and *sui generis* system.

India and so many other countries like Thailand do not protect plants by strict patenting system. But there was a mandate in the TRIPs agreement that plant varieties must be protected. In pursuance to the TRIPs agreement, India has enacted "Protection of Plant Varieties and Farmers' Rights" (PPV&FR) Act, 2001, a *sui*

*generis* system of plant variety protection. The model for this was the UPOV Act.

In US Plant Patent Act of 1930 (PPA) granted property rights for privately developed plant varieties of asexually reproducing plants. These rights were extended to new and distinct asexually varieties for a period of seventeen years. This legislation departed from the US Patent law because living things could receive a plant patent under a more lenient standard than the traditional utility patent requirements of being useful, non-obvious and novel. The protection provided by PPA encouraged the privatization of seed industry, even though seeds were not included under the PPA. Advances in breeding technology provided the momentum for the 1970 Plant Variety Protection Act (PVPA). The PVPA provided protection for sexual reproducing plants, including seed germination. With this Act most commercial crops were protected by patent laws for seventeen years but it was limited by two major exemptions: seed saving by farmers and for research purposes. Under the PVPA "brown bag" exemption, farmers could continue to save, replant and resell protected seeds to other farmers (Chawla, 2007). In 1980 Supreme Court decision on Diamond vs. Chakrabarty case set in motion the trend towards the legal acceptance of the commodification of germplasm and held that the modified organisms (*Pseudomonas*) were patentable. In 1985 USPTO granted a patent on tryptophane overproducing maize obtained through tissue culture (*Hibberd* patent). After *Hibberd,* the PTO granted over 1800 utility patents for germplasm. The common law right of saving seeds was further eroded by Asgrow Seed vs. Winterboer. Thus the seed saving exemption was limited to farmers for replanting in their own farms. Another court decision in a case of J.E.M. AG Supply v. Pioneer Hi-Bred International, Pioneer Hi-Bred, a large seed company, sued a small Iowa seed supply company, Farm Advantage, for violating patents on hybrid corn seed. Justice Thomas, writing for the majority, concluded that newly developed plant breeds are covered by utility patents and that neither the PPA nor the PVPA can limit the scope of a utility patent.

The International Organization for Plant Variety Protection (ASSINSEL) was established in 1938 with the objective to persuade different governments to enact laws for the protection of plant varieties. Several countries especially Europe developed their own system of PBR through UPOV [Union for the Protection of New Varieties of Plants; original in French 'Union International Pour la Protection des Obtentions Vegetales']. It is the only intergovernmental organization in the world that is concerned exclusively with the protection of new plant varieties which have been developed by plant breeders. UPOV protects the rights of plant breeders, mainly via a form of IPR called Plant Breeders' Rights (PBRs). Prior to the signing of the WTO Agreement on TRIPs, in 1995, IPRs legislation was a matter of domestic policy, where member nation states were free to choose how best to protect the rights of innovators in all the areas, including agriculture and breeding of plant varieties. The majority of developing countries, as opposed to developed countries, excluded all life-forms such as plant varieties and animals from patentability as it was considered that no-one can 'own' or be granted 'intellectual property' over what already exists in nature. However, Article 27.3(b) of the TRIPs agreement has now made it mandatory for developing countries to provide patent protection for micro-organisms, non-biological and micro-biological processes as well as providing protection for plant varieties either by patents, or by an effective *sui generic* system', which is essentially a form of Plant Breeders rights or by both patents and *sui generic* system.

To grant plant variety protection, there was only one operational *sui generic* model available as UPOV model. However, The UPOV model is not suitable for developing countries. There are enough arguments to demonstrate that UPOV does not address the agricultural concerns of developing country farmers, where agriculture is largely a livelihood for most and agriculture research is largely publicly funded and therefore publicly owned.

## UPOV

UPOV was signed in Paris in 1961 and it entered into force in 1968. The UPOV aimed to ensure protection of Plant Breeder's Rights (PBRs) by the grant of an exclusive right on the protected new plant variety on the basis of a set of uniform and clearly defined principles. The UPOV Act was revised in 1972, 1978 and 1991. Basically UPOV has two Acts of 1978 and 1991. The 1978 Act entered into force in 1981 while the 1991 Act came into force on April 24, 1998. Some countries have ratified the 1991 Act whereas others the 1978 Act. UPOV Acts of 1978 and 1991 have been summarized in Table 30.2.

### Functions of UPOV

The convention has two main functions:

1. Prescribe minimum rights that must be granted to plant breeders by the member countries, in other words, it specifies a minimum scope of protection.
2. Establish standard criteria for grant of protection.

The purpose of the UPOV convention is to ensure that the member states of union acknowledge the achievement of breeders of new plant varieties by making available to them an exclusive property right on the basis of a set of clearly defined principles. Under the UPOV a plant variety qualifies for protection when it meets the following essential criteria:

i. Distinctness (D)—distinct from existing, commonly known varieties by one or more identifiable morphological, physiological or other characteristics.

ii. Uniformity (U)—sufficiently homogeneous or uniform in appearance under specified environment

iii. Stability (S)—must be stable in appearance and its clonal characteristics over successive generations under the specified environment

iv. Novelty (N)—new in the commercial sense that it must not have been commercialized

**Table 30.2:** A comparison between the two UPOV Acts on plant variety protection

| Sr. No. | Issue | 1978 Act | 1991 Act |
|---|---|---|---|
| 1 | Membership | Only a state can be the party | Intergovernmental organization competent for enacting and implementing legislation with binding upon on all its member states can also be the party |
| 2 | Discovery | Breeder is entitled to protection as discoverer irrespective of the origin, artificial or natural of the variation | A mere discovery is not sufficient. The breeder must also have developed the variety |
| 3 | National treatment | Member state may limit the right on a new variety to national of states which also apply that Act. A similar reciprocity rule may also be applied by a member state granting more extensive rights | Reciprocity rule does not apply. Operation of the principle of national treatment to one and all without qualification |
| 4 | Scope | Authorization of breeder is required for<br>i. the production for purposes of commercial marketing of the propagating material;<br>ii. the offering for sale of the propagating material;<br>iii. the marketing of such material;<br>iv. the repeated use of the new plant variety for the commercial production of another variety (e.g. hybrids);<br>v. the commercial use of the ornamental plants or parts thereof as propagating material in the production of ornamental plants or cut flowers;<br>vi. does not require authorization for use of the material for further research;<br>vii. farmers can use/reuse his produce as seed and can dispose off his farm produce. | Authorization of breeder is essential for<br>i. production or reproduction;<br>ii. conditions for the purpose of propagation;<br>iii. offering for sale;<br>iv. selling or other marketing<br>v. exporting;<br>vi. importing;<br>vii. stocking.<br>Furthermore, the 1991 Act specified four subject matters to which the breeder's right extends<br>i. the protected variety itself<br>ii. varieties which are not clearly distinguishable from the protected variety;<br>iii. varieties which are essentially derived from the protected variety;<br>iv. varieties whose production requires repeated use of protected variety. |
| 5 | Minimum number or species to be covered | Atleast 5 to start with<br>Atleast 10 within 3 years<br>Atleast 18 within 6 years<br>Atleast 24 within 8 years | UPOV 1978 member states, all after 5 years transitional period. All after 10 years if only bound by 1991 Act. To start with while acceding 15 plant genera/species |
| 6 | Period of protection | 18 years for grapevines and trees including rootstocks15 years for all others | 25 years for grapevines and tress including rootstocks20 years for all others |
| 7 | Special title/patent protection | Each state is free for any of the two forms of protection except those where it was a practice before October 31, 1979 for providing protection to the same genera or species by both | No alternative forms of protection |

prior to certain date established by reference to the date of application for protection.

v. A unique and unambiguous denomination (name of the new variety).

Application for its protection can be filed in the country where developed or in any other UPOV member country.

As per the 1991 Act of UPOV, such protection is to be granted to new varieties of all genera and species of plants, for a period of 25 years for all trees and vines and for 20 years for all other plants. The protection granted for the new variety authorizes the breeder with the exclusive right to commercially exploit the variety by direct sale or by licensing to others for sale.

**Breeders exemption:** Under PBR regime, the use of material of protected variety (**the initial variety**) for the development of new varieties is exempted from protection. The PBR for these new varieties will be of the breeder who developed them, and the holder of PBR title of the initial variety will have no claim to it. This provision is called **breeders' exemption**, Under the UPOV 1978 Act, all new varieties evolved using a protected variety were exempted from protection under this provision. But UPOV 1991 Act has somewhat limited the scope of breeder's exemption by bringing 'essentially derived varieties' under the cover of PBR protection granted to the initial variety. An essential derived variety is defined as a variety predominantly derived from another initial variety, which retains the expression of essential characteristics from the genotype or combinations of genotypes of the initial variety except for one or few distinguishable characteristics. As a result of this modification, a breeder who inserts a single gene (e.g. disease resistant gene) in to a protected variety will now have to obtain the permission from the original right holder before marketing the new variety.

**Farmers' Privilege:** PBR system generally allow the farmers to use the material of a protected variety harvested on their farm for planting of their new crop without any obligation to the PBR title holder. This exemption is usually referred to as *farmers' privilege*. Under the UPOV 1978 Act, there was explicit provision for farmer's privilege. But in the UPOV 1991 Act, this privilege has been made 'optional' and each UPOV member state can either allow or disallow this privilege. UPOV 1991 Act deprives the farmers of its rights to use, reuse their produce as a seed. Although farmers are broadly exempted from the breeders monopoly for non-commercial use of their produce from a protected variety including propagating another crop from harvested material on their own farm. It should be clearly understood that farmers privilege applies to the use of seed produced by a farmer for sowing 'his own' fields, PBR, however, does not allow farmers to exchange seeds of protected varieties produced on their farms.

At times there are confusions between patent protection and plant variety protection by PBR. Hence a comparison of the two is presented in Table 30.3. UPOV 1991 Act strengthened the position of PBR holder by eliminating the breeder's exemption from an essentially derived variety and more comparable to a patent. Plant breeders rights are related to the rights available to a patentee, where the model of protection is plant patent based. Under the TRIPs agreement, a patent shall confer on its owner the exclusive right to prevent third parties not having owner's consent from the acts of making, using, offering for sale or importing.

Several developing countries are following the UPOV system. Those that joined UPOV before 1999 could join under the less strict UPOV Act of 1978 (e.g. China) but countries that join now have to do that under the UPOV 1991 Act (e.g. Korea, Singapore, Vietnam). Most of the developing countries have opted for the development of their own *sui generis* system. India and Thailand have enacted a *sui generis* system of plant variety protection known as "Protection of Plant Varieties and Farmers' Rights" (PPV&FR) Act, 2001 and Plant Variety Protection (PVP) Act, 1999 respectively. Acts of both the countries are TRIPS compliant. The model for this was the UPOV Act. Most of the PVP laws in Asia have taken the UPOV Act of 1978 as a basis and many have added articles to cater for Farmers' Rights (e.g. India), or for the protection of new, domestic and wild varieties which are in public domain (e.g. Thailand). The Acts of both the countries grants plant breeders protection against unauthorized use of their new plant varieties and other plant varieties which are under cultivation. Regulation and criteria used in Thailand's PVP Act and Indian PPV&FR Act for registering new varieties in order to provide plant breeder's protection is in line with UPOV guidance.

**Table 30.3:** A comparison between patent protection and plant variety protection

| Sr. No. | Issue | Patent protection | Plant variety protection |
|---|---|---|---|
| 1 | Object | Invention | Plant variety |
| 2 | Documentary examination | Required | Required |
| 3 | Field examination | Not required | Required |
| 4 | Plant material for testing | Not necessary | Required |
| 5 | Conditions for protection | a. Novelty<br>b. Industrial applicability<br>c. Non-obviousness<br>d. A disclosure | a. Commercial novelty<br>b. Distinctiveness<br>c. Uniformity<br>d. Stability |
| 6 | Determination of scope of protection | Determined by claims of the patent | Fixed by national legislation |
| 7 | Use of protected variety for breeding further varieties– Breeder's exemption | No; Require authorization of the patentee of the right holder | Yes: No authorization required except for essentially derived varieties (UPOV Act 1991, PPV&FR Act 2001) |
| 8 | Use of propagating material of the protected variety grown by a farmer for subsequent planting on the same farm | Require authority of the patentee | Generally no authorization required |
| 9 | Term of protection | 20 years | 25 years for tree and vines; 20 years for other (1991 UPOV) while 18 years for tree and vines and 15 years for other (PPV&FR Act) |

For most members of UPOV, agriculture is a commercial activity and supply of seeds is under the control of the corporate sector. For the majority of Indians and also in countries like Pakistan, Thailand, Indonesia, Philippines etc. agriculture is a livelihood and farmers are the principal source of seed supply.

### Plant variety protection in India

The Protection of Plant Varieties and Farmers' Rights (PPV&FR) Act 2001

As stated India is signatory to WTO agreements and ratified the agreement on TRIPS. As per article 27.3(b) of the TRIPS which demand that member countries should protect their plant varieties either by patent, or an effective system of *sui generis* protection, or a combination of these two. In this context India chose a *sui generis* system for protection of plant varieties. An Act named as Protection of Plant Varieties and Farmers' Rights (PPV&FR) Act 2001 has been passed by Parliament on Aug. 8, 2001 and Rules have been framed in 2003. PPV&FR Authority has been constituted with its Head Office located at Delhi. The PPV&FR Act is TRIPs compliant and compatible with UPOV system of plant variety protection. India has the observer status in the UPOV.

The PPV&FR Act 2001 provides protection to the following types of plant varieties (Anonymous, 2007):

i. Newly bred varieties.

ii. Extant varieties–The varieties which were released under section 5 of the Indian Seeds Act, 1966 and have not completed 15 years as on the date of application for their protection.

iii. Farmers' varieties–The varieties which have been traditionally cultivated, including landraces and their wild relatives which are in common knowledge, as well as those evolved by farmers.

iv. Essentially derived varieties (EDV)–An EDV is defined as a variety that is predominantly derived from an initial variety and retains the essential characteristics of the latter while having limited difference(s) to qualify distinctiveness from the initial variety.

v. Transgenic varieties.

An application for registration can be made by any person claiming to be the breeder of the variety, successor of the breeder, assignee, any farmer or group of community of farmers, any person authorized for the above mentioned categories or any University or publicly funded agricultural institution claiming to be the breeder of the variety. It is pertinent to note that Act recognizes the farmer as a cultivator, conserver and breeder. This embraces all farmers, landed or landless, male and female.

To qualify for registration under the act, a new variety has to conform to the criteria of novelty (N), distinctiveness (D), uniformity (U) and stability (S). Besides, a denomination has to be given for the registration of variety. Denomination refers to the label or title of the variety. It is the denomination that is registered.

*Novelty*: A variety is deemed novel when it has not been placed for commercial use in India for a period not exceeding 12 months if bred in India. The variety bred outside India is deemed to have novelty if its protection is sought within a period not exceeding 4 years in the case of annuals and six years in case of vines and trees from the date of its registration in another country.

*Distinctiveness*: It means that that the candidate variety should be different from all other varieties of common knowledge in India and outside for at least one essential character.

*Uniformity*: It means phenotypic similarity among plants subject to the variation that may be expected and allowed due to specific nature of the reproduction of the crop such as vegetative, self and cross-pollinated. Hybrids are to be treated like self-pollinated plants.

*Stability*: Its essential characteristics remain unchanged after repeated propagation.

For extant and farmers' varieties which are in public domain the DUS features will be considered while the novelty feature will not be taken because these varieties are not new and are in public domain. In this act a special clause has been put which states that any variety with terminator gene sequences will not be registered. Thus any transgenic material with GURT (Genetic use restriction technology) sequences will not be registered.

All the varieties will be registered with PPV&FR authority. DUS guidelines for 35 crops have been prepared by ICAR while DUS guidelines for 18 crop species have been notified by PPV&FR Authority in the Gazette as Plant Variety Journal of India. PPV&FR Authority has established testing centres for each and every crop species. Authority has started registration of varieties for 12 crop species of cereals and legumes and 6 crop species of jute and cotton. The registration will then extend to 35 crops that include cereals, pulses, oilseeds, vegetable and two flower species. DUS guidelines are also being prepared for medicinal and aromatic plants, spices, ornamentals and forest trees for which task forces have been constituted by the PPV&FR Authority.

**Researchers' rights:** The researchers have been provided access to protected varieties for conducting experiments or research and use of a variety as an initial source of a variety for the purpose of creating other varieties. This is also refereed to as Researcher's exemption. In case a registered variety is required as a parental line for commercial production of newly developed variety then authorization from the breeder of the registered variety is required.

**Duration of validity of certificate of registration for a variety:** It shall have the validity of nine years initially in case of trees and vines with renewal up to a period of 18 years. For other crops, certificate of registration will be issued for six years initially with renewal up to 15 years. In case of extant varieties the validity period is 15 years from the date of notification of that variety by the Central Government under section 5 of the Seeds Act 1966.

**Breeders' rights:** The certificate of registration for a variety issued under this Act shall confer an exclusive right on the breeder or his successor or his agent or licensee, to produce, sell, market, distribute, import or export of the variety. The IPR accorded with a registration of a plant variety is called plant breeder's right (PBR).

## Farmers' rights

Farmers' rights is a concept developed and adopted in the Food and Agriculture Organization (FAO) conference in 1993 and adopted as a resolution 7/93 and endorsed by all member countries. It recognizes that farming communities have greatly contributed to the creation, selection, and preservation of genetic material and knowledge. The indigenous knowledge systems are similar to general scientific information in that they are part of public knowledge. The usual criteria for recognizing IPR, that is, novelty and non-obviousness, tend to ignore the knowledge systems of rural and tribal families. While the knowledge itself may not be patentable, the products of this knowledge namely "folk" varieties and races and genetic diversity provide the basic raw material for modern plant breeding and biotechnology research (Swaminathan, 1994). The plant breeding and seed industry of the developed countries has used this knowledge for development of new varieties without compensating the farmers who had preserved the genetic stock. It is against this background that the concept of farmers' rights emerged in international forum. Therefore, it is the obligation of world community to help these farmers.

Indian PPV&FR Act is unique in the world in the sense that it has given so many rights to the farmers, for their contribution in the economy, conservation of genetic resources and also in the development of new varieties.

i. Farmer's right to register traditional varieties: A farmer who has bred or developed new variety can get his variety registered in the same manner as the breeder of the variety. The farmer's variety qualifies for registration if the application contains a declaration that the genetic material or parental material acquired for breeding the variety has been lawfully acquired. This grants the farmers the exclusive legal right to produce and market its seeds. Further, the farmers do not have to pay any fee either to register their varieties or to renew these registrations.

ii. Farmer's right to sell seed: A farmer is entitled to save, use, sow, resow, exchange, share and sell seeds produced by him from his own harvest. Seeds saved from the harvest of a crop grown with a variety registered under the Act will also be subject to the same and thus the act allows the farmer to reuse and sell seed as has been the traditional practice in farming community.

iii. The farmer is however, not entitled to sell branded seed of a variety protected under the Act. Branded seed means any seed put in a package or any other container and labeled in a way that would suggest the seed was that of the breeder's registered variety.

iv. Farmer's right for reward and recognition: The contribution of the farming community to the conservation of genetic resources and the use of farmer's varieties in breeding is acknowledged and monetary compensation in this respect is to be paid into a specially created Gene Fund. Revenues collected in the Gene Fund will be used to reward the farmers engaged in conservation and improvement of genetic resources. PPV&FR Authority has constituted Kissan Puraskar awards and their recognition will be called as "Plant Gene Saviour Community Recognition" from 2007 onwards.

v. Farmer's right to claim compensation for under performance of a protected variety from the promised level under given conditions. The Authority, after giving notice to the breeder and after hearing the parties, may direct the breeder to pay such compensation as it deems fit.

vi. Protection against innocent infringement: The Act provides farmers from being prosecuted for innocent infringements of the provisions of the Act. A farmer cannot be prosecuted for infringement of rights specified in the Act if he can prove in court that he was unaware of the existence of such a right. This is a significant departure from the general rule that ignorance of law is no exception. The reason for providing such protection is that farmers in general reuse a bag or container carrying label to sell or share seeds with another farmer. A farmer should not be prosecuted for using the name of a breeder, or title, just because it happened to be in the bag.

vii. Farmer's right to information: The Act exempts farmers or group of farmers or members of a village community from payment of any fees in any proceedings before the Authority or Registrar or Tribunal or the High Court under this Act.

viii. Benefit sharing for use of biodiversity conserved by farming community. According to the concept of benefit sharing, whenever a variety submitted for protection is bred with the possible use of a landrace, extant variety or farmer's variety, a claim can be referred either on behalf of the local community or institution for a share of the royalty.

**Compulsory license**: In the Act a provision of compulsory license has also been put. According to this, after the expiry of three years from the date of issue of certificate of registration of a variety, any person interested can claim in an application to the Authority alleging that reasonable requirements of the public for seeds or other propagating material have not been satisfied or that the seed or other propagating material is not available to the public at a reasonable price and pray for the grant of a compulsory license to undertake production, distribution and sale of the seed or other propagating material of that variety.

The PPV&FR Act is notable for farmer's rights and integration of the some of the important elements of CBD, such as prior informed consent for the use of Indian genetic resources, the concept of benefit sharing and creation of gene fund for strengthening agrobiodiversity conservation.

The Central Government has established the Protection of Plant Varieties and Farmers' Rights Authority. It shall be the duty of the authority to promote the development of new varieties of plants and to protect the rights of the farmers and breeders. It will establish a Plant Varieties Registry for the registration of plant varieties. The registry shall maintain a 'national register of plant varieties' with following details which would be kept at Registry Head Office and its branches:

1. Name of the registered plant variety
2. Name and address of breeder(s)
3. The right of breeder(s) in respect of variety
4. Denomination particulars of variety
5. Seed or other propagating material
6. Specification of salient features of seed/variety
7. Other prescribed matters

The breeder shall be required to deposit specified quantities of seeds/propagules of the registered variety as well as its parental lines in the National Gene Bank as specified by the PPV&FRA.

Citizens of convention countries will have the same rights as citizens of India under the Act. A convention country is a country that is member of such an international convention for protection of plant varieties to which India is also a member, or a country with which India has agreed to grant PBR to citizens of both the countries.

*Advantages of PBR*

1. It will encourage private research resulting in an increase in number of good varieties available to farmers.
2. It will increase competition.
3. Indian farmers and the seed industry will get an access to foreign plant varieties from the countries where IPR laws are already enforced.
4. More genetic variability will be available to Indian plant breeders, as a result of which they can develop better varieties.

*Disadvantages of PBR*

1. It will encourage monopolies in genetic material for specific traits.
2. The holder of PBR may produce less seed than the demand to increase prices for making more profit.

3. Genetic diversity will be reduced because land races will go out of use.
4. It will inhibit free exchange of material and foster unhealthy practices.
5. Farmer's privilege to use his own seed will be threatened.
6. PBR may increase cost of the seed, which the poor farmer may not be able to afford. Thus, it will only help rich farmers.

### Convention on Biological Diversity

In June 1992, 170 countries met in Rio-de-Janerio to discuss the details of a proposed biodiversity treaty under the biodiversity convention. Although USA has not signed this treaty, this Convention on Biological Diversity (CBD) dealt with the sovereign rights of the nations over their genetic resources. This CBD has come into force in December 1993 as the first legal mechanism dealing with biological resources. In respect of biological resources, earlier status according to FAO International Undertaking on Plant Genetic Resources, a non-legally binding mechanism that had adhered to since inception in 1983 that "plant genetic resources are the heritage of mankind" and should be made available without restriction. But as per the CBD, which came into force in 1993 as the first legal mechanism dealing with biological resources, these plant genetic resources are deemed not to be the heritage of mankind, but are the properties of the countries and are therefore tradable commodities. This is in contrast to protection of IPRs, which decrees that natural products are the God's gift to mankind, and hence cannot be exploited exclusively by any one. The objective of the CBD, to be pursued in accordance with its relevant provisions, included (i) the conservation of biological diversity, (ii) the sustainable use of its components, and (iii) the fair and equitable sharing of the benefits arising out of the utilization of genetic resources. The latter may include appropriate (a) access to rightfully owned/possessed genetic resources, (b) transfer of relevant rightfully held technologies, and (c) funding.

Countries need to legislate biodiversity bills (protecting nature's treasures), the basic principle being that no foreigner or foreign organization can take away any biological resources for research or commercial use without permission of the country of origin. No local organization will be allowed to transfer even research results on biological resources to any foreigner without permission. The legislation will also ensure that benefits between conservers of the resources and users are equitably shared. Even though all the signatories had confirmed such approaches for protection of their sovereign rights, most are yet to enact appropriate legislations in this regard. A weak point in the CBD is its ambiguous treatment for the equity. It may be seen that whereas patenting of products of biotechnology is clearly recognized, there are no effective guidelines and conditions defined to recognize and reward the contributions of indigenous communities and the other informal innovators who have been responsible for nurturing, using, and developing biodiversity worldwide.

The key developments in India concerning legislative and regulatory provisions include the enactment of two legislations, one on the protection of plant varieties and farmer's rights (PPVFR) and the other Biological Diversity Act 2002.

Biodiversity is not equally distributed all over the globe. Certain countries are characterized by high species richness and more number of endemic species. These countries are known as Mega biodiversity countries. Twelve such countries have been identified. Together, these countries harbor 70% of the world's recorded vascular plants biodiversity. These countries are: Brazil, Colombia, Ecuador, Peru, Mexico, Madagascar, Zaire, Australia, China, India, Indonesia and Malaysia. India is one of the 12-mega biodiversity countries of the world. With only 2.4% of the land area, India already accounts for 7–8% of the recorded species of the world. Over 46,000 species of plants and 81,000 species of animals have been recorded in the country so far by the Botanical Survey of India, and the Zoological Survey of India, respectively. India is an acknowledged centre of crop diversity, and harbors many wild relatives and breeds of domesticated animals. Thus there is an urgent need to protect the biodiversity and prevent biopiracy.

*Indian Biological Diversity Act 2002*

The main features of Act are to provide for conservation of biological diversity, sustainable use of its components and fair and equitable sharing of the benefits arising out of the use of biological resources, knowledge and for matters connected therewith or incidental thereto. The Biodiversity Act, 2002 primarily addresses access to genetic resources and associated knowledge by foreign individuals, institutions or companies, to ensure equitable sharing of benefits arising out of the use of these resources and knowledge to the country and the people.

A three tiered structure at the national, state and local level has been established for efficient functioning of Biodiversity Authority:

**National Biodiversity Authority (NBA):** All matters relating to requests for access by foreign individuals, institutions or companies, and all matters relating to transfer of results of research to any foreigner.

**State Biodiversity Boards (SBBs):** All matters relating to access by Indians for commercial purposes will be under the purview of the State Biodiversity Boards (SBB).

**Biodiversity Management Committees (BMCs):** Institutions of local self government will be required to set up Biodiversity Management Committees in their respective areas for conservation, sustainable use, and documentation of biodiversity and chronicling of knowledge relating to biodiversity.

NBA and SBBs are required to consult the concerned BMCs on matters related to use of biological resources and associated knowledge within their jurisdiction.

As per the Act only Indian nationals can use the biological resources freely. The persons of the following categories shall be required to take the approval of the NBA for the use of biological resources:

1. a person who is not a citizen of India;
2. a citizen of India, who is a non-resident as defined in section 2 of the Income-Tax Act, 1961;
3. a body corporate, association or organization—
   i. not incorporated or registered in India; or
   ii. incorporated or registered in India under any law for the time being in force which has any non-Indian participation in its share capital or management

However, there is no requirement under the legislation for seeking permission of the NBA for carrying out research, if it is carried out in India by Indians. The only situations that would require permission of the NBA are: (i) when the results of any research which has made use of the country's biodiversity is sought to be commercialized, (ii) when the results of research are shared with a foreigner or foreign institution, and (iii) when a foreign institution/individual wants access to the country's biodiversity for undertaking research. When any person who intends to transfer any biological resource or knowledge, has to make an application in the prescribed form of the NBA. On receipt of an application The NBA may, after making such enquiries as it may deem fit and if necessary after consulting an expert committee constituted for this purpose, by order, grant approval subject to such terms and conditions as it may deem fit, including the imposition of charges by way of royalty or it may reject the application. However, Indian researchers neither require prior approval nor need to give prior intimation to SBB for obtaining biological resource or for conducting research in India.

## Some case studies on plant patents

### Patenting of Basmati Rice in USA

A US patent No. 5663484 on "Basmati Rice lines and grains" was granted to RiceTec, Inc., Texas, USA on September 2, 1997. This patent application was filed on July 8, 1994. This patent allowed 20 claims on three new Basmati rice lines (Bas 867, RT 1117, and RT1121) and grains, which was very broad in nature. Basically this patent claims right on all the Basmati rice varieties if it enters the USA. The Government of India challenged the patent for three claims from 15 to 17 were challenged, which had defined rice grains in the patent without limitation of territory or growing seasons (photoperiod insensitivity). These claims were broadly worded and had wider implications in terms of posing a threat of infringement of patent even by traditional basmati lines being exported to USA from India. The challenged claims were as follows:

15. A rice grain which has:
   i. a starch index of about 27 to 35
   ii. a 2 acetyl-L-pyrroline content of about 150 ppb to about 2000 ppb.
   iii. a length of about 6.2 mm to about 8.0 mm, a width of about 1.6 mm to about 1.9 mm and a length / width ratio of about 3.5 to about 4.5.
   iv. a whole grain index of about 41 to 63.
   v. a lengthwise increase of about 75% to about 150% when cooked and
   vi. a chalk (starch?) index of less than about 20.
16. The rice grain of claim 15, which has a 2 acetyl-L-pyrroline content of about 350 ppb to about 600 ppb.
17. The rice grain of claim 15, which has a burst index of about 4 to about 1.

The challenge was filed on 28 April 2001, requesting for reexamination of patent. In response, M/S Rice Tech Inc. surrendered the three claims 15–17 along with another claim, which also concerned grains. On the basis of withdrawl of these claims by M/S RiceTec, India made a new representation for further reexamination of claims. USPTO has issued a notice on 14 August 2001 confirming claims 8, 9 and 11 and accepting amendments in claims 12 and 13, while all other claims stand cancelled. Eventually the patent is to be restricted to only three rice strains developed by M/S Rice Tec, Inc. The title of the patent has also to be modified accordingly by excluding words basmati rice lines and grains.

### Revocation of the Turmeric Patent

USPTO granted a patent US Patent No 5401504 on "Use of Turmeric in Wound Healing" on 28 March 1995 to University of Mississippi Medical Center, Jackson, USA. This patent generated great concern and public debate on creating private ownership to traditional knowledge.

The Council of Scientific and Industrial Research, Govt. of India, challenged this patent on the basis that prior knowledge on the wound healing property of turmeric existed and patent claim is not novel. Based on printed references predated to the date of patent application, the patent was revoked because it lacks novelty with respect to *prior art*. This patent was revoked on 21 April 1998.

### Revocation of the neem patent

EPO granted European Patent No. 436257B1 to W.R. Grace and Co. and the US Department of Agriculture in 1995 on "Method of controlling fungi on plants by the aid of a hydrophobic extracted neem oil."

Legal opposition to the patent was that the invention lacked novelty and originality as there was an Indian testimony on prior knowledge of insecticidal and fungicidal properties of neem. Moral issues were also raised as grounds for this opposition.

On the basis of representation by filing affidavits and counter affidavits, ultimately in May 2000, EPO concluded that there was adequate evidence in *prior art*, and therefore lack inventive step in the patent claim. The patent was therefore revoked.

The two patents cited for neem and turmeric demonstrate the importance of authentic documentation of traditional knowledge and data base and the ability to provide convincing evidence of *prior art*. Thus, it is essential that for countries having tremendous biodiversity and where common practices as traditional knowledge are employed by tribal and local peoples, there must be proper documentation of information so that no one can misutilize that information for his/her own benefit.

## REFERENCES

Anonymous 2007. *The Protection of Plant Varieties and Farmers' Rights Act, 2001* and Rules, Universal Law Publishing Co., Delhi, 2007.

Chawla, H.S. 2005. Patenting of biological material and biotechnology. *J Intellectual Property Rights* **10:** 44–51.

Chawla, H.S. 2007 Managing intellectual property rights for better transfer and commercialization of agricultural technologies. *J Intellectual Property Rights* **12:** 330–340.

Mishra, J.P. 1999. Biotechnology and intellectual property rights. *Yojana* May Issue 15–20.

Swaminathan, M.S. 1994. Draft Plant Varieties Recognition and Protection Act: Rational and Structure. In: *GATT Accord: India's Strategic Response* (Ramachandria, V., ed.). Commonwealth Publishers, New Delhi, pp. 189–243.

# APPENDIX I

## Metric prefixes

| Factor | Prefix | Symbol |
|---|---|---|
| $10^{18}$ | exa | E |
| $10^{15}$ | peta | P |
| $10^{12}$ | tetra | T |
| $10^{9}$ | giga | G |
| $10^{6}$ | mega | M |
| $10^{3}$ | kilo | k |
| $10^{-3}$ | milli | m |
| $10^{-6}$ | micro | $\mu$ |
| $10^{-9}$ | nano | n |
| $10^{-12}$ | pico | p |
| $10^{-15}$ | femto | f |
| $10^{-18}$ | atto | a |

Examples: 1 ng (nanogram) = $10^{-9}$ gram
1 pmol (picomole) = $10^{-12}$ mol

# APPENDIX II

## Greek alphabets

| | | | | | |
|---|---|---|---|---|---|
| A | $\alpha$ | alpha | N | $\nu$ | nu |
| B | $\beta$ | beta | X | $\chi$ | xi |
| Γ | $\gamma$ | gamma | O | o | omicron |
| Δ | $\delta$ | delta | Π | $\pi$ | pi |
| E | $\epsilon$ | epsilon | P | P | rho |
| Z | $\xi$ | zeta | Σ | $\sigma$ | sigma |
| H | $\eta$ | eta | T | $\tau$ | tau |
| Θ | $\theta$ | theta | Y | $\upsilon$ | upsilon |
| I | $\iota$ | Iota | Φ | $\varphi$ | phi |
| K | $\kappa$ | kappa | X | $\chi$ | chi |
| Λ | $\lambda$ | lambda | Ψ | $\psi$ | psi |
| M | $\mu$ | mu | Ω | $\omega$ | omega |

# APPENDIX III

## Nucleic acid conversion factors

Average MW of a DNA base pair        = 660 Da
1pmol of 1000 bp DNA        = 0.66 µg
1 µg of 1000 bp DNA fragment = 1.5 pmol = 3.0 pmol ends
To calculate picomoles of ends per microgram of linear double strand DNA:

$$(2 \times 10^6)/ 660 \times (\text{number of bases}) = \text{pmol ends}/ \text{µg DNA}$$

In solution:
1 $A_{260}$ unit = ~ 50 µg/ml of double strand DNA = 0.15 mM
1 $A_{260}$ unit = ~ 40 µg/ml of double strand RNA = 0.11 mM
1 $A_{260}$ unit = ~ 33 µg/ml of single strand DNA = 0.10 mM
To calculate the concentration of plasmid DNA in solution using absorbance at 260 nm:
(observed $A_{260}$) × (dilution factor) × (0.05) = DNA concentration in µg/µl
1000 bp DNA open reading frame = 330 amino acids = 37,000 Da protein

# APPENDIX IV
## Agarose gel percentages for resolution of linear DNA

| Gel percentage (%) | DNA size range (bp) |
|---|---|
| 0.5 | 1,000–30,000 |
| 0.7 | 800–12,000 |
| 1.0 | 500–10,000 |
| 1.2 | 400–7,000 |
| 1.5 | 200–3,000 |
| 2.0 | 50–2,000 |

The agarose concentration does not significantly alter the migration of tracking dye, xylene cyanol and bromophenol blue, relative to the molecular weight of DNA fragments. Therefore, over the range of agarose gel percentages shown in this table, xylene cyanol will co-migrate with ~5 kb DNA fragments and bromophenol blue will co-migrate with ~0.5 kb DNA fragments.

# APPENDIX V

## Polyacrylamide gel percentages for resolution of DNA

| Gel percentage (%) | DNA size range (bp) |
|---|---|
| 3.5 | 100–1,000 |
| 5.0 | 75–500 |
| 8.0 | 50–400 |
| 12.0 | 35–250 |
| 15.0 | 20–150 |
| 20.0 | 5–100 |

# APPENDIX VI

## Formulae for commonly used electrophoresis buffers

| Resolution of nucleic acids | | Resolution of proteins | |
|---|---|---|---|
| Buffer | Formula | Buffer | Formula |
| Tris-acetate (TAE) | 40 mM Tris-acetate<br>1 mM EDTA | Separating gel buffer | 1.5 M Tris.HCl, pH 8.8<br>0.4% SDS |
| Tris-phosphate (TPE) | 90 mM Tris-phosphate<br>2 mM EDTA | Stacking gel buffer | 0.5 M Tris.HCl, pH 6.8<br>0.4% SDS |
| Tris-borate (TBE) | 45 mM Tris-borate<br>1 mM EDTA | Tris-glycine running buffer | 25 mM Tris-base<br>192 mM glycine, 0.1% SDS |
| Alkaline | 50 mM NaOH<br>1 mM EDTA | Sample buffer | 10% glycerol, 2.3% SDS<br>62 mM Tris.HCl, pH 6.8 |

# APPENDIX VII
## IUPAC NUCLEOTIDE AMBIGUITY CODES

| Symbol | Meaning | Nucleic Acid |
|---|---|---|
| A | A | Adenine |
| C | C | Cytosine |
| G | G | Guanine |
| T | T | Thymine |
| U | U | Uracil |
| M | A or C | |
| R | A or G | |
| W | A or T | |
| S | C or G | |
| Y | C or T | |
| K | G or T | |
| V | A or C or G | |
| H | A or C or T | |
| D | A or G or T | |
| B | C or G or T | |
| X | G or A or T or C | |
| N | G or A or T or C | |

# Glossary

| | |
|---|---|
| Abiotic stress | The effect of non-living factors which can harm living organisms. These non-living factors include drought, extreme temperature, salt, minerals, etc. |
| Acclimatization | The adaptation of a living organism (plant, animal or microorganism) to a changed environment that subjects it to physiological stress. |
| Adaptor | A synthetic single-stranded oligonucleotide that, after self-hybridization, produces a molecule with cohesive ends and an internal restriction endonuclease site. When the adaptor is inserted into a cloning vector by means of cohesive ends, the internal sequence provides a new restriction endonuclease site. |
| A-DNA | A right handed DNA double helix that has 11 base pairs per turn. DNA exists in this form when partially dehydrated. |
| Adventitious | Adjective used to describe organs developing from positions on the plant from which they would not normally be derived, e.g. shoots from callus, leaves, roots, and embryos from any cell other than a zygote. |
| Agarose gel electrophoresis | A process in which a matrix composed of a highly purified form of agar is used to separate larger DNA and RNA molecules. *See* electrophoresis. |
| Algorithm | A set of rules for calculating or problem solving carried out by a computer program. |
| Alignment | Arrangement of two or more nucleotides of protein sequence to maximize the member of matching monomers. |
| Allele | (Gr. *Allelon*, of one another, mutually each other); allelomorph (adj: allelic, allelomorphic). One of a pair, or series, of variant forms of a gene that occur at a given locus in a chromosome. |
| Alu sequences | A family of 300 bp sequences occurring nearly a million times in the human genome. |
| Amino acid | An organic compound containing an amino ($-NH_2$) and an acid carboxyl group ($-COOH$). |
| Amphidiploid | A species derived by doubling of chromosomes in the $F_1$ hybrid of two species. |
| Amplification | 1. Replication of a gene library in bulk.<br>2. Duplication of gene(s) within a chromosomal segment.<br>3. Creation of many copies of a segment of DNA by polymerase chain reaction. |
| Androgenesis | Development of plants from male gametophytes. |

| | |
|---|---|
| Aneuploidy | The loss or gain of chromosomes from the normal euploid number by various processes. |
| Annealing | The process of heating (denaturing step) and slowly cooling (renaturing step) double stranded DNA to allow the formation of hybrid DNA or complementary strands of DNA or of DNA and RNA. |
| Annotation | Finding genes and other important elements in raw sequence data (structural annotation). Adding pertinent information such as description of the gene, amino acid sequence, or other commentary to the database entry of a raw sequence of DNA nucleotides. |
| Antiauxin | A chemical that interferes with auxin response. Antiauxin may or may not involve prevention of auxin transport or movement in plants. Some antiauxins are said to promote morphogenesis in vitro, such as 2,3,5, tri-iodobenzoic acid (TIBA) or 2,4,5-trichlorophenoxyacetate (2,4,5-T), which stimulate the growth of some cultures. |
| Anticoding strand | The strand of DNA double helix that is actually transcribed. Also known as antisense or template strand. |
| Anticodon | Triplet of bases carried by the tRNA molecule which are complementary to the codon in the mRNA. |
| Antigen | A molecule that is capable of stimulating the production of neutralizing antibody proteins when injected into an organism. |
| Antioxidants | A group of chemicals which prevent oxidation. e.g. ascorbic acid, citric acid. They retard senescence and browning of tissue |
| Antiparallel strands | The two strands of the double helix are organized in opposite orientation, so that the 5′ end of one strand is aligned with the 3′ end of the other strand. |
| Apical dominance | The phenomenon of suppression of growth of an axillary bud in the presence of the terminal bud on the branch. |
| ARLEQUIN v. 2.0 | An online freely downloadable software tool for conducting analysis of molecular variance, population genetic and boot strapping analysis. (For free down loads contact: Schneider, Genetics and Biometrics lab, University of Geneva, Geneva). |
| ARS | (autonomous replicating sequence) Any eukaryotic DNA sequence that initiates and supports chromosomal replication. They have been isolated in yeast cells. |
| Aseptic | Free from all microorganisms. |
| Asexual embryogenesis | The sequence of events whereby embryos develop from somatic cells, also known as somatic embryogenesis. |
| Autoradiography | A technique for the detection of radioactively labeled molecules by overlaying the specimen with photographic film. When the film is developed an image is produced which corresponds to the location of the radioactivity. |
| Autotrophic | Characteristic of plants that are capable of manufacturing their own food, as in photosynthesis. |
| Auxins | A class of plant growth regulators that stimulates cell division, cell elongation, apical dominance and root initiation. 2,4-D, IAA, IBA, NAA are some of the auxins commonly used in plant cell and tissue culture. |

| | |
|---|---|
| Auxotroph | A mutant organism that will not grow on a minimal medium; it requires the addition of some growth regulators. |
| Avirulent | Lacking virulence; a microorganism lacking the properties that normally cause disease. |
| Axenic culture | Free of external contaminants and internal symbionts, which is generally not possible with surface sterilization alone and incorrectly used to indicate aseptic culture. |
| Axillary | An adjective describing the relative position of a bud, i.e. axillary bud, in the axils of leaves. |
| Axillary bud proliferation | A technique of micropropagation of plants in culture, which is achieved primarily through hormonal inhibition of apical dominance and stimulation of lateral branching. |
| BAC | Bacterial artificial chromosome; a cloning vector system for isolating genomic DNA based on the F-factor plasmid, the bacterial sex or fertility plasmid, which has a low copy number because of the strict control of replication. They accept an insert size of 200 to 500 kb. |
| Bacteriophage | A virus that infects and replicates in bacteria. Also called simply phage. |
| Base pair (bp) | It is a partnership of A with T or C with G in a DNA double helix; other pairs can be formed in RNA under certain circumstances. |
| Batch culture | A cell suspension culture in which cells are grown in a finite volume of nutrient medium and follow a sigmoid pattern of growth. |
| B-DNA | The normal form of DNA found in biological systems. It exists as a right handed helix. |
| Binary vector system | A two plasmid system in *Agrobacterium tumefaciens* for gene transfer in plant cells. One plasmid contains the virulence gene (responsible for transfer of the T-DNA), and another plasmid contains the T-DNA borders, the selectable marker and the DNA to be transferred. |
| Bioinformatics | 1. It is a science associated to biomolecular database like activities involving persistent sets of data that are maintained in a consistent state over essentially indefinite periods of time. |
| | 2. The branch of information technology that deals with the storage and analysis of biological data. |
| BIOS | Basis Input-Output Systems. The lowest-level operating system (q.v.) of a computer, controlling elementary functions such as reading and writing to disks, responding to input and displaying readable characters on the monitor. Normally required to run the *operating systems* (q.v.) |
| Biotic stress | Stress resulting from living organisms, which can harm plants, such as viruses, fungi, bacteria, parasitic weeds and harmful insects. |
| Biotransformation | The conversion of a small part of molecule in the structure and composition by means of biological systems. |
| Bit | Binary digit. A bit can represent one of two values (0 and 1—sometimes refereed as *'off* 'or *'on'* |
| BLAST | Basis Local Alignment Search Tool. A program for sequence database similarity searching (see also *FASTA).* |
| Blunt end | The end of a DNA duplex molecule in which neither strand extends beyond the other, also known as flush end. |

| | |
|---|---|
| Boolean network | A network with one or more inputs, whose outputs are constrained by an explicit set of logical rules. |
| Browning | Discoloration due to phenolic oxidation of freshly cut surfaces of explant tissue. In later stages of culture, such discoloration may indicate a nutritional or pathogenic problem, generally leading to necrosis. |
| Browser | An application that process documents written in *HTML*. (q.v.) and interprets the markup language to display formatted *web pages (q.v.)* |
| C | A procedural compiled *programming language* (q.v.) widely used for bioinformatics software. C is part of the UNIX standard. |
| C++ | An object-oriented programming language (q.v.) compatible with C. |
| C value | It is the total amount of DNA in a haploid genome. |
| CAAT box | It is a part of conserved sequence located upstream of the startpoints of eukaryotic transcription units; it is recognized by a large group of transcription factors. It has the consensus sequence GGCCAATCT; it occurs around 75 bases prior to the transcription initiation site. |
| Callus | 1. A tissue consisting of dedifferentiated cells generally produced as a result of wounding or of culturing tissues in the presence of an auxin in particular. |
| | 2. Actively dividing non-organized masses of undifferentiated and differentiated cells often developing from injury (wounding) or in tissue culture in the presence of growth regulators. |
| Cap | The structure found on the 5´- end of eukaryotic mRNA which consist of an inverted methylated guanosine residue. |
| Capsid | External protein shell or coat of a virus particle. |
| cDNA | Complementary DNA, a fragment of DNA which has been produced from an RNA sequence by reverse transcription. Messenger RNA is commonly used to synthesize cDNA. |
| cDNA clone | A double stranded DNA molecule that is carried in a vector and was synthesized in vitro from an mRNA sequence by using reverse transcriptase and DNA polymerase. |
| cDNA cloning | A method of cloning the coding sequence of a gene, starting with its mRNA transcript. It is normally used to clone a DNA copy of a eukaryotic mRNA. |
| cDNA library | A collection of cDNA clones that were generated in vitro from the mRNA sequences isolated from an organism or a specific tissue or cell type or population of an organism. |
| Caulogenesis | Stem organogenesis; induction of shoot development from callus. |
| Cell culture | The growing of cells *in vitro* derived from multicellular organisms. |
| Cell line | Cells (originating from a primary culture) successfully subcultured for the first, second, etc. time. A cell line arises from a primary culture. |
| Cell suspension | Cells in culture in moving or shaking liquid medium, often used to describe suspension cultures of single cells and cell aggregates. |
| centiMorgan | One percent recombination between two loci (cM). |
| Central dogma | The basic concept that in nature genetic information generally can flow only from DNA to RNA to protein. It is now known that information contained in RNA molecules of certain viruses (called retroviruses) can also flow back to DNA. |

| | |
|---|---|
| Chelate | Complex organic molecule that can combine with cations and does not ionize. Chelates can supply micronutrients to plants at slow, steady rates. Usually used to supply iron to plant cells. |
| Chemiluminescence | The emission of light from a chemical reaction. |
| Chemostat | An open continuous culture in which cell growth rate and cell density are held constant by a fixed rate of input of a growth limiting nutrient |
| Chimeric DNA | A recombinant DNA molecule containing unrelated genes. |
| Chimeric gene | A semi-synthetic gene, consisting of the coding sequence from one organism, fused to promoter and other sequences derived from a different gene. Most genes used in transformation are chimeric. |
| Chromosome jumping | A technique that allows two segments of duplex DNA that are separated by thousands of bp (about 200 kb) to be cloned together. |
| Chromosome walking | Isolation of sequential pieces of genomic DNA in a set of clones to form a contiguous overlapping set (a contig). This is usually accomplished by isolating the end fragment of one clone, which is used to rescreen the library for the next overlapping clone i.e. sequential isolation of overlapping molecular clones so as to span large chromosomal intervals. This process (walking) is repeated along the chromosome until the gene of interest is reached. |
| Circularization | A DNA fragment generated by digestion with a single restriction endonuclease will have complementary 5′ and 3′ extensions (sticky ends). If these ends are annealed and ligated, the DNA fragment will have been converted to a covalently closed circle or circularized. |
| Cistron | It is the genetic unit defined by the cis/trans test. A DNA fragment or portion that specifies or codes for a particular polypeptide. |
| Clone | A group of cells, tissues or plants which are in principle genetically identical. A clone is not necessarily homogeneous. |
| Cloning vector | A small, self replicating DNA molecule—usually a plasmid or phage into which foreign DNA is inserted, transferred into an organism and replicated or reproduced. A cloning vector is used to introduce foreign DNA into host cells, where the DNA can be reproduced in large quantities. Examples are plasmids, cosmids, and yeast artificial chromosomes; vector are often recombinant molecules containing DNA sequences from several sources. |
| Coding sequence | That portion of a gene which directly specifies the amino acid sequence of its protein product. Non-coding sequences of genes include control regions, such as promoters, operators and terminators as well as intron sequences of certain eukaryotic genes. |
| Coding strand | The strand of duplex DNA which contains the same base sequence (after substituting U for T) found in the mRNA molecule resulting from transcription of that segment of DNA. Also known as sense strand. The mRNA molecule is transcribed from the other strand known as template or antisense strand. |
| Codon | A group of three nucleotides coding for an amino acid. |
| Cohesive ends | DNA with single stranded ends which are complementary to each other, enabling the different molecules to join each other. Also known as protruding ends, sticky ends, overhangs. Bacteriophage lambda has cohesive ends (the 'cos' site) which allows the formation of concatamers. |

| | |
|---|---|
| Co-integrate vector | A two plasmid system for transferring cloned genes to plant cells. The cloning vector has a T-DNA segment that contained cloned genes. After introduction into *A. tumefaciens*, the cloning vector DNA undergoes homologous recombination with a resident disarmed Ti plasmid to form a single plasmid carrying the genetic information for transferring the genetically engineered T-DNA region to plant cells. |
| Colony hybridization | A technique for using *in situ* hybridization to identify bacteria carrying chimeric vectors whose inserted DNA is homologous with some particular sequence. |
| Comparative genomics | The practice of comparing gene or protein sequence with each other across the genomes of whole organisms, in the hope of elucidating functional and evolutionary significance. |
| Complementary | DNA DNA synthesized by reverse transcriptase from RNA template. See cDNA. |
| Composite transposon | A transposable element formed when two identical or nearly identical transposons insert on either side of a transposable segment of DNA, such as the bacterial transposon Tn5. |
| Concatamer | DNA segment made up of repeated sequences linked end to end. |
| Consensus sequence | The nucleotide sequence that is present in the majority of genetic signals or elements that perform a specific function. |
| Constitutive promoter | An unregulated promoter that allows for continuous transcription of its associated gene. |
| Contig | A set of overlapping clones that provide a physical map of a portion of a chromosome. It refers to contiguous map. |
| Contig map or contiguous map | The alignment of sequence data from large, adjacent regions of the genome to produce a continuous nucleotide sequence across a chromosomal region. |
| Continuous culture | A suspension culture continuously supplied with nutrients by the inflow of fresh medium but the culture volume remains constant. |
| Copy number | The average number of molecules of a plasmid or gene per genome contained in a cell. |
| Copyright | It is a form of intellectual property protection granted to the creators of original works of authorship such as literary, novels, poems, plays, dramatic, musical, artistic, paintings, sculptures, architecture, software, maps, technical documentation, drawings and certain other intellectual works. |
| Cos ends | The 12-base, single strand, complementary extensions of bacteriophage lambda DNA. Also known as *cos* sites. |
| Cosmids | Plasmid vector into which phage lambda *cos* sites have been inserted, as a result the plasmid DNA can be packaged *in vitro* in the phage coat. This vector can accommodate around 40kb fragment of DNA. Cosmids replicate as plasmids. |
| Cos sites | See cos ends. |

| | |
|---|---|
| Cot curve | When duplex DNA is heated, it dissociates into single strands. When temperature is lowered, complementary strands tend to anneal or re-nature. The extent of renaturation depends on the product of DNA concentration in moles of nucleotides per liter, and time in seconds. A graph showing the proportion of renatured DNA against cot is known as cot curve. The cot at which half the DNA has renatured is the half-cot, a parameter indicating the degree of complexity of the DNA. |
| Crown gall | A disease of plants in which tumor forms. The causal agent is the bacterium *Agrobacterium tumefaciens*. A tumor inducing portion of the bacterial genome (Ti plasmid) may be used experimentally as a genetic tool to incorporate (vector) genetic information into plant cells. |
| Cryopreservation | Storage and preservation at ultralow temperature of –196°C. |
| Cryoprotectant | An agent able to prevent freezing and thawing damage to cells. |
| Cybrid | A somatic hybrid in which nucleus is derived from one parent and cytoplasm is derived from both the parents. |
| Cytokinins | A class of plant growth regulators which cause cell division, cell differentiation, shoot differentiation, and the breaking of apical dominance. Some of the cytokinins are BAP, kinetin, zeatin, 2-iP. |
| Cytomic | Cytological approach which integrates genomics and proteomics. |
| DAF, DNA Fingerprinting Amplification | Operates on the same principle as RAPD but uses shorter starters (5 to 8 bases). The electrophoretic profiles of the amplified fragments of DNA are often complex. |
| Database | On a computer, a collection of data records either in a single file or as multiple files. The central component of a database management system. |
| Database Management Systems (DBMS) | A software suite including a database (q.v.) and the utilities required to organize, search, and update maintain data security and control access. |
| DBGET/LinkDB | On the data retrieval tool developed by the Institute of Chemical Research the Human Genome Center in Japan. See also *Entrez*, SRS. |
| Dedifferentiation | Reversion of differentiated to non-differentiated cells. |
| Denaturation | Conversion of DNA or RNA from double stranded to single stranded form; Mostly accomplished by heating at high temperature of 94°C. |
| *de novo* | It means from the beginning, arising, anew, afresh. |
| Design | A design is an idea or a conception pertaining to the shape, configuration, pattern, ornamentation, composition of lines or colors or any other related feature applied to any article in 2D or 3D or both forms. |
| Differentiation | A process in which unspecialized cells develop structures and functions characteristic of a particular type of cell. The development of cells or tissues with a specific function and/or the regeneration of organs or organ like structures (roots, shoots, etc.) or proembryos. |
| Dihaploid | An individual (denoted by 2n = 2x) which arises from a tetraploid (2n = 4x). |
| Diploidization | Process of doubling the chromosome complement (genome) of a cell. |
| Direct repeat | Two or more stretches of DNA within a single molecule which have the same nucleotide sequence in the same orientation. For example: TATTA...TATTA ATAAT...ATAAT |

| | |
|---|---|
| Disarm | To delete from a plasmid or virus genes that are cytotoxic or tumor inducing |
| Disease free | This should be interpreted to mean free from any known disease. |
| Disinfection | Full elimination of internal microorganisms from a culture. |
| DNA amplification | Multiplication of a piece of DNA in a test tube into many thousands of millions of copies. The most commonly used process is PCR. |
| DNA chip | See DNA microarray |
| DNA fingerprinting | See genetic fingerprinting |
| DNA microarray | A small glass surface to which an array of DNA fragments, each with a defined location, is fixed. DNA fragments can be cDNA or short synthesized oligonucleotides. A solution of fluorescently labeled DNA fragments is hybridized to the chip, spots to which hybridization occurs are visible as fluorescence. |
| DNA polymerase | Enzyme that catalyzes the phosphodiester bonds in the formation of DNA. |
| DOS | Disk operating system. A lower-level operating system used to read and copy files and carry out other basic functions. MS-DOS was developed by Microsoft now integrated with the *Windows* |
| Doubling time | Term used in tissue culture and shoot propagation for the time necessary to double the number of cells/shoots *in vitro*. |
| Download | Transfer files from a computer network (or the *Internet,* q.v.) onto a local computer. Cf. *upload*. |
| Downstream | A term used to describe the relative position of sequences in a nucleic acid pertaining to the direction in which the nucleic acid is synthesized i.e. towards the 3′ end. For example, the coding region is downstream of the initiation codon. In molecular biology, site of initiation of transcription, is designated as +1. Thus downstream nucleotides are marked with plus signs, e.g. +2, +20. |
| Electrophoresis | A technique that separates charged molecules (DNA, RNA, proteins) on the basis of relative migration in a appropriate matrix (such as agarose, polyacrylamide) subjected to an electric field. |
| Embryogenesis | Process by which an embryo develops from a fertilized egg cell or asexually from a group of cell(s). |
| Embryo culture | The culture of embryos on nutrient medium. |
| Embryoids | Embryo like structures produced as a consequence of differentiation processes such as embryogenesis and androgenesis. |
| Endomitosis | Duplication of chromosomes without division of the nucleus, resulting in increased chromosome number within a cell. Chromosome strands separate but the cell does not divide. |
| Endonuclease | An enzyme that catalyzes the cleavage of DNA at internal positions, normally cutting it at specific sites. |
| Enhancer sequence | A sequence found in eukaryotes and certain eukaryotic viruses which can increase transcription of a gene when located (in either orientation) up to several kilobases from the gene concerned. These sequences usually act as enhancers when on the 5′ side (upstream) of the gene in question. However, some enhancers are active when placed on the 3′ side (downstream) of the gene. |

| | |
|---|---|
| Entrez | Online data retrieval tool developed by the National Center for Biotechnology Information (NCBI). |
| Epicotyl | The upper portion of the axis of a plant embryo or seedling above the cotyledons. |
| Epigenesis | Describes the developmental process whereby each successive stage of normal development is built on the foundations created by the preceding stages of development. For example, an embryo is built up from a zygote, a seedling from an embryo, and so on. |
| Epigenetic | A term referring to the non-genetic causes of a phenotype. |
| Epigenetic variation | Non-hereditary variations which is at the same time reversible; often the result of a changed gene expression. |
| Episome | It is a plasmid, a genetic extrachromosomal element which replicates within a cell independently of the chromosome and can integrate in to the host bacterial DNA (For example F-fertility factor in *E. coli*). Plasmids and F factors are episomes. |
| Eukaryotes | Cellular organisms having a membrane bound nucleus within which the genome of the cell is stored as chromosome composed of DNA and proteins. For example algae, fungi, plants and animals. |
| Excise | Cut out (with knife, scalpel etc.) and prepare a tissue, organ, etc. for culture. |
| Exon | The coding regions of a gene that are presented in the final mRNA |
| Exonuclease | An enzyme that digests DNA or RNA from the ends of strands. 5′ exonucleases requires a free 5′ end and degrades the molecule in 5′→3′ direction. 3′ exonucleases require a free 3′ end and degrade the molecule in opposite direction. |
| Exonuclease III | An *E. coli* enzyme that removes nucleotides from the 3′ hydroxyl ends of double stranded DNA. Also known as exodeoxyribonuclease III. |
| Explant | A piece of tissue used to initiate tissue culture. |
| Exponential phase | A phase in culture in which cells undergo maximum rate of cell division. It follows the lag phase and preceeds the linear growth phase in most batch propagated suspension cultures. |
| Expressed sequence tag | Short cDNA sequence. So called because it represents part of the sequence of an expressed gene. |
| Expression vector | A vector that has been constructed in such a way that inserted DNA molecule is put under appropriate promoter and terminator sequences for proper transcription and translation. |
| FASTA | (1) A sequence alignment algorithm. (2) FASTA format. A flat file format for the representation of sequence data. |
| F factor | See episome |
| File | On a computer, a discrete collection of bytes |
| Filter sterilization | Process of sterilizing a liquid by passage through a filter, with pores so small that they are impervious to micro-organisms. |
| Flaming | A technique for sterilizing instruments by heating on a flame after dipping in alcohol. |
| Flat File | A plain text file. A file with no markup characters *(see markup language)*. |

| | |
|---|---|
| Flow cytometry | A technique used to sort cells, nuclei or other biological materials by means of flow through apertures of defined size. |
| Forward genetics | The traditional approach to genetics, which starts with a phenotyope and then identifies the gene(s) or mutation that control or cause that trait. This is in contrasts to reverse genetics, which starts from a gene/ sequence in order to determine the phenotype. |
| Friable | A term commonly used to describe a crumbling or fragmenting callus. A friable callus is easily dissected and readily dispersed into single cells or clumps of cells in solution. |
| Fusogen | A fusion inducing agent used for agglutination of protoplasts in somatic hybridization. |
| Gametophyte | Cells and tissues of the haploid stage of the life cycle of plants. |
| Gapped DNA | A duplex DNA molecule with one or more internal single stranded regions. |
| Gel electrophoresis | *See* electrophoresis. |
| Gene | A unit of inheritance which has a specific biological function. It is a specific sequence of nucleotides. |
| Genbank | A public database where DNA sequences are deposited and made freely available. |
| Gene cloning | The process of synthesizing multiple copies of a particular DNA sequence using a bacterial cell or another organism as a host. Used in genetic engineering as cloning. *See* cloning. |
| Gene library | Random collection of cloned fragments in a vector that ideally includes all the genetic information of that species; e.g. wheat, rice, *Arabidopsis*; sometimes called shot gun collection. |
| Genetically modified organism (GMO) | An organism that has been modified by the application of recombinant DNA technology. |
| Genetic engineering | Changes in the genetic constitution of cells by introduction or elimination of specific genes using molecular biology techniques. This a non sexual method of gene transfer. *See* gene cloning. |
| Genetic fingerprinting | A technique in which an individual DNA is analyzed to reveal the pattern of repetition of particular nucleotide sequences throughout the genome. The unique pattern of DNA fragments are identified by Southern hybridization or polymerase chain reaction. |
| Genetic map | The linear array of relative positions of genetic loci on a chromosome based on recombination frequencies (linkage map). |
| Genetic marker | A DNA sequence used to mark or track a particular location (locus) on a particular chromosome. |
| Genetics | The science of heredity and variation. |
| Genetic transformation | Transfer of extracellular genetic information (DNA) among and between species with the use of vectors (e.g. bacterial or viral). |
| Genome | 1. A complete set of chromosomes (hence of genes) inherited as a (haploid) unit from one parent. |
| | 2. The entire complement of genetic material (genes + non-coding sequences) present in each cell of an organism or in a virus or organelle. |
| Genomic library | A collection of clones containing the genomic DNA sequences of an organism. Typically these molecules are propagated in bacteria or virus. |

| | |
|---|---|
| Gibberellins | Group of plant growth regulators, which induce among other things, cell elongation and cell division. |
| Habituation | The acquired ability of cells to grow and divide independently of growth regulators. |
| Hairpin loop | A region of double helix formed by base -pairing within a single strand of DNA or RNA which has folded back on itself. |
| Haploid | Plant with half the number of chromosomes (denoted by 'n') due to reduction division of the diploid (= 2n). |
| Hardening off | Gradual acclimatization of *in vitro* grown plants to *in vivo* conditions. |
| Heteroduplex | A DNA molecule formed by base-pairing between two strands that do not have completely complementary nucleotide sequences. |
| Hetrokaryon | A cell in which two or more nuclei of unlike genetic make-up are present. |
| Heteroplasmic | Refers to the presence of different plastid genomes. Heteroplasmy may be observed at organelle, cellular or clonal levels. |
| High Throughput Biology | An experimental approach that generates massive amounts of raw data using highly automated technologies such as genome sequencing technology or microarray technology. The data is processed by a batch method using computational and other information management tools. |
| Homokaryon | Cell with two or more identical nuclei as a result of fusion. |
| Homoplasmic | Refers to a pure population of plastid genomes. |
| HTML | Hypertext markup language. The markup language that controls how text and multimedia objects are displayed on web pages |
| HTTP | Hypertext transfer protocol. The protocol used to exchange information over the World Wide Web |
| Hyperlink | A word or object in hypertext document (q.v.) usually highlighted, which acts as a link to another |
| Immobilized cells | Cells entrapped in matrices such as alginate, agarose, polyacrylamide designed for use in bioreactors for production of metabolites. |
| Inoculation | Placing of an explant in or on a nutrient medium. |
| Intellectual property rights (IPRs) | It can be defined as the rights given to people over the creation of their minds. They usually give the creator an exclusive right over the use of his/her creations for a certain period of time. |
| Intergeneric | A cross between two different genera. |
| Internet | An international computer network governed by a set of protocols called TCP/IP. |
| Internet Explorer | A popular browser available from Microsoft. See also Netscape Navigator. |
| Interspecific | A cross between two different species of a genus. |
| Insertion sequences | Generic term for DNA insertion sequences found in bacteria capable of genome insertion. Insertion sequence (IS) carries only the genetic functions involved in transposition. |
| Intron | Sequence of DNA interrupting the coding sequence of a gene. This is the portion of a gene that is transcribed but does not appear in the final mRNA transcript. |
| IP | Internet Protocol. See TCP/IP. |
| IP Address | An address to which data is sent or from which it is received over the Internet. Conventionally four integers separated by dots, e.g. 195. 172.6.15. |

| | |
|---|---|
| *In vitro* | Literally means "in glass". Now applied for the growth of a tissue in any type of culture container. |
| Isoenzymes (Isozymes) | Multiple molecular forms of an enzyme exhibiting similar or identical catalytic properties. |
| Java | An object-oriented programming language designed to run on most platforms. Java programs can be run alone or form within an HTML document, in which case they are supplied as an applet. |
| Juvenile phase | The period in the life of a plant during which no flowering can be induced. During the juvenile phase a plant has often very special characteristics (morphology, physiology etc.) which are different from the adult phase. |
| kb | Kilobase, 1000 bases (in RNA) or base pairs (in DNA). DNA sizes are often expressed in kilobases. |
| Klenow fragment | Piece obtained from DNA polymerase I by proteolytic cleavage; it lacks the 5′- 3′ exonuclease activity |
| Lag phase | It describes the first of the five growth phases of most batch propagated cell suspension/callus cultures in which inoculated cells in fresh medium adapt to the new environment and prepare to divide. |
| Laminar air flow cabinet | Cabinet for inoculation which is kept sterile by a continuous non-turbulent flow of sterilized air. |
| Leader | It is the non-translated sequence at the 5′ end of mRNA that precedes the initiation codon. |
| Library | A collection of cloned DNA fragments. The library may consist of cDNAs or genomic clones (fragments cloned directly from cellular DNA). |
| Ligation | Joining of DNA fragments to produce a single DNA molecule. Ligases are enzymes which perform this function. |
| Linker | A small fragment of synthetic DNA that has a restriction site useful for gene splicing. |
| LINUX | A UNIX-like operating system named after its inventor, Linus Thorvald. |
| Liposomes | Artificial phospholipid vesicles. |
| M13 | A single stranded DNA bacteriophage used as a vector for DNA sequencing. |
| Macronutrient | An essential element required by plants in relatively large quantities. |
| MALDI | Matrix-assisted laser desorption/ionization. A technique for regenerating ions in mass spectrometry (q.v.), which is suitable for the analysis of large proteins without significant degradation. |
| MALDI-TOF/MS | Matrix assisted laser desorption ionisation-time of flight/mass-spectrometry, mass-spectrometric method for the determination of molecular mass (e.g., proteins); employed, among other things for quality control in DNA synthesis. |
| MAPMARKER | A software package for conducting Genetic map (QTL and Linkage) analysis. |
| Mass spectrometer | An instrument used to identify chemicals in a substance by their mass and charge. |

| | |
|---|---|
| Meristem | Localized region of active cell division in plants from which permanent tissue is derived. The principal meristems in the flowering plants occur at the tips of the stems and roots (apical meristems), between xylem and phloem of vascular bundles (cambium) in the cortex (cork cambium), in young leaves and, in many grasses, at the bases of internodes (intercalary meristems). |
| Messenger RNA | The RNA that specifies the amino acid sequence for a particular polypeptide chain. |
| Metabolic fingerprinting | An approach which aim to classify samples by global and fast analysis of large number of metabolites |
| Metabolic profiling | Analysis of all metabolites of a metabolic pathway or a given chemical class of metabolites. Metabolic profiling aims at a quantitative assessment of a predefined number of target metabolites. |
| Metabolome | It encompasses all the metabolites that represent the catabolic and anabolic activities being performed by proteins at any given time, plus compounds that can be produced in tissue through non enzymatic reactions. |
| Metabolomics | It denotes comprehensive, nonbiased, high-throughput analyses of the complex metabolite mixtures typical of plant extracts. |
| Micronutrient | An essential element required by plants in relatively small quantity. |
| Monoploid | A cell or individual which has one genome denoted by x and is the lowest number of chromosomes of a polyploid series. e.g. barley has n = x = 7, *Triticum aestivum* has x = 7 and n = 21. |
| Morphogenesis | Developmental pathways in differentiation which results in the formation of recognizable tissues. |
| Multicopy plasmids | These plasmids are present in bacteria at amounts greater than one per chromosome. |
| Mutant | A cell which has undergone a heritable change which has resulted due to a change in its genes or chromosomes. |
| NBRF/PIR | National Biomedical Research Foundation/protein Information Resource. A flat file format for representing sequence information. |
| Netscape Navigator | A popular internet browser (q.v.) |
| Nick translation | A procedure for radiolabeling DNA in vitro using DNA polymerase I. |
| Northern blotting | A technique for the transfer of RNA from an agarose gel to a nitrocellulose filter on which it can be hybridized to a complementary DNA. |
| Okazaki fragments | The short stretches of 1000–2000 bases produced during discontinuous replication which are later joined into a covalently intact strand. |
| Open reading frame | The sequence of DNA or RNA located between the start- code sequence (initiation codon) and the stop-code sequence (termination codon). Sequence is (potentially) translatable into protein. |
| Operator | A site on DNA at which a repressor protein binds to prevent transcription from initiating at the adjacent promoter. |
| Operon | A number of contiguous genes under coordinate control (usually in prokaryotes). |
| Organ culture | Culture of an organ *in vitro* in a way that allows development and/or preservation of the originally isolated organ. |

| | |
|---|---|
| Organogenesis | Type of morphogenesis which results in the formation of organs viz. shoots, roots. |
| Origin of replication | Site of initiation of DNA synthesis. |
| Orthologous genes | Genes coding for the same function that are derived from the same ancestral gene. Two orthologous genes thus belong to two distinct species resulting from a common ancestor and these genes may indicate a position preserved in the genome of two species. |
| Osmoticum | Reagents that increase the osmotic pressure of a liquid. |
| PAC | P1 derived artificial chromosome; a cloning system for isolating genomic DNA based on the F factor plasmid, as in BACs, but also containing some of the elements of the bacteriophage P1 cloning system. |
| P1 clones | A cloning system for the isolating genomic DNA that uses elements from bacteriophage P1 (i.e. recombination sites, loxP and a packaging site, pac, etc.). |
| Packed cell volume (PCV) | A test for determining the viability of cells. It is the percentage volume of cells in a set volume of culture after sedimentation (packing) by means of low speed centrifugation. |
| Palindrome | It is a sequence of DNA that is the same when one strand is read left to right or the other is read right to left; consists of adjacent inverted repeats. |
| Paralogous genes | Two homologous genes found at different chromosomal locations in the same organism that have structural similarities indicating that they are derived from a common ancestral gene |
| Parthenogenesis | Production of an embryo from a female gamete without the participation of a male gamete. |
| Patent | A patent is a Government granted exclusive intellectual property right to an inventor to prevent others from making/using, manufacture and market the invention, provided the invention satisfies certain conditions stipulated in the law. A Patent is territorial in nature i.e. within the boundaries of a particular country, which has given the patent. |
| pBR322 | A standard plasmid cloning vector. |
| pH | The negative logarithm of the hydrogen ion concentration of any solution for measuring acidity or alkalinity. |
| Phytohormone | The generic name for all classes of hormones produced by plants, especially those that elevate plant growth. |
| Picogram | $10^{-12}$ grams, a unit commonly used to express the DNA content per cell or per nuclear genome in a plant. |
| Plasmid | DNA which is not part of the regular genome of an organism and self replicates during growth. Plasmids commonly occur as linear or circular elements, and may be present in one to many copies per cell. |
| Plastome | The genome of plastids |
| Plating efficiency | The percentage of cells plated which give rise to cell colonies. |
| Polyadenylation | Post-transcriptional addition of polyadenylic acid tail to the 3′ end of eukaryotic mRNA. |
| Polycistronic mRNA | A mRNA that includes coding regions of more than one gene. |
| Polyembryony | When two or more embryos are formed after fertilization. |

| | |
|---|---|
| Polysomes | Complexes of ribosomes bound together by a single mRNA molecule; also known as polyribosomes. |
| Positional cloning | A technique used to identify genes, based on their location on a chromosome. |
| Primary culture | Culture resulting from cells, tissues, or organisms taken from an organism. |
| Primary database | A database for primary sequence data. The primary nucleotide databases are NCBI GenBank, the European Molecular biology Laboratory (EMBL) Nucleotide Sequence Database, and the DNA Database of Japan. The primary protein databases are SWISS-PROT and TrEMBL. |
| Primer | It is a short sequence (often of RNA) that is paired with one strand of DNA and provides a free 3′ OH end at which a DNA polymerase starts synthesis of a deoxyribonucleotide chain. |
| Primordium | The earliest detectable stage of an organ. |
| Probe | A radiolabeled nucleic acid molecule used to detect the presence of its complementary strand by hybridization. |
| Proembryo | Early embryonic stage. |
| Promoter | Region of DNA which is recognized by RNA polymerase in order to initiate transcription. |
| Promoter –10 sequence | It is the consensus sequence TATAATG centered about 10 bp before the startpoint of a bacterial gene. It is involved in the initial melting of DNA by RNA polymerase. |
| Promoter –35 sequence | It is the consensus sequence centered about 35 bp before the startpoint of a bacterial gene. It is involved in initial recognition by RNA polymerase. |
| Protoplast | A plasmalemma bound vesicle consisting of a naked cell formed as a consequence of the removal of cell wall by mechanical or enzymatic means. |
| Pseudogene | A sequence of DNA similar to a gene but which is not transcribed; probably the remnant of a once functional gene that accumulated mutations. |
| PSI-BLAST | An iterative version of the BLAST (q.v.) algorithm. |
| Recombinant DNA | DNA molecules in which sequences that are not naturally contiguous to each other are placed next to each other by *in vitro* manipulation. |
| Regeneration | Development and formation of new organs. |
| Regulatory gene | A gene that codes for an RNA or protein product whose function is to control the expression of other genes. |
| Relaxed replication control | It refers to the ability of some plasmids to continue replicating after bacteria cease dividing. |
| Repeated DNA sequence | A sequence of nucleotides which occurs more than once in a genome. Repeated sequences may be present in a few to many millions of copies. The individual repeated sequence may be only a few nucleotides in length up to several kb. |
| Replicon | The segment of the genome in which DNA is replicated and by, definition, contains an origin of replication. |
| Reporter gene | It is the coding unit whose product is easily assayed during genetic transformation (e.g. GUS, chloramphenicol transacetylase etc.). It may be connected to any promoter of interest so that expression of the gene can be used to assay promoter function. |

| | |
|---|---|
| Restriction enzyme | An endonuclease with the ability to cleave DNA at the point where a certain base sequence occurs. |
| Restriction site | A DNA base sequence recognized by a restriction enzyme. |
| RFLP | Restriction fragment length polymorphism, a difference between samples of DNA detected as differing fragment sizes produced after treatment with a restriction enzyme. |
| Restriction map | It is a linear array of sites on DNA cleaved by various restriction enzymes. |
| Reverse engineering | Any modeling process based on the principle of developing the model to fit observed data (cf. simulation). |
| Reverse genetics | Analysis that starts with a gene/ DNA sequence and eventually leads to its biological function. Reverse genetic methods are more amenable to whole genome, high-throughput analysis than those used in forward genetics. |
| RIL, Recombinant Inbred Line | An RIL is formed by crossing two inbred strains, followed by repeated selfing or sibling mating to create a new inbred line whose genome is a mosaic of the parental genomes. |
| RNA interference | This technique involves the transfer of double-stranded RNA into organisms, thereby specifically inactivating genes containing homologous sequences. RNAi has arisen from the observation that sense and antisense RNA are equally effective in suppressing specific gene expression. These result presented a paradox that was resolved by finding that small amounts of double-stranded (ds) RNA contaminating sense and antisense preparations, suppressed gene expression. |
| SAGE (Serial Analysis of Gene Expression) | A method for the comprehensive analysis of gene expression patterns in a given tissue or treatment. It is based on the sequencing of short sections of 9–17 nucleotides ("tags") of cDNAs that have been ligated together into large concatamers. SAGE assumes that the abundance of a particular tag for a gene correlates with the abundance of the transcript in a sample. |
| Satellite DNA | It consists of many tandem repeats (identical or related) of a short basic repeating unit. |
| Search engine | A *server-side* (q.v.) facility that allows the *World Wide Web* (q.v.) to be searched for pages containing particular words, phrases or multimedia objects. |
| Secondary database | A database of sequence information derived from the data in *Primary databases* (q.v.). Examples include PROSITE, BLOCKS, Pfam and PRINTS. |
| Secondary metabolites | Metabolites synthesized by plants besides primary metabolites, which are not essential for its survival (e.g. alkaloids, steroids). |
| Semigamy. | A cross wherein the nucleus of the egg-cell and generative nucleus of the geminated pollen grain divide independently, resulting in a haploid chimera. |
| Sequencing | Determination of the order of nucleotides (base sequences) in a DNA or RNA molecule, or the order of amino acids in a protein. |
| Shoot tip | Terminal (0.5–2.0 mm) portion of a shoot comprising the meristem dome and subjacent leaf and stem tissues. |
| Shot gun collection | Cloning of an entire genome in the form of randomly generated fragments. |

| | |
|---|---|
| Shotgun method | A sequencing method that involves randomly sequencing cloned pieces of the genome, with no foreknowledge of where the piece originally came from. This can be contrasted with "directed" strategies, in which pieces of DNA from known chromosomal location are sequenced. Because there are advantages to both strategies, researchers have used both random shotgun and directed strategies in combination to sequence genomes. |
| Shuttle vector | A plasmid constructed to have origins for replication for two hosts (e.g. *E. coli* and *S. cerevisiea*) so that it can be used to carry a foreign sequence in either prokaryotes or eukaryotes. |
| SIMOCOAL2 | Program for the simulation of complex recombination patterns over large genomic regions under various demographic models. This public domain on-line software tool to conduct has marker, maps, gene, proteins, QTL, literature, diversity analysis. |
| Single copy sequence | A sequence of nucleotides that occurs only once in a genome. |
| Single node culture | Culture of separate lateral buds, each carrying a piece of stem tissue. |
| Smith-Waterman-algorithm | A dynamic programming algorithm (q.v.) for local sequence alignment. |
| Somaclonal variation | The variability generated by the use of a tissue culture cycle. |
| Somatic | Referring to the vegetative or non-sexual stages of a life-cycle. |
| Somatic hybridization | A technique of fusing protoplasts from two contrasting genotypes for production of hybrids or cybrids which contain various mixtures of nuclear and/or cytoplasmic genomes, respectively. |
| Southern transfer (blot) | Transfer of DNA from an electrophoresis gel into a nitrocellulose or nylon membrane. The membrane is then used in hybridization experiments. |
| Splicing | Removal of introns from the primary transcript during the maturation of eukaryotic mRNA. |
| SRS | Sequence retrieval system. A data retrieval tool similar to *Entrez* (q.v.) and DBGET/*Link* DB (q.v.) but it can be used as a stand-alone program as well as over the Internet. |
| Sterile | Medium or object with no perceptible or viable microorganisms. |
| Sterilization | Procedure for the elimination of microorganisms. |
| Sticky ends | Single stranded ends left on a restriction fragment by many type II restriction endonucleases. |
| Structural gene | A gene codes for any RNA or protein product other than a regulator. |
| Subculture | Subdivision of a culture and its transfer to fresh medium. |
| Surface sterilization | Destruction of microorganisms from the surface of the plant parts with the help of chemical agent. |
| Suspension culture | A culture consisting of cells or cell aggregates initiated by placing callus tissues or sometimes seedlings in an agitated liquid medium. |
| SWISS-PROT | Database of confirmed protein sequence with extensive annotations, Maintained by the Swiss Bioinformatics Institute. |
| Synchronous culture | A culture in which the cell cycles (or a specific phase of the cycle) of the majority of cells are synchronous. |
| Synteny | The colinear relationships between the genomes of different organisms. Often chromosomal regions from related organisms, such as various grass species, contain corresponding genetic information and similar gene order. |

| | |
|---|---|
| TATA box | Eukaryotic promoter region analogous to the Pribnow box. It is a conserved AT rich septamer found about 25 bp before the startpoint of each eukaryotic RNA polymerase II transcription unit. |
| TCP/IP | Transmission Control Protocol/Internet Protocol, Protocols that control how data is packaged and reassembled (TCP) and addressed and routed (IP) over the Internet |
| Tilling | Tilling (targeting induced local lesions in genomes. The end product of tilling process is a plant (and its offspring)) that has been identified with a change in a specific gene of interest. The plant line is then useful for determining the overall effect/role of that gene on the characteristics of the plant. The tilled population can be produced and then screened as mutant alleles in specific gene. TILLING is a high throughput method for identifying specific gene knockouts in mutant populations and is useful as a tool of reverse genetics. |
| Ti plasmid | Tumor inducing plasmid often responsible for crown gall (tumor) induction in plants. |
| Tissue | A group of cells that perform a collective function. |
| Tissue culture | The culture of protoplasts, cells, tissues, organs, embryos or seeds *in vitro.* |
| Totipotency | The ability of individual cells to express the phenotype of the whole plant from which it is derived. |
| Trademark | It is a mark capable of being represented graphically and which is capable of identifying and distinguishing the goods or services of one person from those of others and may include any word, name, symbol, design or device, or any combination, shape of goods, their packaging and combination of colors, used, or intended to be used in commerce. In short, a trade mark is a brand name |
| Trailer | Non-translated sequence at the 3′ end of an messenger RNA following the termination codon. |
| Transcription | Copying of a gene into RNA by a DNA dependent RNA polymerase. |
| Transcriptome | The full complement of activated genes, mRNAs or transcripts in a tissue at a particular time in response to a specific stimulus |
| Transformation | Introduction of DNA into the cells of an organism by a method other than conventional sexual crossing. |
| Transformation of plastids | Stable integration of transforming DNA into the plastid genome that is transmitted to the seed progeny. |
| Translation | Copying of mRNA into polypeptide. |
| Transposase | Enzyme activity involved in insertion of transposon at a new site. |
| Transplastome | A transformed plastid genome. |
| Transposon | A DNA element which can insert at random into plasmids or the bacterial chromosome independently of the host cell recombination system. |
| TrEMBL | Translated EMBL, Database of protein sequences translated from the EMBL nucleotide sequence database. Not as extensively annotated as SWISS-PROT. |
| Turbidostat | An open continuous culture into which fresh medium flows in response to an increase in the turbidity of culture. |

| | |
|---|---|
| Two- dimensional polyacrylamide gel electrophoresis | A method of separating proteins as a function of their isoelectirc point (pI) in the first dimension and as a function of their molecular mass in the second dimension. This technique, widely used in plant proteomics, combines isoelectric focusing (IEF) in the first dimension and SDS-PAGE (Sodium Dodecyl Sulfate-Poly Acrylamide Gel Electrophoresis) in the second dimension. |
| Upstream | It identifies sequences proceeding in the opposite direction from expression. For example, the promoter is upstream from the transcription unit, the initiation codon is upstream of the coding region. |
| URL | Uniform Resource Locator. A text address for a web page. |
| Vector | A plasmid, phage, virus or bacterium carrying foreign DNA. |
| Vegetative propagation | The asexual propagation of plants by detachment of some parts of the plant, e.g. a cutting, and its subsequent development into a complete plant. |
| Virion | The physical virus particle (irrespective of its ability to infect cells and reproduce). |
| Vitrification | An undesirable condition of *in vitro* tissues characterized by succulence, brittleness and glassy appearance. |
| Wide hybridization | When individuals from two different species of the same genus or from two different genera are crossed. |
| YAC | Yeast artificial chromosome; a cloning system in yeast comprising an autonomously replicating linear vector containing an exogenous DNA insert, flanked by a yeast centromere and two telomere seeding sequences. |
| Zip | Compressed *archive* format (q.v.) popular on PCs. |
| Zygote | A product of the fusion of male and female gametes. |

# Author Index

# Subject Index